D1674417

Lignin and Lignans as Renewable Raw Materials

Wiley Series in Renewable Resources

Series Editor
Christian V. Stevens – Faculty of Bioscience Engineering, Ghent University, Ghent, Belgium

Titles in the Series
Wood Modification – Chemical, Thermal and Other Processes
Callum A. S. Hill

Renewables – Based Technology – Sustainability Assessment
Jo Dewulf & Herman Van Langenhove

Introduction to Chemicals from Biomass
James H. Clark & Fabien E.I. Deswarte

Biofuels
Wim Soetaert & Erick Vandamme

Handbook of Natural Colorants
Thomas Bechtold & Rita Mussak

Surfactants from Renewable Resources
Mikael Kjellin & Ingegärd Johansson

Industrial Application of Natural Fibres – Structure, Properties and Technical Applications
Jörg Müssig

Thermochemical Processing of Biomass – Conversion into Fuels, Chemicals and Power
Robert C. Brown

Biorefinery Co-Products: Phytochemicals, Primary Metabolites and Value-Added Biomass Processing
Chantal Bergeron, Danielle Julie Carrier & Shri Ramaswamy

Aqueous Pretreatment of Plant Biomass for Biological and Chemical Conversion to Fuels and Chemicals
Charles E. Wyman

Bio-Based Plastics: Materials and Applications
Stephan Kabasci

Introduction to Wood and Natural Fiber Composites
Douglas Stokke, Qinglin Wu & Guangping Han

Cellulosic Energy Cropping Systems
Douglas L. Karlen

Introduction to Chemicals from Biomass, Second Edition
James Clark & Fabien Deswarte

Forthcoming Titles
Sustainability Assessment of Renewables-Based Products: Methods and Case Studies
Jo Dewulf, Steven De Meester & Rodrigo A. F. Alvarenga

Cellulose Nanocrystals: Properties, Production and Applications
Wadood Hamad

Biorefinery of Inorganics: Recovering Mineral Nutrients from Biomass and Organic Waste
Erik Meers and Gerard Velthof

Bio-Based Solvents
François Jerome and Rafael Luque

Lignin and Lignans as Renewable Raw Materials

Chemistry, Technology and Applications

FRANCISCO G. CALVO-FLORES, JOSÉ A. DOBADO, JOAQUÍN ISAC-GARCÍA

Department of Organic Chemistry, University of Granada, Spain

and

FRANCISCO J. MARTÍN-MARTÍNEZ

Department of Civil and Environmental Engineering, Massachusetts Institute of Technology, USA

WILEY

This edition first published 2015
© 2015 John Wiley & Sons, Ltd

Registered office
John Wiley & Sons Ltd, The Atrium, Southern Gate, Chichester, West Sussex, PO19 8SQ, United Kingdom

For details of our global editorial offices, for customer services and for information about how to apply for permission to reuse the copyright material in this book please see our website at www.wiley.com.

The right of the author to be identified as the author of this work has been asserted in accordance with the Copyright, Designs and Patents Act 1988.

All rights reserved. No part of this publication may be reproduced, stored in a retrieval system, or transmitted, in any form or by any means, electronic, mechanical, photocopying, recording or otherwise, except as permitted by the UK Copyright, Designs and Patents Act 1988, without the prior permission of the publisher.

Wiley also publishes its books in a variety of electronic formats. Some content that appears in print may not be available in electronic books.

Designations used by companies to distinguish their products are often claimed as trademarks. All brand names and product names used in this book are trade names, service marks, trademarks or registered trademarks of their respective owners. The publisher is not associated with any product or vendor mentioned in this book.

Limit of Liability/Disclaimer of Warranty: While the publisher and author have used their best efforts in preparing this book, they make no representations or warranties with respect to the accuracy or completeness of the contents of this book and specifically disclaim any implied warranties of merchantability or fitness for a particular purpose. It is sold on the understanding that the publisher is not engaged in rendering professional services and neither the publisher nor the author shall be liable for damages arising herefrom. If professional advice or other expert assistance is required, the services of a competent professional should be sought

The advice and strategies contained herein may not be suitable for every situation. In view of ongoing research, equipment modifications, changes in governmental regulations, and the constant flow of information relating to the use of experimental reagents, equipment, and devices, the reader is urged to review and evaluate the information provided in the package insert or instructions for each chemical, piece of equipment, reagent, or device for, among other things, any changes in the instructions or indication of usage and for added warnings and precautions. The fact that an organization or Website is referred to in this work as a citation and/or a potential source of further information does not mean that the author or the publisher endorses the information the organization or Website may provide or recommendations it may make. Further, readers should be aware that Internet Websites listed in this work may have changed or disappeared between when this work was written and when it is read. No warranty may be created or extended by any promotional statements for this work. Neither the publisher nor the author shall be liable for any damages arising herefrom.

Library of Congress Cataloging-in-Publication Data applied for.

A catalogue record for this book is available from the British Library.

ISBN: 9781118597866

Set in 9/11pt TimesLTStd by SPi Global, Chennai, India

To our families

Contents

Series Preface	xv
Preface	xvii
List of Acronyms	xix
List of Symbols	xxiii

Part I Introduction   1

1. Background and Overview   3
     1.1   Introduction   3
     1.2   Lignin: Economical Aspects and Sustainability   4
     1.3   Structure of the Book   5
         References   7

Part II What is Lignin?   9

2. Structure and Physicochemical Properties   11
     2.1   Introduction   11
     2.2   Monolignols, The Basis of a Complex Architecture   12
     2.3   Chemical Classification of Lignins   16
     2.4   Lignin Linkages   17
     2.5   Structural Models of Native Lignin   20
         2.5.1   Softwood Models   21
         2.5.2   Hardwood Models   28
         2.5.3   Herbaceous Plant Models   28
     2.6   Lignin–Carbohydrate Complex   34
     2.7   Physical and Chemical Properties of Lignins   39
         2.7.1   Molecular Weight   39
         2.7.2   Dispersity Index (D)   40
         2.7.3   Thermal Properties   40
         2.7.4   Solubility Properties   41
         References   42

3. Detection and Determination   49
     3.1   Introduction   49
     3.2   The Detection of Lignin (Color-Forming Reactions)   49
         3.2.1   Reagents for Detecting Lignins   50

3.3	Determination of Lignin	55
3.4	Direct Methods for the Determination of Lignin	55
	3.4.1 Methods for Lignin as a Residue	57
	3.4.2 Lignin in Solution Methods	59
3.5	Indirect Methods for the Determination of Lignin	60
	3.5.1 Chemical Methods	61
	3.5.2 Spectrophotometric Methods	61
	3.5.3 Methods Based on Oxidant Consumption	64
3.6	Comparison of the Different Determination Methods	66
	References	68

4. Biosynthesis of Lignin — 75
- 4.1 Introduction — 75
- 4.2 The Biological Function of Lignins — 75
- 4.3 The Shikimic Acid Pathway — 76
- 4.4 The Common Phenylpropanoid Pathway — 78
- 4.5 The Biosynthesis of Lignin Precursors (the Monolignol-Specific Pathway) — 80
  - 4.5.1 The Biosynthesis of Other Monolignols — 82
  - 4.5.2 The Transport of Monolignols — 83
- 4.6 The Dehydrogenation of the Precursors — 85
- 4.7 Peroxidases and Laccases — 86
- 4.8 The Radical Polymerization — 87
  - 4.8.1 Dimerization — 88
  - 4.8.2 Quinone Methides — 89
  - 4.8.3 Lignification — 90
  - 4.8.4 Interunit Linkage Types — 91
  - 4.8.5 Dehydrogenation Polymer (DHP) — 97
- 4.9 The Lignin–Carbohydrate Connectivity — 97
- 4.10 Location of Lignins (Cell Wall Lignification) — 99
- 4.11 Differences Between Angiosperm and Gymnosperm Lignins — 101
- References — 103

Part III Sources and Characterization of Lignin — 113

5. Isolation of Lignins — 115
- 5.1 Introduction — 115
- 5.2 Methods for Lignin Isolation from Wood and Grass for Laboratory Purposes — 116
  - 5.2.1 Lignin as Residue — 118
  - 5.2.2 Lignin by Dissolution — 120
- 5.3 Commercial Lignins — 127
  - 5.3.1 Kraft Lignin — 128
  - 5.3.2 Sulfite Lignin (Lignosulfonate Process) — 131
  - 5.3.3 Soda Lignin (Alkali Lignin) — 133
  - 5.3.4 Organosolv Pulping — 134
  - 5.3.5 Other Methods of Separation of Lignin from Biomass — 136
- References — 136

6. Functional and Spectroscopic Characterization of Lignins — 145

- 6.1 Introduction — 145
- 6.2 Elemental Analysis and Empirical Formula — 146
- 6.3 Determination of Molecular Weight — 147
  - 6.3.1 Gel-Permeation Chromatography (GPC) — 148
  - 6.3.2 Light Scattering — 149
  - 6.3.3 Vapor-Pressure Osmometry (VPO) — 150
  - 6.3.4 Ultrafiltration (UF) — 151
- 6.4 Functional Group Analyses — 151
  - 6.4.1 Methoxyl Group (MeO) — 152
  - 6.4.2 Phenolic Hydroxyl Group (OH_{ph}) — 152
  - 6.4.3 Total and Aliphatic Hydroxyl Groups (R–OH) — 154
  - 6.4.4 Ethylenic Groups (>C=C<) — 157
  - 6.4.5 Carbonyl Groups (>C=O) — 158
  - 6.4.6 Carboxyl Groups (–COO–) — 158
  - 6.4.7 Sulfonate Groups and Total Sulfur Composition ($R\text{-}SO_2O-$ and S) — 158
- 6.5 Frequencies of Functional Groups and Linkage Types in Lignins — 159
  - 6.5.1 β-O-4′ Linked Units — 159
  - 6.5.2 β-5′ Linked Units — 160
  - 6.5.3 β-1′ Linked Units — 160
  - 6.5.4 α-O-4′ Linked Units (benzyl ethers) — 162
  - 6.5.5 Condensed and Uncondensed Units — 162
  - 6.5.6 Biphenyl Structures — 163
  - 6.5.7 4-O-5′ Linked Units — 163
  - 6.5.8 β-2 and β-6 Linked Units — 163
  - 6.5.9 β-β Linked Units — 163
  - 6.5.10 Dibenzodioxocin Units — 164
- 6.6 Characterization by Spectroscopic Methods — 164
  - 6.6.1 Fourier Transform Infrared (FTIR) Spectroscopy — 164
- 6.7 Raman Spectroscopy — 166
  - 6.7.1 Ultraviolet (UV) — 167
  - 6.7.2 NMR Spectroscopy — 168
  - 6.7.3 Other Spectroscopic Methods — 174
  - References — 175

7. Chemical Characterization and Modification of Lignins — 189

- 7.1 Introduction — 189
- 7.2 Characterization by Chemical Degradation Methods — 189
  - 7.2.1 Oxidation with Nitrobenzene — 190
  - 7.2.2 Oxidation with Cupric Oxide — 193
  - 7.2.3 Permanganate Oxidation — 195
  - 7.2.4 Mild Hydrolysis — 196
  - 7.2.5 Acidolysis — 198
  - 7.2.6 Thioglycolic Acid Hydrolysis (Mercaptolysis) — 200
  - 7.2.7 Thioacetolysis — 200

	7.2.8	Thioacidolysis	201
	7.2.9	Hydrogenolysis	203
	7.2.10	Derivatization Followed by Reductive Cleavage (DFRC)	206
	7.2.11	Nucleus-Exchange Reaction (NE)	209
	7.2.12	Ozonolysis	212
	7.2.13	Pyrolysis	215
7.3	Other Chemical Modifications of Lignins		216
	7.3.1	Acylation	216
	7.3.2	Alkylation	218
	7.3.3	Halogenation	219
	7.3.4	Nitration	221
	7.3.5	Sulfonation	222
	7.3.6	Oxidation	222
	7.3.7	Other Modifications of Lignins	225
7.4	Thermolysis (Pyrolysis) of Lignins		227
7.5	Biochemical Transformations of Lignins		227
	7.5.1	Biodegradation of Lignin	228
	7.5.2	Enzyme-Based Oxidation of Lignin	229
	References		230

Part IV Lignins Applications — 247

8. Applications of Modified and Unmodified Lignins — 249

8.1	Introduction		249
8.2	Lignin as Fuel		252
	8.2.1	Combustion in the Paper Industry	252
	8.2.2	Heating and Power	252
8.3	Lignin as a Binder		253
	8.3.1	Coal Briquettes	253
	8.3.2	Packing	254
	8.3.3	Pelleted Feeds	254
8.4	Lignin as Chelating Agent		254
8.5	Lignin in Biosciences and Medicine		256
8.6	Lignin in Agriculture		257
8.7	Polymers with Unmodified Lignin		258
	8.7.1	Phenol–Formaldehyde Binders	258
	8.7.2	Polyolefin–Lignin Polymers	260
	8.7.3	Polyester–Lignin Polymers	260
	8.7.4	Acrylamide–Lignin Polymers	261
	8.7.5	Polyurethane–Lignin Polymers	261
	8.7.6	Bioplastics (Liquid Wood)	263
	8.7.7	Hydrogels	264
	8.7.8	Foams and Composites	265
	8.7.9	Conducting Polymers	265

8.8	Other Applications of Unmodified Lignins		267
	8.8.1	Lignin in Lead-Acid Batteries	267
	8.8.2	Lignin-Based Nanoparticles and Thin Films	267
	8.8.3	Lignin in Dust Control	268
	8.8.4	Lignin in Concrete Admixtures	269
	8.8.5	Lignin as a Dispersant, Emulsifier, and Surfactant	269
	8.8.6	Lignin as Floating Agent	271
8.9	New Polymeric Materials Derived from Modified Lignins and Related Biomass Derivatives		271
	8.9.1	Modified Lignin in Phenol–Formaldehyde Wood Adhesives	271
	8.9.2	Modified Lignins for Epoxy Resin Synthesis	274
	8.9.3	Polyurethanes	275
	8.9.4	Lignin–Polybutadiene Copolymers	278
8.10	Polymers Derived from Chemicals Obtainable from Lignin Decomposition		278
8.11	Other Applications of Modified Lignins		279
	8.11.1	Nanoparticles (NPs)	279
	8.11.2	Cationic Amphiphilic Lignin Derivatives	281
	8.11.3	Soil Preservation	281
	8.11.4	Fertilizers	281
	References		281

9. High-Value Chemical Products — **289**

9.1	Introduction		289
9.2	Gasification: Syngas from Lignin		291
9.3	Thermolysis of Lignin		291
9.4	Hydrodeoxygenation (Hydrogenolysis)		294
9.5	Hydrothermal Hydrolysis		295
9.6	Chemical Depolymerization		295
	9.6.1	Acid Media Depolymerization	296
	9.6.2	Base Media Depolymerization	296
	9.6.3	Ionic Liquid-Assisted Depolymerization	297
	9.6.4	Supercritical Fluids-Assisted Depolymerization	298
9.7	Oxidative Transformation of Lignin		299
	9.7.1	Oxidation with Chlorinated Reagents	299
	9.7.2	Oxidation with Ozone	300
	9.7.3	Oxidation with Hydrogen Peroxide	300
	9.7.4	Oxidation with Peroxy Acids	302
	9.7.5	Catalytic Oxidation	302
9.8	High-Value Chemicals from Lignin		302
	9.8.1	Vanillin	302
	9.8.2	Dimethyl Sulfide and Dimethylsulfoxide	306
	9.8.3	Active Carbon	306
	9.8.4	Carbon Fiber	308
	References		308

Part V Lignans   313

10. Structure and Chemical Properties of Lignans   315
    10.1 Introduction   315
    10.2 Structure and Classification of Lignans   315
           10.2.1 Lignans   316
           10.2.2 Hybrid Lignans   319
    10.3 Nomenclature of Lignans   319
    10.4 Lignan Occurrence in Plants   325
    10.5 Methods of Determination and Isolation of Lignans from Plants   328
           10.5.1 Lipid Extraction   328
           10.5.2 Solvent Extraction   330
           10.5.3 Separation by Precipitation   330
           10.5.4 Chromatographic Methods   330
           10.5.5 Extraction of Polar Lignans from Biological Materials   331
    10.6 Structure Determination of Lignans   331
    10.7 The Chemical Synthesis of Lignans   332
           10.7.1 Generalities on the Asymmetric Total Synthesis of Lignans 332
           10.7.2 Dibenzylbutane Lignans   335
           10.7.3 Dibenzylbutyrolactone Lignans   337
           10.7.4 Cyclolignans (Aryltetralin Lignans)   338
           10.7.5 Dibenzocyclooctadiene Lignans   343
           References   353

11. Biological Properties of Lignans   369
    11.1 Introduction   369
    11.2 Biosynthesis of Lignans   370
           11.2.1 Pinoresinol Synthase (Dirigent Protein)   370
           11.2.2 The General Biosynthetic Pathway of Lignans   371
           11.2.3 Other Biosynthetic Pathways of Lignans   372
           11.2.4 Biosynthetic Pathways for Neolignans   377
           11.2.5 Biosynthetic Pathways for Norlignans   377
    11.3 Metabolism of Lignans   377
    11.4 Plant Physiology and Plant Defense   383
    11.5 Podophyllotoxin   386
           11.5.1 Extraction, Synthesis, and Biotechnological Approaches   388
           11.5.2 Biological Activities of PPT   389
           11.5.3 The Action Mechanism of PPT   392
           11.5.4 Congeners and Derivatives (Other Aryltetralin Lactones)   393
    11.6 Biological Activity of Different Lignan Structures   397
           11.6.1 Dibenzylbutane Lignans   398
           11.6.2 Dibenzylbutyrolactone Lignans   401
           11.6.3 Dibenzylbutyrolactol Lignans   404
           11.6.4 Furanoid Lignans   404
           11.6.5 Furofuranoid Lignans   407

	11.6.6	Aryltetralin Lignans (No Lactones)	411
	11.6.7	Arylnaphthalene Lignans	413
	11.6.8	Dibenzocyclooctadiene Lignans	418
	11.6.9	Neolignans	422
	11.6.10	Hybrid Lignans	423
	References		423

Part VI Outcome and Challenges   455

12. Summary, Conclusions, and Perspectives on Lignin Chemistry   457

12.1	Sources of Lignin	457
12.2	Structure of Lignin	458
12.3	Biosynthesis and Biological Function	459
12.4	Applications of Lignin	459
12.5	Lignans	461
12.6	Perspectives	462
	References	462

Glossary   465

Index   467

Series Preface

Renewable resources and their modification are involved in a multitude of important processes with a major influence on our everyday lives. Applications can be found in the energy sector, chemistry, pharmacy, the textile industry, paints and coatings, to name but a few fields.

The broad area of renewable resources connects several scientific disciplines (agriculture, biochemistry, chemistry, technology, environmental sciences, forestry ...), but it is very difficult to take an expert view on their complicated interactions. Therefore, the idea to create a series of scientific books, focussing on specific topics concerning renewable resources, has been very opportune and can help to clarify some of the underlying connections in this area.

In a very fast-changing world, trends do not only occur in fashion and politics, hype and buzzwords occur in science too. The use of renewable resources is more important nowadays, however, it is not hype. Lively discussions among scientists continue about how long we will be able to use fossil fuels, opinions ranging from 50 years to 500 years, but they do agree that the reserve is limited and that it is essential to search not only for new energy carriers but also for new material sources.

In this respect, renewable resources are a crucial area in the search for alternatives to fossil-based raw materials and energy. In the field of the energy supply, biomass and renewable-based resources will be part of the solution alongside other alternatives such as solar energy, wind energy, hydraulic power, hydrogen technology and nuclear energy.

In the field of material sciences, the impact of renewable resources will probably be even bigger. Integral utilisation of crops and the use of waste streams in certain industries will grow in importance leading to a more sustainable way of producing materials.

Although our society was much more (almost exclusively) based on renewable resources centuries ago, this disappeared in the Western world in the nineteenth century. Now it is time to focus again on this field of research. However, it should not mean a *retour à la nature*, but it does require a multidisciplinary effort at a highly technological level to perform research on new opportunities, to develop new crops and products from renewable resources. This will be essential to guarantee a level of comfort for a growing number of people living on our planet. The challenge for coming generations of scientists is to develop more sustainable ways to create prosperity and to fight poverty and hunger in the world. A global approach is certainly favoured.

This challenge can only be met if scientists are attracted to this area and are recognized for their efforts in this interdisciplinary field. It is therefore also essential that consumers recognize the fate of renewable resources in a number of products.

Furthermore, scientists do need to communicate and discuss the relevance of their work so that the use and modification of renewable resources may not follow the path of the genetic engineering concept in terms of consumer acceptance in Europe. In this respect, the series will certainly help to increase the visibility of the importance of renewable resources.

Being convinced of the value of the renewables approach for the industrial world, as well as for developing countries, I was myself delighted to collaborate on this series of books focussing on different aspects of renewable resources. I hope that readers become aware of the complexity, interactions and interconnections, and challenges of this field and that they will help to communicate the importance of renewable resources.

I would like to thank the staff from Wiley's Chichester office, especially David Hughes, Jenny Cossham and Lyn Roberts, in seeing the need for such a series of books on renewable resources, for initiating and supporting it and for helping to carry the project to the end.

Last, but not least I want to thank my family, especially my wife Hilde and children Paulien and Pieter-Jan for their patience and for giving me the time to work on the series when other activities seemed to be more inviting.

Christian V. Stevens, Faculty of Bioscience Engineering
Ghent University, Belgium
Series Editor "Renewable Resources"
June 2005

Preface

This book has grown from a mini-review on lignin that we had published in 2010, by invitation from Sarah Higginbotham (nee Hall) at John Wiley & Sons Publishing Group. From the very outset, it was clear to us that tackling the project as authors of a complete work was the most challenging but nevertheless the most robust way of addressing the issue. Conceiving a whole book appeared to be more complete than the common compilation books where the monograph results from the contribution of various authors coordinated by an editor. In our opinion, although these compilation-type books often result in a series of very specific chapters that provide a collection of review articles of high scientific level, they usually lack a strength thread to unify the entire work.

The specific case of lignin is particularly challenging due to the enormous amount of information available, the abundance of undefined concepts, and the diverse areas of knowledge involved in the topic. Native lignin is studied by botanists for its role in plants and plant cells, by biochemists regarding biosynthesis, by chemists concerned with its structure, and even by engineers dealing with lignin coming from paper mill or biorefineries.

A similar situation involves lignans, where these secondary plant metabolites are studied also by botanists, chemists, and even by professionals in biomedical sciences for the biological properties of these molecules in living organisms. A fairly complete description of the nature, structure, properties, synthetic processes, and applications of this family of compounds is provided.

Given such a complex and multidisciplinary outlook, a thorough review was needed of the existing literature, together with classical references, the brainchild of pioneering authors, as well as recent contributions to the topic in order to provide the reader with a broad view of the most comprehensive knowledge on lignin and lignans. As is inevitable with projects of this scope, the final work might not be as complete as it could have been, but we nevertheless trust that the result is thorough enough to be useful to the scientific community interested in the subject.

Throughout the text, lignin is explained from different perspectives, including its role as a structural component of plants, and how it is produced as a by-product of paper industry and a product of biorefineries. Structural models of this biopolymer are disclosed, as well as the developing process that these models have undergone through the years, parallel to the improvement of structural determination methods, both instrumental and chemical ones. This information will provide the reader with an overall idea of the structure of lignin, its origin, its function, its applications, and its potential. The reader will also learn how to appropriately use the term "lignin," as the actual lignin depends on the origin of this material.

During the preparation of the book, special effort was made to review the applications and the potential uses of different lignins, with emphasis on the word "potential." So far, there has been ample academic work on the subject, but the actual results are still relatively modest. Therefore, many topics remain to be developed in the coming years, and they definitely will be, considering the growing importance of renewable raw materials in taking over those of limited availability.

Given our input on lignin, and our experience as authors of the present work, we conclude that this is a highly promising biomaterial, which, in terms of science and technology, still presents many unresolved issues that continue to be investigated. In the literature, terms such as "potential" and "promising" constantly appear, alongside "difficult," "complex," and "underutilized." These modifiers reflect lignin's state of the art. In the coming years, great effort must be needed to ensure lignin the central role as source of raw materials, consumer goods, and much more relevant applications that it deserves. We deeply hope that this book will stimulate further interest and research in this promising biopolymer in its various forms.

Finally, we repeat our appreciation to John Wiley & Sons Publishing Group and its staff for their incalculable help, support, and feedback over the course of the project. Last but not least, we would like to give our special thanks to Dr Ángel Sánchez-González for the design of the front cover, and Mr David Nesbitt for his invaluable work on the revision of the English version of the manuscript and his contribution with the "*Podophyllum peltatum*" illustration.

Francisco G. Calvo-Flores
José A. Dobado
Joaquín Isac-García
Francisco J. Martín-Martínez
January 2015

List of Acronyms

2D	two dimensional
3D	three dimensional
4CL	4-coumarate CoA ligase
Ac-CW	acetylated cell wall
ADF	acid detergent fiber
AFEX	ammonia fiber explosion
AOAC	Association of Analytical Communities (formerly Association of Official Agricultural Chemists)
AOP	advanced oxidation process
ARP	ammonia recycle percolation
ASAM	alkaline sulfite, anthraquinone, methanol
ASL	alkali sulfite lignin
BADGE	bisphenol A diglycidyl ether
BTX	benzene, toluene, and xylene
C4H	cinnamate 4-hydroxylase
CAGT	coniferyl alcohol glucosyltransferase
CBG	coniferin-β-glucosidase
CD	circular dichroism
CEHPL	chain-extended hydroxypropyl lignin
CEL	cellulolytic enzyme lignin
CK	cytokinins
DAD	photodiode array detector
DAHP	3-deoxy-D-arabinose heptulosonic acid-7-phosphate
DBDO	dibenzodioxocin
DCC	N,N'-Dicyclohexylcarbodiimide
DCCC	droplet counter-current chromatography
DCG	dehydrodiconiferyl alcohol-4-β-D-glucoside
DDQ	2,3-dichloro-5,6-dicyano-1,4-benzoquinone
DFRC	derivatization followed by reductive cleavage
DHP	dehydrogenation polymer
DIR	dirigent protein
DMAc	N,N-dimethylacetamide
DMF	dimethylformamide
DMS	dimethyl sulfide
DMSO	dimethyl sulfoxide
DPPH	1,1-diphenyl-2-picrylhydrazyl
EDXA	energy-dispersive X-ray analysis
EMAL	enzymatic mild acidolysis lignin
END	enterodiol
ENL	enterolactone
EPSP	5-enolpyruvylshikimate-3-phosphate

ESR	electron spin resonance
FDA	Food and Drug Administration
FTIR	Fourier transform infrared
G	coniferyl alcohol
GC-MS	gas chromatography-mass spectrometry
GC	gas chromatography
GHG	greenhouse gas
GPC	gel permeation chromatography
H	*p*-coumaryl alcohol
HBS	high-boiling solvents
HDO	hydrodeoxygenation
HDPE	high-density polyethylene
HMR	7-hydroxymatairesinol
HMTA	hexamethylenetetramine
HPA	Heteropoly acids (e.g., $H_3PWO_{12}O_{40}$)
HPL	hydroxypropyl lignin
HPLC	high-performance liquid chromatography
HPSEC	high-pressure size exclusion chromatography
HPSECI	high-pressure size exclusion chromatography infrared
HRMS	high-resolution mass spectrometry
HRP	hydroxyproline-rich protein
HSCCC	high-speed counter-current chromatography
IAT	indulin AT
IM	interference microscopy
IOR	improved oil recovery
IPTES	3-(triethoxysilyl)propylisocyanate
IR	infrared
KL	Klason lignin
LALLS	laser light scattering
LBS	low-boiling solvents
LC-NMR	liquid chromatography-nuclear magnetic resonance
LDPE	low-density polyethylene
LEM	*Lentinus edodes* mycelia
LPF	lignin-modified phenolic resin
LPS	lignin process system
LSA	lignin sulfonic acid, lignosulfonic acid
MAE	microwave-assisted extraction
MDF	medium-density fiberboards
MDI	methylene diphenyl isocyanate
MEKC	micellar electrokinetic capillary chromatography
MOF	metal-organic framework
MPP	mesophase pitch
MSn	multiple-stage mass spectrometry
MWL	milled wood lignin
NADH	nicotinamide adenine dinucleotide
NDGA	nordihydroguaiaretic acid
NE	nucleus exchange
NMR	nuclear magnetic resonance
NOESY	nuclear overhauser effect spectroscopy
NP	nanoparticle

OSB	oriented strand boards
PA66	polyamide 66
PAL	L-Phenylalanine ammonia lyase
PAN	polyacrylonitrile
PE	polyethylene
PEG	polyethylene glycol
PF	phenol formaldehyde
PLCG1	phospholipase C γ1
PLPW	pressurized low-polarity water
PLR	pinoresinol/lariciresinol reductases
PNNL	Pacific Northwest National Laboratory
POM	polyoxometalate
PP	polypropylene
PPG	polypropylene glycol
PPT	podophyllotoxin
PS	polystyrene
PTSA	p-toluenesulfonic acid
PU	polyurethane
PVC	polyvinyl chloride
RP	reverse phase
S	sinapyl alcohol
SAA	soaking in aqueous ammonia
SAR	structure–activity relationship
SDG	secoisolariciresinol diglucoside
SEC	size exclusion chromatography
SECO	secoisolariciresinol
SEL	swelled enzyme lignin
SEM	scanning electron microscopy
SHS	switchable hydrophilicity solvent
SIRD	secoisolariciresinol dehydrogenase
TAL	tyrosine ammonia lyase
TAPPI	Technical Association of the Pulp and Paper Industry
TDMP	2-chloro-4,4,5,5-tetramethyl-1,3,2-dioxaphospholane
TEM	transmission electron microscopy
THF	tetrahydrofuran
TLC	thin-layer chromatography
TOC	total organic carbon
TPA	tonns per annum
TTFA	thallium(III) trifluoroacetate
UDP	uridine diphosphate
UV/Vis	ultraviolet/visible
VCD	vibrational circular dichroism
VPO	vapor pressure osmometry
WG	water-dispersible granules

List of Symbols

δ	NMR chemical shift
$Đ$	dispersity index
ED_{50}	median effective dose
g	contraction factor
IC_{50}	half maximal inhibitory concentration
M_n	average molecular weight
M_w	molecular weight
λ_{max}	The wavelength at which the largest amount of absorption occurs
Log P	partition coefficient
pH	acidity or basicity of an aqueous solution
ppm	parts *per* million
ppu	parts *per* unit
T_g	temperature range of glass transition
rt	room temperature

Part I
Introduction

Part I

1
Background and Overview

1.1 Introduction

Surviving on a small planet with limited resources to support our increasing global population is probably the greatest challenge humanity has faced so far. A large part of the problem is that our economy is driven by many technologies that are not sustainable at all. This necessity of developing sustainable technologies capable of addressing such challenges, together with the increasing concern over environmental protection and questions about future availability of petrochemical feedstock have spurred research and development toward new degradable materials from renewable resources, which are more environmentally friendly and sustainable than the currently used petroleum-based materials. Within this context, lignin, which appears as one of the polymeric components in plants, arises as a promising candidate for some of the desirable applications due to its rich chemical structure and its versatility.

For more than 100 years, scientists and engineers have made efforts to effectively remove lignin from wood when extracting cellulose in the pulping process.[1]

In 1819, the term "lignin," from the Latin word *lignum* meaning "wood" [1], was used for the first time by the Swiss botanist A. P. Candolle (1778–1841). Later, in 1839, A. Payen first described this "encrusting material" in wood. It took, however, about 20 years to accept the term "lignin" to refer to a material as it is currently understood [2].

An understanding of its chemical composition began in 1875, when Bente [3] demonstrated that the noncellulosic constituent of wood, namely lignin, was aromatic in nature. It was further characterized by Benedikt and Bamberger [4] in 1890, who described the methoxy group as typical of lignin chemical structure. Later in 1960, Brauns [5] stated: '*the lignin building stone has a phenyl propane structure that may be regarded as proven, but how the stones are linked together in proto-lignin is still a mystery*'. In addition, in 1920, Klason [6] postulated that lignin was an oxidation product of coniferyl alcohol, which was demonstrated in 1968 by Freudenberg [7].

---

[1] Lignocellulosic fibrous material prepared by chemically or mechanically separating cellulose fibers from wood, fiber crops, or waste paper.

Lignin and Lignans as Renewable Raw Materials: Chemistry, Technology and Applications, First Edition.
Francisco G. Calvo-Flores, José A. Dobado, Joaquín Isac-García and Francisco J. Martín-Martínez.
© 2015 John Wiley & Sons, Ltd. Published 2015 by John Wiley & Sons, Ltd.

Table 1.1 The most common plant phenolic compounds listed according to the count (content) of carbon atoms[a]

Composition	Count of carbons	Types of phenolic substances
C_6	6	Simple phenols, benzoquinones
C_6-C_1	7	Phenolic acids/aldehydes
C_6-C_2	8	Acetophenones, benzofurans
C_6-C_3	9	Phenylpropanoids, benzopyranes (coumarins)
C_6-C_4	10	Naphthoquinones
C_6-C_5	11	Ageratochromenes (prekocens)
$(C_6)_2$	12	Dibenzofurans, dibenzoquinones, biphenyls
$C_6-C_1-C_6$	13	Dibenzopyranes, benzophenones, xanthones
$C_6-C_2-C_6$	14	Stilbenes, anthraquinones, phenanthrenes
$C_6-C_3-C_6$	15	Flavonoids, isoflavones, chalcones, aurones
$C_6-C_4-C_6$	16	Norlignans (diphenylbutadienes)
$C_6-C_5-C_6$	17	Norlignans (conioids)
$(C_6-C_3)_2$	18	Lignans, neolignans
$(C_6-C_3-C_6)_2$	30	Biflavonoids
$(C_6-C_3-C_6)_n$	n	Condensed tannins (flavolans)
$(C_6-C_3)_n$	n	Lignins
$(C_6)_n$	n	Catecholmelanines

[a]Adapted from refs [10–12].

Beyond this historical perspective on early years of lignin research, the rising interest on lignin today has made this natural polymer to go from a waste-side product to a promising source for chemicals, polymers, and many other applications. Lignin is the second most abundant natural polymer together with cellulose, and hemicellulose [8], which are the major sources of nonfossil carbon that make a special contribution to the carbon cycle [9]. Lignin is by far the most abundant substance composed of aromatic moieties in nature (see Table 1.1), and the largest contributor to soil organic matter.

Furthermore, lignin is an important component of secondary cell walls in plant cells, and it helps to maintain the integrity of the cellulose/hemicelluloses/pectin matrix that provides rigidity to the plant. Also, it provides internal transport of nutrients and water, and protects against attack by microorganisms. Apart from this key role in plants, lignin is also obtained from paper industry and other methods. Actually, the many different sources and types of lignin makes it more accurate to refer generically to "lignins" when referring to this multifaceted material. Its diversity also implies that interest in lignin arises from fields of knowledge as diverse as botany, chemistry, chemical engineering, economy, ecology, and so on. Therefore, a general vision about lignin should come from a multidisciplinary approach.

From an ecological viewpoint, lignins are of general significance to the global carbon cycle, since they represent an enormous reservoir of bound organic carbon. However, despite this potential, lignins are a fairly unused renewable raw material that is now gaining the attention of industry, which will make them materials of immense economical importance [13].

1.2 Lignin: Economical Aspects and Sustainability

One-third of the world's land surface is covered by forest, accounting for 3×10^5 million m³ of timber, of which some 2.6×10^9 m³ are harvested annually. Just for comparison, such a vast amount is twice the world production of steel. Cellulose and hemicelluloses form around half of this, and lignin is the remaining bulk constituent that stands as the second largest natural source of organic material.[2] Additionally, the paper industry and, more recently, biorefineries produce large quantities of lignin that today is almost considered a by-product.

Not only from an economic standpoint but also from a sustainability perspective, the misuse or even nonuse of lignin appears to be a colossal mistake. Lignin, in its different versions available, whether native lignin from plants or partially transformed by industrial separation procedures, is a complex material with a great potential by itself or as a source of chemicals. In a world where raw materials are in constant demand, having a renewable source as lignin should be considered as a gift from nature to technology.

Today, many common goods are produced from nonrenewable sources such as oil or carbon. On the contrary, lignin is a middle-term alternative for the production of chemicals, polymers, carbon fibers, or new materials. Also, lignin has several technical applications in many fields, with the great advantage of its biodegradability, low toxicity, antioxidant properties, and low cost. Despite all these properties, lignin in its different forms is currently underutilized, and a great deal of work is pending for the coming years.

1.3 Structure of the Book

As mentioned earlier, the structure, reactions, and applications of lignins have been studied for more than a century. Introducing the term "lignin" in any scientific search engine such as Scifinder®[3] generates more than 90 000 entries. About a half of them, (48 740), have been published in the past 14 years, reflecting the rise of lignin. A closer analysis of the literature shows that 69 500 are scientific articles, 15 200 are patents, 380 are PhD dissertations, 4500 are reviews, and 95 are books. Simple observation of Figure 1.1 shows the exponential growth of interest on lignin. Only in the past year, more than 4000 papers have been published on this topic, and on June 25th, 2012, lignin was nominated by the American Chemical Society as "molecule of the week".[4] Thus, lignin is what is popularly called a hot topic.

In the same way, the term "lignan" yields 6800 entries, from which 6000 are articles, 577 are patents, and only 4 correspond to books.

This book examines the science and technology of lignin, using a multidisciplinary approach. More than 2300 bibliographical references have been compiled to provide the reader with a complete collection of material on the broad and complex field of lignin. To handle such a vast amount of information, the book has been divided into several parts that give a wide vision on lignin's science and technology.

The first part is dedicated to the study of the structure, morphology, composition, and biochemistry, including a review and update on the techniques used in the detection and determination of lignins. In this part, a historical review shows the evolution of different models according to the development of structural identification methods. Also in this first part, great effort is dedicated to the description of the biosynthesis of lignin. This biochemical aspect is basic to explain the final structure of lignin and how different spices are able to produce it. Additionally, it helps to understand how the structural differences lead to different properties of hardness and flexibility in plants. This understanding provides a picture on structure–function relationships that would guide the development of future applications based on lignin.

---

[2] The chemical composition of wood varies among species, but plants are composed of approximately 25% lignin and 75% cellulose and hemicellulose.
[3] Scifinder, version 2014; Chemical Abstracts Service: Columbus, OH, 2014; http://scifinder.cas.org (accessed December 13, 2014).
[4] Lignin – June 25, 2012. http://www.cas.org/motw/lignin (accessed December 4, 2014).

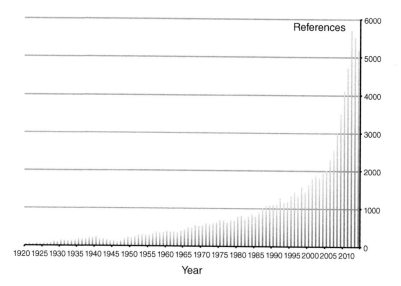

Figure 1.1 Number of entries retrieved using "lignin" as topic in Scifinder® in the 1920–2014 period

Also, the basic constituents of lignins, monolignols are discussed. Monolignols are phenolic compounds produced as secondary metabolites in plants, which present a large variety of biological functions and activities. Monolignols are inherently involved in biochemical processes of the plant itself due to their remarkably rich structural variations. Some of these activities are derived from such structural versatility, which makes them involved even in symbiotic or defensive interactions with other organisms. Despite such varieties, the biosynthetic pathways for their formation differ only in some fractional details, and therefore the types of these phenolic compounds are usually classified only according to the number of carbon atoms and their mutual correlation in the structure (see Table 1.1).

Table 1.1 shows phenylpropanoids and most of other types of plant phenolic compound. The phenylpropanoids are generated only by a limited number of basic biogenetic pathways where a restricted number of two or three key intermediates are involved. Such intermediates are able to generate up to thousands of so-called periphery derivatives formed by very simple specific enzymatic transformations. On the other hand, from the biogenetic standpoint, phenylpropanoids are formed by the shikimate pathway [14].

In general, as shown in Table 1.1, phenylpropanoids contain C_6-C_3 units, or combined $C_6-C_3-C_6$ ones, while their dimers or oligomers contain (9, 15, 18, 30, and n) carbon atoms. It has to be noted that this selection is focused on lignans and lignins.

The second part of the book is dedicated to the different methodologies for the isolation, purification, and chemical characterization of lignins. Many of these procedures can be considered classical methods and are dated accordingly on the first years of development of lignin studies. Throughout these years, a great effort has been made to make a correct detection and characterization of functional groups in lignin, as well as their proportion in the polymer. This characterization implies the determination of the simple units present in lignin and the way that these are linked. This early work has provided a valuable background that contributes to an understanding of the structural complexity of lignin depending on its origin. Part of the description in the present book is also dedicated to the so-called industrial lignins. These types of lignins can be isolated and, therefore, classified from the procedures used by the paper industry. These lignins are obtained in bulk amounts with a relatively high grade of purity, and they present great potential for many applications.

The third part of the monograph is dedicated to the industrial applications of lignin, either native or from the paper industry. Firstly, direct applications of the different types of industrial lignin are described. Additionally, bibliographical sources have been extensively reviewed to offer data on the chemical modification of lignin that improves its properties and/or reactivity when it is used, for example, as a macromonomer in the preparation of other polymers. A remarkable aspect of the industrial use of lignin for the fabrication of goods is its low toxicity and biodegradability, making it a prime candidate for these uses. Another aspect treated in this part of the book concerns the state of the art on the techniques and methodologies for the degradation of lignin macromolecules into high-value chemicals, with a special attention to simple aromatic molecules. The properties of lignin as a raw material for the preparation of aromatic compounds make lignin unique in the natural world.

Finally, an update on chemistry structure, biosynthesis, chemical synthesis, and biological properties of lignans is also provided. These kinds of natural products were included for the common biosynthetic origin and similar structure of both lignans and basic lignin components (see Table 1.1). An exhaustive review of the literature available on lignans led to the development of the last two chapters of the book to complete the work.

References

[1] Jouanin L, Lapierre C. Lignins: Biosynthesis, Biodegradation and Bioengineering, Advances in Botanical Research. vol. 61. Academic Press; 2012.
[2] Schulze F. Beitrage zur kenntniss des lignins. Chem Zentralbl. 1857;21:321–325.
[3] Bente F. Über die constitution des tannen- und pappelholzes. Ber Dtsch Chem Ges. 1875;8(1): 476–479.
[4] Benedikt R, Bamberger M. Über eine quantitative reaction des lignins. Monatsh Chem. 1890; 11(1):260–267.
[5] Brauns FE, Brauns DA. The Chemistry of Lignin: Supplement. New York: Academic Press; 1960.
[6] Klason P. Constitution of the lignin of pine wood. Ber Dtsch Chem Ges B. 1920;53:1864–1873.
[7] Freudenberg K. The constitution and biosynthesis of lignin. In: Freudenberg K, Neish AC, editors. Constitution and Biosynthesis of Lignin, Molecular Biology, Biochemistry and Biophysics. vol. 2. Berlin: Springer-Verlag; 1968. p. 45–122.
[8] Vanholme R, Demedts B, Morreel K, Ralph J, Boerjan W. Lignin biosynthesis and structure. Plant Physiol. 2010;153(3):895–905.
[9] Austin AT, Ballaré CL. Dual role of lignin in plant litter decomposition in terrestrial ecosystems. Proc Natl Acad Sci U S A. 2010;107:4618–4622.
[10] Harmatha J, Zídek Z, Kmoníčkova E, Šmidrkal J. Immunobiological properties of selected natural and chemically modified phenylpropanoids. Interdiscipl Toxicol. 2011;4(1):5–10.
[11] Mann J. Natural Products: Their Chemistry and Biological Significance. Longman Scientific & Technical; 1994.
[12] Harmatha J. Structural abundance and biological significance of lignans and related plant phenylpropanoids. Chem Listy. 2005;99(9):622–632.
[13] Hofrichter M, Steinbuchel A, editors. Biopolymers: Lignin, Humic Substances and Coal. vol 1. Weinheim: John Wiley & Sons, Ltd; 2001.
[14] Umezawa T. Diversity in lignan biosynthesis. Phytochem Rev. 2003;2(3):371–390.

Part II
What is Lignin?

2

Structure and Physicochemical Properties

2.1 Introduction

Lignin is a plant-derived biopolymer, basic structural constituent of wood and plants, which is formed mainly by three phenolic units, known as monolignols, and a few carbohydrate moieties. This lignin is so-called native lignin. Also, the name lignin is applied to a by-product from the separation of different components of plant biomass in paper industry and biorefineries. In this case, the size and structure of this kind of lignin substantially differ from native lignin.

Notwithstanding its importance in nature and despite the abundance of studies on its role as one of the main components in plants, a single definition of lignin has not been established, not only because of its intrinsic molecular complexity but also because of its diverse structural composition. These two factors, molecular complexity and structural diversity, often make it more accurate to refer to as lignins when discussing this peculiar material.

For decades, the concept of lignin has evolved according to the advances in research on this area. Thus, in the subsequent sections, it will be considered how the scientific community has described lignin throughout the years. The first thorough description was provided in 1960 by Brauns, who described lignins as polymers with the following characteristics [1]:

- Lignins are plant polymers made from phenylpropanoid building units.
- Lignins contain most of the wood methoxyl content.
- Lignins are resistant to acid hydrolysis, readily oxidized, soluble in hot alkaline and bisulfite, and readily condensed with phenols or thiols.
- In reaction with nitrobenzene in hot alkaline solution, lignins yield mainly vanillin, syringaldehyde, and *p*-hydroxybenzaldehyde depending on the origin of the lignins.
- When boiled in HCl/EtOH solution, lignins give a mixture of aromatic ketones resulting from cleavage of lignins' major interunit ether linkages (β-*O*-4).

Lignin and Lignans as Renewable Raw Materials: Chemistry, Technology and Applications, First Edition.
Francisco G. Calvo-Flores, José A. Dobado, Joaquín Isac-García and Francisco J. Martín-Martínez.
© 2015 John Wiley & Sons, Ltd. Published 2015 by John Wiley & Sons, Ltd.

12   Lignin and Lignans as Renewable Raw Materials

This definition was generally accepted [2], but later extended by Brunow et al. [3], who provided the most precise and comprehensive definition to date [4]. This author also prefers to employ the term protolignin,[1] and defines protolignins and lignins as natural polymers with the following features:

- Protolignins are biopolymers consisting of phenylpropanoid units with an oxygen atom at the para-position (i.e., –OH or –O–C) and with none, one, or two methoxyl groups in the para-position to this oxygen atom.
- The phenylpropanoid building units are connected to one another by a series of characteristic linkages. There are a series of characteristic end-groups.
- All the types of structural elements detected in protolignins are consistent with those formed by oxidation of the p-hydroxycinnamyl alcohols in vitro.
- The structural units in protolignin are not linked to each other in any particular order.
- Lignins are not optically active.
- Protolignins are blanched and cross-linked to other cell wall components. There are strong indications of the occurrence of linkages between lignins and carbohydrates. There are esters that exist in some types of lignins.

From these definitions, it is clear that lignin is not a constitutionally defined compound, but on the contrary it is a class of phenolic natural polymers with a broad composition and a variety of linkages between building units. Therefore, in this chapter, we show the most remarkable structural characteristics of lignins together with some common physicochemical properties.

2.2   Monolignols, The Basis of a Complex Architecture

Lignin is not a constitutionally defined compound, but rather a physically and chemically heterogeneous material. Its structural diversity arises mainly from the combination of three phenylpropane derivatives that are the main building blocks of its complex architecture. These phenolic compounds are hydroxycinnamyl alcohols (see Figure 2.1) or monolignols that share the most abundant phenylpropane unit (M1) and differ in the phenyl functionalization. These are commonly known as p-coumaryl ($M1_H$),[2] coniferyl ($M1_G$),[3] and sinapyl ($M1_S$)[4] alcohols, where the subindices H, G, and S denote the specific monolignol within this M1 type. When these three alcohol moieties are in the polymer, each monolignol is constituent p-hydroxyphenyl (H), guaiacyl (G), and syringyl (S) residues, respectively [6, 7].

Despite these three most abundant types, other less-abundant monomers of lignin, such as $M1_{SH}$ or some other with different phenylpropane units (M2 to M12), have been reported as well (see Figure 2.2). All of them "conjugate variously" in the biosynthesis process of lignin to form a 3D polymer, which does not have an ordered and regular macromolecular structure. These processes of formation of phenylpropanoid macromolecules termed lignin is called lignification [5], which includes the biosynthesis of monolignols, their transport to the cell wall, and the polymerization into the final macromolecule.

Monolignols in plants are not abundant in free forms, but rather exist as 4-O-β-D-glucosides. These glycosylated derivatives of the main monolignols, called p-glucocoumaryl alcohol glucoside, coniferin and syringin, are transportable and stored in lignifying tissues. They are formed by monolignol UDP-glucose coniferyl-alcohol glucosyltransferase [8, 9], and have been isolated from all gymnosperms,[5] as well as from a limited number of Angiosperms [10] (see Figure 2.3).

---

[1] Immature form of lignin that can be extracted from the plant cell wall with EtOH or dioxane.
[2] (E)-4-(3-hydroxyprop-1-enyl)phenol.
[3] 4-[(1E)-3-hydroxy-1-propenyl]-2-methoxyphenol.
[4] (E)-4-(3-hydroxyprop-1-enyl)-2,6-dimethoxyphenol.
[5] A group of seed-producing plants that includes conifers, cycads, Ginkgo, and Gnetales.

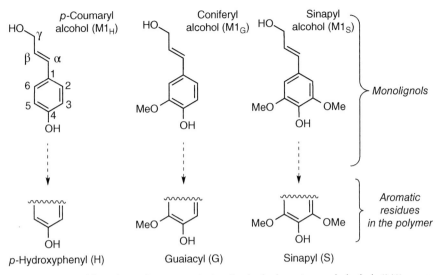

Figure 2.1 *Chemical formula, and atom numbering for the hydroxycinnamyl alcohols (M1) monomers and residues of lignin.*

In comparison to their corresponding monolignols, these glucosides are also more soluble. These glucoconjugates of monolignols are possibly moved first from the cytosol to the vacuole and then transported from the vacuole to the cell wall through an yet-unknown mechanism [11].

Besides, other nonconventional monolignols that have not been discussed yet might be found also in smaller amounts, as it occurs in some grassy and herbaceous species. Figure 2.4 shows some of these unconventional monolignols that have been found as end-groups in some specific plants [12].

Monolignols are at the bottom level of the hierarchy found in lignin structure. Starting from monolignols, an extremely complex architecture is developed. Lignins are synthesized by peroxidase-mediated ether linkages with aryl-glycerol and β-aryl ether. From these linkages, many stereocenters are formed, although the final polymer found in nature is an optically inactive form. The polymerization of monolignols and their ratio within the polymer varies depending on plants, woody tissues, and cell wall layers. Cellulose, hemicellulose, and lignin form structures called microfibrils, which are organized into macrofibrils that mediate structural stability in the plant cell wall (see Figures 2.5 and 2.6) [14].

The cell wall is composed by different layers, as shown in Figure 2.5. These layers are ordered from the outer section to the inner one in the following manner: the middle lamella (ML),[6] the primary wall (P), and the secondary wall. These are called S1, S2, and S3, respectively, with lignins being located principally in the ML (S1) and the secondary wall (S3). The relative amount of lignin in ML is higher than in the other two layers because it is thinner than the others. Nevertheless, it contributes only a minor fraction to the total lignin content. On the other hand, the secondary wall presents lower values in percentage, but it accounts as the major lignin container of the wall.

In spite of its relatively simple constituents described so far, the variety of linkages and arrangements end up in a complex structure that shows many different forms depending on the origin of lignin. Thus, the composition, structure, and lignin ratio in plants depend on the plant species. For example, in softwood, lignin represents about 30% of the total mass, while in hardwood this share falls to 20–25%. For herbaceous species, the average content of lignin reaches even lower values. Also, the variable M_w found in all these cases is a consequence of the random cross-linked polymerization of phenolic moieties, originating from radical coupling reactions between phenolic radicals.

---

[6] The middle lamella is a layer that forms between adjacent cells and holds them together.

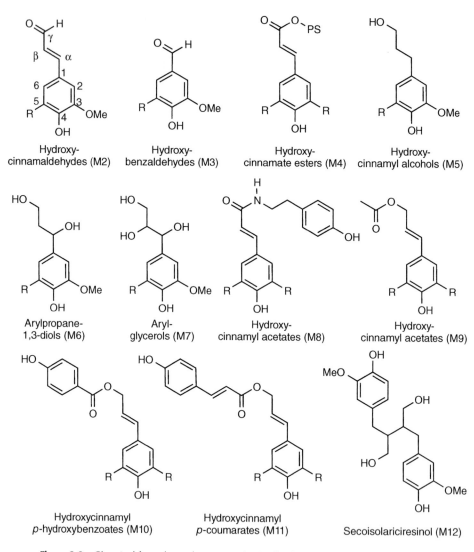

Figure 2.2 Chemical formula, and atom numbering for the M2–M12 monomers of lignin

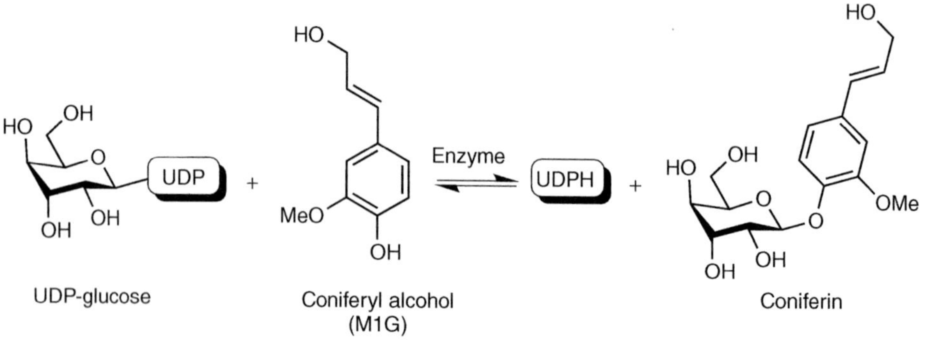

Figure 2.3 Scheme for the formation of lignoglucoside derivatives by means of the enzyme coniferyl-alcohol glucosyltransferase

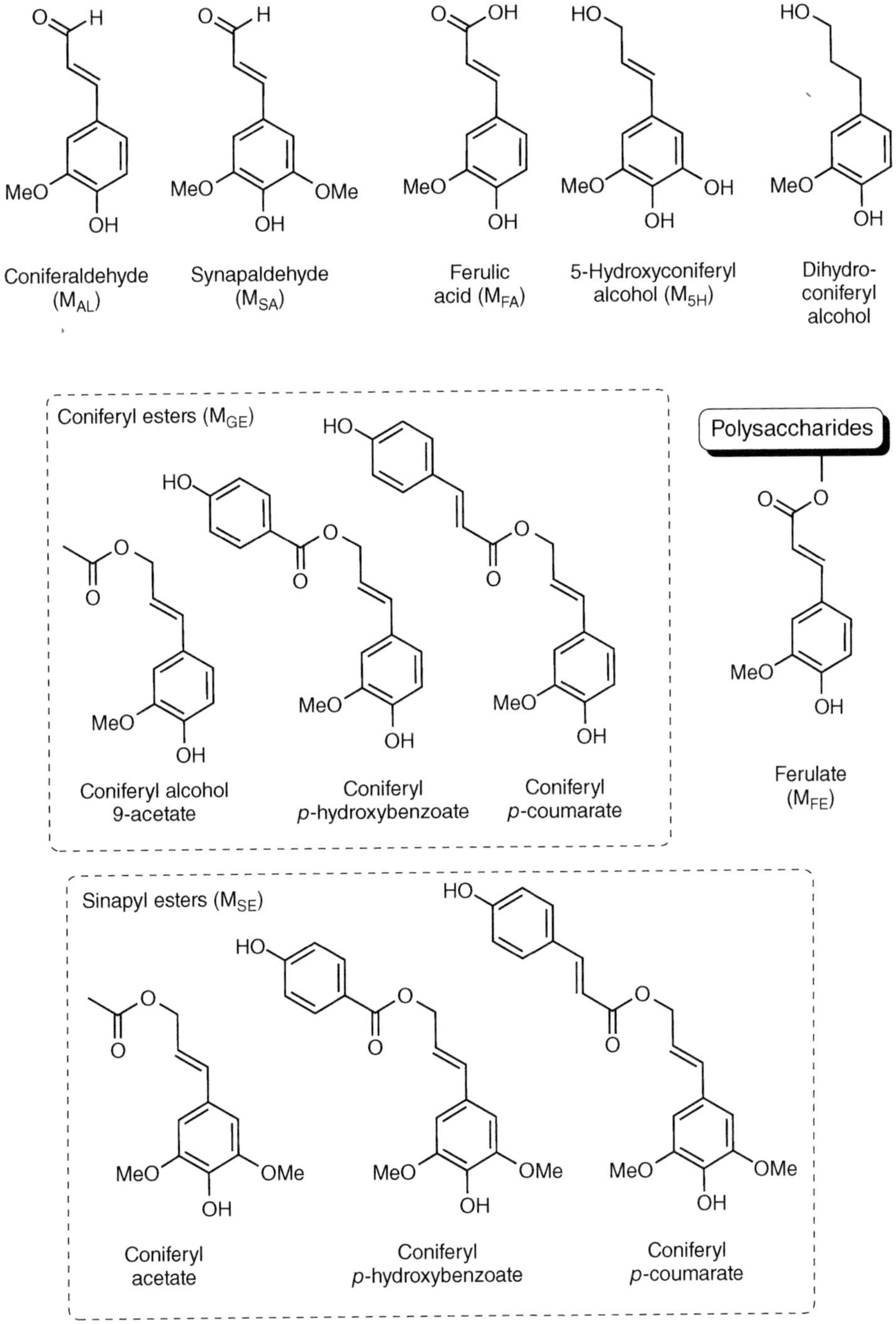

Figure 2.4 Chemical formula for nonconventional monolignols [12, 13]

16 Lignin and Lignans as Renewable Raw Materials

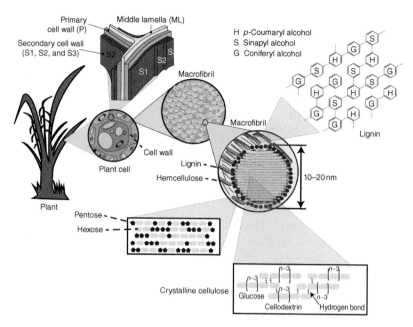

Figure 2.5 Scheme of the microfibril and structure of the plant cell wall [14]

2.3 Chemical Classification of Lignins

Lignin can be classified according to two different criteria. A general classification is based, for instance, on plant taxonomy, depending on which three different categories are considered:

- Gymnosperm lignins (Softwood)
- Angiosperm lignins (Hardwood)
- Grass lignins.

However, this classification has many exceptions, and, therefore, a more robust criterion has been proposed based on a chemical approach. The underlying chemical composition of lignin is closely related to the taxonomy to which it belongs, so that both classifications are somehow related. Hence, lignin from gymnosperms presents more guaiacyl residues, lignin from angiosperms contains a mixture of guaiacyl and syringyl residues, and lignin from grass bears a mixture of all three aromatic residues. On the contrary, Bryophyta[7] species do not contain any lignin.

Within the framework of a chemical classification, "the abundance of the basic phenol units" in the polymer, namely guaiacyl (G), syringyl (S), and *p*-hydroxyphenyl (H), enable lignin to be classified into four main group types known as

- Type-G
- Type-G-S
- Type-H-G-S
- Type-H-G.

Hence, lignin from softwood is composed mainly of moieties derived from coniferyl alcohol (type-G), hardwood lignin contains residues derived from both coniferyl and sinapyl alcohols (type-G-S)

---

[7] The simplest plants that grow on land, composed by three types, for example: mosses, liverworts, and hornworts.

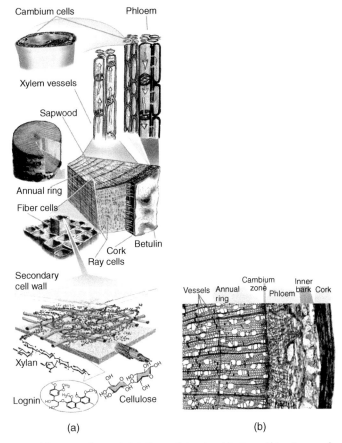

Figure 2.6 Structural features of wood. (a) General structural features;(b) micrograph of birch surface structures. Copyright from ref. [15]. (See insert for color representation of this figure.)

[16], whereas Lignins derived from grasses and herbaceous crops[8] contain the three basic phenol units (type-H-G-S). Among these three, hardwood lignin has a higher content of methoxyl groups on average, which makes this lignin less condensed and more amenable to chemical conversion. Furthermore, the ratio of monolignols in every lignin group is also variable depending on the plant species [17] (see Table 2.1).

2.4 Lignin Linkages

The structural diversity of lignin arises not only from the existence of different monolignols that act as building blocks, but also from the different ways in which these building blocks connect with each other to produce the complex architecture of lignin. It is indeed possible to find several linkages

---

[8] 'Crops' refer to plants that are grown on a large scale for food, clothing, and other human uses. They are nonanimal species or varieties grown to be harvested as food, livestock fodder, fuel, or for any other economic purpose (for example, for use as dyes, medicinal, and cosmetic use). Major crops include sugarcane, pumpkin, maize (corn), wheat, rice, cassava, soybeans, hay, potatoes, and cotton.

18 Lignin and Lignans as Renewable Raw Materials

Table 2.1 Amount of the different monolignols in lignin from various plant types[a]

Scientific name	Common name	p-Coumaryl alcohol ($M1_H$) (%)	Coniferyl alcohol ($M1_G$) (%)	Sinapyl alcohol ($M1_S$) (%)
Coniferous	Softwoods	< 5[b]	> 95	0[c]
Eudicotyledonous	Hardwoods	0–8	25–50	45–75
Monocotyledonous	Grasses	5–35	35–80	20–55

[a]Data from ref. [17, p. 203].
[b]Higher amount in compression wood.
[c]Some exceptions exist (see ref. [18]).

among the monolignol units, which lead to different C–O and C–C intermonomeric bonds in the polymer, as depicted in Figures 2.7–2.9.

These linkages are usually noted as follows:

- The monolignol "drawn" on the right is represented with ′ (prime).
- Carbon linked to the carbon chain has locant number 1.
- Carbons of side chains are described with Greek letters (α, β, and γ) starting with the closest to the aromatic ring.

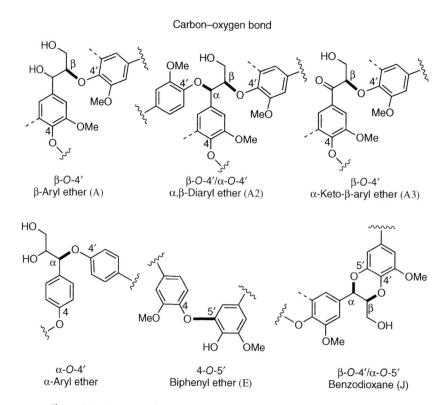

Figure 2.7 Common phenylpropane linkages in lignin (carbon–oxygen bond)

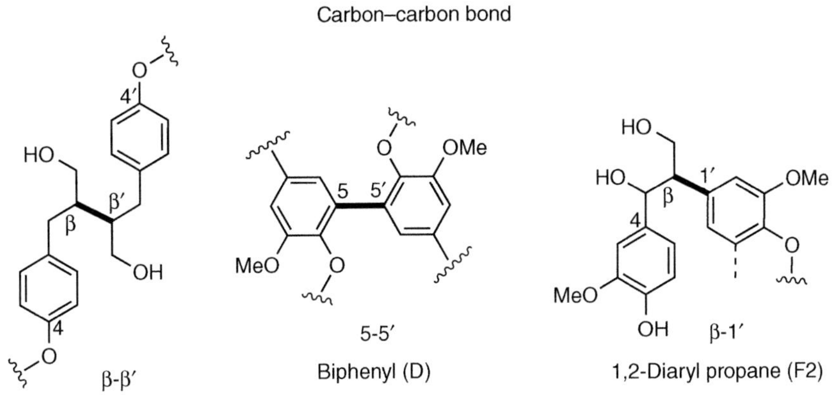

Structure and Physicochemical Properties 19

Carbon–carbon bond

Figure 2.8 Common phenylpropane linkages in lignin (carbon–carbon bond)

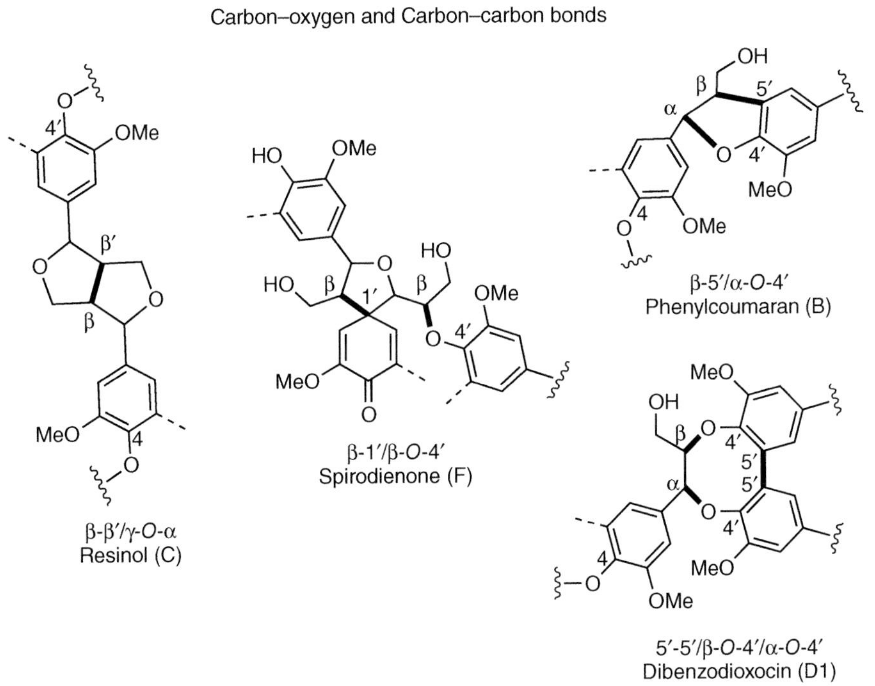

Carbon–oxygen and Carbon–carbon bonds

Figure 2.9 Common phenylpropane linkages in lignin (carbon–oxygen and carbon–carbon bonds)

According to this nomenclature, eight types of bond arrangements are considered:

- Only carbon–carbon bonds: β-β′, β-1′, and 5-5′.
- Only carbon–oxygen bonds: β-O-4′, α-O-4′, and 4-O-5′.
- Carbon–carbon and carbon–oxygen bonds: β-5′/α-O-4′, β-β′/α-O-γ′.

Table 2.2 Percentage of total linkages present in softwood and hardwood lignins

Linkage type	Dimer structure	Percentage of total linkages (%)[a]	
		Softwood	Hardwood
β-O-4'	Phenylpropane β-aryl ether	45–50	60
5-5'	Biphenyl and dibenzodioxocin	18–25	5
β-5'	Phenylcoumaran	9–12	6
β-1'	1,2-Diaryl propane	7–10	7
α-O-4'	Phenylpropane α-aryl ether	6–8	7
4-O-5'	Diaryl ether	4–8	7
β-β'	β-β-linked structures	3	3

[a]Values from ref. [22].

Table 2.3 Types and frequencies of linkages and main functional groups in softwood and hardwood lignins (dilignol/functional groups per 100 ppu) [23]

Dilignol linkages[a]	Softwood	Hardwood	Functional groups[a]	Softwood	Hardwood
β-O-4' (1)	43–50	50–65	Methoxyl (a)	92–96	132–146
β-5/α-O-4' (2)	9–12	4–6	OH$_{phenolic}$ (b)	20–28	9–20
α-O-4'(2')	6–8	4–8	OH$_{benzyl}$ (c)	16	–
β-β' (3)	2–4	3–7	OH$_{aliphatic}$ (d)	120	–
5-5' (4)	10–25	4–10	Carbonyl (e)	20	3–7
4-O-5' (5)	4	6–7	Carboxyl (f)	–	11–13
β-1' (6)	3–7	5–7			
C-6,C-2 (7)	3	2–3			

[a]See Figure 2.10 for letters and numbers meaning.

Apart from these common monomeric linkages, there is another one where three phenolic units are involved. It leads to the so-called dibenzodioxocin structure (α-O-4'/β-O-4"/3'-3") and was discovered in plant lignins by a Finnish group in the mid-1990s. Dibenzodioxocins represent a new lignin structure that was not discovered in many decades of research in softwood cells, and today it is proposed to be the main branching point in softwood lignin [19–21].

The percentage of intermonomeric linkage types in softwood and hardwood lignins has been described by Sjöström [22]. Sjöström, E. In both cases, the mayor linkage is the β-O-4 bond (see Tables 2.2, and 2.3).

An alternative, more abstract description of lignin and its linkages, showing average units and bonds, has been presented by Rodrigues Pinto et al. [23]. This description is less intuitive but closer to the real nature of lignin (see Table 2.3 and Figure 2.10).

2.5 Structural Models of Native Lignin

For many years, researchers have been looking for an accurate model to describe native lignin.[9] However, its complex nature and inherent difficulties in its analysis have made it difficult to identify

---

[9] A lignin isolated in such a way that the solvent does not react with the lignin or alter it in any way.

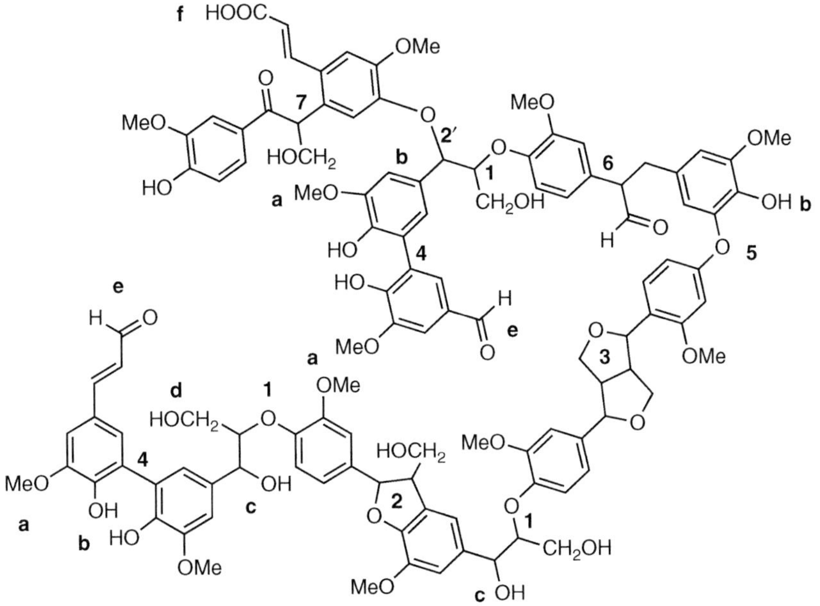

Figure 2.10 Scheme of a hardwood lignin fragment, with the notation of linkages and functional groups of Table 2.3 [23–27]

a complete structure of a lignin molecule, hampering the development of an accurate model. The most common methodology employed for that purpose is based on the hydrolysis of lignin and the subsequent analysis of the fragments obtained from the degradation process. The methods used for the detection and isolation of the fragments, and the diversity of native lignin depending on the vegetable species, contribute to the uncertainty concerning its structure. Nevertheless, several models have been proposed over the years, being revisited when new analytical methods are implemented. For many decades, lignin has been considered a cross-linked network polymer, and lately this perception has slightly changed.

2.5.1 Softwood Models

In softwood[10] lignin gymnosperms the dominant linkage is the β-O-4′ one. Several models for this kind of lignin have been described. Neish [28] proposed the first one for softwood lignin in 1968 based on the experimental data available at that time for spruce lignin. In this model, 18 units of monolignols are represented assuming that more than 100 units formed the native state [16] (see Figure 2.11).

In 1974, Glasser and Glasser [29] developed a structural model of softwood lignin by means of a mathematical simulation of the oxidative coupling reactions of main monolignols. They provided a model for the polymer with an M_w of approximately 14 000. Figure 2.12 shows a sketch of the structure where 70 monolignols units are included.

Later, in 1977, Adler described a new model with 16 C_9 units derived from the oxidative degradation of spruce lignin, in which these units were distributed according to the most reliable analytical data. It was a partially limited model because certain units and linkages were not considered exactly due to the arbitrary choice of the sequence of units, in which proportions of certain structural details

---

[10] Botanically softwoods are gymnosperms or conifers. Anatomically, softwoods are nonporous and do not contain vessel. Typically, softwoods are cone-bearing plants with needle- or scale-like evergreen leaves.

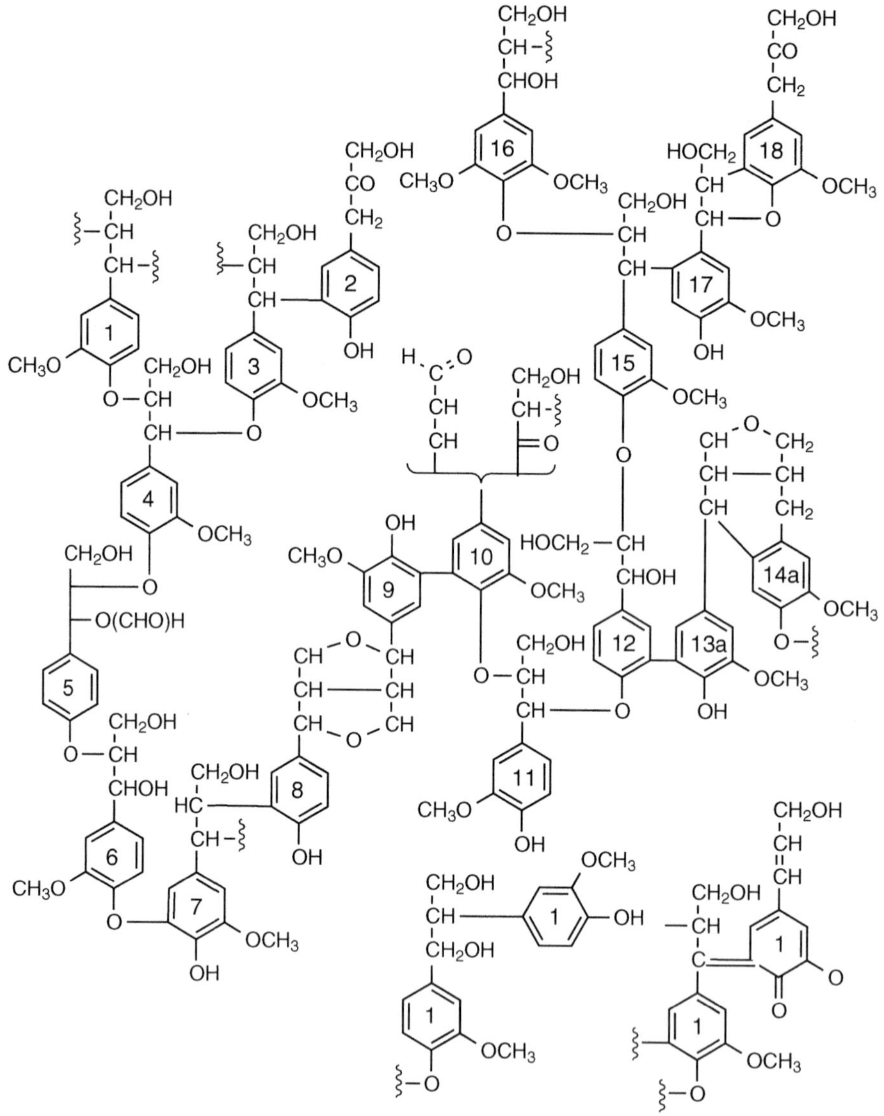

Figure 2.11 Lignin according to Neish [28]

were not accounted for in a complete way [27]. Despite this limitation, Adler's model is among the most extensive lignin structural ones and it has been extensively used (see Figure 2.13).

In 1980, Sakakibara [30] proposed a structural model of softwood lignin (see Figure 2.16) according to degradation products resulting from hydrolysis with dioxane/water and catalytic hydrogenolysis. In this work, 39 hydrolysis products were identified (see Figures 2.14–2.17).

In 1995, Karhunen et al. [19, 20] reported a new eight-membered ring (dibenzodioxocin) linkage in softwood lignin. This linkage was found by 2D NMR techniques and is now proposed to be the main branching point in softwood lignin [21].

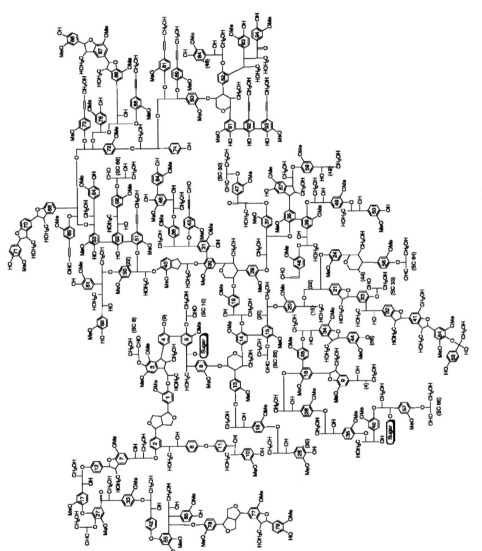

Figure 2.12   Glasser and Glasser lignin model [29]

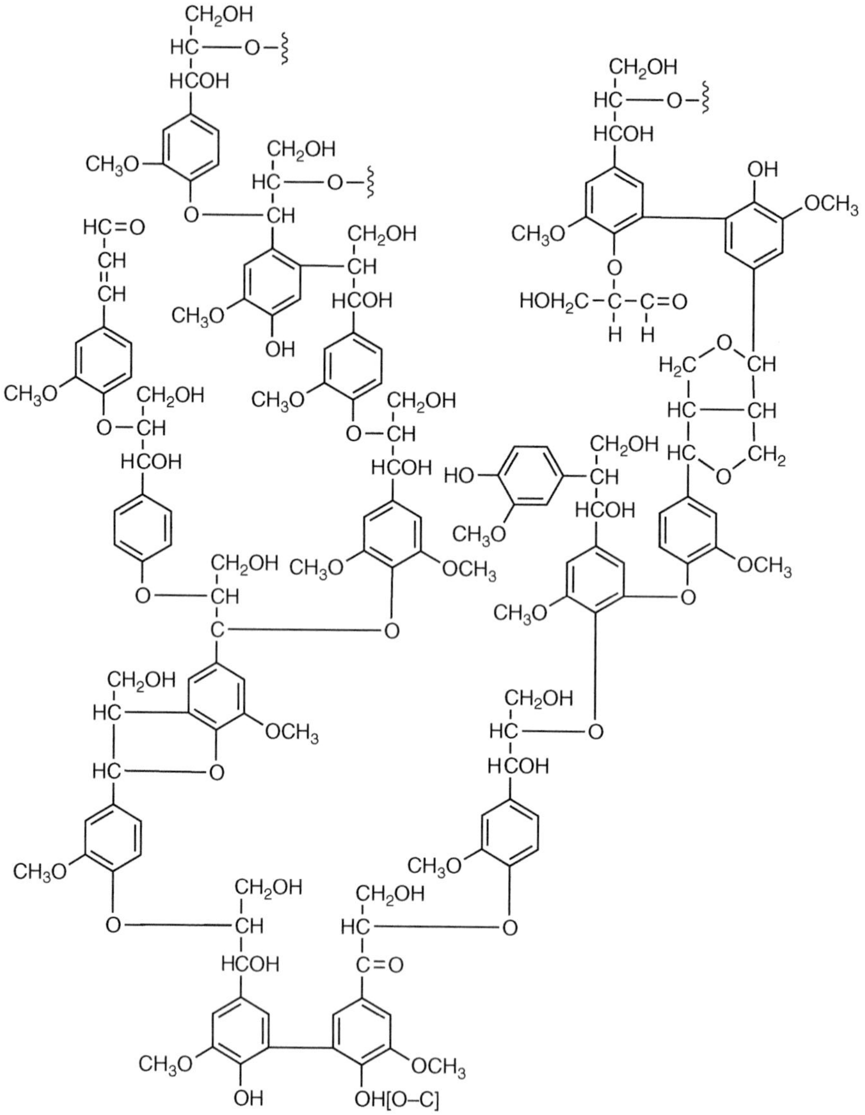

Figure 2.13 *Lignin according to Adler [27]*

More recently, in 2001, Brunow [31] developed a softwood-lignin model based on spruce wood data, formed with 25 units of monolignols, using a color code for every unit, bold black bonds for radical coupling linkages, and gray bonds for post-coupling internal re-aromatization reactions (see Figure 2.18). This model attempts to define more than the primary structure related to the coupling sequence and idea of the main linkages and the average of them. Softwood lignin is relatively branched by 4-*O*-5′ units, and dibenzodioxocin units.

This model also includes glycerol and cinnamaldehyde end units. Initially, it was thought that glycerol units in particular were formed during the isolation of lignin in the milling step, but it has been found in biomimetically synthesized polymers, which have not undergone this physical treatment.

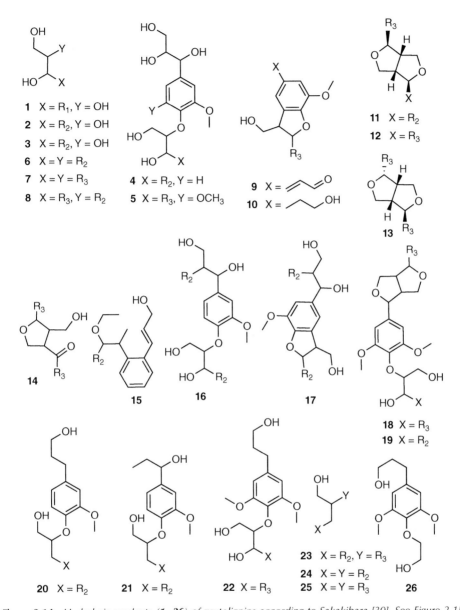

Figure 2.14 Hydrolysis products (**1–26**) of protolignins according to Sakakibara [30]. See Figure 2.15 for the meaning of R1-R3.

Another additional problem that arises with these models is that although the native lignin is an achiral material, the structure shown in Figure 2.18 has 46 chiral centers, and the optical centers must be generated randomly from the 17 billion physically distinct isomers due to the relative stereochemistries of pairs of centers in ring structures such as phenylcoumarans, resinols, and dibenzodioxocins [32]. Therefore, the proposed softwood models increase in complexity, given the possibility of an enormous numbers of possible isomers.

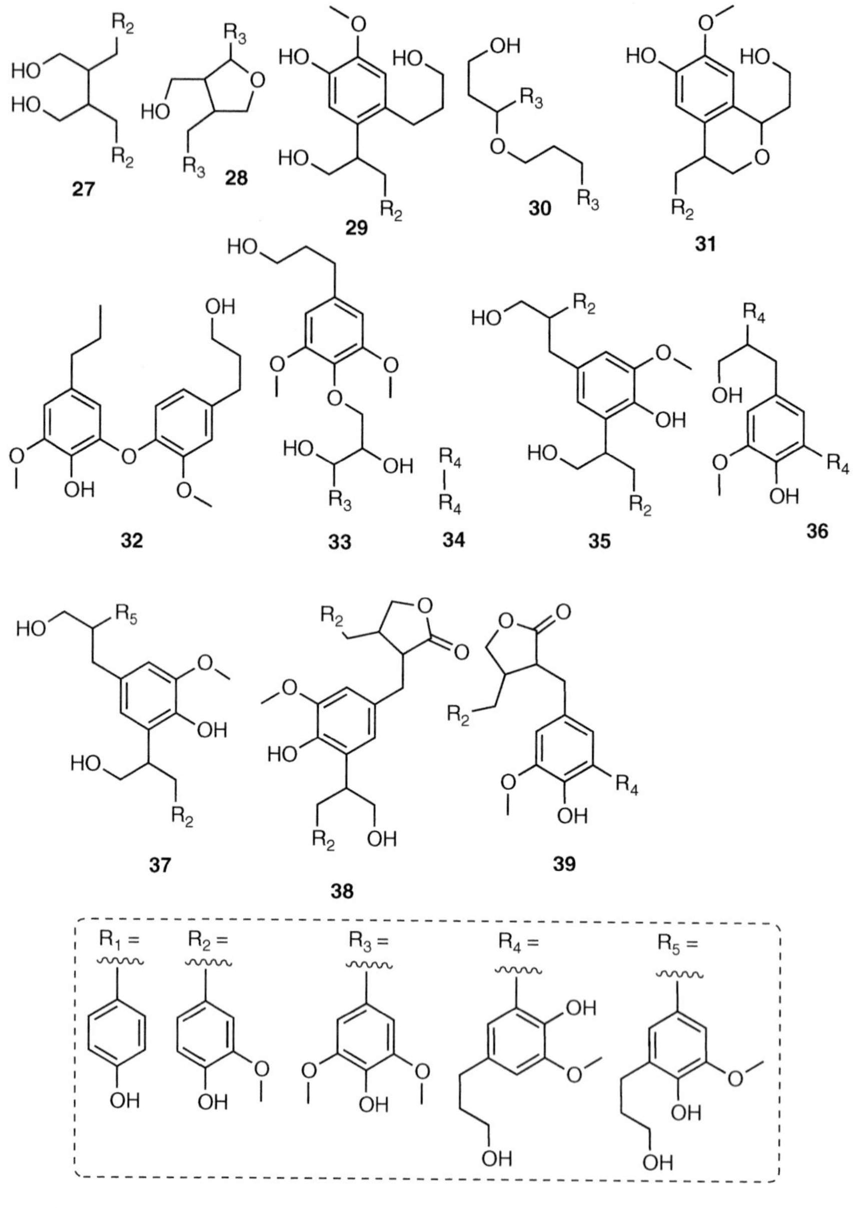

Figure 2.15 Hydrolysis products (27–39) of protolignins according to Sakakibara [30]

For milled softwood lignin, Crestini et al. [33], according to their experimental data, in 2011 proposed a model based on supramolecular aggregates of linear oligomers rather than a network polymer (see Figure 2.19). This model was previously suggested by Wayman and Obiaga [34], in 1974, but on that occasion, the model was supported with the current analytical techniques.

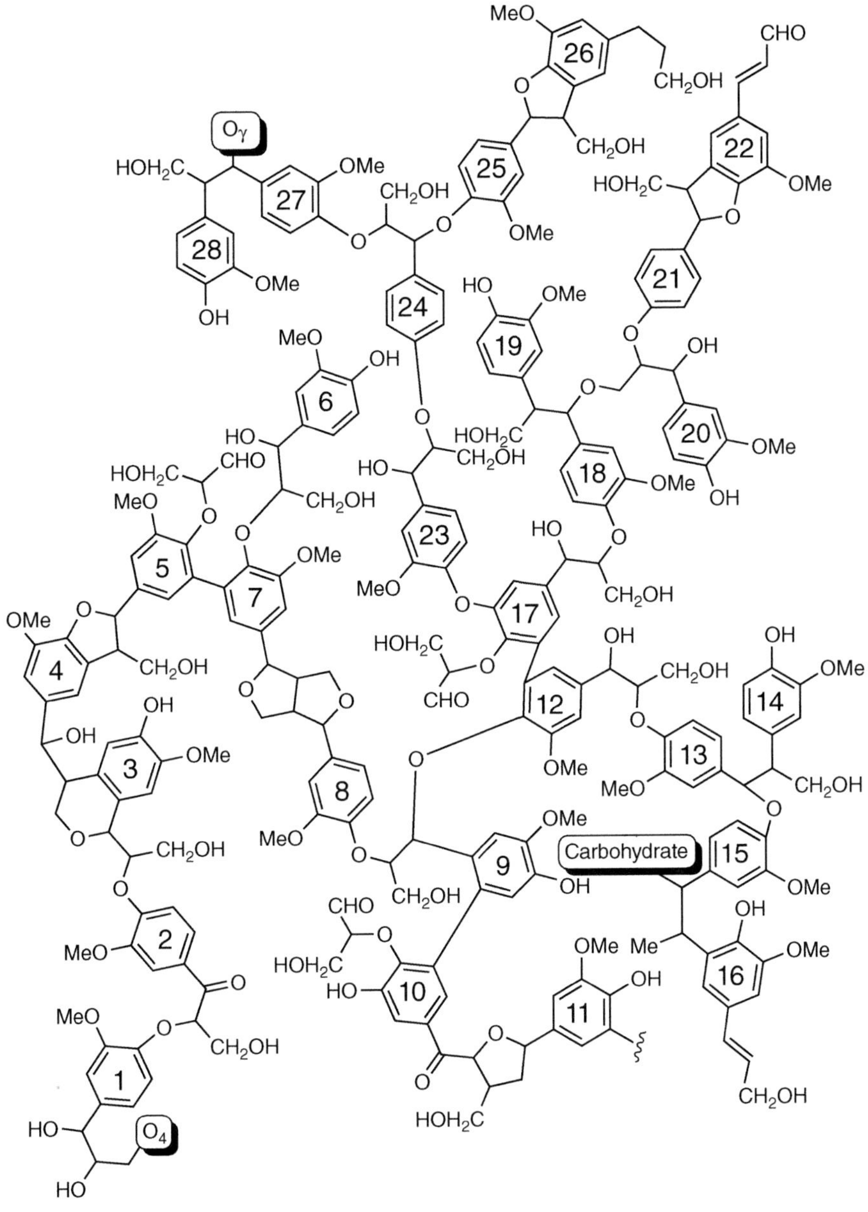

Figure 2.16 Sakakibara structural model for softwood lignin [30]

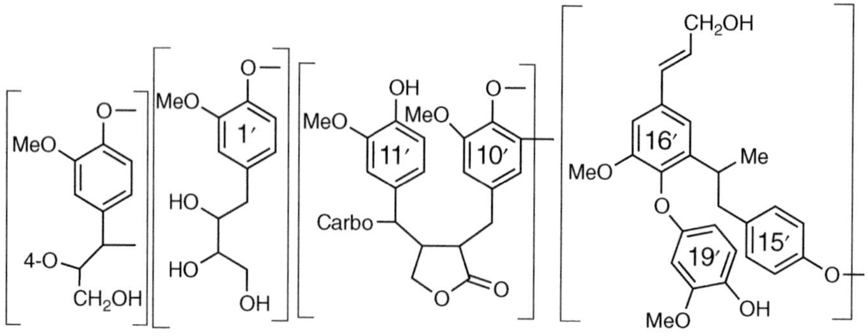

Figure 2.17 Other fragments in the Sakakibara structural model for softwood lignin [30]

2.5.2 Hardwood Models

The structure of hardwood[11] lignins varies greatly from one species to another. The major difference is the ratio of syringylpropane to guaiacylpropane units (G/S ratio). Hardwood lignin contains a higher proportion of sinapyl units, which results in a considerable percentage of unevenly distributed linear lignin. Typical composition of hardwood lignin is about a 100:70:7 ratio of coniferyl alcohol, sinapyl alcohol, and *p*-coumaryl alcohol, respectively.

One of the first models of hardwood lignin was proposed by Nimz [35], in 1974, for beech wood (see Figure 2.20). Later, Boerjan *et al.* [36] suggested, in 2003, a model for hardwood lignin based on poplar wood. This model is formed by 20 units of monolignols less branched than the model proposed for softwood lignin (see Figure 2.21). On the other hand, several acylated units have been detected in many hardwoods, for example, *p*-hydroxybenzoates (PB) in poplars [37] or acetylated units.

All these data agree with the complexity and variability of lignin, and therefore, the softwood and hardwood lignin models are only approximations to the linkage types and their approximate relative nature and frequencies.

2.5.3 Herbaceous Plant Models

The first studies on lignin in herbaceous plants have shown that it presents greater variability according to the species, the part of the plant, and also the isolation method. In 1970, Simionescu and Anton [38] proposed a scheme of a fragment of the Brauns lignin of reeds consisting of 14 phenylpropane units (see Figure 2.22).

The main type of bond present were alkyl-aryl ether (α-*O*-4 and β-*O*-4 bonds) [39]. Semiempirical formulas of lignin in herbaceous show this diversity (see Table 2.4).

Some remarkable facts can be pointed out. For example, one difference of lignin from many herbaceous plants such as sisal, kenaf, abaca, or curaua is that these types of lignins are extensively acylated, especially at the Cγ of the lignin side chain (up to 80% acylation) with acetate and/or *p*-coumarate groups and preferentially over syringyl units. The structure of these acetylated lignins can be essentially regarded as syringyl units linked mostly through β-*O*-4' ether bonds. The lignin polymer for these herbaceous plants is rather linear and unbranched [40] (see Figure 2.23).

---

[11] Botanically hardwoods are angiosperms. Anatomically, hardwoods are porous; that is, they contain vessel elements. Typically, hard woods are plants with broad leaves that lose their leaves in autumn or winter.

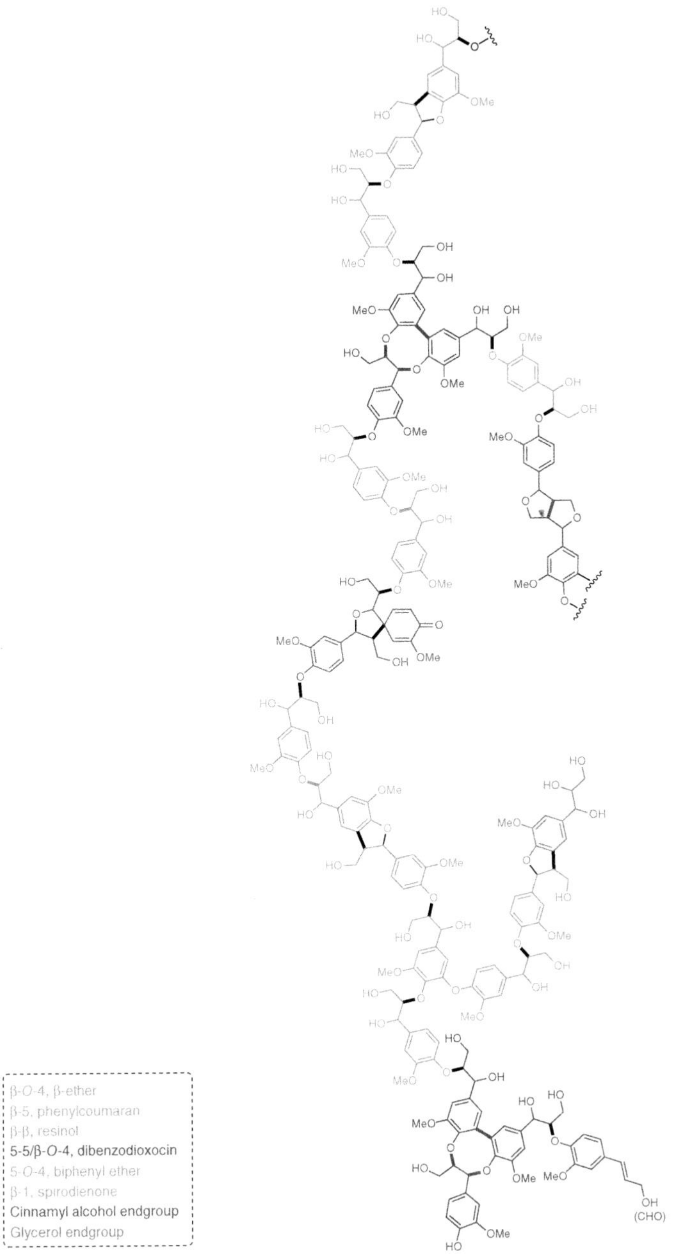

Figure 2.18 Spruce lignin model proposed by Brunow [31]. (See insert for color representation of this figure.)

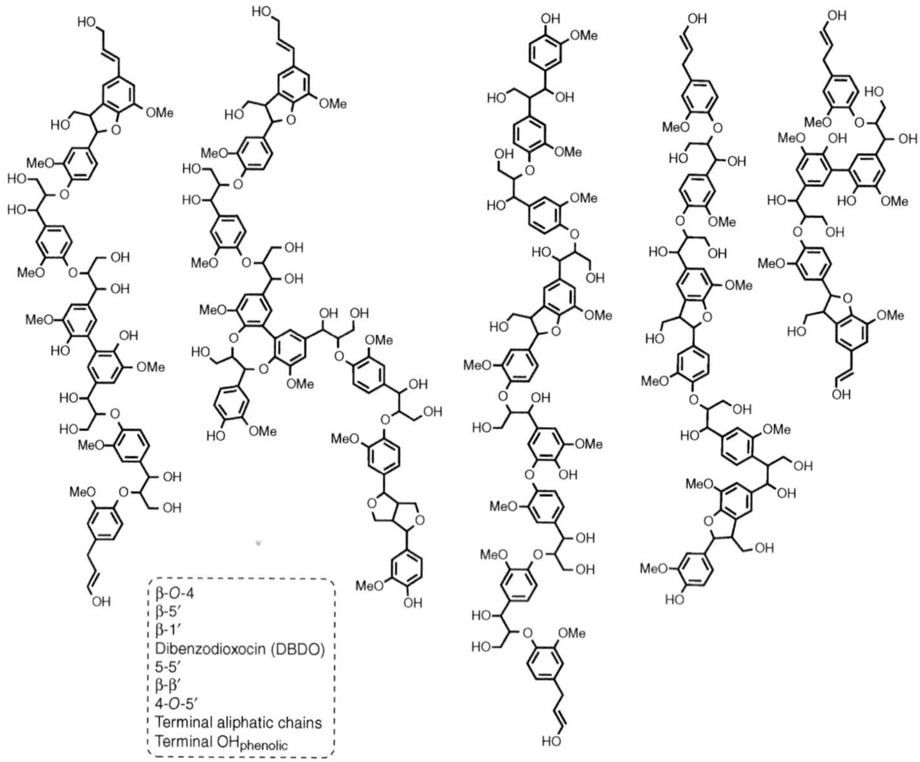

Figure 2.19 Milled softwood-lignin model proposed by Crestini et al. [33]. (See insert for color representation of this figure.)

In wheat straw, Banoub and Delmas [41] have characterized lignin moieties extracted using the AVIDEL methodology[12] by combined techniques such as atmospheric pressure chemical ionization mass spectrometry (APCI-MS), tandem mass spectrometry (MS/MS), and matrix-assisted laser desorption/ionization time-of-flight mass spectrometry (MALDI-TOFMS). With these techniques, fragments that range from dimeric structures to octameric ones were detected (see Figure 2.24).

In conclusion, unlike softwood and hardwood lignins, for herbaceous plants, there is no commonly accepted structural model that accurately describes these lignins, mainly due to the aforementioned diversity of lignin types in herbaceous plants. The most detailed structural data come from the composition analysis of the fragments and from the study of some isolated fragments.

2.5.3.1 Genetically Modify Species

In recent years, a research line has been opened in order to develop genetically modified plant species with several purposes. One of the most remarkable goals is to produce lignins that are easily separable from cellulose and other components from original vegetable biomass for paper industry or for biorefineries and biofuels factories. Today, this difficult separation is a limiting factor for the development of this area to compete under advantageous conditions with the petrochemical industry. Transgenic species show some differences in lignin structure with respect to natural plants. These differences

---

[12] Selective separation of the main constituents of vegetable matter by treating with a formic acid / acetic acid / water mixture at atmospheric pressure and slightly raised temperature.

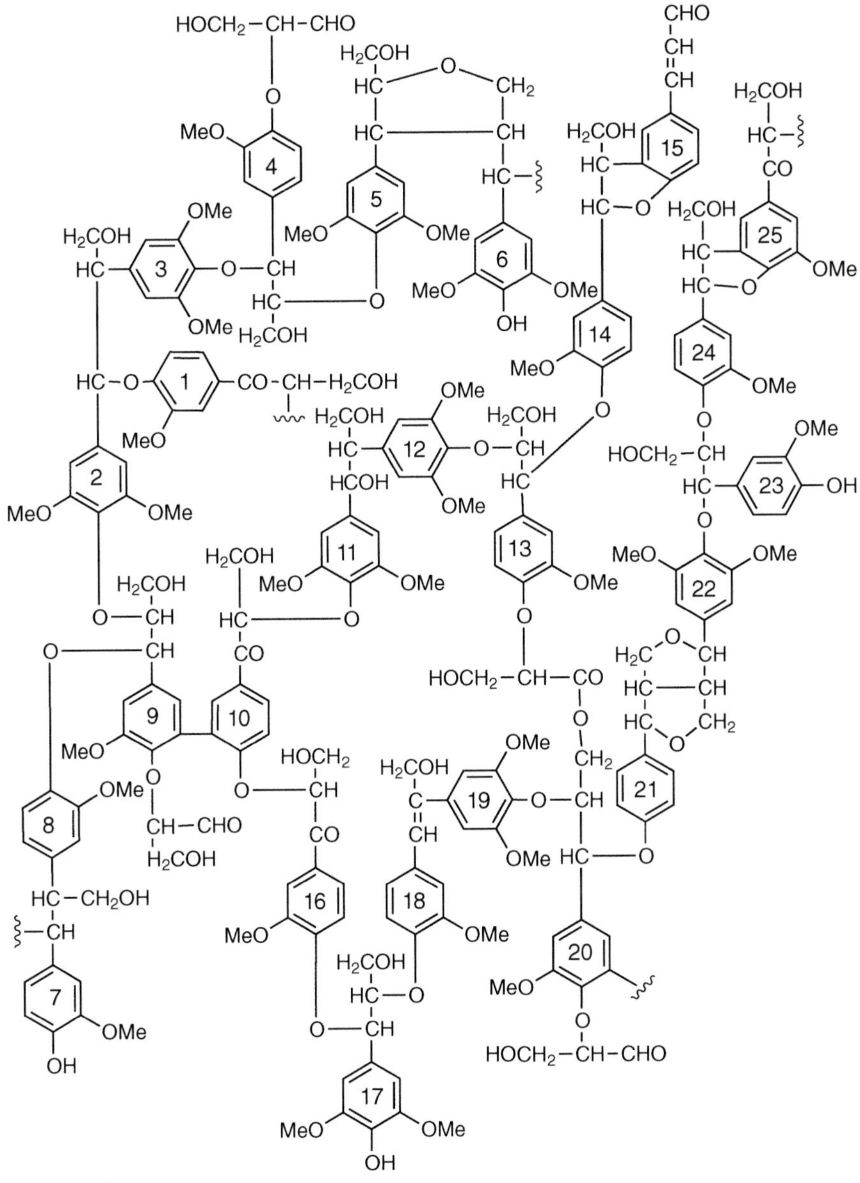

Figure 2.20 Structure of beech wood lignin proposed by Nimz [35]

consist of several average basic monolignols and in some cases of bond types between monolignols and the way lignin is linked to carbohydrates (see Figure 2.25).

One example of these lignins has been modeled by such authors as Stewart et al. [43] (see Figure 2.26). Poplar lignins, extremely rich in S units, are produced by cinnamate-4-hydroxylase/5-hydroxylase (C4H::F5H) overexpression. The plants analyzed appear to have the highest S content on record, with some 97.5% of the lignin deriving from sinapyl alcohol. The S/G average of

32 Lignin and Lignans as Renewable Raw Materials

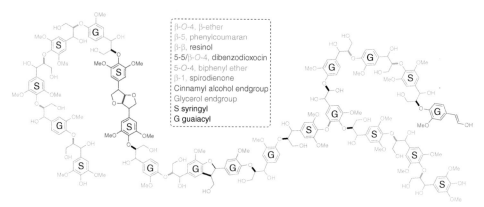

Figure 2.21 Hardwood lignin model proposed by Boerjan et al. [36]. (See insert for color representation of this figure.)

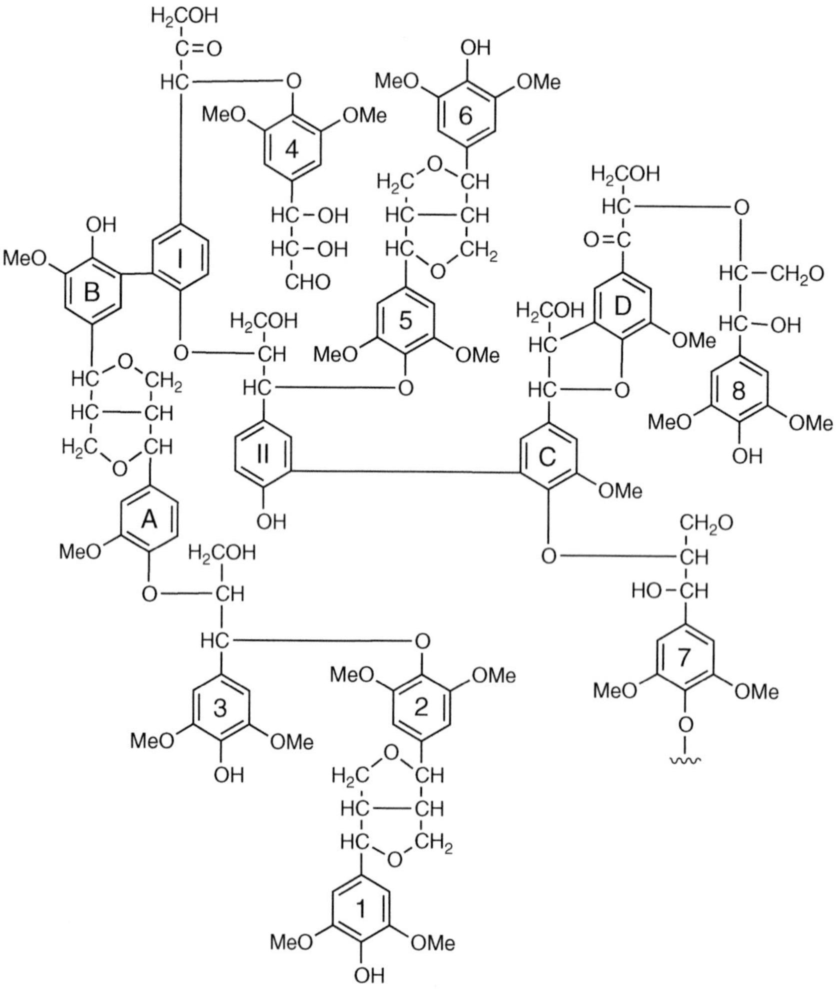

Figure 2.22 Structure of a fragment macromolecule of reed stem lignin [38]

Table 2.4 Semiempirical formulas of the lignins of some herbaceous plants[a]

Sample (family)	Semiempirical formula $C_9H_xO_y$							
	x	y	OCH_3	OH_{ph}	OH_{al}	O_{CO}	O_{ar-al}	OOH
Gramineae								
Rice husk DLA	7.57	1.13	0.93	0.33	0.89	0.31	0.67	
Rice stem DLA	8.03	1.53	0.97	0.39	0.84	0.32	0.61	
Reed stem DLA	7.82	1.52	1.11	1.21		0.21		
Reed stem DLA	6.46		1.19	0.64	0.32	0.25	0.52	0.14
Wheat straw MGL	6.99	2.76	1.11					
Rye straw DLA	11.15	1.76	0.99	0.16	1.12	0.20		
Malvaceae								
DLA of Tashkent-1 cotton plant stems, early period	7.67	2.01	0.53	0.66	0.33	0.44	0.34	0.03
DLA of Tashkent-1 cotton plant stems, early period	6.24	1.59	0.78	0.73	0.7	0.18	0.37	0.08
DLA of Tashkent-1 cotton plant stems, ripe stems	6.22	1.21	0.82	0.63	0.88	0.46	0.37	0.17
DLA of Tashkent-60 cotton plant	7.96	0.62	0.99	0.28	1.33	0.31	0.72	0.06
DLA of C-6030	8.7	0.59	1.18	0.29	1.19	0.32	0.71	0.24
Althea rhyticarpa DLA	7.6	1.0	1.07	0.43	0.93	0.52	0.57	
Hollyhock DLA	6.55	0.99	1.15	0.31	0.85	0.3	0.69	0.03
Althea nudiflora	6.84	1.21	1.12	0.38	0.93	0.41	0.62	0.08
Kenaf stem DLA	6.9	0.9	1.34	0.25	1.01	0.17	0.75	0.02
Kenaf phloem DNA	6.97	0.73	1.37	0.34	0.88	0.18	0.66	0.026
Kenaf tow DLA	7.14	0.67	1.39	0.27	0.92	0.18	0.73	0.02
Linaceae								
Flax tow DLA	10.74	1.29	0.87	0.14	1.09	0.19		
Tiliaceae								
Jute stem DLA	7.78	2.96	1.27					
Jute stem DLA	8.33	3.52	1.26					
Castor-oil plant stem DLA	7.8	0.68	1.18	0.19	1.05	0.53	0.81	0.03

[a] Data obtained from refs [39, and references therein].
[b] OH value.
[c] From COOH group.

approximately 38 suggests that the lignin chains are predominantly linear, and this is supported by comparative NMR determination of isolated lignin fractions. As might be anticipated from the substantial compositional shifts documented between the control and the transgenic trees, the structure of the polymer can be dramatically altered with this methodology, with substantial differences from natural native lignins of nontransgenic species. Whereas the β-ether link remains predominant,

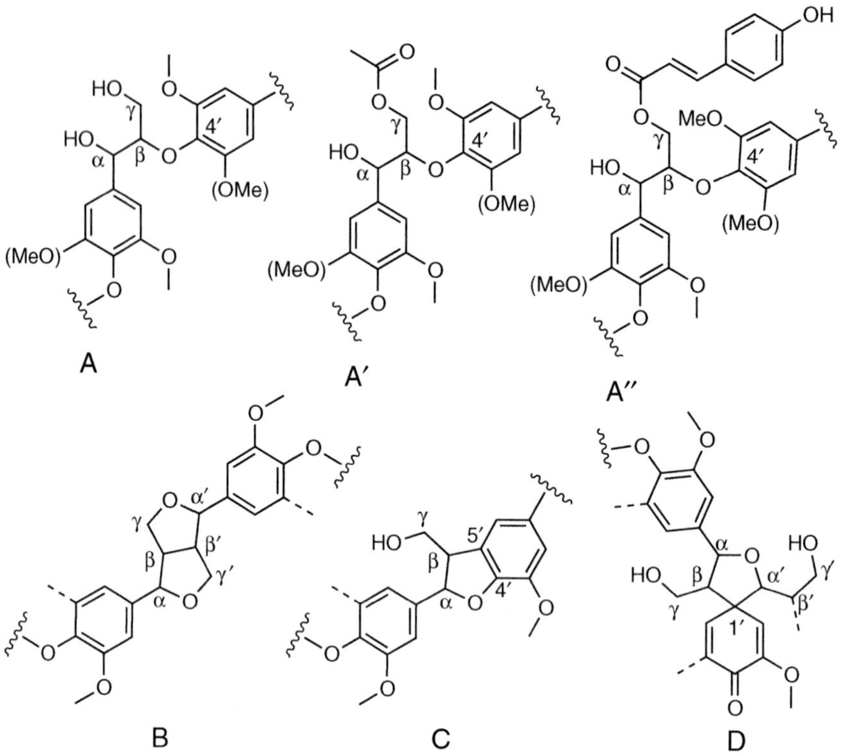

Figure 2.23 *Main structures present in the highly acylated lignins [40]. (A) β-O-4' linked substructures; (A') β-O-4' linked substructures with acetylated γ-carbon; (A') β-O-4' linked substructures; with p-coumaroylated γ-carbon; (B) phenylcoumaran structures formed by β-5' and α-O-4' linkages; (C) resinol structures formed by β-β', α-O-γ', and γ-O-α' linkages; and (D) spirodienone structures formed by β-1', and α-O-α' linkages*

phenylcoumarans units become rare in this lignin type (at high S levels) and spirodienones S increase. An unexpected finding was that β-ether A levels were not higher at the extreme S levels (compared the native S levels common to wild-type trees). The contribution from β-β' of resinols units was compensatorily higher. This fact suggests more dihydrodimerization in the polymerization process so that consequently shorter chain lengths result in the modified lignin, as confirmed by a fuller NMR analysis. These results remain consistent with the accepted theory of lignification based on combinatorial radical coupling reactions under simple chemical control.

2.6 Lignin–Carbohydrate Complex

The term lignin–carbohydrate complex (LCC) was first introduced by Björkman (1954). Lignin and polysaccharides interact in vascular plant cell walls, forming a stable aggregate that has been denominated LCC consisting of sugar chains and relatively small lignin fragments attached as pendant side chains [44]. The occurrence of stable lignin-carbohydrate bonds creates significant problems

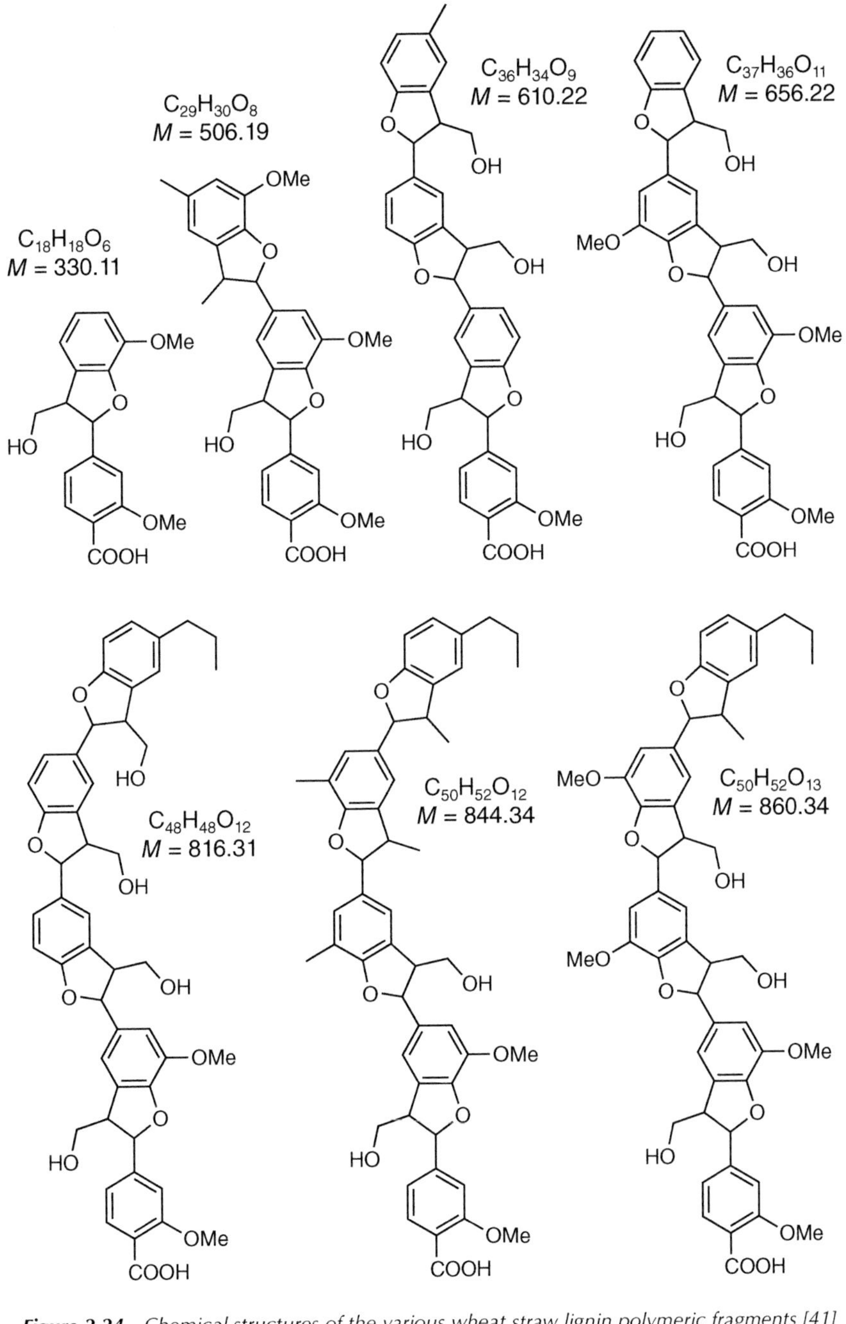

Figure 2.24 Chemical structures of the various wheat straw lignin polymeric fragments [41]

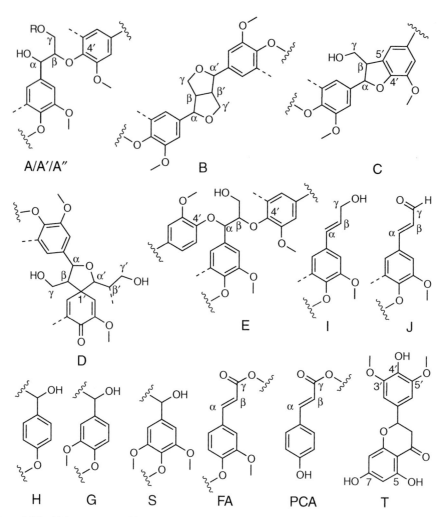

Figure 2.25 Main structures of lignin fractions of Arundo donax, involving different side-chain linkages, and aromatic units identified by 2D HSQC NMR [42]. (A) β-O-4' linkages; (A') β-O-4' linkages with acetylated γ-carbon; (A') β-O-4' linkages with p-coumaroylated γ-carbon; (B) resinol structures formed by β-β', α-O-γ', and γ-O-α' linkages; (C) phenylcoumaran structures formed by β-5' and α-O-4' linkages; (D) spirodienone structures formed by β-1' and α-O-α' linkages; (E) α,β-diaryl ether substructures; (H) p-hydroxyphenyl unit; (G) guaiacyl unit; (S) syringyl unit; (I) cinnamyl alcohol end-groups; (J) cinnamyl aldehyde end-groups; (FA) ferulate; (PCA) p-coumarate; (T) tricin

in selective separation and isolation of lignin and carbohydrate preparations from biomass[13]. Three factors determine the formation of LCC linkages [45]:

- The functional groups present in lignin and carbohydrate polymers (the primary and secondary alcohols, carboxyl and carbonyl groups).

---

[13] All purified holocellulose materials contain a certain amount of lignin, and all purified lignin fractions contain a certain amount of monosaccharides.

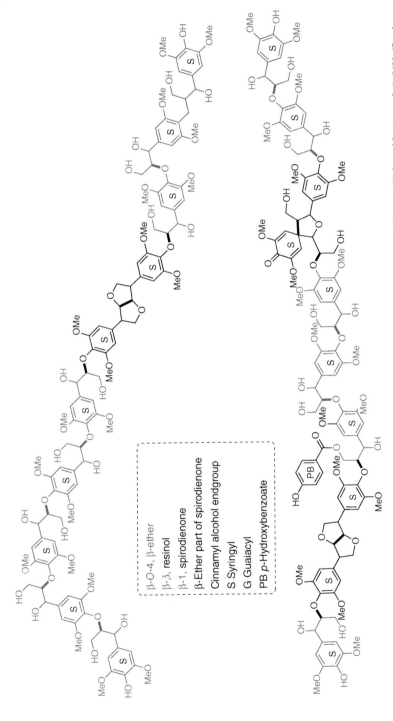

Figure 2.26 Representation of lignin polymers from the C4H:F5H-up-regular transgenic trees, as predicted from NMR-based lignin analysis [43]. (See insert for color representation of this figure.)

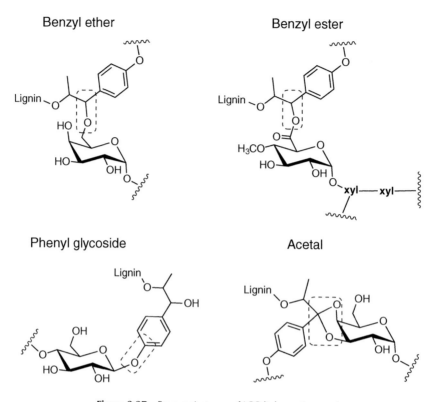

Figure 2.27 Four main types of LCC linkages in wood

- The enzymes involved in cell wall architecture (mainly glycosyl-transferases, laccases, and peroxidases).
- The free radicals formed in lignin polymerization that are so reactive that they can react with carbohydrates.

Although there are many possibilities for the formation of LCC linkages in the cell walls, investigation on wood yield of lignin fragments covalently bound to sugars [46–50], and the experimental findings support that LCC linkages in wood are mainly phenyl glycoside bonds [51], benzyl esters [47], benzyl ethers [52], and acetal bonds [53]. Thus, carbohydrates from LCC are basically hemicellulose constituents (see Figure 2.27).

Hemicellulose is a heteropolymer formed by pentoses, mainly xylose and arabinose, and hexoses such as glucose, galactose, mannose, and sugar acid derivatives. Hemicellulose, unlike cellulose, is easily hydrolyzed by a diluted acid or base as well as a myriad of hemicellulase enzymes. However, cellulose is crystalline, strong, and resistant to hydrolysis, while hemicellulose has a random, amorphous structure with little strength (see Figures 2.5 and 2.6). Hemicelluloses can be classified into four types based on main type of sugar residue [54]:

- Xylans present in secondary cell walls of hardwood and herbaceous plants.
- Xyloglucans present in primary cell walls of higher plants and bound to cellulose.

- Mannans present in secondary cell walls of conifers (softwood) and Leguminosae.
- Mixed linkage β-glucans that are common in poales[14] and Pteridophytes.[15]

The information about LCC linkages is usually gained by indirect methods, such as isolation by enzymatic treatments with hemicellulases or cellulases and with model compound studies. Recently, Balakshin *et al.* [55] have quantified the nature of LCC linkages with high-resolution NMR spectroscopy in crude milled wood lignin extracted with acetic acid (LCC-AcOH) and cellulolytic enzyme lignin (CEL) from loblolly pine (*Pinus taeda*) and white birch (*Betula pendula*). With this methodology, different amounts of benzyl ether, benzyl ester, and phenyl glycoside LCC bonds have been characterized, but acetal units have not been detected, probably because of the hydrolysis of such moieties under acidic conditions.

In certain species, this LCC has been investigated. For example, Du *et al.* [56] suggested several groups of LCC in spruce wood, and Terashima *et al.* [57] have also proposed these supramolecular structures for lignin associated with carbohydrates in softwood.

LCCs are very difficult entities to separate into their respective components by chemical treatments or special purification techniques. Therefore, LCCs use to accompany cellulose or lignin in paper industrial processes. This is a technical problem when a high-purity material is needed. The knowledge of the structure and properties of LCC is key to avoid this problem or to improve the applications of this wood derivative.

On the other hand, lignin molecules in solution are approximately spherical particles and slightly solvated with solvent [58]. In living plant tissue, lignin exits in an excess of water, forming complex structures to carbohydrate polymers. Lignin is hydrophobic compared to carbohydrate polymers in plant tissue. However, the number of hydroxyl groups attaching to lignin is sufficient to act as a reaction site for hydrogen bond formation with water molecules.

2.7 Physical and Chemical Properties of Lignins

In this section, we briefly discuss those properties of lignins that are intimately related to structure and not those related to the detection, isolation, or characterization of lignins, which are discussed in Chapters 3, 5, and 6.

2.7.1 Molecular Weight

For the determination of lignin structures, the most commonly used parameter is the weight-average molar mass, also called molecular weight (M_w). Despite that these are conclusive data for most natural lignins, the use of M_w to characterize the lignin type is still under debate. Thus, for example, it is common to refer to values of "number-average molar mass" (M_n) and the M_w, but these do not have an univocal acceptance. This controversy comes from the different aspects, such as the origin of lignins, depending on the plant species, the isolation methods, and the diverse analytical techniques used by researchers. Therefore, in many cases the results are discordant (see Section 6.3).

The M_w of isolated lignin has been determined by the usual techniques. The main problem, when the solution properties of isolated lignin are measured, is the solubility in organic solvents. The M_w of soluble portion of lignin is smaller than that of the truly isolated sample. The different techniques

---

[14] Large order of flowering plants in the monocotyledons, which includes grasses.
[15] Vascular plants that do not have flowers and seeds and reproduce by spores.

suggest that the M_w of isolated lignin is in the 10^3–10^5, range depending on the plant species, the processing method, and also the measuring method.

The average M_w for softwood is greater than 2×10^4, and less for hardwoods. For lignosulfonates, the measured M_w is as high as 10^6.

2.7.2 Dispersity Index (Đ)

The dispersity index[16] is a measure of the width of molecular weight distributions (MWD) [59] and is expressed as

$$Đ_M = \frac{M_w}{M_n}; \qquad (2.1)$$

where M_w is the weight-average molar mass (or molecular weight) and M_n is the number-average molar mass [60].

A considerable amount of experimental data support the idea that soluble lignins and their derivatives show remarkable polydispersity [61–67]. The $Đ_M$ may vary from 2.5–3.0 to over 10–11 for different lignins types [68], indicating that lignins are extremely polydisperse materials. Fractionation studies on lignosulfonates [69] and alkali lignins [63, 70] have shown that they contain M_w ranging from 1000 to more than a million [71, p. 189]. Glasser et al. [68] have produced a universal plot of $Đ_M$ versus M_w for a series of prototype lignins supplying a linear expression with some predictive power.

2.7.3 Thermal Properties

The glass transition temperature (T_g) is one of the most widely used thermal properties for describing amorphous polymers. T_g curves for most lignins have indicated that mass slightly decreases at temperatures between 0 and 150 °C. This loss can be attributed to vaporization of residual water in the samples that are submitted to this test. The solid-state molecular motion for lignins has been evaluated using several experimental techniques such as dilatometry [72], IR spectroscopy [73], viscoelastic measurements [74–77], thermal analysis [75, 78–81], nuclear magnetic relaxation in both broadline and pulsed NMR [75, 82], and dielectric [83], piezoelectric, [84] and acoustic measurements [85] (see Table 6.6 in page 149).

As examples of the evaluation of this property, it has been found that the thermal decomposition of samples starts at approximately 200 °C. The extrapolated decomposition temperatures, for Kraft lignin, alkaline lignin and lignins from sawdust sample, have been observed at approximately 321 °C, 294 °C, 324 °C, and 331 °C, respectively, for pulping lignins. By contrast, T_g of the wood powder was observed at approximately 370 °C and 390 °C. Accordingly, in the derivative T_g curves of the wood powder, the peak at 390°C was attributed to lignin and a shoulder at 370 °C was attributed to cellulose. The residual mass of lignin at 500 °C was higher than that of wood powder [86].

The results from the aforementioned measurements suggest that the molecular motion of lignin is similar to those of synthetic polymers with phenyl groups in the main chain. However, because of the presence of hydroxyl groups in the molecular chain of lignins, the molecular motion of lignin is markedly influenced by hydrogen bonds to water molecules. It has also been suggested that the molecular properties of *in situ* lignin are affected by coexisting polysaccharides through the formation of LCCs.

These results are in accordance with the variety of chemical components and complex higher order structure of lignins, this being broader than in another synthetic amorphous polymers with similar M_w. When polar ionic groups are introduced into the molecular chain, molecular motion of polyelectrolyte lignin is markedly restricted. Molecular motion of lignin is induced by the phenylpropane units in

---

[16] IUPAC has deprecated the use of the term polydispersity index having replaced it with the term dispersity, represented by the symbol Đ (pronounced D-stroke).

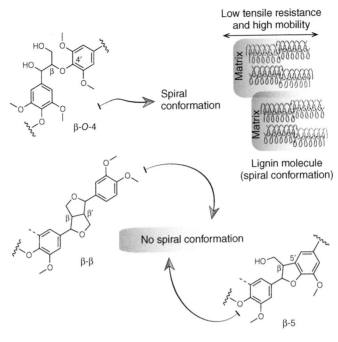

Figure 2.28 Sketch of a structural spiral motif of lignin molecule with β-O-4, β-β, and β-5 [90]

the molecular chain and bulky side-chain molecules and slight cross-linked bonds establish intra- and intermolecular chain. Rigid groups in the main chain and cross-linking restrict molecular motion and have the effect of increasing the T_g.

In the case of acetylated organosolv lignin, there is a weak change of T_g at 122 °C. On the other hand, in the nonacetylated organosolv lignin, a clear change in the curve has not been detected. These T_g values are in the glass transition region reported in the literature for several types of lignins, between 110 and 150 °C [87, 88]. As described in the literature, polymer plastification decreases the T_g value. This study shows that the acetylated lignins exhibit lower T_g values probably due to the reduction in the degree of intermolecular association occurring within the lignin. Lignin-esterified derivative decreases the availability of OH groups in the lignin structure and the addition of nonreactive acetyl groups can also make the chains stand off from one another easily, reducing T_g [89].

2.7.4 Solubility Properties

In 1962, Goring [92] pointed out the two broad aspects to the physical behavior of lignin. The first aspect concerns the solubility of lignin in several solvents. One remarkable fact is that unmodified native lignins from biomass are insoluble in most common solvents. For proper solubilization, some physicochemical transformations must be made. It is necessary to obtain lignins fragments with lower M_w in order to solubilize them. The study of the solubilized aggregates with different techniques yields information on the size and configuration of the macromolecule in solution. The second deals with the bulk properties of solid or solvent-swollen polymers.

Native lignin is insoluble in most of conventional organic solvents [93]. Therefore, solubility properties of lignins can be altered if they undergo a pretreatment. In this sense, Björkman [94] found in 1956 that when spruce wood is milled thoroughly, more than 50% of the lignin residue could be

Table 2.5 Guaiacyl/syringyl ratio (G/S) in the lignins of birch cells and their structural and rigidity level, respectively[a]

Morphological regions	G/S ratio	Probable types of lignin structures	Molecular rigidity level
Fiber S2 (F)	12:88	No condensed	Low rigidity
Vessel S2 (V)	88:12	Condensed	Rigid
Ray parenchyma	49:51	Semi-condensed	Moderated rigidity
MLcc (F/F)	91:9	Condensed	Rigid
MLcc (F/V)	80:20	Condensed	Rigid
MLcc (F/R)	100:0	Condensed	High rigidity
MLcc (R/R)	88:12	Condensed	Rigid

[a]Values from refs [90, 91]

extracted in aqueous dioxane because it is partially degraded by this physical transformation. During the milling process, some covalent bonds are broken [95], giving low M_w lignin fractions that can be solubilized in the solvent mixture.

The predominant physicochemical properties of lignin macromolecules in solution are related to properties of the isolated polymer as the intrinsic viscosity, branching parameter, and the degree of polydispersity. Their determination for every lignin type provided useful structural information related to the overall architecture of the former protolignin [96].

Lignin molecules in solution are spherical particles and slightly solvated with solvent. Dong and Fricke [97] and suggested that the lignin molecules (Kraft lignin) in DMF have a compact spherical structure with a constant density by measure of viscosity.

The intrinsic viscosities [98] of native lignins at comparable M_w were found to be 1/40 of those of polysaccharides and about 1/40 of other synthetically produced polymers [70, 99, 100]. The low intrinsic viscosities of several native and pulping lignins in dioxane [101], such as Kraft lignin [102], lignosulfonate [103], or alkali lignin [70, 100] and in other solvents, may provide some conclusive facts, as Goring stated [99, 101]; for example, that these molecules in solution show compact spherical microgel particles. Similar conclusions on how lignins remain in solution can be drawn when other parameters are measured, such as "sedimentation coefficients" and "diffusion constants" [99, p. 705] [104]. The "branching parameter," or contraction factor (g) introduced by Zimm and Stockmayer [105], when measured on various alkali lignin fractions, decrease with an increases in the M_w of the microparticle solution [70], as might be expected for such molecular configurations. In general, the chains of the lignin macromolecules in solution are more densely packed than those of linear flexible synthetic commercial polymers such as PS.

Abreu et al. [90] have proposed a relationship between lignin structure and composition, G/S ratio in the lignins of birch cells and wood properties (structural and rigidity level) (see Table 2.5 and Figure 2.28).

References

[1] Lu F, Ralph J. Chapter 6 - lignin. In: Cereal Straw as a Resource for Sustainable Biomaterials and Biofuels. Amsterdam: Elsevier; 2010. p. 169–207.

[2] Bolker HI. Lignin. In: Natural and Synthetic Polymers: An Introduction. Marcel Dekker Incorporated; 1974. p. 577–621.

[3]   Brunow G, Lundquist K, Gellerstedt G. Lignin. In: Sjöström E and Alén R, editors. Analytical Methods in Wood Chemistry, Pulping, and Papermaking, Springer Series in Wood Science. Springer-Verlag; 1999. p. 77–124.
[4]   Ralph J, Lundquist K, Brunow G, Lu F, Kim H, Schatz PF, et al. Lignins: natural polymers from oxidative coupling of 4-hydroxyphenyl-propanoids. Phytochem Rev. 2004;3(1-2):29–60.
[5]   Hatfield R, Vermerris W. Lignin formation in plants: the dilemma of linkage specificity. Plant Physiol. 2001;126(4):1351–1357.
[6]   Rolando C, Monties B, Lapierre C. Thioacidolysis. In: Lin SY, Dence CW, editors. Methods in Lignin Chemistry, Springer Series in Wood Science. Springer Berlin Heidelberg; 1992. p. 334–349.
[7]   Ralph J, Lu F. The DFRC method for lignin analysis. 6. A simple modification for identifying natural acetates on lignins. J Agric Food Chem. 1998;46(11):4616–4619.
[8]   Ibrahim RK, Grisebach H. Purification and properties of udp-glucose: Coniferyl alcohol glucosyltransferase from suspension cultures of paul's scarlet rose. Arch Biochem Biophys. 1976;176(2):700–708.
[9]   Whetten R, Sederoff R. Lignin biosynthesis. Plant Cell. 1995;7(7):1001–1013.
[10] Fukushima K. Regulation of syringyl to guaiacyl ratio in lignin biosynthesis. J Plant Res. 2001;114(4):499–508.
[11] Liu CJ, Miao YC, Zhang KW. Sequestration and transport of lignin monomeric precursors. Molecules. 2011;16(1):710–727.
[12] Gandini A, Belgacem MN. Lignins as components of macromolecular materials. In: Belgacem MN, Gandini A, editors. Monomers, Polymers and Composites from Renewable Resources. Elsevier Science; 2008. p. 243–272.
[13] Vanholme R, Morreel K, Ralph J, Boerjan W. Lignin engineering. Curr Opin Plant Biol. 2008;11(3):278–285.
[14] Rubin EM. Genomics of cellulosic biofuels. Nature. 2008;454(7206):841–845.
[15] Leisola M, Pastinen O, Axe DD. Lignin-designed randomness. J Med Chem. 2012;2012(3):1.
[16] Fengel D, Wegener G. Wood: Chemistry, Ultrastructure, Reactions. W. de Gruyter; 1984.
[17] Gellerstedt G, Henriksson G. Lignins: major sources, structure and properties. In: Belgacem MN, Gandini A, editors. Monomers, Polymers and Composites from Renewable Resources. Amsterdam: Elsevier; 2011. p. 201–224.
[18] Donaldson LA. Lignification and lignin topochemistry - an ultrastructural view. Phytochemistry. 2001;57(6):859–873.
[19] Karhunen P, Rummakko P, Sipilä J, Brunow G, Kilpeläinen I. Dibenzodioxocins; a novel type of linkage in softwood lignins. Tetrahedron Lett. 1995;36(1):169–170.
[20] Karhunen P, Rummakko P, Sipilä J, Brunow G, Kilpeläinen I. The formation of dibenzodioxocin structures by oxidative coupling. A model reaction for lignin biosynthesis. Tetrahedron Lett. 1995;36(25):4501–4504.
[21] Karhunen P, Mikkola J, Pajunen A, Brunow G. The behaviour of dibenzodioxocin structures in lignin during alkaline pulping processes. Nord Pulp Pap Res J. 1999;14(2):123–128.
[22] Sjöström E. Wood Chemistry: Fundamentals and Applications. Academic Press; 1981.
[23] Rodrigues Pinto PC, Borges da Silva EA, Rodrigues AE. Insights into oxidative conversion of lignin to high-added-value phenolic aldehydes. Ind Eng Chem Res. 2011;50(2):741–748.
[24] Sjöström E. Wood Chemistry: Fundamentals and Applications. 2nd ed. Academic Press; 1993.
[25] Evtuguin DV, Pascoal Neto C, Silva AMS, Domingues PM, Amado FML, Robert D, et al. Comprehensive study on the chemical structure of dioxane lignin from plantation eucalyptus globulus wood. J Agric Food Chem. 2001;49(9):4252–4261.

[26] Pinto PC, Evtuguin DV, Pascoal Neto C. Effect of structural features of wood biopolymers on hardwood pulping and bleaching performance. Ind Eng Chem Res. 2005;44(26):9777–9784.
[27] Adler E. Lignin chemistry-past, present and future. Wood Sci Technol. 1977;11:169–218.
[28] Neish AC. Monomeric intermediates in the biosynthesis of lignin. In: Freudenberg K, Neish AC, editors. Constitution and Biosynthesis of Lignin, Molecular Biology, Biochemistry and Biophysics. vol. 2. Berlin: Springer-Verlag; 1968. p. 1–43.
[29] Glasser WG, Glasser HR. Simulation of reactions with lignin by computer (SIMREL). II. Model for softwood lignin. Holzforschung. 1974;28(1):5–11.
[30] Sakakibara A. A structural model of softwood lignin. Wood Sci Technol. 1980;14(2):89–100.
[31] Brunow G. Methods to reveal the structure of lignin. In: Hofrichter M, Steinbuchel A, editors. Biopolymers: Lignin, Humic Substances and Coal. vol 1. John Wiley & Sons, Ltd; 2001. p. 89–116.
[32] Ralph J, Brunow G, Boerjan W. Lignins. In: eLS. John Wiley & Sons, Ltd; 2007.
[33] Crestini C, Melone F, Sette M, Saladino R. Milled wood lignin: a linear oligomer. Biomacromolecules. 2011;12(11):3928–3935.
[34] Wayman M, Obiaga TI. The modular structure of lignin. Can J Chem. 1974;52(11):2102–2110.
[35] Nimz H. Beech lignin-proposal of a constitutional scheme. Angew Chem Int Ed. 1974;13(5):313–321.
[36] Boerjan W, Ralph J, Baucher M. Lignin biosynthesis. Annu Rev Plant Biol. 2003;54:519–546.
[37] Mansfield SD, Kim H, Lu F, Ralph J. Whole plant cell wall characterization using solution-state 2D NMR. Angew Chem Int Ed. 2012;7(9):1579–1589.
[38] Simionescu CI, Anton I. A study of some representative types of reed lignin. Cellul Chem Technol. 1970;4:589–611.
[39] Dalimova GN, Abduazimov KA. Lignins of herbaceous plants. Chem Nat Compd. 1994;30(2):146–159.
[40] del Río JC, Rencoret J, Marques G, Gutiérrez A, Ibarra D, Santos JI, et al. Highly acylated (acetylated and/or *p*-coumaroylated) native lignins from diverse herbaceous plants. J Agric Food Chem. 2008;56(20):9525–9534.
[41] Banoub JH, Delmas M. Structural elucidation of the wheat straw lignin polymer by atmospheric pressure chemical ionization tandem mass spectrometry and matrix-assisted laser desorption/ionization time-of-flight mass spectrometry. J Mass Spectrom. 2003;38(8):900–903.
[42] You TT, Mao JZ, Yuan TQ, Wen JL, Xu F. Structural elucidation of the lignins from stems and foliage of arundo donax Linn. J Agric Food Chem. 2013;61(22):5361–5370.
[43] Stewart JJ, Akiyama T, Chapple C, Ralph J, Mansfield SD. The effects on lignin structure of overexpression of ferulate 5-hydroxylase in hybrid poplar. Plant Physiol. 2009;150(2):621–635.
[44] Yuan TQ, Sun SN, Xu F, Sun RC. Characterization of lignin structures and lignin-carbohydrate complex (LCC) linkages by quantitative ^{13}C and 2D HSQC NMR spectroscopy. J Agric Food Chem. 2011;59(19):10604–10614.
[45] Cornu A, Besle JM, Mosoni P, Grenet E. Lignin-carbohydrate complexes in forages: structure and consequences in the ruminal degradation of cell-wall carbohydrates. Reprod Nutr Dev. 1994;34(5):385–398.
[46] Björkman A. Isolation of lignin from finely divided wood with neutral solvents. Nature. 1954;174(4440):1057–1058.
[47] Koshijima T, Watanabe T, Azuma J. Existence of benzylated carbohydrate moiety in lignin-carbohydrate complex from pine wood. Chem Lett. 1984;10:1737–1740.

[48] Azuma J, Takahashi N, Koshijima T. Isolation and characterization of lignin-carbohydrate complexes from the milled-wood lignin fraction of *Pinus densiflora* Sieb. *et* Zucc. Carbohydr Res. 1981;93(1):91–104.
[49] Lawoko M, Henriksson G, Gellerstedt G. Characterization of lignin-carbohydrate complexes (LCCs) of spruce wood (Picea abies L.) isolated with two methods. Holzforschung. 2006;60(2):156–161.
[50] Lawoko M, Henriksson G, Gellerstedt G. Characterization of lignin-carbohydrate complexes from spruce sulfite pulp. Holzforschung. 2006;60(2):162–165.
[51] Kondo R, Sako T, Iimori T, Hiroyuki I. Formation of glycosidic lignin-carbohydrate complex in the dehydrogenative polymerization of coniferyl alcohol. Mokuzai Gakkaishi. 1990; 36(4):332–338.
[52] Watanabe T, Koshijima T. Evidence for an ester linkage between lignin and glucuronic acid in lignin-carbohydrate complexes by DDQ-oxidation. Agric Biol Chem. 1988; 52(11):2953–2955.
[53] Xie Y, Yasuda S, Wu H, Liu H. Analysis of the structure of lignin-carbohydrate complexes by the specific ^{13}C tracer method. J Wood Sci. 2000;46(2):130–136.
[54] Schädel C, Blöchl A, Richter A, Hoch G. Quantification and monosaccharide composition of hemicelluloses from different plant functional types. Plant Physiol Biochem. 2010;48(1):1–8.
[55] Balakshin M, Capanema E, Gracz H, Chang H, Jameel H. Quantification of lignin-carbohydrate linkages with high-resolution NMR spectroscopy. Planta. 2011;233(6): 1097–1110.
[56] Du X, Gellerstedt G, Li J. Universal fractionation of lignin-carbohydrate complexes (LCCs) from lignocellulosic biomass: an example using spruce wood. Plant J. 2013;74(2):328–338.
[57] Terashima N, Yoshida M, Hafren J, Fukushima K, Westermark U. Proposed supramolecular structure of lignin in softwood tracheid compound middle lamella regions. Holzforschung. 2012;66(8):907–915.
[58] Hatakeyama H, Hatakeyama T. Lignin structure, properties, and applications. In: Abe A, Dusek K, Kobayashi S, editors. Biopolymers, Advances in Polymer Science. vol. 232. Springer Berlin Heidelberg; 2010. p. 1–63.
[59] Rogošić M, Mencer HJ, Gomzi Z. Polydispersity index and molecular weight distributions of polymers. Eur Polym J. 1996;32(11):1337–1344.
[60] Gilbert RG, Hess M, Jenkins AD, Jones RG, Kratochvil P, Stepto RFT, *et al.* Dispersity in polymer science (*IUPAC* recommendations 2009). Pure Appl Chem. 2009;81(2):351–353.
[61] Matron J, Marton T. Molecular weight of Kraft lignin. Tappi. 1964;47(8):471–476.
[62] Peniston QP, McCarthy JL. Lignin. I. Purification of lignin sulfonic acids by continuous dialysis. J Am Chem Soc. 1948;70(4):1324–1328.
[63] Gardon JL, Mason SG. Physicochemical studies of ligninsulfonates. I. Preparation and properties of fractionated samples. Can J Chem. 1955;33:1477–1490.
[64] Sjostrom E, Haglund P, Janson J. Changes in cooking liquor composition during sulfite pulping. Sven Papperstidn. 1962;65:855–859.
[65] Moacanin J, Felicetta VF, Haller W, McCarthy JL. Lignin. VI. Molecular weights of lignin sulfonates by light scattering. J Am Chem Soc. 1955;77(13):3470–3475.
[66] Felicetta VF, Ahola A, McCarthy JL. Lignin. VII. Distribution in molecular weight of certain lignin sulfonates. J Am Chem Soc. 1956;78(9):1899–1904.
[67] Szabo A, Goring DAI. Degradation of a polymer gel: application to delignification of sprucewood. Tappi. 1968;51(10):440–444.
[68] Glasser WG, Dave V, Frazier CE. Molecular weight distribution of (semi-) commercial lignin derivatives. J Wood Chem Technol. 1993;13(4):545–559.

[69] Rezanowich A, Goring DAI. Polyelectrolyte expansion of a lignin sulfonate microgel. J Colloid Sci. 1960;15:452–471.
[70] Gupta PR, Goring DAI. Physicochemical studies of alkali lignins: III. Size and shape of the macromolecule. Can J Chem. 1960;38(2):270–279.
[71] Brauns FE, Brauns DA. The Chemistry of Lignin: Supplement. New York: Academic Press; 1960.
[72] Ramiah MV, Goring DAI. The thermal expansion of cellulose, hemicellulose, and lignin. J Polym Sci, Part C: Polym Symp. 1965;11:27–48.
[73] Hatakeyama H, Nakano J, Hatano A, Migita N. Variation of infrared spectra with temperature of lignin and lignin model compounds. Tappi. 1969;52(9):1724–1728.
[74] Kimura M, Hatakeyama H, Nakano J. Torsional braid analysis of soft wood lignin and its model polymers. Mokuzai Gakkaishi. 1975;21(11):624–628.
[75] Hatakeyama H, Nakamura K, Hatakeyama T. Studies on factors affecting the molecular motion of lignin and lignin-related polystyrene derivatives. Trans Tech Sect (Can Pulp Pap Assoc). 1980;6(4):105–110.
[76] Yano S, Hatakeyama T, Hatakeyama H. Primary dispersion region of dioxane lignin. Rep Prog Polym Phys Jpn. 1981;24:273–274.
[77] Yano S, Hatakeyama H, Hatakeyama T. Temperature dependence of the tensile properties of lignin/paper composites. Polymer. 1984;25(6):890–893.
[78] Hatakeyama H, Kubota K, Nakano J. Thermal analysis of lignin by differential scanning calorimetry. Cellul Chem Technol. 1972;6(5):521–529.
[79] Hatakeyama H, Iwashita K, Meshitsuka G, Nakano J. Effect of molecular weight on the glass transition temperature of lignin. Mokuzai Gakkaishi. 1975;21(11):618–623.
[80] Tinh N, Zavarin E, Barrall EMI. Thermal analysis of lignocellulosic materials. I. Unmodified materials. J Macromol Sci, Rev Macromol Chem. 1981;20(1):1–65.
[81] Tinh N, Zavarin E, Barrall EMI. Thermal analysis of lignocellulosic materials. Part II. Modified materials. J Macromol Sci, Rev Macromol Chem. 1981;21(1):1–60.
[82] Hatakeyama H, Nakano J. Nuclear magnetic resonance studies on lignin in solid state. Tappi. 1970;53(3):472–475.
[83] Zhao G, Norimoto M, Yamada T, Morooka T. Dielectric relaxation of water adsorbed on wood. Mokuzai Gakkaishi. 1990;36(4):257–263.
[84] Suzuki Y, Hirai N, Ikeda M. Piezoelectric relaxation of wood I. Effects of wood species and fine structure on piezoelectric relaxation. Mokuzai Gakkaishi. 1992;38(1):20–28.
[85] Sasaki T, Norimoto M, Yamada T, Rowell RM. Effect of moisture on the acoustical properties of wood. Mokuzai Gakkaishi. 1988;34:794–803.
[86] Hatakeyama H, Hatakeyama T. Thermal properties of isolated and *in situ* lignin. In: Heitner C, Dimmel D, Schmidt J, editors. Lignin and Lignans: Advances in Chemistry. CRC press; 2010. p. 301–319.
[87] Nada AMA, Yousef HA, El-Gohary S. Thermal degradation of hydrolyzed and oxidized lignins. J Therm Anal Calorim. 2002;68(1):265–273.
[88] Bouajila J, Dole P, Joly C, Limare A. Some laws of a lignin plasticization. J Appl Polym Sci. 2006;102(2):1445–1451.
[89] Lisperguer J, Perez P, Urizar S. Structure and thermal properties of lignins: characterization by infrared spectroscopy and differential scanning calorimetry. J Chil Chem Soc. 2009; 54(4):460–463.
[90] Abreu HS, Latorraca JVF, Pereira RPW, Monteiro MBO, Abreu FA, Amparado KF. A supramolecular proposal of lignin structure and its relation with the wood properties. An Acad Bras Cienc. 2009;81(1):137–142.
[91] Higuchi T. Biosynthesis and Biodegradation of Wood Components. Elsevier Science; 1985.
[92] Goring DAI. Physical chemistry of lignin. Pure Appl Chem. 1962;5:233–254.

[93] Schuerch JC. The solvent properties of liquids and their relation to the solubility, swelling, isolation, and fractionation of lignin. J Am Chem Soc. 1952;74:5061–5067.
[94] Björkman A. Finely divided wood. I. Extraction of lignin with neutral solvents. Sven Papperstidn. 1956;59:477–485.
[95] Chang HM, Cowling EB, Brown W. Comparative studies on cellulolytic enzyme lignin and milled wood lignin of sweetgum and spruce. Holzforschung. 1975;29(5):153–159.
[96] Argyropoulos DS, Menachem SB. Lignin. In: Eriksson KEL, Babel W, Blanch HW, Cooney CL, Enfors SO, Eriksson KEL, et al., editors. Biotechnology in the Pulp and Paper Industry, Advances in Biochemical Engineering/Biotechnology. vol. 57. Springer Berlin Heidelberg; 1997. p. 127–158.
[97] Dong D, Fricke AL. Intrinsic viscosity and the molecular weight of Kraft lignin. Polymer. 1995;36(10):2075–2078.
[98] Meister JJ, Richards EG. Molecular weight distribution of aspen lignins estimated by universal calibration. In: Glasser WG, Sarkanen S, editors. Lignin. Washington, DC: American Chemical Society; 1989. p. 82–99.
[99] Goring DAI. Polymer properties of lignin and lignin derivatives. In: Sarkanen KV, Ludwig CH, editors. Lignins: Occurrence, Formation, Structure and Reactions. New York: Wiley-Interscience; 1971. p. 695–768.
[100] Gupta PR, Robertson RF, Goring DAI. Physicochemical studies of alkali lignins. II. Ultracentrifugal sedimentation analysis. Can J Chem. 1960;38:259–269.
[101] Rezanowich A, Yean WQ, Goring DAI. The molecular properties of milled wood and dioxane lignins: sedimentation, diffusion, viscosity, refractive index increment, and ultraviolet absorption. Sven Papperstidn. 1963;66:141–149.
[102] Lindberg JJ, Tylli H, Majani C. Notes on the molecular weight and the fractionation of lignins with organic solvents. Pap Puu. 1964;46(9):521–526.
[103] Yean WQ, Goring DAI. Simultaneous sulphonation and fractionation of spruce wood by a continuous flow method. Pulp Pap Mag Can. 1964;65(6):T127–T132.
[104] Pla F, Robert A. Study of extracted lignins by GPC, viscosimetry and ultracentrifugation: determination of the degree of branching. Holzforschung. 1984;38(1):37–42.
[105] Zimm BH, Stockmayer WH. The dimensions of chain molecules containing branches and rings. J Chem Phys. 1949;17:1301–1314.

3

Detection and Determination

3.1 Introduction

Lignin can be defined chemically[1] or functionally to stress what lignin does within the plant. It has been recognized for over 50 years that lignin is a polymeric material composed of phenylpropanoid units derived primarily from three cinnamyl alcohols (monolignols): *p*-coumaryl, coniferyl, and sinapyl alcohols [1]. Functionally, lignins strengthen cell walls, facilitate water transport, and impede the degradation of wall polysaccharides, thus acting as a major line of defense against pathogens, insects, and other herbivores.

Ever since Bente, in 1875, stated that the noncellulosic constituent of wood, or lignin, was aromatic in nature, color reactions have found widespread use in the detection and identification of lignin and several have also been adapted for use in the quantitative determination of lignin.

Because lignin is fairly resistant to both chemical and biological degradations, it might be thought that it would be relatively easy to measure [1]. However, lignin has been defined, at least in general chemical terms, for more than 50 years, and several well-defined procedures to quantify its content in plant tissues have been developed and approved such as Association of Official Agricultural Chemists (AOAC) International or standard wood chemistry methods.

Innumerable methods exist for the study of lignins, including several meant to detect their presence or to determine their location and others to determine them quantitatively or to study in depth their structure and composition.

3.2 The Detection of Lignin (Color-Forming Reactions)

Due to the widespread number of functional groups present in lignin structure, lignins can react with diverse chemical reagents, giving rise to a widespread number of color reactions, which have been

---

[1] For example, its chemical composition and structure.

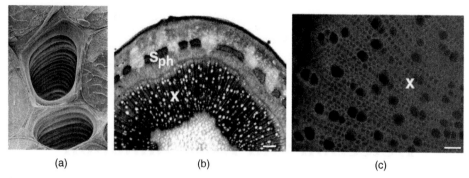

Figure 3.1 *Lignification pattern in* Populus *tissues. (a) Scanning electron micrograph of xylem elements in a Zinnia stem. Courtesy of Kim Findlay and Copyright from ref. [7]. (b) Transverse section of stem segment. Lignin deposition, visualized under the light microscope after phloroglucinol–HCl staining (red color) (x-xylem, ph-phloem, s-sclerenchyma. Bars = 100 μm). Copyright from ref. [8]. (c) Secondary xylem from stem. Lignin distribution by fluorescent microscopy (autofluorescence). Copyright from ref. [8]. (See insert for color representation of this figure.)*

used in its cyto- and histochemical detection. The detection of lignin can be performed using more than 150 color reactions, most of them developed by botanists and plant physiologists [2, 3].

Prior to performing the color reaction, the lignified material should be extracted with EtOH/benzene (1:2, v/v) and then with water to remove products that might interfere with the tests.

Among the most commonly used histochemical tests, the reaction with toluidine blue, the Wiesner test[2] [4], and the Mäule reaction[3] [5] deserve mention. One of the characteristics of lignin is its blue autofluorescence when illuminated with light of 330–380 nm wavelength (see Figure 3.1). All these characteristics have been used to detect and study the presence of lignins in different plant tissues [6].

3.2.1 Reagents for Detecting Lignins

Brauns, [14, 15] in 1952, classified the reagents used in these reactions into the following five groups: aliphatic compounds, phenols, aromatic amines, heterocyclic compounds, and inorganic reagents.

3.2.1.1 Aliphatic Compounds

Several alcohols and ketones in the presence of mineral acids give color reactions with lignified tissues. A sampling of the colors produced by different reagents was described by Brauns in 1952 as follows:

- MeOH/HCl gives a red color.
- Amyl alcohol/H_2SO_4 gives a blue color.
- 6-Methyl-hept-5-en-2-one gives a purple color.

3.2.1.2 Phenols and Aromatic Amines

The reactions are carried out by first moistening the plant material with diluted HCl and then applying to it an aqueous solution of the phenol derivatives listed in Table 3.1.[4]

---

[2] Specific for cinnamaldehyde groups.
[3] Specific for syringyl groups.
[4] When the phenols are insoluble in water, an alcoholic solution of the corresponding phenol derivatives is used instead.

Table 3.1 Color reactions of lignin with phenol derivatives[a]

Reagent	Coloration	Author [Ref.]
Phenol	Greenish-blue	Runge [9]
o-Cresol	Greenish-blue	Grandmougin [10]
m-Cresol	Blue	Grandmougin [10]
p-Cresol	Green	Grandmougin [10]
Thymol	Green	Czapek [11]
Catechol	Blue	Wiesner [12]
Resorcinol	Blue	Wiesner [12]
Hydroquinone	Olive	Grandmougin [10]
Phloroglucinol	Violet-red	Wiesner [12]
Pyrogallol	Green	Wiesner [12]
1,2,3,5-Tetrahydroxybenzene	Green	Fuchs [13]
Orcinol	Red	Wiesner [12]
o-Nitrophenol	Yellowish olive	Grandmougin [10]
m-Nitrophenol	Yellowish olive	Grandmougin [10]
p-Nitrophenol	Yellowish olive	Grandmougin [10]
o-Aminophenol	Yellow	Grandmougin [10]
m-Aminophenol	Yellow	Grandmougin [10]
p-Aminophenol	Brownish-yellow	Grandmougin [10]
4-methylquinoline	Red	Ihl (1890)
β-Naphthol	Greenish-blue	Grandmougin [10]
β-Naphthol	Pink	Grandmougin [10]
1,2-Dihydroxynaphthalene	Light green	Fuchs [13]
1,4-Dihydroxynaphthalene	Flesh color	Fuchs [13]
1,5-Dihydroxynaphthalene	Dirty green	Fuchs [13]

[a] Data taken from ref. [3].

The reaction most used is the Wiesner test (Wiesner tinction). Phloroglucinol[5] reacted in HCl/EtOH with coniferyl aldehyde, sinapyl aldehyde, vanillin, and syringaldehyde to yield either pink pigments (in the case of hydroxycinnamyl aldehydes) or red-brown pigments (in the case of hydroxybenzaldehydes)[6] [16].

The Wiesner test has been used to determine the amount of cinnamaldehyde units in native lignin. Adler et al. [4] have identified the coniferyl aldehyde units as the group responsible for color formation based on the similarity of UV/vis absorption spectra (λ_{max} = 550 nm). Many other color reactions are attributable to the same structural unit [15, 17–19] (see Figure 3.2).

With aromatic amines, the reactions are performed by adding a 1% solution of the base in dilute HCl or H_2SO_4. A sampling of the colors produced is listed in Table 3.2.

3.2.1.3 Heterocyclic Compounds

A comprehensive list of heterocyclic compounds giving rise to colored products on reactions with lignin was compiled by Brauns in 1960 [15]. A sampling of the colors produced is presented in Table 3.3.

---

[5] 1,3,5-Trihidroxybenzene.
[6] The histological sections are dyed with a solution containing phloroglucinol (1%, p/v) on HCl/EtOH (25:75, v/v), during 10–15 min.

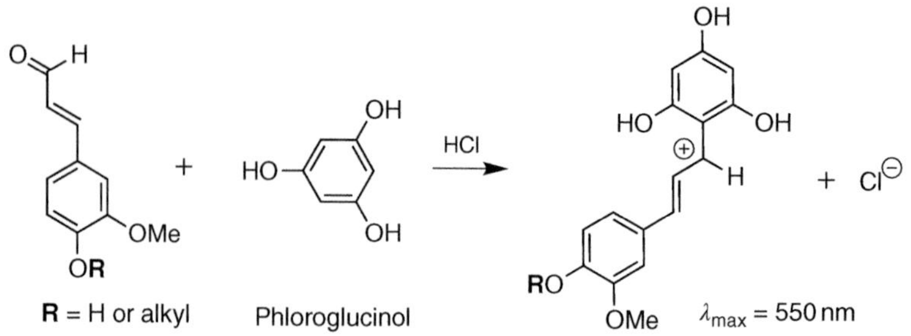

Figure 3.2 *UV/vis detection of cinnamaldehyde units based on the Wiesner reaction with phloroglucinol in acid media*

Table 3.2 Color reactions of lignin with amines[a]

Reagent	Coloration	Author [Ref.]
Aniline	Yellow	Runge [9]
o-Toluidine	Yellow	Grandmougin [10]
m-Toluidine	Yellow	Grandmougin [10]
p-Toluidine	Yellow	Grandmougin [10]
o-Nitroaniline	Yellow	Grandmougin [10]
m-Nitroaniline	Orange	Grandmougin [10]
p-Nitroaniline	Orange	Berge [20]
1-Methyl-2-amino-6-nitrobenzene	Yellow	Grandmougin [10]
o-Phenylenediamine	Orange-brown	Grandmougin [10]
m-Phenylenediamine	Yellow	Molisch (1887)
p-Phenylenediamine	Orange-brown	Grandmougin [10]
1-Methyl-2,5-diaminobenzene	Orange	Hegler (1890)
Dimethyl-p-phenylenediamine	Red	Grandmougin [10]
p-Aminodiphenylamine	Bordeaux-brown	Grandmougin [10]
Benzidine	Orange	Schneider [21]
β-Naphtylamine	Orange-yellow	Nickel [22, p. 51]
β-Naphtylamine	Orange-yellow	Nickel [22, p. 51]
Diphenylamine	Golden orange	Ellram [23]

[a] Data taken from ref. [3].

Table 3.3 Color reactions of lignin with heterocycles[a]

Reagent	Coloration	Author(s) (Year)
Furans-type	Green	Brauns (1952)
Pyrrole	Red	Ihl (1885)
Indole	Red	Niggl (1881)
Skatole	Red	Mattirolo (1885)
Carbazolo	Red	Mattirolo (1885)
Thalline	Orange	Hegler (1889)
Acridine	Green fluorescence	Ko (2004)
Acriflavine	Green fluorescence	Christiernin et al. (2005)

[a] Data taken from ref. [3, and references therein] [24, 25].

In addition to these historical methods, two tinctions based in fluorescence were reported in the last years (see Table 3.3 and Figure 3.3):

"Acriflavine tinction": In order to show the presence of lignins in cell walls, cells were dyed with acriflavine (0.1%, p/v), then washed with water, and later examined with a confocal laser microscope (Leica DM IRE2). The samples were excited with argon-laser light of 488 nm, giving the green fluorescence of the images with a filter in the rank of wavelengths of 520–530 nm [24].

"Acridine orange tinction": Similarly to the acriflavine method, this tinction shows the presence of lignins in the cell walls. The cell cultures were dyed with acridine 0.1% for 15 min, washed with water, and later examined with a confocal laser microscope (Leica DM IRE2). The samples were excited with argon-laser light of 488 nm, causing the green fluorescence of the images with a filter in the rank of wavelengths 505–530 nm [25].

3.2.1.4 Inorganic Reagents

Inorganic compounds are also used to produce a color reaction, and a selection of them is summarized in Table 3.4. It is noteworthy that some of these reactions were investigated in connection with studies aimed to determine the structural units of lignin.

The oldest known color reaction with inorganic reagents, which dates from 1895 was the Cross–Bevan, in which chlorine is used to detect the syringyl units in the lignin by the reaction of Figure 3.4 in order to give a product that absorb at $\lambda_{max} = 550$ nm.

Figure 3.3 Structures of acridine and acriflavine heterocycles

Table 3.4 Color reactions of lignin with inorganic compounds

Reagent	Coloration	Author (Year)
Cobalt thiocyanate	Blue	Casparis (1921)
Chlorine/sodium sulfite	Pink-purple red	Cross and Bevan (1895)
$KMnO_4$/HCl/ammonia (sol.)	Deep red	Mäule (1900)
Hypochlorite/Zn–water/ hydrogen sulfite/H_2SO_4	Red-orange-brown	Combes (1906)
HCl/potassium chlorate	Orange	Podbreznik (1929)
Hypochlorite/phloroglucinol/ iodine-IK	Bright red	Morquer (1929)
Vanadium pentoxide/ phosphoric acid	Yellow-brown	Brauns (1952)
HCl (concentrated)/H_2SO_4 (concentrated)	Green	Brauns (1952)
Bisulfite/H_2SO_4 (concentrated)	Red	Brauns (1952)

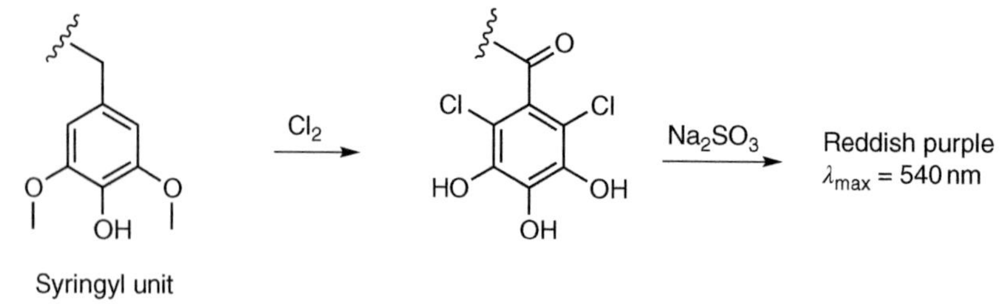

Figure 3.4 Cross–Bevan color reaction of syringyl lignin [26]

The steps comprising the Mäule test may be portrayed as follows: the syringyl nucleus in lignin is converted to the chloromethoxycatechol and hence to a methoxy-*o*-quinone [27, 28] (see Figure 3.5):

3.2.1.5 Unclassified Reagents

In addition to the aforementioned reactions, other reactions have been used in the detection of lignin. For example, the lignin transformations into quinonemethides (yellow compounds), developed in 1956 by Adler and Stenemur [30], which consists of successive treatments with hydrogen bromide in chloroform and sodium bicarbonate; or in indophenols by displacement of the lignin side chain upon reaction with quinone monochlorine or the oxidation of lignin with potassium nitrosodisulfonate (Fremy's salt) to yield *o*-quinones (red-brown compounds) was developed in 1961 by Adler and Lundquist [31], (see Figures 3.6 and 3.7).

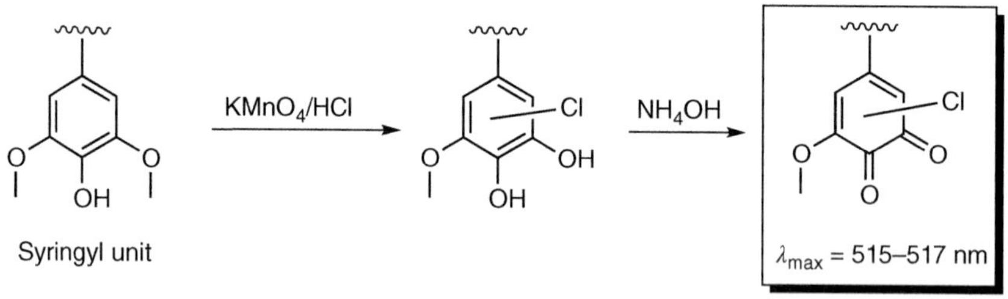

Figure 3.5 Mäule test reaction steps [29, p. 26]

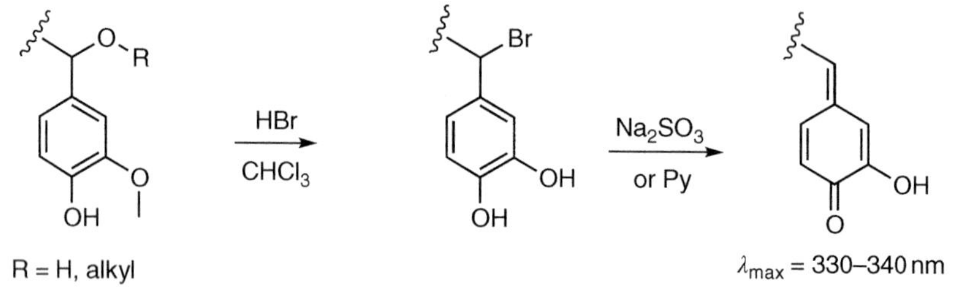

Figure 3.6 Adler and Stenemur color reaction of guaiacyl lignin [30]

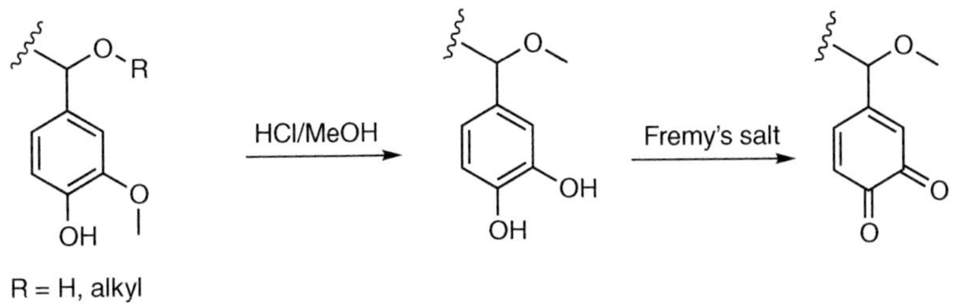

Figure 3.7 Adler and Lundquist color reaction of guaiacyl lignin [31]

3.3 Determination of Lignin

There is no main applicable method for the quantitative determination of total lignin in lignocellulosics. Nevertheless, the determination of lignin is an analysis performed routinely for characterizing materials; assessing the effects of chemical, physical, and biological treatments of wood and pulp; monitoring effluents in wood-processing industries; and, in the case of chemical pulps, estimating bleach chemical requirements [32].

Numerous methods have been developed and modified through the years to quantitatively determine the amount of lignin in a given type of plant tissue. The methods for determining lignin in samples can be divided into two basic categories: those that rely on a gravimetric determinations and those based on noninvasive methods. Some of the aforementioned methods require that the plant sample be subjected to some kind of extraction to remove potentially interfering compounds. This is the step that has most changed in terms of adapting procedures developed for different plant species.

Tables 3.5 and 3.6 summarize the lignin content of various types of plants and wood fibers determined by some of the procedures that will be explained in the following sections.

The various protocols described in the literature for the quantitative estimation of lignin can be grouped in the following two categories:

- Direct methods: The lignin is separated from the other plant constituents and weighed as such. The oldest and most common method is based on gravimetry. These methods were also used for the isolation of lignin (see Chapter 5).
- Indirect methods: Some characteristic groups of lignin, for example, the methoxyl group, are determined, and, by multiplication with a suitable factor, the percentage of lignin is computed. In some cases, advantage is taken of the fact that lignin produces a color reaction with certain reagents or by means of other characteristic chemical properties of lignin.

3.4 Direct Methods for the Determination of Lignin

The direct methods for the quantitative estimation of lignin may be split into two subcategories:

- Those that depend upon dissolving cellulose and other carbohydrates, leaving the lignin as an insoluble residue.
- Those that dissolve out lignin, thus separating it from cellulose and other carbohydrate derivatives.

Examples of the first subclass include 72% sulfuric acid or the fuming hydrochloric acid methods, with the method of Mehta being an example of the second subcategory.

Table 3.5 Lignin content in various types of plants[a]

Plant scientific	Plant common name	Lignin content	[Ref.]
Gymnosperms			
Picea abies	Norway spruce	28	[33]
Picea abies	Norway spruce (compression wood)	39	[34]
Pinus radiata	Monterey pine	27	[35, p. 249]
Pinus sylvestris	Scots pine	28	[35, p. 249]
Pseudotsuga menziesii	Douglas fir	29	[35, p. 249]
Tsuga canadensis	Eastern hemlock	31	[35, p. 249]
Angiosperms–eudicotyledons			
Acacia mollissima	Black wattle	21	[35, p. 249]
Betula verrucosa	Silver birch	20	[36]
Eucalyptus globulus	Blue gum eucalyptus	22	[35, p. 249]
Eucalyptus grandis	Rose eucalyptus	25	[37]
Populus tremula	European aspen	19	[36]
Corchorus capsularis	Jute	13	[38]
Hibiscus cannabinus	Kenaf	12	[38]
Linum usitatissimum	Flax	2.9	[38]
Angiosperms–monocotyledons			
Oryza species	Rice straw	6.1	[39]
Saccharum species	Bagasse	14	[39]

[a] Data obtained from ref. [39].

Table 3.6 Chemical lignin content of common natural and wood fibers[a]

Fiber type	Lignin (%)	Fiber type	Lignin (%)	Fiber type	Lignin (%)	Fiber type	Lignin (%)
Stalk fiber		**Cane**		**Grass**		**Wood**	
Rice	12–16	Bagasse	19–24	Esparto	17–19	Seed flax	21–23
Wheat	16–21	Bamboo	21–31	Sabai	22	Kenaf	15–19
Barley	14–15					Jute	21–22
Oat	16–19					Hemp	9–13
Rye	16–19					Rami	–
Leaf		**Seed hull**		**Reed**		**Bast**	
Abaca (Manila)	7–9	Cotton	0.7–1.6	Coniferous	26–34	Reed	22–24
Sisal (agave)	7–9	Deciduous	23–30				

[a] Data obtained from ref. [40].

3.4.1 Methods for Lignin as a Residue

These consist of the hydrolysis and solubilization of the carbohydrate component of the lignified material, leaving the lignin as a residue, which is determined gravimetrically.

3.4.1.1 Sulfuric Acid Method: Klason Method

The most widely applied variation employs 72% sulfuric acid to promote carbohydrate hydrolysis. Then, it is diluted with water and later the lignin is filtered, washed, and dried. The isolated lignin is referred to as "Klason lignin" [41, 42].

Many slightly modified conditions in the sulfuric acid method have been employed: for example, preliminary hydrolysis with HCl and/or hot water, concentration of H_2SO_4, hydrolysis time, temperature of the acid. Also, for the appropriate use of this method, several corrections must be made. Thus, this procedure is essentially the TAPPI official test methods (T222 om-83, T222 om-02, and T249 cm-00). Among the main factors affecting this determination are as follows:

- Concentration of sulfuric acid used.
- Time and temperature of the treatment.
- The pre-extraction treatment.
- The acid-soluble lignin in the filtrate.

For example, Table 3.7 shows the influence of the concentration of H_2SO_4 in the determination of lignin content.

This method gives correct values for softwoods, but hardwoods contain variable amounts of "acid-soluble lignin", which must be estimated and added to the Klason lignin to determine the total lignin content. The method is also not applicable for herbaceous or annual plants[7] or for bleached chemical pulp where the amounts of lignin are too small to be accurately weighed.

In 2011, Aldaeus et al. [45] published a simplified method for the determination of lignin in Kraft lignin and black liquors[8] with reduced or omitted acid pretreatment, which is faster and yields results equivalent to the standard method.

In most samples, a small portion of the total lignin may be soluble in the diluted acid solution from the second stage of the Klason lignin procedure. The acid-soluble lignin in wood and pulp can be determined by means of UV spectroscopy. Two problems have been addressed with this method, one in the determination of the extinction coefficient and other in the choice of the absorption maximum to be used [46].

The acid-soluble lignin content, generally determined by UV spectroscopy, is affected by diverse factors [14]:

- The reliability of the absorptivity value.
- The formation of carbohydrate degradation products that absorb at $\lambda_{max} = 280$ nm.[9]
- The partial hydrolysis of ester groups in the lignin.
- Some loss of methoxyl groups such as MeOH [47].

3.4.1.2 Method of Ellis

This method, adopted by AOAC International for the determination of lignin in annual plants, developing wood, and forage consists of a pre-extraction of the sample with EtOH/benzene, followed by successive treatments with a proteolytic enzyme, diluted H_2SO_4, and finally 72% H_2SO_4. This is the

---

[7] Because such materials contain variable amounts of proteins.
[8] Aqueous solution composed by lignin, hemicellulose, and inorganic chemicals that is used in the Kraft pulping process.
[9] Same as the furfural and hydroxymethylfurfural absorption range.

Table 3.7 Lignin determination with 72% and 80% sulfuric acid methods[a]

Material	Source (variety)	Lignin (%) 72%, 16 h at 10°C	Lignin (%) 80%, 2 h at 5°C
Bagasse	Florida	19.6	19.4
	Hawaii	21.3	21.8
	Mexico	18.3	18.7
	Louisiana	20.3	20.3
	Florida	18.8	19.7
	Philippines	20.1	19.8
Bamboo[b]	Mississippi (*Arwzdinaria gigalltea*)	22.2	19.5
Cornstalks	Iowa	14.7	14.7
	Israel (hybrid)	15.3	13.7
	Israel (yellow dent)	13.0	16.3
Broom cornstalks	Illinois	17.3	18.5
Crotalaria	South Carolina	22.4	23.0
Kenaf fiber	Florida	10.9	10.1
Ramie fiber	Florida	20.6	21.4
Barley straw	Nebraska (Exond)	15.4	17.1
Oat straw	Illinois (Clinton)	17.7	17.4
Rice straw	Louisiana (Zenith)	12.7	14.2
Rye straw	Minnesota (commercial)	18.0	19.1
Wheat straw	Illinois (Kawvale)	15.9	15.7
	Nebraska (Pawnee)	18.2	18.5
	Illinois (Pawnee)	20.0	20.1
	North Dakota (Premier)	16.3	17.4
	Washington (Rex)	16.4	17.8
	North Dakota (Stewart Durum)	16.2	16.5
	Kansas (Tenmarq)	17.0	16.3
Leaf or hard fiber			
Abaca	Philippines (commercial)	9.6	10.1
Henequen	Yucatan (commercial)	8.8	9.7
Sisal	Africa (commercial)	7.1	7.8
Forage fiber[c]			
Fescue	Kentucky (K34)	5.6	5.4
		5.4	5.6
Sudan grass	Illinois (Piper)	6.8	6.7
		6.8	6.9
		7.0	6.7
Sorghum	Illinois (Atlas)	7.9	7.9

[a] See refs [43, 44].
[b] This material ground to pass a US standard 40 screen.
[c] Acid–pepsin pretreatment used with all forage materials.

preferred method in cases where the protein content of the material is high, as in herbaceous and annual plants [48].

The chief limitations of this method are as follows:

- The possible need of correction based on nitrogen analysis because the proteins condense with lignin, thereby adding to its weight [49, pp. 190–195].
- The possible condensation of lignin with the carbohydrate-degradation products.
- In grass, the ash content may be as high as 5–10% [50].

3.4.1.3   Acid Detergent

In 1970, Goering and Van Soest [51] proposed to use an acid detergent extraction step to produce an acid detergent fiber (ADF). The procedure consists of a treatment with cetyltrimethylammonium bromide in 0.5 M H_2SO_4, before the treatment with 72% H_2SO_4. This treatment removes many of the potentially contaminating substances from the lignin residue. Goering and Van Soest recommended using a permanganate oxidation, similar to the procedure discussed earlier, to remove the lignin and leave all other materials [52, 53]. Theander et al. [54, 55] proposed a simplified scheme (Uppsala method) to quickly produce cell wall preparations (alcohol-insoluble residues).

3.4.1.4   Fuming Hydrochloric Acid Method

In 1913, Willstätter and Zechmeister [56] observed that fuming HCl,[10] will completely hydrolyze cellulose in the cold. This method has been applied to the determination of lignin content in straw and feedstuffs. The method is subject to possible error by incomplete hydrolysis of proteins when applied to the determination of lignin content in feedstuffs.

3.4.1.5   Method of König and Rump

In this method, the cellulosic material is hydrolyzed by heating the sample with HCl under a pressure of 5–6 atm [57].

3.4.1.6   Precipitate Formation Method

Browning [50] developed a method based on the reaction of lignosulfonates with organic bases to produce a precipitate, which is determined gravimetrically or turbidimetrically.

3.4.2   Lignin in Solution Methods

The methods used to determine lignin in solution may be classified according to whether the procedure is based on derivative formation, light absorbance, or modification by a specific chemical reaction. The dissolved lignin sources include pulping and bleaching[11] liquors and ground, surface, and sea waters containing discharges from pulp and paper mills and other wood-processing industries [32].

3.4.2.1   Spectrophotometric Methods

These depend on the capacity of lignin to absorb radiation corresponding to various regions of the electromagnetic spectrum. Under proper conditions, the magnitude of the response at a selected frequency is proportional to the concentration of lignin.

---

[10] HCl at over 40% with $d = 1.212–1.223$ g/ml at 15 °C.
[11] Removal of colored residual lignin from chemical pulp to increase its brightness, cleanliness, and other desirable properties, while preserving the strength (cellulose integrity) and carbohydrate yield (cellulose and hemicellulose) of the unbleached fiber.

UV/vis absorption spectrophotometry has been employed to determine the lignin concentration by incorporating a finely ground sample of the material in a KCl pellet and measuring the absorbance at λ_{max} values of 210 or 280 nm [58].

Other spectrophotometric methods used to determine lignin in wood and pulps are as follows:

- UV-microspectrophotometric method [59, 60]
- IR spectroscopy [61]
- FTIR [62]
- Multiple internal reflectance IR spectrometry [63]
- Diffuse reflectance Fourier transform spectrometry [64]
- Near-IR spectroscopy [65, 66] and near-IR reflectance spectroscopy (NIRS) [67]
- ^{1}H-NMR spectroscopy [68–70]
- Solid-state ^{13}C-NMR spectrometry (^{13}C CP/MAS/NMR, cross-polarization/magic angle spinning NMR) [71, 72].

In most samples, several problems have been addressed with these methods, the main one being the choice of an appropriate "standard lignin" with which the instrument to use is calibrated.

The ^{13}C-NMR method has the advantage of being applicable in the solid state, avoiding the risk of concurrent chemical modifications.

Totally, dissolving the sample in a suitable solvent and measuring the UV absorbance of the solution at λ_{max} values characteristic for lignin (usually 280 nm). Among the solvents used to dissolve the lignocellulosic material are: ionic liquids [73], sulfuric acid [74], phosphoric acid [75], nitric acid [76, 77], sodium chlorite solution [78], cadoxene (cadmium oxide/ethylenediamine) [79], acetyl bromide/acetic acid [80], and thioglycolic acid [50].

Some of these spectrophotometric methods use UV absorption spectrophotometry to determine the concentration of lignin in sulfite and Kraft pulping liquors. Moreover, another method based on fluorescence spectrophotometry has been proposed [81]. One of the advantages of this fluorescence method is that it can detect lignin in concentrations two to three orders of magnitude less than the required for absorption spectroscopy.

3.4.2.2 Alkali Method or Method of Mehta

The method of Mehta [82] is based on the fact that the lignin fraction in a lignified plant material can be removed by heating the latter with alkali under pressure. The alkaline solution is then filtered and acidified with HCl. The lignin is filtered off, washed with water, and redissolved in boiling EtOH. The alcoholic solution is filtered, and the filtrate evaporates to dryness. The lignin values are considerably lower than those reached by the aforementioned methods, owing probably to the partial degradation of the lignin by the NaOH solution.

3.5 Indirect Methods for the Determination of Lignin

Indirect methods do not involve the isolation of a lignin residue. Instead, the lignin contents are calculated as the difference between 100% and the polysaccharide content or, more commonly, by measuring some characteristic structural functionality, property, or chemical response and by relating the result to concentration.

The indirect methods for the quantitative estimation of lignin may be subdivided into two classes: (I) Those that depend upon a reaction of the lignin and (II) those that dissolve out the lignin and/or derivative, and then this is determined spectroscopically.

3.5.1 Chemical Methods

Chemical methods are the most extended indirect ones for the determination of lignin structure. In the following subsections, the majority of these procedures are described as well as the basic information that can be obtained from them.

3.5.1.1 Method of Schulze

The oldest indirect method for the determination of lignin is that of Schulze [83], who oxidized lignified plant material with HNO_3 and potassium chlorate and assumed that the weight loss represented the presence of lignin. This method only bears historical interest, as the results are too imprecise, because not only is the lignin completely oxidized by the drastic oxidizing reagent employed but also the hemicelluloses and perhaps some of the cellulose are destroyed.

3.5.1.2 Method of Benedikt and Bamberger

Benedikt and Bamberger [84] determined the percentage of methoxyl in a large number of woods and showed that there was a close relationship between the percentage of methoxyl and the lignin content. The method of Benedikt and Bamberger is based on the assumption that in lignified plant materials lignin is the only substance containing the methoxyl group and that the methoxyl content in lignin from various sources is the same.

3.5.1.3 Method of Cross, Bevan, and Briggs

This method is based on a reaction between lignin and phloroglucinol [26] (see Section 3.2.1.2 on page 50 for details).

3.5.1.4 Method of Seidel

This method is based on the fact that when HNO_3 is added to lignin, oxides of nitrogen are given off [85]. For the determination, a special apparatus is required, and the quantity of nitrogen oxides produced is determined by titration with potassium permanganate. By the use of an empirical factor, the percentage of lignin in the sample can be calculated.

3.5.1.5 Method of Waentig and Gierisch

This method is based on the fact that lignin takes up a considerable quantity of chlorine, and by determining the amount of chlorine absorbed, it is possible to estimate the percentage of lignin [86]. The percentage increase in weight corresponds to the "chlorine number." This number multiplied by 0.71 gives the approximate percentage of lignin in the sample.

3.5.1.6 Method of Mehta

Taking advantage of the fact that a solution of phosphotungstic and phosphomolybdic acids in phosphoric acid is a very sensitive reagent for detecting extremely small amounts of aromatic substances containing hydroxyl groups, Mehta [82] developed a micromethod to estimate lignin.

3.5.2 Spectrophotometric Methods

These methods depend on the capacity of lignin derivatives to absorb radiation corresponding to various regions in the electromagnetic spectrum.

In a procedure developed by Saka *et al.* [87], in 1978, the lignin is brominated and the bromine uptake, which is proportional to the lignin content, is determined by a combination of scanning

62 *Lignin and Lignans as Renewable Raw Materials*

electron microscopy or transmission electron microscopy (SEM or TEM), and energy-dispersive X-ray analysis (EDXA).

Totally dissolving the sample (chemically modified lignin) in a suitable solvent and measuring the UV absorbance of the solution at λ_{max} values characteristic for lignin (usually 280 nm). Among the solvents employed to "dissolve" the lignocellulosic material are sodium chlorite solution [78], cadoxene (cadmium oxide/ethylenediamine) [79], acetyl bromide/acetic acid [80], and thioglycolic acid [50].

3.5.2.1 The Acetyl Bromide Method

The method is based in solubilization of lignins into a solution. This is the most widely accepted method, and has the advantage that it is quick and simple, is adaptable to small samples, does not require correction, provides precise absorbance values, involves minimal chemical modification of the sample, and is accompanied by less interference from nonlignin products [32] (see Figure 3.8). The original method of Johnson [80] has been developed and/or modified several times [88–96].

As main limitation is the sample size, which is not uniform in the different modifications of the procedure, it requires the use of a proper absorptivity value depending on the sample. Marton [88] found that bleached pulps dissolved incompletely in the acetyl bromide/acetic acid solution.

Nevertheless, it presents some problems such as the degradation of xylans by acetyl bromide, which increases the absorbance in the region of 270–280 nm, wavelength zone where lignins are quantified. This causes an overvaluing of the amount of lignins present in the sample [97]. This overvaluation can be avoided by lowering the temperature of reaction to 50 °C and increasing the reaction time to 2–4 h.

3.5.2.2 The Thioglycolate Method

This method involves the formation of thioethers of benzyl alcohol groups (see Figure 3.9) by treatment of the lignin with thioglycolic acid under acidic conditions [98]. This chemically modified lignin

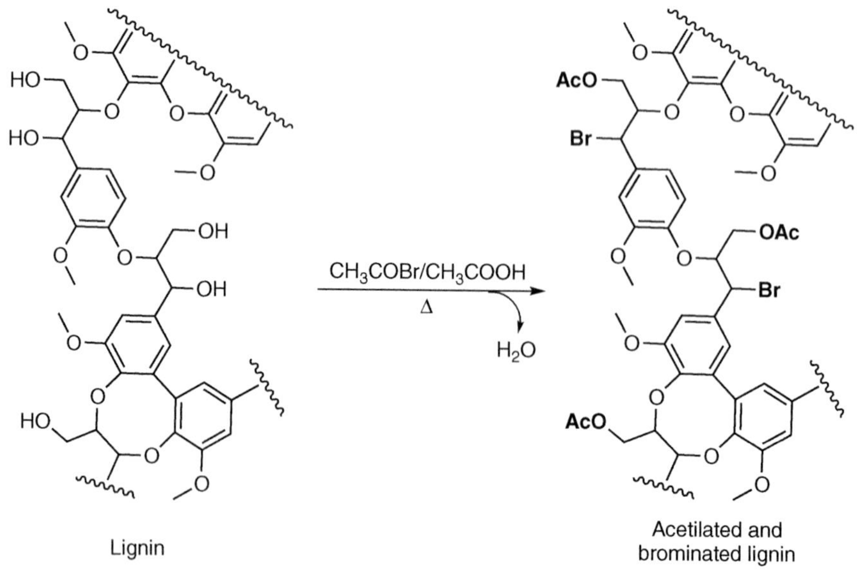

Figure 3.8 *Acetyl derivatization of lignin by acetyl bromide reagent to yield lignin soluble under acidic conditions [46]*

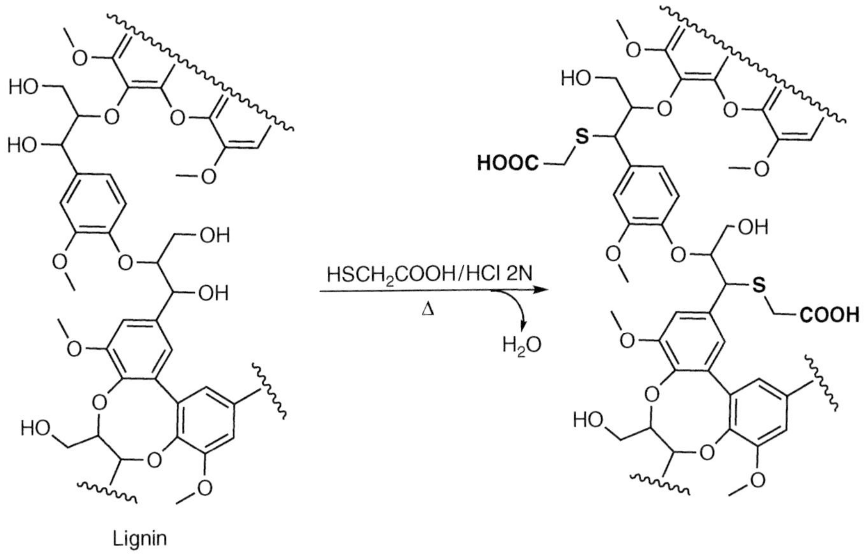

Figure 3.9 Thio derivatization of lignin by thioglycolate reagent to yield lignin soluble under alkaline conditions [46]

contains acid groups, rendering lignin soluble under alkaline conditions. The original procedure of Browning [50] has been modified recently for small samples [46, 99, 100]. The quantification of the lignin was based on the absorbance values at 280 nm and it required a lignin standard for calibration. Although originally proposed [50] in 1940, this method has not been widely used perhaps because of the lack of lignin standards required for calibration [46].

3.5.2.3 Pearl and Benson Method: Nitrosation Method

This method, which involves a specific chemical reaction, that is, the nitrosation reaction, has found the widest application [101].

Reaction of the phenolic units in lignin with acidified sodium nitrite leads to the formation of a nitrosophenol which, upon the addition of alkali, is tautomerized to a colored quinone mono-oxime. The absorbance of the latter structure is measured at 430 nm and related to lignin concentration by calibration with a standard lignin.

The sensitivity of the method is such that lignosulfonate concentrations as low as 0.2–0.5 ppm in water can be determined [102].

In comparison to other colorimetric and to UV spectrophotometric procedures, the Pearl and Benson method has been found to be less affected by interfering impurities [103].

The main drawback of the method is its lack of specificity for phenols having the characteristic lignin structure (e.g., tannins).

3.5.2.4 Method of Morrison

This method consists of digesting of the plant material with methyl bromide [90]. After that, the three monolignols building blocks are formed again. Thus, due to their aromatic nature, and by assuming that these building blocks represent a certain proportion of the total mass of lignin, an estimate of the lignin concentration in the original material can be made measuring the monolignol absorption in the UV spectra.

3.5.3 Methods Based on Oxidant Consumption

These methods are used exclusively in the analysis of unbleached pulps. Thus, all the procedures are based on the common principle that the lignin will consume the applied oxidant at a much higher rate than the carbohydrate component of the pulp. Hence, the oxidant consumption can be regarded as a measure of lignin concentration in the pulp. The lignin concentration determined by such procedures is usually expressed as the amount of oxidant consumed per unit weight of pulp (the Roe chlorine number or the Kappa number). These numbers can be converted into Klason lignin by applying a conversion factor. The two most commonly used oxidants are chlorine and potassium permanganate.

3.5.3.1 Chlorine: The Roe Chlorine Number

Chlorine reacts rapidly and extensively with **lignin in unbleached pulp at room temperature**. Chlorine consumption not only provides a method for estimating the lignin content but also indicates the total bleach requirement. "The Roe chlorine number" determination consists of measuring the uptake of gaseous chlorine by a known weight of the unbleached pulp. This method was modified by Johansson [104] in 1935, and later adopted by TAPPI as a standard method T202 ts-66.

3.5.3.2 Hypochlorite: The Hypo Number

Developed by Colombo *et al.* [105] and McLean [106], and eventually adopted by TAPPI as revised official test method T253 am-86. In this procedure, the pulp is reacted with acidified sodium or calcium hypochlorite, and the chlorine consumption is measured by titration. The ratio of hypo number to the Roe number is 1.052.

3.5.3.3 The Methanol Number

Ni *et al.* [107] developed a procedure based on the conversion of methoxyl groups in the lignin to MeOH when the pulp is chlorinated under a specified set of conditions.

3.5.3.4 The Permanganate Number Test

Several adaptations of the basic procedure have been made, but all are based on the addition of an excess of 0.1 N potassium permanganate to an aqueous suspension of the pulp sample and measurement of the amount consumed by titration of the residual permanganate. The permanganate number is the number of milliliters of permanganate consumed by 1 g of oven-dried pulp.

3.5.3.5 The Kappa Number

The Kappa number indicates the lignin content or bleachability of pulp. It is a standardized analysis method based on the milliliters of 0.1N potassium permanganate solution consumed when 1 g of moisture-free pulp is treated according to the ISO 302:2004 specification. It is applicable to all kind of chemical and semichemical pulps and shows a value range of 1–100.

The permanganate number is affected by the size of the pulp sample and the amount of permanganate applied. This problem has been solved by Tasman and Berzins [108–111], using a procedure, which includes an adjustment of the sample size to ensure that approximately half of the applied permanganate is consumed. This corrected procedure has been termed the Kappa number and has been adopted as a standard procedure by the technical pulp and paper organizations of several countries and specifically by TAPPI as a historical method T236 hm-85. The original method has been developed and/or modified several times [50, 112, 113]. In 1989, Birkett and Gambino [114] developed a method for estimating the Kappa number in unbleached pulps, based on near-IR spectroscopy (see Table 3.8). Today, the Kappa number is determined by ISO 302:2004. ISO 302 is applicable

Table 3.8 Relationships for the interconversion of lignin content and chlorine (hypo), permanganate, and Kappa numbers[a]

Material/relationship	Author [Ref.]
Kraft pulp	
% Klason lignin = Kappa $N°$ × 0.15	Kyrklund [115], Tasman [111]
% Klason lignin = Roe chlorine $N°$ × 0.84	Brauns [14]
Permanganate $N°$ = Roe chlorine $N°$ × 2.87	Brauns [14]
Chlorine (hypo) $N°$ = Roe chlorine $N°$ × 1.052	Kyrklund [116]
Kraft and polysulfide pulps	
Kappa $N°$ = chlorine $N°$ × 5.91	Kyrklund [115]
Kappa $N°$ = Chlorine $N°$ × 5.0	Tasman [111]
Sulfide pulps	
% Klason lignin = Kappa $N°$ × 0.165	Loras [117]
% Klason lignin = Kappa $N°$ × 0.187	Kyrklund [115]
% Total lignin = Kappa $N°$ × 0.252	Kyrklund [118]
% Total lignin = chlorine (hypo) $N°$ × 0.90	Kyrklund [115]
Softwood pulps	
log(Kappa $N°$) = 0.958 + 0.0253 permanganate $N°$	Hatton [119]
Hardwood pulps	
log(Kappa $N°$) = 0.727 + 0.0421 permanganate $N°$	Hatton [119]

[a] Data obtained from ref. [32, p. 49].

Table 3.9 Typical Kappa numbers range for representative pulps[a]

Pulp	Kappa $N°$ range
Kraft (bleachable grade). Softwood	25–35
Kraft (bleachable grade). Hardwood	14–18
Kraft softwood liner. Primary	80–95
Kraft softwood liner. Secondary	40–50
Neutral sulfite semichemical. Softwood	80–100
Bisulfite. Softwood	30–50
Acid sulfite. Softwood	16–22
Acid sulfite. Hardwood	14–20
Kraft (chlorinated and alkali-extracted). Softwood	5–8
Kraft (chlorinated and alkali-extracted). Hardwood	3–6
Acid sulfite (chlorinated and alkali-extracted). Softwood	3–5
Acid sulfite (chlorinated and alkali-extracted). Hardwood	2–4

[a] Data obtained from ref. [32, p. 53].

to all kind of chemical and semichemical pulps and gives a Kappa number in the range of 1–100 (see Table 3.9).

The method is applicable to all types and grades of pulps. Typical values are listed in Table 3.9. The main drawback is that it may not reflect the lignin content. However, its advantage is that it is applicable to very small samples with low permanganate consumption [113]. This method has been adopted by TAPPI and designated as an Useful Method UM-246.[12] Berzins [120] introduced "the Rapid Kappa number test" reducing the reaction time to 5–10 min.

The Kappa number is calculated as follows:

$$K = p\frac{f}{W};\qquad(3.1)$$

where

$$p = \frac{(b-a)N}{0.1};\qquad(3.2)$$

and where K = Kappa number; f = factor for correction to a 50% permanganate consumption, depending on the volume of p varies from 0.958 for $p = 30$ up to 1.044 for $p = 70$); W = weight of moisture-free pulp in the specimen g, p = amount of 0.1N permanganate actually consumed by the sample, ml; a = amount of thiosulfate consumed by the sample, ml; b = amount of thiosulfate consumed at the blank determination, ml; and N = normality of thiosulfate.

Table 3.8 shows relationships for the interconversion of lignin content and chlorine (hypo), permanganate, and Kappa numbers.

3.5.3.6 Total Organic Carbon (TOC) Method

Measurement of total organic carbon (TOC) content in pulping liquors has been related to lignin concentration. In this approach, a liquor sample is combusted to CO_2, which is measured gravimetrically [121]. The experimental data on TOC were correlated with the Kappa number and yield. The TOC content of black liquor appears to be a useful parameter for batch digester control.

3.5.3.7 Other Methods

Recently, Lourenço et al. [122] reported the total lignin content obtained with analytical pyrolysis gas chromatography and mass spectrometry (Py-GC/MS). The sample is thermally degraded in the absence of oxygen, forming volatile fragments that can be separated by GC and identified by MS. The lignin-derived pyrolysis products were added together to give the total lignin (Py-lignin). The results were similar to those found with wet chemistry.

3.6 Comparison of the Different Determination Methods

Since lignin is fairly resistant to both chemical and biological degradation, it might be thought that it would be relatively easy to measure [1].

Lignin has been defined, at least in general chemical terms, for more of 50 years, and several well-defined procedures to quantify lignin in plant tissues have been developed and approved as AOAC International or standard wood chemistry methods. Of the several types of methods available to determine the lignin in plant samples, none can be considered a standard unambiguous method for all samples. With the increasing interest in altering lignin quantity and composition for a variety of reasons (increased ease of pulping, digestibility, etc.), plus availability of molecular techniques to accomplish this task, the following question arises: What method should be used to measure lignin? [46] (see Tables 3.10 and 3.11). The extraction with acetyl bromide is one of the most commonly used methods. It allows quick and easy detection of lignin in small samples of cell walls (3–6 mg). Nevertheless, it presents some problems, such as the degradation of xylans by acetyl bromide, which increases the absorbance in the region of 270–280 nm, the wavelength zone where lignins are quantified. This causes an overvaluing of the amount of lignins in the sample [97]. This overestimate can be

---

[12] Technical Association of the Pulp and Paper Industry, Atlanta, 1985, Useful Method UM-246.

Table 3.10 Lignin contents of woods and pulps given by different methods[a]

Sample[b]	Standard method			Acetyl bromide method	
	Klason lignin	Acid-soluble lignin	Total lignin	Conventional	Modified
P. radiata WM	28.1	0.5	28.6	26.3	29.0
Eucalyptus regnans WM	23.0	4.8	27.8	24.1	27.7
P. radiata RMP	28.1	0.4	28.5	25.6	28.3
Mixed E CSP	22.4	5.3	27.7	24.1	27.6
E. regnans EP	33.4	2.4	35.8	-	37.6
E. regnans AEEP	15.9	1.4	17.3	-	17.9
P. radiata BP	1.7	3.5	5.2	4.6	5.4
P. radiata KP1	9.1	0.4	9.5	10.0	8.9
P. radiata KP2	5.0	0.4	5.4	6.4	5.4
E. regnans KP1	8.0	1.8	9.8	6.1	10.7
E. regnans KP2	3.3	1.1	4.4	4.1	5.6

[a]Data taken from ref. [32, p. 47] and ref. [92].
[b]WM, wood meal; RMP, refiner mechanical pulp; CSP, cold soda pulp; EP, exploded pulp; AEEP, alkali-extracted exploded pulp; BP, bisulfite pulp; KP, kraft pulp.

Table 3.11 Lignin concentrations given by four analytical procedures[a,b]

Sample	g/kg of Dry matter			
	Acid detergent lignin (ADL)	Permanganate lignin (PerL)	Klason lignin (KL)	Acetyl bromide soluble lignin (ABSL)
Bromegrass Y	28.5	56.0	102.2	123.3
Bromegrass M1	30.4	64.1	100.4	127.5
Bromegrass M2	36.5	69.8	109.8	144.6
Bromegrass M3, W	45.6	67.0	130.1	139.0
Setaria M3, W	72.5	80.9	135.3	136.0
Setaria M3, W (tiller)	61.6	65.8	126.0	130.95
Oat straw (stem)	83.3	111.4	171.1	186.3
Oat straw (leaf)	106.9	71.3	138.0	123.5
Wheat straw (stem)	89.1	122.0	184.2	213.0
Wheat straw (leaf)	103.4	74.3	141.5	149.9
Corn stalk PA	24.8	45.2	76.7	91.9
Alfalfa Y	83.6	134.6	123.0	116.5
Alfalfa FB, lower 30	92.5	157.5	144.8	134.7
Alfalfa FB, upper 30	59.3	95.3	111.4	71.3
Alfalfa PSD	90.6	153.7	138.8	117.2
Red clover FB	41.7	115.5	71.2	90.4
Aspen wood	69.5	190.5	158.6	181.5
Pine wood	245.5	255.3	249.7	401.2
Mean	75.9	107.3	134.1	148.8

[a]Y, young; M, mature (1–3 refer to three different maturity stages); W, wild; PA, past anthesis; FB, full bloom; PSD, past seed development.
[b]Data taken from ref. [46, 124].

avoided by lowering the temperature of reaction to 50 °C and increasing the reaction time to 2–4 h. The method is quite suitable for the routine study of small samples due to its speed and simplicity.

Iiyama and Wallis [92] determined the lignin contents of several pine and eucalyptus materials and compared the results with those using the Klason method (see Table 3.10).

Brinkmann et al. [67] made a comparison of different methods (thioglycolic acid, acetyl bromide, and ADF lignin) as a basis for calibrating near-IR reflectance spectroscopy. Jung et al. [124] compared Klason lignin and acid detergent lignin methods in ten diverse forage samples. The yield of the Klason lignin proved substantially higher than that of acid detergent lignin.

Fukushima and Hatfield [46, 123] comparatively evaluated the four most popular methods for measuring lignin (in forage samples). Clearly, the "acid detergent lignin procedure" gave the lowest lignin values, and the permanganate lignin method registered the much higher values (see Table 3.11).[13]

Of the aforementioned methods, none is clearly superior to all the others in providing an accurate measure of the total lignin in a given sample. Thus, there is no single method that can be classified in general as rapid, noninvasive, apt for large sample numbers, and accurate for measuring of cell wall lignin.

Many methods based on gravimetric or spectrophotometric analysis have been developed to quantitatively determine lignin content in plants [32, 125, 126] with disadvantages including a relatively large sample size and time-consuming procedures (gravimetric method), and the difficulty in finding an appropriate calibration standard (spectrophotometric methods). The method of Ellis is preferred for the determination of lignin in annual plants, developing wood and forage, where the protein content of the material is high.

Recently, a near-infrared (NIR) spectroscopy method was modified to improve the precision of determining the lignin content and assessing the lignin syringyl/guaiacyl (S/G) ratio [46, 95, 127–130], and high-throughput screening of plant cell wall compositions *via* pyrolysis molecular beam MS to analyze the lignin content and S/G ratio has also been reported [131]. The method can be used to analyze ball- or Wiley-milled samples at a microscale *via* direct dissolution and ^{1}H-NMR analysis of biomass using perdeuterated pyridinium chloride/[D_6]DMSO bisolvent system.

References

[1] Sarkanen KV, Ludwig CH. Lignins: Occurrence, Formation, Structure and Reactions. Wiley-Interscience; 1971.
[2] Phillips M. The chemistry of lignin. Chem Rev. 1934;14(1):103–170.
[3] Nakano J, Meshitsuka G. The detection of lignin. In: Dence CW, Lin SY, editors. Methods in Lignin Chemistry, Springer Series in Wood Science. Berlin: Springer-Verlag; 1992. p. 23–32.
[4] Adler E, Björkquist J, Häggroth S. Über die ursache der farbreaktionen des holzes. Acta Chem Scand. 1948;2:93–94.
[5] Srivastava LM. Histochemical studies on lignin. Tappi. 1966;49(4):173–183.
[6] Novo Uzal E. Lignificación en cultivos celulares de gimnospermas basales. Universidade da Coruña. A Coruña, España; 2008.
[7] Roberts K, McCann MC. Xylogenesis: the birth of a corpse. Curr Opin Plant Biol. 2000; 3(6):517–522.
[8] Barakat A, Bagniewska-Zadworna A, Choi A, Plakkat U, DiLoreto DS, Yellanki P, et al. The cinnamyl alcohol dehydrogenase gene family in *Populus*: phylogeny, organization, and expression. BMC Plant Biol. 2009;9:26–40.
[9] Runge FF. Über einige produkte der steinkohlendestillation. Ann Phys. 1834;107(5):65–78.
[10] Grandmougin E. Zusammenstellungen von ligninreaktionen. Z Farben- Text-Chem. 1906; 5:321–323.

---

[13] This method is also based on using the acid detergent fiber pretreatment.

[11] Czapek F. Biochemie der Pflanzen. Jena: Fischer; 1913.
[12] Wiesner J. Note über das verhalten des phloroglucius und einiger verwandter körper zur verholzten zellmembran. Sitzungsber Akad Wiss Wien, Math-Naturwiss Kl, Abt 2B. 1878; 77(I):60–66.
[13] Fuchs W. Die Chemie des Lignins. Berlin: J. Springer; 1926.
[14] Brauns FE. The Chemistry of Lignin. New York: Academic Press; 1952.
[15] Brauns FE, Brauns DA. The Chemistry of Lignin: Supplement. New York: Academic Press; 1960.
[16] Pomar F, Merino F, Barceló AR. O–4–Linked coniferyl and sinapyl aldehydes in lignifying cell walls are the main targets of the Wiesner (phloroglucinol-HCl) reaction. Protoplasma. 2002;220(1–2):0017–0028.
[17] Harada T, Nikuni Z. On a new bacterium which colors lignin red in urine. Nippon Nogeikagaku Kaishi. 1950;23(10):415–421.
[18] Hachihama Y, Jodai M. Lignin no Kagaku (Chemistry of Lignin). Nippon Hyoron Sha; 1946.
[19] Nakamura T, Kitaura S. Lignin color reactions. Ind Eng Chem. 1957;49(9):1388–1388.
[20] Berge A. New reagent on wood pulp. Bull Soc Chim Belg. 1906;20:158–159.
[21] Schneider H. The unna methods for establishing the oxidizing and reducing regions and their use on plant objects. Benzidine as a reagent for lignification. Z Wiss Mikrosk Mikrosk Tech. 1914;31:51–69.
[22] Nickel E. Die Farbenreaktionen der Kohlenstoffverbindungen. Berlin: Hermann Peters; 1890.
[23] Ellram W. New reactions of vanadic acid, molybdic acid, and thiocyanates. Sitzungsber Naturforsch Ges Univ Dorpat, xi 1895:28.
[24] Christiernin M, Ohlsson AB, Berglund T, Henriksson G. Lignin isolated from primary walls of hybrid aspen cell cultures indicates significant differences in lignin structure between primary and secondary cell wall. Plant Physiol Biochem. 2005;43(8):777–785.
[25] Ko JH, Han KH, Park S, Yang J. Plant body weight-induced secondary growth in Arabidopsis and its transcription phenotype revealed by whole-transcriptome profiling. Plant Physiol. 2004;135(2):1069–1083.
[26] Cross CF, Bevan EJ, Briggs JF. Lignone-phloroglucid formation without a color reaction. Chem Ztg. 1907;31:725–727.
[27] Meshitsuka G, Nakano J. Studies on the mechanism of lignin color reaction (XII). Mäule color reaction. Mokuzai Gakkaishi. 1978;24(8):563–568.
[28] Iiyama K, Pant R. The mechanism of the Mäule colour reaction introduction of methylated syringyl nuclei into softwood lignin. Wood Sci Technol. 1988;22(2):167–175.
[29] Lin SY, Dence CW, editors. Methods in Lignin Chemistry. Berlin: Springer; 1992.
[30] Adler E, Stenemur B. Lignin model studies. Quinonemethides. Chem Ber. 1956;89:291–303.
[31] Adler E, Lundquist K. Estimation of uncondensed phenolic units in spruce lignin. Acta Chem Scand. 1961;15:223–224.
[32] Dence CW. The determination of lignin. In: Dence CW, Lin SY, editors. Methods in Lignin Chemistry, Springer Series in Wood Science. Berlin: Springer-Verlag; 1992. p. 33–61.
[33] Önnerud H, Gellerstedt G. Inhomogeneities in the chemical structure of spruce lignin. Holzforschung. 2003;57(2):165–170.
[34] Önnerud H. Lignin structures in normal and compression wood. Evaluation by thioacidolysis using ethanethiol and methanethiol. Holzforschung. 2003;57(4):377–384.
[35] Sjöström E. Wood Chemistry: Fundamentals and Applications. 2nd ed. Academic Press; 1993.
[36] Önnerud H, Gellerstedt G. Inhomogeneities in the chemical structure of hardwood lignins. Holzforschung. 2003;57(3):255–265.
[37] Bassa A, Sacon VM, da Silva Junior FG, Barrichelo LEG. Polpação kraft convencional e modificada para madeiras de *Eucalyptus grandis* e hibrido (*E. grandis* x *E. urophylla*). In: 35°

Congresso e Exposicao Anual de Celulose e Papel; (Sao Paulo, Brazil; October 14–17); 2002. p. 5.

[38] del Rio JC, Rodriguez IM, Gutierrez A. Chemical characterization of fibers from herbaceous plants commonly used for manufacturing of high quality paper pulps. 9th European Workshop on Lignocellulosics and Pulp (Vienna, Austria; August, 27–30); 2006. p. 109–112.

[39] Gellerstedt G, Henriksson G. Lignins: major sources, structure and properties. In: Belgacem MN, Gandini A, editors. Monomers, Polymers and Composites from Renewable Resources. Amsterdam: Elsevier; 2011. p. 201–224.

[40] Rowell RM, Han JS, Rowell JS. Characterization and factors affecting fibre properties in natural polymers and agro fibres based composites. In: Frollini E, Leao AL, Mattoso LHC, editors. Natural Polymers and Agrofibers Based Composites. Brazil: Embrapa Agricultural Instrumentation; 2000. p. 115–134.

[41] Klason P. Chemical composition of deal (fir wood). Ark Kemi, Mineral Geol. 1908;3(Art. 5):1–20.

[42] Klason P. Determination of lignin in sulfite wood pulp. Papierfabrikant. 1910;8:1285–1286.

[43] Clark TF, Wolff IA. Search for new fiber crops. XI. Compositional characteristics of Illinois kenaf at several population densities and maturities. Tappi. 1969;52(11):2111–2116.

[44] Bagby MO, Nelson GH, Helman EG, Clark TF. Determination of lignin in non-wood plant fiber sources. Tappi. 1971;54:1876–1878.

[45] Aldaeus F, Schweinebarth H, Törngren P, Jacobs A. Simplified determination of total lignin content in Kraft lignin samples and black liquors. Holzforschung. 2011;65(4):601–604.

[46] Hatfield R, Fukushima RS. Can lignin be accurately measured? Crop Sci. 2005;45(3): 832–839.

[47] Bland DE. Methanol lignin from *Eucalyptus regnans* F. Muell. and its purification by countercurrent distribution. Biochem J. 1960;75:195–201.

[48] Ellis GH. Report on [the determination of] lignin and cellulose in plants. J Assoc Off Agric Chem. 1949;32:287–291.

[49] Lai YZ, Sarkanen KV. Isolation and structural studies. In: Sarkanen KV, Ludwig CH, editors. Lignins: Occurrence, Formation, Structure and Reactions. New York: Wiley-Interscience; 1971. p. 165–240.

[50] Browning BL. Methods of Wood Chemistry. vol. 2. Wiley Interscience; 1967.

[51] Goering HK, Van Soest PJ. Forage fiber analyses (apparatus, reagent, procedures, and some applications). Washington, DC, 20402: ARS/USDA Handbook No. 379, Superintendent of Documents, US Government Printing Office; 1970. http://trove.nla.gov.au/version/46035777 (accessed March 11, 2015).

[52] Van Soest PJ. Use of detergents in analysis of fibrous feeds. II. A rapid method for the determination of fiber and lignin. J Assoc Off Agric Chem. 1963;46(5):829–835.

[53] Rowland AP, Roberts JD. Lignin and cellulose fractionation in decomposition studies using acid-detergent fiber methods. Commun Soil Sci Plant Anal. 1994;25(3-4):269–277.

[54] Theander O. Advances in the chemical characterisation and analytical determination of dietary fibre components. In: Birch GG, Parker KJ, editors. Dietary Fibre. Applied Science Publishers; 1983. p. 77–93.

[55] Theander O, Westerlund EA. Studies on dietary fiber. 3. Improved procedures for analysis of dietary fiber. J Agric Food Chem. 1986;34(2):330–336.

[56] Willstätter R, Zechmeister L. Zur kenntnis der hydrolyse von cellulose I. Ber Dtsch Chem Ges. 1913;46(2):2401–2412.

[57] König J, Rump E. Chemie und struktur der pflanzen-zellmembran. Z Unters Nahr- Genussm Gebrauchsgegenstaende. 1914;28(4):177–222.

[58] Bolker HI, Somerville NG. Ultraviolet spectroscopic studies of lignins in the solid state. I. Isolated lignin preparations. Tappi. 1962;45:826–829.

[59] Fergus BJ, Goring DAI. The distribution of lignin in birch wood as determined by ultraviolet microscopy. Holzforschung. 1970;24(4):118–124.

[60] Boutelje JB, Jonsson U. Ultraviolet microscope photometry of pulp fibers. UV-absorbance and its relationship to chlorine number, kappa number and lignin content. Cellul Chem Technol. 1980;14(1):53–67.

[61] Kolboe S, Ellefsen O. Infrared investigations of lignin. A discussion of some recent results. Tappi. 1962;45:163–166.

[62] Ibarra D, del Rio JC, Gutierrez A, Rodriguez IM, Romero J, Martinez MJ, et al. Chemical characterization of residual lignins from eucalypt paper pulps. J Anal Appl Pyrolysis. 2005;74(1–2):116–122.

[63] Marton J, Sparks HE. Determination of lignin in pulp and paper by infrared multiple internal reflectance. Tappi. 1967;50(7):363–368.

[64] Schultz TP, Templeton MC, McGinnis GD. Rapid determination of lignocellulose by diffuse reflectance Fourier transform infrared spectrometry. Anal Chem. 1985;57(14):2867–2869.

[65] Casler MD, Jung HJG. Selection and evaluation of smooth bromegrass clones with divergent lignin or etherified ferulic acid concentration. Crop Sci. 1999;39(6):1866–1873.

[66] Tsuchikawa S. A review of recent near infrared research for wood and paper. Appl Spectrosc Rev. 2007;42(1):43–71.

[67] Brinkmann K, Blaschke L, Polle A. Comparison of different methods for lignin determination as a basis for calibration of near-infrared reflectance spectroscopy and implications of lignoproteins. J Chem Ecol. 2002;28(12):2483–2501.

[68] Fort DA, Remsing RC, Swatloski RP, Moyna P, Moyna G, Rogers RD. Can ionic liquids dissolve wood? Processing and analysis of lignocellulosic materials with 1-n-butyl-3-methylimidazolium chloride. Green Chem. 2007;9(1):63–69.

[69] Swatloski RP, Spear SK, Holbrey JD, Rogers RD. Dissolution of cellulose with ionic liquids. J Am Chem Soc. 2002;124(18):4974–4975.

[70] Jiang N, Pu Y, Ragauskas AJ. Rapid determination of lignin content *via* direct dissolution and ^{1}h NMR analysis of plant cell walls. ChemSusChem. 2010;3(11):1285–1289.

[71] Haw JF, Maciel GE, Schroeder HA. Carbon-13 nuclear magnetic resonance spectrometric study of wood and wood pulping with cross polarization and magic-angle spinning. Anal Chem. 1984;56(8):1323–1329.

[72] Rencoret J, Marques G, Gutierrez A, Ibarra D, Li J, Gellerstedt G, et al. Structural characterization of milled wood lignins from different eucalypt species. Holzforschung. 2008;62(5):514–526.

[73] Kline LM, Hayes DG, Womac AR, Labbe N. Simplified determination of lignin content in hard and soft woods *via* UV-spectrophotometric analysis of biomass dissolved in ionic liquids. BioResources. 2010;5(3):1366–1383.

[74] Giertz HW. Determination of residual lignin. A colorimetric method for the control of the bleaching process. Sven Papperstidn. 1945;48:485–489.

[75] Bethge PO, Gran G, Ohlsson KE. Determination of lignin in chemical wood pulp. I. Principles and methods. Sven Papperstidn. 1952;55:44–48.

[76] Bartunek R. A nonhydrolytic spectrophotometric method for determining lignin in high-purity pulps. Tappi. 1959;42:553–556.

[77] Henriksen A, Kesler RB. The Nu-number, a measure of lignin in pulp. Tappi. 1970;53(6):1131–1140.

[78] Schadenböck W, Prey V. New quantitative lignin determination by ultraviolet spectrophotometry. Papier. 1972;26(3):116–118.

[79] Sjöström E, Enström B. Spectrophotometric determination of residual lignin in pulp after dissolution in cadoxene. Sven Papperstidn. 1966;69(15):469–476.

[80] Johnson DB, Moore WE, Zank LC. The spectrophotometric determination of lignin in small wood samples. Tappi. 1961;44:793–798.
[81] Bublitz WJ, Meng TY. The fluorometric behavior of pulping waste liquors: a valuable tool for lignin and pulping research. Tappi. 1978;61(2):27–30.
[82] Mehta MM. Biochemical and histological studies on lignification: Part I. The nature of lignin: its physiological significance and its estimation in timbers. Biochem J. 1925;19(6):958–978.
[83] Schulze F. Beitrage zur kenntniss des lignins. Chem Zentralbl. 1857;21:321–325.
[84] Benedikt R, Bamberger M. Über eine quantitative reaction des lignins. Monatsh Chem. 1890;11(1):260–267.
[85] Seidel F. Studien über den Zellulosedarstellungsprozess und eine Methode für Bestimmung des Reinheitsgrades von Zellulosen. Techn. Hochsch. Dresden, Germany; 1907.
[86] Waentig P, Gierisch W. Determination of the degree of lignification of vegetable fibers. Angew Chem. 1919;32, I:173–175.
[87] Saka S, Thomas RJ, Gratzl JS. Lignin distribution: determination by energy dispersive analysis of X-rays. Tappi. 1978;61(1):73–76.
[88] Marton J. Determination of lignin in small pulp and paper samples by using the acetyl bromide method. Tappi. 1967;50(7):335–337.
[89] Bagby MO, Cunningham RL, Maloney RL. Ultraviolet spectral determination of lignin. Tappi. 1973;56:162–163.
[90] Morrison IM. A semi-micro method for the determination of lignin and its use in predicting the digestibility of forage crops. J Sci Food Agric. 1972;23(4):455–463.
[91] Morrison IM. Improvements in the acetyl bromide technique to determine lignin and digestibility and its application to legumes. J Sci Food Agric. 1972;23(12):1463–1469.
[92] Iiyama K, Wallis AFA. An improved acetyl bromide procedure for determining lignin in woods and wood pulps. Wood Sci Technol. 1988;22(3):271–280.
[93] Iiyama K, Wallis AFA. Determination of lignin in herbaceous plants by an improved acetyl bromide procedure. J Sci Food Agric. 1990;51(2):145–161.
[94] Fukushima RS, Hatfield RD. Extraction and isolation of lignin for utilization as a standard to determine lignin concentration using the acetyl bromide spectrophotometric method. J Agric Food Chem. 2001;49(7):3133–3139.
[95] Chang XF, Chandra R, Berleth T, Beatson RP. Rapid, microscale, acetyl bromide-based method for high-throughput determination of lignin content in *Arabidopsis thaliana*. J Agric Food Chem. 2008;56(16):6825–6834.
[96] Fukushima RS, Kerley MS. Use of lignin extracted from different plant sources as standards in the spectrophotometric acetyl bromide lignin method. J Agric Food Chem. 2011; 59(8):3505–3509.
[97] Hatfield RD, Grabber J, Ralph J, Brei K. Using the acetyl bromide assay to determine lignin concentrations in herbaceous plants: Some cautionary notes. J Agric Food Chem. 1999;47(2):628–632.
[98] Holmberg B. Thioglycolic acid lignin in spruce wood. Sven Papperstidn. 1930;33:679–686.
[99] Bruce RJ, West CA. Elicitation of lignin biosynthesis and isoperoxidase activity by pectic fragments in suspension cultures of castor bean. Plant Physiol. 1989;91(3):889–897.
[100] Lange BM, Lapierre C, Sandermann H Jr. Elicitor-induced spruce stress lignin. Structural similarity to early developmental lignins. Plant Physiol. 1995;108(3):1277–1287.
[101] Pearl IA, Benson HK. The determination of lignin in sulphide pulping liquor. Pap Trade J. 1940;111:35–36.
[102] Goldschmid O. Ultraviolet spectra. In: Sarkanen KV, Ludwig CH, editors. Lignins: Occurrence, Formation, Structure and Reactions. New York: Wiley-Interscience; 1971. p. 241–266.
[103] Jurkiewicz S. Effect of disturbing factors in the quantitative determination of lignin compounds in aqueous solutions. Przegl Papier. 1977;33(1):30–35.

[104] Johansson D. Testing methods for evaluation of unbleached chemical pulp. Pap Trade J. 1935;101(13):101–104.
[105] Colombo P, Corbetta D, Pirotta A, Ruffini G. Chlorine number as a method for evaluation of lignin content of a pulp. Pulp Pap Mag Can. 1962;63:T126–T140.
[106] McLean JD. Hypo number (chlorine number using sodium hypochlorite solution). Pulp Pap Mag Can. 1965;66(3):T103–T106.
[107] Ni Y, Kubes GJ, Van Heiningen ARP. Methanol number: a fast method to determine lignin content of pulp. J Pulp Pap Sci. 1990;16(3):J83–J86.
[108] Tasman JE, Berzins V. The permanganate consumption of pulp materials. I. Development of a basic procedure. Tappi. 1957;40:691–695.
[109] Tasman JE, Berzins V. The permanganate consumption of pulp materials. II. the KAPPA number. Tappi. 1957;40:695–699.
[110] Tasman JE, Berzins V. The permanganate consumption of pulp materials. III. The relationship of the KAPPA number to the lignin content of pulp materials. Tappi. 1957;40:699–704.
[111] Tasman JE. The permanganate consumption of pulp materials. IV. The KAPPA correction coefficient. Pulp Pap Mag Can. 1959;60(No. C):231–235.
[112] Jayme G, Jerratsch HJ. The Kj-method, a modification of the Kappa number determination, as applied to semichemical pulps obtained in yields over 70 per cent. Papier. 1960;14:718–723.
[113] Berzins V. Micro kappa numbers. Pulp Pap Mag Can. 1966;67(3):T206–208.
[114] Birkett MD, Gambino MJT. Estimation of pulp kappa number with near-infrared spectroscopy. Tappi J. 1989;72(9):193–197.
[115] Kyrklund B, Strandell G. Applicability of the chlorine number for evaluation of the lignin content in pulp. Pap Puu. 1969;51:299–305.
[116] Kyrklund B, Strandell G. A modified chlorine number for evaluation of cooking degree of high yield pulps. Pap Puu. 1967;49:99–106.
[117] Loras V, Loschbrandt F. Determination of lignin in sulfite pulps. Nor Skogind. 1961;15:302–309.
[118] Kyrklund B, Palenius I. The total lignin content of sulphite pulps and its relation to the kappa number. Pap Puu. 1964;46:513–520.
[119] Hatton JV. Kappa number-permanganate number relations for softwoods and hardwoods. Tappi. 1975;58(10):150–151.
[120] Berzins V. A rapid procedure for the determination of kappa number. Pulp Pap Mag Can. 1965;48(1):15–20.
[121] Genco JM, Hassler JC, Busayasakul N. Total organic carbon as an indicator of wood delignification. Tappi J. 1984;67(7):92–95.
[122] Lourenco A, Gominho J, Marques AV, Pereira H. Comparison of Py-GC/FID and wet chemistry analysis for lignin determination in wood and pulps from *Eucalyptus globulus*. BioResources. 2013;8(2):2967–2980.
[123] Fukushima RS, Hatfield RD. Comparison of the acetyl bromide spectrophotometric method with other analytical lignin methods for determining lignin concentration in forage samples. J Agric Food Chem. 2004;52(12):3713–3720.
[124] Jung HJG, Varel VH, Weimer PJ, Ralph J. Accuracy of klason lignin and acid detergent lignin methods as assessed by bomb calorimetry. J Agric Food Chem. 1999;47(5):2005–2008.
[125] Iiyama K, Wallis AFA. Effect of acetyl bromide treatment on the ultraviolet spectra of lignin model compounds. Holzforschung. 1989;43(5):309–316.
[126] Bose SK, Wilson KL, Francis RC, Aoyama M. Lignin analysis by permanganate oxidation. Part 1. Native spruce lignin. Holzforschung. 1998;52(3):297–303.
[127] Yeh TF, Chang HM, Kadla JF. Rapid prediction of solid wood lignin content using transmittance near-infrared spectroscopy. J Agric Food Chem. 2004;52(6):1435–1439.

[128] Dang VQ, Bhardwaj NK, Hoang V, Nguyen KL. Determination of lignin content in high-yield kraft pulps using photoacoustic rapid scan Fourier transform infrared spectroscopy. Carbohydr Polym. 2007;68(3):489–494.

[129] Yao S, Wu G, Xing M, Zhou S, Pu J. Determination of lignin content in acacia spp. using near-infrared reflectance spectroscopy. BioResources. 2010;5(2):556–562.

[130] Parkås J, Brunow G, Lundquist K. Quantitative lignin analysis based on permanganate oxidation. BioResources. 2007;2(2):169–178.

[131] Sykes R, Yung M, Novaes E, Kirst M, Peter G, Davis M. High-throughput screening of plant cell-wall composition using pyrolysis molecular beam mass spectroscopy. Methods Mol Biol. 2009;581(Biofuels):169–183.

4
Biosynthesis of Lignin

4.1 Introduction

Lignin is the defining constituent of wood and the second most abundant natural polymer on earth after cellulose. It is produced mainly by the oxidative coupling of three monolignols: *p*-coumaryl alcohol, coniferyl alcohol and sinapyl alcohol [1] (see Figure 4.1). The process of lignin formation is called lignification.

Plant lignification is a tightly regulated complex cellular process that occurs *via* three sequential steps:

- The synthesis of monolignols.
- The transport of monomeric precursors across plasma membrane.
- The oxidative polymerization of monolignols to form lignin macromolecules within the cell wall.

Although there is a reasonable understanding of monolignol biosynthesis, some aspects of lignin assembly remain elusive [2]. Many of these issues have already been raised by Neish [3], some of which remain unresolved today.

Two major hypotheses for a unified mechanism of lignin biosynthesis are frequently debated in the literature. The first one is the combinatorial random coupling [3, 4] between monolignol radicals. This is the most widely accepted mechanism for lignin polymerization, but there is a second hypothesis on protein-directed synthesis involving dirigent proteins (DIR) [5, 6].

The lack of knowledge on the process is mainly due to the difficulty of its study. Thus, Sederoff *et al.* [7] pointed out that there is no tissue or plant species in which the biosynthetic pathway of lignin has been completely characterized.

4.2 The Biological Function of Lignins

One basic question concerning lignin biosynthesis is whether it is possible to imagine the nature of lignin in plants without polyphenols. They have been a basic feature of plants ever since their

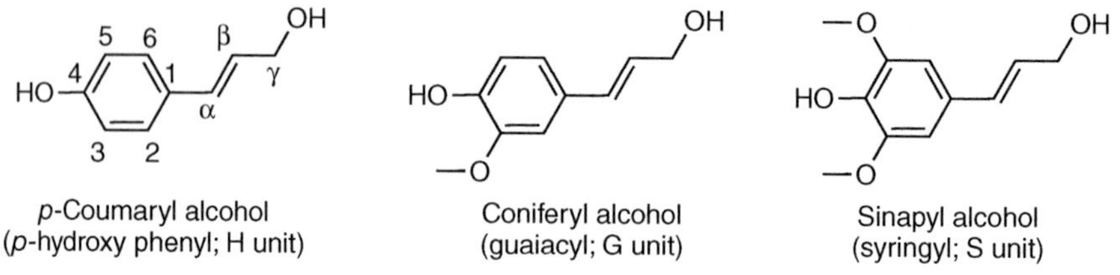

Figure 4.1 Structure of monolignols

early colonization of land. Polyphenolic molecules can be considered secondary metabolites and they are indeed relevant for many key functional aspects of plant life. Several roles can be attributed to polyphenols in plants, including structural roles in different supporting or protective tissues. Another role may be related to defense strategies and signaling properties, including the interactions between plants and their close environment. Higher plants synthesize several thousand different known phenolic compounds, and the number of these fully characterized is continually increasing [8].

Lignin is primarily synthesized and deposited in the secondary cell wall of specialized cells such as xylem vessels, tracheids, and fibers. It is also deposited in minor amounts in the periderm [4].

Lignin has at least the following four key roles in plants [9]:

- Lignin gives rigidity to cell walls: Lignin cements and fixes polymers in the woody plant cell walls, in close association with polysaccharides, so that it makes the fibers relatively stiff, and able to serve as a mechanical support for stems and branches.
- Lignin glues different cells together in woody tissues: In wood, the middle lamella consists mainly of lignin, which works as an efficient and resistant glue, keeping together the different cells.
- Lignin makes the cell wall hydrophobic: The polymer inhibits the swelling of the cell walls in water, and thereby that water leaks from a woody cell wall, that is, it makes the cell wall waterproof. In nonwoody plants, this is the main function of lignin [10].
- Lignin protects against microbial degradation of wood: A lignified woody tissue is simply so compact that the polysaccharide-degrading proteins excreted by microorganisms cannot penetrate the cell wall [10]. Thus, it serves as a barrier against microorganisms. Some specialized fungi and bacteria can, however, degrade lignin efficiently, and they have potential applications in pulp and paper industry, as will be further discussed in Chapter 7. Nevertheless, wood is degraded much slower than nonlignified plant materials.

The structure of lignin makes it very suitable for these functions. An additional role for lignin has been reported [11] involving complexes of lignin phenolic acids in forage legumes and grasses. The presence of lignin phenolic acids is thought to inhibit the digestion of potentially digestible carbohydrates by ruminants [12].

4.3 The Shikimic Acid Pathway

Lignins and lignans are both biosynthesized through the same pathway in the earlier steps (see Figure 4.2). Thus, the lignin biosynthesis is developed through a long sequence of reactions involving the following:

- The shikimate pathway, which produces L-phenylalanine and L-tyrosine.
- Phenylpropanoid common pathway from L-phenylalanine (and/or L-tyrosine) to *p*-hydroxycinnamoyl-CoA.
- The specific pathway of the biosynthesis of lignins, which leads to the *p*-hydroxycinnamoyl-CoA synthesis of cinnamyl alcohols.

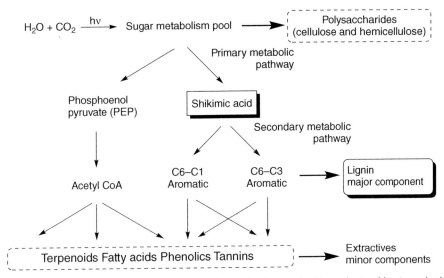

Figure 4.2 *Primary and secondary metabolic pathways leading to the biosynthesis of lignin and other wood components [13]*

Table 4.1 The eight core monolignol biosynthetic enzymes[a]

Enzyme	Enzyme family
Phenylalanine ammonia-lyase (PAL)	Lyase, class I-like
Cinnamic acid 4-hydroxylase (C4H)	Cytochrome P450 monooxygenase
p-Coumaroyl-CoA ligase (4CL)	Acyl-CoA synthetase
Hydroxycinnamoyltransferase (HCT)	BAHD acyltransferase
p-Coumarate 3-hydroxylase (C3H)	Cytochrome P450 monooxygenase
Caffeoyl-CoA O-methyltransferase (CCoAOMT)	SAM-dependent methyltransferase
Cinnamoyl-CoA reductase (CCR)	Dehydrogenase
Cinnamyl alcohol dehydrogenase (CAD)	Dehydrogenase

[a]Data obtained from ref. [43].

- The oxidation of cinnamyl alcohols to yield the corresponding 4-O-phenoxy radicals, which polymerize spontaneously to give a growing lignin polymer.

The shikimate pathway, which is present in bacteria, yeasts, and plants, but not in animals, is the entry pathway toward an overabundance of phenolic compounds. This plastid-localized[1] pathway is highly transcriptionally controlled [14, 15]. In seven enzymatic steps, the glycolytic intermediate phosphoenol pyruvate and the pentose phosphate pathway intermediate erythrose-4-phosphate are metabolized into chorismate *via* 3-dehydroshikimate as an intermediate [16]. These enzymes are 3-deoxy-D-arabinose heptulosonic acid-7-phosphate (DAHP) synthase; dehydroquinate synthase; 3-dehydroquinate dehydratase; 3-dehydroshikimate reductase; shikimate kinase; 5-enolpyruvylshikimate-3-phosphate (EPSP) synthase; and chorismate synthase (see Figure 4.3 and Table 4.1).

---

[1] The plastid is a major organelle found in the cells of plants and algae.

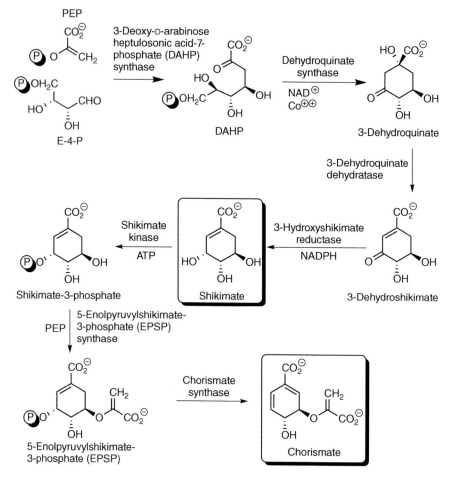

Figure 4.3 The shikimate–chorismate pathway [17]

Although the biosynthetic pathway is not yet fully elucidated, 3-dehydroshikimate also serves as a precursor for gallic acid and, thus, gallotannin biosynthesis [18, 19]. Chorismate serves as the precursor for p-aminobenzoate, which is an intermediate in tetrahydrofolate biosynthesis [20], and the aromatic amino acids phenylalanine, tyrosine, and tryptophan [21].

Phenylalanine and tyrosine are produced from chorismate via arogenate by the action of three enzymes: chorismate mutase; prephenate aminotransferase; arogenate dehydrogenase; or arogenate dehydratase (see Figure 4.4). Phenylalanine is produced from chorismate by the action of two enzymes, that is, a dehydratase and an aminotransferase, through the exact order of action is unknown [22–24].

4.4 The Common Phenylpropanoid Pathway

In 1971, Higuchi [25] demonstrated that lignin is synthesized from L-phenylalanine and cinnamic acids [26]. These acids are derived from carbohydrates through the shikimic and cinnamic acid

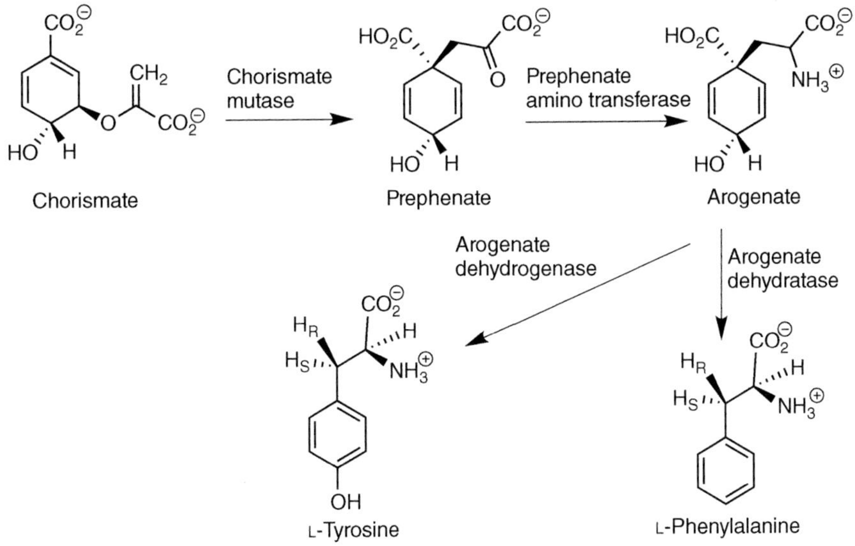

Figure 4.4 Biosynthetic pathway from chorismate to L-tyrosine and L-phenylalanine via arogenate [17]

pathways. Supporting evidence for this pathway was gained when radioactive glucose (^{14}C labeled) was administered into plants, producing shikimic acid [27], and radioactive lignins [28, 29].

The general phenylpropanoid pathway uses phenylalanine as an entry substrate and results, after seven steps, in feruloyl-CoA [1, 30]. Following the deamination of phenylalanine to cinnamate, this deamination process is catalyzed by L-phenylalanine ammonia-lyase (PAL), a key enzyme found only in plants that can synthesize lignin [28, 29, 31] and some cinnamic acid derivatives (see Figure 4.5).

In grasses, a PAL isozyme catalyzing the deamination of both phenylalanine (PAL activity) and tyrosine (tyrosine ammonia-lyase activity) *in vitro* might be involved in the direct conversion of tyrosine to *p*-coumarate *in vivo* [32].

It bears mentioning that an additional enzyme, tyrosine ammonia-lyase (TAL), which catalyzes the formation of *p*-coumaric acid from L-tyrosine, is characteristically found only in grasses [33–35]. The presence of this enzyme may account for the presence of *p*-coumaryl alcohol as an additional lignin monomer as well as esterified *p*-coumaric acid present mainly in grasses.

Hydroxylation of the aromatic ring of the cinnamate leads to *p*-coumarate, a reaction catalyzed by cinnamate 4-hydroxylase (C4H). Activation of the acid to a thioester by 4-coumarate CoA ligase (4CL) yields *p*-coumaroyl-CoA. The subsequent 3-hydroxylation of *p*-coumaroyl-CoA to caffeoyl-CoA involves three enzymatic steps, at least in dicots. First, *p*-coumaroyl-CoA is transesterified to its quinic or shikimic acid ester derivative by hydroxycinnamoyl-CoA: shikimate/quinate hydroxycinnamoyltransferase (HCT). *p*-Coumaroyl shikimate or quinate is then hydroxylated by *p*-coumarate 3-hydroxylase (C3H, named when it was assumed that *p*-coumarate was the direct substrate) and then transesterified again by HCT to caffeoyl-CoA.

As lignification proceeds, cinnamic acid is hydroxylated to *p*-coumaric and caffeic acids by specific hydroxylase enzymes [36, 37]. The caffeic acid thus formed is then methylated to ferulic acid by *O*-methyltransferase (OMT) [38–40]. Up to this point, the biosynthetic pathways for softwood and hardwood lignins are believed to be common [25].

Recently, an alternative 3-hydroxylation pathway has been found; the poplar (*Populus trichocarpa*) heterodimeric C4H/C3H protein complex efficiently converts *p*-coumaric acid into caffeic acid [41], after which 4CL might convert caffeic acid into caffeoyl-CoA. Further methylation of the

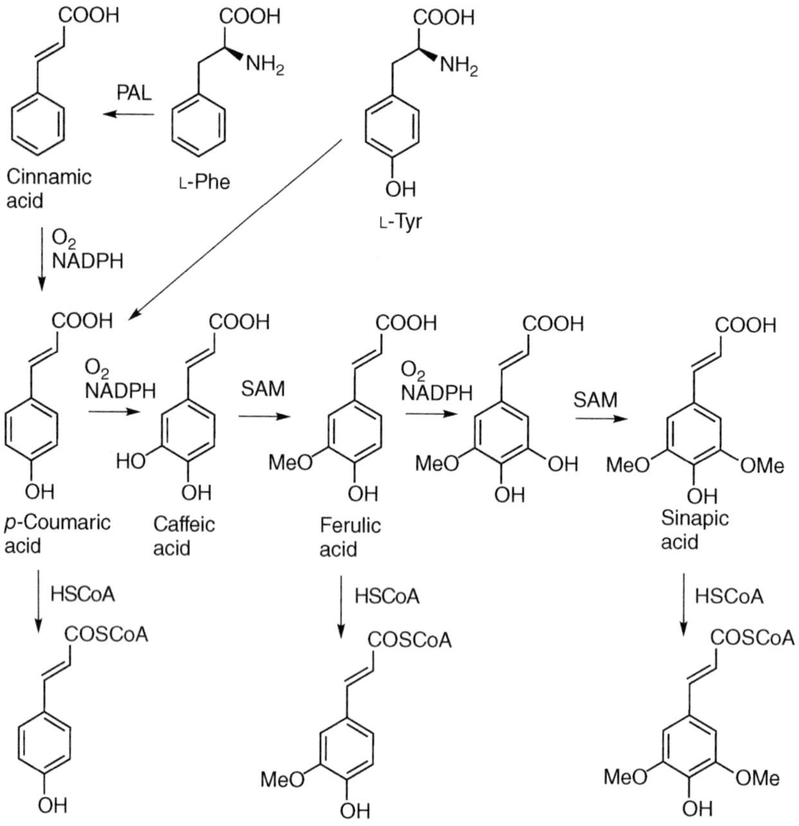

Figure 4.5 *Common phenylpropanoid pathway. PAL, Phenylalanine ammonia-lyase; TAL, tyrosine ammonia-lyase; COMT, CoA O-methyltransferase; C4H, cinnamic acid 4-hydroxylase; C3H, emphp–coumarate 3-hydroxylase; NADPH, nicotinamide adenine dinucleotide phosphate; CCoAOTM, caffeoyl–CoA O-methyltransferase; F6H, flavonol 6-hydroxylase; 4CL, emphp-coumaroyl-CoA ligase; HSCoA, coenzyme A (CoA, CoASH, or HSCoA); CCoA3H, 4-hydroxycinnamoyl-CoA 3-hydroxylase [42]*

3-hydroxyl group by caffeoyl-CoA O-methyltransferase (CCoAOMT) yields feruloyl-CoA. The eight core monolignol biosynthetic enzymes are summarized in Table 4.1.

Various pathways branch off from the general phenylpropanoid pathway, including the monolignol-specific pathway and the pathways toward flavonoids, benzenoids, and coumarins as well as sinapate and ferulate esters (see Figure 4.6).

4.5 The Biosynthesis of Lignin Precursors (the Monolignol-Specific Pathway)

The lignin biosynthetic pathway can be subdivided into four steps starting with the following:

- The biosynthesis of monolignols.
- Transport of monolignols from the site of synthesis to the site of polymerization.
- Dehydrogenation.
- Polymerization of monolignols to give the final product.

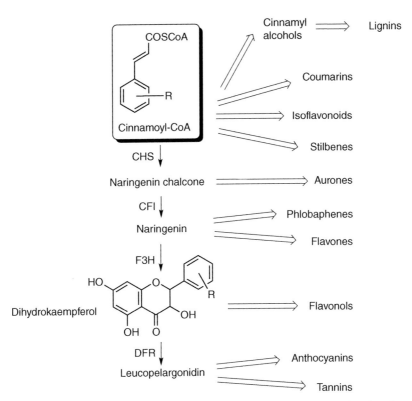

Figure 4.6 Schematic view of the central role of cinnamoyl-coenzyme A in the phenylpropanoid metabolism [48]

The hydroxycinnamyl alcohol monomers serve as the precursor of lignans, according to Dewick [42], and are produced from L-phenylalanine and L-tyrosine, mediated by a series of cinnamic acid derivatives. The reduction of these acids leads to three alcohols (p-coumaryl alcohol, coniferyl alcohol, and sinapyl alcohol), which are the main precursors for all lignins and lignans (see Figure 4.6).

The monolignol-specific pathway includes four well-studied enzymatic steps that convert feruloyl-CoA into the monolignols coniferyl alcohol and sinapyl alcohol [1, 30].

First, feruloyl-CoA is reduced to coniferaldehyde by cinnamoyl-CoA reductase (CCR). The hydroxylation at the 5-position is catalyzed by ferulate 5-hydroxylase (F5H), which is also now often called coniferaldehyde 5-hydroxylase (CAld5H) to reflect its preferred substrate, and 5-hydroxyconiferaldehyde is produced [44, 45] (see Figure 4.7). The subsequent methylation of the newly introduced 5-hydroxyl group is catalyzed by caffeic acid O-methyltransferase (COMT), the preferred substrate of which is also now known to be the aldehyde [46, 47], to provide sinapaldehyde. Further reduction to its corresponding alcohols, coniferyl alcohol and sinapyl alcohol, is catalyzed by cinnamyl alcohol dehydrogenase (CAD).

Data demonstrate that CAD, not SAD, is the enzyme responsible for S lignin biosynthesis in woody angiosperm xylem [49].

The monolignol biosynthetic pathway envisages a metabolic grid leading to G and S units, through which the O-methylation and hydroxylation reactions may occur at different levels. Dixon et al. [51] in 2001, postulated an alternative model suggesting independent pathways to G and S lignins.

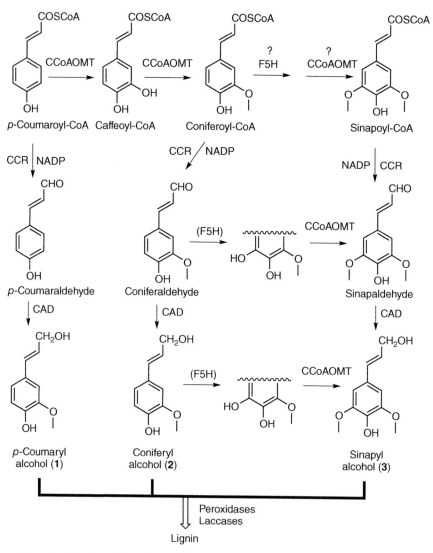

Figure 4.7 Monolignol-specific pathway to lignins. CCoAOMT, caffeoyl-CoA O-methyltransferase; CCR, cinnamoyl-CoA reductase; NADP oxidoreductase; CAD, cinnamyl alcohol dehydrogenase; F5H, ferulate 5-hydroxylase [50]

In 2003, Amthor [52] made a quantitative analysis on the efficiency of lignin biosynthesis,[2] and found that the biosynthesis *via* tyrosine (in monocots) is higher than that *via* phenylalanine, in all plants.

4.5.1 The Biosynthesis of Other Monolignols

The general phenylpropanoid- and monolignol-specific pathways also provide hydroxycinnamic acids, which include *p*-coumaric, caffeic, ferulic, and sinapic acids. Also, dihydro-hydroxycinnamyl

---
[2] Carbon percent (%) of sucrose retained in lignin.

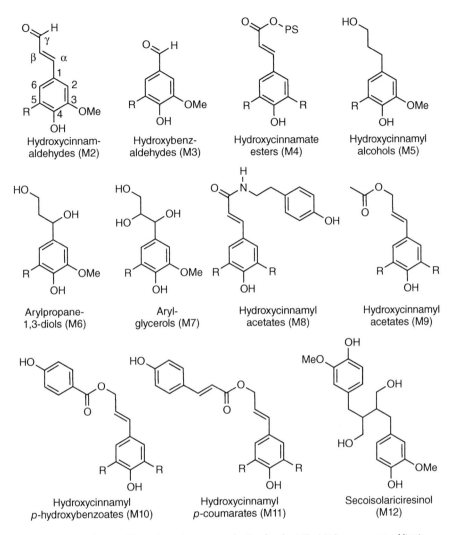

Figure 4.8 Chemical formula and atom numbering for the M2–M12 monomers of lignin

alcohols, hydroxybenzaldehydes, and hydroxycinnamic acids and other products from an incomplete monolignol biosynthesis, such as hydroxycinnamaldehydes, are found in lignins of wild-type plants [1, 53–57] (see Figure 4.8).

For instance, traditional monolignols are often acylated at their C-position with acetate, *p*-hydroxybenzoate or *p*-coumarate groups [58–64]. Such acylated units can even be highly abundant; for example, coniferyl and sinapyl acetate may constitute 50% or more of the units in lignin from kenaf (*Hibiscus cannabinus*) [63, 65].

4.5.2 The Transport of Monolignols

The synthesis of monolignols occurs in the cytoplasm. To reach the cell wall, they must be transported across the cell membrane (see Figure 4.9). The transport of these precursors can occur by passive

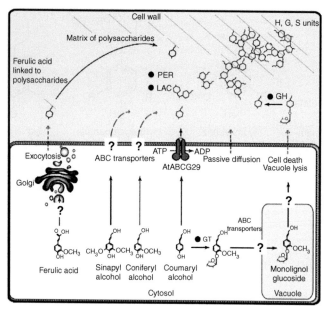

Figure 4.9 Monolignol biosynthesis and transport in the plant cell. Three main lignin building blocks (p-coumaryl alcohol, coniferyl alcohol, and sinapyl alcohol) are synthesized in the cytosol. Monolignols are transported from the cytosol to different locations: to the cell wall for oxidative cross-linking by apoplastic peroxidases (PER) and laccases (LAC) into lignins; into the vacuole for storage as glucoconjugates and, for ferulic acid, into the Golgi apparatus for incorporation into polysaccharides (pectins or arabinoxylans). Most of these transport routes still remain to be discovered. In principle, these hydrophobic molecules may passively diffuse through membranes, undergo active transport through membrane transporters or through Golgi-derived vesicles or simply be released from dying cells. p-Coumaryl alcohol is exported across the plasma membrane by the AtABCG29 transporter, whereas free coniferyl alcohol and sinapyl alcohol are exported by other, so far unknown ABC transporters. Monolignols are selectively imported into the vacuole as glucoconjugates by unknown ABC transporters. Monolignol glucoconjugates are generated by cytosolic glucosyl transferases (GT) and need to be deglucosylated by the cell wall-associated glucosylhydrolases (GH) before incorporation into lignin polymers. A hypothetical transport route of ferulic acid into Golgi vesicles is also represented. Copyright from ref. [67]. (See insert for color representation of this figure.)

diffusion or by an energized active process. Miao and Liu [66] demonstrated that the transport across plasma membrane and their sequestration into vacuoles involved ATP-binding cassette-like transporters, that recognize different chemical forms (aglycones or glycoconjugates) in conveying them to distinct sites.

Monolignols, such as those synthesized by plants, are toxic for the same plants [68]; this is a mechanism to protect plants themselves for the intrinsic toxicity of monolignols and to provide an efficient transport along the plants, and their conversion into glycosides, which are easily transported to the cell *via* the cell membrane [1, 69] because the monolignols are insoluble in plants. Most of these glycosides are glucose derivatives. Monolignols are converted into their respective glucosides by the action of enzyme UDP-glycosyltransferase and later are hydrolyzed to the corresponding H, G, and S units, respectively, in the cell wall. This conversion again renders the monolignols soluble and less toxic to the plant cells.

These former glucosides can be stored in plant vacuoles, which are transported to the cell wall as the need arises. In gymnosperms and some angiosperms, monolignol 4-O-β-D-glucosides accumulate

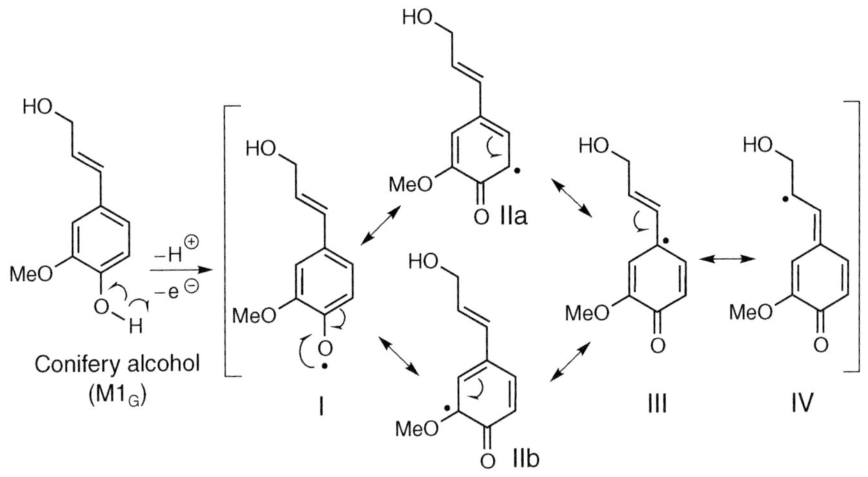

Figure 4.10 Resonance Lewis structures for the coniferyl alcohol radical [42]

to high levels in the cambial tissues [70]. After the synthesis process, the lignin precursors or monolignols are transported to the cell wall and, on arriving, they are oxidized and polymerized. It has been hypothesized that these monolignol glucosides are in fact storage or transport forms of the monolignols. It has been proposed that an UDP-glucose, the so-called coniferyl alcohol glucosyltransferase (CAGT) [70], together with coniferin-β-glucosidase (CBG), may regulate storage and mobilization of monolignols for lignan and/or lignin biosynthesis [71].

4.6 The Dehydrogenation of the Precursors

After their biosynthesis, monolignols are translocated to the apoplast *via* a largely unsolved mechanism [66]. The dehydrogenation process of monolignols to produce the corresponding radicals has been attributed to enzymes of several classes, such as peroxidases,[3] laccases,[4] polyphenol oxidases, and coniferyl alcohol oxidases. These enzymes, which act alone or in combination with several of them, are responsible for the dehydrogenation of the monolignols in plants, but whether monolignol oxidation occurs through redox shuttle-mediated oxidation remains unclear [72]. Upon entering the cell wall matrix, monolignols are oxidized either by peroxidases or laccases to monolignol radicals.

The dehydrogenative polymerization of lignin monomers in plants is caused by a class of enzymes called peroxidases or the peroxidase-H_2O_2 system. These enzymes are capable of abstracting a proton from the phenolic hydroxyls of the precursor molecules, creating resonance-stabilized free radicals [12]. These radicals can be detected and accurately quantified using ^{31}P NMR when they react with a nitroxide phosphorus compound [73, 74].

Peroxidase induces one-electron oxidation of the phenol group, allowing the delocalization of the unpaired electron through resonance forms (see Figure 4.10). Direct contact between the peroxidase/laccase and the substrate is not needed. Radical-transfer reactions can also pass the radical from one molecule to another.

---

[3] A group of enzymes containing the heme group, which use H_2O_2 for activation.
[4] Copper-containing enzymes that use molecular oxygen for activation.

4.7 Peroxidases and Laccases

Usually, peroxidases are considered the main enzymes involved in the polymerization step to form lignin, since both have a spatial and temporal correlation with the cell wall lignification (see Figure 4.11). However, there is a resurgence of those theories in which other enzymes participate in the polymerization pathway, such as laccases [75, 76]. From these results, the role that these other enzymes play in polymerization processes of cinnamyl alcohols became relevant.

In 1990, Lewis and Yamamoto [79] set four basic criteria that should be met by those enzymes involved in the polymerization of cinnamyl alcohols in the cell wall. These criteria are the following:

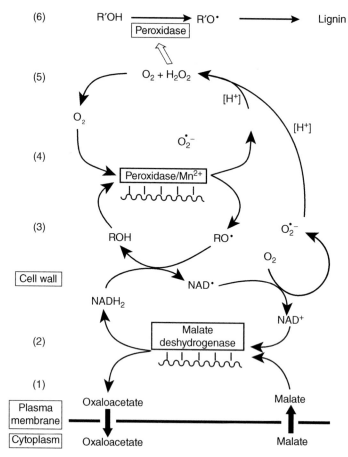

Figure 4.11 *Proposed scheme of reactions generating H_2O_2 by wall-bound/associated malate dehydrogenase and peroxidase in the plant cell walls [77]. In summary, malate is oxidized by O_2 to yield oxaloacetate and H_2O_2 that can be utilized for the formation of monolignol radicals to initiate lignin formation. (1) Well-documented malate–oxaloacetate shuttles transport reducing equivalents across the plasma membrane; (2) bound malate dehydrogenase (MDH) in lignifying the cell walls of Forsythia differs from other MDHs; (3) most efficient monophenol (ROH) stimulation of H_2O_2 formation in the cell walls was affected by coniferyl alcohol; (4) high concentrations of Mn in wood have been reported; (5) H_2O_2 production via the superoxide radical ($O_2^{\bullet -}$) cycle is driven by O_2, ROH, and wall-bound peroxidase; (6) H_2O_2 formation largely depends on the availability of monolignols in the cell walls [78]*

- Display substrate specificity of cinnamyl alcohols.
- Have present a subcellular localization in the cell wall.
- Have a known primary structure.
- Present a temporal correlation with the phases of active cell wall lignification.

To these four criteria, Ros Barceló [80] added two new ones:

- This enzyme should be widely distributed in all vascular plants known hitherto.
- The enzyme should have a high affinity for alcohols cinnamyl compounds during oxidation of lignin type.

Very few enzymes currently believed to be involved in lignification meet all these aforementioned requirements.

For example, laccases have low affinity to cinnamyl alcohols, are clearly unable to form highly polymerized lignin-type compounds, and, in many cases, lack a quantitative correlation between the levels of the enzyme and lignification [81].

In the case of peroxidases, it is difficult to imagine that a single isozyme is exclusively responsible for such a complex and organized pathway such as the polymerization of cinnamyl alcohols to lignin. However, peroxidases meet many of the previously defined requirements. Thus, these enzymes exhibit a high affinity toward cinnamyl alcohol, during oxidation of lignin-type compounds, they show a quantitative correlation with the cell wall lignification and, finally, are present in high levels in all tissues of vascular plants.

Two laccases involved in lignification have been identified [82]. A review of the literature regarding plant laccases and selected fungal laccases and their role during lignin biosynthesis has been made by Dean and Eriksson [83].

Peroxidases (H_2O_2: hydrogen donating: H_2O_2 oxyreductases) are hemoproteins that catalyze the monoelectronic oxidation of different substrates by H_2O_2:

$$2RH + H_2O_2 \longrightarrow 2R + 2H_2O \tag{4.1}$$

Peroxidases are widely distributed in the plant kingdom and are classified into two major superfamilies: one that includes plant, fungal, and bacterial peroxidases; and another constituted by animal peroxidases [84, 85]. At the same time, within the superfamily of plant, fungal, and bacterial peroxidases, there are three classes that can be defined based on the structural differences between them:

- Class I: These are glycoprotein in nature and show moderate substrate specificity for ascorbic acid. Additionally, these class I peroxidases are inhibited by thiol reagents such as p-chloromercuribenzoate, and are considered generally very thermolabile.
- Class II: These include all the fungal secretory peroxidases (manganese peroxidases).
- Class III: These are all the plant secretory peroxidases. Glycoprotein nature, they are located in vacuoles and the cell walls. These peroxidases show low substrate specificity, with a modest, but significant, affinity for coniferyl alcohol, in addition to presenting an unusual thermal stability. All these features distinguish plant ascorbate peroxidases (class I).

4.8 The Radical Polymerization

Decades ago, Freudenberg and Neish [86] determined an average elementary composition for spruce lignin of $C_9H_{7.15}O_2-(H_2O)_{0.4}(OCH_3)_{0.92}$. This formula closely resembles that of coniferyl aldehyde (**5**), $C_9H_7O_2(OCH_3)$, indicating that the lignin of conifers roughly corresponds to the multiple of a slightly oxidized coniferyl alcohol unit. In other words, lignins are polymerized oxidation products of monolignols [78].

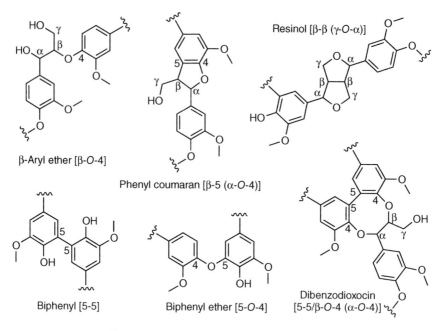

Figure 4.12 *Major linkages present in lignins [43]*

The monolignols that result from the above-described monolignol-specific pathway are used for at least three different product classes: oligolignols/lignin, monolignol 4-*O*-hexosides, and neolignans.

Oligolignols are racemic radical coupling products of monolignols that arise during lignin polymerization [62, 87, 88] (see Figure 4.12).

Lignans are formed by the initial stereospecific β–β coupling of two monolignol radicals [89] (see Figure 4.13). Secondary metabolites arising from two monolignol radicals that are stereospecifically β-*O*-4- or β-5-coupled are called neolignans [89]. Stereospecific coupling reactions in neolignan biosynthesis appear to be assisted by DIR [5, 89–91]. Because of their antioxidant properties, neolignans are believed to be involved in defense responses [5]. Some may also have a hormonal function; dehydrodiconiferyl alcohol-4-β-D-glucoside (DCG) has been associated with cell division-promoting activities [92–94].

The third metabolic class derived from monolignols includes the 4-*O*-glucosylated monolignols.[5] Several glucosyltransferases involved in their biosynthesis have been described in Arabidopsis,[6] as well as β-glucosidases that convert the monolignol 4-*O*-glucosides back to their respective aglycones [95–97]. The biological role of monolignol 4-*O*-hexosides has not been unequivocally defined, but they could serve as storage forms for their aglycones [95, 98]. This hypothesis is supported by the finding that monolignol 4-*O*-glucosides are sequestered into the vacuoles of *Arabidopsis*, whereas monolignol aglycones are transported to the apoplast [66, 99].

4.8.1 Dimerization

A monolignol free radical can then undergo radical coupling reactions, producing a variety of dimers, termed dilignols. The combination of monomeric radicals through only β-*O*-4 and β-5 coupling would

---

[5] For example, coniferin and syringin.
[6] They are small flowering plants related to cabbage and mustard.

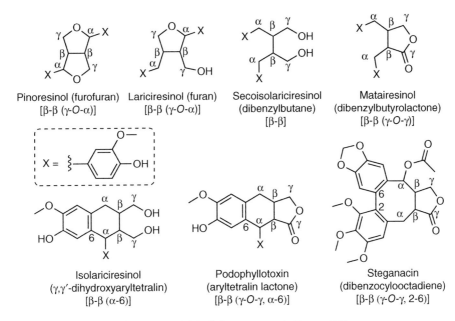

Figure 4.13 Major linkages present in lignans [43]

lead to a linear polymer. However, branching of the polymer may take place through subsequent nucleophilic attack by water, alcohols, or phenolic hydroxyl groups on the benzyl carbon of the quinone methide intermediate. The dilignols then undergo further endwise polymerization, instead of combining with one another (see Figures 4.14 and 4.15).

Monolignol radicals are formed, which can react in different ways to form dimers. For example, when two coniferyl alcohol radicals react, they can form 15 possible dimers, of which only five are stable enough. These dimers can be oxidized to form new radicals that can react either with a new monolignol radical or with another dimeric radical, and so on. This is the basis for the very large resulting complex polymeric species [101].

The three monolignols differ in the degree of O-methylation (see Figures 4.14 and 4.15), each one with five mesomeric radical forms, because it is possible for many substructures to form during dimerization [102].

$$C_{m,n} = \frac{m(m+1)}{n(n-1)} = 120; \qquad (4.2)$$

where m is the number of mesomeric forms ($m = 3 \times 5 = 15$), and n is the degree of polymerization ($n = 2$).

In lignins, the phenylpropanoid units are interconnected through ether bonds and carbon-carbon bonds [103], leading to major substructures: aryl-glycerol-β-aryl ether, phenylcoumaran, diarylpropane, resinol, biphenyl and diphenyl ether, and other minor ones (see Figure 4.12).

Monolignol dimerization and lignification are substantially different processes [104], explaining why lignification produces frequencies of the various units that are different from those produced by dimerization or bulk polymerization *in vitro* [105].

4.8.2 Quinone Methides

The polymerization process proceeds with the coupling of the different resonance structures. Among the products, highly reactive quinone methides are formed and further react by addition to

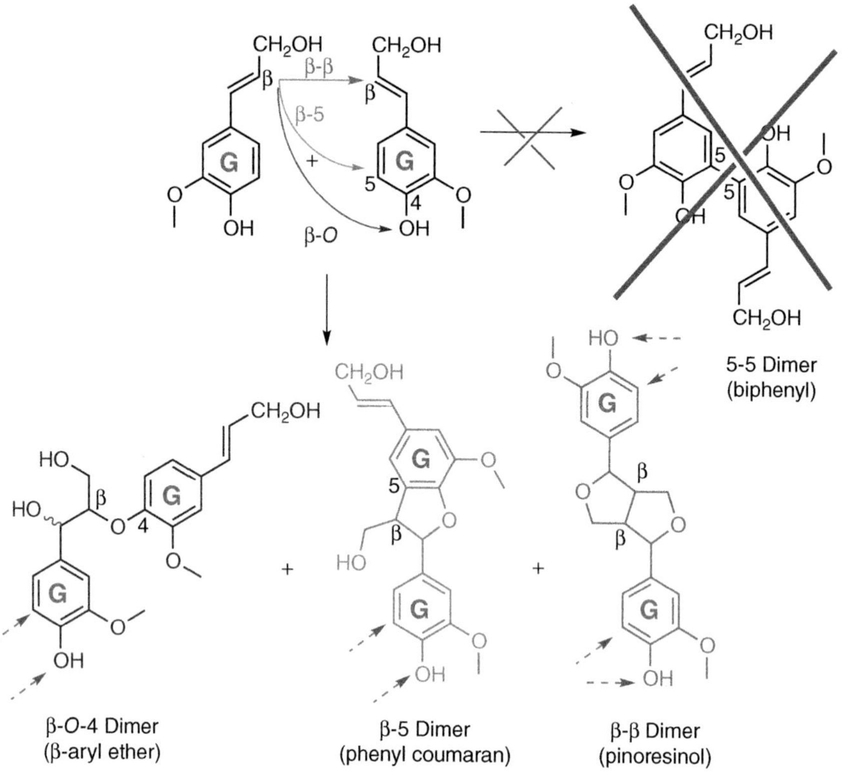

Figure 4.14 *Dimerization of monolignols: coniferyl alcohol dehydrodimerization [100]. (See insert for color representation of this figure.)*

various nucleophiles. Free-radical-exchange reactions, such as hydrogen abstraction from suitable portions of other molecules, are also possible [106, 107] (see Figure 4.16).

Addition reactions to a quinone methide lead to the formation of the various interunit linkages in lignin (see Figure 4.16). Recently Ralph et al. [108] published a review on the role of quinone methides in lignification.

The first dilignol to be isolated and identified was a phenylcoumaran structure analogous to that of dehydrodiconiferyl alcohol [109]. This structure, which comprises about 10% of the lignin units, is formed by the coupling of two free coniferyl alcohol radicals centered at C_5 and C_β positions of the phenylpropanoid (structures II and IV in Figure 4.10), giving rise to a dimeric quinone methide. This coupling followed by aromatization of the ring allows the creation of an α-O-4 linkage (see Figure 4.16).

4.8.3 Lignification

The polymerization step is the last step in the biosynthesis of lignins from cinnamyl alcohol. After the oxidation of the monomeric alcohols to phenoxy radicals, the reaction changes dramatically.

The reactions are no longer subjected to enzymatic control, but to a random polymerization process [110–113].

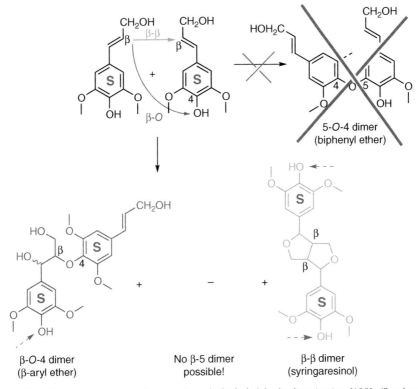

Figure 4.15 Dimerization of monolignols: sinapyl alcohol dehydrodimerization [100]. (See insert for color representation of this figure.)

Because radical coupling is a purely chemically driven process, independent of control by any proteinaceous agent, any phenolic molecule having the proper chemical kinetic and thermodynamic radical-generation and cross-coupling propensities can couple into the lignin polymer [1, 58, 59, 61, 62, 88, 114] (see Figures 4.17 and 4.18).

The monolignol radical is resonance-stabilized, having various sites of enhanced single-electron density in the molecule. Mutual coupling of monolignols (dimerization) and cross-coupling with the growing polymer lead not only to the characteristic H, G, and S units but also to various interunit linkage types.

These monomers are linked together *via* endwise and radical coupling reactions to produce H, G, and S lignin, respectively [4, 86]. This process is known as lignification.

This reaction produces an optically inactive hydrophobic heteropolymer [115], composed of H (*p*-hydroxyphenyl), G (guaiacyl), and S (syringyl) units.

4.8.4 Interunit Linkage Types

At least 60% of the interunit linkages in dicots are β-aryl ethers arising from β-*O*-4 coupling of a monolignol at its β-position to the 4-*O*-position of the growing oligomer (see Figure 4.19). These β-aryl ether linkages can, unlike other prevalent interunit linkages, be cleaved by harsh alkaline or acidic pretreatment of the lignocellulosic biomass [4].

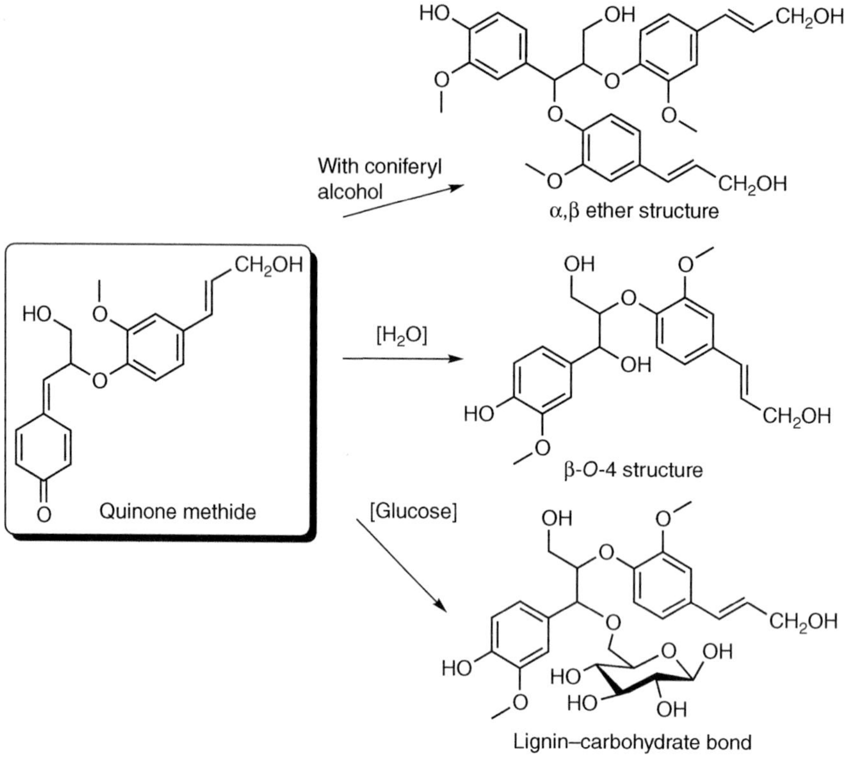

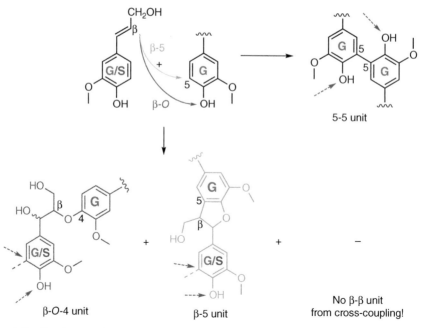

Figure 4.16 Formation of different interunit linkages in lignin via addition reactions to a quinone methide [12]

Figure 4.17 Lignification of monolignols: monolignol cross-coupling with a G-end unit [100]. (See insert for color representation of this figure.)

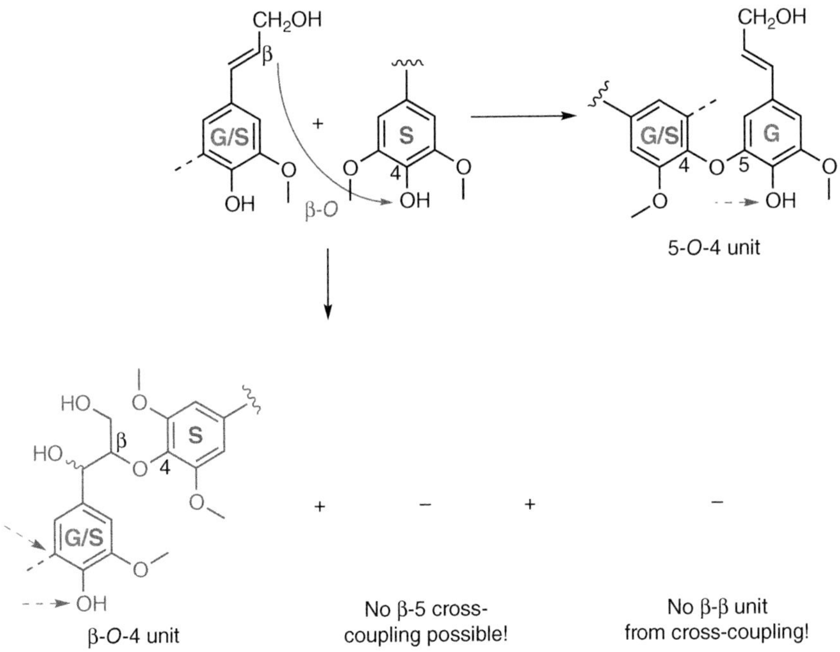

Figure 4.18 *Lignification of monolignols: monolignol cross-coupling with an S-end unit [100]. (See insert for color representation of this figure.)*

The two other major interunit linkages in lignin are phenylcoumarans and resinols formed by β-5 and β–β coupling, respectively (see Figure 4.19). Both are carbon–carbon linkages (among the so-called "condensed linkages") that can be broken only under extremely harsh conditions that would also degrade polysaccharides.

For the three major types of linkages, the incoming monolignol radical reacts exclusively at its β-position, enabling the resulting 4-O-phenolic function produced after rearomatization of the quinone methide intermediate to enter another coupling reaction. Although the β-O-4, β-5, and β–β couplings yield a linear polymer, branching can occur whenever the 4-O- or 5-position of one lignin oligomer or polymer couples with the 5-position of another lignin oligomer or polymer, producing 5-5 and 4-O-5 linkages. The coupling reactions involved in lignification have been previously reviewed in detail [88](see Figure 4.20).

Alternatively, the phenoxy oxygen and the β-carbon are considered the most reactive species, readily coupling into aryl ether linkages [104, 117]. This may account for the high abundance of the β-O-4 interunit linkages in lignin, estimated to be as high as 50% in softwoods and almost 60% in hardwoods [104].

Other bonding patterns that occur in lignin are the so-called "diaryl propane" units or β-1 (see Figure 4.21). They are present in about 5–10% of the total phenylpropane units in lignin, and they are thought to be relatively stable under alkaline pulping conditions.

Ether and ester linkages are also common in lignin. With respect to ether linkages, several types have been identified to be present, namely biaryl ethers, noncyclic benzyl alkyl ethers, and diphenyl ethers. They comprise only about 6% of the phenylpropanoid units in spruce lignin [118, 119]. Non-cyclic benzyl alkyl ethers, present in relatively small amounts in lignin (2–3%), have beneficial effects during pulping. The combination of C_β and phenoxy radicals (see Figure 4.22) results in the formation of β-O-4 linkages [118–120], linking approx. 50% of the lignin units [104]. Structures that

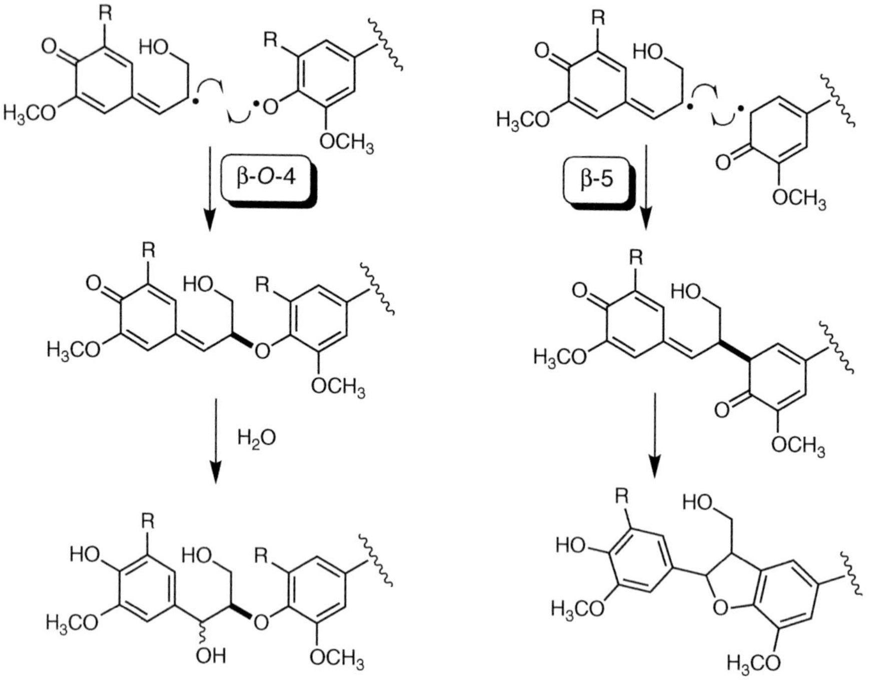

Figure 4.19 Coupling reactions (endwise polymerization or dimerization) for the major structural units of lignin polymer (β-O-4 and β-5 linkages) [116]

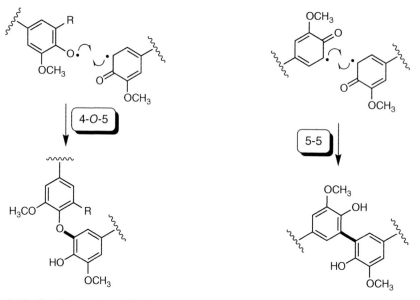

Figure 4.20 Coupling reactions (oligolignol cross-coupling) for the major structural units of lignin polymer (4-O-5 and 5-5 linkages) [116]

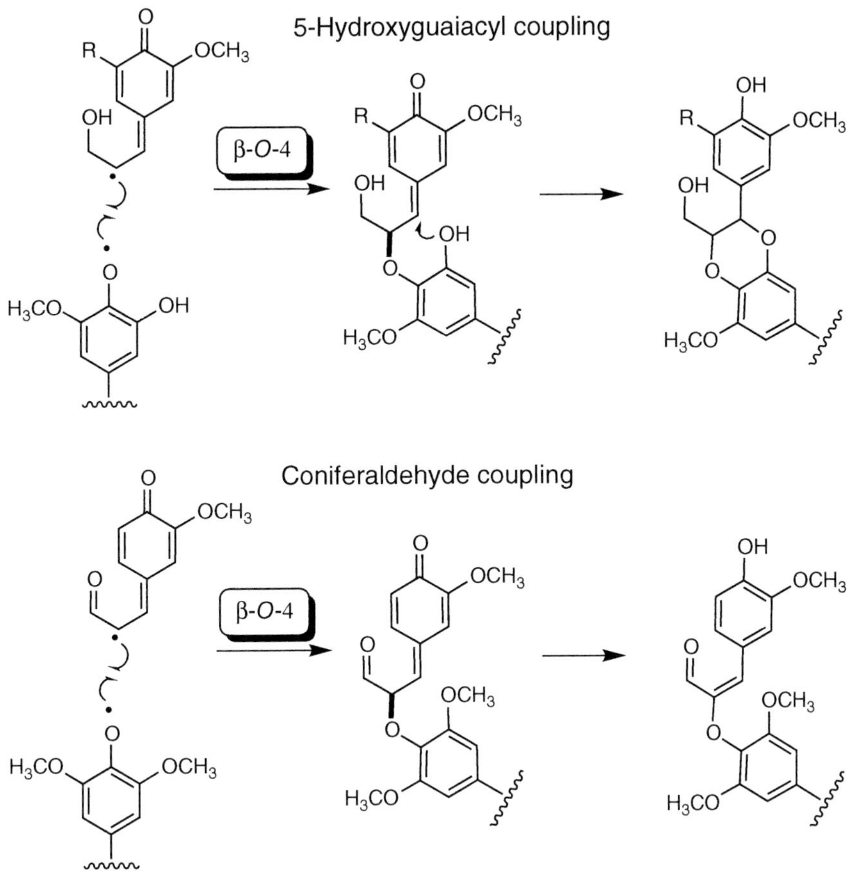

Figure 4.21 The diaryl propane (β-1) interunit linkages of lignin [12]

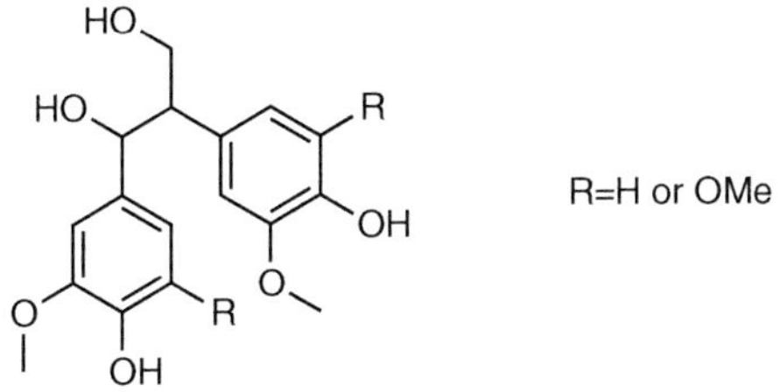

Figure 4.22 Coupling reactions (5-hydroxyguaiacyl and coniferaldehyde ones) for the major structural units of lignin polymer (β-O-4 linkage) [116]

contain β-ethers are relatively infrequent and have been estimated to account for less than 1% of the interunit linkages in lignin. Another lignin structure, containing two ether groups is the pinoresinol (present in less than 5% of the units) [121–125]. This structure is formed by the β–β coupling of two coniferyl alcohol radicals followed by a double-ring closure as illustrated in Figure 4.23.

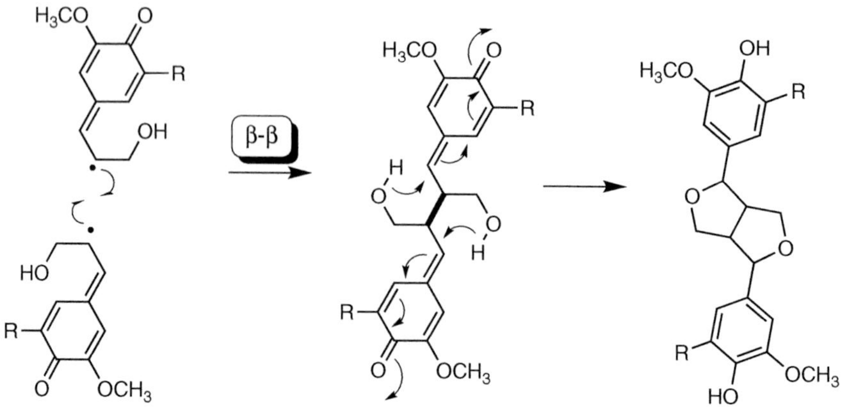

Figure 4.23 Coupling reaction (dimerization) for the major structural units of lignin polymer (β–β linkage) [12]

Benzyl alcohol groups are also common structures in lignin. These, which account for more than 30% of the phenylpropane units, are formed by the addition of a water molecule to a quinone methide. Karhunen et al. [126, 127] reported another bonding pattern present in softwood lignin. This involves the formation of α-, β-ethers on the same 5-5′ biphenyl structures. The new octagonal moiety has been identified as the dibenzodioxocin of Figure 4.24.

Table 4.2 shows the types and frequencies of linkages in softwood and hardwood lignins (Figure 4.21 and 4.22).

Based on these structural details, a number of lignin models have been proposed by a variety of investigators (see Chapter 2), including those of Freudenberg [119], Brauns and Brauns [129, p. 173], Erdtman [130, 131], Adler [132], Forss and Fremer [133], and Glasser and Glasser [134]. Among the models proposed for softwood lignin, Adler's is still the most widely cited [119, 135].

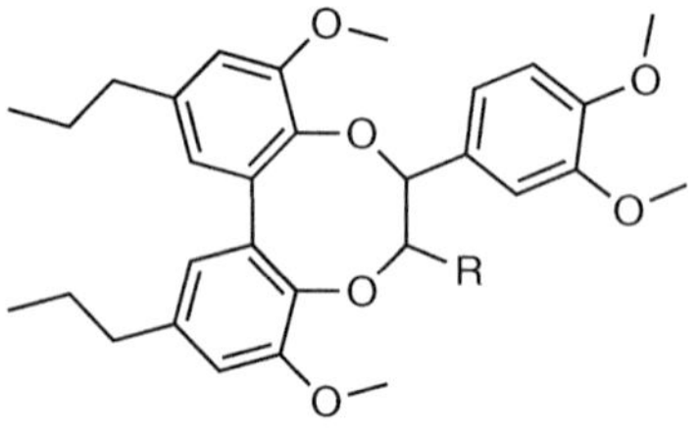

Figure 4.24 Dibenzodioxocin present in softwood lignin

Table 4.2 Types and frequencies of linkages in softwood and hardwood lignins[128][a]

Dilignol linkages	Softwood	Hardwood	Dilignol linkages	Softwood	Hardwood
β-O-4 (**1**)	43–50	50–65	5 – 5′ (**4**)	10–25	4–10
β-5 / α-O-4 (**2**)	9–12	4–6	4-O-5′ (**5**)	4	6–7
α-O-4 (**2′**)	6–8	4–8	β-1 (**6**)	3–7	5–7
β-β (**3**)	2–4	3–7	C-6,C-2 (**7**)	3	2–3

[a] See Figure 2.10 for numbers meaning

The first example of DIR activity was for (+)-pinoresinol synthase activity, which was obtained from *Forsythia intermedia* [136] and was then purified and named as DIR [5] (see Figure 4.25). Notably, the DIR controlling this transformation lacks oxidative catalytic capacity by itself, but in the presence of an appropriate oxidase it is able to confer absolute specificity to the coupling reaction [137].

In lignin, there are some types of reduced structures that cannot be explained by oxidative coupling. Holmgren *et al.* [138] pointed out that NADH reduced a β-aryl ether quinone methide to its benzyl derivative.

4.8.5 Dehydrogenation Polymer (DHP)

A major milestone in lignin chemistry was Freudenberg's success in providing experimental evidence for the enzyme-initiated dehydrogenative polymerization theory. This became possible by polymerizing coniferyl alcohol to a lignin-like dehydrogenation polymer (DHP) [76]. Further attempts to synthesize lignin DHPs using the peroxidase/H_2O_2 system were successfully accomplished by Freudenberg *et al.* [135], [86, p. 78] and Sarkanen and Ludwig [4]. Yet, as Sarkanen and Ludwig pointed out, "these lignin polymer models", despite the fact that structurally they were lignin-like, they were not identical to lignins formed *in vivo* [4].

In the synthesis of DHP, two different modes have been employed, the "Zulauf" and the "Zutropf" methods. The former involves a stepwise addition of enzyme to a solution of coniferyl alcohol, whereas in the latter, the reverse is used. From a structural standpoint, the two methods give slightly different DHPs but the Zutropf–DHP is considered to be somewhat more "lignin-like" [139].

Recently, Van Parijs *et al.* [140] presented a model of *in vitro* lignin polymerization that allowed them to predict the reaction conditions controlling the primary structure of lignin polymers. Kishimoto *et al.* [141] produced DHP by the Zutropf method[7] from mixtures of various ratios of coniferyl and sinapyl alcohols and studied the relationships of the S/G ratio and the interunit linkage types of lignin. Grabber [142] showed in a review how the lignin composition, structure, and cross-linking of different DHPs affect its degradability.

4.9 The Lignin–Carbohydrate Connectivity

Polysaccharides and lignins are the major plant polymers, carbohydrates such as cellulose being the majority; however, some other polysaccharides such as hemicellulose can also be found. In wood species, approximately 40–50% of the biomass corresponds to cellulose and 20–30% to hemicellulose [143]. Lignin is known to be the most abundant aromatic plant polymer on the earth and the lignin content in typical wood species ranges from approximately 20% to 35% [4, 104, 143, 144]. These averages may differ depending on whether the species is hardwood or softwood.

Cellulose is a structural component of the cell wall in green plants and certain microorganisms. The cellulose structure is a long, linear homopolymer composed of D-glucopyranose units linked by β-(1→4)-glycosidic bonds [143].

Under the denomination of hemicellulose, a diverse and complex material is described. The composition of hemicellulose depends on the nature of the wood. Galactoglucomannan (see Figure 2.6 in page 17) is the major hemicellulose constituent of softwood species, representing approximately 16% of the wood [143]. Another major hemicellulose component in softwood is arabino-(4-*O*-methylglucurono)-xylan (see Figure 2.6 in page 17), which represents about 7–15% of the wood.

---

[7] Gradual monolignol addition.

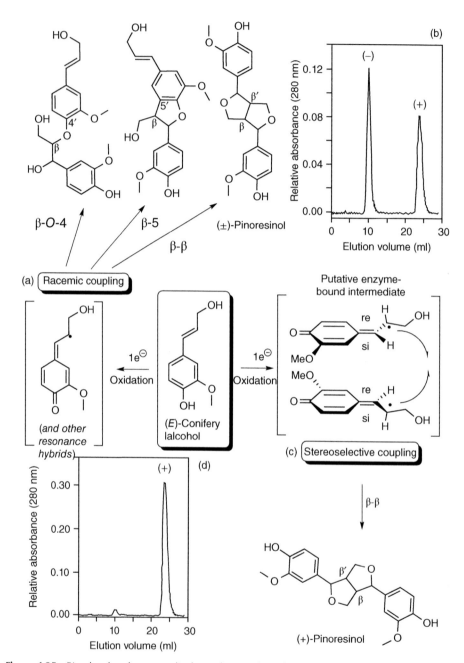

Figure 4.25 Bimolecular phenoxy radical coupling products from E-coniferyl alcohol. (a) Dimeric lignans formed via "random" coupling. (c) Stereoselective coupling to give (+)-pinoresinol. (b and d) HPLC profiles show chirality of pinoresinol obtained for each case, respectively [5]

Xylan is the main component of hardwood hemicelluloses (10–35% of the wood). In contrast to softwood xylans, hardwood xylans do not contain α-D-arabinofuranosyl units and, therefore, are glucuronoxylans [143].

According to studies on wood and grasses [12], it is known that lignin is closely linked to carbohydrate moieties and not only deposited on the cell walls [145–148]. For many years, the nature of these bonds and links has been studied [149–151]. It is assumed that this lignin–carbohydrate links take place by the reaction between the intermediate quinone methides with lignols and carbohydrates during the biosynthesis process (see Figure 4.16 on page 92).

Lignins can be attached to the cell wall polysaccharides by nucleophilic addition of the hydroxyl groups of the polysaccharide to the resulting quinone methide structure from the β-O-4 coupling of two monolignol radicals. Through this mechanism, lignins strengthen to the plant cell walls, facilitating the transport of water and preventing the degradation of the wall polysaccharides of the wall, thus acting as a better line of defense against herbivorous vertebrates, insects, and fungi.

There is no perfect method for isolating LCC preparations and analyzing them [152]. The quantification of LCC linkages with traditional wet chemistry methods is limited mostly to relative quantification of carbohydrate sites linked to lignin [153–160]. Quantitative information on various types of linkages between lignin and carbohydrates is very scarce. In fact, we have found only two reports evaluating the amounts of lignin–carbohydrate linkages in LCC preparations isolated from wood [151, 161].

4.10 Location of Lignins (Cell Wall Lignification)

Lignin is first synthesized and then deposited in the secondary cell wall of specialized cells such as xylem vessels, tracheids, and fibers.

The deposition of lignin proceeds in different phases, each proceeded by the deposition of the polysaccharide matrix in the S2 (middle layer of secondary wall) layer. The bulk of lignin is deposited after cellulose and hemicellulose have been deposited in the S3 (innermost layer of secondary wall) layer. Generally, the lignin concentration is higher in the middle lamella and cell corners than in the S2 layer of secondary wall [162–164].

The first step in the lignification process involves the synthesis of the necessary monolignols by the plant cells. This is a process mediated by Golgi apparatus [165]. Once monolignols are synthesized, they are oxidized by the corresponding enzymes, peroxidases, and laccases, and, later, polymerization takes place. The lignification begins in the cell corners, ML, and the secondary wall, S1, extending to the rest of the secondary wall to the lumen. Lignification in the ML and P wall begins later than the forming of the second wall, but the lignification in the secondary wall generally occurs when the secondary wall formation has finished, as can be deduced by the presence of a S3 lamella [163].

Because lignification begins at the farthest region of protoplast, it has been suggested that there might be starting points linked to specific regions of the cell wall, where the polymerization begins [166]. Enzymes linked to specific points in the cell wall could act as initiation sites, but that peroxidases and oxidases are restricted to the lignifying and lignified cells walls during xylem formation, indicating that these enzymes are not really starting sites [167, 168]. Ideally, starting sites should be on the wall during its formation, rather than relying on the diffusion of the enzymes involved toward lignification sites. Although these enzymes are known to be present in the cell wall, which is lignified, there is information on its exact location and how its distribution varies during lignification. The ultrastructural location of the enzymes is required to confirm their role in the control of the chemistry of the cell surface. Other candidates for starting points are the DIR, hydroxyproline-rich proteins (HRPs), and polysaccharides of the cell wall proteins [169–172].

Terashima and Fukushima [173] have shown that lignification and cellulose deposition in the plant cell wall proceeds in three distinct phases (see Figure 4.26):

- Initially lignification occurs at the cell corner and middle lamella, after the deposition of pectins is complete and the formation of the secondary wall S1 has been initiated.
- During the second phase, an extensive deposition of cellulose microfibrils, mannan and xylan in the S2 layer takes place. However, the lignification process proceeds very slowly during this stage.
- Finally, during the third phase, lignification proceeds extensively. This occurs after the deposition of cellulose microfibrils in the S3 layer of the secondary wall has taken place.

Recently, Terashima has proposed a supramolecular structure of lignin: a globular lignin as an aggregate of oligolignols folded at the β-O-4 bond is deposited into the network composed of thin cellulose microfibrils, hemicelluloses, and pectin [174].

Terashima et al. [174] has also shown that the structure of lignin with regard to the content of condensed structures differs in the compound middle lamella from that in the secondary wall.

There are variations in the nature of lignins found in the same cell wall, this being observed heterogeneity during the lignin deposition in the secondary cell wall, with differences not only in the content but also in the main monomeric composition. Recently, these differences have been elucidated using solution-state 2D NMR [175].

Related to the variations in the content, in 1985, Donaldson [176] found that the lignin concentrations are in the 16–27% range for the secondary cell wall. However, in the same main lamella of conifers values are higher, in the 38–88% range. With respect to the monomeric composition, lignins deposited in the main lamella and cell corners are rich in H units, the most predominant units being those of G type in the secondary enlargement [173] (see Figure 4.27).

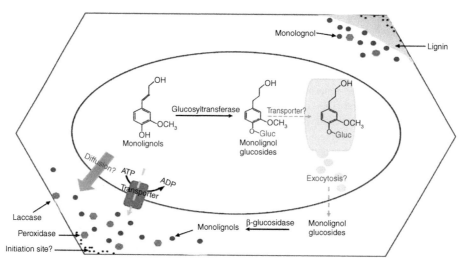

Figure 4.26 *The scheme for monolignol deposition and the subsequent initiation of lignin polymerization within the cell wall. Symplastic transport of monolignols may export them to the cell wall through active transport or by passive diffusion. Alternatively, they may be sequestrated and stored as glucoconjugates into the vacuoles in gymnosperms, before their subsequent transport to the cell wall and hydrolysis to free monolignols for polymerization. The deposited monolignols in the cell wall diffuse to initiation sites where the polymerization process begins. The polymerization to form different bond-linkages of lignin is known to be a random chemical process. However, the nature of initiating sites and the way in which the amount and type of lignin formation is controlled across the cell wall are poorly understood. Copyright from ref. [2]. (See insert for color representation of this figure.)*

4.11 Differences Between Angiosperm and Gymnosperm Lignins

It has been observed differences in the content of lignin, monomers composition (G/S ratio), bond types, and xylem function in plants belonging to the angiosperm and gymnosperm groups.

The lignin content, as is known, differs for hardwood and softwood species and herbaceous plants. In hardwood trees, it is 27–33%, softwood trees 18–25%, and for herbaceous plants it is 17–24%. As might be expected, the highest amounts of lignin occur in the compressed wood on the lower part of branches and leaning stems [4]. On the other hand, lignin corresponding to bark trees differs from wood lignin structure [177]. By contrast, lignin does not occur in algae, lichens, or mosses [178].

The composition of wood lignin is composed basically of S and G monolignol units linked by carbon-carbon and ether-type bonds [104]. Hardwood plants belong to dicotyledonous angiosperm group and lignin in them have certain common properties related to composition and structure. In this case, G and S units are the most abundant, whereas H units can be found as traces. For plants with softwood, which are classified as gymnosperms, lignin is composed mostly of G units and some H units [26]. In the case of grasses and other herbaceous plants, lignin is constituted by G and S units in a similar average for monocots and higher proportions of H units for dicots [162]. For angiosperms, lignin is composed mainly of S and G units while total lignin is a combination of S and G lignin in varied proportion. However, there are gymnosperms with predominant S lignins and angiosperms with predominant G lignins [79]. In monocotyledonous, lignins contain significant amounts of p-coumaric acid [103, 179].

In the case of grass family, Ralph et al. [179] have detected significant amounts of p-coumaric acid linked by ether bonds to lignins. Billa et al. [180] have proposed that lignin in herbaceous plants have a structure with a G–S nucleus similar to the one found in angiosperms, when the H units formed from p-coumaric acid are an essential part of the linking points with polysaccharides of the cell wall. Also, ferulic acid is reportedly linked by ether bonds to hemicellulose polymer, and these moieties can provide starting points for lignin polymer growing [181].

It has also been found that the presence of acetylated lignin is higher in herbaceous than in other angiosperms, and additionally those not shown in the gymnosperms. This acetylation occurs more frequently in S units, although the role of these acetylated lignins remains unknown [63].

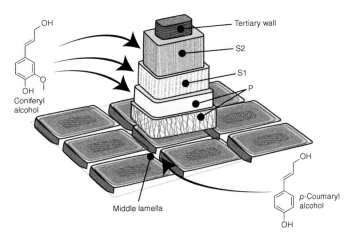

Figure 4.27 Monolignols and differential cell wall targeting. Telescopic representation of a conifer tracheid. p-Coumaryl alcohol is preferentially deposited in the middle lamella whereas secondary alcohols in the secondary wall [6]

The low S/G ratio in lignins has remarkable economic importance in pulp-paper production. The presence on great amounts of G make lignin more compact, so that the improvement of S monomers in lignin is more economical and less environmentally damaging because less aggressive conditions for pulp production are necessary. S and G monolignols differ basically in the number of methyl groups attached to the aromatic core to the monomers and in the number of free hydroxyl groups. This plays a major role in the compactness of the resulting lignin. Syringyl units have less potential radical formation sites than do guaiacyl units, so that lignin formed from S units is less compact than that formed when G monomers predominate.

Lapierre *et al.* [182] showed that the lignins in angiosperms differ from those in gymnosperms in the ratio of bond types: the β-*O*-4 bond is about 60–65% in angiosperms, and only 30% in gymnosperms, in grass the ratio is lower. This also determines the compactness of the resulting lignin.

Santos *et al.* [183] observed significant variations in lignin structures for hardwood species, such as the S/G ratio, predominant bond types, degree of condensation, and elemental and methoxy contents.

Three basic monolignols (S, G, and H) are incorporated in different stages during the cell wall formation, depending on the plant family or other factors such as the plant component. For example, in angiosperms, H units are deposited first, followed by G and S units [163]. In vessels, lignin is generally rich in G units, whereas lignin in fibers is typically rich in S units [164], while in the secondary walls of ray parenchyma a large proportion of S units is also found. On the other hand, in gymnosperms, the lignin deposited in compressed wood is rich in H units [184]. Some differences in timing of monolignol deposition have been found to be associated with variations in lignin condensation in the individual cell wall layers. This can be shown by immunocytochemistry with antibodies raised against pure H, pure G, or mixed G–S synthetic lignins [185]. The final composition of lignin in plants and the amount that is formed in each case varies among taxa, cell types, and individual cell wall layers and is influenced by developmental and environmental cues [186].

Xylem cells show differences in behavior also depending on plant type. Thus, gymnosperms, tracheid xylem involved in both mechanical support and water transports and secondary xylem of woody angiosperms contains two specialized cell types: the elements of the vessels, for water transport, and the fibers, which provide structural support. This function specialization in the xylem of angiosperms produces greater efficiency and economy. Large vessel elements exhibit a reduced lignin content (about 20% dry matter) compared to gymnosperms (30% dry matter). This low lignin content and high carbohydrate proportion requires significantly less energy because it consumes less carbon in the plant growing "sinks" [52]. In the case of angiosperms, primary xylem cells and water conducting vessels of secondary xylem are richer in lignin-type G, as the most basal tracheid gymnosperms, while the nonconducting cells of the xylem fibers are rich in S units [187].

These cells have specialized in transporting water in seed plants, showing a rich lignin-type G composition. It has been suggested that, to conserve the biosynthetic pathway of these lignins, a strong selective pressure and its regulation govern the transporting cells of the xylem during the evolution of seed plants [187]. On the other hand, there is molecular evidence to suggest that an ancient pathway, predominantly for the synthesis of coniferyl alcohol, was conserved among seed plants, and another new side pathway for the biosynthesis of sinapyl alcohol evolved more recently in plants of the family of angiosperms [187]. This divergence apparently occurred in basal angiosperms, before the segregation the two branches known as monocots from eudicots.

Over the evolution of seed in plants, coevolution differences have arisen in the set of substrate for lignification not only with the segregation of the functions of water transport and support, but also with the shape/structure of the transporting cells. The xylem of gymnosperms is composed only of tracheids with an estimated diameter of less than 30 μm, while the xylem of angiosperms group is typically formed by vessels with large lumens[8][188]. Thus, xylem vessels of angiosperms are more water drivers, thus maintaining high photosynthetic capacity [188]. However, this system of large

---

[8] In the diameter range of 100–200 μm.

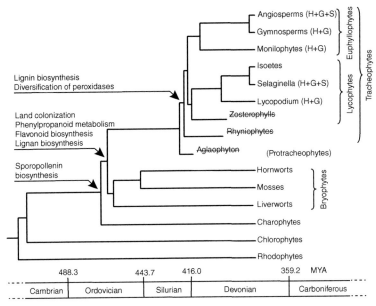

Figure 4.28 *Plant phylogenetic tree marked with the major milestones of evolution of lignin biosynthesis. Strikethrough text means extinct lineage [43]*

vessels can also become a disadvantage due to frequent strokes that occur during periods of freezing and thawing of xylem vessels [189]. This is one of the reasons why gymnosperms still dominate many regions with warm climate of the Northern Hemisphere, and why the subtropical rainforests constitute the only habitat where the gymnosperms tend to be absent [188]. Thus, it can be concluded that the chemical complexity of lignins has increased during the course of plant evolution from ancestral species, such as Pteridophytes and gymnosperms, to the most evolved species, such as monocots (see Figure 4.28).

References

[1] Boerjan W, Ralph J, Baucher M. Lignin biosynthesis. Annu Rev Plant Biol. 2003;54:519–546.
[2] Liu CJ. Deciphering the enigma of lignification: precursor transport, oxidation, and the topochemistry of lignin assembly. Mol Plant. 2012;5(2):304–317.
[3] Neish AC. Monomeric intermediates in the biosynthesis of lignin. In: Freudenberg K, Neish AC, editors. Constitution and Biosynthesis of Lignin, Molecular Biology, Biochemistry and Biophysics. vol. 2. Berlin: Springer-Verlag; 1968. p. 1–43.
[4] Sarkanen KV, Ludwig CH. Lignins: Occurrence, Formation, Structure and Reactions. Wiley-Interscience; 1971.
[5] Davin LB, Wang HB, Crowell AL, Bedgar DL, Martin DM, Sarkanen S, *et al.* Stereoselective bimolecular phenoxy radical coupling by an auxiliary (dirigent) protein without an active center. Science. 1997;275(5298):362–367.
[6] Davin LB, Lewis NG. Lignin primary structures and dirigent sites. Curr Opin Biotechnol. 2005;16(4):407–415.
[7] Ellis BE, Kuroki GW, Stafford HA, Sederoff R, Campbell M, O'Malley D, *et al.* Genetic regulation of lignin biosynthesis and the potential modification of wood by genetic engineering

in loblolly pine. In: Ellis BE, editor. Genetic Engineering of Plant Secondary Metabolism, Recent Advances in Phytochemistry. vol. 28. US: Springer-Verlag; 1994. p. 313–355.

[8] Boudet AM. Evolution and current status of research in phenolic compounds. Phytochemistry. 2007;68(22–24):2722–2735.

[9] Henriksson G. Lignin. In: Ek M, Gellerstedt G, Henriksson G, editors. Wood Chemistry and Biotechnology, Pulp and Paper Chemistry and Technology. vol. 1. Walter de Gruyer; 2009. p. 121–145.

[10] Falkehag SI. Lignin in materials. Appl Polym Symp. 1975;28(Proc. Cellul. Conf., 8th, 1975, Vol. 1):247–257.

[11] Jeffries TW. Biodegradation of lignin-carbohydrate complexes. Biodegradation. 1990;1(2-3): 163–176.

[12] Argyropoulos DS, Menachem SB. Lignin. In: Eriksson KEL, Babel W, Blanch HW, Cooney CL, Enfors SO, Eriksson KEL, et al., editors. Biotechnology in the Pulp and Paper Industry, Advances in Biochemical Engineering/Biotechnology. vol. 57. Springer Berlin Heidelberg; 1997. p. 127–158.

[13] Hu Z. Elucidation of the Structure of Cellulolytic Enzyme Lignin from Loblolly Pine (*Pinus taeda*). North Carolina State University, USA. North Carolina; 2006.

[14] Chen Y, Zhang X, Wu W, Chen Z, Gu H, Qu LJ. Overexpression of the wounding-responsive gene *AtMYB15* activates the shikimate pathway in *Arabidopsis*. J Integr Plant Biol. 2006;48(9):1084–1095.

[15] Tzin V, Galili G. New insights into the shikimate and aromatic amino acids biosynthesis pathways in plants. Mol Plant. 2010;3(6):956–972.

[16] Herrmann KM, Weaver LM. The shikimate pathway. Annu Rev Plant Physiol Plant Mol Biol. 1999;50:473–503.

[17] Lewis NG, Sarkanen S. Lignin and lignan biosynthesis: developed from a symposium sponsored by the Division of Cellulose. Paper, and Textile at the 211th National Meeting of the ACS, New Orleans, LA; March 24–29, 1996. vol. 697. Washington, DC: American Chemical Society; 1998.

[18] Dewick PM, Haslam E. Phenol biosynthesis in higher plants; gallic acid. Biochem J. 1969; 113(3):537–542.

[19] Werner RA, Rossmann A, Schwarz C, Bacher A, Schmidt HL, Eisenreich W. Biosynthesis of gallic acid in Rhus typhina: discrimination between alternative pathways from natural oxygen isotope abundance. Phytochemistry. 2004;65(20):2809–2813.

[20] Basset GJC, Quinlivan EP, Ravanel S, Rebeille F, Nichols BP, Shinozaki K, et al. Folate synthesis in plants: The *p*-aminobenzoate branch is initiated by a bifunctional PabA-PabB protein that is targeted to plastids. Proc Natl Acad Sci U S A. 2004;101(6):1496–1501.

[21] Knaggs AR. The biosynthesis of shikimate metabolites. Nat Prod Rep. 2003;20(1):119–136.

[22] Cho MH, Corea ORA, Yang H, Bedgar DL, Laskar DD, Anterola AM, et al. Phenylalanine biosynthesis in *Arabidopsis thaliana*: identification and characterization of arogenate dehydratases. J Biol Chem. 2007;282(42):30827–30835.

[23] Yamada T, Matsuda F, Kasai K, Fukuoka S, Kitamura K, Tozawa Y, et al. Mutation of a rice gene encoding a phenylalanine biosynthetic enzyme results in accumulation of phenylalanine and tryptophan. Plant Cell. 2008;20(5):1316–1329.

[24] Corea ORA, Ki C, Cardenas CL, Kim SJ, Brewer SE, Patten AM, et al. Arogenate dehydratase isoenzymes profoundly and differentially modulate carbon flux into lignins. J Biol Chem. 2012;287(14):11446–11459.

[25] Higuchi T. Formation and biological degradation of lignins. Adv Enzymol Relat Areas Mol Biol. 1971;34:207–283.

[26] Higuchi T. Lignin biochemistry: biosynthesis and biodegradation. Wood Sci Technol. 1990; 24(1):23–63.

[27] Hasegawa M, Higuchi T. Formation of lignin from glucose in Eucalyptus tree. Nippon Rin Gakkaishi. 1960;42:305–308.
[28] Kratzl K, Zauner J. Incorporation of radioactivity in wood studied by means of xylitol-C14. Holzforsch Holzverwert. 1962;14:108–111.
[29] Schubert WJ, Acerbo SN. Conversion of D-glucose into lignin in Norwegian spruce. A preliminary report. Arch Biochem Biophys. 1959;83:178–182.
[30] Humphreys JM, Chapple C. Rewriting the lignin roadmap. Curr Opin Plant Biol. 2002;5(3): 224–229.
[31] Young MR, Towers GHN, Neish AC. Taxonomic distribution of ammonia lyases for L-phenylalanine and L-tyrosine in relation to lignification. Can J Bot. 1966;44(3):341–349.
[32] Rosler J, Krekel F, Amrhein N, Schmid J. Maize phenylalanine ammonia-lyase has tyrosine ammonia-lyase activity. Plant Physiol. 1997;113(1):175–179.
[33] Higuchi T, Ito Y, Kawamura I. p-Hydroxyphenylpropane component of grass lignin and role of tyrosine ammonia-lyase in its formation. Phytochemistry. 1967;6(6):875–881.
[34] Neish AC. Formation of m- and p-coumaric acids by enzymic deamination of the corresponding isomers of tyrosine. Phytochemistry. 1961;1:1–24.
[35] Brown SA. Studies of lignin biosynthesis using isotopic carbon. IX. Taxonomic distribution of the ability to utilize tyrosine in lignification. Can J Bot. 1961;39:253–258.
[36] Vaughan PFT, Butt VS. Action of o-dihydric phenols in the hydroxylation of p-coumaric acid by a phenolase from leaves of spinach beet (Beta vulgaris). Biochem J. 1970;119(1):89–94.
[37] Russell DW. Metabolism of aromatic compounds in higher plants. X. Properties of the cinnamic acid 4-hydroxylase of pea seedlings and some aspects of its metabolic and developmental control. J Biol Chem. 1971;246(12):3870–3388.
[38] Kutsuki H, Higuchi T. Activities of some enzymes of lignin formation in reaction wood of Thuja orientalis, Metasequoia glyptostroboides and Robinia pseudoacacia. Planta. 1981;152(4):365–368.
[39] Kutsuki H, Shimada M, Higuchi T. Regulatory role of cinnamyl alcohol dehydrogenase in the formation of guaiacyl and syringyl lignins. Phytochemistry. 1982;21(1):19–23.
[40] Luederitz T, Schatz G, Grisebach H. Enzymic synthesis of lignin precursors. Purification and properties of 4-coumarate: CoA ligase from cambial sap of spruce (Picea abies L.). Eur J Biochem. 1982;123(3):583–586.
[41] Chen HC, Li Q, Shuford CM, Liu J, Muddiman DC, Sederoff RR, et al. Membrane protein complexes catalyze both 4- and 3-hydroxylation of cinnamic acid derivatives in monolignol biosynthesis. Proc Natl Acad Sci USA. 2011;108(52):21253–21258.
[42] Dewick PM. Medicinal Natural Products: A Biosynthetic Approach. John Wiley & Sons, Ltd; 2002.
[43] Weng JK, Chapple C. The origin and evolution of lignin biosynthesis. New Phytol. 2010; 187(2):273–285.
[44] Humphreys JM, Hemm MR, Chapple C. New routes for lignin biosynthesis defined by biochemical characterization of recombinant ferulate 5-hydroxylase, a multifunctional cytochrome P450-dependent monooxygenase. Proc Natl Acad Sci USA. 1999;96(18): 10045–10050.
[45] Osakabe K, Tsao CC, Li L, Popko JL, Umezawa T, Carraway DT, et al. Coniferyl aldehyde 5-hydroxylation and methylation direct syringyl lignin biosynthesis in angiosperms. Proc Natl Acad Sci USA. 1999;96(16):8955–8960.
[46] Li L, Popko JL, Umezawa T, Chiang VL. 5-hydroxyconiferyl aldehyde modulates enzymatic methylation for syringyl monolignol formation, a new view of monolignol biosynthesis in angiosperms. J Biol Chem. 2000;275(9):6537–6545.
[47] Parvathi K, Chen F, Guo D, Blount JW, Dixon RA. Substrate preferences of O-methyltransferases in alfalfa suggest new pathways for 3-O-methylation of monolignols. Plant J. 2001;25(2):193–202.

[48] Weisshaar B, Jenkins GI. Phenylpropanoid biosynthesis and its regulation. Curr Opin Plant Biol. 1998;1(3):251–257.

[49] Barakate A, Stephens J, Goldie A, Hunter WN, Marshall D, Hancock RD, et al. Syringyl lignin is unaltered by severe sinapyl alcohol dehydrogenase suppression in tobacco. Plant Cell. 2011;23(12):4492–4506.

[50] Li HR, Feng YL, Yang ZG, Wang J, Daikonya A, Kitanaka S, et al. New lignans from *Kadsura coccinea* and their nitric oxide inhibitory activities. Chem Pharm Bull. 2006;54(7):1022–1025.

[51] Dixon RA, Chen F, Guo D, Parvathi K. The biosynthesis of monolignols: a metabolic grid, or independent pathways to guaiacyl and syringyl units? Phytochemistry. 2001;57(7):1069–1084.

[52] Amthor JS. Efficiency of lignin biosynthesis: a quantitative analysis. Ann Bot. 2003;91(6):673–695.

[53] Baucher M, Chabbert B, Pilate G, Van Doorsselaere J, Tollier MT, Petit-Conil M, et al. Red xylem and higher lignin extractability by down-regulating a cinnamyl alcohol dehydrogenase in poplar. Plant Physiol. 1996;112(4):1479–1490.

[54] Ralph J, MacKay JJ, Hatfield RD, O'Malley DM, Whetten RW, Sederoff RR. Abnormal lignin in a loblolly pine mutant. Science. 1997;277(5323):235–239.

[55] Ralph J, Brunow G, Harris P, Dixon RA, Schatz PF, Boerjan W. Lignification: are lignins biosynthesized via simple combinatorial chemistry or via proteinaceous control and template replication? In: Daayf F, Lattanzio V, editors. Recent Advances in Polyphenol Research. vol. 1. Oxford, UK: Wiley-Blackwell; 2008. p. 36–66.

[56] Ralph J. Hydroxycinnamates in lignification. Phytochem Rev. 2010;9(1):65–83.

[57] Sibout R, Baucher M, Gatineau M, Van Doorsselaere J, Mila I, Pollet B, et al. Expression of a poplar cDNA encoding a ferulate-5-hydroxylase/coniferaldehyde 5-hydroxylase increases S lignin deposition in *Arabidopsis thaliana*. Plant Physiol Biochem. 2002;40(12):1087–1096.

[58] Lu F, Ralph J. Detection and determination of *p*-coumaroylated units in lignins. J Agric Food Chem. 1999;47(5):1988–1992.

[59] Lu F, Ralph J. Preliminary evidence for sinapyl acetate as a lignin monomer in kenaf. Chem Commun. 2002;(1):90–91.

[60] Lu F, Ralph J, Morreel K, Messens E, Boerjan W. Preparation and relevance of a cross-coupling product between sinapyl alcohol and sinapyl *p*-hydroxybenzoate. Org Biomol Chem. 2004;2(20):2888–2890.

[61] Lu F, Ralph J. Novel tetrahydrofuran structures derived from β-; β-coupling reactions involving sinapyl acetate in Kenaf lignins. Org Biomol Chem. 2008;6(20):3681–3694.

[62] Morreel K, Ralph J, Kim H, Lu F, Goeminne G, Ralph S, et al. Profiling of oligolignols reveals monolignol coupling conditions in lignifying poplar xylem. Plant Physiol. 2004;136(3):3537–3549.

[63] Del Rio JC, Marques G, Rencoret J, Martinez AT, Gutierrez A. Occurrence of naturally acetylated lignin units. J Agric Food Chem. 2007;55(14):5461–5468.

[64] Withers S, Lu F, Kim H, Zhu Y, Ralph J, Wilkerson CG. Identification of grass-specific enzyme that acylates monolignols with *p*-coumarate. J Biol Chem. 2012;287(11):8347–8355.

[65] Ralph J. An unusual lignin from kenaf. J Nat Prod. 1996;59(4):341–342.

[66] Miao YC, Liu CJ. *ATP*-binding cassette-like transporters are involved in the transport of lignin precursors across plasma and vacuolar membranes. Proc Natl Acad Sci USA. 2010;107(52):22728–22733, S22728/1–S22728/5.

[67] Sibout R, Höfte H. Plant cell biology: the ABC of monolignol transport. Curr Biol. 2012;22(13):R533–R535.

[68] Srivastava S. Molecular Characterization of Cinnamoyl CoA Reductase (CCR) Gene in *Leucaena leucocephala*. Pune, India: University of Pune; 2009.

[69] Whetten R, Sederoff R. Lignin biosynthesis. Plant Cell. 1995;7(7):1001–1013.
[70] Steeves V, Forster H, Pommer U, Savidge R. Coniferyl alcohol metabolism in conifers - i. Glucosidic turnover of cinnamyl aldehydes by UDPG: coniferyl alcohol glucosyltransferase from pine cambium. Phytochemistry. 2001;57(7):1085–1093.
[71] Dharmawardhana DP, Ellis BE, Carlson JE. A β-glucosidase from lodgepole pine xylem specific for the lignin precursor coniferin. Plant Physiol. 1995;107(2):331–339.
[72] Onnerud H, Zhang L, Gellerstedt G, Henriksson G. Polymerization of monolignols by redox shuttle-mediated enzymatic oxidation: a new model in lignin biosynthesis. Plant Cell. 2002;14(8):1953–1962.
[73] Zoia L, Perazzini R, Crestini C, Argyropoulos DS. Understanding the radical mechanism of lipoxygenases using ^{31}P NMR spin trapping. Bioorg Med Chem. 2011;19(9):3022–3028.
[74] Zoia L, Argyropoulos DS. Phenoxy radical detection using ^{31}P NMR spin trapping. J Phys Org Chem. 2009;22(11):1070–1077.
[75] O'Malley DM, Whetten R, Bao W, Chen CL, Sederoff RR. The role of laccase in lignification. Plant J. 1993;4(5):751–757.
[76] Dean JFD, Eriksson KEL. Laccase and the deposition of lignin in vascular plants. Holzforschung. 1994;48(Suppl.):21–33.
[77] Gross GG, Janse C, Elstner EF. Involvement of malate, monophenols, and the superoxide radical in hydrogen peroxide formation by isolated cell walls from horseradish (*Armoracia lapathifolia Gilib.*). Planta. 1977;136(3):271–276.
[78] Gross GG. From lignins to tannins: forty years of enzyme studies on the biosynthesis of phenolic compounds. Phytochemistry. 2008;69(18):3018–3031.
[79] Lewis NG, Yamamoto E. Lignin: occurrence, biogenesis and biodegradation. Annu Rev Plant Physiol Plant Mol Biol. 1990;41:455–496.
[80] Ros Barceló A. Lignification in plant cell walls. Int Rev Cytol. 1997;176:87–132.
[81] Alba CM, De Forchetti SM, Tigier HA. Phenoloxidase of peach (*Prunus persica*) endocarp: its relationship with peroxidases and lignification. Physiol Plant. 2000;109(4):382–387.
[82] Berthet S, Demont-Caulet N, Pollet B, Bidzinski P, Cezard L, Le Bris P, et al. Disruption of LACCASE4 and 17 results in tissue-specific alterations to lignification of *Arabidopsis thaliana* stems. Plant Cell. 2011;23(3):1124–1137.
[83] Dean JFD, Sterjiades R, Eriksson KEL. Purification and characterization of an anionic peroxidase from sycamore maple (*Acer pseudoplatanus*) cell suspension cultures. Physiol Plant. 1994;92(2):233–240.
[84] Welinder KG. Superfamily of plant, fungal and bacterial peroxidases. Curr Opin Struct Biol. 1992;2(3):388–393.
[85] Hiraga S, Sasaki K, Ito H, Ohashi Y, Matsui H. A large family of class III plant peroxidases. Plant Cell Physiol. 2001;42(5):462–468.
[86] Freudenberg K, Neish AC. Constitution and Biosynthesis of Lignin, Molecular Biology, Biochemistry and Biophysics. vol. 2. Berlin: Springer-Verlag; 1968.
[87] Morreel K, Dima O, Kim H, Lu F, Niculaes C, Vanholme R, et al. Mass spectrometry-based sequencing of lignin oligomers. Plant Physiol. 2010;153(4):1464–1478.
[88] Ralph J, Lundquist K, Brunow G, Lu F, Kim H, Schatz PF, et al. Lignins: natural polymers from oxidative coupling of 4-hydroxyphenyl- propanoids. Phytochem Rev. 2004;3(1-2):29–60.
[89] Umezawa T. Diversity in lignan biosynthesis. Phytochem Rev. 2003;2(3):371–390.
[90] Beejmohun V, Fliniaux O, Hano C, Pilard S, Grand E, Lesur D, et al. Coniferin dimerisation in lignan biosynthesis in flax cells. Phytochemistry. 2007;68(22-24):2744–2752.
[91] Pickel B, Constantin MA, Pfannstiel J, Conrad J, Beifuss U, Schaller A. An enantiocomplementary dirigent protein for the enantioselective laccase-catalyzed oxidative coupling of phenols. Angew Chem Int Ed. 2010;49(1):202–204.

[92] Binns AN, Chen RH, Wood HN, Lynn DG. Cell division promoting activity of naturally occurring dehydrodiconiferyl glucosides: do cell wall components control cell division? Proc Natl Acad Sci USA. 1987;84(4):980–984.
[93] Teutonico RA, Dudley MW, Orr JD, Lynn DG, Binns AN. Activity and accumulation of cell division-promoting phenolics in tobacco tissue cultures. Plant Physiol. 1991;97(1):288–297.
[94] Li X, Bonawitz ND, Weng JK, Chapple C. The growth reduction associated with repressed lignin biosynthesis in *Arabidopsis thaliana* is independent of flavonoids. Plant Cell. 2010;22(5):1620–1632.
[95] Lim EK, Jackson RG, Bowles DJ. Identification and characterisation of *Arabidopsis* glycosyltransferases capable of glucosylating coniferyl aldehyde and sinapyl aldehyde. FEBS Lett. 2005;579(13):2802–2806.
[96] Escamilla-Trevino LL, Chen W, Card ML, Shih MC, Cheng CL, Poulton JE. *Arabidopsis thaliana* β-glucosidases BGLU45 and BGLU46 hydrolyse monolignol glucosides. Phytochemistry. 2006;67(15):1651–1660.
[97] Lanot A, Hodge D, Jackson RG, George GL, Elias L, Lim EK, *et al.* The glucosyltransferase UGT72E2 is responsible for monolignol 4-*O*-glucoside production in *Arabidopsis thaliana*. Plant J. 2006;48(2):286–295.
[98] Vanholme R, Storme V, Vanholme B, Sundin L, Christensen JH, Goeminne G, *et al.* A systems biology view of responses to lignin biosynthesis perturbations in *Arabidopsis*. Plant Cell. 2012;24(9):3506–3529.
[99] Alejandro S, Lee Y, Tohge T, Sudre D, Osorio S, Park J, *et al.* AtABCG29 is a monolignol transporter involved in lignin biosynthesis. Curr Biol. 2012;22(13):1207–1212.
[100] Ralph J, Brunow G, Boerjan W. Lignins. In: eLS. John Wiley & Sons, Ltd; 2007.
[101] Leisola M, Pastinen O, Axe DD. Lignin-designed randomness. J Med Chem. 2012;2012(3):1.
[102] Ros Barcelo A, Ros Gomez LV, Gabaldon C, Lopez-Serrano M, Pomar F, Carrion JS, *et al.* Basic peroxidases: the gateway for lignin evolution? Phytochem Rev. 2004;3(1–2):61–78.
[103] Ralph J, Bunzel M, Marita JM, Hatfield RD, Lu F, Kim H, *et al.* Peroxidase-dependent cross-linking reactions of *p*-hydroxycinnamates in plant cell walls. Phytochem Rev. 2004;3(1–2):79–96.
[104] Adler E. Lignin chemistry-past, present and future. Wood Sci Technol. 1977;11:169–218.
[105] Syrjänen K, Brunow G. Regioselectivity in lignin biosynthesis. The influence of dimerization and cross-coupling. J Chem Soc [Perkin 1]. 2000;(2):183–187.
[106] Freudenberg K, Schluter H. Intermediates in the formation of lignin. Chem Ber. 1955;88:617–625.
[107] Freudenberg K, Geiger H. Pinoresinolide and other intermediates of lignin formation. Chem Ber. 1963;96:1265–1270.
[108] Ralph J, Schatz PF, Lu F, Kim H, Akiyama T, Nelsen SF. Quinone methides in lignification. In: Rokita SE, editor. Quinone Methides, Reactive Intermediates in Chemistry and Biology. John Wiley & Sons, Inc.; 2009. p. 385–420.
[109] Freudenberg K, Harkin JM, Reichert M, Fukuzumi T. Enzymes participating in lignification. Dehydrogenation of sinapin alcohol. Chem Ber. 1958;91:581–590.
[110] Lüdemann HD, Nimz H. ^{13}C-kernresonanzspektren von ligninen, 2. Buchen- und fichtenbjörkman-lignin. Makromol Chem. 1974;175(8):2409–2422.
[111] Lüdemann HD, Nimz H. ^{13}C-kernresonanzspektren von ligninen, 1. Chemische verschiebungen bei monomeren und dimeren modellsubstanzen. Makromol Chem. 1974;175(8):2393–2407.
[112] Lüdemann HD, Nimz H. Carbon-13 nuclear magnetic resonance spectra of lignins. Biochem Biophys Res Commun. 1973;52(4):1162–1169.
[113] Lapierre C, Monties B, Guittet E, Lallemand JY. Photosynthetically carbon-13 labeled poplar lignins: carbon-13 NMR experiments. Holzforschung. 1984;38(6):333–342.

[114] Vanholme R, Demedts B, Morreel K, Ralph J, Boerjan W. Lignin biosynthesis and structure. Plant Physiol. 2010;153(3):895–905.
[115] Ralph J, Peng J, Lu F, Hatfield RD, Helm RF. Are lignins optically active? J Agric Food Chem. 1999;47(8):2991–2996.
[116] Bonawitz ND, Chapple C. The genetics of lignin biosynthesis: connecting genotype to phenotype. Annu Rev Genet. 2010;44:337–363.
[117] Lyr H. Detoxification of heartwood toxins and chlorophenols by higher fungi. Nature. 1962;195:289–290.
[118] Freudenberg K, Harkin JM. Supplement to the constitutional scheme for spruce lignin. Holzforschung. 1964;18(6):166–168.
[119] Freudenberg K. A schematic constitutional formulation for spruce lignin. Holzforschung. 1964;18(1–2):3–9.
[120] Adler E, Pepper JM, Eriksoo E. Action of mineral acid on lignin and model substances of guaiacylglycerol β-aryl ether type. Ind Eng Chem. 1957;49:1391–1392.
[121] Adler E, Marton J, Smith-Kielland I, Sömme R, Stenhagen E, Palmstierna H. Zur kenntnis der carbonylgruppen im lignin. I. Acta Chem Scand. 1959;13:75–96.
[122] Adler E, Marton J. Carbonyl groups in lignin. II. Catalytic hydrogenation of model compounds containing aryl carbinol, aryl carbinol ether, ethylene, and carbonyl groups. Acta Chem Scand. 1961;15:357–369.
[123] Marton J, Adler E. Carbonyl groups in lignin. III. Mild catalytic hydrogenation of björkman lignin. Acta Chem Scand. 1961;15:370–383.
[124] Marton J, Adler E, Persson KI, Dam H, Sjöberg B, Toft J. Carbonyl groups in lignin. IV. Infrared absorption studies and examination of the volumetric borohydride method. Acta Chem Scand. 1961;15:384–392.
[125] Adler E, Lundquist K. Spectrochemical estimation of phenylcoumaran elements in lignin. Acta Chem Scand. 1963;17:13–26.
[126] Karhunen P, Rummakko P, Sipilä J, Brunow G, Kilpeläinen I. Dibenzodioxocins; a novel type of linkage in softwood lignins. Tetrahedron Lett. 1995;36(1):169–170.
[127] Karhunen P, Rummakko P, Sipilä J, Brunow G, Kilpeläinen I. The formation of dibenzodioxocin structures by oxidative coupling. A model reaction for lignin biosynthesis. Tetrahedron Lett. 1995;36(25):4501–4504.
[128] Rodrigues Pinto PC, Borges da Silva EA, Rodrigues AE. Insights into oxidative conversion of lignin to high-added-value phenolic aldehydes. Ind Eng Chem Res. 2011;50(2):741–748.
[129] Brauns FE, Brauns DA. The Chemistry of Lignin: Supplement. New York: Academic Press; 1960.
[130] Erdtman H. Chemical nature of lignin. Tappi. 1949;32:71–74.
[131] Erdtman H. Outstanding problems in lignin chemistry. Ind Eng Chem. 1957;49:1385–1386.
[132] Adler E. The chemical structure of lignin. Sven Kem Tidskr. 1968;80(9):279–290.
[133] Forss K, Fremer KE. The repeating unit in spruce lignin. Pap Puu. 1965;47(8):443–454.
[134] Glasser WG, Glasser HR. Evaluation of lignin's chemical structure by experimental and computer simulation techniques. Pap Puu. 1981;63(2):71–74, 77–80, 82–83.
[135] Freudenberg K, Chen CL, Cardinale G. Die oxydation des methylierten natürlichen und künstlichen lignins. Chem Ber. 1962;95(11):2814–2828.
[136] Pare PW, Wang HB, Davin LB, Lewis NG. (+)-pinoresinol synthase: a stereoselective oxidase catalyzing 8,8′-lignan formation in *Forsythia intermedia*. Tetrahedron Lett. 1994;35(27):4731–4734.
[137] Lewis NG, Davin LB, Sarkanen S. Lignin and lignan biosynthesis: distinctions and reconciliations. ACS Symp Ser. 1998;697(Lignin and Lignan Biosynthesis):1–27.
[138] Holmgren A, Brunow G, Henriksson G, Zhang L, Ralph J. Non-enzymatic reduction of quinone methides during oxidative coupling of monolignols: implications for the origin of benzyl structures in lignins. Org Biomol Chem. 2006;4(18):3456–3461.

[139] Nimz H, Mogharab I, Lüdemann HD. Carbon-13-NMR spectra of lignins. 3. Comparison of spruce lignin with synthetic lignin according to freudenberg. Makromol Chem. 1974; 175(9):2563–2575.

[140] van Parijs FRD, Morreel K, Ralph J, Boerjan W, Merks RMH. Modeling lignin polymerization. I. Simulation model of dehydrogenation polymers. Plant Physiol. 2010; 153(3):1332–1344.

[141] Kishimoto T, Chiba W, Saito K, Fukushima K, Uraki Y, Ubukata M. Influence of syringyl to guaiacyl ratio on the structure of natural and synthetic lignins. J Agric Food Chem. 2010;58(2):895–901.

[142] Grabber JH. How do lignin composition, structure, and cross-linking affect degradability? A review of cell wall model studies. Crop Sci. 2005;45(3):820–831.

[143] Fengel D, Wegener G. Wood: Chemistry, Ultrastructure, Reactions. W. de Gruyter; 1984.

[144] Sakakibara A, Sano Y. Chemistry of lignin. In: Hon DNS, editor. Wood and Cellulosic Chemistry. New York: Marcel Dekker; 2001. p. 109–174.

[145] Mereweather JWT. Lignin-carbohydrate complex in wood. Holzforschung. 1957;11:65–80.

[146] Koshijima T, Watanabe T, Yaku F. Structure and properties of the lignin-carbohydrate complex polymer as an amphipathic substance. In: Glasser WG, Simo S, editors. Lignin. Washington, DC: American Chemical Society; 1989. p. 11–28.

[147] Leary GJ, Newman RH. Cross polarization/magic angle spinning nuclear magnetic resonance (CP/MAS NMR) spectroscopy. In: Lin SY, Dence CW, editors. Methods in Lignin Chemistry, Springer Series in Wood Science. Springer Berlin Heidelberg; 1992. p. 146–161.

[148] Grabber JH, Hatfield RD, Ralph J, Zon J, Amrhein N. Ferulate crosslinking in cell walls isolated from maize cell suspensions. Phytochemistry. 1995;40(4):1077–1082.

[149] Das NN, Das SC, Dutt AS, Roy A. Lignin-xylan ester linkage in jute fiber (*Corchorus capsularis*). Carbohydr Res. 1981;94(1):73–82.

[150] Joseleau JP, Gancet C. Selective degradations of the lignin-carbohydrate complex from aspen wood. Sven Papperstidn. 1981;84(15):R123–R127.

[151] Obst JR. Frequency and alkali resistance of lignin-carbohydrate bonds in wood. Tappi. 1982;65(4):109–112.

[152] Balakshin M, Capanema E, Berlin A. Isolation and analysis of lignin-carbohydrate complexes preparations with traditional and advanced methods: a review. Stud Nat Prod Chem. 2014;42:83–115.

[153] Eriksson O, Goring DAI, Lindgren BO. Structural studies on the chemical bonds between lignins and carbohydrates in spruce wood. Wood Sci Technol. 1980;14(4):267–279.

[154] Karlsson O, Ikeda T, Kishimoto T, Magara K, Matsumoto Y, Hosoya S. Isolation of lignin-carbohydrate bonds in wood. Model experiments and preliminary application to pine wood. J Wood Sci. 2004;50(2):142–150.

[155] Minor JL. Chemical linkage of pine polysaccharides to lignin. J Wood Chem Technol. 1982;2(1):1–16.

[156] Minor JL. Chemical linkage of polysaccharides to residual lignin in loblolly pine kraft pulps. J Wood Chem Technol. 1986;6(2):185–201.

[157] Iversen T. Lignin-carbohydrate bonds in a lignin-carbohydrate complex isolated from spruce. Wood Sci Technol. 1985;19(3):243–251.

[158] Iversen T, Waennstroem S. Lignin-carbohydrate bonds in a residual lignin isolated from pine kraft pulp. Holzforschung. 1986;40(1):19–22.

[159] Watanabe T. Structural studies on the covalent bonds between lignin and carbohydrate in lignin-carbohydrate complexes by selective oxidation of the lignin with 2,3-dichloro-5,6-dicyano-1,4-benzoquinone. Wood Res. 1989;76:59–123.

[160] Watanabe T, Karina M, Sudiyani Y, Koshijima T, Kuwahara M. Lignin-carbohydrate complexes from *Albizia falcata* (L.) back. Wood Res. 1993;79:13–22.

[161] Kosikova B, Joniak D, Kosakova L. The properties of benzyl ether bonds in the lignin-saccharidic complex isolated from spruce. Holzforschung. 1979;33(1):11–14.
[162] Baucher M, Monties B, Van Montagu M, Boerjan W. Biosynthesis and genetic engineering of lignin. Crit Rev Plant Sci. 1998;17(2):125–197.
[163] Donaldson LA. Lignification and lignin topochemistry - an ultrastructural view. Phytochemistry. 2001;57(6):859–873.
[164] Saka S, GDAI. Localization of lignin in wood cell walls. In: Higuchi T, editor. Biosynthesis and Biodegradation of Wood Components. New York: Academic Press; 1985. p. 51–62.
[165] Samuels AL, Rensing KH, Douglas CJ, Mansfield SD, Dharmawardhana DP, Ellis BE. Cellular machinery of wood production: differentiation of secondary xylem in *Pinus contorta* var. latifolia. Planta. 2002;216(1):72–82.
[166] Donaldson LA. Mechanical constraints on lignin deposition during lignification. Wood Sci Technol. 1994;28(2):111–118.
[167] Fukuda H, Komamine A. Lignin synthesis and its related enzymes as markers of tracheary-element differentiation in single cells isolated from the mesophyll of *Zinnia elegans*. Planta. 1982;155(5):423–430.
[168] Deighton N, Richardson A, Stewart D, McDougall GJ. Cell-wall-associated oxidases from the lignifying xylem of angiosperms and gymnosperms. Monolignol oxidation. Holzforschung. 1999;53(5):503–510.
[169] Davin LB, Lewis NG. Dirigent proteins and dirigent sites explain the mystery of specificity of radical precursor coupling in lignan and lignin biosynthesis. Plant Physiol. 2000;123(2):453–461.
[170] Müsel G, Schindler T, Bergfeld R, Ruel K, Jacquet G, Lapierre C, *et al*. Structure and distribution of lignin in primary and secondary cell walls of maize coleoptiles analyzed by chemical and immunological probes. Planta. 1997;201(2):146–159.
[171] Terashima N, Atalla RN, Ralph SA, Landucci LL, Lapierre C, Monties B. New preparation of lignin polymer models under conditions that approximate cell wall lignification. I. Synthesis of novel lignin polymer models and their structural characterization by ^{13}C NMR. Holzforschung. 1995;49(6):521–527.
[172] Terashima N, Atalla RH, Ralph SA, Landucci LL, Lapierre C, Monties B. New preparations of lignin polymer models under conditions that approximate cell wall lignification. II. Structural characterization of the models by thioacidolysis. Holzforschung. 1996;50(1):9–14.
[173] Terashima N, Fukushima K. Biogenesis and structure of macromolecular lignin in the cell wall of tree xylem as studied by microautoradiography. In: Lewis NG, Paice MG, editors. Plant Cell Wall Polymers. Washington, DC: American Chemical Society; 1989. p. 160–168.
[174] Terashima N, Yoshida M, Hafren J, Fukushima K, Westermark U. Proposed supramolecular structure of lignin in softwood tracheid compound middle lamella regions. Holzforschung. 2012;66(8):907–915.
[175] Mansfield SD, Kim H, Lu F, Ralph J. Whole plant cell wall characterization using solution-state 2D NMR. Angew Chem Int Ed. 2012;7(9):1579–1589.
[176] Donaldson LA. Critical assessment of interference microscopy as a technique for measuring lignin distribution in cell walls. N Z J For Sci. 1985;15(3):349–360.
[177] Zimmermann W, Nimz H, Seemueller E. Proton and nitrogen-13 NMR spectroscopic study of extracts from corks of rubus idaeus, solanum tuberosum, and *Quercus suber*. Holzforschung. 1985;39(1):45–49.
[178] Nimz HH, Tutschek R. Carbon-13 NMR spectra of lignins, 7. The question of the lignin content of mosses (*Sphagnum magellanicum* Brid.). Holzforschung. 1977;31(4):101–106.
[179] Ralph J, Hatfield RD, Quideau S, Helm RF, Grabber JH, Jung HJG. Pathway of *p*-coumaric acid incorporation into maize lignin as revealed by NMR. J Am Chem Soc. 1994;116(21):9448–9456.

[180] Billa E, Tollier MT, Monties B. Characterization of the monomeric composition of *in situ* wheat straw lignins by alkaline nitrobenzene oxidation: effect of temperature and reaction time. J Sci Food Agric. 1996;72(2):250–256.

[181] Jacquet G, Pollet B, Lapierre C, Mhamdi F, Rolando C. New ether-linked ferulic acid-coniferyl alcohol dimers identified in grass straws. J Agric Food Chem. 1995;43(10):2746–2751.

[182] Lapierre C, Pollet B, Rolando C. New insights into the molecular architecture of hardwood lignins by chemical degradative methods. Res Chem Intermed. 1995;21(3-5):397–412.

[183] Santos RB, Capanema EA, Balakshin MY, Chang HM, Jameel H. Lignin structural variation in hardwood species. J Agric Food Chem. 2012;60(19):4923–4930.

[184] Timell TE, editor. Compression Wood in Gymnosperms. Springer-Verlag; 1986.

[185] Chabannes M, Ruel K, Yoshinaga A, Chabbert B, Jauneau A, Joseleau JP, *et al.* In situ analysis of lignins in transgenic tobacco reveals a differential impact of individual transformations on the spatial patterns of lignin deposition at the cellular and subcellular levels. Plant J. 2001;28(3):271–282.

[186] Campbell MM, Sederoff RR. Variation in lignin content and composition: mechanisms of control and implications for the genetic improvement of plants. Plant Physiol. 1996;110(1):3–13.

[187] Peter G, Neale D. Molecular basis for the evolution of xylem lignification. Curr Opin Plant Biol. 2004;7(6):737–742.

[188] Brodribb TJ, Holbrook NM, Hill RS. Seedling growth in conifers and angiosperms: impacts of contrasting xylem structure. Aust J Bot. 2005;53:749–755.

[189] Sperry JS, Hacke UG, Wheeler JK. Comparative analysis of end wall resistivity in xylem conduits. Plant Cell Environ. 2005;28(4):456–465.

Part III

Sources and Characterization of Lignin

Part IV

5
Isolation of Lignins

5.1 Introduction

Isolation of lignin is a nontrivial initial step that determines the scale-up and optimization of many of the industrial processes concerning this natural polymer, as well as most of the final applications of any lignin derived from them. One of the fundamental questions underlying such isolation relies on whether to choose quality or quantity when finding a reproducible isolation method. On the one hand, it is desirable to obtain a lignin that is very similar to native lignin, and only the smallest changes are made. The structure of this almost-unaltered lignin is easier to determine. On the other hand, there is a need to produce high amounts of lignin upon separation from wood components in order to improve any industrial process by using the lignin produced (paper, bioethanol, high-value products, etc.). These two premises, quality and quantity, may look simple, but fulfilling both of them at the same time in a satisfying manner is not straightforward.

The pretreatment is an essential technique to provide a robust and reproducible process for biomass conversion because it overcomes some physical and chemical barriers that make lignin resistant to later bioprocesses such as hydrolysis and fermentation [1]. A recent critical analysis of pretreatment technologies was published by da Costa Sousa et al. [2], in which the various pretreatment technologies are divided in four categories: physical pretreatment (e.g., ball milling), solvent fractionation (including the organosolv process, together with phosphoric acid fractionation and the use of ionic liquids), chemical pretreatment (acidic, alkaline, and oxidative), and biological treatment (using mainly fungi).

The process toward a controlled isolation of lignin has been extended for years. In 1967, Pearl [3] wrote that "It is practically impossible to isolate two lignin preparations with identical properties, even by the same procedure." Things have changed since Pearl made his statement but, although many methods are available nowadays for the isolation and purification of lignin, none of them indeed provides a 100% yield and structural authenticity. One of the reasons for the lack of perfect isolation is the aggressive treatment often performed during the process. For lignin to be isolated in a way

---

Lignin and Lignans as Renewable Raw Materials: Chemistry, Technology and Applications, First Edition.
Francisco G. Calvo-Flores, José A. Dobado, Joaquín Isac-García and Francisco J. Martín-Martínez.
© 2015 John Wiley & Sons, Ltd. Published 2015 by John Wiley & Sons, Ltd.

that it closely represents the native material, reactive chemicals and elevated temperatures must be avoided. Also, it is important to remark at this point that all these preparations are distinct from the so-called protolignin,[1] which is a term used for the material as it occurs in the plant tissue [4, 5].

Despite the inherent difficulty for isolating lignin, it is clear that working with pure isolated samples of lignins is desirable when characterizing these materials. Many reviews have been published in this regard, such as the critical review by Lai and Sarkanen [6] in 1971, and more recently by Monteil Rivera *et al.* [7]. Moreover, several monographs such as Brauns, [8, 9] Fengel and Wegener [10], Gellerstedt and Henriksson [11], or more recently Lu and Ralph [12] complete the extensive literature on the topic.

In a general perspective, isolation methods are usually classified according to the scale of the process, which also implies the target application of the isolated lignin. Accordingly, two main groups are formed: procedures for laboratory purposes and those for commercial application, this later implying a larger scale.

The laboratory-scale methods can be in turn classified by lignin as residue or by dissolution. Some of the methods under this classification include, for example, isolating lignin for studying its structure and properties. These methods isolate small to moderate amounts of lignin, and only in some specific cases have these been extended to industrial or semiindustrial applications. Another more practical approach is also considered, for example, industrial lignins produced as by-products of pulping and bioethanol production processes [13]. These are the aforementioned commercial applications of lignin. All these, that is, procedures for laboratory purposes and those for commercial applications, will be discussed in this chapter.

Furthermore, it is common to categorize isolated lignins from biomass according to the corresponding isolation procedure, and two main approaches are generally used when the isolation is performed from wood: acidolysis and enzymatic treatments of lignocellulosics.

Tables 3.5 and 3.6 on p. 56 summarize the lignin content of various types of plants and wood fibers determined by some of the procedures that will be explained in the following sections.

5.2 Methods for Lignin Isolation from Wood and Grass for Laboratory Purposes

Several methods have been used so far for isolating lignin from wood or herbaceous plants, but none of them leave lignin unchanged after isolation. Whatever method is employed, the resulting lignin preparation is no longer identical to native lignin [14], and it is not a measurement of the total lignin in the wood or grass either (see Chapter 3). To distinguish the different lignins obtained, these are usually called after the author's name of the procedure.

Nevertheless, regardless of the method used to isolate lignin, the final ground plant material is usually set free from various "extractives," such as fatty substances, resins, and volatile oils. This is generally accomplished by extracting the lignified material with an organic solvent or a mixture of these (preferably an alcohol/benzene solution).

The methods described in the literature for the isolation of lignin can be divided into two main classes depending on whether the lignin is the object or residue of the separation:

- Removal of cellulose and other components (by hydrolysis), leaving lignin as an insoluble residue.
- Removal of lignin from the cellulose and other components.

A more detailed description of these methods is provided in Table 5.1.

---

[1] Immature form of lignin that can be extracted from the plant cell wall with EtOH or dioxane.

Table 5.1 Main methods used for the isolation of lignin

Process	Reagents	Lignin name	Yield (% lignin)
Lignin as residue			
Acid hydrolysis of polysaccharides		Acid hydrolysis (AHL)	> 80
	H_2SO_4	Klason	
	HCl	Willstäter	
	HCl/H_3PO_4	Urban	
	HCl/H_2SO_4	Halse	
	HF	Fredenhagen	
	HBr/H_2SO_4	Runkel	
	CF_3COOH	Fengel	
Polysaccharides oxidation			> 80
	$Na_3H_2IO_6$	Purves	
	$Cu(NH_3)_4(OH)_2$	Freudenberg/Cuoxam	
Lignin by dissolution			
	Aqueous-dioxane	Brauns	< 10
Mechanical disintegration and cellulolytic enzymes	Dioxane/H_2O	Björkman (MWL)	20–30
	Dioxane/H_2O	Pew (CEL)	80
	Ionic liquids		< 80
Mild acid hydrolysis	Alcohol/dioxane, phenol, acetic acid with acid catalyst, alcohols	Organosolv	25–50
	Enzymes	EMAL	60
Steam hydrolysis	Steam/acetic acid	SEL	50–90
Alkaline hydrolysis	NaOH/EtOH or H_2O	Alkali (soda)	> 80
	$NaOH/Na_2S$	Kraft	90–95
	$NaOH/O_2$ or H_2O_2		> 80
Sulfonation	$HSO_3^-/SO_3^=$ and base	Sulfite	> 80
Biological treatment	EtOH or dioxane/H_2O	Rot fungi	20

Note that among all these methods, only three yield a chemically almost unchanged lignin:

- 'Brauns lignin': extraction with organic solvents [8, 9, 15].
- "Milled-wood lignin" (MWL): Björkman introduced the use of a preliminary, extensive grinding, followed by solvent extraction [16–18].
- "Cellulolytic enzyme lignins," (CELs): Later, it was found that the amount of solubilized lignin could be increased when the finely ground wood meal is treated with hydrolytic enzymes prior to solvent extraction [19–21].

5.2.1 Lignin as Residue

Most of the methods described here have been discussed in Chapter 3 in relation to the gravimetric determination of total lignin content. Therefore, in this section, only some particular aspects related to isolation are addressed (see Table 5.1).

5.2.1.1 Klason Lignin (Sulfuric Acid)

The most widely applied variation of the Klason (sulfuric acid) (see Chapter 3) method employs 72% of H_2SO_4 to promote carbohydrate hydrolysis. Then, it is diluted with water, heated at high temperature (100–125 °C) to promote a second hydrolysis and finally the lignin is filtered, washed, and dried. The lignin isolated is referred to as Klason lignin (KL) [22, 23]. Many analogous versions of the sulfuric acid method have been published, recently reviewed by Sluiter *et al.* [24]. Different factors affect the method, such as pretreatment of the sample, temperature and time of hydrolysis, H_2SO_4 concentration, dilution, temperature, and time in the second hydrolysis. The "acid-insoluble lignin" contents of a variety of representative lignified materials are recorded in Table 5.2 [25–31].

These values reflect generally recognized trends based on the source,[2] on the botanical classification,[3] on anatomical differences,[4,5] in the case of pulps, on the process used to delignify the wood (acid sulfite, bisulfite, Kraft), and on fiber classification[6] [32].

5.2.1.2 Willstätter Method (Fuming Hydrochloric Acid)

Wood meal is extracted and hydrolyzed with concentrated HCl. In 1913, Willstätter and Zechmeister [33] recorded the observation that fuming HCl (42–43% at 15 °C) will completely hydrolyze cellulose at low temperature, yielding an insoluble lignin residue.

5.2.1.3 Urban Lignin

Urban [34] avoids the use of fuming HCl and strong H_2SO_4 for the hydrolysis of the polysaccharides. Thus, the wood is treated with a mixture consisting of HCl and phosphoric acid (3:1, v/v).

5.2.1.4 Freudenberg Lignin: Cuoxam or Cuproxam Lignin

In this method, lignin is isolated by alternately hydrolyzing wood with boiling 1% H_2SO_4 and cuprammonium hydroxide solution (Schweizer's reagent, tetraamine copper(II)) [35, 36]. All the carbohydrate components in extracted wood meal may be dissolved by complexation giving, lignin in high yields of the so-called Freudenberg, cuoxam, or cuproxam lignins [37]. This lignin is totally insoluble in organic solvents and is known to retain the morphological features of wood [38, 39], being the cuoxam lignin the most suitable for model delignification experiments.

5.2.1.5 Halse, Runkel, and Fengel Lignins

The residue that does not dissolve after heating wood meal with HCl and H_2SO_4 is called Halse lignin [40, 41]. The residue that does not dissolve after heating with HBr and H_2SO_4 is named Runkel lignin [42] and the residue that does not dissolve after heating with trifluoroacetic acid is named Fengel lignin [43].

---

[2] Woods compared to annual plants.
[3] Gymnosperms or softwoods compared to angiosperms or hardwoods.
[4] Normal compared to reaction wood, earlywood compared to latewood, sapwood.
[5] Compared to heartwood (the dead, inner wood, of a woody stem or branch, which often comprises the majority of the stem's cross section).
[6] Fine compared to fiber fractions.

Table 5.2 Acid-insoluble (Klason) lignin contents of lignified materials [32, p. 37][a]

Material	Lignin (%)	Author [Ref.]
Softwood		
Norway spruce	28.6	Timell [25, p. 294–311]
Norway spruce compression wood	38.8	"
Larch sapwood	26.0	"
Larch hardwood	28.6	"
Pine earlywood	28.8	"
Pine latewood	27.4	"
Hardwood		
White birch normal wood	22.0	Timell [27]
White birch tension Wood	16.1	"
Birch fibers	12.6	Obst [26]
Birch ray cells	26.7	"
Nonwood Fiber		
Bagasse	19.6	Bagby et al. [28][a]
Bamboo	22.2	"
Wheat straw	17.0	"
Kenaf	10.9	"
Sorghum	7.9	"
Pulp		
Pine kraft	4.8	Kyrklund and Strandell [29]
Birch kraft	5.0	"
Spruce kraft	2.8	"
Birch acid sulfite	3.2	"
Spruce bisulfite (Mg base)	9.9	"
Birch bisulfite (Mg base)	4.0	"
Scots pine kraft (> 50 mesh)	5.2	Lindström and Glad-Nordmark [30]
Scots pine kraft(< 300 mesh)	23.4	"
Norway spruce TMP(> 150 mesh)	23.5	Robinson [31]
Norway spruce TMP(< 150 mesh)	35.8	"

[a] 80% of the acid method was used.

5.2.1.6 Fredenhagen Lignin

This method uses, as reagent, hydrogen fluoride in order to hydrolyze cellulose [44]. Defaye et al. [45] showed the almost quantitative and rapid (30 min, rt) solubilization of the carbohydrates from the lignocellulosic matrix when this is treated with HF. Also, the insoluble lignin was analyzed indicating that aryl ether linkages are not cleaved.

5.2.1.7 Periodate Lignin: Purves Lignin

This method is based on the conversion of monosaccharide units in polysaccharide to dialdehydes by sodium paraperiodate (see Figure 5.1). These dialdehydes are susceptible to hydrolysis by boiling water (see Figure 5.1) [46, 47].

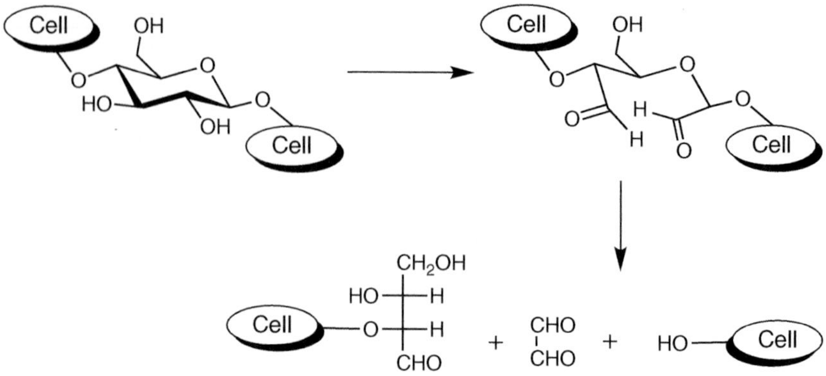

Figure 5.1 Degradation of monosaccharide units to dialdehydes

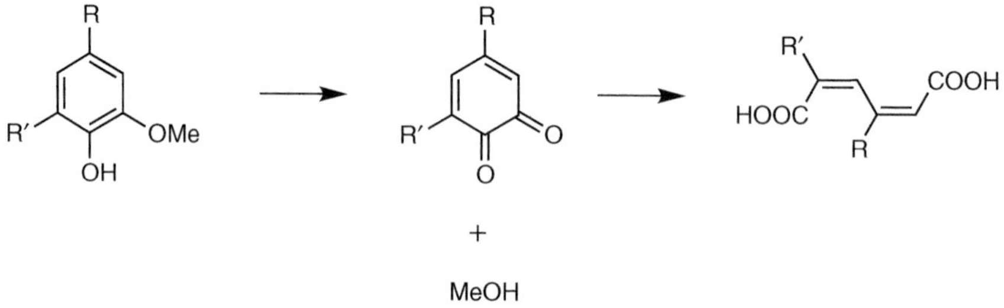

Figure 5.2 Oxidative cleaving of the aromatic ring, via o-quinone, to muconic acid

Mild oxidation of extracted wood meal with periodic acid (HIO_3) dissolved nonlignin components by hot water hydrolysis, there is less alteration in lignin structure. Degraded carbohydrates are dissolved, finally giving an insoluble periodate or Purves lignin.

This method avoids the condensation processes that cause changes in the acid lignins, but oxidative modification occurs instead, notably the units with free phenolic hydroxyl groups, which are converted into o-quinones and then into muconic acid structures (see Figure 5.2) [48].

This method is tedious and requires at least several successive treatments with periodate (4.5% aqueous $Na_3H_2IO_6$, 24 h at 20 °C), each followed by a treatment with boiling water (3 h).

5.2.2 Lignin by Dissolution

Whatever the process to isolate lignin, the first stage of preparation of the sample will consist of a mechanical treatment. As almost all the methods discussed require ball milling of wood, it is relevant to understand how lignin is degraded during the milling [49].

At the moment, most of these methods are considered as pretreatments for the enhancement of EtOH production potential or pulping from biomass [50, 51].

5.2.2.1 Brauns Lignins (Native Lignin)

Brauns lignins are obtained by extraction with organic solvents (EtOH/dioxane), which are removed by evaporation [8, 9, 15].

When sprucewood meal is allowed to stand for 3–4 days in 96% EtOH and is then filtered and the clear light-brown filtrate evaporated under reduced pressure, a resinous residue remains, which is washed with distilled water and ether. With the removal of the resins by the ether, the native lignin[7] is obtained as a very finely divided light cream-colored powder, which is precipitated repeatedly from its solution in dioxane into ether until a constant methoxyl content results [15].

This procedure has the drawback of a very low yield (2–10% of the lignin present in wood) and the contamination with extractives [6, 52].

5.2.2.2 Milled-Wood Lignin (MWL)

Björkman [16–18, 53] introduced the so-called MWL method, which uses a preliminary, extensive grinding, with the subsequent solvent extraction. It is the most commonly used method and involves thorough milling of the plant material, followed by extraction with dioxane/water (96:4, v/v) for 24 h. Lignin is heat-sensitive, so extractions should not be at the boiling point of the solvents. Solvents are removed by evaporation [17, 52]. The yields are usually low, due possibly to the chemical changes occurring during the isolation process (see Table 5.3). Moreover, a possible variation in lignin composition can occur in different parts of the plant. To prepare a lignin sample representative of a certain species of wood, it is customary to choose sapwood.[8]

The milling is carried out either in a nonswelling medium such as toluene, or in the dry state. Ball milling to some extent modifies the structure of lignin [55, 56]. Figure 5.3 shows the fragmentation reaction suggested during the milling of wood, and the representative data on MWL isolated from different wood species are recovered in Table 5.3.

Hardwood MWLs are found to contain more carbohydrates than do softwood MWLs.[9]

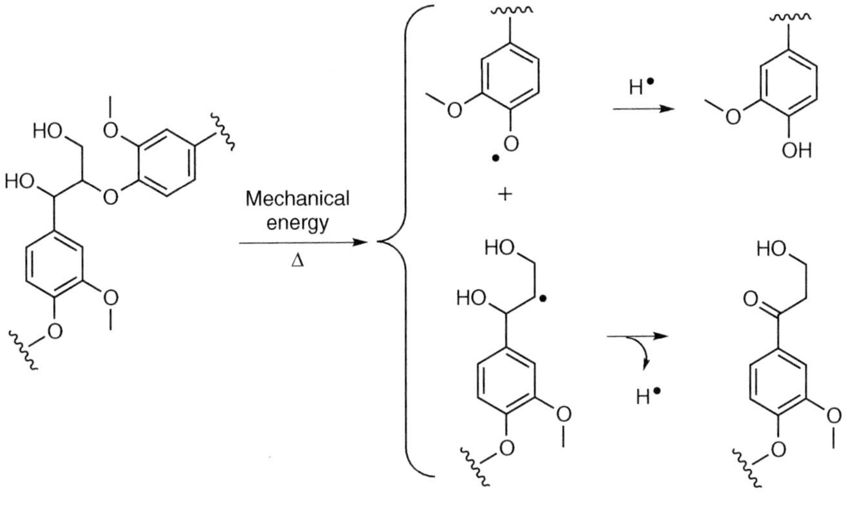

Figure 5.3 *Lignin fragmentation reaction suggested to take place during the milling of wood [11]*

---

[7] A lignin isolated in such a way that the solvent does not react with the lignin or alter it in any way.
[8] The relative amounts of sapwood and heartwood in any stem can vary significantly among individuals, species, and growing conditions.
[9] MWL usually yields about 25 % of the lignin in wood.

Table 5.3 Representative data on MWL isolated from different wood species[a]

Scientific name	Common name	Lignin (%)[b]	–OCH$_3$ (%)	–OH$_p$ [c] (%)	Carbohydrate (%)	[Ref.]
P. abies	Norway spruce	19	15.45	30	1.9	[16]
P. abies	Norway spruce	17	15.2	20	4.1	[21]
P. abies	Norway spruce	28	15.2	20	4.3	[21]
Picea mariana	Black spruce	–	15.41	28	–	[18]
P. mariana	Black spruce	25	15.3	23	< 9.6	[54]
P. mariana	Black spruce	69	13.7	21	< 7.7	[54]
Pinus silvestris	Scots pine	–	15.74	27	–	[18]
Betula verrucosa	Silver birch	–	21.51	–	7.5	[18]
Liquidambar styraciflua	Sweetgum	17	21.4	14	3.6	[21]
Liquidambar styraciflua	Sweetgum	43	21.5	13	3.8	[21]
Populus tremuloides	Trembling aspen	24	19.6	21	< 8.0	[54]
Populus tremuloides	Trembling aspen	61	20.9	21	< 4.3	[54]

[a] Data obtained from ref. [11].
[b] % Content total.
[c] Phenolic hydroxyls.

Many purification procedures have been used to reduce contaminants in isolated lignins. The first, developed by Björkman in 1954, for the MWL includes two precipitations [16]:

- The native lignin dissolved in 90% acetic acid was precipitated into water.
- The precipitation product is then dissolved in 1,2-dichloroethane/EtOH (2:1) and precipitated into ethyl ether.

Sakakibara and Nakayama [57, 58] found that heating wood in neutral dioxane/water (1:1) at high temperatures (175 °C, 2 h) dissolves lignins to about 50% yield.

Recently, Zhang et al. [59] reported the extraction and purification of a lignin from sugarcane bagasse (lignin from herbaceous plants is known to differ from those from softwoods and hardwoods). These researchers proposed a three-step extraction of lignin fractions with 96% dioxane, 50% dioxane, and 80% dioxane containing 1% NaOH at boiling temperature. The total yield in lignin was 72% based on dried material.

5.2.2.3 Cellulolytic Enzyme Lignins (CEL)

Pew [19–21] found that the amount of solubilized lignin could be increased if the finely ground wood meal was treated with hydrolytic enzymes prior to solvent extraction. The treatment of the finely ground wood meal with cellulolytic enzymes prior to solvent extraction removes part of the polysaccharides and increases the yield of lignin [21]. After extraction with dioxane/water, the crude extract still contains carbohydrates. To overcome this problem, Chen et al. [60] developed the swelled enzyme lignin (SEL) method. Swelling loosens and reduces the associations between lignin and hemicelluloses and as a result not only shortens the digestion time but also leads to a more intensive enzymatic degradation of the polysaccharides.

CEL is purified by precipitation into water from a solution in acetic acid. An alternative method based on liquid–liquid extraction has been proposed, yielding lignins with less carbohydrate contamination [52].

For grass plants [61], Jung and Himmelsbach [62] developed a lignin isolation method from wheat straw by ball milling (8 days) and enzyme treatment (4 days). Extraction with 50% dioxane/water produced lignin with the highest yield compared to the lignin extracted by standard methods (96% dioxane).

CEL is probably more representative of the total lignin in wood than is MWL, with higher yields and less chemical changes, but its preparation procedure is more tedious [63]. Recently, Hu et al. [64] were able to obtain CEL preparations with yields of up to 86%. Imai et al. [65] reported a boost of 200% in the reaction rate of the enzymatic hydrolysis by previous ultrasonic pretreatment of the sample.

5.2.2.4 Enzymatic Mild Acidolysis Lignin (EMAL)

Wu and Argyropoulos [54] have proposed a procedure using the combination of enzymatic and mild acidolysis of lignin. Guerra et al. [66, 67] extensively studied the effect of milling (vibratory, rotatory, or ball milling) on the structure of EMALs, and reported that the yield of EMAL is about fourfold greater than that of the corresponding MWL, and about twofold greater than CEL method isolated from the same batch of milled wood. Furthermore, the isolated lignin samples are more representative of the total lignin in milled wood.

5.2.2.5 Dioxane Acidolysis Lignins

Freudenberg and Stumpf [68, 69] also showed that lignin could be extracted with dioxane/HCl (9:1) or with tetrahydrofuran and HCl from wood swollen by water, at rt.

5.2.2.6 Organosolv Lignin[10]

The delignification of wood using organic solvents (organosolv pulping) has generated interest since it was introduced toward the end of the last century [70]. Organosolv lignins are a group of lignins that are isolated directly from biomass extracting lignin with mixtures of organic solvents and some additives or catalysts. Either low-boiling-point solvents (LBSs) such as EtOH, MeOH, or acetone can be used, or high-boiling-point solvents (HBSs) such as glycols. In many cases, organic solvents are used together with water. The most widely used solvents are primary alcohols such as MeOH and EtOH, but carboxylic acids as acetic and formic have been used. Organic solvents can be mostly recovered by distillation, this being a great advantage. Once the biomass is treated, Organosolv lignins are recovered by precipitation in water after evaporating organic solvents. Catalysts (acids or bases) are commonly used although "uncatalyzed" pulping can be performed at higher temperatures by acetic acid. In alkaline systems, the lignification rate is governed by the cleavage of β-ether bonds; under acidic conditions, α-ether cleavage occurs to a great extent [71].

The original technique was developed by Kleinert [72], in 1971, using mixtures of EtOH/water. In short, lignin is extracted from biomass with an organic solvent/water mixture at a high temperature and pressure [73, 74] and later isolated by precipitation from the solution.

HBSs such as ethylene glycol and tetrahydrofurfuryl alcohol [75, 76] were used for pulping. Gast and Puls [77] found the pulping efficiency of ethylene glycol to be improved by the addition of aluminum sulfate or chloride as a catalyst. Nakamura and Takauti [78], in 1941, were the first to use ethylene glycol as a pulping agent. Kishimoto and Sano [79, 80] showed that butane-1,4-diol, at 220 °C, and butane-1,3-diol, at 200 °C, are very effective even without acid catalyst. The same authors [81] demonstrated that the addition of a reducing sugar to the fresh solvent drastically accelerated the lignification rate.

These environmentally benign pulping processes [82] are able to solubilize and isolate lignin from other biomass component to produce a polymer more similar to native lignin that lignins isolated from traditional paper industry pulping.

The main factors that affect the delignification of organosolv pulp are extraction temperature, time, solvent composition, and hydrogen ion concentration [73].

Lawther et al. [83] presented yields of organosolv lignins extracted with different mixtures under various conditions (see Table 5.4).

---
[10] For further information see Section 5.3, p. 127.

Table 5.4 Yield of organosolv lignin under various extraction conditions[a,b]

Solvents mixture	Yield (%)	Solvents mixture	Yield (%)
EtOH/water	27.7	n-Butanol/water	23.9
MeOH/water	24.2	Dioxane/water	24.2
Propan-1-ol/water	27.1	Acetone/water	26.4

[a] Organic solvent/water (60:40, v/v) in 0.02 N H_2SO_4 at 75 °C for 2 h.
[b] Values obtained from ref. [83].

Table 5.5 Yield of organosolv lignin in EtOH/water at 75 °C under various extraction conditions[a,b]

EtOH/water (v/v)[a]	Yield (%)	H_2SO_4 conc. (N)[a]	Yield (%)	Time (h)[a]	Yield (%)
40/60	24.8	0.00	20.0	1.0	24.1
50/50	24.9	0.01	19.9	1.5	26.9
60/40	27.7	0.02	27.7	2.0	27.7
70/30	31.2	0.03	27.6	2.5	27.8
80/20	29.8	0.1	27.9	3.0	28.0
90/10	24.1	0.5	32.5	5.0	28.3
		1.0	43.4	12.0	28.7

[a] Default experimental conditions: EtOH/water (60:40) in 0.02 N H_2SO_4 at 75 °C for 2 h.
[b] Values obtained from ref. [83].

They found that EtOH/water is the most efficient extracting solvent and they studied other different parameters, such as the acid concentration and time (see Table 5.5), affecting the yield of the extraction.

Finally, they found that treatment of wheat straw with 0.5 M NaOH at 75 °C for 2 h released 52% lignin (double yield obtained from organosolv systems).

5.2.2.7 From Ionic Liquid

Zakzeski et al. [13] give a table of the lignin solubility data in different ionic liquids. Ionic liquids are salts with melting points below an arbitrary set point of 100 °C. They often have tunable physical properties based on the choice of cation and anion pair, a negligible vapor pressure, and good thermal stability [84].

Lee et al. [85] observed that ethyl-methylimidazolium acetate [EMIM][OAc], was able to selectively extract lignin from wood. Fort et al. [84] found that 1-butlyl-3 methylimidazolium chloride, [BMIM][Cl] was capable of dissolving both cellulose and lignin, and they were able to use precipitation solvents to isolate cellulose from the remaining biomass components.

The great challenge associated with the use of ionic liquid is the separation of substrate and product after the reaction. Nevertheless, typical analytical methods include solute extraction using organic solvents [86]. Another is the necessary recyclability of ionic liquid, given its high price [87].

Pu et al. [88] specifically investigated the effect of various anions on the solubility of lignin from Kraft pulp. They discovered that up to 20 wt% of lignin was soluble using $CF_3SO_3^-$ or $MeSO_4^-$ anions, and for 1-butyl-3-methylimidazolium salts, the order of solubility was $MeSO_4^- > Cl^- \approx Br^- > [PF_6^-]$ [88]. The noncoordinating PF_6^- anion was essentially ineffective in dissolving lignin [88]. Tan et al. [89] observed that 1-ethyl-3-methylimidazolium cation with a mixture of alkylbenzene sulfonate anions (mainly xylenesulfonate), [EMIM][BMS], dissolved lignin, and an extraction yield of 93% was attained.

Wen et al. [90] studied the chemical transformations of lignin during ionic liquid pretreatment. Results revealed that a decrease of aliphatic OH and an increase in phenolic hydroxyl groups occurred in lignin as the pretreatment proceeded. The increased phenolic OH was mainly as a result of cleavage of β–O–4′ linkages, while the reduced aliphatic OH is probably attributed to the dehydration reaction.

5.2.2.8 Alcoholysis

Lignin is isolated by heating wood with various hydroxylic compounds in the presence of a catalyst. Gruss [91], in 1921, recorded the observation that lignin can be isolated by heating the wood with alcohol and HCl. Instead of EtOH, other hydroxyl compounds have been used, such as butyl and amyl alcohols [92], ethylene glycol [93], and monomethyl ether of ethylene glycol [94].

Recently, Monteil Rivera et al. [7, 95] extracted lignin from ground biomass (see Table 5.6) using microwave irradiation of a suspension in EtOH/water (81:19, v/v) and H_2SO_4 (0.5 N).

5.2.2.9 Phenol Lignin

The fact that lignin can be extracted, at temperatures above 150 °C from wood with phenols was firstly reported by Bühler. As in the preceding case, the phenols also combine with the lignin by heating wood with phenol and HCl [96].

5.2.2.10 Alkali Method

Both aqueous [97] and alcoholic [97–99] sodium hydroxide solutions have been employed. The ease with which lignin can be obtained by this method depends somewhat on the character of the lignified plant material: wood lignin requires aggressive treatment (5% NaOH, 130–170 °C) to become soluble in aqueous media; grass lignins can be isolated in good yields by mild alkaline treatments, even at room temperature [100]. The large proportion of free phenolic groups presented in grass lignins seems to play a key role [101].

5.2.2.11 Steam-Exploded Lignin

One method that it has gained much attention is the so-called steam-exploded delignification [11]. Although it has not yet commercially available, it has a promising future as a possible means for simple and cheap separation of wood components.

When wood is treated with steam at high temperature under pressure (i.e., about 1.38–3.00 MPa and 180–230 °C for 1–20 min) [102], followed by a sudden decompression in the presence of some chemicals, a hemicellulose degradation and a partial hydrolysis of lignin occurs.

These conditions result in a water-insoluble lignin material with a low level of carbohydrate and wood extractive impurities. With this procedure, the lignin itself has a somewhat reduced molecular

Table 5.6 Lignin yields after microwave extraction[a] of various biomass[b]

Biomass	Klason lignin	Acid-soluble lignin	Total lignin	Isolated lignin (%)
Triticale straw	17.8	3.34	21.1	12.7
Wheat straw	17.5	3.10	20.5	12.4
Corn residues	18.1	3.89	22.0	19.2
Flax shives	23.2	2.11	25.3	18.2
Hemp hurds	22.7	1.86	24.6	13.9

[a] (81/19 EtOH/H_2O; 0.5 N H_2SO_4; 140 °C).
[b] Data obtained from ref. [7, p. 358].

mass because certain acid hydrolysis reactions take place. This type of separation process is often combined with enzymatic hydrolysis to produce carbohydrates for fermentation [103]. With this procedure, the fibrous mass of wood or herbaceous plants actually explode and lignin fibers are liberated. Depending on time and temperature, different degrees of polymer modification and or degradation can be achieved. Once the delignification occurs, the lignin fraction can be extracted, to a large extent, by treatment with either aqueous alkali or organic solvents. A residue highly enriched in cellulose remains separate from lignin [104, 105]. Li et al. [106] recently applied a steam-explosion pretreatment process to both softwoods and hardwoods. They noted that the process with SO_2 pre-impregnation allowed efficient extraction of lignin from hardwood, but only low fractionation efficiencies were observed with softwoods [13].

Brownell and Saddler [107] compared different pretreatment methods with steam explosion, and differing views on the relative importance of mechanical and chemical effects were outlined. Hydrolysis was desirable; pyrolysis was undesirable. The effects of the initial moisture content on steam consumption, mechanism and rate of heat transfer, and subsequent glucose yield were summarized. They proposed the use of steam explosion as a pretreatment for woody biomass prior to enzymatic hydrolysis [107–109]. In this study, the effect of SO_2 impregnation (1.6% SO_2 on dry wood input) and H_2SO_4 impregnation (0.58% H_2SO_4 on dry wood input) on the exploded substrates produced are compared with the case where no acid was added. Both acid catalysts substantially improve the survival of pentose sugars when treatments of equal severity were compared. However, H_2SO_4 reduces the extent to which lignin may be extracted from the water-washed exploded substrates with caustic [108].

According to Garrote et al. [110], the general advantages of steam-explosion processes compared to other pretreatment technologies for chemical utilization of lignocellulose are as follows:

- No chemicals are used except water.
- Good yield of hemicelluloses is achieved with low degraded by-products.
- Equipment corrosion is minimal due to a mild pH of reaction media when compared to acid hydrolysis processes.
- Stages of acid handling and acid recycling are avoided.
- Disruption of the solid residues from bundles to individual fibers occurs due to the explosion effect.

Recently, Bauer et al. [111] reviewed the steam-explosion pretreatment.

5.2.2.12 Brown-Rot Lignin (Schubert-Nord, Enzymatically Released Lignin)

Schubert and Nord et al. [112–114] attempted to isolate lignin by degrading the carbohydrate material with brown rot fungi. It was determined that the amount of Brauns or MWL lignin that could be obtained increased with the fungal treatments.[11]

Advantages of biological pretreatment (with white-, brown-, and soft-rot fungi) are low-energy requirement and mild operation conditions. Nevertheless, the rate of biological hydrolysis is very low and requires long pretreatment times [115–117].

5.2.2.13 Other Methods

Friedrich [118] obtained a lignin preparation by treating wood meal with a mixture of 3% HCl in glacial acetic acid at 100 °C.

In 2003, Lu and Ralph [119] suggested a method for the complete solubilization of milled wood by acetylation and dissolving the residue in N-methylimidazole/DMSO. The preparation obtained

---

[11] A wood-decay fungus is a variety of fungus that digests moist wood, causing it to rot. Some wood-decay fungi, such as brown rot, attack dead wood, while some are parasitic and colonize living trees. Fungi that not only grow on wood but also actually cause it to decay, are called lignicolous fungi. Various lignicolous fungi consume wood in various ways; for example, some attack the carbohydrates in wood, and some others decay lignin. Brown-rot fungi break down hemicellulose and cellulose. Cellulose is broken down by H_2O_2 that is produced during the break down of hemicellulose.

was called acetylated cell wall (Ac-CW). This allowed the characterization of wood components by NMR [49].

Liquid hot water (LHW) processes are biomass pretreatments with water at high temperature and pressure. Other terms are hydrothermolysis, hydrothermal pretreatment, aqueous fractionation, solvolysis, or aquasolv [120]. Between 40% and 60% of the total biomass is dissolved in the process, with 4–22% of the cellulose, 35–60% of the lignin and all of the hemicellulose being removed. In addition, acetic acid is formed during the treatment and acts as a catalyst for polysaccharide hydrolysis.

5.3 Commercial Lignins

The objective of any chemical pulping processes is to remove lignin in order to separate cellulosic fibers, producing a pulp suitable for the manufacture of paper and other related products. Every year, the paper industry produces an enormous amount of lignin as a by-product. In 1995, only from Kraft pulping, the amount exceeded 20 million metric tons. Today, these materials are the main source of lignin available, and as has been pointed out elsewhere in the text, the main use of this huge amount of a very complex and structured polymer has been as a fuel for the same factory. Table 5.4 shows a schematic description of Kraft and sulfite processes.

Depending on the pulping procedure [121], these commercial lignins can be divided into two major categories. The first one comprises sulfur-containing lignins. In this group, Kraft lignin and lignosulfonates are included (see Figure 5.4). The second category comprises nonsulfur lignins produced from other different processes, which do not include soda lignins, Organosolv lignins, steam-explosion lignins, steam-explosion hydrolysis lignins, or oxygen delignification lignins. In

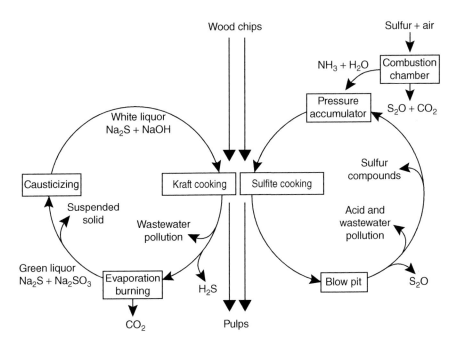

Figure 5.4 Kraft and sulfite processes of pulp formation [82]

the following paragraphs, a short review on the most common isolation methods of commercial lignins will be addressed.

The techniques may be based on the differences between lignin and contaminants in solubility and M_w. The criteria used for the selection methods are as follows:

- The lignin should be isolated in a quantitative yield.
- The isolated lignin should be free from contaminants.
- The procedure should be simple and easy to perform.

In addition to differences in structure and botanic origin, commercial lignins also vary in M_w. Most such lignins are of weight-average molar mass (M_w) ranging between 3000 and 20 000 and dispersity indexes ($Ð_M = M_w/M_n$) between 2 and 12 [122] (see Section 2.7.2, p. 40).

5.3.1 Kraft Lignin

Most of the paper pulp produced worldwide is generated by the Kraft process [123], which converts wood into wood pulp when wood is treated with $Na_2S/NaOH$ solution ("white liquor") at a temperature of 155 – 175 °C range for several hours. The method is able to dissolve about 90–95% of all lignin present in the starting material. Thus, cellulose is separated as a solid from a rich lignin fluid fraction called "black liquor." This lignin remains in this solution [124], but as a fragmented polymer compared to native lignin.

Kraft lignin (thiolignin, sulfate lignin) is isolated in the so-called delignification process, from black liquor by precipitation and neutralization with an acid solution (pH = 1–2), and subsequently dried to a solid form. Kraft lignin has been recovered from black liquors by acid precipitation and heat coagulation. This is a consequence of the low solubility of Kraft lignin in acidic aqueous media. Several methods have been carried out to obtain acid media in order to precipitate Kraft lignin from black liquors (CO_2, acetic acid, H_3PO_4, H_2SO_4, and HCl). In the case of CO_2, a low precipitation yield (60–80%) of pine Kraft lignin [125, 126] results, but a higher recovery of lignin (up to 95%) can be achieved with acetic or strong mineral acids.

Theliander [127] showed that the yield in the precipitation step is influenced mainly by pH, ion strength, and temperature. Furthermore, the composition of the precipitated solid was almost independent of the conditions used in the precipitation and subsequent separation stage [127].

This procedure yields a polymer with a lower M_w, but which is highly modified from the former native lignin. During digestion, several reactions take place. In addition to the cleavage of most of lignin–carbohydrate linkages, depolymerization of the lignin occurs, and reactions with HO⁻ and HS⁻ ions and new polymerization reactions of the fragments also occur. During the cooking process, quinone methides are formed, and these very reactive structures are involved as intermediates for repolymerization and many other processes [128, 129]. Figure 5.5 depicts a schematic representation of the chemical transformations that are carried out during Kraft pulping.

Chakar and Ragauskas [130] have detailed the process chemistry surrounding Kraft pulping including a description of the ways in which the primary linkages in lignin are disrupted during the Kraft process. Delignification is controlled by two main reactions: degradation and condensation ones. The prevalent degradation reactions are the cleavage of the α- and β-aryl ether bonds [128]. Condensation reactions proceed *via* Michael addition of external (sulfide and hydroxy anions) and internal (carbanions from phenolic structures) to an acceptor (quinone methides or formaldehyde) (see Figure 5.5).

A significant percentage of the commercial Kraft lignin is in sulfomethylated, water-soluble form, which involves the reaction of the alkaline lignin with sodium sulfite and formaldehyde, which has been formed *in situ*, during pulping [131]. The heterogeneity of lignin produced, depending on the wood origin, is demonstrated by the analytical data of lignin fractions resulting from fractional precipitation. The purity of lignin can be measured by its ash content, for example, an ash content of less than 1% indicates high purity. A model of pine Kraft lignin is described in Figure 5.6.

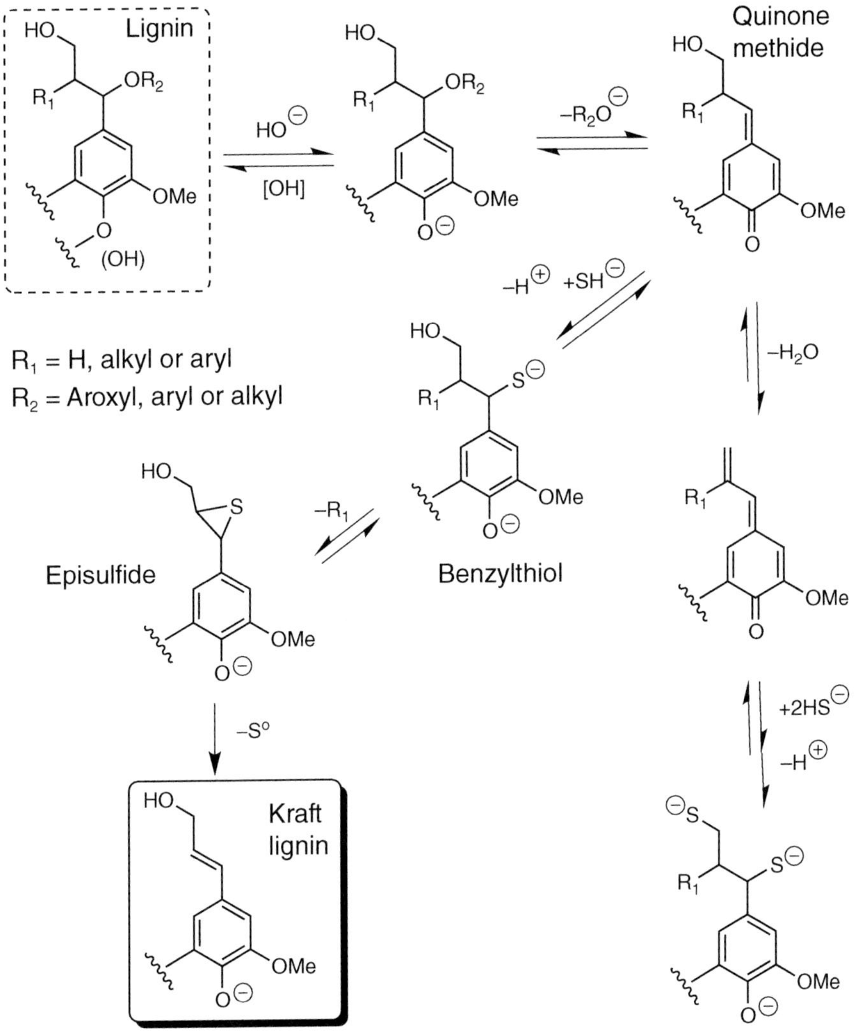

Figure 5.5 Main reactions involved in the formation of Kraft lignin during pulping [128]

Organic solvents or mixtures can be used to fractionate lignins (from a typical black liquor of pine) to produce fractions with high purity and narrow dispersity (see Table 5.7) [132–134].

Lin [135] recovered the bulk of the Kraft lignin (> 80%) at pH = 10.5 with a uniform chemical composition (see Table 5.8).

Although the Kraft process is the most predominant pulping process worldwide, the recovery of Kraft lignin for chemical uses is not very extended, and worldwide "Lignotech Borregaard" is the main company that is currently practicing it on an industrial basis [136]. Kraft lignins are currently produced commercially, for instance, by MeadWestvaco, the world's largest producer of Kraft lignin, and by the LignoBoost[12] technology, a process owned by Metso Corporation, in which lignin is extracted from pulp mill black liquor [13].

---

[12] Metso LignoBoost: Lignin from black liquor, Metso: Helsinki, Finland, 2009.

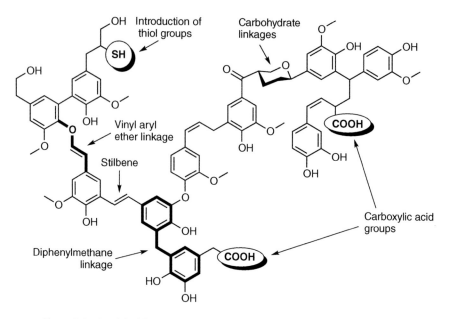

Figure 5.6 Model addressing the structural characteristics of pine Kraft lignin [13]

Table 5.7 Yield and analytical data of Kraft lignins obtained after solvent fractionation of isolated black liquor lignin[a]

Fraction	solvent	Yield (%)	M_n	M_w	$Đ_M = M_w/M_n$
1	CH_2Cl_2	9	450	620	1.4
2	n-Propanol	22	900	18 290	1.4
3	Methanol	26	18 710	28 890	1.7
4	MeOH/CH_2Cl_2 (7:3)	28	38 800	82 000	22.0
5	Residue	14	5800	180 000	31.0

[a]Data taken from ref. [11].

Table 5.8 Yield of Kraft lignin isolated from pine Kamyr black liquor at different pH[a]

Fraction no.	pH of precipitation	Yield (% of total)
1	11.0	55
2	10.5	26
3	9.5	5
4	8.5	3
5	8.0	3
6	2.0	5

[a]Data taken from ref. [135, p. 78].

5.3.2 Sulfite Lignin (Lignosulfonate Process)

The sulfite process is a pulping method, which can use several salts of sulfurous acid such as sulfites or bisulfites. Tilghman, in 1866, was granted the first patent on the sulfite-pulping process, by using an aqueous solution of calcium hydrogen sulfite and sulfur dioxide in pressurized systems; although Ekman in Sweden in 1874 was the principal initiator of the sulfite pulp industry. Depending on the combination of salts employed, there are different versions of the method: the cooking may be acidic, neutral, or alkaline (see Table 5.9).

The waste liquid formed during the separation of cellulose, generally by the treatment of softwood, is rich on lignins. These lignins, which cannot be isolated by precipitation from this pulping system and which contain about 4–8% sulfur, are called lignosulfonates because of the presence of sulfonate groups in the structure (see Figure 5.7).

Table 5.9 Sulfite-pulping procedures used to extract lignin from wood[a]

Process	Reactive agent(s)	pH	Temperature [°C]
Acid sulfite	SO_2/HSO_3^-	1–2	125–145
Bisulfite	HSO_3^-	3–5	150–175
Neutral sulfite	$HSO_3^-/SO_3^=$	6–7	150–175
Alkaline sulfite/ anthraquinone	$NaSO_3$	9–13	50–175

[a]Data taken from ref. [137].

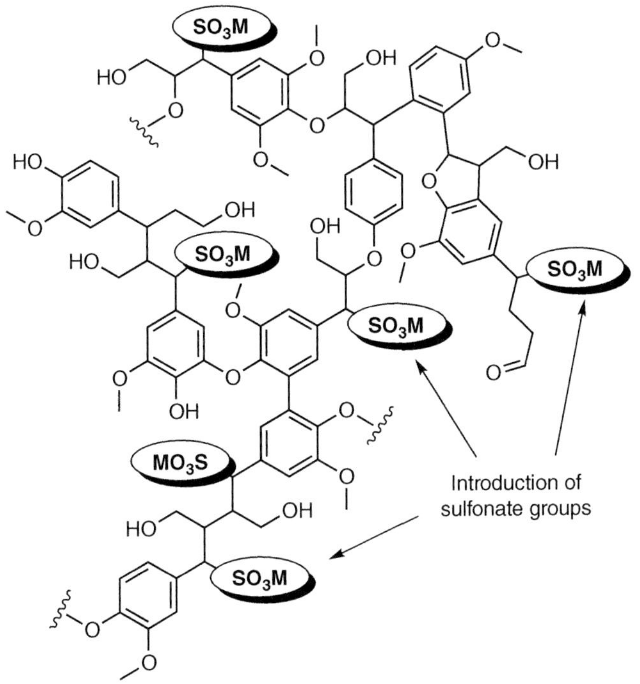

Figure 5.7 Model depicting the structural characteristics of lignosulfonate lignin [13]

Lignosulfonates present both hydrophobic and hydrophilic properties. They are water soluble and chemically modified with sulfonate groups, carbohydrates, and small amounts of wood extracts and inorganic compounds. Lignosulfonates are commercially available mainly as calcium, magnesium, and ammonium salts. The production procedure is less aggressive than Kraft pulping, and the M_w of the final product is higher than that of Kraft lignin, rendering a structure more similar to that of the original lignin.

In 1990, Lin and Lin [138] published a summary of lignosulfonate lignin processes. Normally, the sulfite process is carried out at 140–170 °C with an aqueous solution of a sulfite or bisulfite salt of sodium, ammonium, magnesium, or calcium for wood digestion. The type of salt determines the pH of the digestion. During the digestion process of wood, several chemical transformations take place, such as the cleavage of linkages between the lignin and the cellulose and other carbohydrates, and some linkages of the lignin polymer as ether bonds, which are responsive of interconnection lignin units. In addition, the sulfonation of the lignin aliphatic chain take place in the C_α-positions *via* the intermediate carbocation [136]. Lignin in this pulping process remains in a form of water-soluble sulfonic acids.

Lignin isolation requires it to be isolated from the liquor. This task can be carried out by diverse techniques: alcoholic fermentation of the sugars in the sulfite liquor, sugar removal by chemical destruction, ultrafiltration to remove sugars, and precipitation. The last technique is performed in three steps: firstly, by treatment with a long-chain alkyl amine (for example, octyl amine, N,N-dimethylhexadecylamine), (see Table 5.10) then by extraction with an organic solvent (for example, octan-1-ol), and finally by regeneration of sulfonate with alkaline water. The purity of isolated lignosulfonates can be measured by the amounts of reducing sugars and inorganic salts that remain in the solids it contains.

Like the lignin in black liquor, lignosulfonate in the spent sulfite liquor is heterogeneous with respect to chemical composition. The main problems encountered in isolating lignosulfonates involve the formation of an emulsion and foam. These problems can be solved by proper selection of the alkyl alcohol and amine used in the isolation procedure. Table 5.10 lists the amines and solvents used in extraction studies [139–143].

Fractionation of a spent sulfite liquor by stepwise addition of octylamine give five fractions of lignosulfonate (see Table 5.11) [135].

Other methods are used for purification of lignins, such as ultrafiltration, gel-permeation chromatography (GPC), and ion-exclusion techniques. The main properties of ligninsulfonates summarized in Table 5.12.

Table 5.10 Amines and solvents used in the extraction of lignosulfonates[a]

Amine	Solvent	[Ref.]
Dodecylamine	Amyl alcohol	[139]
Dicyclohexylamine	Butanol	[140]
Tri-*n*-hexylamine	Butanol	[141]
Dodecylamine	Butanol	[142]
Dodecylamine, dioctylamine, trioctylamine	Butanol, Pentanol, Metylisobutyl ketone, 1,2-Dichloroethane, cyclohexane	[143]

[a]Data taken from ref. [135, p. 80].

Table 5.11 *Lignosulfonate fractions isolated from spruce spent sulfite liquor by addition of octylamine*[a]

Fraction No.	Quantity of amine (% eq. of octilamine)	Yield (% of total)
1	5	6
2	15	20
3	25	30
4	25	23
5	30	18

[a]Data taken from ref. [135, p. 79].

Table 5.12 *Chemical properties of lignin sulfonate*[a]

Chemical or Physical property	Value
Color	Light tan to dark brown[b,c,d]
Physical state	Solid (powder)
Odor	Odorless
Melting point	Decomposes at > 200 °C
Boiling point	Exists in a solid state
Solubility	Soluble in water[d]
Stability	Soil half-life up to 1 year depending on chemical species and soil flora[b,d]
Reactivity	Lignin sulfonates are corrosive to aluminum and aluminum alloys in the absence of calcium carbonate
Oxidizing or reduction action	Reacts as acids to neutralize bases. Usually do not react as reducing or oxidizing agents[c]
Flammability/flame extension	Flash point data not available although probably combustible[c]
Exploitability	Dust explosions may occur if fine particles are mixed with air in the presence of an ignition source. Ignition temperature is approximately 250–300 °C[b]

[a] Values taken from [144, and references therein].
[b] Sodium lignosulfonate.
[c] Ammonium lignosulfonate.
[d] Calcium lignosulfonate.

Lignotech Borregaard is the largest producer of lignosulfonates worldwide [136], and Tembec is the second. Other producers are La Rochette Venizel, Nippon Paper chemicals, Cartiere Burgo, and Domsjö Fabriker AB (MoDo group).

5.3.3 Soda Lignin (Alkali Lignin)

Alkali lignin is produced in the soda pulping process, industrialized since 1853, and traditionally used for nonwood fibers, such as straw, sugarcane bagasse, and flax. Such raw materials still remain a notable source for many types of papers in developing countries [145]. Biomass is digested by the

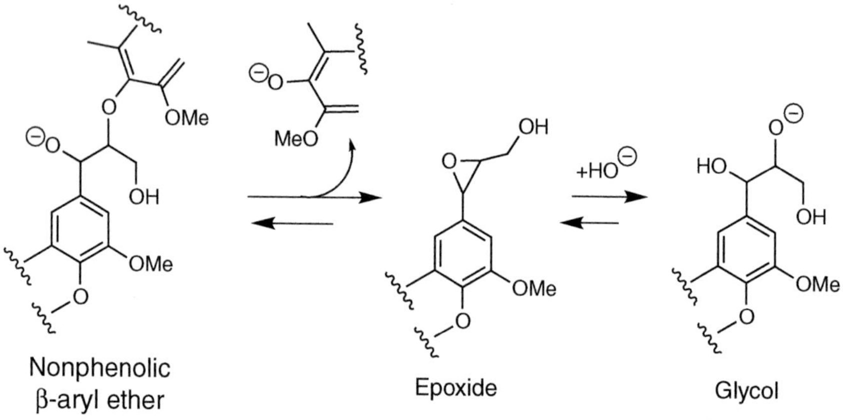

Nonphenolic β-aryl ether → **Epoxide** → **Glycol**

Figure 5.8 Main reactions involved in the formation of soda lignin [128]

treatment with an aqueous solution of NaOH (or lime, calcium hydroxide) at temperatures of about ≤ 160 °C. Lignin depolymerization occurs principally by the cleavage of α- and β-aryl ether bonds and saponification of intermolecular ester bonds cross-linking hemicelluloses and lignin. Quinone methide structures are formed in soda pulping, and contributes to depolymerization. Under these reaction conditions, stilbene or styryl aryl moieties are formed, which are alkali stable structures (see Figure 5.8).

The lignin is recovered lowering the pH until precipitation and the solid is isolated by decantation or filtration, as in the case of Kraft lignin. This lignin-process system (LPS) is used industrially to recover nonwood lignins from soda black liquors. The main producer in the world of sulfur-free soda lignins is Green Value SA.

5.3.4 Organosolv Pulping

Delignification of wood using organic solvents (organosolv pulping) has generated interest ever since they were introduced (see Section 5.2.2.6, p. 123). Benefits of organosolv pulping include: the production of a high-quality lignin, which might facilitate higher value applications of lignin and potentially lowering the enzyme cost by the separation of lignin before the enzymatic hydrolysis of the cellulose fraction.

Eight procedures for the commercial production of organosolv lignins have been the subject of major research efforts (see Table 5.13), which summarize the main methods for lignin isolation with organic solvents. The Acetosolv[13] and Formacell processes[14] remove both lignin and hemicelluloses under mild conditions, without cellulose degradation [146–148].

One of the more popular procedures is the Alcell© method. Lignin is extracted with a mixture of water and a low M_w alcohol, such as EtOH. In this case, biomass is treated with a 1:1 mixture of water and EtOH at 175–195 °C for 1 h as the cooking medium and a liquid/solid biomass ratio about 4–7. The lignins produced are sulfur-free and show very high purity. As by-product, a considerable amount of furfural is produced [149]. This furfural by-product can be recovered.

The acetocell [150] and acetosolv [151] procedures are quite similar, and both use a mixture of acetic acid/water to isolate lignin. In case of the acetosolv method, a mineral acid is added as

---

[13] HCl-catalyzed and acetic acid media.
[14] Addition of formic acid to aqueous acetic acid.

Table 5.13 Organosolv lignins [137]

Lignin type	Solvent
Alcell© [149]	EtOH/water
Acetocell [150]	Acetic acid/water
Acetosolv [151]	Acetic acid/HCl pulping
ASAM [152]	Alkaline sulfite/anthraquinone/MeOH
Batelle/Geneva phenol [153]	Phenol /acid/water
Formacell [154]	Acetic acid/formic acid/water
Milox [155]	Formic acid/ H_2O_2
Organocell [74]	MeOH pulping, followed by MeOH, NaOH, and anthraquinone pulping

catalyst [156], and thus for 300 g of wood chips, 1.5 l of 95% aqueous acetic acid, containing 0.1 or 0.16% HCl are required. For optimal results, the mixture must be refluxed for 3–5 h. Afterwards, the excess acetic acid is decanted, and the residue is extracted with hot (105 °C) 95% acetic acid in a column with water. Acetic and HCl can be recovered by evaporation of the extract. This procedure is economically inadvisable because it is a costly process for pulp mills.

ASAM pulping [152] is a method to produce lignin from wood chips. Starting wood chips are pre-steamed for 15 min and then sodium hydroxide, sodium sulfite, MeOH, and anthraquinone as catalyst are added. The mixture is cooked for 90–180 min at 170–180 °C. The cooking process can be modified depending on the desired pulp grading if it is going to be used for making papers or boards.

The Battelle-Geneva process [153], uses a mixture of phenol and water for the treatment of biomass, with an average of phenol about 40% of the total amount of solvent. The pulping is carried out at temperatures near 100 °C and at atmospheric pressure. The phase separation of the homogeneous mixture of phenol/water, at the pulping reaction condition, takes place by cooling down the spent liquor. In some cases, acids or alkalis can be added to water to facilitate the degradation of the lignin [157].

The Formacell pulping process [154] employs a mixture of water, formic acid, and acetic acid. The presence or formic acid in the pulping liquor provides an easier cleavage of lignin α-ether linkages and so lignin fragments are dissolved at lower temperatures and generally at atmospheric pressure.

The Milox procedure [155] uses formic acid and H_2O_2 to produce *in situ* performic acid. The method is performed in three pathways, when H_2O_2 is added in the first and last one. The temperature for these stages is about 80 °C, while for the middle pathway is increased to 100 °C. The method can be applied to softwood, hardwood, and nonwoody plants, and the lignin produced by this procedure has a lower M_w than others.

The Organocell process is a procedure based on a treatment of biomass with an alkali and an organic solvent. In the first stage, the wood chips are initially impregnated at a temperature of around 120 °C with a mixture of alcohol (MeOH or EtOH) and water. The wood chips are softened and then transported into the digester and treated with alkali and catalytic amounts of anthraquinone. The pulping liquor is heated to 165–170 °C, under pressure, approximately 13 bar. Then, the wood chips are cooked for a period, depending on the type of wood and the later use of the pulp. The amount of alkali used in this method is related, too, with the final degree of delignification of the wood chips. By this technique, a wide range of pulps from hardwoods, softwoods, and annual pants can be prepared with properties very similar to Kraft pulping with the advantage that bleaching can be done without chlorine chemicals.

5.3.5 Other Methods of Separation of Lignin from Biomass

In addition to the steam-exploded lignin (see Section 5.2.2.11), several other methods for pretreating and isolating lignins are available, including the ammonia fiber explosion (AFEX) process [158], the CO_2 explosion [116], and the "hot water process" (see Section 5.2.2.13, p. 126). In the AFEX process, the biomass is treated with liquid ammonia at high temperatures and pressures. After a few seconds, the pressure is swiftly reduced. In the CO_2 explosion treatment, high-pressure CO_2 is injected in the reactor and then released by an explosive decompression. Yields of CO_2 explosion are lower than those with steam or AFEX. An alternative use of CO_2 is extraction with supercritical CO_2 (SC-CO_2) [159].

The dilute-acid process provides an effective separation of the lignin from the other biomass streams but suffers from low yields and also corrosion of equipment from the acidic environment. "The alkaline oxidation process" uses O_2 or H_2O_2 to degrade lignin, which is then easy to recover. The process suffers from slow delignification rates, which is the main reason that the process is not used extensively [13].

The pretreatment of biomass with aqueous ammonia at elevated temperatures reduces the lignin content. This ammonia pretreatment includes the ammonia recycle percolation (ARP) and soaking in aqueous ammonia (SAA). Biomass can be delignified also by treatment with an oxidizing agent such as H_2O_2, O_3, O_2, or air.

Pyrolysis lignin could also be considered a possible feedstock. The pyrolysis process typically involves relatively high temperatures (450 °C) for short times, typically 2 s. No waste except flue gas and ash is produced during the process. At higher temperatures, more than 300 °C, cellulose is decomposed into gaseous products and residual char. At lower temperature, the decomposition rate is very slow [13]. Compared with the native lignin of wood, pyrolytic lignin consists of relatively low M_w components due to fragmentation during pyrolysis [160]. Mild acid hydrolysis[15] of the products from the pyrolysis pretreatment can convert 80–85% of cellulose into reducing sugars [161, p. 57].

References

[1] Ramos LP. The chemistry involved in the steam treatment of lignocellulosic materials. Quim Nova. 2003;26(6):863–871.
[2] da Costa Sousa L, Chundawat SPS, Balan V, Dale BE. 'Cradle-to-grave' assessment of existing lignocellulose pretreatment technologies. Curr Opin Biotechnol. 2009;20(3):339–347.
[3] Pearl IA. The Chemistry of Lignin. New York: M. Dekker; 1967.
[4] Glasser WG, Barnett CA, Muller PC, Sarkanen KV. The chemistry of several novel bioconversion lignins. J Agric Food Chem. 1983;31(5):921–930.
[5] Glasser WG, Barnett CA, Sano Y. Classification of lignins with different genetic and industrial origins. J Appl Polym Sci: Appl Polym Symp. 1983;37(Proc. Cellul. Conf., 9th, 1982, Part 1):441–460.
[6] Lai YZ, Sarkanen KV. Isolation and structural studies. In: Sarkanen KV, Ludwig CH, editors. Lignins: Occurrence, Formation, Structure and Reactions. New York: Wiley-Interscience; 1971. p. 165–240.
[7] Monteil-Rivera F, Phuong M, Ye M, Halasz A, Hawari J. Isolation and characterization of herbaceous lignins for applications in biomaterials. Ind Crops Prod. 2013;41(1):356–364.
[8] Brauns FE. The Chemistry of Lignin. New York: Academic Press; 1952.
[9] Brauns FE, Brauns DA. The Chemistry of Lignin: Supplement. New York: Academic Press; 1960.
[10] Fengel D, Wegener G. Wood: Chemistry, Ultrastructure, Reactions. W. de Gruyter; 1984.

---

[15] 1 NH_2SO_4, 97 °C, 2.5 h.

[11] Gellerstedt G, Henriksson G. Lignins: major sources, structure and properties. In: Belgacem MN, Gandini A, editors. Monomers, Polymers and Composites from Renewable Resources. Amsterdam: Elsevier; 2011. p. 201–224.
[12] Lu F, Ralph J. Chapter 6 - lignin. In: Cereal Straw as a Resource for Sustainable Biomaterials and Biofuels. Amsterdam: Elsevier; 2010. p. 169–207.
[13] Zakzeski J, Bruijnincx PCA, Jongerius AL, Weckhuysen BM. The catalytic valorization of lignin for the production of renewable chemicals. Chem Rev. 2010;110(6):3552–3599.
[14] Phillips M. The chemistry of lignin. Chem Rev. 1934;14(1):103–170.
[15] Brauns FE. Native lignin I. Its isolation and methylation. J Am Chem Soc. 1939;61(8): 2120–2127.
[16] Björkman A. Finely divided wood. I. Extraction of lignin with neutral solvents. Sven Papperstidn. 1956;59:477–485.
[17] Björkman A. Finely divided wood. V. The effect of milling. Sven Papperstidn. 1957;60: 329–335.
[18] Björkman A. Finely divided wood. II. The properties of lignins extracted with neutral solvents from softwoods and hardwoods. Sven Papperstidn. 1957;60:158–169.
[19] Pew JC. Properties of powdered wood and isolation of lignin by cellulolytic enzymes. Tappi. 1957;40:553–558.
[20] Pew JC, Weyna P. Fine grinding, enzyme digestion, and the lignin-cellulose bonds in wood. Tappi. 1962;45(3):247–256.
[21] Chang HM, Cowling EB, Brown W. Comparative studies on cellulolytic enzyme lignin and milled wood lignin of sweetgum and spruce. Holzforschung. 1975;29(5):153–159.
[22] Klason P. Chemical composition of deal (fir wood). Ark Kemi, Mineral Geol. 1908; 3(Art. 5):1–20.
[23] Klason P. Determination of lignin in sulfite wood pulp. Papierfabrikant. 1910;8:1285–1286.
[24] Sluiter JB, Ruiz RO, Scarlata CJ, Sluiter AD, Templeton DW. Compositional analysis of lignocellulosic feedstocks. 1. Review and description of methods. J Agric Food Chem. 2010;58(16):9043–9053.
[25] Timell TE, editor. Compression Wood in Gymnosperms. Springer-Verlag; 1986.
[26] Obst JR. Guaiacyl and syringyl lignin composition in hardwood cell components. Holzforschung. 1982;36(3):143–152.
[27] Timell TE. The chemical composition of tension wood. Sven Papperstidn. 1969;72:173–181.
[28] Bagby MO, Nelson GH, Helman EG, Clark TF. Determination of lignin in non-wood plant fiber sources. Tappi. 1971;54:1876–1878.
[29] Kyrklund B, Strandell G. Applicability of the chlorine number for evaluation of the lignin content in pulp. Pap Puu. 1969;51:299–305.
[30] Lindström T, Glad-Nordmark G. Chemical characterization of the fines fraction from unbleached Kraft pulps. Sven Papperstidn. 1978;81(15):489–492.
[31] Robinson DH. A Comprehensive Study of the Bleaching of Mechanical Pulps with Hydrogen Peroxide and Sodium Dithionite. Syracuse, NY: SUNY College of Environmental Science and Forestry; 1977.
[32] Dence CW. The determination of lignin. In: Dence CW, Lin SY, editors. Methods in Lignin Chemistry, Springer Series in Wood Science. Berlin: Springer-Verlag; 1992. p. 33–61.
[33] Willstätter R, Zechmeister L. Zur kenntnis der hydrolyse von cellulose I. Ber Dtsch Chem Ges. 1913;46(2):2401–2412.
[34] Urban H. Fir wood. Cellul-Chem. 1926;7:73–78.
[35] Freudenberg K, Harder M, Markert L. Bemerkungen zur chemie des lignins. (VII. Mitteilung über lignin und cellulose). Ber Dtsch Chem Ges. 1928;61(8):1760–1765.
[36] Freudenberg K, Zocher H, Dürr W. Lignin and cellulose. XI. Further experiments with lignin. Ber Dtsch Chem Ges B. 1929;62:1814–1823.

[37] Glennie DW, McCarthy JL. Chemistry of lignin. In: Libby CE, editor. Pulp and Paper Science and Technology. McGraw-Hill; 1962. p. 82–107.
[38] Fengel D, Przyklenk M. Supermolecular structure of cell wall components. 4. composition of the fractions from ion exchange chromatography. Sven Papperstidn. 1975;78(17):617–620.
[39] Fengel D. Fractionation experiments with the alkali extract from spruce holocellulose. Part 2. Optimizing the ion exchange chromatographic fractionation. Holzforschung. 1976;30(5):143–148.
[40] Halse OM. Determination of cellulose and wood fiber in paper. Pap J. 1926;14:121.
[41] Zadrazil F, Brunnert H. The influence of ammonium nitrate supplementation on degradation and *in vitro* digestibility of straw colonized by higher fungi. Eur J Appl Microbiol Biotechnol. 1980;9(1):37–44.
[42] Runkel ROH, Wilke KD. Thermoplastic properties of wood. II. Holz Roh- Werkst. 1951;9:260–270.
[43] Fengel D, Wegener G. Wood: Chemistry, Ultrastructure, Reactions. Walter De Gruyter Incorporated; 1989.
[44] Fredenhagen K, Cadenback G. The decomposition of cellulose by hydrofluoric acid and a new method of wood saccharification by means of highly concentrated hydrofluoric acid. Angew Chem. 1933;46:113–117.
[45] Defaye J, Gadelle A, Papadopoulos J, Pedersen C. Hydrogen fluoride saccharification of cellulose and lignocellulose materials. J Appl Polym Sci: Appl Polym Symp. 1983;37(Proc. Cellul. Conf., 9th, 1982, Part 2):653–670.
[46] Ritchie PF, Purves CB. Periodate lignins, their preparation and properties. Pulp Pap Mag Can. 1947;48(12):74–82.
[47] Wald WJ, Ritchie PF, Purves CB. Elementary composition of lignin in northern pine and black spruce woods, and of the isolated klason and periodate lignins. J Am Chem Soc. 1947;69:1371–1377.
[48] Adler E, Hernestam S. Estimation of phenolic hydroxyl groups in lignin. I. Periodate oxidation of guaiacol compounds. Acta Chem Scand. 1955;9(2):319–334.
[49] Balakshin MY, Capanema EA, Chang HM. Recent advances in the isolation and analysis of lignins and lignin–carbohydrate complexes. In: Hu TQ, editor. Characterization of Lignocellulosic Materials. Blackwell Publishing Ltd.; 2008. p. 148–170.
[50] Haghighi Mood S, Hossein Golfeshan A, Tabatabaei M, Salehi Jouzani G, Najafi GH, Gholami M, et al. Lignocellulosic biomass to bioethanol, a comprehensive review with a focus on pretreatment. Renewable Sustainable Energy Rev. 2013;27:77–93.
[51] Singh R, Shukla A, Tiwari S, Srivastava M. A review on delignification of lignocellulosic biomass for enhancement of ethanol production potential. Renewable Sustainable Energy Rev. 2014;32:713–728.
[52] Lundquist K. Wood. In: Lin SY, Dence CW, editors. Methods in Lignin Chemistry, Springer Series in Wood Science. Springer Berlin Heidelberg; 1992. p. 65–70.
[53] Björkman A. Isolation of lignin from finely divided wood with neutral solvents. Nature. 1954;174(4440):1057–1058.
[54] Wu S, Argyropoulos DS. An improved method for isolating lignin in high yield and purity. J Pulp Pap Sci. 2003;29(7):235–240.
[55] Ikeda T, Holtman K, Kadla JF, Chang HM, Jameel H. Studies on the effect of ball milling on lignin structure using a modified DFRC method. J Agric Food Chem. 2002;50(1):129–135.
[56] Fujimoto A, Matsumoto Y, Chang HM, Meshitsuka G. Quantitative evaluation of milling effects on lignin structure during the isolation process of milled wood lignin. J Wood Sci. 2005;51(1):89–91.
[57] Sakakibara A, Nakayama N. Hydrolysis of lignin with dioxane and water. I. Formation of cinnamic alcohols and aldehydes. Mokuzai Gakkaishi. 1961;7:13–18.

[58] Sakakibara A, Nakayama N. Hydrolysis of lignin with dioxane and water. II. Identification of hydrolysis products. Mokuzai Gakkaishi. 1962;8:153–156.
[59] Zhang AP, Liu CF, Sun RC, Xie J. Extraction, purification, and characterization of lignin fractions from sugarcane bagasse. BioResources. 2013;8(2):1604–1614.
[60] Chen JY, Shimizu Y, Takai M, Hayashi J. A method for isolation of milled-wood lignin involving solvent swelling prior to enzyme treatment. Wood Sci Technol. 1995;29(4):295–306.
[61] Higuchi T, Ito Y, Shimada M, Kawamura I. Chemical properties of milled wood lignin of grasses. Phytochemistry. 1967;6(11):1551–1556.
[62] Jung HJG, Himmelsbach DS. Isolation and characterization of wheat straw lignin. J Agric Food Chem. 1989;37(1):81–87.
[63] Lapierre C, Monties B, Rolando C. Thioacidolysis of lignin: comparison with acidolysis. J Wood Chem Technol. 1985;5(2):277–292.
[64] Hu Z, Yeh TF, Chang HM, Matsumoto Y, Kadla JF. Elucidation of the structure of cellulolytic enzyme lignin. Holzforschung. 2006;60(4):389–397.
[65] Imai M, Ikari K, Suzuki I. High-performance hydrolysis of cellulose using mixed cellulase species and ultrasonication pretreatment. Biochem Eng J. 2004;17(2):79–83.
[66] Guerra A, Filpponen I, Lucia LA, Saquing C, Baumberger S, Argyropoulos DS. Toward a better understanding of the lignin isolation process from wood. J Agric Food Chem. 2006;54(16):5939–5947.
[67] Guerra A, Filpponen I, Lucia LA, Argyropoulos DS. Comparative evaluation of three lignin isolation protocols for various wood species. J Agric Food Chem. 2006;54(26):9696–9705.
[68] Stumpf W, Freudenberg K. Lösliches lignin aus fichten- und buchenholz. Angew Chem Int Ed. 1950;62(22):537–537.
[69] Freudenberg K. Neuere ergebnisse auf dem gebiete des lignins und der verholzung. In: Zechmeister L, editor. Progress in the Chemistry of Organic Natural Products. vol. 11. Vienna: Springer; 1954. p. 43–82.
[70] McDonough TJ. The chemistry of organosolv delignification. Tappi J. 1993;76(8):186–193.
[71] Sarkanen KV, Hoo LH. Kinetics of hydrolysis of *erythro*-guaiacylglycerol β-(2-methoxyphenyl) ether and its veratryl analogue using HCl and aluminum chloride as catalysts. J Wood Chem Technol. 1981;1(1):11–27.
[72] Kleinert TN. Organosolv pulping and recovery process [US 3,585,104]; 1971.
[73] Lora JH, Aziz S. Organosolv pulping: a versatile approach to wood refining. Tappi. 1985;68(8):94–97.
[74] Lindner A, Wegener G. Characterization of lignins from organosolv pulping according to the Organocell process. Part 1. Elemental analysis, nonlignin portions and functional groups. J Wood Chem Technol. 1988;8(3):323–340.
[75] Bogomolov BD, Groshev AS, Popova GI, Vishnyakova AP. Delignification of wood with tetrahydrofurfuryl alcohol. 1. Comparison of the delignifying capacity of tetrahydrofurfuryl alcohol and other organic solvents. Koksnes Kim. 1979;4:21–24.
[76] Bogomolov BD, Groshev AS, Popova GI, Vishnyakova AP. Delignification of wood with tetrahydrofurfuryl alcohol. 2. Study of the delignification of birchwood. Koksnes Kim. 1979;6:43–46.
[77] Gast D, Puls J. Improvement of the ethylene glycol water system for the component separation of lignocelluloses. In: Palz W, Coombs J, Hall DO, editors. Energy from Biomass, 3rd E. C. Conference, EUR 10024. Commission European Communities. London: Elsevier Applied Science Publishers; 1985. p. 949–952.
[78] Nakamura H, Takauti E. Zellstoffherstellung mittels aethylenglykol. Celluloid-Ind. 1941;17(3):19–26.
[79] Kishimoto T, Sano Y. Delignification mechanism during high-boiling solvent pulping. Part 1. Reaction of guaiacylglycerol-β-guaiacyl ether. Holzforschung. 2001;55(6):611–616.

[80] Kishimoto T, Sano Y. Delignification mechanism during high-boiling solvent pulping: Part 2. Homolysis of guaiacylglycerol-β-guaiacyl ether. Holzforschung. 2002;56(6):623–631.
[81] Kishimoto T, Sano Y. Delignification mechanism during high-boiling solvent pulping. V. Reaction of nonphenolic β-*O*-4 model compounds in the presence and absence of glucose. J Wood Chem Technol. 2003;23(3 & 4):279–292.
[82] Sridach W. The environmentally benign pulping process of non-wood fibers. Suranaree J Sci Technol. 2010;17(2):105–123.
[83] Lawther JM, Sun RC, Banks WB. Extraction and comparative characterization of ball-milled lignin (LM), enzyme lignin (LE) and alkali lignin (LA) from wheat straw. Cellul Chem Technol. 1996;30(5-6):395–410.
[84] Fort DA, Remsing RC, Swatloski RP, Moyna P, Moyna G, Rogers RD. Can ionic liquids dissolve wood? Processing and analysis of lignocellulosic materials with 1-n-butyl-3-methylimidazolium chloride. Green Chem. 2007;9(1):63–69.
[85] Lee SH, Doherty TV, Linhardt RJ, Dordick JS. Ionic liquid-mediated selective extraction of lignin from wood leading to enhanced enzymatic cellulose hydrolysis. Biotechnol Bioeng. 2009;102(5):1368–1376.
[86] Holbrey JD, Rogers RD, Mantz RA, Trulove PC, Cocalia VA, Visser AE, et al. Physicochemical properties. In: Wasserscheid P, Welton T, editors. Ionic Liquids in Synthesis. Wiley-VCH; 2008. p. 57–174.
[87] Zhu S. Use of ionic liquids for the efficient utilization of lignocellulosic materials. J Chem Technol Biotechnol. 2008;83(6):777–779.
[88] Pu Y, Jiang N, Ragauskas AJ. Ionic liquid as a green solvent for lignin. J Wood Chem Technol. 2007;27(1):23–33.
[89] Tan SSY, MacFarlane DR, Upfal J, Edye LA, Doherty WOS, Patti AF, et al. Extraction of lignin from lignocellulose at atmospheric pressure using alkylbenzenesulfonate ionic liquid. Green Chem. 2009;11(3):339–345.
[90] Wen JL, Yuan TQ, Sun SL, Xu F, Sun RC. Understanding the chemical transformations of lignin during ionic liquid pretreatment. Green Chem. 2014;16(1):181–190.
[91] Gruss J. A new reagent for wood and vanillin. Ber Dtsch Bot Ges. 1921;38:361–368.
[92] Hägglund E, Urban H. Lignin acetals. Cellul-Chem. 1927;8:69–71.
[93] Hibbert H, Rowley HJ. Lignin and related compounds. I. New method for the isolation of spruce-wood lignin. Can J Res. 1930;2:357–363.
[94] Fuchs W. Extraction of lignin with hydrochloric acid in methylglycol. Ber Dtsch Chem Ges B. 1929;62:2125–2132.
[95] Monteil-Rivera F, Huang GH, Paquet L, Deschamps S, Beaulieu C, Hawari J. Microwave-assisted extraction of lignin from triticale straw: optimization and microwave effects. Bioresour Technol. 2012;104:775–782.
[96] Kalb L, Schoeller V. Cellulose determination by means of phenol. Cellul-Chem. 1923;4:37–40.
[97] Beckmann E, Liesche O, Lehmann F. Lignin aus winterroggenstroh. Angew Chem. 1921;34(50):285–288.
[98] Phillips M. The chemistry of lignin. I. Lignin from corn cobs. J Am Chem Soc. 1927;49(8):2037–2040.
[99] Powell WJ, Whittaker H. Chemistry of lignin. Part I. Flax lignin and some derivatives. J Chem Soc, Trans. 1924;125:357–364.
[100] Liu X, Lee Z, Tai D. Fractional studies on the characteristics of high alkali-soluble lignins of wheat straw. Cellul Chem Technol. 1989;23(5):559–572.
[101] Lapierre C, Jouin D, Monties B. On the molecular origin of the alkali solubility of gramineae lignins. Phytochemistry. 1989;28(5):1401–1403.

[102] Focher B, Marzetti A, Crescenzi V. Steam Explosion Techniques: Fundamentals and Industrial Applications: Proceedings of the International Workshop on Steam Explosion Techniques: Fundamentals and Industrial Applications, Milan, Italy; 20–21 October 1988. Gordon and Breach Science Publishers; 1991.

[103] San Martin R, Blanch HW, Wilke CR, Sciamanna AF. Production of cellulase enzymes and hydrolysis of steam-exploded wood. Biotechnol Bioeng. 1986;28(4):564–569.

[104] De Long EA. A method of rendering lignin separable from cellulose and hemicellulose in lignocellulosic material and the product so produced [Patent]; 1981. CA1096374.

[105] Josefsson T, Lennholm H, Gellerstedt G. Steam explosion of aspen wood. Characterisation of reaction products. Holzforschung. 2002;56(3):289–297.

[106] Li J, Gellerstedt G, Toven K. Steam explosion lignins; their extraction, structure and potential as feedstock for biodiesel and chemicals. Bioresour Technol. 2009;100(9):2556–2561.

[107] Brownell HH, Saddler JN. Steam pretreatment of lignocellulosic material for enhanced enzymatic hydrolysis. Biotechnol Bioeng. 1987;29(2):228–235.

[108] Mackie KL, Brownell HH, West KL, Saddler JN. Effect of sulfur dioxide and sulfuric acid on steam explosion of aspenwood. J Wood Chem Technol. 1985;5(3):405–425.

[109] Brownell HH. The lignin-carbohydrate bond. II. Isolation of milled wood lignin and lignin-carbohydrate complex. Tappi. 1965;48(9):513–519.

[110] Garrote G, Dominguez H, Parajo JC. Hydrothermal processing of lignocellulosic materials. Holz Roh- Werkst. 1999;57(3):191–202.

[111] Bauer A, Lizasoain J, Theuretzbacher F, Agger JW, Rincon M, Menardo S, et al. Steam explosion pretreatment for enhancing biogas production of late harvested hay. Bioresour Technol. 2014;166:403–410.

[112] Schubert WJ, Nord FF. Investigations on lignin and lignification. I. Studies on softwood lignin. J Am Chem Soc. 1950;72(2):977–981.

[113] Schubert WJ, Nord FF. Investigations on lignin and lignification. II. The characterization of enzymatically liberated lignin. J Am Chem Soc. 1950;72(9):3835–3838.

[114] Nord FF, Schubert WJ. Enzymic studies on cellulose, lignin, and the mechanism of lignification. Holzforschung. 1951;5:1–9.

[115] Cardona CA, Sanchez OJ. Fuel ethanol production: Process design trends and integration opportunities. Bioresour Technol. 2007;98(12):2415–2457.

[116] Sun Y, Cheng J. Hydrolysis of lignocellulosic materials for ethanol production: a review. Bioresour Technol. 2002;83(1):1–11.

[117] Tengerdy RP, Szakacs G. Bioconversion of lignocellulose in solid substrate fermentation. Biochem Eng J. 2003;13(2-3):169–179.

[118] Friedrich A. Lignin. IV. Soluble pine lignin prepared by various methods. Z Physiol Chem. 1928;176:127–143.

[119] Lu F, Ralph J. Non-degradative dissolution and acetylation of ball-milled plant cell walls: high-resolution solution-state NMR. Plant J. 2003;35(4):535–544.

[120] Mosier N, Wyman C, Dale B, Elander R, Lee YY, Holtzapple M, et al. Features of promising technologies for pretreatment of lignocellulosic biomass. Bioresour Technol. 2005;96(6):673–686.

[121] Biermann CJ. Essentials of Pulping and Papermaking. Academic Press; 1993.

[122] Glasser WG, Dave V, Frazier CE. Molecular weight distribution of (semi-) commercial lignin derivatives. J Wood Chem Technol. 1993;13(4):545–559.

[123] Gierer J. Chemical aspects of kraft pulping. Wood Sci Technol. 1980;14(4):241–266.

[124] Gellerstedt G, Zhang L. Chemistry of TCF-Bleaching with oxygen and hydrogen peroxide. In: Argyropoulos DS, editor. Oxidative Delignification Chemistry. Washington, DC: American Chemical Society; 2001. p. 61–72.

[125] Loutfi H, Blackwell B, Uloth V. Lignin recovery from Kraft black liquor: preliminary process design. Tappi J. 1991;74(1):203–210.

[126] Alen R, Patja P, Sjostrom E. Carbon dioxide precipitation of lignin from pine Kraft black liquor. Tappi. 1979;62(11):108–110.

[127] Theliander H. The lignoboost process: solubility of lignin. In: International Chemical Recovery Conference. vol. 2. 2010 International Chemical Recovery Conference; (Williamsburg, VA; USA; 29 March - 1 April); 2010. p. 33–42.

[128] Gierer J. Chemistry of delignification. Part 1: general concept and reactions during pulping. Wood Sci Technol. 1985;19(4):289–312.

[129] Gierer J. Reactions of lignin during pulping. description and comparison of conventional pulping processes. Sven Papperstidn. 1970;73(18):571–596.

[130] Chakar FS, Ragauskas AJ. Review of current and future softwood kraft lignin process chemistry. Ind Crops Prod. 2004;20(2):131–141.

[131] Adler E, Karl MHE. Method of producing water-soluble products from black liquor lignin; 1954.

[132] Lindberg JJ, Tylli H, Majani C. Notes on the molecular weight and the fractionation of lignins with organic solvents. Pap Puu. 1964;46(9):521–526.

[133] Mörck R, Yoshida H, Kringstad KP, Hatakeyama H. Fractionation of Kraft lignin by successive extraction with organic solvents. I. Functional groups, carbon-13 NMR-spectra and molecular weight distributions. Holzforschung. 1986;40(Suppl.):51–60.

[134] Mörck R, Reimann A, Kringstad KP. Fractionation of Kraft lignin by successive extraction with organic solvents. III. Fractionation of Kraft lignin from birch. Holzforschung. 1988;42:111–116.

[135] Lin SY. Commercial spent pulping liquors. In: Lin SY, Dence CW, editors. Methods in Lignin Chemistry, Springer Series in Wood Science. Springer Berlin Heidelberg; 1992. p. 75–80.

[136] Lora J. Industrial commercial lignins: sources, properties and applications. In: Belgacem MN, Gandini A, editors. Monomers, Polymers and Composites from Renewable Resources. Elsevier Science; 2008. p. 225–241.

[137] Calvo-Flores FG, Dobado JA. Lignin as renewable raw material. ChemSusChem. 2010;3(11):1227–1235.

[138] Lin SY, Lin IS. Lignin. In: Arpe HJ, editor. Ullmann's Encyclopedia of Industrial Chemistry, Ullmann's Encyclopedia of Industrial Chemistry. vol. 15, 5th ed. Weinheim: VCH; 1990. p. 305–315.

[139] Harris EE, Hogan D. Isolation and fractionation of lignosulfonates in sulfite spent liquors. Ind Eng Chem. 1957;49:1393.

[140] Felicetta VF, McCarthy JL. Pulp mills research program at the University of Washington. Tappi. 1957;40:851–866.

[141] Eisenbraun E. Separation and fractionation of lignosulfonic acids from spent sulfite liquor with trihexylamine in organic solvents. Tappi. 1963;46:104–107.

[142] Forss KG, Pirhonen IMJ. Fractionating lignosulfonic acids by amine-alcohol extraction [Patent]; 1972. DE2055292A.

[143] Kontturi AK, Sundholm G, Nielsen KM, Zingales R, Vikholm I, Urso F, et al. The extraction and fractionation of lignosulfonates with long chain aliphatic amines. Acta Chem Scand, Ser A. 1986;40:121–125.

[144] for the USDA National Organic Program II. Lignin sulfonate. technical evaluation report. Washington, DC: The U.S. Department of Agriculture's Agricultural Marketing Service; 2011. http://www.ams.usda.gov/AMSv1.0/getfile?dDocName=STELPRDC5089351 (accessed March 12, 2015).

[145] Kocurek MJ, Ingruber OV, Wong A. Sulfite Science and Technology, Pulp and Paper Manufacture. Joint Textbook Committee of the Paper Industry, TAPPI; 1985.

[146] Nimz HH, Casten R. Chemical processing of lignocellulosics. Holz Roh- Werkst. 1986;44(6):207–212.
[147] Sano Y, Nakamura M, Shimamoto S. Pulping of wood at atmospheric pressure. II. Pulping of birch wood with aqueous acetic acid containing a small amount of sulfuric acid. Mokuzai Gakkaishi. 1990;36(3):207–211.
[148] Pan XJ, Sano Y. Comparison of acetic acid lignin with milled wood and alkaline lignins from wheat straw. Holzforschung. 2000;54(1):61–65.
[149] Pye EK, Lora JH. The Alcell process. A proven alternative to kraft pulping. Tappi J. 1991;74(3):113–118.
[150] Neumann N, Balser K. Acetocell - an innovative process for pulping, totally free from sulfur and chlorine. Papier. 1993;47(10A):16–24.
[151] Parajo JC, Alonso JL, Vazquez D, Santos V. Optimization of catalyzed acetosolv fractionation of pinewood. Holzforschung. 1993;47(3):188–196.
[152] Black NP. ASAM alkaline sulfite pulping process shows potential for large-scale application. Tappi J. 1991;74(4):87–93.
[153] Johansson A, Aaltonen O, Ylinen P. Organosolv pulping - methods and pulp properties. Biomass. 1987;13(1):45–65.
[154] Saake B, Lummitsch S, Mormanee R, Lehnen R, Nimz HH. Production of pulps using the Formacell process. Papier. 1995;49(10A):V1–V7.
[155] Kauliomaki SV, Laamanen LA, Poppius KJ, Sundquist JJ, Wartiovaara IYP. Process for bleaching organic peroxyacid cooked material with an alkaline solution of hydrogen peroxide [US 4,793,898]; 1988.
[156] Nimz HH, Granzow C, Berg A. Acetosolv pulping. Holz Roh- Werkst. 1986;44(9):362.
[157] Azadi P, Inderwildi OR, Farnood R, King DA. Liquid fuels, hydrogen and chemicals from lignin: A critical review. Renewable Sustainable Energy Rev. 2013;21:506–523.
[158] Teymouri F, Laureano-Perez L, Alizadeh H, Dale BE. Optimization of the ammonia fiber explosion (AFEX) treatment parameters for enzymatic hydrolysis of corn stover. Bioresour Technol. 2005;96(18):2014–2018.
[159] Demessie ES, Hassan A, Levien KL, Kumar S, Morrell JJ. Supercritical carbon dioxide treatment: effect on permeability of Douglas fir heartwood. Wood Fiber Sci. 1995;27(3):296–300.
[160] Bayerbach R, Meier D. Characterization of the water-insoluble fraction from fast pyrolysis liquids (pyrolytic lignin). Part IV: structure elucidation of oligomeric molecules. J Anal Appl Pyrolysis. 2009;85(1–2):98–107.
[161] Fan LT, Gharpuray MM, Lee YH. Cellulose Hydrolysis. vol. 3. Springer Berlin Heidelberg; 1987.

6

Functional and Spectroscopic Characterization of Lignins

6.1 Introduction

Although lignin is one of the most abundant biopolymers, it has received relatively little attention as a polymer *per se*. "Lignins should be regarded as a family of 3D polymers, spherical in solution, containing a variety of functional sites and capable of a surprising selection of modifying reactions." [1]

The constitutional model of a lignin (see Chapter 2) gives the broad picture of the reactive groups available in native lignin. These consist of various types of ethers, primary and secondary hydroxyl groups, phenolic, carbonyl, and carboxyl groups, ester functions, ethylenic linkages, and sulfur-containing groups (in some pulps). There also exists a number of typical aromatic sites and activated aliphatic location capable of involvement in modification reactions [1].

One of the greatest challenges in the structural biochemistry of the lignified cell wall is to determine the nature and proportion of the building units and interunit linkages in native lignin structures. Before the advent of powerful NMR methods, chemical degradation reactions of lignins were the only viable ways to gain structural information [2, 3].

As mentioned previously, in the structure of lignins (see Chapter 2), different types of bonds exist, when joining the different subunits. Among them, β-*O*-4′ bonds are very numerous but quite labile, making them the target of many of the depolymerization processes. Nevertheless, the C–C bonds are highly resistant, especially the 5-5′ biphenyl one, which is characteristic of guaiacyl dimers. This explains why coniferous wood, rich in guaiacyl units, is hardly depolymerized [2]. Also, other bonds such as β-5′, β-1′, β-β′, or 4-*O*-5′ are similarly very resistant to degradation.

6.2 Elemental Analysis and Empirical Formula

The elementary composition of lignin slightly varies with the source and the isolation method. Lignin is composed of 60–65% carbon, 5–7% hydrogen, and 28–35% oxygen. Unmodified lignin is free of other elements. Given the variety of materials and methods for their isolation, the observed agreement is fairly good (see Table 6.1).

Based on the assumption that the phenylpropane units remain intact,[1] lignin can be expressed in terms of a semiempirical C_9-unit structure. On the basis of elemental analysis and methoxy content, empirical C_9-unit structures can be calculated in accordance with the procedure developed by Lenz [17], as outlined in Table 6.2.

Table 6.1 Elementary composition of various lignin preparations[a]

Source	Method of isolation	Carbon (%)	Hydrogen (%)	Author(s) [Ref.]
Flax	NaOH/H$_2$O	63.9	5.8	Powell and Whittaker [4]
Larch		63.8	5.2	Powell and Whittaker [5]
Pine		63.4	5.6	Powell and Whittaker [5]
Spruce		64.0	5.5	Powell and Whittaker [5]
Ash		63.2	5.6	Powell and Whittaker [5]
Birch		63.2	5.5	Powell and Whittaker [5]
Poplar		63.3	5.8	Powell and Whittaker [5]
Corn cobs		62.5	5.0	Phillips [6]
Oat hulls		64.4	5.2	Phillips [7]
Rye straw	NaOH/alcohol	63.0	5.6	Beckmann et al. [8]
Corn cobs		61.1	5.8	Phillips [9]
Spruce	HCl (fuming)	62.4	6.4	Heuser et al. [10]
Spruce		64.0	5.3	Hägglund and Malm [11]
Spruce	H$_2$SO$_4$	63.9	5.3	Klason [12]
Spruce	Urban	63.9	6.0	Urban [13]
Wood		64.7	5.5	Freudenberg and Harder [14]
Spruce	Freudenberg	63.6	5.7	Freudenberg et al. [15]

[a]Data taken from ref. [16].

Table 6.2 Example of calculation of empirical C_9 formula[a]

Element/ fragment	Average (%)	M_w	M_w ratio	From OCH$_3$	From acetyl	Net ratio[b]	C_9 formula[c]
C	64.30	12.010	5.35	0.35	1.20	3.80	9.00
H	5.88	1.008	5.83	1.05	1.80	2.98	7.05
O	29.82	16.000	1.86	0.35	0.60	0.91	2.15
OCH$_3$	10.86	31.034	0.35			0.35	0.83
S	0.93	32.064	0.03			0.03	0.07
Acetyl	25.86	43.044	0.60			0.60	1.42

[a]Data taken from ref. [17].
[b]Molecular ratio-methoxy-acetyl.
[c]Net ratio × 2.368, for example normalized to 9.00.

---

[1] Not absolutely correct.

Table 6.3 Formulas for milled wood lignins (MWLs)[a]

Wood species	C_9 formula	OMe free formula	[Ref.]
Spruce	$C_9H_{8.83}O_{2.37}(OMe)_{0.96}$	$C_9H_{9.05}O_2(H_2O)_{0.37}$	[19]
	$C_9H_{7.92}O_{2.40}(OMe)_{0.92}$	$C_9H_{8.04}O_2(H_2O)_{0.40}$	[21, p. 113]
	$C_9H_{7.92}O_{2.88}(OMe)_{0.96}$	$C_9H_{7.12}O_2(H_2O)_{0.88}$	[20]
Beech	$C_9H_{8.50}O_{2.86}(OMe)_{1.43}$	$C_9H_{8.21}O_2(H_2O)_{0.86}$	[19]
	$C_9H_{7.49}O_{2.53}(OMe)_{1.39}$	$C_9H_{7.82}O_2(H_2O)_{0.53}$	[21, p. 113]
	$C_9H_{7.93}O_{2.95}(OMe)_{1.46}$	$C_9H_{7.49}O_2(H_2O)_{0.95}$	[20]
Birch	$C_9H_{9.03}O_{2.77}(OMe)_{1.58}$	$C_9H_{9.07}O_2(H_2O)_{0.77}$	[19]

[a] Data taken from ref. [23, p. 123].

Table 6.4 Elemental analysis, methoxyl content, and calculated C_9 formula of various microwave extracted grass lignins[a]

Biomass	Elemental analysis (%)					OCH_3	C_9 formula
	C	H	N	O	S	(%)[b]	
Triticale straw	65.96	6.48	0.81	26.47	0.28	17.76	$C_9H_{8.70}N_{0.11}S_{0.02}O_{1.98}(OMe)_{1.05}$
Wheat straw	66.60	6.29	0.62	26.03	0.46	17.52	$C_9H_{8.29}N_{0.08}S_{0.03}O_{1.92}(OMe)_{1.02}$
Corn residues	65.49	6.31	0.78	27.14	0.28	17.29	$C_9H_{8.51}N_{0.10}S_{0.02}O_{2.09}(OMe)_{1.02}$
Flax shives	65.63	6.25	0.13	27.77	0.22	20.28	$C_9H_{8.01}N_{0.02}S_{0.01}O_{2.02}(OMe)_{1.22}$
Hemp hurds	66.86	5.81	0.00	27.19	0.14	22.62	$C_9H_{6.73}N_{0.00}S_{0.01}O_{1.80}(OMe)_{1.36}$

[a] Data taken from ref. [22].
[b] % w/w.

These formulas highlight the fact that the monomeric units comprising lignins have a C_9 carbon skeleton, and they underscore the significance of the methoxyl group both for identifying the lignin type and for indicating the contributions of component monomer units [18]. Usual methoxyl content is 13–16% (softwoods) or 18–22% (hardwoods).

It is a common practice to express the chemical composition of lignins in terms of a C_9 unit structure in order to evaluate and compare the complex structural characteristics of these materials. Thus, the formula of a typical milled wood lignin (MWL) from softwood is reported as $C_9H_{8.3}O_{2.7}(OCH_3)_{0.97}$ and that from a typical hardwood such as $C_9H_{8.7}O_{2.9}(OCH_3)_{1.58}$.

Some representative empirical C_9 formulas of softwood and hardwood MWLs are shown in Table 6.3. These values vary according to the source and nature of the lignin preparations [19, 20] [21, p. 113]. In Table 6.4, calculated C_9 formula of various grass lignins are shown [22].

6.3 Determination of Molecular Weight

For determining the lignin structure, the most commonly used parameter is molecular weight. However, today the use of molecular weight to characterize the lignin type is still under debate. Thus, for example, values of number average molecular weight (M_n) and average molecular weight (M_w) are not universally accepted, because different analytical techniques yield discordant results.

Depending on the statistical method applied for the determination of the molecular weight, it is possible to obtain different average values. Four parameters are extensively used: mole fraction, weight fraction, and two other parameters that can be related to measured quantities, the so-called

Table 6.5 Average molar mass data (M_w, M_n) and dispersity index ($Đ$) of organosolv lignins extracted from various types of herbaceous biomass[a]

Source	M_w	M_n	($Đ$)
Triticale straw	2320	798	2.9
Wheat straw	2015	780	2.6
Corn residues[b]	2609	842	3.1
Flax shives	4500	1408	3.2
Hemp hurds	2418	1054	2.3

[a]Data taken from ref. [22, p. 361].
[b]Stalk and stover.

number average molar mass or M_n, that is sometimes referred to as number average molecular weight (NAMW).[2] Finally, the mass average molar mass (M_w), where w is weight, and frequently referred to as weight average (weight average molecular weight (WAMW)).

The M_n is the ordinary arithmetic mean or average of the molecular masses of individual macromolecules. Some significant properties are dependent on molecular size, so a larger molecule will have in this parameter a larger contribution than a smaller one. The M_w is calculated by the following equation:

$$M_w = \frac{\sum_i N_i M_i^2}{\sum_i N_i M_i}, \tag{6.1}$$

where N_i corresponds to the number of molecules of molecular mass M_i.

The understanding of the macromolecular properties of lignins requires information on M_n and M_w, together with their distributions. These parameters are useful in studying the behavior of macromolecules in solution as well as their conformation and size [24].

M_n is determined mainly by the measurement of colligative properties, for example, cryoscopy, ebulliometry, and membrane and vapor-pressure osmometry; meanwhile light scattering is the absolute method to determine M_w.

The major problem in measuring the physical properties of isolated lignin is its solubility in organic solvents. It is thought that unavoidable uncertainty comes from the fact that the molecular mass of the soluble portion of lignin is comparably smaller than that of a truly isolated sample. This point becomes more complex since the solution properties of lignin are thought to depend not only on plant species but also on the isolation process [25, 26].

Recently, Monteil-Rivera et al. [22] determined the M_w of various extracted lignins, without acetylation, by gel-permeation chromatography (GPC) analysis using THF as the eluent and polystyrene standards (see Table 6.5).

6.3.1 Gel-Permeation Chromatography (GPC)

Also referred to as gel-filtration chromatography (GFC) or size-exclusion chromatography (SEC), GPC fractionates macromolecules according to molecular size [27–30].

The stationary phase used involved cross-linked polydextran gels having varying pore size (Sephadex) [31]. Such gels in water form a 3D network that acts as a molecular sieve. When an

---
[2] This is the ordinary arithmetic mean or average of the molecular masses for the individual macromolecules, and it is determined by measuring the molecular mass of n polymer molecules, summing the masses, and dividing by n.

Table 6.6 Molecular weight (M_w), dispersity index ($Đ$), and glass-transition temperatures (T_g) of selected lignins[a]

Isolation method	Species	M_w	$Đ$[b]	$T_g(°C)$[c]
Milled wood	Pine	11 400	8.8	160
Milled wood	Alder	7000	6.4	110–130
Kraft	Pine	4300	3.3	169
Kraft	Hardwood mix	3000	2.9	–
Lignosulfonate	Mix	1000–10 000	6–8	–
Organosolv	Pine	1400	2.8	91
Organosolv	Aspen	2100	3.5	97
Acid hydrolysis	Pine	40 000	50	96
Acid hydrolysis	Aspen	10 100	15.3	95
Steam explosion	Aspen	2300	2.9	139

[a] Data taken from refs [49–51].
[b] Dispersity index determined by GPC (gel-permeation chromatography).
[c] Determined by DSC (differential scanning calorimetry).

aqueous solution of macromolecules is allowed to move through a column containing the gel, the molecules of low molecular size penetrate the gel particle pores but large molecules are excluded and pass directly through the column (elute first). Detection is usually done by UV or by refractive index, although detectors based on low-angle laser light scattering (LALLS) [32] and differential viscometry [33, 34] are also available.

The application of the method requires that the lignin sample be completely soluble in THF; consequently, the vast majority of lignin preparations should be undertaken only after appropriate derivatization (acetylation or methylation).

This technique has been widely applied to lignin materials since the early 1960s, such as lignosulfonates [35–38] and Kraft lignins [37–46]. Hatakeyama et al. [27] reported M_w in a range from 2000 to 7500 for fractionated lignin. Lindner and Wegner [47], by fractionation of organosolv lignins (spruce and pine lignin), reported that M_w ranges from 2700 to 11 000, and M_w/M_n ranges from 1.8 to 2.4. Recently, Lebo et al. [48] reported results of detailed characterization of lignosulfonates: M_w 64 000, M_w/M_n 8.8, and fractionated samples show M_w from 4600 to 40 000. Other examples of M_w, T_g, and dispersity index ($Đ$) values for different lignins are presented in Table 6.6.

The introduction of cross-linked, semirigid polyethylene (PE) gels of uniform particle size has made it possible to perform GPC in high-pressure systems using small columns. With such systems, the experimental time can be substantially reduced and the resolution power enhanced. A variety of solvents can be used. The technique is referred to as high-performance size-exclusion chromatography (HPSEC) [52].

6.3.2 Light Scattering

This physicochemical method has been considered one of the most useful for deriving the M_w of polymers in solution. It is also probably the most suitable for determining the actual shape of molecules [53, 54].

The determination of M_w is based on the principle that the intensity of light scattered at some angle is a function of the size of the scattering particles.

This has been used to determine the M_w of soda and Kraft lignins [55, 56] and lignosulfonates [57–59]. See Table 6.7.

Table 6.7 M_w of different lignins determined by light scattering[a]

Lignin	M_w range	[Ref.]
Soda	28 000–146 000	[56]
Kraft	5600–25 000	[60]
	2000–17 500	[60]
	13 700–48 300	[61]
Lignosulfonates	7900–126 000	[58]

[a] Data taken from ref. [53, p. 506].

Table 6.8 Molecular weight (M_w) of spruce dioxane lignin[a]

Method	Fraction				
	D3	D4	D5	D6	D7
Laser light scattering (LALLS)	1600	2100	3350	5400	7530
Size-exclusion chromatography (SEC)	1550	2100	3100	5150	7250

[a] Data taken from ref. [53, p. 506].

With lignins, several difficulties must be overcome to ensure reproducible results, since the relatively small dimensions of the molecules give low-intensity scattered light. These problems have recently been overcome by the use of new and sophisticated instruments[3] [32, 43, 60–64].

Table 6.8 shows very similar values of M_w from fractions of spruce dioxane lignin determined by laser light scattering (LALLS) and SEC.

6.3.3 Vapor-Pressure Osmometry (VPO)

Several methods based on colligative properties were used to determine M_n of lignins; for example, cryoscopy, ebulliometry, osmotic pressure, and vapor-pressure osmometry (VPO). These methods, except VPO, have not been used because of many experimental difficulties. The latter method is used to determine M_n in the range of 100 to 10 000. The method is based on the measurement, at a given temperature, of the vapor-pressure depression of the solvent for diluting polymer solutions [54, 65, 66].

Table 6.9 shows very similar M_n values for spruce dioxane lignin fractionated by preparative SEC and by VPO [66]. Table 6.10 shows that M_n of black cottonwood alkali lignins measured in

Table 6.9 M_n of spruce dioxane lignin fractionated by preparative SEC[a]

Method	Solvent	Fraction								
		1	2	3	4	5	6	7	8	9
VPO	THF	820	970	1250	1650	2250	3100	4000	5000	6050
VPO	Dioxane	800	970	1300	1700	2250	3100	4000	5000	6050
SEC	THF	820	980	1250	1650	2300	3150	4300	5100	5900

[a] Data taken from refs [53, p. 516, 67, 68].

---

[3] The use of a laser source, an interference filter, and an analyzing polarizer.

Table 6.10 M_n of black cottonwood lignins determined by VPO measured at various temperatures in 2-methoxyethanol[a]

Fraction	Temperature (°C)				Fraction	Temperature (°C)			
	25	37	45	60		25	37	45	60
1	3700	33 750	33 700	33 700	3	53 900	53 950	53 860	53 900
2	53 000	53 000	53 050	53 000	4	63 400	63 400	63 400	63 450

[a] Data taken from refs [63, 66, p. 516].

Table 6.11 M_n of different Kraft lignins determined by VPO[a]

Wood species	Range of M_n	[Ref.]	Wood species	Range of M_n	[Ref.]
Pine	13 600–23 700	[69]	Pine	13 400–23 500	[60]
Hardwood	13 000–13 100	[69]	Southern slash pine	23 300–63 100	[61]

[a] Data taken from ref. [66, p. 516].

Table 6.12 Comparison of M_n from different methods[a,b]

Sample	VPO	GPC/LALLS	GPC/DV	Sample	VPO	GPC/LALLS	GPC/DV
Red oak HPL	1416	6708	1535	Aspen	1393	4004	1591
RO:PO[c]	1108	5433	1567	Westvaco	1499	3711	1597

[a] Data taken from ref. [70].
[b] Experimental time of 24 h. See text for the meaning of acronyms.
[c] Red Oak:Propylene Oxide; Identical to the red oak HPL, but was made in larger quantities to allow preparative fractionation.

2-methoxyethanol determined by VPO does not depend on temperature. Table 6.11 lists the M_n ranges for Kraft lignins determined by VPO.

Siochi et al. [70] reported the M_n and distribution of hydroxypropylated lignins from various trees using GPC, static low-angleLALLS, and VPO. Comparisons of the M_n of hydroxypropylated lignins determined by different methods: VPO and GPC, using an LALLS (GPC/LALLS) or a differential viscosity (DV) detector (GPC/DV) are listed in Table 6.12.

6.3.4 Ultrafiltration (UF)

Ultrafiltration (UF) is a pressure-driven membrane-separation technique, based on molecular size [71], which has separation capabilities from 20 to 100 000 Å. UF has been successfully applied to fractionation of Kraft lignin [72, 73] and lignosulfonates [74]. The most distinctive advantage over classical physical techniques is that it is not influenced by impurities, as the other techniques are.

The aforementioned methods suggest that the molecular mass of isolated lignin lies in the range of 10^3–10^5, depending on the plant species, processing method, and measuring method.

6.4 Functional Group Analyses

The expression of analysis data based on C_9 units is widely used but is erroneous in relation to the loss of C_3 side chain features and the incorporation of nonlignin components [75]; analysis data based on functional group contents is an alternative. Table 6.13 shows frequencies of the principal functional groups in spruce MWL lignin.

Table 6.13 Functional group of spruce MWL[a]

Functional groups	Per C_6C_3	[Ref.]	Functional groups	Per C_6C_3	[Ref.]
Aliphatic OH	1.09, 0.93	[76, 77]	Ar–CH = CH = CHO	0.03–0.04	[78]
Phenolic OH	0.26, 0.33	[77, 79]	Ar–CH = CH–CH$_2$OH	0.03	[80]
Total carbonyl	0.20	[81]	Phenolic C_α–OH	0.05–0.06	[82, 83]
$C_\alpha = O$	0.06–0.07	[81]	Nonphenolic C_α–OH	0.15, 0.10	[81, 83]
Nonconjugated C = O	0.10	[81]			

[a]Data taken from refs [2, 23].

6.4.1 Methoxyl Group (MeO)

The presence of the methoxyl group has been established in the lignin preparations isolated from various sources and by different methods. According to Freudenberg et al. [84], the methoxy groups present in lignin are attached to aromatic nuclei.

The treatment of MeO- or EtO-containing compound with concentrated HI at refluxing temperature leads to the formation of methyl or ethyl iodide (see Figure 6.1). When these pass through a solution of silver nitrate, the process yields a precipitate of silver iodide, which is used to determine quantitatively the methoxyl or ethoxyl content of the compound [85–88]. This procedure is not appropriate for sulfur-containing products. Viebock and Brecher [89], and Viebock and Schwappach [90] eliminated the interference by conversion of the alkyl iodide into iodic acid by oxidation with bromine, while simultaneously oxidizing H_2S to H_2SO_4.

The methoxyl content of the lignin slightly varies with the source and the method used for its isolation, as shown in Table 6.14.

6.4.2 Phenolic Hydroxyl Group (OH$_{ph}$)

The phenolic hydroxyl group is one of the most remarkable functionalities affecting the physical and chemical properties of lignin polymers. Physical and chemical methods, or a combination of both, have been used to estimate the phenolic hydroxyl content of lignin [2].

The method of Aulin-Erdtman is based on the difference in absorption at 300 nm of phenolic units in neutral and in alkaline solutions [99, 100]. A simple and rapid method for determining phenolic hydroxyl groups in lignin preparations was developed by Goldschmid [101] based on these procedures.

The titration method is based on the acidity of phenolic hydroxyl groups.[4] The determination of the phenolic hydroxyl content of lignin by conductometric or potentiometric titration in aqueous

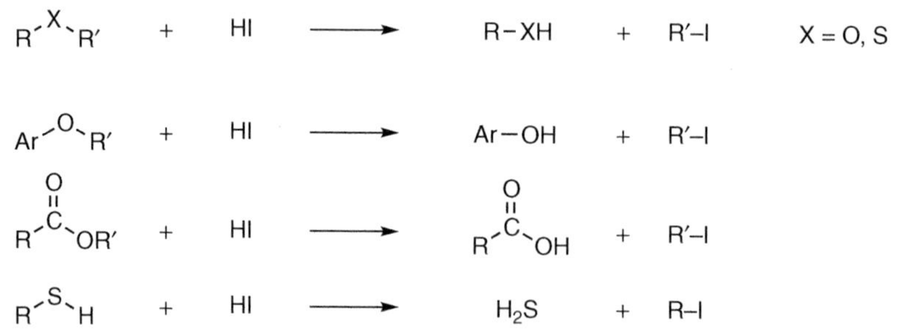

Figure 6.1 Main reactions of HI with functional groups present in lignin [91]

---
[4] Interference of carboxyl and/or sulfonate groups.

Table 6.14 Percentage of methoxyl in various lignin preparations[a]

Source	Method of isolation	OCH_3 (%)	Author(s) [Ref.]
Spruce	H_2SO_4	14.47	Klason [92, p. 27]
Fir	H_2SO_4	13.95	König and Rump [93]
Sugar maple	H_2SO_4	21.00	Sherrard and Harris [94]
Spruce	HCl (fuming)	14.39	Hägglund [95]
Spruce	HCl (fuming)	13.10	Fischer and Schrader [96]
Spruce	HCl (fuming)	11.60	Kürschner [97, p. 34]
Flax	$NaOH/H_2O$	14.90	Powell and Whittaker [4]
Winter rye straw	NaOH/alcohol	14.85	Beckmann et al. [8]
Corn cobs	NaOH/alcohol	12.10	Phillips [9]
Corn cobs	$NaOH/H_2O$	14.30	Phillips [98]
Oat hulls	$NaOH/H_2O$	15.80	Phillips [7]
Oat hulls	NaOH/alcohol	15.61	Phillips [7]

[a] Data taken from ref. [16, p. 126].

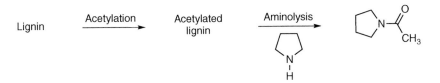

Figure 6.2 The aminolysis method [112]

[102, 103] or nonaqueous [104, 105] solutions. It has also been claimed that the potentiometric titration with sodium colamine in ethylenediamine could be used to determine all free phenolic groups [104]. Later, Pobiner [106] developed a nonaqueous titration method using tetra-*n*-butylammonium hydroxide as the titrant, with dimethylformamide (DMF) as the solvent and *p*-hydroxybenzoic acid as internal standard.

On the other hand, different authors [17, 107, 108] have used ^{1}H NMR spectroscopy for the determination of the phenolic hydroxyl content of lignin. Also, the estimation of the phenolic hydroxyl content of lignin by ^{13}C-NMR, both in solution and in the solid state, is generally based on the methyl or carbonyl signals of phenolic acetyl groups [75], or on the signals corresponding to C_3, C_4, and C_5 in the aromatic ring [109, 110]. Robert and Brunow [76] have estimated the phenolic hydroxyl groups in MWL by ^{13}C NMR.

Chemical methods include the determination of the increase in methoxyl content resulting from diazomethane methylation [111], or the increase in phenolic acetyl group after acetylation [17, 77].

The aminolysis method is based on the finding that the rate of deacetylation of aromatic acetates in pyrrolidine is higher than that of aliphatic acetates [77] (see Figure 6.2).

Adler *et al.* [113], in 1958, developed a method based on the oxidation of simple phenolic compounds (**1**) with aqueous sodium periodate to *o*-quinones (**2**) in which 1 mol of MeOH per mole of phenolic hydroxyl group is released (see Figure 6.3).

The structural units with a free phenolic group react with 1-nitroso-2-naphthol to give compounds with a maximum absorption at 505 nm. Olcay [114] has determined free phenolic hydroxyl groups in lignins by applying this reaction. Also, free phenolic groups have been determined by reaction with dinitrofluorobenzene.

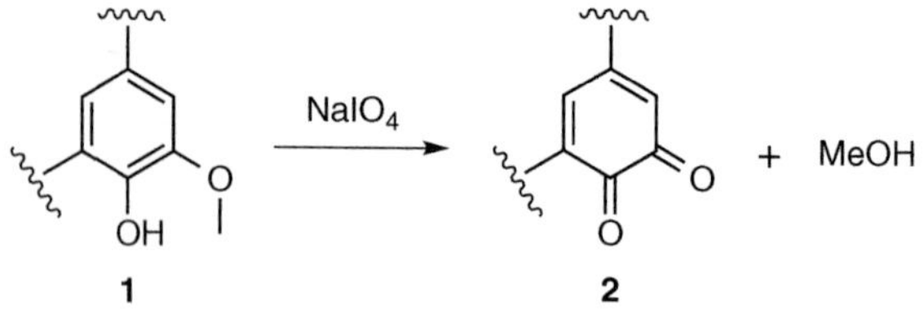

Figure 6.3 Oxidation of guaiacyl moiety with sodium periodate [23]

Table 6.15 Phenolic hydroxyl groups of spruce MWLs[a]

Method	Phenol–OH/OCH$_3$	[Ref.]	Method	Phenol–OH/OCH$_3$	[Ref.]
^{1}H NMR	0.29, 0.27	[17, 107]	Titration	0.33–0.34	[104]
^{1}H NMR	0.27, 0.24	[108]	Aminolysis	0.33	[77]
^{13}C NMR	0.20	[76]		0.15–0.20[b]	
Periodate	0.30	[83]			

[a] Data taken from ref. [23].
[b] Cellulolytic enzyme lignin.

Two degradation methods have been used to estimate phenolic hydroxyl content. The first, developed by Gellerstedt et al. [115–117], is based on the sequential oxidation of lignin after ethylation with potassium permanganate and H_2O_2; the second one, employed by Lapierre and Rolando [118], is based on the thioacidolysis after diazomethane methylation. The phenolic hydroxyl group contents found so far are summarized in Table 6.15.

Table 6.16 shows the phenolic hydroxyl content of a variety of lignins. From an inspection of the values, it is clear that the phenolic content of lignin decreases in the following order: "Kraft lignin >> Kraft pulp > wood," a trend, which underscores the significance of aryl ether cleavage in delignification and of the effect of the phenolic hydroxyl groups on lignin solubility [112].

6.4.3 Total and Aliphatic Hydroxyl Groups (R–OH)

Lignins are produced by polymerization of three monomeric precursors (monolignols). These compounds all have a terminal aliphatic hydroxyl in addition to a phenolic hydroxyl group at C_4 of the aromatic ring. The presence of hydroxyl groups in lignin is indicated by the fact that it can be acetylated and alkylated (see Chapter 7). Alcoholic and phenolic hydroxyl groups seem to be present (see Table 6.17).

In general, the aliphatic hydroxyl content of a lignin is determined by subtraction of the phenolic hydroxyl content from the total hydroxyl content of the lignin. The total hydroxyl content can be determined by acetylation with acetic anhydride-pyridine reagent followed by saponification of the acetate and titration of the resulting products in acetic acid with a standard solution of NaOH [88].

Recently, Monteil-Rivera et al. [22] determined quantitatively the content of aliphatic, phenolic, and carboxylic OH units in five isolated herbaceous lignins by means of phosphitylation with 2-chloro-4,4,5,5 tetramethyl-1,3,2-dioxaphospholane (TDMP) and using ^{31}P NMR (see Table 6.18). In addition, they determined the molar H/G/S ratio.

Various benzyl alcohols, dilignols, and tri- and tetralignols have been isolated by mild hydrolysis and hydrogenolysis of lignin.

Glycerol side chains can exist in the native lignin macromolecules, but the presence in lignins is rather insignificant because of formation of only small amounts of formaldehyde by periodate oxidation [23, 125, 126].

Table 6.16 Phenolic hydroxyl contents of some lignins[a]

Species (preparation)	Phenolic content		Method	[Ref.]
	mmol g^{-1} lignin	No (100 C$_9$ units)		
Spruce (wood)		10	UV	[119]
Spruce (wood meal)	0.70	13	Periodate	[120]
	0.65	12	Aminolysis	[120]
Spruce (MWL)		18	Titration	[121]
		20	Periodate	[122]
		20	UV	[122]
		20	Periodate	[119]
		20	^{13}C NMR	[76]
		26	Periodate	[17]
		27	^{1}H NMR	[17]
		30	Periodate	[113]
		33	Aminolysis	[77]
Spruce (CEL)		20	Periodate	[122]
		20	UV	[122]
Pine (wood meal)	0.73	13	Aminolysis	[123]
Pine (Kraft pulp)	0.84–1.47	15–26	Aminolysis	[123]
Pine (Kraft lignin)	2.9–4.1	55–70	Aminolysis	[75]
Aspen (wood meal)	0.49	10	Periodate	[120]
	0.45	9	Aminolysis	[120]
Sweetgum (MWL)		14	Periodate	[122]

[a]Data taken from ref. [112, 432].

Table 6.17 Total, phenolic, and aliphatic hydroxyl contents of some representative milled wood and bamboo lignins (MWL and MBL)[a]

Lignin preparation	Species	C$_9$ formulae	OH groups[b]		
			Total	Phenolic	Aliphatic
MWL	Spruce	C$_9$H$_{7.66}$O$_2$(H$_2$O)$_{0.48}$(OMe)$_{0.94}$	1.46	0.28	1.18
MWL	Zhong-Yang Mu	C$_9$H$_{6.49}$O$_2$(H$_2$O)$_{0.92}$(OMe)$_{1.13}$	1.33	0.22	1.11
MBL	Chinese bamboo	C$_9$H$_{5.52}$O$_2$(H$_2$O)$_{0.84}$(OMe)$_{1.25}$	1.49	0.36	1.13
Willstatter (HCl) lignin	Sweetgum	C$_9$H$_{4.86}$O$_2$(H$_2$O)$_{1.15}$(OMe)$_{1.60}$	1.51	0.12	1.39
Pine Kraft lignin		C$_9$H$_{7.42}$O$_2$(H$_2$O)$_{0.28}$S$_{0.08}$(OMe)$_{0.77}$	1.35	0.58	0.77
Bamboo Kraft lignin		C$_9$H$_{5.52}$O$_2$(H$_2$O)$_{1.60}$S$_{0.08}$(OMe)$_{1.00}$	1.00	0.44	0.56

[a]Data taken from ref. [124, p. 410].
[b](mol/C$_9$ unit).

Table 6.18 Hydroxyl group content (determined by ^{31}P NMR analysis) of organosolv lignins isolated from various types of plant biomass[a]

Biomass	OH (mmol/g)						H/G/S molar ratio
	H phenolic	G phenolic	S phenolic	Total phenolic	Total aliphatic	Total –COOH	
Triticale straw	0.42	0.79	0.69	2.02	1.50	0.17	22/42/36
Wheat straw	0.38	1.12	0.77	2.50	1.38	0.15	17/49/34
Corn wastes[b]	0.94	0.88	0.78	2.71	1.39	0.23	36/34/30
Flax shives	0.02	1.20	0.56	1.97	2.95	0.02	1/67/32
Hemp hurds	0.17	1.20	1.53	3.39	1.19	0.01	6/41/53
Softwood (Indulin AT)	0.20	2.00	–	3.57	2.43	0.35	9/91/0
Hardwood (Organosolv)	0.12	0.98	1.84	3.39	1.25	0.21	4/33/63

[a] Data taken from ref. [22, p. 360].
[b] Stalk and stover.

Benzyl alcohol units with free phenolic hydroxyls (**3**) have been determined with the quinone monochloroimide (**4**) color reaction by Gierer [127, 128] (see Figure 6.4).

Adler et al. [79, 129] found that p-alkoxybenzyl alcohol (**6**) is oxidized to the corresponding aryl ketone (**8**) with DDQ (**7**), which is reduced to **9**. The α-ketone formed was determined spectrophotometrically, giving a value of 0.16/OCH$_3$ (see Figure 6.5).

The benzyl alcohol units have also been determined semiquantitatively from NMR spectra as 0.33/OCH$_3$ [130], 0.32 [17], 0.31 [76], or 0.29 in spruce MWL and 0.45 in birch MWL.

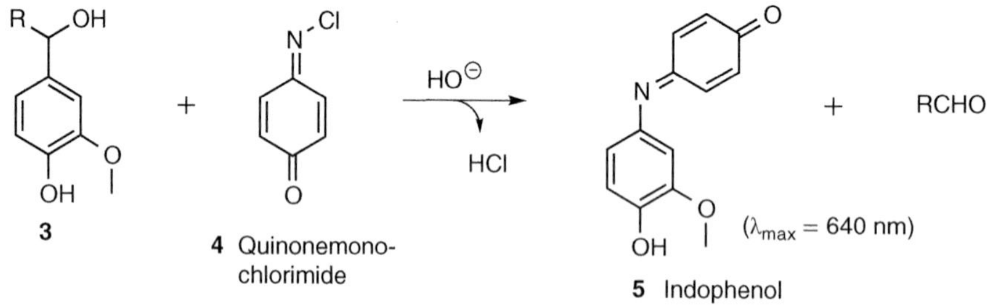

Figure 6.4 Reaction of quinone monochloroimide (**4**) with p-hydroxybenzyl alcohol groups (**3**) [23]

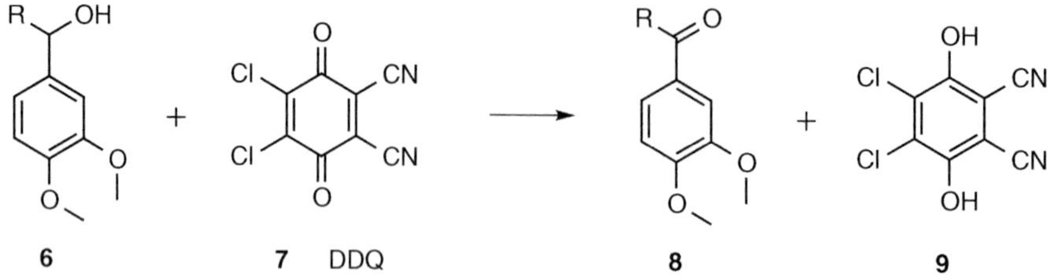

Figure 6.5 Reaction of DDQ with p-hydroxybenzyl alcohol groups (**6**) [23]

6.4.4 Ethylenic Groups (>C=C<)

Small amounts of ethylenic groups are present *in situ* in lignin, mainly as components of unattached cinnamaldehyde and cinnamyl alcohol end groups [131, pp. 195–200]. Stilbene and other conjugated structures are generated on the treatment of lignin with alkali or acid. The ethylenic group content of several lignins is listed in Table 6.19.

The presence of small amounts of coniferyl alcohol groups in the lignin molecules was demonstrated with a color reaction reported by Lindgren and Mikawa [80]. Coniferyl alcohol (**10**) and its 4-*O*-methyl ethers react with nitrosodimethylnitrile after tosylation, giving the *p*-dimethylaminoanilide of styrylglyoxalnitrile (**11**), which is red in color (λ_{max} = 475 nm), *via* an intermediate. Coniferyl alcohol groups in spruce lignin were estimated to have the same content as cinnamaldehyde moieties (2%) by this color reaction. (See Figure 6.6)

Table 6.19 *Ethylene group contents of some lignins*[a]

Material (softwood)	Structure	Method	Number/C_9 units	[Ref.]
Brauns native lignin	Cinnamaldehyde	Colorimetric	2.5	[78]
Lignosulfonate	Cinnamaldehyde	Colorimetric	2	[78]
MWL from birch wood	Cinnamaldehyde	UV spectroscopy	1–2	[132]
MWL	Coniferyl alcohol	UV spectroscopy	3	[133]
MWL	Coniferaldehyde	UV spectroscopy	3–4	[133, 134]
Wood meal	Coniferaldehyde + coniferyl alcohol	Hydroformylation	8	[135]
Alkali lignin	Unspecified	Peracid oxidation	10	[136]
Kraft MWL	Stilbenes	UV spectroscopy	7–8	[137]
Refiner mechanical pulp	*o*-Quinones	Oxyphosphorane	10–12	[138]

[a] Data taken from ref. [139, p. 442].

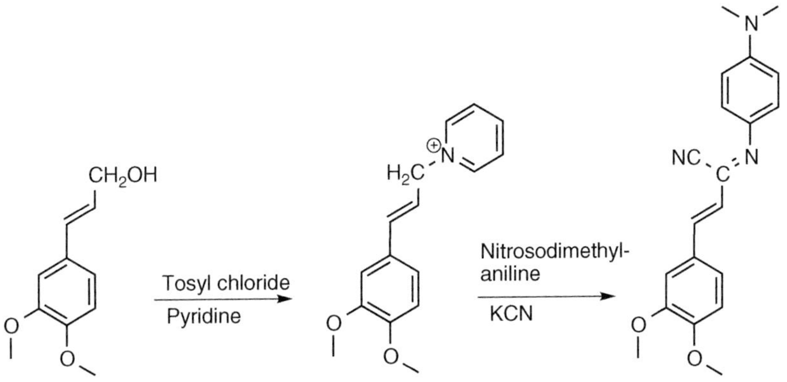

(**10**)

(**11**) *p*-Dimethylaminoanilide of styrylglyoxylnitrile (red, λ_{max} = 475 nm)

Figure 6.6 *Color reaction of coniferyl alcohol groups [23]*

6.4.5 Carbonyl Groups (>C=O)

The evidence of the presence of carbonyl and carboxyl groups in lignin preparations was reported by Adler et al. [78, 140, 141] and by Pew [142] during an investigation into the mechanism of the Wiesner reaction. In lignin detections, plant tissues treated with phloroglucinol and concentrated HCl give a purple color (see Figure 3.2 on p. 52). In a procedure based on this reaction, the total content of carbonyl groups in spruce lignin was estimated by UV spectroscopy to be $(2-3)/(100C_9$ units) [78, 143].

The carbonyl content increases with ball-milling of the wood meal during the preparation of MWL, and it is also produced during the isolation process. Technical lignins, in particular alkali lignins, contain appreciable amounts of carbonyl groups [144]. Furthermore, the carbonyl content of lignin preparations has been determined using common reagents such as phenylhydrazine [145], thiobenzylhydrazide [111], and hydroxylamine hydrochloride [81, 146–148].

Other methods used consist of reduction with either sodium or potassium borohydride [147, 149], catalytic hydrogenation [133], and UV spectroscopy [81].

6.4.6 Carboxyl Groups (–COO–)

Carboxyl groups have been detected in native lignin by IR spectroscopy [150]. The most widely accepted methods for determining carboxyl groups in lignin are based directly or indirectly on the acidity of this group. The analysis involves the neutralization of the carboxylic acid by using potentiometry or conductivity to detect the end point [151–155].

The carboxyl groups have also been determined using the so-called calcium acetate method, which is an ion-exchange method [156].

The carboxyl contents of different lignins are shown in Table 6.20. In general, the carboxyl content is less than 1% for lignins isolated without strong oxidation.

6.4.7 Sulfonate Groups and Total Sulfur Composition (R-SO$_2$O– and S)

In 1892, Lindsey and Tollens [160] deduced that sulfur found in sulfite pulping was chemically bound to lignin in the form of sulfonic acid.

Methods of determining the sulfonate content of lignin fall into two main categories: those typified by conductometric titration, in which sulfonate content is measured directly, and those that measure the sulfur content and assume that all the sulfur is present in sulfonate form.

Table 6.20 Carboxyl contents of some lignins[a]

Material (softwood)	COOH, mEq/g	Method	[Ref.]
Hardwood Kraft	1.44	Potentiometric titration	[106]
Hardwood native	0.92	Potentiometric titration	[106]
Hardwood + softwood LSA	0.31–2.08	Potentiometric titration	[106]
Wheat straw MW	0.81	Potentiometric titration	[106]
Spruce MW	0.12	Potentiometric titration	[155]
Decayed spruce	0.55	Potentiometric titration	[157]
Birch hydrolysis	0.41	CaCH$_3$COO	[132]
Spruce LSA	0.61–1.26	Conductometric titration	[158]
Softwood Kraft	0.8	Conductometric titration	[28]
Softwood Kraft	0.89	Methylation	[159]

[a] Data taken from ref. [139, p. 463].

Table 6.21 Sulfonic acid contents in lignin determined by different methods[a]

Material (softwood)	Yield (%)	Sulfur (mmol/kg) A[b]	Sulfonic acid, (mmol/kg) B[b]	C[b]	[Ref.]
Wood, sulfonated[c] for 12.5 min	99	128	120	107	[152]
Wood, sulfonated[c] for 35 min	97	216	181	179	[152]
Wood, sulfonated[c] for 110 min	95	250	227	213	[152]
CMP, sulfonated[c]	92	269	236	246	[152]
High-yield bisulfite	70	219	208	196	[152]
Low-yield bisulfite	47	40	28	20	[152]
Low-yield acid sulfite	42	24	6	6	[152]
Oxidatively sulfonated Kraft lignin (KMnO$_4$)	–	1250	1250	–	[161]
LSA	–	1811[d]	1503	–	[162]

[a] Data taken from ref. [163, p. 481].
[b] A = gravimetric, B = conductometric, C = magnesium elution methods.
[c] At pH = 7.
[d] Sulfur content measured by combustion.

Such methods are precipitation of barium sulfate,[5] X-ray fluorescence spectroscopy, and combustion of pulp followed by the analysis of sulfur as SO_2 or as sulfate. Table 6.21 shows a comparison of sulfonic acid contents determined by various methods.

6.5 Frequencies of Functional Groups and Linkage Types in Lignins

Figure 6.7 shows types and frequencies of linkages and main functional groups in softwood and hardwood lignins. This abstract description of lignin showing linkages and average units and bonds is less expected but closer to the real nature of lignin [164] (see also Table 2.3 on p. 20).

Figure 6.8 summarizes the main types of lignin structural units [23].

Tables 6.22 and 6.23 list the frequencies of the typical linkage units in hardwood and softwood lignins.

The most remarkable linkage types in the lignin molecule are B (β-O-4) and then D (β-5), E(5-5), $C(\beta$-1), and A (α-O-4).

6.5.1 β-O-4′ Linked Units

Arylglycerol-β-aryl ether units (B) belong to the most notable substructures in lignin molecules. Adler et al. [170] indicated that β-O-4′ units may be 25–30% of all phenylpropanes. From ^{1}H NMR spectra, Lundquist [176] estimated later that 40–50% of birch lignin units [177] and 30–50% of spruce lignin units are attached to an adjacent unit by a β-O-4′ linkage.

---

[5] Canadian Pulp and Paper Association Standard G28 1970.

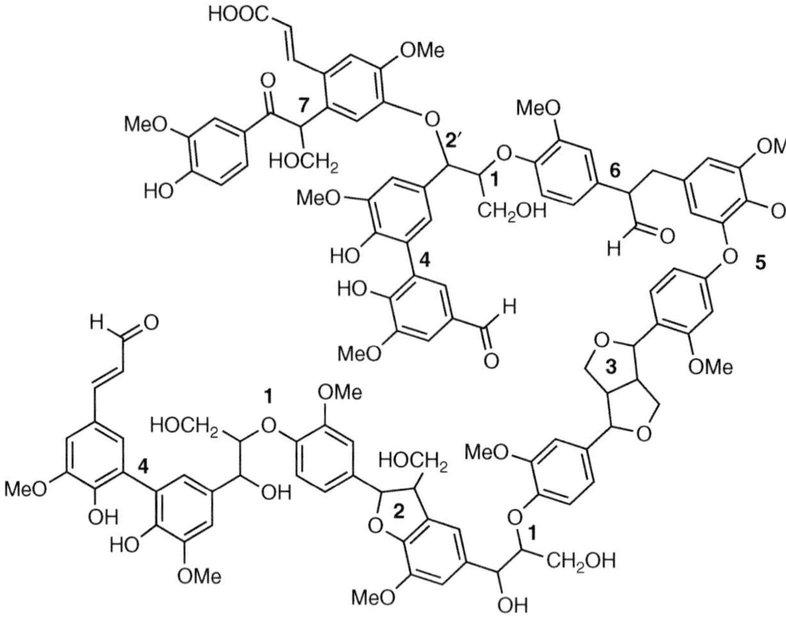

Dilignol linkages		Softwood	Hardwood
β-O-4	(1)	43–50	50–65
β-5 / α-O-4	(2)	9–12	4–6
α-O-4	(2′)	6–8	4–8
β-β	(3)	2–4	3–7
5-5′	(4)	10–25	4–10
4-O-5′	(5)	4	6–7
β-1	(6)	3–7	5–7
C-6,C-2	(7)	3	2–3

Figure 6.7 Scheme of a hardwood lignin fragment, with the frequency of linkages [164]

6.5.2 β-5′ Linked Units

β-5′ linkage units are represented by the phenylcoumaran structure (D). Adler et al. [171] indicated that spruce MWL contained 0.11 of these β-5′ structures per methoxyl after acidolysis, but their 0.03 are β-5′ not ring-closed. Freudenberg et al. [178] also isolated β-5 linkage units after acid methanolysis of spruce wood. Phenylcoumaran lignols were isolated through dioxane/water hydrolysis [82, 179].

6.5.3 β-1′ Linked Units

Nimz [180] first isolated diarylpropanediols (C) from the degradation products of spruce and beech lignins by mild hydrolysis.

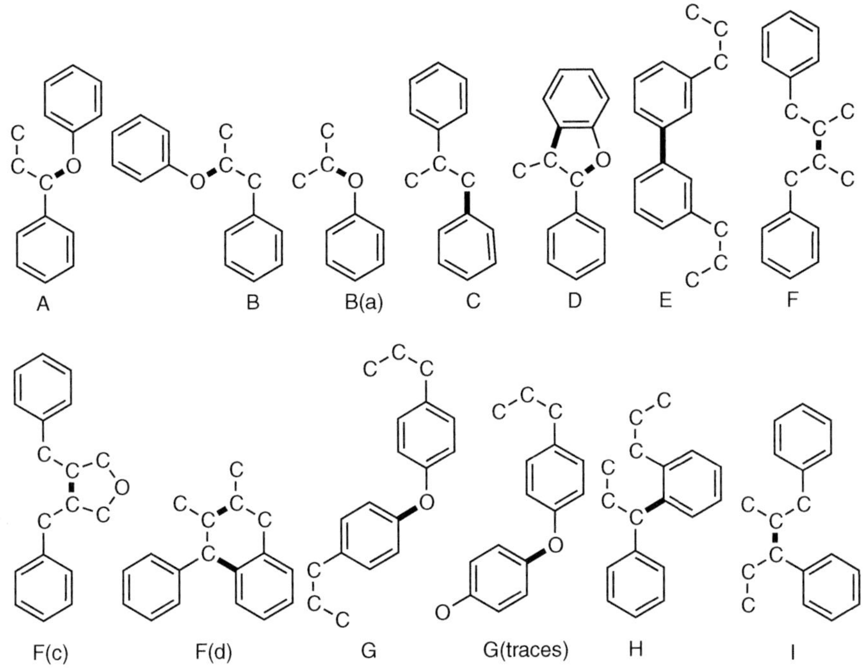

Figure 6.8 Typical linkage units in lignin [23]

Table 6.22 Structural units of spruce MWL[a]

Structural units	Per C_6C_3	[Ref.]
A: α-O-4′ (open)	0,12, 0.07, 0.06–0.08	[165–167]
Phenolic	0.04, 0.02	[129, 168]
Nonphenolic	0.05–0.09, 0.06	[129, 165]
B: β-O-4′	0.49–0.51, 0.50	[166, 167]
	(0.25–0.30, 0.3–0.5)[b]	[169, 170]
	0.02[c]	[126]
C: β-5′	0.14, 0.9–0.12	[171] [172, p. 154]
Noncyclic	0.03	[171]
D: β-1′	0.15, 0.02, 0.07	[166, 173] [172, p. 154]
E: 5-5′, 5-6′	0.19–0.22, 0.10–0.11	[166, 173]
F: β-β′	0.13	[172, p. 154]
Pinoresinol units	0.05–1.0, 0.02–0.03	[169, 174]
G: 4-O-5′, 4-O-1′	0.07–0.08, traces	[166, 173]

[a]Data taken from refs [2, 23].
[b]Except displaced side chain units.
[c]Arylglyceraldehyde-β-aryl ether.

Table 6.23 Structural units (per 100 C_6C_3 units) of birch MWL and beech lignin[a]

Units	Birch[b] G	Birch[b] S	Birch[b] Total	Beech[c] Total
A: α-O-4′ (open)			6	
B: β-O-4′	22–28	34–39	60	65[e]
β-O-4′[d]			2	
C: β-1′			7	15
D: β-5′			6	6
E: 5-5′		4.5	4.5	2.3
F: β-β′			3	5
β-β′[f]				2
β-β′, α-6′[g]				0.5
G: 4-O-5′, 4-O-1′	1	5.5	6.5	1.5
H: C(α)-2, C(α)-6	1–1.5	0.5–1	1.5–2.5	
I: α-β				2.5

[a] Data taken from ref. [23].
[b] Data taken from ref. [2].
[c] Data taken from ref. [175].
[d] In glyceraldehyde-β-aryl ether.
[e] A + B.
[f] In dibenzyl-THF units.
[g] In tetralin units.

6.5.4 α-O-4′ Linked Units (benzyl ethers)

Noncyclic benzyl ethers (A) are very difficult to isolate from the degradation products of lignin because of the labile nature of these ether linkages. Adler and Gierer [146] estimated that the total amount of benzyl alcohol and noncyclic benzyl ether was about 0.43/OCH_3 in spruce MWL after treatment of lignin with MeOH/HCl.

6.5.5 Condensed and Uncondensed Units

The guaiacyl units possessing an unsubstituted 5-position are called "uncondensed," and the units that carry C–C or ether bonds at this position are called "condensed".[6]

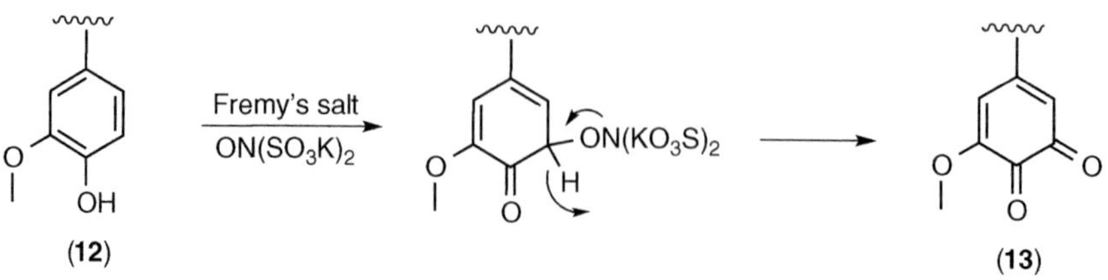

Figure 6.9 Determination of condensed units by oxidation of guaiacyl nucleus with Fremy's salt [23]

---

[6] The condensed units here are aromatic rings in which at least one carbon atom, not necessarily the five-carbon alone, is linked directly to another carbon atom outside the ring.

Fremy's salt[7] oxidizes p-substituted phenols to o-quinones (see Figure 6.9). For example, 4-propylguaiacol (**12**) is oxidized to methoxy-5-propyl-o-quinone (**13**).

Adler and Lundquist [181] applied this oxidation to estimate uncondensed units in lignin. The o-quinone (**13**) formed can be quantitatively determined by means of spectrophotometry. In this way, it was found that 0.15–0.18 units per methoxyl in MWL were uncondensed phenolic units, corresponding to 50–60% of the 0.30 phenolic units present. However, as the oxidation with Fremy's salt can be applied only to the units with a free phenolic hydroxyl, no information is available on etherified units.

By means of ^{1}H NMR spectroscopy, about 45–50% condensed units [130, 182] and lower amounts of them (35% and 30% for spruce and birch MWLs) have been estimated.

6.5.6 Biphenyl Structures

Aulin-Erdtman [134] found that the most obvious effect of increasing the pH of lignin solutions was a higher absorption in the UV spectrum above 300 nm. This absorption band is characteristic of biphenyl structures (E). The number of hydroxy-biphenyl units in black spruce BL was estimated at 0.05/OCH_3. Different biphenyl units (5-5, 5-6, 5-1, and 6-6) are presumed to be present in lignin from the permanganate oxidation products.

6.5.7   4-O-5' Linked Units

Freudenberg and Chen [183] first isolated 4-O-5' and 1-O-4' (G) type biphenyl ether compounds by permanganate oxidation. The estimated frequency of 4-O-5' linked units vary from 0.015/C_6C_3 for beech lignin [175] to 0.065/C_6C_3 for birch lignin [184].

6.5.8   β-2 and β-6 Linked Units

Metahemipinic (**15**) and hemipinic (**14**) acids, detected in the products of permanganate oxidation, are presumed to exist in β-6 and β-2 linked units in the lignin molecule [183] (see Figure 6.10).

Larsson et al. [184], and Erickson et al. [166] estimated the content of β-2 and β-6 linked units to be 0.015–0.025/C_6C_3 for birch lignin [184] and 0.025–0.03/C_6C_3 for spruce lignin [166], respectively.

6.5.9   β-β Linked Units

β-β type structures (F) are involved in lignans represented by pinoresinol and syringaresinol. β-β linked units are involved much more in hardwood lignin than softwood lignin. Lundquist [176] estimated a low content for pinoresinol units (0.02–0.03/OCH_3) in spruce MWL from its

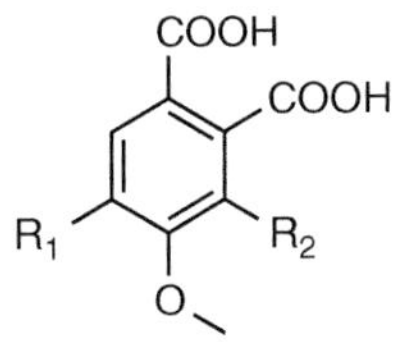

14 (R_1 = H, R_2 = OMe)
15 (R_1 = OMe, R_2 = H)

Figure 6.10   *The hemipinic and metahemipinic acids*

---

[7] Potassium nitrosodisulfonate, $K_4[ON(SO_3)_2]_2$.

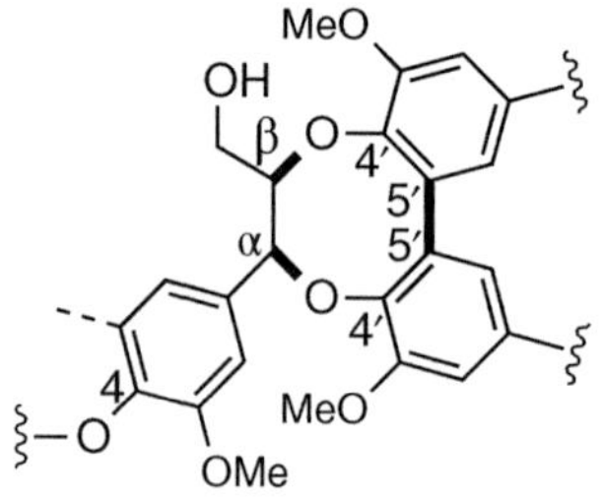

5'-5' / β-O-4' / α-O-4'
Dibenzodioxocin (D1) (**16**)

Figure 6.11 *Dibenzodioxocin (DBDO)*

^{1}H NMR spectrum. Larsson *et al.* [184] also estimated 0.03–0.05/C_6C_3 units for birch lignin and Erickson *et al.* [166] 0.02/C_6C_3 for spruce lignin.

Nimz and Das [175] isolated cyclolignan-bearing β-β and α-6 linkages from beech lignin by thioacetic acid degradation. Nimz and Das also isolated α-β type compounds and tetrahydrofuran dilignols involving γ-O-γ and β-β linkages (see Figure 7.9 in page 202).

6.5.10 Dibenzodioxocin Units

To these monomeric linkages, another one is added in which three phenolic units are involved. It is the so-called dibenzodioxocin structure (α-O-4' / β-O-4''/3'-3''). In softwood cells, they represent the main branching point [185–187] (see Figure 6.11).

Argyropoulos *et al.* [188] have estimated the presence of 3.7 dibenzodioxocin rings/100 C_9 units in softwood lignin by ^{31}P NMR measurements and DFRC method.

6.6 Characterization by Spectroscopic Methods

Nondegradative methods involve mostly modern spectroscopic and other nonevasive techniques and yield structural information about the structure of lignin without the need of subjecting it to harsh chemical environments. Thus, they have been used for the characterization of lignin in solid state as well as in solution.

Other methods used to determine lignin in wood and pulps include UV spectrophotometry [189], FTIR [190], NIR [191], and NMR [192].

6.6.1 Fourier Transform Infrared (FTIR) Spectroscopy

In IR spectroscopy, a vibrational transition that involves a change in dipole moment results in absorption of an IR photon. The energy of the absorbed photon is equal to the energy difference between two vibrational states of a molecule.

IR spectroscopy has been used often to characterize lignin because the technique is simple and the sample to be studied does not need to be dissolved in any solvent and is required only in very small quantities. Typical FTIR spectra of lignins is shown in Figure 6.12.

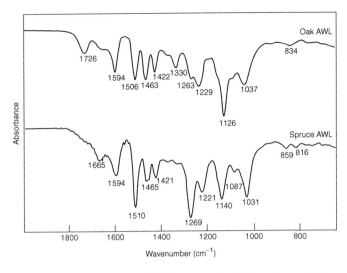

Figure 6.12 FTIR spectra of MWLs from oak, birch, and spruce [23]

Table 6.24 IR-absorption bands of lignins[a]

Band	Absorption (cm^{-1})
OH stretching	3450–3400
OH stretching (in CH- and CH$_2$-)	2940–2820
C=O stretching (nonconjugated)	1715–1710
C=O stretching (conjugated to aromatic ring)	1675–1660
Aromatic ring vibrations	1605–1600
	1515–1505
C–H deformations, asymmetric	1470–1460
Aromatic ring vibrations	1430–1425
C–H deformations, symmetric	1370–1365
Syringyl ring breathing	1330–1325
Guaiacyl ring breathing	1270–1275
C–H and C–O deformations	1085–1030

[a]Data taken from ref. [193, p. 272].

Hergert [193] reviewed the early research on FTIR spectroscopy of lignin and summarized the IR band assignments (see Table 6.24).

Table 6.25 shows the assignments of FTIR peaks for softwood and hardwood lignins, and there is some significant difference in the two types of lignins.

Recently, Monteil-Rivera et al. [22] summarized the main FTIR peaks for extracted herbaceous lignins (see Table 6.26). In this case, not only remarkable differences appeared with respect to wood lignins, but also between the various herbaceous-type of lignins.

Kolboe and Ellefsen [194] used IR spectroscopy as an method for estimating lignin content. The absorption at 1515 cm^{-1} was chosen for the determination of lignin. The lignin content was estimated in 28–29%, in agreement with the generally accepted value.

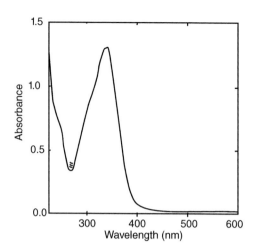

Figure 6.13 Example of a UV/vis spectra of lignin [197]

Table 6.25 Assignments of FTIR peaks in lignins[a]

Assignment[b]	Band (cm^{-1})	
	Softwood	Hardwood
O–H (s)	3425–3400	3450–3400
C–H (s) in CH$_3$ and CH$_2$	2920	2940
C=O (s) in nonconjugated ketone and conjugated COO	1715	1715–1710
C=O (s) in conjugated ketone	1675–1660	1675–1660
Aromatic (sv)	1605	1595
Coumaran ring	1495	
C–H (d) (asymmetric)	1470–1460	1470–1460
Aromatic (sv)	1430	1425
C–H (d) (symmetric)	1370	1370–1365
Syringyl (rb) with C–O (s)		1330–1325
Guaiacyl (rb) with C–O (s)	1270	1275
C–H (id) in guaiacyl	1140	1145
C–H (id) in syringyl		1130
C–O (d) in secondary alcohol and aliphatic ether	1085	1085
C–H (id) in guaiacyl, and C–O (d) in primary alcohol	1035	1030

[a] Data taken from refs [23, 193].
[b] s = stretching; sv = skeletal vibrations; d = deformations; rb = ring breathing; id = inplane deformation; d = deformation.

It has been demonstrated that the IR spectrum is also useful as an indicator of the ratio of syringyl-propane to guaiacylpropane units in lignins [195].

6.7 Raman Spectroscopy

In Raman spectroscopy, the electromagnetic field induces a dipole moment in the molecule, resulting in an exchange of energy and vibrational transition simultaneously. The energy of the exciting

Table 6.26 Assignments of FTIR peaks (cm^{-1}) of lignins extracted from various herbaceous biomass[a,b]

Assignment	Lignin			
	T	W	C	F
O–H stretching	3350	3363	3350	3398
C–H stretching (CH$_3$ and CH$_2$)	2922, 2850	2922, 2850	2924, 2850	2933, 2848
C=O stretching, nonconjugated	1708	1708	1701	1716
C=O stretching, conjugated	1653	1653	1655	1659
Aromatic skeletal vibrations (S > G)	1597	1597	1603	1593
Aromatic skeletal vibrations (G > S)	1512	1512	1512	1508
C–H deformations (CH$_3$ and CH$_2$)	1458	1458	1456	1462
Aromatic skeletal vibrations	1423	1425	1425	1421
Aliphatic C–H stretch in CH$_3$, not in OMe	1356	1354	1365	1369
S ring (or G ring condensed) breathing	1328	1331	1327	1327
G ring breathing	1261	1261	1263	1265
C–C and C–O stretch, G condensed > G	1215	1211	1207	1217
C–O stretch in ester groups (HGS)	1165	1161	1167	–
Aromatic C–H in plane deformation (S)	1117	1115	1119	1122
Aromatic C–H in plane deformation (G > S)	1030	1030	1032	1030
–CH=CH– out-of-plane deformation (trans)	–	–	984	–
Aromatic C–H out-of-plane deformation (G)	816	816	–	854, 816
Aromatic C–H out-of-plane deformation (S+H)	837	839	833	–

[a] Data taken from ref. [22].
[b] T = triticale straw; W = wheat straw; C = corn waste (stalk + stover); F = flax shives; and H = hemp hurds.

photons is higher than the energy difference between the two vibrational states, and the exchange with the field results in a scattered photon, shifted in frequency from the incident photon by an amount equal to the energy difference between the vibrational states. Although Raman spectroscopy is complementary to IR one, only scant information has been retrieved by this method in the analytical chemistry of lignin [196].

6.7.1 Ultraviolet (UV)

Lignin shows characteristic π-π* transitions of the aromatic nuclei that make it a strong absorber of UV light. In general, softwood lignin shows a maximum at 280–285 nm and hardwood lignin at

274–276 nm. In soluble lignins, maxima occur at about 205 nm. In Figure 6.13, an example of a UV/vis spectra of lignin is shown.

UV spectrophotometric investigations are used extensively to characterize lignin preparations. Lignin ultraviolet (UV)–visible spectroscopy was comprehensively reviewed by Goldschmid [198]. More recently, detailed descriptions of measurement techniques for solution absorbance [189] and diffuse reflectance have appeared.

As cited earlier, Aulin-Erdtman [199–202] applied UV spectra to estimate the amounts of certain functional groups, especially the phenolic hydroxyls of lignin.

The absorbance at either 205 or 280 nm is the basis of several techniques for the quantitative determination of lignin (see Chapter 3). The characteristic UV absorption spectra of lignin (at either 205 or 280 nm) have been used for the quantitative and qualitative determination of native lignins [203, 204], and lignosulfonates [205], as well as the study of delignification of wood during pulping [206, 207].

Schöning and Johansson [208] studied the UV absorption of lignin from pulp-waste liquor and concluded that acid-soluble lignin components should be determined at 205 nm. Wegener *et al.* [209] found that hexafluoropropanol is an excellent solvent for UV and IR spectroscopy of lignins.

Fergus and Goring [210, 211] pioneered the use of UV microscopy for studying the distribution of lignin across the cell wall, in different morphological regions within woody tissue. Similar efforts by Boutelje and Eriksson [212] and Yang and Goring [119] firmly established the use of UV microscopy for such endeavors.

6.7.2 NMR Spectroscopy

The application of NMR spectroscopy to the characterization, classification, and detailed structural elucidation of lignins has seen widespread utility, because it gives information on the lignin molecule that cannot be gained by chemical analysis.

6.7.2.1 Proton Spectra (^{1}H NMR)

The fact that the proton nucleus is of 100% natural abundance and of high sensitivity made the ^{1}H NMR experiment rather popular in the early days of applying this technique to lignin. Ludwig *et al.* [107, 213] were the first to determine the ^{1}H NMR spectra of lignin. They estimated the free benzylic hydroxyls, aliphatic and aromatic hydroxyls, and total aliphatic hydrogens in the lignin preparations.

Almost invariably, ^{1}H NMR is used on acetylated lignins [107, 177, 213–215], since this affords better signal resolution. ^{1}H NMR signal assignment has been based on model compound [216–218], decoupling [219], and 2D NMR experiments [220]. Therefore, ^{1}H NMR is able to quantify a number of notable lignin structural features. ^{1}H NMR signals from acetylated lignins are summarized in Table 6.27.

Nevertheless, there are some essential limitations to ^{1}H NMR spectroscopy of lignins. These include the rather limited range of chemical shifts (12 ppm), extensive signal overlapping and proton-coupling effects.

Lenz [17] studied ^{1}H NMR spectra of both underivatized and acetylated lignins. He observed that lignin preparations from both hardwoods and softwoods showed marked differences in the degree of condensation of aromatic rings, phenolic and aliphatic hydroxyl groups, and the number of highly shielded aliphatic protons.

Ede *et al.* [221] studied ^{1}H–^{1}H correlation spectroscopy (COSY) and *J*-resolved spectra of acetylated MWL, concluding that the results confirm the presence of most of the known lignin structural units.

Table 6.27 Main ^{1}H chemical shifts (δ) assignment from TMS of acetylated spruce MWL and beech MWL[a]

Assignment	δ (ppm)
Hydrocarbon contaminant	1.26
Aliphatic acetate	1.95, 2.01, 2.02
Aliphatic acetate[b]	2.13
Aromatic acetate	2.28–2.29
Unknown[c]	2.94
H_β in β-β	3.08
OCH_3	3.76–3.81
H_γ	4.18, 4.27–4.28
H_γ primarily in β-O-4 (erythro) and β-5	4.39
H_γ	4.43
H_β in β-O-4 (birch)	4.60
H_β in β-O-4 (spruce) including CH_2 in cinnamyl alcohol	4.65
H_α in β-β (birch) including CH_2 in cinnamyl alcohol	4.70
H_α in β-5 and noncyclic benzyl aryl ethers (birch)	5.44
H_α in β-5 and noncyclic benzyl aryl ethers and H_β in aryloxypropiophenones (spruce)	5.49
H_α in β-O-4 and β-1 and vinyl protons	6.01–6.06
Aromatic and vinyl protons	6.93–6.94
Aromatics in benzaldehyde, vinyl protons on the carbon adjacent to aromatic rings in cinnamaldehyde (spruce)	7.41
Aromatics ortho to C=O (birch)	7.50
Aromatics ortho to C=O (spruce)	7.53
Formyl protons in cinnamaldehyde	9.64
Formyl protons in benzaldehyde	9.84–9.86

[a]Data taken from refs [23, 176, 177].
[b]Including aromatic acetate in biphenyl structure.
[c]The peak is absent in the acetate of lignin reduced with $NaBH_4$.

Further developments have claimed that, after appropriate derivatization, it is possible to quantitatively determine the phenolic hydroxyl groups in lignin [108].

6.7.2.2 Carbon Spectra (^{13}C NMR)

Lüdemann and Nimz [222–225] were the first to study ^{13}C NMR spectra of lignins. This early work has allowed the ^{13}C NMR spectroscopy to become an indispensable tool for the structural elucidation of lignin. Compared to proton NMR, the ^{13}C NMR spectra of lignin offer considerably better resolution, with no coupling effects, over a much wider chemical shift range (200 ppm).

Convenience, speed, and a wealth of even quantitative information [226–228] have contributed to the widespread use of ^{13}C NMR for the structural analysis of lignins. Furthermore, a variety of newly developed pulse sequences such as distortionless enhancement by polarization transfer (DEPT) [229], incredible natural abundance double quantum transfer experiment (INADEQUATE) [230], heteronuclear multiple-quantum correlation (HMQC) [231], as well as acquisitions on ^{13}C-enriched lignin samples [228, 232, 233] have improved our understanding of lignin structure and reactivity.

Table 6.28 Main ^{13}C chemical shifts (δ, ppm) assignment from TMS of acetylated spruce MWL and beech MWL[a]

Assignment		Spruce (ppm)	Beech (ppm)
(structure: CHOR$_2$ on ring with R$_1$O, OMe, OR)	A: R$_1$ = H, R$_2$ = Ac B: R$_1$ = H, R$_2$ = alkyl C: R$_1$ = OMe, R$_2$ = Ac D: R$_1$ = OMe, R$_2$ = alkyl		
α-CO and γ-CO in cinnamaldehyde		194.9	195.2
α-CO		192.3	192.6
CO in primary acetoxyl		171.4	171.7
CO in secondary acetoxyl		170.5	171.0
CO in aromatic acetoxyl		169.5	170.0
B(3), C3/5, C(3,5), D3/5, D(3/5), C$_\alpha$ in cinnamaldehyde		153.5	153.8
A(3), A4, B4		151.4	152.0
B3		149.1	148.7
A3		148.3	
C-4 in B-ring in cyclic β-5 (dehydro diconiferyl-alcohol acetate)		145.4	144.9
A(4), B(1), D(1)		140.8	140.4
A(1), D4		137.7	137.9
C(4), C(1)		136.7	136.4
D1			136.0
B1, C1		134.1	133.8
A1, Cβ in cinnamaldehyde		132.4	132.6
C(4), D(4), C-1 in cyclic β-5, C-2/6 in p-hydroxyphenyl ring		129.3	129.3
A6, B6		120.7	120.5
A5, A(6), B(6)		118.8	118.8
B5		116.5	116.5
A2, A(2)		112.9	112.4
C(2/6), D(2/6)			106.9
C2/6, D2/6			105.0
C$_\alpha$ in cyclic β-5		88.5	88.7
C$_\alpha$ in cyclic β-β (pinoresinol acetate)		86.3	86.7
C$_\alpha$ in GOA[b], C$_{\alpha/\beta}$ in α,β-diaryl ether		80.7	81.4
C$_\alpha$ in β-1(1,2-disyringylpropane-1,3-diol acetate)		76.5	77.1
C$_\alpha$ in GOA[b]		75.5	75.5
C$_\alpha$ in GOA[b] (diastereomer)		74.8	
C$_\gamma$ in open β-β (dibenzyltetrahydrofuran)		73.3	73.3
C$_\gamma$ in β-β		72.6	72.6
C$_\gamma$ in cyclic β-5 and cinnamyl alcohol acetate		66.2	66.0
C$_\gamma$ in β-1 and α,β-diaryl ether			64.3
C$_\gamma$ in GOA[b]		64.1	63.3
OCH$_3$		56.4	56.4
C$_\beta$ in β-β		55.4	55.4
C$_\beta$ in β-1 and β-5		51.4	51.3
C$_\alpha$ in open β-β		41.0	41.2
CH$_3$ in acetoxyl		20.5	20.5

[a] Data taken from refs [23, 234].
[b] GOA = guaiacylglycerol-β-aryl ether acetate.

Table 6.29 Main ^{13}C chemical shifts (δ, ppm) assignment from TMS of nonacetylated lignin[a]

Assignment	δ_C	Assignment	δ_C
C_9 in p-coumarate (PCE)	166.5	C_6, FE ester	123.0
C_4 in PCE	160.0	C_1 and C_6 in Ar–C(=O)C-C units	122.6
C_7 in PCE	144.7	C_6 in G units	119.4
C_2/C_6 in PCE	130.3	C_6 in G units	118.4
C_1 in PCE	125.1	C_5 in G units	115.1
C_3/C_5 in PCE	116.0	C_5 in G units	114.7
C_8 in PCE	115.0	C_2 in G units	111.1
C_3/C_5, etherified S units	152.5	C_2 in G units	110.4
C_3, etherified G units	149.7	C_2/C_6, S with α-CO	106.8
C_3, G units	148.4	C_2/C_6, S units	104.3
C_4, etherified G	148.0	Cα in G type β-5 units	86.6
C_4, nonetherified G	146.8	Cβ in G type β-O-4 units (threo)	84.6
C_4, etherified 5–5	145.8	Cβ in G type β-O-4 units (erythro)	83.8
C_4, nonetherified 5–5	145.0	Cγ in β-β; Cγ, β-aryl ether	72.4
C_4, S etherified	143.3	Cα in G type β-O-4 units (erythro)	71.8
C_1, S etherified	138.2	Cα in G type β-O-4 units (threo)	71.2
C_1, G etherified	134.6	Cγ in G type β-O-4 units with α-C=O	63.2
C_1, (S and G) nonetherified	133.4	Cγ in G type β-5, β-1 units	62.8
C_5, etherified 5-5	132.4	Cγ in G type β-O-4 units	60.2
C_1, nonetherified 5-5	131.1	C in Ar–OCH$_3$	55.6
Cβ in Ar–CH=CH–CHO	129.3	Cβ in β-β units	53.9
Cα and Cβ in Ar–CH=CH–CH$_2$OH	128.0	Cβ in β-5 units	53.4
C_2/C_6, in H units	128.1	CH$_3$ group, ketones (conj) or in aliphatic	36.8
C_5/C_5' in nonetherified 5-5	125.9	CH$_2$ in aliphatic side chain	29.2
C_1 and C_6 in Ar–C(=O)C-C	122.6	CH$_3$ or CH$_2$ group in saturated side chains	26.7
C_5, nonetherified 5-5	125.9	γ-CH$_3$ in n-propyl side chain	14.0

[a] Data taken from ref. [235].

Despite all these advantages, some drawbacks remain concerning the efficient use of ^{13}C NMR. These stem from the fact that the ^{13}C nucleus is not sensitive enough (1/5800 compared to proton) and that it is only 1.1% naturally abundant. Consequently, relatively large sample sizes and a long acquisition times are essential. An extensive compilation of ^{13}C signals from acetylated lignins are summarized in Table 6.28, and for nonacetylated lignins in Table 6.29.

These chemical shifts can be divided into the following three regions:

- Carbonyl carbons appear at 200–160 ppm.

- C_1-C_6 aromatic carbons, and carbons of double bonds on the side chain, at 160–100 ppm.
- Carbons on saturated side chains at 90 to 20 ppm.
- In addition, methoxyl carbons always appear in the narrow range of 56.3 ± 0.2 ppm.

Obst and Ralph [236] have tried to determine the relative syringyl/guaiacyl ratios for hardwood lignins. Lapierre et al. [237] easily estimated the ratios of syringyl/guaiacyl for hardwood lignins using the signal intensities of C_2 plus C_6 for each from the conventional NMR spectra.

Figure 6.14 depicts an modern example of the application of 2D NMR spectroscopy (^{13}C-^{1}H (HSQC, heteronuclear single-quantum correlation) correlation spectra) to the detailed structural elucidation of lignins (in this case, lignin unit compositions) [238].

6.7.2.3 Solid-State NMR

When standard techniques of ^{13}C NMR are applied, solid samples usually yield weak and broad signals. These problems can be overcome using cross-polarization pulse sequences and spinning samples at the magic angle of 54.7 °C relative to the applied magnetic field. The first CP/MAS spectrum for solid wood was made by Schaefer and Stejskal [239] in 1976.

Advances in instrumentation have made the use of solid-state NMR spectroscopy a routine operation [240, 241]. As such, various solid-state ^{13}C NMR experiments can now be performed routinely on lignin [242, 243] and solid wood or plant samples [244–246]. It is thus possible to gain some information about the lignin within a sample without its prior isolation [247].

Dipolar dephasing when used as part of the acquisition protocol of solid-state spectra of wood and pulps has been documented to yield significant information in relation to the degree of chemical modification occurring in lignin [248].

6.7.2.4 Phosphorus Spectra (^{31}P NMR)

^{31}P NMR spectroscopy has provided a new analytical tool capable of detecting a variety of functional groups in isolated lignins [249, 250]. More specifically, solution ^{31}P NMR has been used to examine soluble lignins [251, 252] and carbohydrate [253] samples after phosphitylation with 1,3,2-dioxaphospholanyl chloride and 2-chloro-4,4,5,5 tetramethyl-1,3,2-dioxaphospholane (TMDP) [254, 255] (see Figure 6.15).

From a single quantitative ^{31}P NMR experiment, it is possible to determine the three principal forms of phenolic hydroxyls present in lignins as well as primary hydroxyls, carboxylic acids, and the two diastereomeric forms of β-O-4 structures [256].

6.7.2.5 Other Active Nuclei in NMR

Efforts to overcome some of the limitations imposed by ^{1}H and ^{13}C NMR spectroscopies have promoted the examination of other NMR-active nuclei which, when covalently linked to lignin by appropriate derivatization procedures, may provide additional structural information for these polymers.

^{19}F NMR has been proposed as a means of detecting different functional groups in lignins [257–261].

Early attempts were made to examine the potential of silylation followed by silicon NMR for the determination of hydroxyl groups lignins [262, 263]. While the method offers resolved signals for aromatic, phenolic, and carboxylic acids, large sample concentrations and long acquisition times are essential due to the low natural abundance, low magnetic moments, and high relaxation times of the ^{29}Si nuclei.

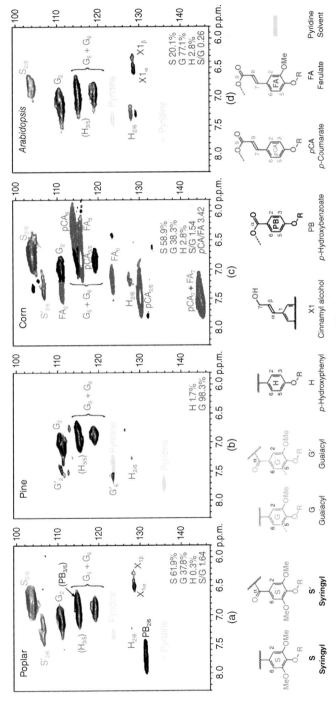

Figure 6.14 2D NMR spectra revealing lignin unit compositions. Partial short-range $^{13}C^{-1}H$ (HSQC) correlation spectra (aromatic regions only) of cell wall gels in DMSO-d_6/pyridine-D_5 (4:1, v/v) from (a) two-year-old greenhouse-grown poplar wood, (b) mature pine wood, (c) senesced corn stalks and (d) senesced Arabidopsis inflorescence stems. Contours in this region are used to measure S/G/H ratios, as well as relative p-hydroxybenzoate (PB, in poplar), p-coumarate (pCA in corn) and ferulate (FA in corn) levels [238]. (See insert for color representation of this figure.)

Figure 6.15 Derivatization of hydroxylic structures with TMPD [254, 255]

6.7.3 Other Spectroscopic Methods

In addition to the methods described earlier, other spectroscopic methods are being used to characterize lignins both in solid state and in solution.

Two analytical techniques are the most valuable for separate and identify low-M_w fragments such as those found in product mixtures after the application of chemical degradation techniques (see Section 7.2, p. 189) and in spent pulping liquors and bleaching[8] effluents:

Gas chromatography-mass spectrometry (GS-MS), the objective of which is to confirm the identity of compounds suggested by GC analysis and to determine the possible structures of unidentified components in the sample [264].

High-performance liquid chromatography (HPLC), which is the fastest-growing technique in chromatography, despite being used for nonpreparative separations [265].

6.7.3.1 Electron Microscopy

One of the techniques in electron microscopy for detecting of lignin involves the use of $KMnO_4$ stains. This technique provides many excellent details of the ultrastructural features of lignin in wood [266, and references therein].

6.7.3.2 Interference Microscopy (IM)

Interference microscopy (IM) according to Jamin [267] and Lebedeff [268] causes interference by splitting a single beam of polarized light into two beams, which are in phase. One beam then passes through the specimen while the other passes through the reference medium. The two beams travel over different optical path lengths because of the difference in optical density between the specimen and the reference medium. The two beams are recombined and interfere with each other to produce an image with color or intensity contrast between the specimen and background. By manipulating the image contrast, it is possible to measure the refractive index of the specimen and hence to gain information on the chemical composition of the sample. Quantitative IM can be used to measure the distribution of lignin in sections of wood or pulp fibers. It is ideally suited to the study of lignification [269].

6.7.3.3 Electron Spin Resonance (ESR)

Electron spin resonance (ESR) has found applications to be a highly sensitive tool for the detection and identification of free-radical species in lignin [270, and references therein].

Using conventional ESR, Wan and Depew [271] could elucidate the initial photochemical reactions of lignin model compounds. ESR technique could be used to monitor the cation radical and other types of radical formation during induced breakdown of model compounds [272]. Several model compounds were irradiated with γ-rays and the radicals formed were studied for their postirradiation behavior at various temperatures, *in vacuo* and in air by ESR, and their behavior was compared with that of isolated lignin radicals [273].

---

[8] See Glossary for a definition.

6.7.3.4   Pyrolysis-Gas Chromatography-Mass Spectrometry (Py-GC/MS)

Recently, Lourenço et al. [274] reported the total lignin content detected with analytical pyrolysis gas chromatography, and mass spectrometry (Py-GC/MS). The sample is thermally degraded in the absence of oxygen-forming volatile fragments that can be separated by gas chromatography and identified by mass spectrometry.

References

[1] Allan GG. Modification reactions. In: Sarkanen KV, Ludwig CH, editors. Lignins: Occurrence, Formation, Structure and Reactions. New York: Wiley-Interscience; 1971. p. 511–573.
[2] Adler E. Lignin chemistry-past, present and future. Wood Sci Technol. 1977;11:169–218.
[3] Glasser WG. Classification of lignin according to chemical and molecular structure. In: Glasser WG, Northey RA, Schultz TP, editors. Lignin: Historical, Biological, and Materials Perspectives, ACS Symposium Series No. 742. Washington, DC: American Chemical Society; 1999. p. 216–238.
[4] Powell WJ, Whittaker H. Chemistry of lignin. Part I. Flax lignin and some derivatives. J Chem Soc, Trans. 1924;125:357–364.
[5] Powell WJ, Whittaker H. XXII.- the chemistry of lignin. Part II. A comparison of lignins derived from various woods. J Chem Soc, Trans. 1925;127:132–137.
[6] Phillips M. The chemistry of lignin. II. Fractional extraction of lignin from corn cobs. J Am Chem Soc. 1928;50(7):1986–1989.
[7] Phillips M. The chemistry of lignin. IV. Lignin from oat hulls. J Am Chem Soc. 1930;52(2):793–797.
[8] Beckmann E, Liesche O, Lehmann F. Lignin aus winterroggenstroh. Angew Chem. 1921;34(50):285–288.
[9] Phillips M. The chemistry of lignin. I. Lignin from corn cobs. J Am Chem Soc. 1927;49(8):2037–2040.
[10] Heuser E, Schmitt R, Gunkel L. Methylation of lignin. Cellul-Chem. 1921;2:81–86.
[11] Hägglund E, Malm CJ. Lignin prepared by the action of hydrochloric acid. Cellul-Chem. 1923;4:73–77.
[12] Klason P. Lignin content of spruce wood. Cellul-Chem. 1923;4:81–84.
[13] Urban H. Fir wood. Cellul-Chem. 1926;7:73–78.
[14] Freudenberg K, Harder M. Zur konstitution des lignins. II. Formaldehyde as a cleavage product of lignin. Ber Dtsch Chem Ges B. 1927;60:581–585.
[15] Freudenberg K, Zocher H, Dürr W. Lignin and cellulose. XI. Further experiments with lignin. Ber Dtsch Chem Ges B. 1929;62:1814–1823.
[16] Phillips M. The chemistry of lignin. Chem Rev. 1934;14(1):103–170.
[17] Lenz BL. Application of nuclear magnetic resonance spectroscopy to characterization of lignin. Tappi. 1968;51(11):511–519.
[18] Dence CW. The determination of lignin. In: Dence CW, Lin SY, editors. Methods in Lignin Chemistry, Springer Series in Wood Science. Berlin: Springer-Verlag; 1992. p. 33–61.
[19] Björkman A. Finely divided wood. II. The properties of lignins extracted with neutral solvents from softwoods and hardwoods. Sven Pappersttdn. 1957;60:158–169.
[20] Fengel D, Wegener G, Feckl J. Characterization of analytical and technical lignins. 1. Chemical analyses. Holzforschung. 1981;35(2):51–57.
[21] Freudenberg K. The constitution and biosynthesis of lignin. In: Freudenberg K, Neish AC, editors. Constitution and Biosynthesis of Lignin, Molecular Biology, Biochemistry and Biophysics. vol. 2. Berlin: Springer-Verlag; 1968. p. 45–122.

[22] Monteil-Rivera F, Phuong M, Ye M, Halasz A, Hawari J. Isolation and characterization of herbaceous lignins for applications in biomaterials. Ind Crops Prod. 2013;41(1):356–364.

[23] Sakakibara A, Sano Y. Chemistry of lignin. In: Hon DNS, editor. Wood and Cellulosic Chemistry. New York: Marcel Dekker; 2001. p. 109–174.

[24] Goring DAI. Polymer properties of lignin and lignin derivatives. In: Sarkanen KV, Ludwig CH, editors. Lignins: Occurrence, Formation, Structure and Reactions. New York: Wiley-Interscience; 1971. p. 695–768.

[25] Hatakeyama H, Hatakeyama T. Lignin structure, properties, and applications. Adv Polym Sci. 2010;232(1):1–63.

[26] Ben-Ghedalia D, Yosef E. Effect of isolation procedure on molecular weight distribution of wheat straw lignins. J Agric Food Chem. 1994;42(3):649–652.

[27] Hatakeyama H, Iwashita K, Meshitsuka G, Nakano J. Effect of molecular weight on the glass transition temperature of lignin. Mokuzai Gakkaishi. 1975;21(11):618–623.

[28] Mörck R, Yoshida H, Kringstad KP, Hatakeyama H. Fractionation of Kraft lignin by successive extraction with organic solvents. I. Functional groups, carbon-13 NMR-spectra and molecular weight distributions. Holzforschung. 1986;40(Suppl.):51–60.

[29] Forss K, Kokkonen R, Sågfors PE. Determination of the molar mass distribution of lignins by gel permeation chromatography. In: Glasser WG, Sarkanen S, editors. Lignin. Washington, DC: American Chemical Society; 1989. p. 124–133.

[30] Gellerstedt G. Gel permeation chromatography. In: Lin SY, Dence CW, editors. Methods in Lignin Chemistry, Springer Series in Wood Science. Springer Berlin Heidelberg; 1992. p. 487–497.

[31] Porath J, Flodin P. Gel filtration: a method for desalting and group separation. Nature. 1959;183(4676):1657–1659.

[32] Kolpak FJ, Cietek DJ, Fookes W, Cael JJ. Analysis of lignins from spent alkaline pulping liquors by gel permeation chromatography/low-angle laser light scattering (GPC/LALLS). J Appl Polym Sci: Appl Polym Symp. 1983;37:491–507.

[33] Meister JJ, Richards EG. Molecular weight distribution of aspen lignins estimated by universal calibration. In: Glasser WG, Sarkanen S, editors. Lignin. Washington, DC: American Chemical Society; 1989. p. 82–99.

[34] Siochi EJ, Haney MA, Mahn W, Ward TC. Molecular weight determination of hydroxypropylated lignins. ACS Symp Ser. 1989;397(Lignin):100–108.

[35] Jensen W, Fremer KE, Forss K. Separation of the components in spent sulfite liquor. Fractionation of the aromatic components by ion exclusion and gel filtration. Tappi. 1962;45(4676):122–127.

[36] Forss K, Fremer KE. The dissolution of wood components under different conditions of sulfite pulping. Tappi. 1964;47(8):485–493.

[37] Alekseev A, Forss K, Johanson M, Stenlund B. On the polymerization of spruce lignosulfonates in acid bisulfite pulping. Pap Puu. 1978;60(4a):195–198.

[38] Forss KG, Stenlund BG, Sagfors PE. Determination of the molecular-weight distribution of lignosulfonates and Kraft lignin. Appl Polym Symp. 1976;3(28):1185–1194.

[39] Forss K, Janson J, Sagfors PE. Influence of anthraquinone and sulfide on the alkaline degradation of the lignin macromolecule. Pap Puu. 1984;66(2):77–79.

[40] Connors WJ. Gel chromatography of lignins, lignin model compounds, and polystyrenes using sephadex LH-60. Holzforschung. 1978;32(4):145–147.

[41] Connors WJ, Lorenz LF, Kirk TK. Chromatography separation of lignin model by molecular weight using sephadex LH-20. Holzforschung. 1978;32(3):106–108.

[42] Connors WJ, Sarkanen S, McCarthy JL. Gel chromatography and association complexes of lignin. Holzforschung. 1980;34(3):80–85.

[43] Dolk M, Pla F, Yan JF, McCarthy JL. Lignin. 22. Macromolecular characteristics of alkali lignin from western hemlock wood. Macromolecules. 1986;19(5):1464–1470.

[44] Kondo R, McCarthy JL. Incremental delignification of hemlock wood and characterization of lignin products. Holzforschung. 1985;39(43):231–234.
[45] Sarkanen S, Teller DC, Abramowski E, McCarthy JL. Lignin. 19. Kraft lignin component conformation and associated complex configuration in aqueous alkaline solution. Macromolecules. 1982;15(4):1098–1104.
[46] Yan JF, Pla F, Kondo R, Dolk M, McCarthy JL. Lignin. 21. Depolymerization by bond cleavage reactions and degelation. Macromolecules. 1984;17(10):2137–2142.
[47] Lindner A, Wegener G. Characterization of lignins from organosolv pulping according to the organocell process. Part 3. Permanganate oxidation and thioacidolysis. J Wood Chem Technol. 1990;10(3):331–350.
[48] Lebo SE, Braten SM, Fredheim GE, Lutnaes BF, Lauten RA, Myrvold BO, et al. Characterization of lignocellulosic materials. In: Hu TQ, editor. Characterization of Lignocellulosic Materials. Blackwell Publishing Ltd.; 2008. p. 188–205.
[49] Glasser WG, Barnett CA, Muller PC, Sarkanen KV. The chemistry of several novel bioconversion lignins. J Agric Food Chem. 1983;31(5):921–930.
[50] Glasser WG, Dave V, Frazier CE. Molecular weight distribution of (semi-) commercial lignin derivatives. J Wood Chem Technol. 1993;13(4):545–559.
[51] Sjöström E. Wood Chemistry: Fundamentals and Applications. 2nd ed. Academic Press; 1993.
[52] Striegel A, Yau WW, Kirkland JJ, Bly DD. Modern Size-Exclusion Liquid Chromatography: Practice of Gel Permeation and Gel Filtration Chromatography. John Wiley & Sons, Inc.; 2009.
[53] Pla F. Light scattering. In: Lin SY, Dence CW, editors. Methods in Lignin Chemistry, Springer Series in Wood Science. Springer Berlin Heidelberg; 1992. p. 498–508.
[54] Dong D, Fricke AL. Effects of multiple pulping variables on the molecular weight and molecular weight distribution of Kraft lignin. J Wood Chem Technol. 1995;15(3):369–393.
[55] Gupta PR, Goring DAI. Physicochemical studies of alkali lignins: III. Size and shape of the macromolecule. Can J Chem. 1960;38(2):270–279.
[56] Lindberg JJ, Tylli H, Majani C. Notes on the molecular weight and the fractionation of lignins with organic solvents. Pap Puu. 1964;46(9):521–526.
[57] Forss K, Schott O, Stenlund B. Light absorption and fluorescence of lignosulfonates dissolved in water and dimethyl sulfoxide. Pap Puu. 1967;49(8):525–530.
[58] Forss K, Stenlund B. Molecular weights of lignosulfonates fractionated by gel chromatography. Pap Puu. 1969;51(1):93–95, 97–105.
[59] Moacanin J, Felicetta VF, Haller W, McCarthy JL. Lignin. VI. Molecular weights of lignin sulfonates by light scattering. J Am Chem Soc. 1955;77(13):3470–3475.
[60] Plastre D. Determination des proprietes macromoleculaires de la lignine, mise a point de techniques d'analyse. Grenoble: University of Grenoble; 1983.
[61] Kim H. The Effect of Kraft Pulping Conditions on Molecular Weights of Kraft Lignins. Orono, MN: University of Maine; 1985.
[62] Pla F. Etude de la estructure macromeleculaire des lignines. Grenoble: University of Grenoble, France; 1980.
[63] Pla F, Dolk M, Yan JF, McCarthy JL. Lignin. 23. Macromolecular characteristics of alkali lignin and organosolv lignin from black cottonwood. Macromolecules. 1986;19(5): 1471–1477.
[64] Pla F, Froment P, Capitini R, Tistchenko AM, Robert A. Study of extraction lignin by light diffusion with a laser source. Cellul Chem Technol. 1977;11(6):711–718.
[65] Froment P, Pla F. Determinations of average molecular weights and molecular weight distributions of lignin. In: Glasser WG, Sarkanen S, editors. Lignin. Washington, DC: American Chemical Society; 1989. p. 134–143.
[66] Pla F. Vapor pressure osmometry. In: Lin SY, Dence CW, editors. Methods in Lignin Chemistry, Springer Series in Wood Science. Springer Berlin Heidelberg; 1992. p. 509–517.

[67] Pla F, Robert A. Hydrodynamic behavior of an extracted lignin. Cellul Chem Technol. 1974;8(1):3–10.
[68] Froment P, Robert A. Importance of standardization in the determination of the average molecular weights of lignin by gel-permeation chromatography. Cellul Chem Technol. 1977;11(6):691–696.
[69] Matron J, Marton T. Molecular weight of Kraft lignin. Tappi. 1964;47(8):471–476.
[70] Siochi EJ, Ward TC, Haney MA, Mahn B. The absolute molecular weight distribution of hydroxypropylated lignins. Macromolecules. 1990;23(5):1420–1429.
[71] Meister JJ, Richards EG. Determination of a polymer's molecular weight distribution by analytical ultracentrifugation. In: Glasser WG, Sarkanen S, editors. Lignin. Washington, DC: American Chemical Society; 1989. p. 58–81.
[72] Forss KG, Fuhrmann AGMU. Adhesive for the manufacture of plywood, particle boards, fiber boards and similar products [Patent]; 1978. US4105606A.
[73] Hill M, Fricke AL. Ultrafiltration studies on a Kraft black liquor. Tappi J. 1984;67(6):100–103.
[74] Collins JW, Torkelson JM, Webb AA. Some viscosity properties of lignosulfonates isolated by ultrafiltration. J Agric Food Chem. 1977;25(4):743–746.
[75] Robert DR, Bardet M, Gellerstedt G, Lindfors EL. Structural changes in lignin during Kraft cooking. Part 3. On the structure of dissolved lignins. J Wood Chem Technol. 1984;4(3):239–263.
[76] Robert DR, Brunow G. Quantitative estimation of hydroxyl groups in milled wood lignin from spruce and in a dehydrogenation polymer from coniferyl alcohol using carbon-13 NMR spectroscopy. Holzforschung. 1984;38(2):85–90.
[77] Månsson P. Quantitative determination of phenolic and total hydroxyl groups in lignins. Holzforschung. 1983;37(3):143–146.
[78] Adler E, Ellmer L, Gjertsen P. Coniferyl aldehyde groups in wood and in isolated lignin preparations. Acta Chem Scand. 1948;2:839–840.
[79] Beck HD, Adle E. Oxidation with quinones. Acta Chem Scand. 1961;15:218–219.
[80] Lindgren BO, Mikawa H. The presence of cinnamyl alcohol groups in lignin. Acta Chem Scand. 1957;11:826–835.
[81] Adler E, Marton J, Smith-Kielland I, Sömme R, Stenhagen E, Palmstierna H. Zur kenntnis der carbonylgruppen im lignin. I. Acta Chem Scand. 1959;13:75–96.
[82] Aoyama M, Sakakibara A. Isolation of a new phenylcoumaran compound from hydrolysis products of hardwood lignin. Mokuzai Gakkaishi. 1976;22(10):591–592.
[83] Adler E, Hernestam S. Estimation of phenolic hydroxyl groups in lignin. I. Periodate oxidation of guaiacol compounds. Acta Chem Scand. 1955;9(2):319–334.
[84] Freudenberg K, Belz W, Niemann C. Lignin and cellulose. X. The aromatic nature of lignin. Ber Dtsch Chem Ges B. 1929;62:1554–1561.
[85] Zeisel S. Über ein verfahren zum quantitativen nachweise von methoxyl. Monatsh Chem. 1885;6(1):989–997.
[86] Zeisel S, Fanto R. Neues verfahren zur bestimmung des glycerins. Z Landwirtsch Versuchswes Dtsch-Oesterr. 1902;5:729–745.
[87] Zeisel S, Fanto R. Bestimmung des rohglycerins im weine mittelst der jodidmethode. Z Anal Chem. 1903;42(9-10):549–578.
[88] Roth H. Analytik der alkoholischen und phenolischen hydroxylgruppe. In: Müller E, Bayer O, Meerwein H, Ziegler K, editors. Methoden der organischen Chemie (Houben-Weyl). vol. 4. Stuttgart, Federal Republic of Germany: George Thieme; 1954. p. 330–380.
[89] Viebock F, Brecher C. New method for the volumetric determination of methoxy- and ethoxy-groups. II. Microanalysis. Ber Dtsch Chem Ges B. 1930;63:3207–3210.
[90] Viebock F, Schwappach A. New method for the volumetric determination of the methoxyl and ethoxyl groups. Ber Dtsch Chem Ges B. 1930;63:2818–2823.

[91] Chen CL. Determination of methoxyl groups. In: Lin SY, Dence CW, editors. Methods in Lignin Chemistry, Springer Series in Wood Science. Springer Berlin Heidelberg; 1992. p. 465–472.

[92] Klason P. Beiträge zur Kenntnis der chemischen Zusammensetzung des Fichtenholzes, Schriften des Vereins der Zellstoff- und Papierchemiker und -Ingenieure. Borntraeger; 1911.

[93] König J, Rump E. Chemie und struktur der pflanzen-zellmembran. Z Unters Nahr- Genussm Gebrauchsgegenstaende. 1914;28(4):177–222.

[94] Sherrard EC, Harris EE. Factors influencing properties of isolated wood lignin. Ind Eng Chem. 1932;24(1):103–106.

[95] Hägglund E. Lignin. Ark Kemi, Mineral Geol. 1918;7:1–20.

[96] Fischer F, Schrader H. The dry distillation of lignin and cellulose. Gesammelte Abh Kennt Kohle. 1922;6:106–116.

[97] Kurschner K. In: Enke F, editor. Zur Chemie der Ligninkorper. Stuttgart; 1925.

[98] Phillips M. The chemistry of lignin. III. The destructive distillation of lignin from corn cobs. J Am Chem Soc. 1929;51(8):2420–2426.

[99] Aulin-Erdtman G. Spectrographic contributions to lignin chemistry. V. Phenolic groups in spruce lignin. Sven Papperstidn. 1954;57:745–760.

[100] Wexler AS. Characterization of lignosulfonates by ultraviolet spectrometry. Direct and difference spectrograms. Anal Chem. 1964;36(1):213–221.

[101] Goldschmid O. Determination of the phenolic hydroxyl content of lignin preparations by ultraviolet spectrophotometry. Anal Chem. 1954;26:1421–1423.

[102] Hachihama Y, Shinra K, Kyogoku Y. Waste sulfite liquor. I. Conductometric titration of digesting liquor in the sulfite method with the addition of organic acids or sugars. II. Conductometric titration of lignosulfonic acid and waste liquor. Kogyo Kagaku Zasshi. 1944;47:209–215.

[103] Mikawa H, Sato K, Takasaki C, Ebisawa K. The cooking mechanism of wood. XII. The nature of the weakly acidic group of lignosulfonic acid. Bull Chem Soc Jpn. 1955;28:653–660.

[104] Freudenberg K, Dall K. Phenol groups in lignin. Naturwissenschaften. 1955;42:606–607.

[105] Sarkanen K, Schuerch C. Conductometric determination of phenolic groups in mixtures such as isolated lignins. Anal Chem. 1955;27:1245–1250.

[106] Pobiner H. Improved inflection points in the non-aqueous potentiometric titration of acid functionalities in lignin chemicals by using internal standardization and ion exchange. Anal Chim Acta. 1983;155(C):57–65.

[107] Ludwig CH, Nist BJ, McCarthy JL. Lignin. XIII. The high resolution nuclear magnetic resonance spectroscopy of protons in acetylated lignins. J Am Chem Soc. 1964;86(6):1196–1202.

[108] Li S, Lundquist K. A new method for the analysis of phenolic groups in lignins by proton NMR spectrometry. Nord Pulp Pap Res J. 1994;9(3):191–195.

[109] Hemmingson JA, Dekker RFH. CP/MAS ^{13}C NMR study of residual lignin structure in autohydrolysis-exploded woods and bagasse. J Wood Chem Technol. 1987;7(2):229–244.

[110] Leary GJ, Lloyd JA, Morgan KR. A ^{13}C CP/MAS NMR study of residual lignin in Kraft pulps. Holzforschung. 1988;42(3):199–202.

[111] Björkman A, Person B. Finely divided wood. IV. Some reactions of the lignin extracted by neutral solvents from *Picea abies*. Sven Papperstidn. 1957;60:285–292.

[112] Lai YZ. Determination of phenolic hydroxyl groups. In: Lin SY, Dence CW, editors. Methods in Lignin Chemistry, Springer Series in Wood Science. Springer Berlin Heidelberg; 1992. p. 423–434.

[113] Adler E, Hernestam S, Wallden I. Estimation of phenolic hydroxyl groups in lignin. Sven Papperstidn. 1958;61:641–647.

[114] Olcay A. Determination of free phenolic hydroxyl content of lignin. Holzforschung. 1970;24(5):172–175.

[115] Gellerstedt G, Gustafsson K. Structural changes in lignin during Kraft cooking. Part 5. Analysis of dissolved lignin by oxidative degradation. J Wood Chem Technol. 1987;7(1):65–80.
[116] Gellerstedt G, Lindfors EL. Structural changes in lignin during Kraft pulping. J Wood Chem Technol. 1984;38(3):151–158.
[117] Gellerstedt G, Gustafsson K, Northey RA. Structural changes in lignin during kraft cooking. Part 8. Birch lignins. Nord Pulp Pap Res J. 1988;3(2):87–94.
[118] Lapierre C, Rolando C. Thioacidolyses of pre-methylated lignin samples from pine compression and poplar woods. Holzforschung. 1988;42(1):1–4.
[119] Yang JM, Goring DAI. The phenolic hydroxyl content of lignin in spruce wood. Can J Chem. 1980;58(23):2411–2414.
[120] Lai YZ, Guo XP, Situ W. Estimation of phenolic hydroxyl groups in wood by a periodate oxidation method. J Wood Chem Technol. 1990;10(3):365–377.
[121] Lindner A, Wegener G. Characterization of lignins from organosolv pulping according to the Organocell process. Part 1. Elemental analysis, nonlignin portions and functional groups. J Wood Chem Technol. 1988;8(3):323–340.
[122] Chang HM, Cowling EB, Brown W. Comparative studies on cellulolytic enzyme lignin and milled wood lignin of sweetgum and spruce. Holzforschung. 1975;29(5):153–159.
[123] Gellerstedt G, Lindfors EL. Structural changes in lignin during kraft cooking. Part 4. Phenolic hydroxyl groups in wood and kraft pulps. Sven Papperstidn. 1984;87(15):R115–R118.
[124] Chen CL. Determination of total and aliphatic hydroxyl groups. In: Lin S, Dence C, editors. Methods in Lignin Chemistry, Springer Series in Wood Science. Springer Berlin Heidelberg; 1992. p. 409–422.
[125] Adler E, Yllner S. Synthesis and reactions of α-(3-methoxy-4-hydroxyphenyl)glycerol (guaiacylglycerol). II. Synthesis. Acta Chem Scand. 1953;7:570–581.
[126] Lundquist K, Lundgren R. Acid degradation of lignin. VII. Cleavage of ether bonds. Acta Chem Scand. 1972;26(5):2005–2023.
[127] Gierer J. Die reaktion von chinonmonochlorimid mit lignin. I. Spezifität der reaktion auf p-oxybenzylalkoholgruppen und deren bestimmung in verschiedenen ligninpräparaten. Acta Chem Scand. 1954;8:1319–1331.
[128] Gierer J. Die reaktion von chinonmonochlorimid mit lignin. II. Isolierung und identifizierung der gebildeten farbstoffe. Chem Ber. 1956;89(2):257–262.
[129] Adler E, Becker HD, Ishihara T, Stamvik A. The benzyl alcohol groups in spruce lignin. Holzforschung. 1966;20(1):3–11.
[130] Ludwig CH, Nist BJ, McCarthy JL. Lignin. XIII. High resolution nuclear magnetic resonance spectroscopy of protons in acetylated lignins. J Am Chem Soc. 1964;86(6):1196–1202.
[131] Lai YZ, Sarkanen KV. Isolation and structural studies. In: Sarkanen KV, Ludwig CH, editors. Lignins: Occurrence, Formation, Structure and Reactions. New York: Wiley-Interscience; 1971. p. 165–240.
[132] Klemola A. Investigations of birchwood (*Betula pubescens*) lignin degraded by steam hydrolysis. Suom Kemistil A. 1968;41(7-8):166–180.
[133] Marton J, Adler E. Carbonyl groups in lignin. III. Mild catalytic hydrogenation of björkman lignin. Acta Chem Scand. 1961;15:370–383.
[134] Aulin-Erdtman G, Hegbom L. Spectrographic contributions to lignin chemistry. VIII. δε-studies on Brauns' 'native lignins' from coniferous woods. Sven Papperstidn. 1958;61:187–210.
[135] Nahum LS. Estimation of double bond content in lignin from the results of the oxo reaction of wood and lignin model compounds. Tappi. 1969;52(4):712–714.
[136] Redinger L. Alkali lignin, its condensation products with phenols, and preparation of curable resins. Monatsber Dtsch Akad Wiss Berl. 1961;3(10):571–578.
[137] Falkehag SI, Marton J, Adler E. Chromophores in kraft lignin. Adv Chem Ser. 1966;59:75–89.

[138] Lebo SE, Lonsky WFW, McDonough TJ, Medvecz PJ, Dimmel DR. The occurrence and light-induced formation of ortho-quinonoid lignin structures in white spruce refiner mechanical pulp. J Pulp Pap Sci. 1990;16(5):J139–J143.

[139] Dence CW. Determination of ethylenic groups. In: Lin SY, Dence CW, editors. Methods in Lignin Chemistry, Springer Series in Wood Science. Springer Berlin Heidelberg; 1992. p. 435–445.

[140] Adler E, Björkquist J, Häggroth S. Über die ursache der farbreaktionen des holzes. Acta Chem Scand. 1948;2:93–94.

[141] Adler E. Sulfite pulping properties of spruce wood from unpeeled, floated logs. I. Catechol tannins in the surface layer of the sapwood. Sven Papperstidn. 1951;54:445–450.

[142] Pew JC. Structural aspects of the color reaction of lignin with phenols. J Am Chem Soc. 1951;73(4):1678–1685.

[143] Adler E, Marton J. Carbonyl groups in lignin. II. Catalytic hydrogenation of model compounds containing aryl carbinol, aryl carbinol ether, ethylene, and carbonyl groups. Acta Chem Scand. 1961;15:357–369.

[144] Chen CL. Determination of carbonyl groups. In: Lin SY, Dence CW, editors. Methods in Lignin Chemistry, Springer Series in Wood Science. Springer Berlin Heidelberg; 1992. p. 446–457.

[145] Schubert WJ, Nord FF. Investigations on lignin and lignification. I. Studies on softwood lignin. J Am Chem Soc. 1950;72(2):977–981.

[146] Adler E, Gierer J. The alkylation of lignin with alcoholic hydrochloric acid. Acta Chem Scand. 1955;9:84–93.

[147] Gierer J, Söderberg S, Smith-Kielland I, Sömme R, Stenhagen E, Palmstierna H. Über die carbonylgruppen des lignins. Acta Chem Scand. 1959;13:127–137.

[148] Gierer J, Lenz B. Reactions of lignin during sulfate cooking. VI. Formation of 1,2-glycol groups in milled wood lignin on treatment with 2n NaOH at 170°. Sven Papperstidn. 1965;68(9):334–338.

[149] Marton J, Adler E, Persson KI, Dam H, Sjöberg B, Toft J. Carbonyl groups in lignin. IV. Infrared absorption studies and examination of the volumetric borohydride method. Acta Chem Scand. 1961;15:384–392.

[150] Ekman KH, Lindberg JJ. Origin of the infrared bands in the 1720 cm^{-1} region in lignins. Pap Puu. 1960;42(1):21–22.

[151] Chum HL, Ratcliff M, Schroeder HA, Sopher DW. Electrochemistry of biomass-derived materials. I. Characterization, fractionation, and reductive electrolysis of ethanol-extracted explosively-depressurized aspen lignin. J Wood Chem Technol. 1984;4(4):505–352.

[152] Katz S, Beatson RP, Scallan AM. The determination of strong and weak acidic groups in sulfite pulps. Sven Papperstidn. 1984;87(6):R48–R53.

[153] Gaslini F, Nahum LZ. Conductometric titration of alkali lignins. Sven Papperstidn. 1959;62:520–523.

[154] Crozier TE, Johnson DC, Thompson NS. Changes in a southern pine dioxane lignin on oxidation with oxygen in sodium carbonate media. Tappi. 1979;62(9):107–111.

[155] Scalbert A, Monties B. Comparison of wheat straw lignin preparations. II. Straw lignin solubilization in alkali. Holzforschung. 1986;40(4):249–254.

[156] Ekman K, Enkvist T. Some determinations of weak and strong acidic groups in various lignin preparations and pulps. Pap Puu. 1955;37:369–377, 382–.

[157] Kirk TK, Chang HM. Decomposition of lignin by white-rot fungi. II. Characterization of heavily degraded lignins from decayed spruce. Holzforschung. 1975;29(2):56–64.

[158] James AN, Tice PA. The presence of carboxyl groups in lignosulfonate preparations. Tappi. 1965;48(4):239–244.

[159] Marton J, Adler E. Reactions of lignin with methanolic hydrochloric acid. A discussion of some structural questions. Tappi. 1963;46:92–98.
[160] Lindsey JB, Tollens B. Wood sulphite-liquor and lignin. Justus Liebigs Ann Chem. 1892;267(2–3):341–366.
[161] Öster R, Kringstad KP, Hirose S, Hatakeyama H. Oxidative sulfonation of Kraft lignin. Nord Pulp Pap Res J. 1988;3(2):68–74.
[162] Samuelson O, Westlin A. A contribution to the chemistry of the sulphite process. Part II. Sven Papperstidn. 1948;51:438–444.
[163] Beatson RP. Determination of sulfonate groups and total sulfur. In: Lin SY, Dence CW, editors. Methods in Lignin Chemistry, Springer Series in Wood Science. Springer Berlin Heidelberg; 1992. p. 473–484.
[164] Rodrigues Pinto PC, Borges da Silva EA, Rodrigues AE. Insights into oxidative conversion of lignin to high-added-value phenolic aldehydes. Ind Eng Chem Res. 2011;50(2):741–748.
[165] Freudenberg K, Harkin JM, Werner HK. The existence of benzyl aryl ethers in lignin. Chem Ber. 1964;97(3):909–920.
[166] Erickson M, Larsson S, Miksche GE, Wiehager AC, Lindgren BO, Swahn CG. Gaschromatographische analyse von ligninoxydationsprodukten. VIII. Zur struktur des lignins der fichte. Acta Chem Scand. 1973;27:903–914.
[167] Lundquist K, Miksche GE, Ericsson L, Berndtson L. Presence of glyceraldehyde-2-aryl ether structures in lignin. Tetrahedron Lett. 1967;(46):4587–4591.
[168] Adler E, Miksche GE, Johansson B. Benzyl aryl ether linkage in lignin. I. Liberation of phenolic hydroxyl in lignin preparations by cleavage of easily hydrolyzable alkyl aryl ether structures. Holzforschung. 1968;22(6):171–174.
[169] Lundquist K, Paterson A, Ramsey L. NMR studies of lignins. 6. Interpretation of the proton NMR spectrum of acetylated spruce lignin in a deuterioacetone solution. Acta Chem Scand, Ser B. 1983;37(8):734–736.
[170] Adler E, Pepper JM, Eriksoo E. Action of mineral acid on lignin and model substances of guaiacylglycerol β-aryl ether type. Ind Eng Chem. 1957;49:1391–1392.
[171] Adler E, Delin S, Lundquist K. Phenylcoumaran elements in spruce lignin. Acta Chem Scand. 1959;13:2149–2150.
[172] Sarkanen KW. Precursors and their polymerization. In: Sarkanen KV, Ludwig CH, editors. Lignins: Occurrence, Formation, Structure and Reactions. New York: Wiley-Interscience; 1971. p. 95–163.
[173] Lundquist K, Miksche GE. A new linkage principle for guaiacylpropane units in spruce lignin. Tetrahedron Lett. 1965;(25):2131–2136.
[174] Ogiyama K, Kondo T. Pinoresinol type structural units in lignin molecules. iv. yields of the dilactone from enzymic dehydrogenation of lignins. Mokuzai Gakkaishi. 1968;14(8):416–420.
[175] Nimz H, Das K. Low molecular weight degradation products of lignin. II. Dimeric phenolic degradation products obtained by degradation of beech lignin with thioacetic acid. Chem Ber. 1971;104(8):2359–2380.
[176] Lundquist K. NMR studies of lignins. 4. Investigation of spruce lignin by proton NMR spectroscopy. Acta Chem Scand, Ser B. 1980;34(1):21–26.
[177] Lundquist K. NMR studies of lignins. 2. Interpretation of the proton NMR spectrum of acetylated birch lignin. Acta Chem Scand, Ser B. 1979;33(1):27–30.
[178] Freudenberg K, Chen CL, Harkin JM, Nimz H, Renner H. Lignin. Chem Commun. 1965;(11):224–225.
[179] Sano Y, Sakakibara A. Hydrolysis of lignin with dioxane and water. XIV. Isolation of a new trimeric compound. Mokuzai Gakkaishi. 1975;21(8):461–465.

[180] Nimz H. Mild hydrolysis of beech lignin. ii. isolation of a 1,2-diarylpropane derivative and its conversion to a hydroxystilbene. Chem Ber. 1965;98(10):3160–3164.
[181] Adler E, Lundquist K. Estimation of uncondensed phenolic units in spruce lignin. Acta Chem Scand. 1961;15:223–224.
[182] Morohoshi N, Sakakibara A. Chemical composition of reaction wood. I. Mokuzai Gakkaishi. 1971;17(9):393–399.
[183] Freudenberg K, Chen CL. Oxidation products of pine lignin. Chem Ber. 1967;100(11): 3683–3688.
[184] Larsson S, Miksche GE, Sood MS, Nielsen BE, Ljunggren H, Ehrenberg L. Gaschromatographische analyse von ligninoxydationsprodukten. IV. Zur struktur des lignins der birke. Acta Chem Scand. 1971;25:647–662.
[185] Karhunen P, Rummakko P, Sipilä J, Brunow G, Kilpeläinen I. Dibenzodioxocins; a novel type of linkage in softwood lignins. Tetrahedron Lett. 1995;36(1):169–170.
[186] Karhunen P, Rummakko P, Sipilä J, Brunow G, Kilpeläinen I. The formation of dibenzodioxocin structures by oxidative coupling. A model reaction for lignin biosynthesis. Tetrahedron Lett. 1995;36(25):4501–4504.
[187] Karhunen P, Mikkola J, Pajunen A, Brunow G. The behaviour of dibenzodioxocin structures in lignin during alkaline pulping processes. Nord Pulp Pap Res J. 1999;14(2):123–128.
[188] Argyropoulos DS, Jurasek L, Kristofova L, Xia Z, Sun Y, Palus E. Abundance and reactivity of dibenzodioxocins in softwood lignin. J Agric Food Chem. 2002;50(4):658–666.
[189] Lin SY. Ultraviolet photospectrometry. In: Lin SY, Dence CW, editors. Methods in Lignin Chemistry, Springer Series in Wood Science. Springer Berlin Heidelberg; 1992. p. 217–232.
[190] Ibarra D, del Rio JC, Gutierrez A, Rodriguez IM, Romero J, Martinez MJ, et al. Chemical characterization of residual lignins from eucalypt paper pulps. J Anal Appl Pyrolysis. 2005;74(1-2):116–122.
[191] Tsuchikawa S. A review of recent near infrared research for wood and paper. Appl Spectrosc Rev. 2007;42(1):43–71.
[192] Rencoret J, Marques G, Gutierrez A, Ibarra D, Li J, Gellerstedt G, et al. Structural characterization of milled wood lignins from different eucalypt species. Holzforschung. 2008;62(5):514–526.
[193] Hergert HL. Infrared spectra. In: Sarkanen KV, Ludwig CH, editors. Lignins: Occurrence, Formation, Structure and Reactions. New York: Wiley-Interscience; 1971. p. 267–297.
[194] Kolboe S, Ellefsen O. Infrared investigations of lignin. A discussion of some recent results. Tappi. 1962;45:163–166.
[195] Sarkanen KV, Chang HM, Allan GG. Species variation in lignins. III. Hardwood lignins. Tappi. 1967;50(12):587–590.
[196] Atalla RH, Agarwal UP, Bond JS. Raman spectroscopy. In: Lin SY, Dence CW, editors. Methods in Lignin Chemistry, Springer Series in Wood Science. Springer Berlin Heidelberg; 1992. p. 162–176.
[197] Pomar F, Merino F, Barceló AR. O–4–Linked coniferyl and sinapyl aldehydes in lignifying cell walls are the main targets of the Wiesner (phloroglucinol-HCl) reaction. Protoplasma. 2002;220(1–2):0017–0028.
[198] Goldschmid O. Ultraviolet spectra. In: Sarkanen KV, Ludwig CH, editors. Lignins: Occurrence, Formation, Structure and Reactions. New York: Wiley-Interscience; 1971. p. 241–266.
[199] Aulin-Erdtman G. Spectrographic contributions to lignin chemistry. II. Preliminary report. Sven Papperstidn. 1952;55:745–749.
[200] Aulin-Erdtman G. Spectrographic contributions to lignin chemistry. III. Investigations on model compounds. Sven Papperstidn. 1953;56:91–101.

[201] Aulin-Erdtman G, Hegbom L. Spectrographic contributions to lignin chemistry. VII. The ultraviolet absorption and ionization δε-curves of some phenols. Sven Papperstidn. 1957;60:671–681.

[202] Aulin-Erdtman G, Sanden R. Spectrographic contributions to lignin chemistry. IX. Absorption properties of some 4-hydroxyphenyl, guaiacyl, and 4-hydroxy-3,5-dimethoxyphenyl type model compounds for hardwood lignins. Acta Chem Scand. 1968;22(4):1187–209.

[203] Alibert G, Boudet A. Lignification of poplar. I. Demonstration of method for the determination and monomeric analysis of lignins. Physiol Veg. 1979;17(1):67–74.

[204] Janshekar H, Brown C, Fiechter A. Determination of biodegraded lignin by ultraviolet spectrophotometry. Anal Chim Acta. 1981;130(1):81–91.

[205] Steinitz YL. Microbial desulfonation of lignosulfonate - a new approach. Eur J Appl Microbiol Biotechnol. 1981;13(4):216–221.

[206] Gadda L. Delignification of Wood Fibre Cell Wall During Alkaline Pulping Processes. Åbo (Turku): Institute of Wood Chemistry and Pulp and Paper Technology; 1981.

[207] Wardrop AB. The phase of lignification in the differentiation of wood fibers. Tappi. 1957;40:225–243.

[208] Schöning AG, Johansson G. Absorptiometric determination of acid-soluble lignin in semi-chemical bisulfite pulps and in some woods and plants. Sven Papperstidn. 1965;68(18):607–613.

[209] Wegener G, Przyklenk M, Fengel D. Hexafluoropropanol as valuable solvent for lignin in UV and IR spectroscopy. Holzforschung. 1983;37(6):303–307.

[210] Fergus BJ, Goring DAI. Location of guaiacyl and syringyl lignins in birch xylem tissue. Holzforschung. 1970;24(4):113–117.

[211] Fergus BJ, Goring DAI. The distribution of lignin in birch wood as determined by ultraviolet microscopy. Holzforschung. 1970;24(4):118–124.

[212] Boutelje JB, Eriksson I. An UV-microscopy study of lignin in middle lamella fragments from fibers of mechanical pulp of spruce. Sven Papperstidn. 1982;85(6):R39–R42.

[213] Ludwig CH, Nist BJ, McCarthy JL. Lignin. XII. The high resolution nuclear magnetic resonance spectroscopy of protons in compounds related to lignin. J Am Chem Soc. 1964;86(6):1186–1196.

[214] Simionescu CI, Draganova R, Kusmanova D. NMR-spectroscopic studies of dioxane-lignin fractions. Cellul Chem Technol. 1981;15(4):455–464.

[215] Brunow G, Lundquist K. Comparison of a synthetic dehydrogenation polymer of coniferyl alcohol with milled wood lignin from spruce, using proton NMR spectroscopy. Pap Puu. 1980;62(11):669–670, 672.

[216] Ralph J, Ede RM, Robinson NP, Main L. Reactions of β-aryl lignin model quinone methides with anthrahydroquinone and anthranol. J Wood Chem Technol. 1987;7(2):133–160.

[217] Brunow G, Sipila J, Makela T. On the mechanism of formation of noncyclic benzyl ethers during lignin biosynthesis. Part 1. The reactivity of β-O-4-quinone methides with phenols and alcohols. Holzforschung. 1989;43(1):55–59.

[218] Gellerstedt G, Gierer J. Reactions of lignin during acidic sulfite pulping. Sven Papperstidn. 1971;74(5):117–127.

[219] Hauteville M, Lundquist K, Von Unge S. NMR studies of lignins. 7. Proton NMR spectroscopic investigation of the distribution of erythro and threo forms of β-O-4 structures in lignins. Acta Chem Scand, Ser B. 1986;40(1):31–35.

[220] Lundquist K, Stern K. Analysis of lignins by proton NMR spectroscopy. Nord Pulp Pap Res J. 1989;4(3):210–213.

[221] Ede RM, Brunow G, Simola LK, Lemmetyinen J. Two-dimensional proton-proton chemical shift correlation and J-resolved NMR studies on isolated and synthetic lignins. Holzforschung. 1990;44(2):95–101.

[222] Lüdemann HD, Nimz H. Carbon-13 nuclear magnetic resonance spectra of lignins. Biochem Biophys Res Commun. 1973;52(4):1162–1169.
[223] Lüdemann HD, Nimz H. ^{13}C-kernresonanzspektren von ligninen, 2. Buchen- und fichten-björkman-lignin. Makromol Chem. 1974;175(8):2409–2422.
[224] Nimz H, Mogharab I, Lüdemann HD. Carbon-13-NMR spectra of lignins. 3. Comparison of spruce lignin with synthetic lignin according to freudenberg. Makromol Chem. 1974;175(9):2563–2575.
[225] Nimz HH, Lüdemann HD. Carbon-13 NMR spectra of lignins. 6. Lignin and DHP acetates. Holzforschung. 1976;30(2):33–40.
[226] Landucci LL. Quantitative carbon-13 NMR characterization of lignin 1. A methodology for high precision. Holzforschung. 1985;39(6):355–359.
[227] Obst JR, Landucci LL. Quantitative carbon-13 NMR of lignins - methoxyl:aryl ratio. Holzforschung. 1986;40(Suppl.):87–92.
[228] Bardet M, Gagnaire D, Nardin R, Robert D, Vincendon M. Use of carbon-13 enriched wood for structural NMR investigation of wood and wood components, cellulose and lignin, in solid and in solution. Holzforschung. 1986;40(Suppl.):17–24.
[229] Bardet M, Foray MF, Robert D. Use of the DEPT pulse sequence to facilitate the carbon-13 NMR structural analysis of lignins. Makromol Chem. 1985;186(7):1495–1504.
[230] Ralph J. NMR of lignin model compounds: application of long-range C-H correlations through oxygen. Holzforschung. 1988;42(4):273–527.
[231] Ede RM, Kilpelaeinen I. Homo- and hetero-nuclear 2D NMR techniques: unambiguous structural probes for non-cyclic benzyl aryl ethers in soluble lignin samples. Res Chem Intermed. 1995;21(3-5):313–328.
[232] Ellwardt PC, Haider K, Ernst L. Investigation of microbial lignin degradation by carbon-13 NMR spectroscopy of specifically carbon-13-enriched DHP-lignin from coniferyl alcohol. Holzforschung. 1981;35(3):103–109.
[233] Lewis NG, Newman J, Just G, Ripmeister J. Determination of bonding patterns of carbon-13 specifically enriched dehydrogenatively polymerized lignin in solution and solid state. Macromolecules. 1987;20(8):1752–1756.
[234] Gagnaire D, Robert D. A polymer model of lignin (D.H.P.) carbon-13 selectively labeled at the benzylic positions: synthesis and NMR study. Makromol Chem. 1977;178(5):1477–1495.
[235] Wen JL, Sun SL, Xue BL, Sun RC. Recent advances in characterization of lignin polymer by solution-state nuclear magnetic resonance (NMR) methodology. Materials. 2013;6:359–391.
[236] Obst JR, Ralph J. Characterization of hardwood lignin: investigation of syringyl/guaiacyl composition by carbon-13 nuclear magnetic resonance spectroscopy. Holzforschung. 1983;37(6):297–302.
[237] Lapierre C, Monties B, Guittet E, Lallemand JY. The quantitative measurements in hardwood lignin carbon-13 NMR spectra. Holzforschung. 1985;39(6):367–368.
[238] Mansfield SD, Kim H, Lu F, Ralph J. Whole plant cell wall characterization using solution-state 2D NMR. Angew Chem Int Ed. 2012;7(9):1579–1589.
[239] Schaefer J, Stejskal EO. Carbon-13 nuclear magnetic resonance of polymers spinning at the magic angle. J Am Chem Soc. 1976;98(4):1031–1032.
[240] Zhang M, Maciel GE. Enhanced signal-to-noise ratios in the nuclear magnetic resonance analysis of solids, using large-sample magic-angle spinners. Anal Chem. 1990;62(6):633–638.
[241] Zhang M, Maciel GE. Large-sample carbon-13 MAS NMR spectroscopy of coal: relaxation and spin counting. Fuel. 1990;69(5):557–563.
[242] Hatfield GR, Maciel GE, Erbatur O, Erbatur G. Qualitative and quantitative analysis of solid lignin samples by carbon-13 nuclear magnetic resonance spectrometry. Anal Chem. 1987;59(1):172–179.

[243] Maciel GE, O'Donnell DJ, Ackerman JJH, Hawkins BH, Bartuska VJ. A carbon-13 NMR study of four lignins in the solid and solution states. Makromol Chem. 1981;182(8):2297–2304.

[244] Haw JF, Maciel GE, Biermann CJ. Carbon-13 nuclear magnetic resonance study of the rapid steam hydrolysis of red oak. Holzforschung. 1984;38(6):327–331.

[245] Haw JF, Maciel GE, Linden JC, Murphy VG. Nuclear magnetic resonance study of autohydrolyzed and organosolv-treated lodgepole pinewood using carbon-13 with cross polarization and magic-angle spinning. Holzforschung. 1985;39(2):99–107.

[246] Argyropoulos DS, Morin FG. Probing the macromolecular structure of wood and pulps with proton spin-lattice relaxation time measurements in the solid state. Wood Sci Technol. 1995;29(1):19–30.

[247] Leary GJ, Newman RH. Cross polarization/magic angle spinning nuclear magnetic resonance (CP/MAS NMR) spectroscopy. In: Lin SY, Dence CW, editors. Methods in Lignin Chemistry, Springer Series in Wood Science. Springer Berlin Heidelberg; 1992. p. 146–161.

[248] Kimura T, Kimura F, Argyropoulos DS, Gray DG. Carbon-13 CP/MAS NMR study of photodegraded stoneground wood pulp. Holzforschung. 1992;46(4):331–336.

[249] Argyropoulos DS. Phosphorus-31 NMR in wood chemistry: a review of recent progress. Res Chem Intermed. 1995;21(3-5):373–395.

[250] Argyropoulos DS, Heitner C, Schmidt JA. Observation of quinonoid groups during the light-induced yellowing of softwood mechanical pulp. Res Chem Intermed. 1995;21(3-5):263–274.

[251] Argyropoulos DS. Quantitative phosphorus-31 NMR analysis of lignins, a new tool for the lignin chemist. J Wood Chem Technol. 1994;14(1):45–63.

[252] Argyropoulos DS. Quantitative phosphorus-31 NMR analysis of six soluble lignins. J Wood Chem Technol. 1994;14(1):65–82.

[253] Archipov Y, Argyropoulos DS, Bolker H, Heitner C. Phosphorus-31 NMR spectroscopy in wood chemistry. phosphite derivatives of carbohydrates. Carbohydr Res. 1991;220:49–61.

[254] Granata A, Argyropoulos DS. 2-chloro-4,4,5,5-tetramethyl-1,3,2-dioxaphospholane, a reagent for the accurate determination of the uncondensed and condensed phenolic moieties in lignins. J Agric Food Chem. 1995;43(6):1538–1544.

[255] Jiang ZH, Argyropoulos DS, Granata A. Correlation analysis of ^{31}P NMR chemical shifts with substituent effects of phenols. Magn Reson Chem. 1995;33(5):375–382.

[256] Saake B, Argyropoulos DS, Beinhoff O, Faix O. A comparison of lignin polymer models (DHPs) and lignins by ^{31}P NMR spectroscopy. Phytochemistry. 1996;43(2):499–507.

[257] Barrelle M, Fernandes JC, Froment P, Lachenal D. An approach to the determination of functional groups in oxidized lignins by fluorine-19 NMR. J Wood Chem Technol. 1992;12(4):413–424.

[258] Barrelle M. A new method for the quantitative fluorine-19 NMR spectroscopic analysis of hydroxyl groups in lignins. Holzforschung. 1993;47(3):261–267.

[259] Ahvazi BC, Argyropoulos DS. Quantitative trifluoromethylation of carbonyl-containing lignin model compounds. J Fluorine Chem. 1996;78(2):195–198.

[260] Ahvazi BC, Argyropoulos DS. 19f nuclear magnetic resonance spectroscopy for the elucidation of carbonyl groups in lignins. 1. Model compounds. J Agric Food Chem. 1996;44(8):2167–2175.

[261] Sevillano RM, Mortha G, Barrelle M, Lachenal D. Fluorine-19 NMR spectroscopy for the quantitative analysis of carbonyl groups in lignins. Holzforschung. 2001;55(3):286–295.

[262] Brezny R, Schraml J. Silicon-29 NMR spectral studies of kraft lignin and related model compounds. Holzforschung. 1987;41(5):293–298.

[263] Nieminen MOJ, Pulkkinen E, Rahkamaa E. Determination of hydroxyl groups in kraft pine lignin by silicon-29 NMR spectroscopy. Holzforschung. 1989;43(5):303–307.

[264] Chen CL. Gas chromatography-mass spectrometry (GC-MS). In: Lin SY, Dence CW, editors. Methods in Lignin Chemistry, Springer Series in Wood Science. Springer Berlin Heidelberg; 1992. p. 527–548.
[265] Lewis NG. High performance liquid chromatography (HPLC). In: Lin SY, Dence CW, editors. Methods in Lignin Chemistry, Springer Series in Wood Science. Springer Berlin Heidelberg; 1992. p. 549–567.
[266] Saka S. Electron microscopy. In: Lin SY, Dence CW, editors. Methods in Lignin Chemistry, Springer Series in Wood Science. Springer Berlin Heidelberg; 1992. p. 133–145.
[267] Jamin MJ. Sur un refracteur differentiel pour la lumiere polarisee. C R Seances Acad Sci, Vie Acad. 1868;LXVII:814–816.
[268] Lebedeff AA. Polarization interferometer and its applications. Rev Opt, Theor Instrum. 1930;9:385–413.
[269] Donaldson LA. Interference microscopy. In: Lin SY, Dence CW, editors. Methods in Lignin Chemistry, Springer Series in Wood Science. Springer Berlin Heidelberg; 1992. p. 122–132.
[270] Hon DNS. Electron spin resonance (ESR) spectroscopy. In: Lin SY, Dence CW, editors. Methods in Lignin Chemistry, Springer Series in Wood Science. Springer Berlin Heidelberg; 1992. p. 274–286.
[271] Wan JKS, Depew MC. Applications of ESR and CIDEP to mechanistic studies of lignin chemistry. Res Chem Intermed. 1998;24(8):831–847.
[272] Kalyanaraman B. Radical intermediates during degradation of lignin model compounds and environmental pollutants: an electron spin resonance study. Xenobiotica. 1995;25(7):667–675.
[273] Hatakeyama H, Nakano J. Electron spin resonance studies on lignin and lignin model compounds. Cellul Chem Technol. 1970;4(3):281–291.
[274] Lourenço A, Gominho J, Marques AV, Pereira H. Comparison of Py-GC/FID and wet chemistry analysis for lignin determination in wood and pulps from *Eucalyptus globulus*. BioResources. 2013;8(2):2967–2980.

7

Chemical Characterization and Modification of Lignins

7.1 Introduction

The term "chemical modification" has been used to designate different processes by different authors over the years. The term may include both degradation and pulping reactions studied in Chapters 3, 5, and 6, such as those involving the formation of new derivatives of lignins in view of their potential applications (see Chapters 8 and 9) and/or study. It should be borne in mind that in one case or the other, the reaction of lignins with a chemical reagent (or simply with temperature) may cause depolymerization and repolymerization reactions, and therefore, can trigger significant changes in the lignin structure. Therefore, this chapter has been divided into separate sections for manageable reading in an otherwise unwieldy topic.

The conversion of lignin polymers into phenols of low M_w by degradation methods allows an analytical procedure to study the composition of lignins, providing useful information on the structure of the original polymers. Nevertheless, these degradation processes offer low yields, due mainly to very strong bonds in the chemical structure of lignin. At this point, lignins differ from other biopolymers such as nucleic acids, proteins, or polysaccharides, which contain weak bonds that can be broken either chemically or enzymatically by means of hydrolase action. Although the degradation methods have been criticized for their low effectiveness, and the potential formation of artifacts, the most complete description of the types and bonding patterns of lignins has been achieved thanks to degradation methods such as thioacidolysis, acidolysis, oxidation with permanganate or hydrogenolysis ones [1].

7.2 Characterization by Chemical Degradation Methods

The literature on lignin describes several different types of degradation methods, which produce mixtures of degradation products. The methods described in this section require the sample to be exposed to certain reagents and specified conditions that degrade the lignin in a prescribed manner, in accordance with documented reaction sequences. The products are then quantitatively analyzed by chromatographic techniques, yielding structural information on the building blocks of the lignin sample [2].

Such techniques suffer from being laborious, involving many steps and complex chemical manipulations, often subjecting the derived quantitative information to large errors and diminished reproducibility. Despite these limitations, these methods have offered significant advances to our knowledge of lignin structure and reactivity.

Among the first pioneering techniques, acidolysis [3, 4], thioacetolysis [5], and hydrogenolysis [6] played an undisputed role in our current knowledge of lignin structure.

Lignin chemical degradation methods can be classified according to the mechanism underlying the depolymerization of the lignin network, namely oxidative, solvolytic, or hydrogenolytic reactions. In addition to these chemical lignin-fragmentation procedures, analytical pyrolysis has been used to evaluate the lignin content and structure in lignocellulosic materials.

Today, a large number of chemical methods can be used to identify lignin on the basis of its pattern of degradation products. Such methods have also been used to gain information on the structure of different types of lignins. Table 7.1 summarizes some commonly used methods.

Probably, the most widely used degradation methods are oxidation with nitrobenzene and with cupric oxide in alkaline medium. Oxidation with nitrobenzene is quite effective, although all alkaline oxidations shorten the side chains of lignins without providing information on its functional groups nor on the type of bonds between subunits. The acidolysis method preserves the skeletons C_6–C_3, providing more information and less interference. However, the yield is relatively low due to side reactions. These problems are solved in the thioacidolysis method, which is a depolymerization process, under acid conditions, which act only to break the β-O-4 bonds. Without a doubt, the great limitation of the thioacidolysis method is its application to lignins with a low proportion of β-O-4 bonds, since the results can hardly be comparable to the polymer assembly. In these cases, the complementary use of the oxidation with the nitrobenzene method constitutes a very useful tool.

7.2.1 Oxidation with Nitrobenzene

The oxidation of lignin with nitrobenzene in an alkaline solution was introduced by Freudenberg et al. [7, 8, 22] in 1939, to find evidence for the aromatic nature of lignins, and was further optimized by Leopold et al. [9–11]. On nitrobenzene oxidation, normal softwoods and their lignins give rise to vanillin as the major product. Vanillin (**2**) was produced in a yield of about 25% on oxidation of spruce wood meal with nitrobenzene in 2 M NaOH solution at 160 °C for 3 h. In addition, insignificant amounts of other minor oxidations products were obtained [23]. Leopold et al. [11] studied these oxidation products from spruce wood in detail, also deducing the presence of p-hydroxyphenyl

Table 7.1 Chemistry methods for the analysis of lignin[a]

Method	Reaction	[Ref.]
Oxidation with nitrobenzene (alkaline)	Oxidative elimination of side chains	[7–11]
Oxidation with $KMnO_4$ or H_2O_2	Oxidative elimination of side chains	[12, 13]
Thioacetolysis	Cleavage of alkyl aryl ethers	[14]
Hydrogenolysis	Reductive cleavage of ethers	[6, 15]
Acidolysis	Hydrolytic cleavage of ethers	[4, 16]
Thioacidolysis	Hydrolytic cleavage of ethers	[17, 18]
Acetyl bromide–zinc/acetic anhydride	Reductive cleavage of ethers	[19, 20]

[a]See ref. [21, p. 211].

units in softwood lignin. Hardwood and their lignins yield vanillin (**2**) and syringaldehyde (**3**) as the major products [24, 25]. Grasses and their lignins afford *p*-hydroxybenzaldehyde (**1**), vanillin (**2**), and syringaldehyde (**3**) as the major products [26].

The mechanism consists of an alkaline two-transfer process as in the conversion of isoeugenol to vanillin (see Figure 7.1). In addition to aldehydes, many other phenylpropanoid-derived compounds were produced.

Figure 7.2 shows all the products identified in nitrobenzene and cupric oxide oxidation of lignins.

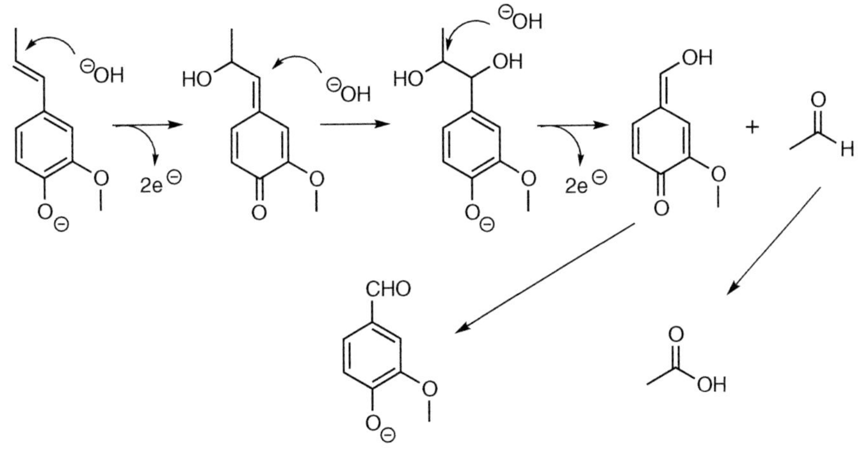

Figure 7.1 Mechanism of lignin oxidation meditated by nitrobenzene [27, p. 434]

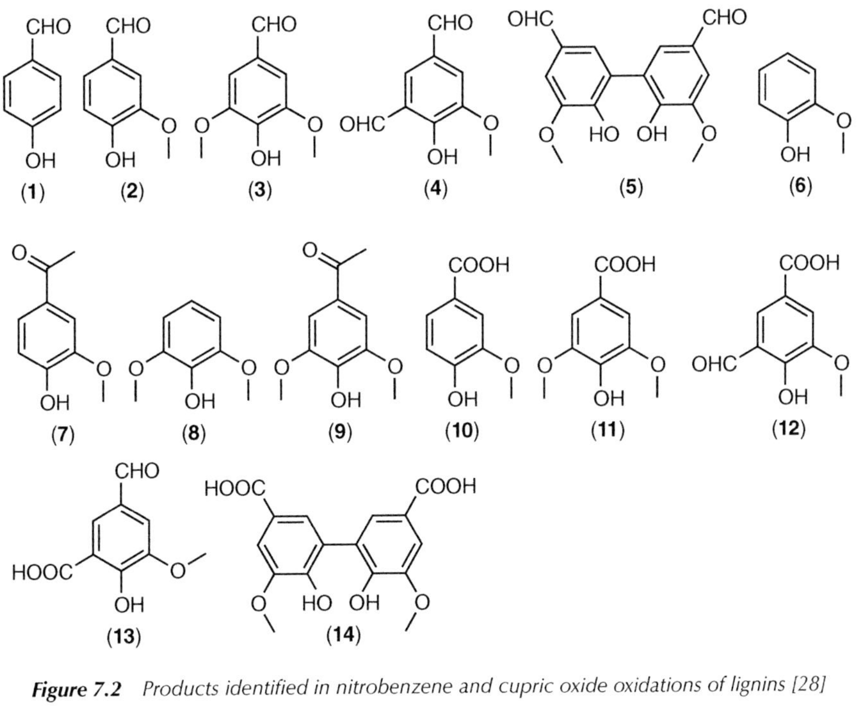

Figure 7.2 Products identified in nitrobenzene and cupric oxide oxidations of lignins [28]

Table 7.2 shows the yields and the compounds obtained in the nitrobenzene oxidation mixture of Norway spruce (*Picea abies*) independently obtained by Leopold *et al.* [11] and Pew [29].

The yield and molar ratio of phenolic aldehydes depend on the plant species being investigated. Thus, the nitrobenzene oxidation is not only relevant in terms of characterization of lignins, by providing information on the minimal quantities and the relative amounts of the uncondensed *p*-hydroxyphenyl-, guaiacyl-, and syringylpropane units present in a lignin, but also in terms of the taxonomy of vascular plants.

The qualitative and quantitative determination of oxidation products has been made by using gas chromatography (GC), gas chromatography-mass spectrometry (GC-MS), or high-performance liquid chromatography (HPLC). Several applications of alkaline nitrobenzene oxidation have been documented by Brauns and Brauns [31] and by Pearl [32, pp. 202–205]. Representative uses to which the procedure has been put include [28]: the botanical classification of plants [24, 25], the characterization of lignin in various morphological regions of wood tissue [33, 34], estimation of the proportion of uncondensed guaiacyl nuclei in softwood lignin [35], the monitoring of changes in lignin during plant growth [36], and the assessment of the effect on lignin of various chemical [37–40], and biological [41, 42] treatments.

Table 7.3 summarizes the yield of phenolic aldehydes obtained from lignins and woods from several plant species by nitrobenzene oxidation.

The syringaldehyde/vanillin (S/V) molar ratio of the products from sweetgum HCl lignin is 3.1 compared for sweetgum wood. This indicates that the uncondensed guaiacylpropane units of the lignins in the wood are more susceptible to acid-catalyzed condensation than are uncondensed syringylpropane ones [43].

As mentioned earlier, nitrobenzene oxidation has been employed as a means of characterizing lignin in different morphological regions of wood cell walls (see Table 7.4).

Table 7.2 Yields and oxidation products from Picea abies wood on nitrobenzene oxidation[a]

Oxidation products	Structure no.[b]	Yield (% of Klason lignin)	
		A[c]	B[d]
p-Hydroxybenzaldehyde	1	0.25	–
Vanillin	2	27.5	25.8
Syringaldehyde	3	0.06	–
5-Formylvanillin	4	0.23	–
Dehydrodivanillin	5	0.80	2.2
Guaiacol	6	–	–
Acetoguaicone	7	0.05	–
Syringol	8	–	–
Acetosyringone	9	–	–
Vanillic acid	10	4.8	1.3
Syringic acid	11	0.02	–
5-Formylvanillic acid	12	0.1	–
5-Carboxyvanillin	13	1.2	0.6
Dehydrodiveratric acid	14	0.03	–

[a] See refs [28, p. 303] and [30].
[b] See Figure 7.2 for structure number.
[c] From ref. [11].
[d] From ref. [29].

Table 7.3 Yields of phenolic aldehydes from several plant species on nitrobenzene oxidation[a,b]

Plant species	Preparation	Total lignin content (%)	(H)	(V)	(S)	Total aldehyde	Molar ratio (H:V:S)
Spruce (*Picea glauca*)	Wood	27.6	+	33.4	+	33.4	–
	MWL	97.8	+	33.9	+	33.9	–
Birch (*Betula papyrifera*)	Wood	22.6	–	14.5	36.2	50.7	0:1:2.5
	MWL	95.6	–	14.1	34.2	49.5	0:1:2.4
Sweetgum (*Liquidambar styraciflua*)	Wood	25.8	–	11.2	32.1	43.3	0:1:2.9
	MWL	96.4	–	10.3	223.4	33.7	0:1:2.3
	HCl lignin	–	–	7.6	28.8	31.4	0:1:23.1
	Kraft lignin	–	–	5.4	11.0	16.4	0:1:2.0
Zhong-yang Mu (*Bischofia polycarpa*)	Wood	33.3	–	27.0	10.6	37.6	0:1:0.39
	MWL	93.8	–	27.1	12.0	39.1	0:1:0.44
Bamboo (*Phyllostachys pubescens*)	MBL	88.1	7.9	19.0	25.7	52.6	0.4:1:1.4
	MBL-S	–	0.8	17.8	24.3	42.9	0.05:1:1.4
	Kraft lignin	–	+	7.2	4.5	11.7	0:1:0.6

[a] Data from ref. [28, p. 316], yield in mol % C_9 unit.
[b] *p*-Hydroxybenzaldehyde (H); vanillin (V); syringaldehyde (S); MBL-S MBL after saponification with alkaline solution; + = trace amount.

Table 7.4 Alkaline nitrobenzene oxidation products of the fiber fraction from birch wood soft xylem[a]

	Total aldehydes	Products Vanillin	Syringaldehyde	S/V molar ratio
Fiber fraction	9.5	7.3	2.2	0.25
Whole wood	42.7	8.0	34.7	3.62

[a] Yield based on lignin content determined by the acetyl bromide method. Data taken from ref. [33].

The effect of several white-rot fungi on the lignin in birch wood [42] (see Table 7.5) reveal that the syringylpropane lignin units were preferentially degraded.

In this method, a high risk of interference exists between the lignins and other phenols from the wall, which derive in similar products.

7.2.2 Oxidation with Cupric Oxide

In 1942, Pearl [44] found that vanillin could be obtained in a yield of about 20% of the lignin when sulfite liquor solids containing about 40% of lignosulfonate were heated with 4 M NaOH solution in

Table 7.5 Nitrobenzene oxidation of birch wood decayed by four ascomycetes[a,b]

	Lignin loss (%)	Vanillin (mg)	Syringaldehyde (mg)	S/V (molar ratio)
Sound birch	–	0.72	3.17	3.68
Decayed by:				
Hypoxylon fuscum	24.0	0.88	2.74	2.60
Libertella betulina	28.8	1.00	2.96	2.47
Hypoxylon multiforme	31.6	1.05	3.24	2.58
Daldinia concentrica	44.0	1.74	3.24	1.55

[a] Data taken from refs [28, 42].
[b] Yield, mg/ca. 40 mg wood simple.

Table 7.6 Yields on nitrobenzene and cupric oxide oxidation products from Picea abies and Populus tremuloides woods[a,b]

Oxidation products	Structure no.	Yield (% of Klason lignin)	
		Nitrobenzene	Cupric oxide
Norway spruce (Picea abies)			
Vanillin	2	27.5a	15.9
Guaiacol	6	–	1.4
Acetoguaiacone	7	–	3.9
Total yield	8	27.5a	21.9
Aspen (Populus tremuloides)			
Acetosyringone	9	–	5.3
p-Hydroxybenzaldehyde	1	–	1.8
Vanillin	2	12.4	7.8
Syringaldehyde	3	30.0	20.0
Acetoguaiacone	7	–	2.0
Total yield		42.4	36.9

[a] Data taken from ref. [28, p. 304].
[b] Data taken from ref. [46] except a from ref. [11].

the presence of cupric sulfate in an autoclave at 160 °C for several hours. Cupric sulfate was later replaced by cupric hydroxide [45]. The optimal conditions for performing the cupric oxidation are heating a mixture of lignin, cupric oxide, and NaOH (1.5–0.2 M) in a ratio 1:6:3 in an autoclave at 170 °C for 2–5 h [45, 46].

Other metal oxides that have been used as oxidants include Cu, Hg, Ag, and Co oxides. The main oxidation products are aromatic aldehydes or aromatic carboxylic acids, depending on the oxidant used. Silver oxide gives acids, whereas cupric oxide yields aldehydes, and Hg and Co produce a mixture of both acid and aldehyde.

As it is listed in Table 7.6, cupric oxide oxidation provides results comparable to those obtained from nitrobenzene oxidation [46], although in less yield.

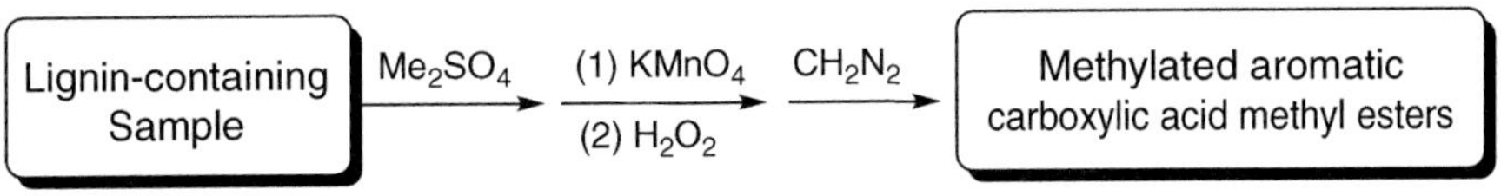

Figure 7.3 Reaction sequence for the oxidative degradation of lignin with KMnO$_4$ [13]

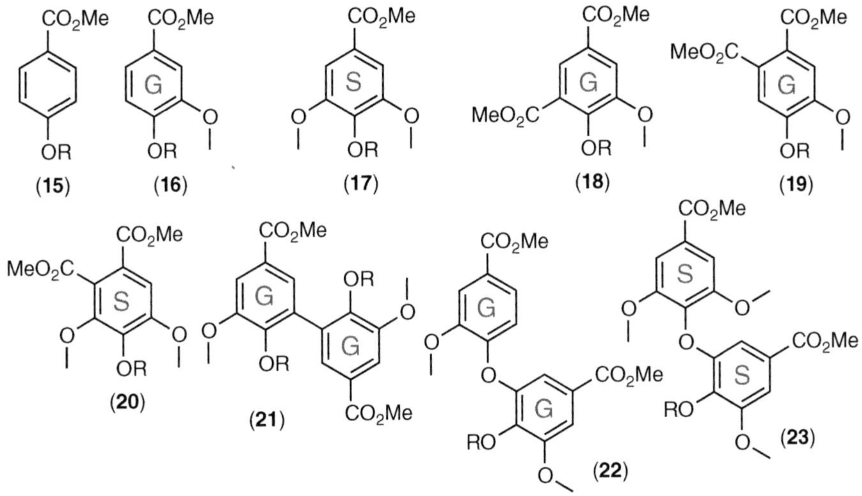

Figure 7.4 Main carboxylic acid methyl esters formed in the oxidation of lignin with K$_2$MnO$_4$ [13]

7.2.3 Permanganate Oxidation

Much of our knowledge about the structure of lignin in wood and pulp is based on results obtained from permanganate oxidation. This method involves the selective degradation of all aliphatic side chains attached to aromatic moieties in lignin, resulting in the formation of a mixture of aromatic carboxylic acid structures. The identity of these as well as the amount of each individual acid provides information about the substitution pattern in a particular lignin and the frequency with which individual linkages between phenylpropane units occur [13].

In the original method, wood was first subjected to alkaline hydrolysis at high temperature followed by methylation with dimethyl sulfate.[1] The oxidation step was carried out by the addition of portions of potassium permanganate at pH = 7 until the purple color of the solution remained [47]. Freudenberg [48] was able to identify nine different acids by this technique (see Figures 7.3, and 7.4).

This procedure was modified later by Larsson and Miksche [49] and Erickson et al. [12] . They found that considerably higher yields of the aromatic carboxylic acids were obtained if oxidation was carried out by a mixture of sodium periodate and permanganate in aqueous *tert*-butyl alcohol with NaOH at 82 °C. Moreover, of the introduction of modern spectroscopic methods, permanganate oxidation conducted under alkaline conditions instead of neutral conditions, introduction of a second oxidation step with alkaline hydrogen peroxide. Finally, the mixture of acids was methylated with diazomethane to obtain derivatives suitable for GC separation [49–59].

---

[1] The free hydroxyl phenolic groups were converted into methoxyl groups stabilized toward oxidation.

196 Lignin and Lignans as Renewable Raw Materials

Table 7.7 Relative frequency of aromatic carboxylic acids formed in the permanganate oxidation of selected woods and pulps

Sample	Relative frequency, mol %					
	15	16	18	20	21	22
Wood (spruce)	6	67	10	8	6	3
Wood (pine)	3	69	10	7	7	4
MWL (spruce)	4	54	12	8	15	7
CTMP (spruce)	6	60	10	15	6	3
Sulfite pulp (spruce)	4	40	19	19	7	11
Kraft pulp (pine)	3	42	16	6	21	12
O_2-Bleached Kraft pulp (pine)	1	38	19	6	24	12

[a]See ref. [13, p. 330].

Finally, Freudenberg et al. [60] developed a four-step standardized procedure. The procedure commences with an alkaline CuO predegradation step, followed by a methylation step, and ends with two oxidation steps involving permanganate and H_2O_2 [13]. The complete reaction sequence is outlined in Figure 7.3.

Oxidative degradation with permanganate has also been used for elucidating the structure of lignins in nonwoody plant materials [61], lignins modified in various technical processes: neutral sulfite [62–64], Kraft pulps [65–69], and lignin of bleaching Kraft pulp with chlorine and chlorine dioxide [70–73]. A total of 40 aromatic acids have been identified in the reaction mixture obtained on oxidation of methylated spruce MWL, which had not been subjected to ether cleavage [51, 54].

In Table 7.7 is listed the relative frequency of aromatic carboxylic acids formed in the permanganate oxidation of selected woods and pulps. These values indicate the relative distribution of types of structural units in lignin [12, 56, 74].

Bose et al. [75, 76] have used the permanganate oxidation method to estimate (analyze) condensed and uncondensed structures in different lignins. They used and alkaline CuO pretreatment to hydrolyze most of the interunitary ether bonds.

Recently, Parkås et al. [77] have reported a quantitative analysis of lignins based on oxidation with permanganate. They review the literature of the method and calculate the distribution of different types of acids according to the percentage of lignin phenolic units.

A new capillary electrophoretic technique was developed for the separation of lignin degradation products after permanganate oxidation, yielding information on the quality and quantity of various linkages in the lignin molecule [78].

7.2.4 Mild Hydrolysis

7.2.4.1 Hydrolysis with Water

Nimz [79–83] percolated wood powder with water at 100 °C for several weeks ("mild hydrolysis"), showing that beech wood loses about 40% of its lignin, while only 20% of lignin in spruce wood goes into solution. From hydrolysis products of spruce, eight dilignols, two diastereoisomeric trilignols, and one tetralignol were isolated (see Figure 7.5). From beech, syringaresinol and three diarylpropane diols were isolated.

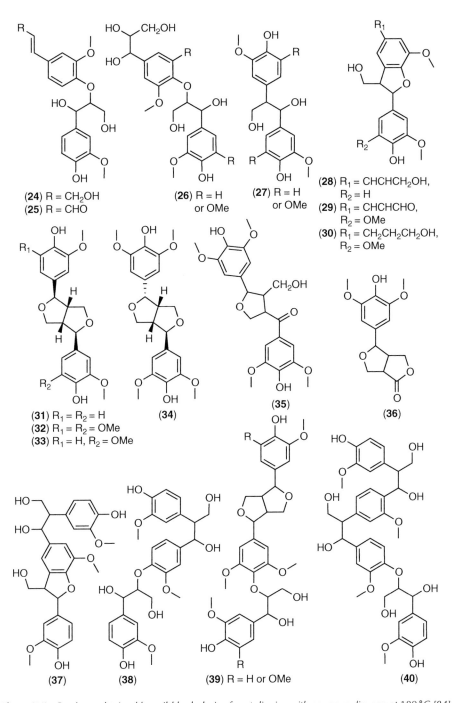

Figure 7.5 Products obtained by mild hydrolysis of protolignins with aqueous dioxane at 180 °C [84]

7.2.4.2 Hydrolysis with Dioxane/Water

Sakakibara et al. [85–94] found that 40–60% of lignin can be dissolved by treating wood powder with a dioxane/water (1:1) mixture at 180 °C. The many degradation products are almost the same as those reported by Nimz et al. [79–83].

7.2.4.3 Solvolysis

To find a method to degrade lignin without involving simultaneous condensation reactions, Nimz [79–83] and Sakakibara et al. [85–94] subjected softwood and hardwood to "mild hydrolysis" under neutral or slightly acidic conditions followed by percolation with hot water and a mild hydrolysis with aqueous dioxane at 180 °C, respectively. Separately, they isolated and identified many monolignol-to-trilignol hydrolysis products in small amounts.

Under the "mild hydrolysis" conditions, the main reaction, probably, is the cleavage of benzyl aryl-ether linkages (see Figure 7.6). This has been used to gain valuable information on the biosynthesis process in lignin and the end groups in lignin.

Sano [96, 97] subjected extractive-free oak wood meal to "solvolysis" in p-cresol/water (1:1) at 180 °C for 30 min in order to explain the delignification mechanism of wood lignin by solvolysis pulping with aqueous phenol at elevated temperatures. From reaction products of the wood, six compounds were isolated and identified [96–98].

7.2.5 Acidolysis

This treatment was introduced in lignin chemistry by Pepper et al. [99] in 1959 as a procedure for isolating lignin [100]. The term acidolysis refers to the refluxing of lignin with 0.2 M HCl in dioxane/water (9:1, v/v), although other similar mixtures have been used [101–103]. Acidolysis reactions in the presence of catalysts other than HCl have also been studied [104, 105]. Since it had been found that this treatment, "acidolysis," results in the formation of an ether-soluble oil in addition to a high M_w product, it was applied to lignin preparations.

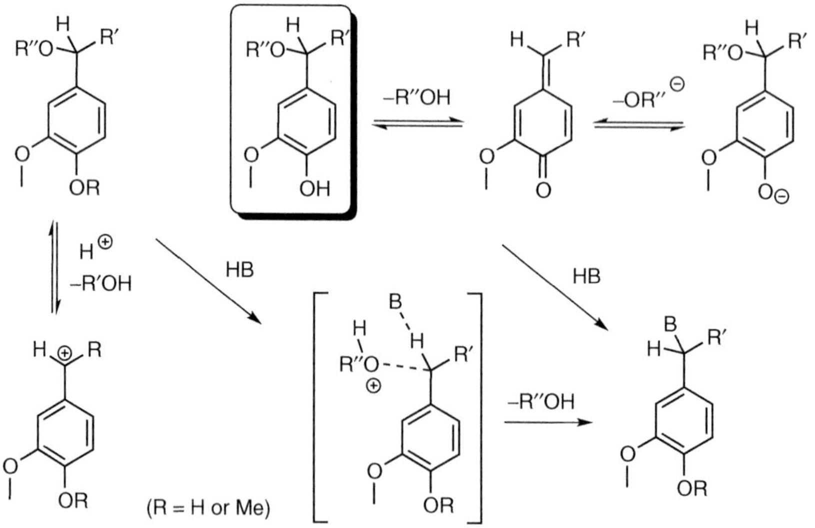

Figure 7.6 Solvolytic reactions of benzyl alcohol and benzyl ether groups [95]

Table 7.8 Monomeric phenols detected in lignin acidolysis reaction mixtures[a]

41	42	43	R
a	a	a	$-CH_3-CO-CH_2OH$
b	b	b	$-CH(OH)-CO-CH_3$
c	c	c	$-CO-CH(OH)-CH_3$
d	d	d	$-CH_2-CO-CH_3$
e	e	e	$-CO-CO-CH_3$
f	f	-	$-CH_2-CHO$
g	g	g	$-CHO$
h	h	h	$-COOH$
i	i	i	$-CH=CH-CHO$
j	-	j	$-CH=CH-COOH$

[a] Data taken from ref. [100].

Acidolysis causes the selective cleavage of aryl ether bonds in the lignin. When spruce MWL was subjected to the acidolysis treatment, the low-molecular portion of the resulting mixture could be resolved by gel filtration into fractions containing monomeric, dimeric, and oligomeric compounds, respectively. Lignin can be characterized by analysis of their low M_w acidolysis products. Acidolysis has a close relationship with "ethanolysis"[2] and solvolysis (see Section 7.2.4.3).

A method for characterizing lignin based on the analysis of acidolysis monomers was developed by Lundquist and Kirk [106]. Monomeric phenolic degradation products were analyzed as trimethylsilyl derivatives by GC and GC-MS. Lignin acidolysis products may also be analyzed by HPLC [16].

The acidolysis monomers detected are shown in Table 7.8. Analysis of these monomers has been used as a tool for the characterization of a variety of isolated lignins and lignin components in plant materials [17, 106–117]. In the monomeric fraction the predominating ketol (**41**) obtained in yields of 5–6% from the lignin. In the acidolysis of birch MWL, a number of syringyl analogs have been detected. The yields of the syringyl monomers were higher than those of the guaiacyl ones, although the S/G ratio is about 1:1 in birch. This is because some of the guaiacyl units are linked to an adjacent unit by 5-5, β-5, and 5-*O*-4 bonds, which cannot occur in syringyl units.

Acidolysis of spruce or birch lignins results in the formation of the dimeric products shown in Figure 7.7, and this finding is evidence for the presence of substantial amounts of β-1 and β-β structures in the original lignins [4, 118]. The quantitative estimation has indicated that about 10% of the C_9 units in the spruce lignin are connected to an adjacent unit by an α-*O*-4 as well as a β-5 linkage.

Recently, Govender *et al.* [119] have systematically studied the acidolysis of many of hardwoods (Eucalypti and poplar) in order to determine the S/G ratio. Several variables were investigated: reaction temperature, reaction time, and effect of the addition of 2,6-di-*tert*-butyl-4-methylphenol (BHT). The S/G ratios documented indicate that the growing environment is a key factor; for example sunshine, rainfall, soil, nutrients, age of the sample, and site.

In acidolysis of Klason lignin in dioxane/water/HCl, a diketone-containing side chain was observed as the major products ($CH_3-CO-CO-Ar$) (entry 'e' in Table 7.8).

---

[2] Heating of lignin with HCl in EtOH.

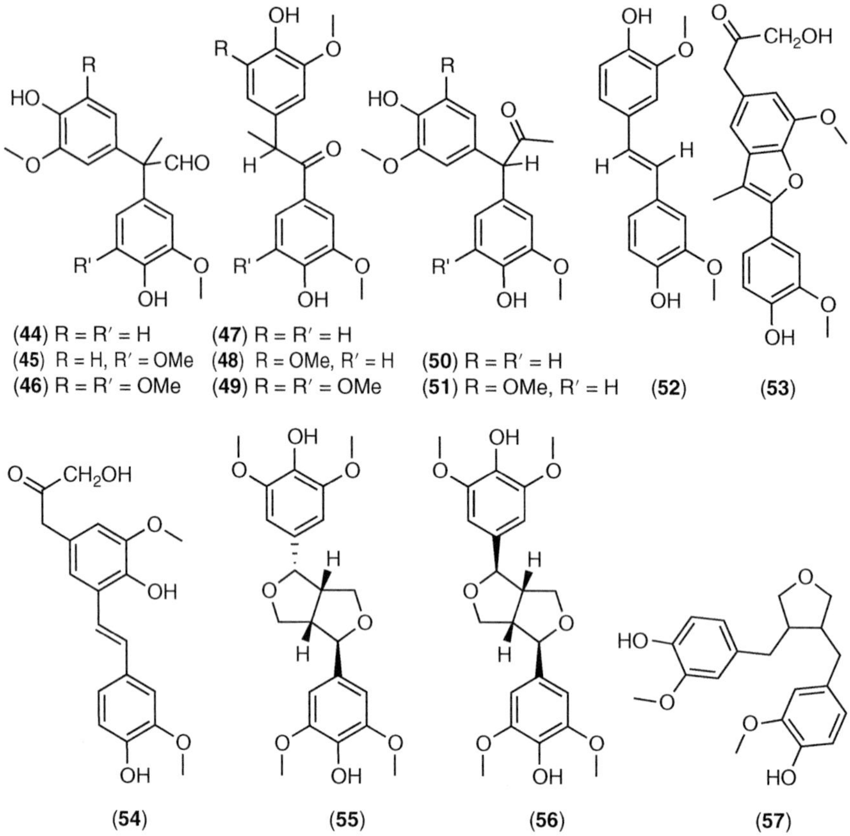

Figure 7.7 Dimeric lignin acidolysis products [100]

The yield of acidolysis monomers reflects the amount of intact lignin in a sample. The main criticism of all methods involving acidolysis has been the fact that almost invariably the product yields are rather low as a result of condensation reactions taking place in acidic media [120].

7.2.6 Thioglycolic Acid Hydrolysis (Mercaptolysis)

In 1930, Holmberg [121] by treating lignin with thioglycolic acid under acidic conditions obtained a modified lignin containing thioglycolic ethers of benzyl alcohol groups. This is the base of the thioglycolate methods for the determination of lignin (see Section 3.5.2.2 on page 62).

7.2.7 Thioacetolysis

The principle of the three-step degradation method has been formulated by Nimz and Das [122] as shown in Figure 7.8 Treatment of wood with thioacetic acid and boron trifluoride converts the arylglycerol-β-aryl ether unit (40.1) *via* the benzylium ion (40.2) into the S-benzyl thioacetate (40.3). Subsequent saponification with 2N NaOH at 60 °C gives a benzyl thiolate ion (40.4), which loses the β-aryloxy group by nucleophilic attack of the neighboring thiolate ion on the C_β atom to give an episulfide (40.5) The latter dimerises to dithianes or polymerizes to thioethers. In a final step,

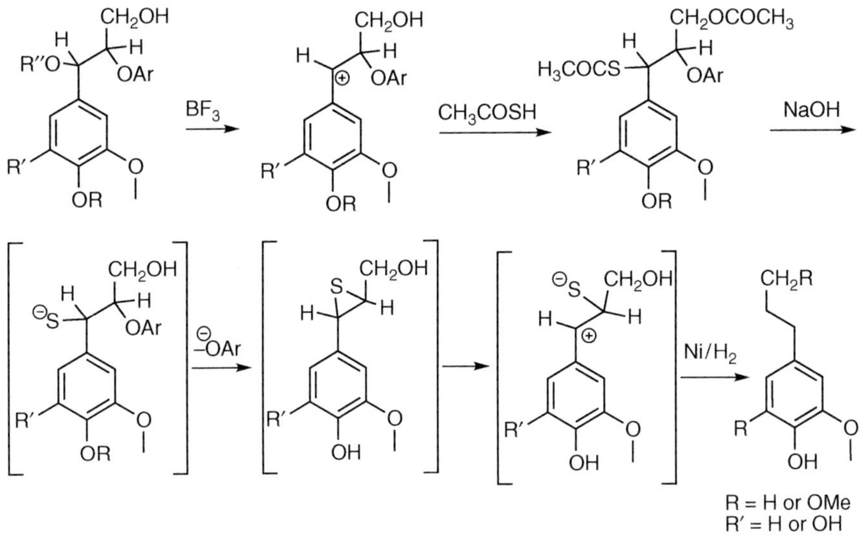

Figure 7.8 Degradation of lignin by thioacetolysis with thioacetic acid [122]

treatment with Raney nickel and alkali at 115 °C removes the sulfur and yields the reduced phenolic reaction products.

The 20 dimers derived from beech wood are shown in Figure 7.9. Most of the bond types exhibited by these dimers are identical to those revealed by other degradation methods, especially the acidolysis and oxidative degradation mentioned earlier.

"Thioacetolysis" causes cleavage of β-O-4 bonds, and brings about a more deep-ground fragmentation of the lignin than does acidolysis [14, 122]. As much as 91% of the lignin of beech wood and 77% of the lignin of spruce wood were degraded to mixtures of monomeric to tetrameric products.

On the basis of the yields of degradation products, Nimz [5] has calculated the frequencies of the bond types and has also proposed a structure for lignin (see Figure 2.20 on page 31).

7.2.8 Thioacidolysis

Thioacidolysis, that is, solvolysis in dioxane/ethanethiol with boron trifluoride etherate, is an acid-catalyzed reaction, which results in the depolymerization of lignins and which can be used to estimate the amount of uncondensed alkyl aryl ether structures[3] [18]. Lapierre et al. [123] have used this principle to develop the technique of thioacidolysis.

Thioacidolysis, as in acidolysis, proceeds mainly by the cleavage of arylglycerol-β-aryl ether linkages. However, thioacidolysis is performed in anhydrous media. The reagent combines a hard Lewis acid, such as boron trifluoride etherate, and a soft nucleophile such as ethanethiol [124, 125]. This combination has been shown to quantitatively and selectively cleave the arylglycerol-β-aryl ether linkages in lignin. Subsequent developments of the technique involving the desulfurization of the products by using Raney nickel have been claimed to offer improvements over earlier protocols [126]. Thioacidolysis is carried out using a few milligrams of sample in dioxane at 100 °C for 4 h instead of thioacetolysis performed at 20 °C for a week. The mechanism by which lignin structures

---
[3] Excluded methyl aryl ethers.

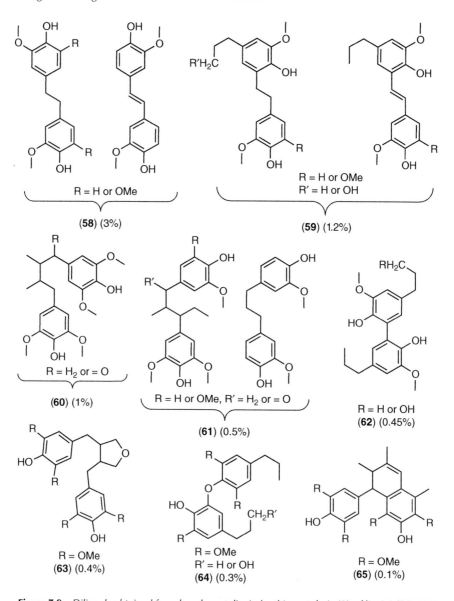

Figure 7.9 Dilignols obtained from beech protolignin by thioacetolysis (% of lignin) [84, 122]

are degraded, which is similar in principle to that proposed for thioacetolysis, has been investigated by Lapierre et al. [127] and is depicted in Figure 7.10.

The various monomeric products recovered reflect the structure of the original unit and included thioethyl derivatives of arylglycerol-β-aryl ether, and so on (see Figure 7.11).

Among the thioethylated monomers from spruce MWL, compounds **66** and **67**, which were formed from uncondensed β-O-4 linked units, gave a total yield of 93% of the monolignols. They reflect the higher content of uncondensed β-O-4 linkages in the MWL products.

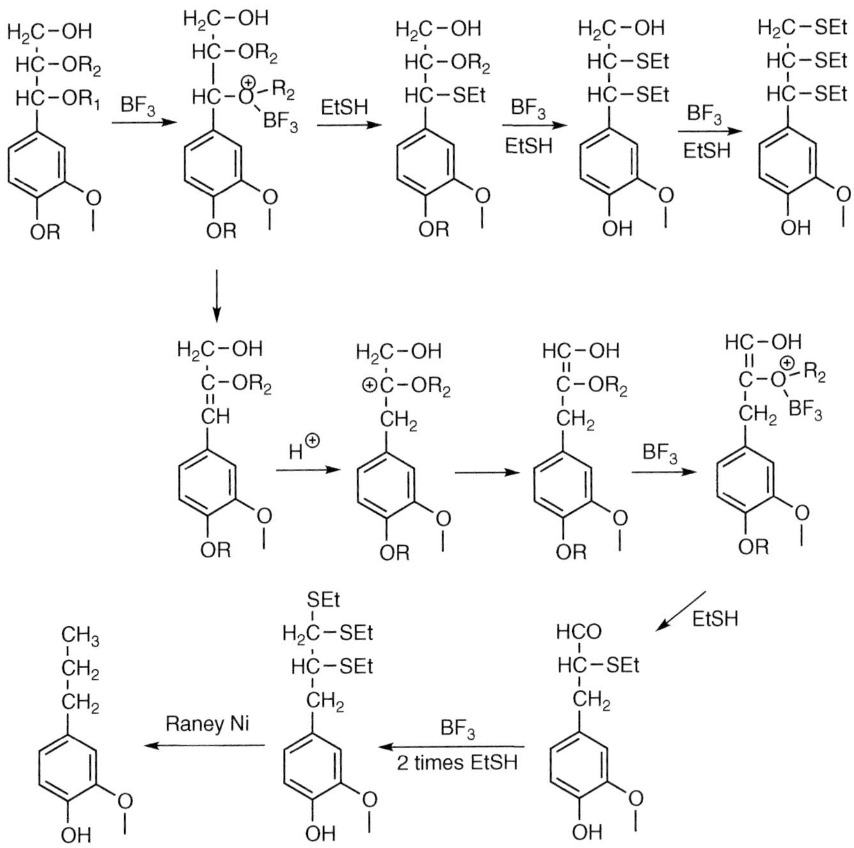

Figure 7.10 Reaction mechanism of β-O-4 substructure units by thioacidolysis and subsequently desulfurization with Raney nickel [17, 84]

The yields of compounds that were characterized as monolignols and dilignols among the thioacidolysis mixture were only 40–50% of lignin [128], which may reflect the limitation of thioacidolysis method to degrade lignin. Table 7.9 lists the yields of monomeric products resulting from thioacidolysis method.

The wide range of values shows the large variation in lignin structure, which can be associated with genetic origin and the procedure of its isolation.

Table 7.10 shows a comparison of different degradation procedures for milled wood lignin (MWL) and enzymatically isolated lignin (EL) from poplar wood.

These values show that nitrobenzene leads to higher product yields, indicating that the nitrobenzene method can degrade phenolic components other than lignin. The low yield in permanganate oxidation indicates that the units are mainly of the guaiacyl type. Table 7.10 indicates also that none of the degradation methods is capable of providing an unequivocal characterization of lignin.

7.2.9 Hydrogenolysis

Another procedure that has significantly contributed to our understanding of the lignin structure is hydrogenolysis. Hydrogenolysis is a process in which a carbon–carbon or carbon–heteroatom single

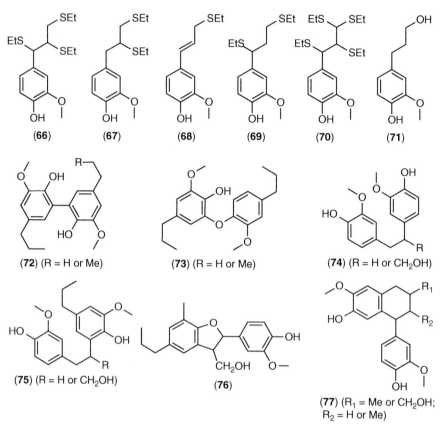

Figure 7.11 *Thioethylated monomers and Raney nickel-desulfurated dimers produced by thioacidolysis of spruce MWL [84, 126]*

bond is cleaved by reaction with hydrogen.[4] Catalytic hydrogenolysis is one of the most effective methods for gaining information on the chemical structure of lignin, since high yields of mono-, di-, tri-, and oligolignols are formed without the occurrence of secondary condensation or polymerization reactions [6].

This method has been reviewed by several authors [32, pp. 202–205] [31, 130] [131, pp. 818–821] [132]. Lindblad [133], in 1931, was the first to subject wood to catalytic hydrogenolysis in order to obtain products of commercial value, for example, low M_w phenols [134–136].

Harris *et al.* [137] isolated several *n*-propylcyclohexane derivatives by reaction of an isolated lignin with hydrogen. Lautsch [138] and Freudenberg *et al.* [139] studied the high-pressure hydrogenolysis of commercial lignin with various catalysts and produced cyclohexanol derivatives in high yield.

Numerous studies have been done on hydrogenolysis of lignin to monomeric phenylpropanoids [140–148]. With shorter reaction times and moderately active catalysts, lignin can be hydrogenolyzed to di- and trilignols [149–171].

The main reaction in the hydrogenolysis of lignins is the cleavage of ether linkages connecting the carbon atoms of the side chain and the *para*-position of a phenolic ring in an adjacent unit.

---

[4] Hydrogenation is the addition of hydrogen to a multiple bond.

Table 7.9 Yields[a] of p-hydroxyphenyl (H), guaiacyl (G), and syringyl (S) monomeric thioacidolysis products[b]

Samples	H	G	S	Total	H/G	S/G
Lignocellulosic:[c]						
Spruce wood (Picea abies)	–	1006	–	1006	–	–
Pine wood (Pinus pinaster)						
Compression wood	207	934	–	1141	0.22	–
Opposite wood	25	992	–	1017	0.03	–
Poplar wood (Populus euramericana)						
Tension wood	–	732	1080	1812	–	1.48
Opposite wood	–	763	1186	1949	–	1.55
Birch wood (Betula verrucosa)	–	525	1341	1866	–	2.55
Wheat straw (Triticum aestivum)						
cv. Champlein	43	457	636	1136	0.08	1.07
cv. Capitole	53	535	650	1238	0.10	1.21
Rice straw (Oryza sativa)						
Normal rice	95	287	251	632	0.33	0.87
Brittle-stem mutant	38	228	235	501	0.17	1.03
Lignin:[d]						
Pine (Pinus pinaster) MWL						
Compression wood	146	627	21	794	0.23	0.03
Opposite wood	15	803	8	826	0.02	0.01
Poplar (Populus euramericana) MWL						
Tension wood	–	497	513	1010	–	1.03
Opposite wood	–	520	601	1021	–	1.16
Aspen (Populus tremuloides) SEL						
	–	254[e]	415[e]	669[e]	–	1.63[e]
	–	123[f]	232[f]	355[f]	–	1.89[f]
Wheat (Triticum aestivum) MWL	27	410	343	780	0.06	0.84

[a] Microm/g lignin.
[b] See ref. [129, p. 347].
[c] Yield, mmol/g klason lignin.
[d] Yield, mmol/g lignin.
[e] At 25 bar and 4.5 s.
[f] At 35 bar and 125 s.

Depending on the reaction conditions, C–C linkages may also be broken during hydrogenolysis (see Figure 7.12). Moreover, hydrogenation accompanies hydrogenolysis.

The factors affecting hydrogenolysis are catalyst activity, catalyst amount, hydrogen pressure, solvent, and pH. Various catalysts have been used in the hydrogenolysis of lignin: Raney nickel, copper chromite, cobalt octacarbonyl, and palladium, rhodium, and ruthenium on different supporting surfaces. Catalysts with high-hydrogenation activity convert lignin mainly into cyclohexyl derivatives,

Table 7.10 Yields[a] of monomeric guaiacyl (G) and syringyl (S) products recovered from poplar lignin fractions after degradation by various procedures[b]

Lignin fractions	Thioacidolysis	Acidolysis	Nitrobenzene oxidation	Permanganate oxidation
Poplar MWL				
S + G	1024	583	1660	274
S/G	1.20	0.70	1.40	0.39
Poplar EL				
S + G	1243	630	1770	106
S/G	1.67	0.91	1.87	0.48

[a] Microm/g lignin.
[b] MWL = milled wood lignin; EL = enzymatically isolated lignin.

while catalysts having lower activity give rise to aromatic derivatives. Although heterogeneous catalysts have been used in hydrogenolysis, a few examples of homogeneous catalysts may be mentioned. For example, Sakakibara et al. [135] used different metal carbonyls.

Hydrogenation and hydrogenolysis reactions compete with each other. If hydrogenation occurs first, the hydrogenolysis is extremely restricted. However, when this sequence is reversed, both cleavage and reduction take place.

The highest yields of monomeric hydrogenolysis products was obtained with hydrogen in the presence of Raney nickel (52.2% of combined yield of eight monomeric products based on original Klason lignin) [142]. Other results with different catalysts are listed in Table 7.11.

In general, yield of monomeric products from hardwood exceeds that from softwood lignin. In hardwood lignins, there is a greater abundance of syringyl-type monomers than guaiacyl-type monomers. More than 50 di- and trilignols have been isolated from the hydrogenolysis product mixtures of softwoods and hardwoods (see Figure 7.13 and Figure 7.14).

Nimz and Das [122] cleaved beech wood with thioacetic acid using boron trifluoride as a catalyst followed by reduction with Raney nickel to produce many dimeric compounds. The products were slightly different from those formed by hydrogenolysis (see Section 7.2.7).

Recently, Barta et al. [172] showed the complete hydrogenolysis of phenyl ether bonds, coupled with hydrogenation of aromatic rings of an organosolv lignin by hydrogen transfer from supercritical MeOH, using a Cu-doped metal oxide as the catalyst, at a relatively mild temperature (300 °C). The reaction produces a complex mixture composed chiefly of monomeric substituted cyclohexyl derivatives.

7.2.10 Derivatization Followed by Reductive Cleavage (DFRC)

In 1997, Lu and Ralph [19] developed a new analytical degradation method, the DFRC[5] procedure, as an alternative to thioacidolysis. It consists of three successive steps (see Figure 7.15):

- Acetyl bromide treatment, which results in bromination of the benzylic position and acetylation of free hydroxyl groups.
- Reductive cleavage of the β-O-4 bonds with zinc dust.
- Acetylation of newly formed free phenolic groups for GC determination of the lignin-derived monomers.

---

[5] Also the abbreviation reflects the Dairy Forage Research Center where the method was developed.

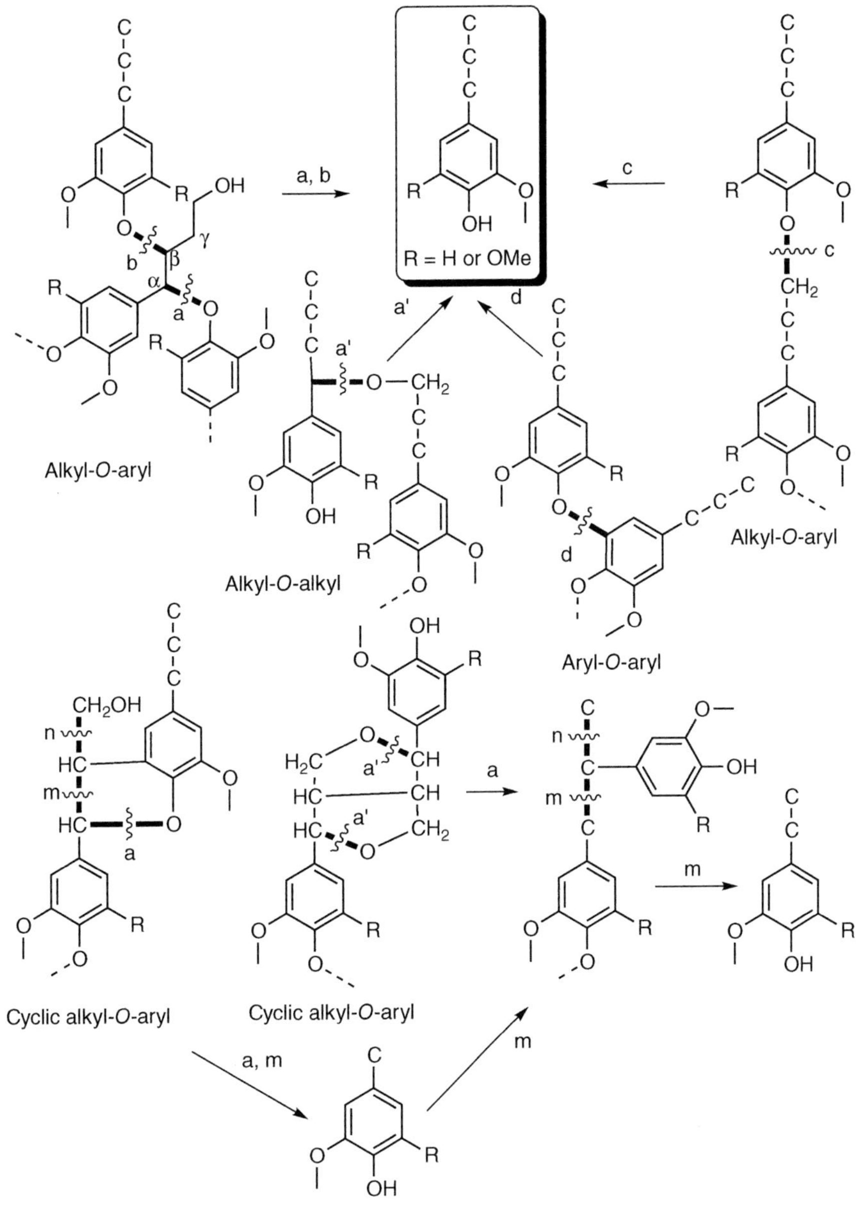

Figure 7.12 Cleavage patterns in the hydrogenolysis of interunitary lignin linkages [6]

Similar to thioacidolysis, this method selectively cleaves β-aryl ethers in lignins and releases analyzable monomers for quantification. The main monomers recovered, originating from lignin units, involved only in arylglycerol-β-ethers structures. Since DFRC does not scramble the β-carbon stereochemistry, the recovery of optically inactive β-5 (see Figure 7.16) and β-β DFRC dimers confirmed the racemic nature of lignins [173].

Table 7.11 Monomeric aromatic products formed in the hydrogenolysis of lignin[a,b]

Product		Aspen wood meal[c]	Spruce wood meal[d]	Birch MWL[e]	Oak MWL[e]	Spruce MWL[f]
	Dihydroconiferyl alcohol (**71**)	10.3	10.6	2.0	0.6	8.1
	4-n-Propylguaiacol (**78**)	0.3	5.6	2.3	2.6	5.9
	4-n-Propylsyringol (**79**)	2.9	–	3.9	7.3	–
	Dihydrosinapyl alcohol (**80**)	31.0	–	7.9	0.8	–
	4-Methylguaiacol (**81**)	–	0.1	1.1	1.0	3.5
	4-Methylsyringol (**82**)	1.7	–	2.0	3.1	–

Table 7.11 (continued)

Product		Aspen wood meal[c]	Spruce wood meal[d]	Birch MWL[e]	Oak MWL[e]	Spruce MWL[f]
[structure: Et, H, OH, OMe]	4-Ethylguaiacol (83)	1.0	0.2	0.9	0.6	2.1
[structure: Et, MeO, OH, OMe]	4-Ethylsyringol (84)	3.6	–	1.1	0.7	–
Total		50.7	16.5	21.2	16.7	19.6
Yield phenylpropane derivatives		44.5	16.2	16.1	11.3	14.0

[a] Yield (% of original lignin) [6, p. 358].
[b] Data taken from refs [140–143].
[c] Raney nickel, 195 °C, 5 h, dioxane–water (1:1).
[d] Rhodium-C, 195 °C, 5 h, dioxane–water (1:1).
[e] Copper chromite, 240–260 °C, 48 h, dioxane.
[f] Copper chromite, 240 °C, 47 h, dioxane

A unique feature of this method is that it depolymerizes lignin back to its component monomers [174, p. 183], [19, 20, 175].

By a modification of the DFRC method, Del Rio *et al.* [176] have examined the occurrence of native acetylated lignin in a large set of vascular plants.

7.2.11 Nucleus-Exchange Reaction (NE)

In 1978, Funaoka and Abe [177] developed a chemical method to degrade lignin in a medium consisting of boron trifluoride and excess phenol. In this method, C_α-aryl carbon–carbon linkages in phenylpropane units are selectively and quantitatively cleaved, following phenolation at the C_α position, releasing phenyl nuclei typified by guaiacol from softwood lignins (see Figure 7.17). Guaiacol subsequently is partially demethylated to catechol. The key step in this sequence, the release of a phenyl nucleus on reaction with phenol, was termed nucleus exchange (NE) by the original authors [178].

NE applied to softwood lignin leads to the quantitative conversion of noncondensed guaiacyl nuclei to guaiacol and catechol. The sum of guaiacol and catechol represents the quantity of noncondensed phenyl nuclei in softwood lignins.

In the NE reaction, lignin is degraded by boron trifluoride in the presence of excess phenol. The reaction has been demonstrated to occur predominantly at the C_α position of the phenylpropane side chain [179–184].

In general, the fraction of condensed guaiacyl nuclei in softwood lignins is in the range 43% to 55%. The NE method provides a rapid and relatively simple means of determining the degree of condensation in protolignins. In Table 7.12, the quantities of noncondensed and condensed guaiacyl nuclei have been determined for various softwood protolignins using the NE reaction. The method is not suitable for grass lignins, which contain a significant amount of *p*-hydroxyphenyl units, which yield phenol in the NE reaction [185].

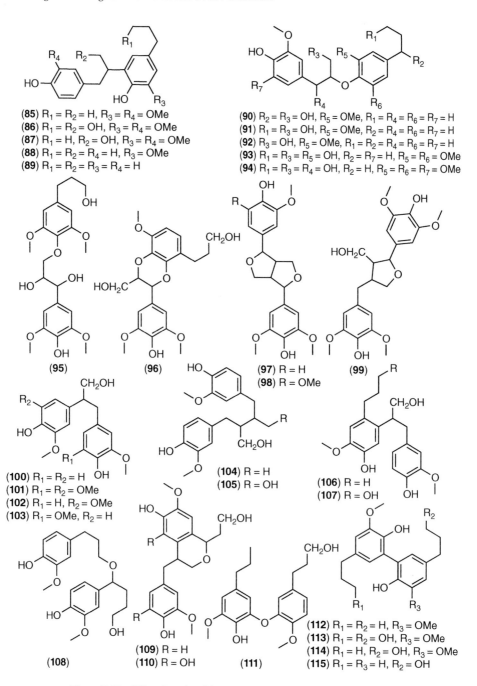

Figure 7.13 Dilignols isolated from the hydrogenolysis of protolignins [6]

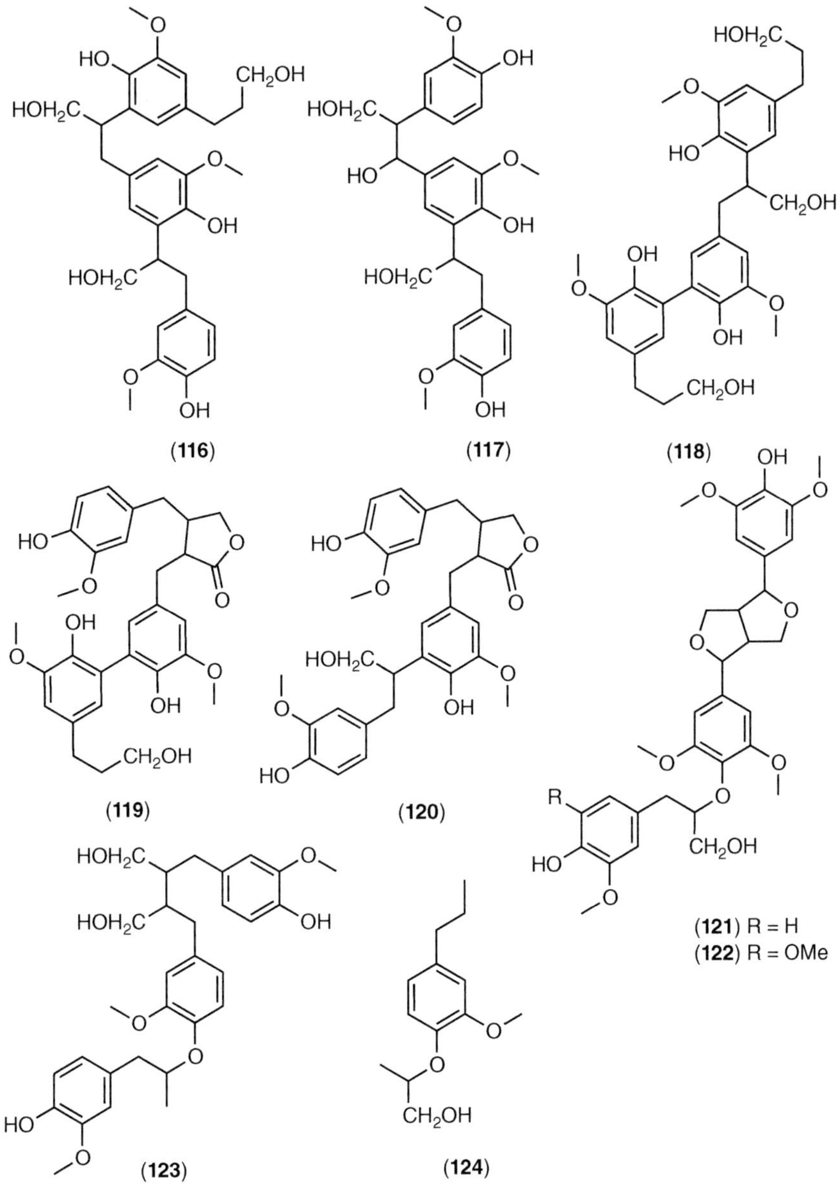

Figure 7.14 Trilignols isolated from the hydrogenolysis of protolignins [6]

In Table 7.13, the ratio of noncondensed and condensed guaiacyl nuclei has been determined for various parts of woody plants using the NE reaction [186].

Finally, in Table 7.14, the quantities of noncondensed and condensed guaiacyl and syringyl nuclei have been determined for hardwood protolignin using the NE reaction [187].

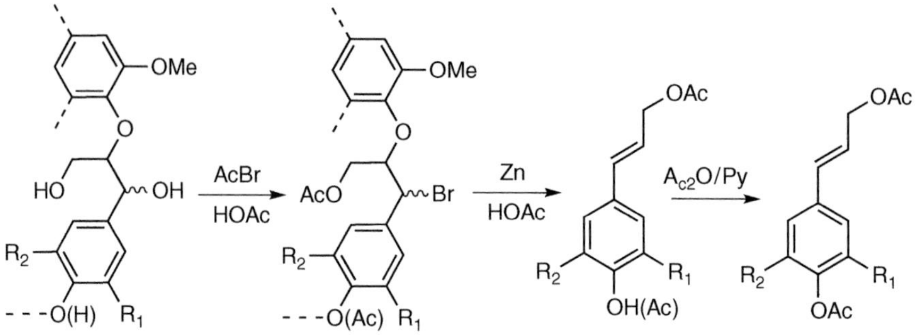

Figure 7.15 Reactions of the DFRC degradation method on lignin structures. Degradation of H ($R_1 = R_2 = H$), G ($R_1 = OMe$, $R_2 = H$), and S ($R_1 = R_2 = OMe$) lignin units involved only in arylglycerol-β-ether structures [19]

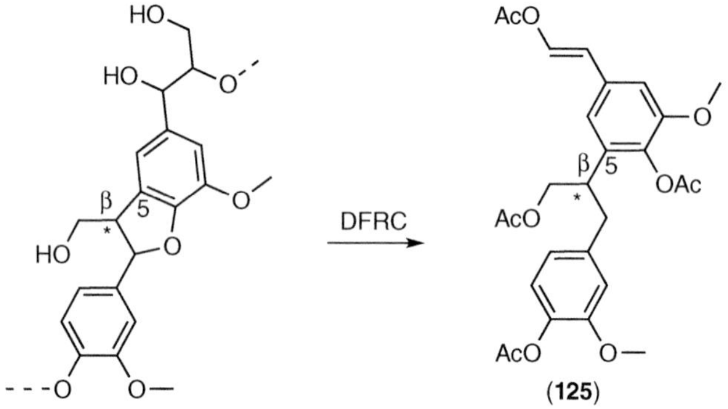

Figure 7.16 Main dimer recovered from G units involved in phenylcoumaran structures. The chiral center at β-carbon is denoted with an asterisk (*) [19]

7.2.12 Ozonolysis

Ozone has long been known to cleave carbon–carbon double and triple bonds, and the products are often useful in elucidating the structure of the parent compound. The mechanism was suggested by Criegee [188] in 1975 and has been revisited using ^{17}O NMR spectroscopy by Geletneky and Berger [189]. The first step is a 1,3-dipolar cycloaddition of ozone to the alkene, leading to the primary ozonide,[6] which decomposes to give a carbonyl oxide and a carbonyl compound (see Figure 7.18 a). Carbonyl oxides are similar to ozone in being 1,3-dipolar compounds and undergo 1,3-dipolar cycloaddition to the carbonyl compounds with the reverse regiochemistry, leading to a mixture of three possible secondary ozonides (1,2,4-trioxolanes)[7] (see Figure 7.18 b). The Criegee mechanism is valid for reactions in hydrocarbons, CH_2Cl_2, or other noninteractive solvents. Alcohols react with carbonyl oxide to give hydroperoxy hemiacetals.

The first ozonolysis on wood and lignins was conducted in 1913 by Doree and Cunningham [190] who found that lignins were highly reactive to ozone becoming degraded to low-molecular weight compounds, while polysaccharides were much more resistant [191].

---

[6] Molozonide, 1,2,3-trioxolane, or Criegee intermediate.
[7] These secondary ozonides are more stable than the primary ozonides.

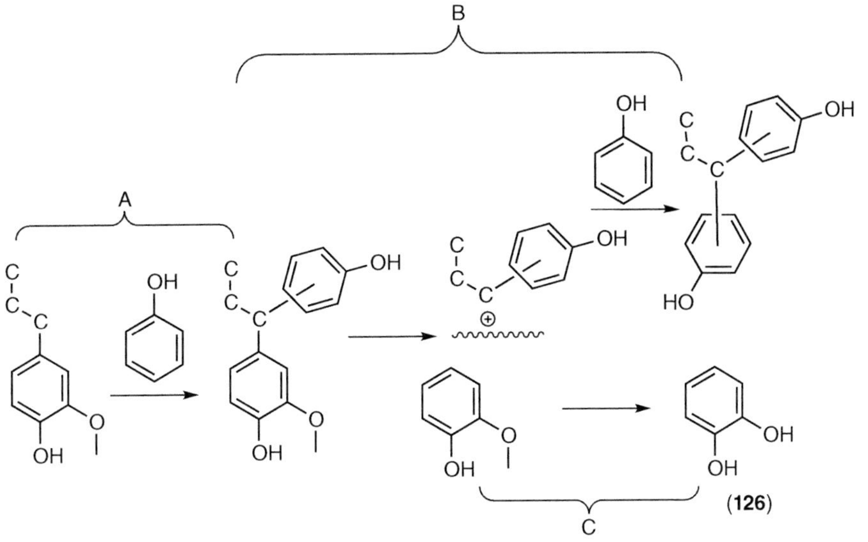

Figure 7.17 Formation of catechol (**126**) from softwood lignin in the NE method [178]

Table 7.12 Fractions of condensed and noncondensed guaiacyl nuclei in softwood protolignins Mol (% of original lignin)[a]

Species		Condensed	Noncondensed
Douglas fir	*Pseudotsuga menziesii*	50	50
Western hemlock	*Tsuga heterophylla*	55	45
Spruce	*Picea abies*	48	52
Yezo spruce	*Picea jezoensis*	47	53
Glehn's spruce	*Picea glehnii*	45	55
Slash pine	*Pinus elliottii*	50	50
Japanese red pine	*Pinus densiflora*	43	57
Japanese black pine	*Pinus thunbergii*	48	52
Japanese larch	*Larix leptolepis*	55	45
Japanese fir	*Abies firma*	49	51
Japanese hemlock	*Tsuga sieboldii*	48	52
Sugi	*Cryptomeria japonica*	45	55
Hinoki cypress	*Chamaecyparis obtusa*	45	55
Japanese torreya	*Torreya nucifera*	53	47

[a]Data taken from ref. [178, p. 382].

Nakano *et al.* [192, 193] discovered that extensive ozonolysis of lignin destroyed the aromatic moieties while leaving the side chains intact. Ozonolysis provides the best opportunity for gathering information on the frequency of the different side chains or for characterizing of their stereochemistry. In general, aromatic rings react more slowly with ozone than do alkenes. Nevertheless, phenols and phenol ethers react rather rapidly with O_3. The reactivity follows the order: syringyl > guaiacyl > 4-hydroxyphenyl [194, 195].

Table 7.13 Characteristics of Douglas fir lignin[a]

	Whole wood	Secondary wall	Middle lamella
Lignin content	29.8[b]	25.9[c]	46.2[d] (% on ML fraction)
Condensed/noncondensed	50/50	48/52	71/29

[a]Data taken from ref. [186] and ref. [178, p. 383].
[b]% on wood.
[c]% on SW fraction.
[d]% on ML fraction.

Table 7.14 Characterization of hardwood protolignin by the NE method[a]

Sample	Guaiacyl, mol % on lignin			Syringyl, mol % on lignin		S/G[b]	S/V[c]
	(NC)[d]	(C)[d]	(T)[d]	(NC)[d]	(T)[d]		
Sweetgum	22	17	39	61	61	1.56	2.82

[a]Data taken from ref. [187] and ref. [178, p. 383].
[b]$S/G = \dfrac{\text{Total syringyl nuclei}}{\text{Total guaiacyl nuclei}}$, from nitrobenzene oxidation.
[c]$S/V = \dfrac{\text{Syringaldehyde} \times \text{syringic acid}}{\text{Vanillin} + \text{vanillic acid}}$, from nitrobenzene oxidation.
[d]Noncondensed (NC), condensed (C), total (T).

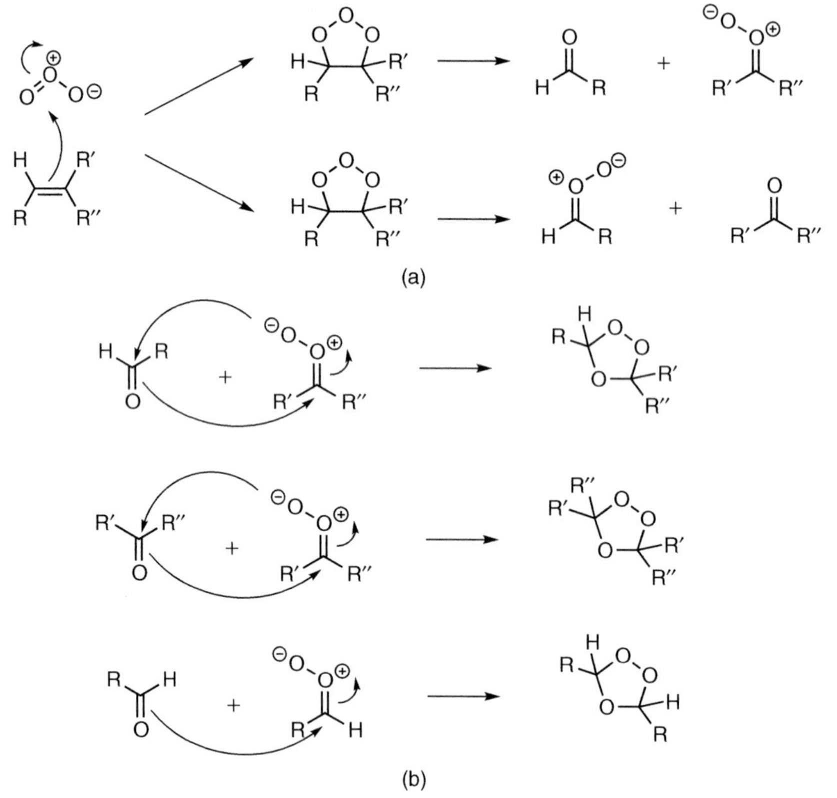

Figure 7.18 Mechanism of Criegee ozonolysis [188, 189]

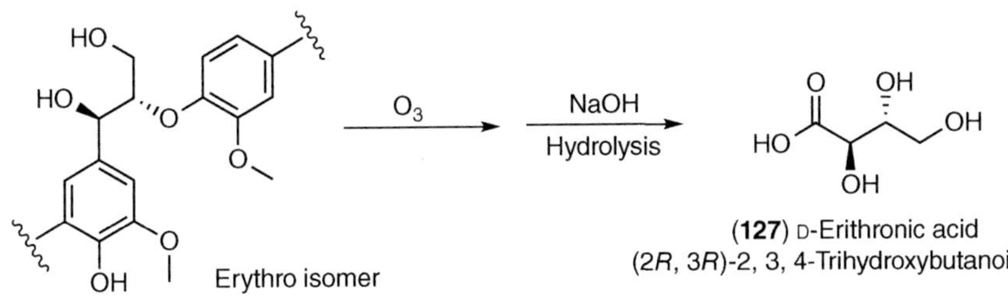

Figure 7.19 Formation of two tetronic acids from side chains of arylglycerol-β-aryl ether structures by ozonolysis [203]

(127) D-Erithronic acid
(2R, 3R)-2, 3, 4-Trihydroxybutanoic acid

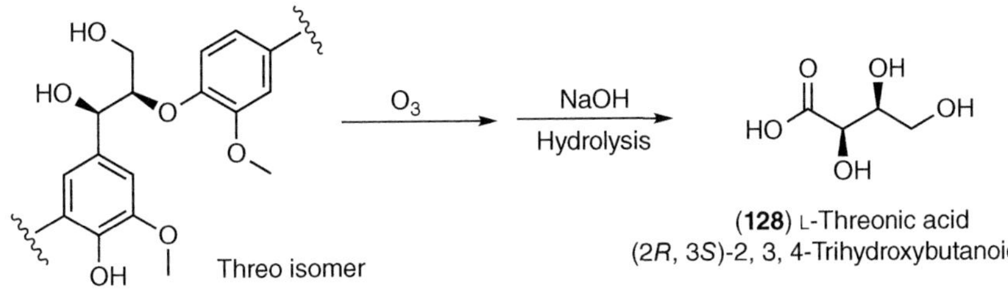

(128) L-Threonic acid
(2R, 3S)-2, 3, 4-Trihydroxybutanoic acid

The ratio of ozone to substrate can be < 10, 10–30, or > 30 moles of O_3 per mole of substrate. The use of intermediate ozone-to-substrate ratios gives products of the more reactive aromatic rings[8] [195–198]. When the highest ratios are used, all aromatic rings are completely destroyed [192–194, 199–202] (see Figure 7.19).

Different solvents have been used: carboxylic acid, alcohol media, aqueous media, and aprotic media (see Table 7.15). When the solvent is or contains alcohol, it is often incorporated in the ozonolysis products. To ensure complete oxidation of aldehydes to carboxylic acids, the ozonolysis products of lignin are treated with H_2O_2 [193].

Table 7.16 lists the E/T ratios for some lignocelluloses and lignin of different origin and processing histories [202].

These data are based on yields of the corresponding threonic and erythronic acids. The E/T ratio for lignin side chains involved in β-O-4 linkages is of practical as well as theoretical interest, since *erythro* structures hydrolyze more rapidly than do threo structures [213]. Any increase in the ratio of threo to erythro serves as an indication of increased diarylpropane content in the lignin [201]. Ozonolysis of α-aryl condensed structures found in Klason lignin gives malonic acid [192].

7.2.13 Pyrolysis

In addition to these chemical lignin-fragmentation procedures, analytical pyrolysis has been used to evaluate the lignin content and structure in lignocellulosic materials [214, 215]. The main advantages of analytical pyrolysis are as follows:

- Its high output, since this procedure does not involve any time-consuming wet chemistry.
- The low sample demand.

The pyrolytic depolymerization of lignins proceeds mainly by cleavage of labile ether bonds and thereby suffers the same limitations as most chemical degradations.

---

[8] Those containing a free phenolic hydroxyl.

Table 7.15 Media used in ozonolysis of lignins[a]

Solvent	Dispersing medium	Author [Ref.]
A: Carboxylic acid	Acetic acid	Phillips and Goss [204]
		Aulin-Erdtman et al. [193]
B: Alcohol	MeOH	Phillips and Goss Kolsaker et al. [205]
	MeOH/ClCH$_2$CH$_2$Cl	Tanahashi et al. [194]
		Haluk and Metche [195]
	MeOH/dioxane (2:1, v/v)	Tomita et al. [206]
	CH$_2$Cl$_2$/MeOH (5:1, v/v)	Kondo et al. [207]
	Methyl cellosolve	Kratzl et al. [196]
C: Aqueous	Water	Dobinson [208]
	Acetone/water (9:1, v/v)	Soteland [209]
	Acetone/water (4:1, v/v)	Balousek et al. [210]
D: Aprotic media	Acetic anhydride[b]	Tishchenko [211]
	Ethyl acetate[b]	Tishchenko et al. [211]
	CHCl$_3$[b]	Kratzl et al. [196]
	CCl$_4$[b]	Kratzl et al. [196]

[a] Data taken from ref. [191].
[b] c.f. ref. [212].

7.3 Other Chemical Modifications of Lignins

The term "chemical modification" has been used to mean different things by different authors over the years. Here chemical modification will be defined as a chemical reaction between some reactive part of a lignocellulosic cell wall polymer and a simple single chemical reagent, with or without catalyst, to form a covalent bond between the two. Moreover, ketone and aldehyde groups, the most reactive functional groups in lignin are the phenolic groups, the benzyl alcohol (α-OH) and noncyclic benzyl ether groups (α-O-4), which under acidic and alkaline conditions are subject to condensation reactions. From an industrial standpoint, the reactions of lignin in pulping processes are the most remarkable (see Chapter 5) since these reactions influence not only the pulping process itself but also the structural features of technical lignins gained as by-products [216].

Acid and base catalyzed solvolytic reactions are relevant in many aspects of lignin chemistry. They are useful for structural determination, for preparing derivatives, and for the degradation of polymers.

Halogenation and nitration are among the most widely studied of all lignin reactions. Lignin is composed of phenolic nuclei and such units undergo halogenation and nitration reactions.

The reactivity of the various functional groups in lignin, compared to those in carbohydrates, is the key to producing chemical pulps (see Chapter 5).

For many years, different chemical oxidations of lignin have been carried out. Among the different kinds of oxidations, those with the strongest oxidizing agents break the aromatic ring, whereas milder oxidizing agents change only the side chain while keeping the aromatic ring intact. At the beginning, these oxidation reactions helped to identify the lignin polymeric structure and the different linkages between the precursors [63]. Moreover, the phenylpropanoid units produced with the degradation of the lignin polymer confirmed these as building blocks of lignin [217].

7.3.1 Acylation

Among the chemical modification procedures that have been studied over the years, acetylation, using acetic anhydride, has been the most thoroughly studied and has given the most consistent results [218, 219].

Chemical Characterization and Modification of Lignins

Table 7.16 Erythro/threo ratios of isolated and in situ lignins[a]

Substrate	Ozonolysis time (min)	E/T ratio	Substrate	Ozonolysis time (min)	E/T ratio
Lignins			**Wood meals**		
Kraft lignin	6	0.75	Norway spruce	40	1.05
Klason lignin[b]	18	0.76	Norway spruce (AcOH)	40	1.03
Klason lignin[c]	18	0.78	Tsuga heterophylla	30	1.02
Soda lignin[c]	6	1.94	T. heterophylla	40	1.27
Native spruce lignin[c]	6	0.75	T. heterophylla	40	1.12
Native spruce lignin	6	0.79	T. heterophylla	60	1.01
Native spruce lignin	18	0.88	T. heterophylla[e]	40	1.36
Native spruce lignin	40	0.82	T. heterophylla[e]	40	1.11
MWL fiber[d]	6	2.18	Pseudotsuga menziesii	40	1.01
MWL wood[d]	6	1.84	P. menziesii[e]	40	1.39
Fiber lignin (enzymatic)[d]	6	2.00	Populus deltoides	20	2.72
Wood lignin (enzymatic)[d]	6	2.06	P. deltoides	40	3.03
Grasses			P. deltoides	60	2.42
Bagasse	40	1.92	Arbutus menziesii tension sapwood	40	3.69
Wheat straw	40	1.88	Liquidambar styraciflua	40	2.86
Wheat straw (saponified)	40	1.94	Eucalyptus gummifera	40	2.60
Pulps			Eucalyptus grandis	40	2.45
Soda pulp (cotton wood)	40	3.05	Corchorus capsularis fiber	40	2.41
Bagasse pulp	40	1.94	Corchorus capsularis wood	40	2.02
Wheat straw pulp	40	2.25			

[a] Data taken from refs [202] and [191, p. 402].
[b] Tsuga heterophylla.
[c] Populus deltoides.
[d] Corchorus capsularis.
[e] Compression wood.

In 1919, Pringsheim and Magnus [220] acetylated Willstatter lignin (see Chapter 5) with acetic anhydride and pyridine. The acetyl content of the product varied considerably with the source of the lignin, from 19.85% in pine wood, 27.20% in straw, and 37.85% in white beech. Later, in 1924, Heuser and Ackermann [221] performed a systematic study of the acetylation of lignin. He employed five different acetylation methods, and the products containing the highest percentages of acetyl resulted when acetic anhydride and pyridine or acetyl chloride were the acetylating agents. The acetylation of wood was first performed in 1928 by Fuchs [222], using acetic anhydride and H_2SO_4 as a catalyst, isolating lignin from pine wood. The acetylation of wood with the subsequent separation of the acetylated lignin fraction has also been reported by Suida and Titsch [223, 224], as well as by Friese [225].

Beckmann et al. [226] prepared a benzoyl derivative of alkali lignin by treating lignin with benzoyl chloride and pyridine. The same authors also prepared *p*-bromobenzoyl and *p*-nitrobenzoyl derivatives. A benzoyl derivative of "primary lignin" has also been prepared by Friedrich and Diwald [227]. A tosyl[9] derivative of Urban lignin has been prepared by Freudenberg and Hess [228]. The product contained 7.0% sulfur. Benzoyl derivatives of ligninsulfonic acid have been prepared by Klason [229, p. 20] and by Dorée and Hall [230].

In 1945, Brauns et al. [231] reported the preparation of lignin esters by reacting the lead salt of alkali lignin with acid chloride. The following esters have been prepared: propionyl, butyryl, caproyl, pelargonyl, stearoyl, and benzoyl. Other metal derivatives of alkali lignin have been prepared and used for the preparation of esters. They include Hg(II), Sn(II), and Cu(II).

In 2005, Thielemans and Wool [232] prepared different Kraft lignin esters for use in unsaturated thermosets. Kraft lignins were esterified with several anhydrides to alter their solubility behavior in nonpolar solvents. Increasing the carbon chain length on the ester group improved the solubility of Kraft lignin in nonpolar solvents, with butyrated lignin being completely soluble in styrene. Esterification with unsaturated groups, such as methacrylic anhydride, improved the solubility to a lesser extent than the saturated analogs. The aromatic hydroxyl groups were found to be threefold more reactive than the aliphatic ones.

In 2012, Chandran et al. [233–235] reported the development of a novel photoresponsive biomaterial by modifying biopolymeric core of lignin by incorporating 2-(5-(4-dimethylamino-benzylidin)-4-oxo-2-thioxo-thiazolidin-3-yl) acetic and 4-[(*E*)-2-(3-hydroxynaphthalen-2-yl) diazen-1-yl] benzoic acids, two chromophoric systems, through esterification by DCC coupling. The products were characterized by spectroscopic methods and their photochemical and photophysical properties have been investigated. The results of the studies show that incorporation of the chromophoric system into the polymeric core enhanced the thermal stability of the chromophoric system and core materials.

7.3.2 Alkylation

The reactions of methylation have been studied to determine the nature and number of hydroxyl groups in the lignin structure (see Chapter 6). Diazomethane has been used to methylate acidic hydroxyl groups; dimethyl sulfate and sodium hydroxide methylate all other hydroxyl groups.

In 1921, Heuser et al. [236] methylated lignin by suspending lignin in a 10% NaOH solution and adding gradually dimethyl sulfate. The original methoxyl content was thereby increased from 14.65% to 20.73%. Freudenberg and Hess [228] treated lignin with an ether solution of diazomethane, and after 2 days the percentage of methoxyl increased from 15.5% to 19.8%. Klason [237] methylated calcium ligninsulfonate. Heuser and Samuelsen [238] methylated ligninsulfonic acid with dimethyl sulfate. The methoxyl content of the product augmented to 25.43% by five successive methylations, thus indicating that this is the maximum percentage of methoxyl that can be introduced in ligninsulfonic acid. The methylated product was light yellow, but when exposed to the air it became darker in color. It was insoluble in water, ether, petroleum ether, and mineral acids, and only slightly soluble in EtOH.

Jones and Brauns [239] described the ethylation of lignin by diazoethane and diethyl sulfate and the preparation of mixed methyl-ethyl ethers.

Using the reaction between the lead salt of lignin and the appropriate halide, Brauns et al. [231] prepared the following ethers of alkali hardwood lignin: methyl, propyl, isopropyl, butyl, amyl, heptyl, decyl, lauryl, stearyl, and benzyl. In general, iodide reacts more readily than bromide, and longer chain halides require a longer time for condensation than shorter chain halides. Stearyl bromide reacts at temperatures of approximately 210 °C. in 5 to 6 h; stearyl iodide, on the other hand, reacts readily at 180 °C.

---

[9] *p*-Toluenesulfonyl.

Treatment of native (Brauns) lignin with 0.5% MeOH/HCl, either for 48 h at rt or for 2.5 h under reflux, resulted in extensive methylation [240]. These conditions were considered mild enough to avoid lignin degradation.

The acid-catalyzed alkylation of lignins with MeOH or EtOH in dioxane[10] has been studied by NMR [241]. Complete alkylation required 4–6 days and resulted in the introduction of approximately one alkoxyl group in every second phenylpropane unit. The alkylation can be explained by etherification of benzyl alcohols, reetherification of benzyl ethers, esterification of carboxylic acids, and the formation of acetals from benzaldehydes, cinnamaldehydes, and glyceraldehyde-2-aryl ethers.

Four hydroxypropyl lignin (HPL) derivatives were obtained by Siochi et al. [242] by reacting an organosolv lignin with propylene oxide, following a procedure described by Wu and Glasser [243]. The molecular weight distribution of these HPLs was determined by using GPC/LALLS and GPC/DV.

Wood liquefaction can be performed by treatment with phenol or polyhydric alcohols, under acid-catalyzed conditions [244–247]. Many efforts have been done to replace phenol with multifunctional alcohols as (poly)ethylene glycol, glycerol and diethylene glycol [248–250]. The addition of glycerol and diethylene glycol under the PTSA catalysis allowed to liquefy wood in a 3:1 (by mass) ratio [248]. The liquefaction of lignin with phenol has also been studied on models [251, 252].

7.3.3 Halogenation

Halogenation forms the basis for remarkable technological and laboratory processes in the wood and wood-pulping industries (see Chapter 3). Chlorine is used both as a pulping agent[11] and as a bleaching agent [255]. Halogen substitution may occur on the aromatic nucleus or on the side chain of the lignin unit. Table 7.17 shows as a large variety of lignins and conditions that have been used in halogenation.

The chlorination of the lignin was first studied by Cross and Bevan [256, and reference therein] in 1880. By the chlorination of jute, they produced a substance, with a 26.8% of chlorine, which they designated "lignone chloride." This substance dissolved in alkaline solutions and in glacial acetic acid and alcohol. In 1913, Heuser and Sieber [257] studied the action of chlorine on spruce wood. After 2 h, the wood absorbed 31% of its weight of chlorine. The chlorinated product was extracted with absolute alcohol and the resulting product contained 22.7% of chlorine. The percentage of chlorine was, therefore, considerably less than that recorded by Cross and Bevan. Hägglund [258] chlorinated Willstatter lignin at 0 °C for a product containing 46% of chlorine. According to Jonas [259], when lignin is chlorinated under carefully controlled temperature conditions, the same chlorolignin derivatives were consistently obtained.

In 1923, Tropsch [260] refluxed lignin with antimony pentachloride containing a small amount of iodine and produced perchloroethane and hexachlorobenzene.

Polčín [261] has studied the reaction of chlorine gas with alkali sulfite lignin (ASL). He concluded that, in aqueous media, methoxyl groups were oxidatively split from aromatic rings and that o-quinones were formed.

For the isolation of halolignins, the procedures employed depend on the starting material. For wood meal halogenated, the procedure consists of an extraction with an organic solvent,[12] followed by a concentration in vacuo, and precipitation into water. For isolated lignin, the halolignin is usually recovered directly by filtration or after precipitation by the addition of water. Halolignins are

---

[10] Alcohol/dioxane, 2:1, catalyst, 0.15 M p-toluenesulfonic acid (PTSA).
[11] The Celdecor-Pomilio process consists of chopping the straw, soaking it in caustic soda, digesting it mildly, and chlorinating the semipulp [253]. The chlorination is the second stage of the digestion process. The chlorolignins formed are dissolved with an alkaline wash, and the pulp screened and bleached [254].
[12] Usually 96% EtOH.

220 Lignin and Lignans as Renewable Raw Materials

Table 7.17 Composition of halolignins prepared under different conditions[a]

Substrate	Reaction medium	Time[b]	T (°C)	C	H	Cl	OCH$_3$
Chlorine reactive							
Spruce wood	H$_2$O	c	c	42.5	4.12	25.0	4.70
Birch wood	H$_2$O	10.0 h	20.0	35.4	5.3	12.1	8.10
White spruce wood	MeOH	d	c	43.1	3.4	35.4	16.95
Poplar wood	CCl$_4$	10.0 h	20.0	40.4	2.6	28.0	9.65
Maple wood	MeOH	2.5 h	65.0	40.8	3.8	29.5	19.80
Spruce HCl lignin	H$_2$O	5.0 h	c	–	–	23.8	–
Spruce HCl lignin	AcOH	10.0 h	c	–	–	31.1	–
Spruce Klason lignin	MeOH	c	10.0	–	–	34.1	16.80
Maple Klason lignin	MeOH	c	10.0	–	–	33.4	16.60
White spruce MeOH lignin	CCl$_4$	c	c	–	–	26.3	12.10
Softwood lignosulfonic acid	H$_2$O	9.0 h	0.0	44.3	4.1	27.1	8.10
Bromine reactive							
Red pine wood	CCl$_4$	4.0 h	2.0	–	–	–	–
Spruce Klason lignin	H$_2$O	c	c	–	–	–	5.40
Spruce Klason lignin	CCl$_4$	c	c	–	–	–	9.50
Spruce EtOH lignin	AcOH	1.0 h	c	38.5	3.35	–	6.60

[a] Data taken from ref. [255].
[b] h = hours, d = days.
[c] Unspecified.
[d] Undeterminable.

commonly soluble in NaOH, Na$_2$CO$_3$, and NH$_4$OH solutions, pyridine, acetic acid, ethyl acetate, alcohol, acetone, dioxane, and DMSO, but insoluble in water, chloroform, CCl$_4$, ether, benzene, and petroleum ether. The M_w lies in the range 1000 to 6000 [255, and references therein].

In 1921, Karrer and Widmer [262] found that wood dissolved completely when treated for several hours with acetyl bromide. From this solution, a bromine derivative of lignin may be separated.

In 1995, Brage et al. [263] proposed a general mechanism for the reactions of conjugated lignin structures with chlorine dioxide (see Figure 7.20). This includes the initial oxidation on the conjugated structures by the oxidant to give radical cations and chlorous acid. The β-radical mesomer of this cation reacts with another molecule of chlorine dioxide affording chlorous esters of the methylene quinonium ↔ benzylium ion type. The latter are considered to be key intermediates in the formation of the final products. The subsequent steps are ionic in nature, all involving an attack on the C_α atom of the quinonium ↔ benzylium ion moiety by internal or external nucleophiles.

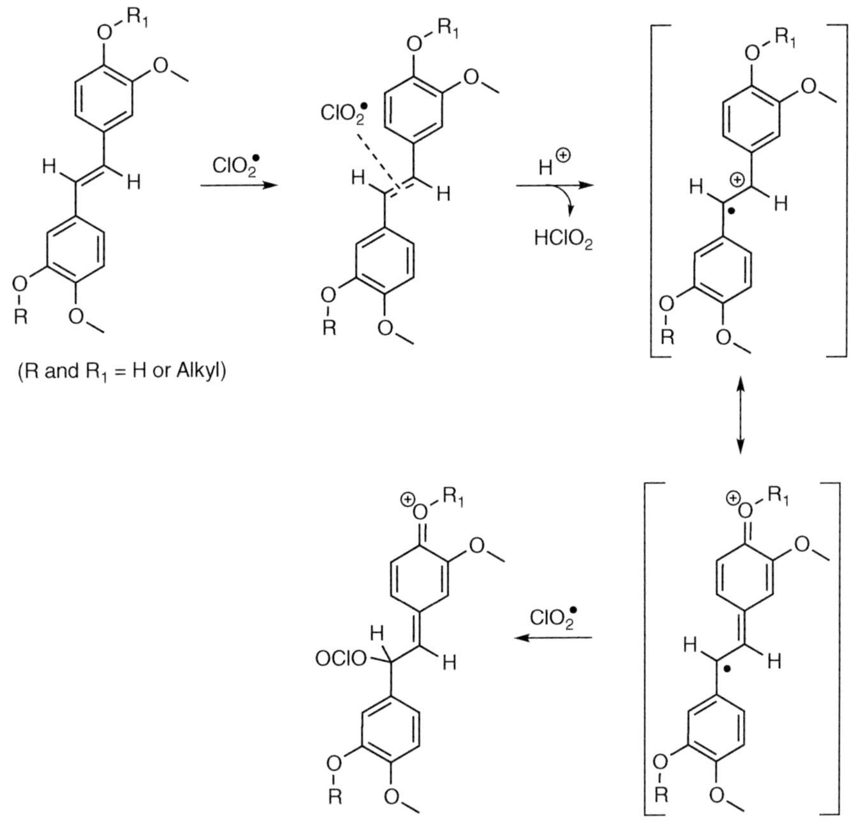

Figure 7.20 Initial radical reactions of conjugated structures with chlorine dioxide (formation of chlorous ester intermediates) [263]

7.3.4 Nitration

Lignin is nitrated very readily with HNO_3, as demonstrated in the 1920s by several authors [264–268], and it was also found to react with gaseous NO_2.

For the nitration of lignin, diluted HNO_3 has been used as a nitrating agent (see Table 7.18). Normally, nitration is conducted in nonaqueous medium, and the nitrolignin produced is isolated, by precipitation, using water addition. The nitrolignin solubility is similar to those observed for halolignins. Nitrolignins present nitrogen as aromatic nitro groups, as has been established by the identification of monomeric nitroaromatic compounds [255, and references therein].

Glycol-lignin[13] was readily nitrated by adding it to a mixture consisting of acetic anhydride/fuming HNO_3 (4:1) at temperature below 0 °C. Acetylation occurred simultaneously. The nitrated lignin could be reduced with sodium amalgam in alkaline solution. The resulting reduction product could be diazotized, and the latter when coupled with β-naphthol disulfonic acid afforded a product that dyed silk and wool a brownish color when mordanted with tannic acid [269].

---

[13] Obtained by digesting wood with ethylene glycol at 110 °C, in the presence of a catalyst, such as iodine or HCl.

Table 7.18 Composition of nitrolignins prepared under different conditions[a]

Substrate	Reaction medium	Time[b]	T (°C)	C	H	Total N	NO_2	OCH_3
		HNO_3 reactive						
Spruce wood	MeOH	2.0 h	65.0	58.1	5.12	3.1	10.0	15.35
Spruce wood	EtOH	2.0 h	78.5	57.1	5.09	3.6	10.9	10.07
Poplar wood	HNO_3	4.0 d	40.0	52.8	5.50	3.7	12.1	8.20
Spruce HCl lignin	HNO_3	c	c	52.4	3.90	4.3	14.1[d]	9.60
Pine wood	HNO_3/H_3PO_4	4.0 h	20.0	–	–	12.4	8.2	3.60
		$AcOH/NO_3$[e] reactive						
Beech wood	$AcOH/NO_3$[e]	6.0 h	0.0	–	–	5.9	8.2	5.00
		NO_2 reactive						
Methylated spruce cuoxam lignin	–	0.5 h	c	58.4	5.70	4.0	13.1[d]	23.50

[a] Data taken from ref. [255].
[b] h = hours, d = days.
[c] Unspecified.
[d] Calculated assuming all nitrogen to be present in the form of nitro groups.
[e] Acetyl nitrate prepared from acetic acid anhydride with nitric acid.

Phelps [270, p. 32] produced a yellow dyestuff by the treatment of ligninsulfonic acid with HNO_3. Other nitro derivatives of ligninsulfonic acid were prepared by Oman [271] and by Doroe and Hall [230].

7.3.5 Sulfonation

When wood is heated with a solution of sulfurous acid and acid sulfites, the lignin is solubilized, leaving cellulose in a more or less pure condition. This is the base in the sulfite process for pulping wood. The solution containing the lignin (waste sulfite liquor) has been the subject of numerous investigations (see Chapter 5).

Lindsey and Tollens [272], in 1892, were the first to demonstrate that the lignin in waste sulfite liquor was present in the form of a calcium salt of a sulfonic acid. Klason [273] employed calcium chloride for the precipitation of the ligninsulfonic acids. The fraction that was precipitated with this reagent, Klason, called α-ligninsulfonic acid, and the portion that remained in solution was called β-ligninsulfonic acid. According to Klason, of the total ligninsulfonic acids in sulfite liquor, roughly two-thirds were found to be in the form of the α-acid and one-third in the form of the β-acid. The most important mechanisms of degradation and condensation during sulfate pulping have been clarified by Gierer and Lindeberg [274] in the 1970s.

7.3.6 Oxidation

Various chemical oxidations of lignin have been investigated over many years. Oxidation reactions were initially used to identify the structure of lignin polymers and the mode of linkage between the precursors [63]. The use of strong oxidizing agents has shown the break-up of the aromatic ring, whereas milder oxidizing agents caused changes in the side chain while keeping the ring intact. Degradation of the lignin polymer yielded various phenylpropanoid compounds, which confirmed these as the building block of lignin [217]. Lignin offers an enormous potential as a feedstock for renewable aromatic compounds [275–277]. In addition, the removal of lignin from lignocellulosic fibers is noticeable for textile and paper applications. Lignin removal required the breakage of chemical linkages, commonly through oxidation with metal salts or metal complexes [278].

According to Chan and Allan [27] the oxidation reactions can be classified in three categories according to the degree of lignin degradation observed:

- Degrading lignin to aromatic carbonyl compounds and carboxylic acids.
- Degrading aromatic rings, normally to CO_2 and dicarboxylic aliphatic acids.
- Oxidation limited to specific groups.

The first category is relevant in the characterization of lignin and the production of commercial aromatic products, and it comprises (includes), as is well known, oxidation with nitrobenzene, metal oxides, molecular oxygen, and permanganate, all in alkaline medium. The second category includes strong oxidants such as peracetic acid, HNO_3, Cl_2, and other chlorine derivatives, O_3 and H_2O_2. The final category includes oxidants employed in functional group analysis as periodic salts, Fremy's salt, DDQ, and alkali peroxides.

7.3.6.1 Molecular Oxygen

In 1921, Fischer et al. [279] subjected lignin to "pressure oxidation," which consisted essentially of heating lignin with 1.253N NaOH solution in an autoclave at 200 °C, under 55 atm pressure; in addition to some simple aliphatic acids, small quantities of aromatic acids resulted.

Oxygen is an oxidizing agent that has also been used in the conversion of lignin into aldehydes. The method requires the use of high temperatures and pressures. Various examples confirming aldehyde formation using catalysts such as ammoniated manganese sulfate and copper sulfate [280], or a perovskite-type oxide catalyst [281]. Recently, Wong et al. [282] showed the formation of vanillin and syringaldehyde in an oxygen delignification process of pine, eucalyptus, and wheat straw.

In oxidation with O_2 of four softwood and three hardwood lignins, *Pinus* spp. Kraft lignin provides the highest production of vanillin, but only in 4.4% w/w[14][283].

7.3.6.2 Hydrogen Peroxide

Hydrogen peroxide (H_2O_2) has found worldwide use as a bleaching agent in the pulp industry. H_2O_2 is a strong oxidizing agent. Its alkaline reaction with model compounds, such as cresol, was found to produce only aliphatic acids. However, at the lowest temperature H_2O_2 was able to degrade only unetherified phenolics [284]. Oxidative studies on lignin using biocatalysts and organometallic catalysts with oxygen and H_2O_2 have shown the formation of various oxidative products [285].

Oxidation of lignin with H_2O_2 in neutral and in alkaline solutions yielded simple organic acids, such as formic, acetic, oxalic, succinic, and malonic acids [27], [286, and references therein].

Two mechanistic types have been proposed: In the first type, the reaction is initiated by the attack of a hydroperoxide anion (HOO^-) at electron-deficient sites within the molecule (see Figure 7.21). In the second type, this is initiated by an attack of a phenolate anion upon unionized H_2O_2 (electrophile) [27].

In Figure 7.22, oxidations of several lignin model compounds are shown.

The lignins from grasses pretreated with alkaline H_2O_2 increases the fraction of lignins involved in condensed linkages from 88–95% to ≈99%. This indicates significant scission of β-*O*-4 bonds by pretreatment and/or induction of lignin condensation reactions [287].

---

[14] Corresponding to 36% of nitrobenzene oxidation yield.

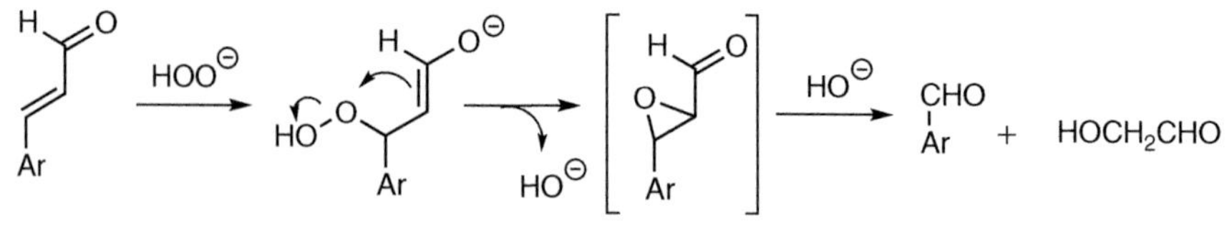

Figure 7.21 Mechanism of the lignin oxidation meditate by H_2O_2 [27, p. 474]

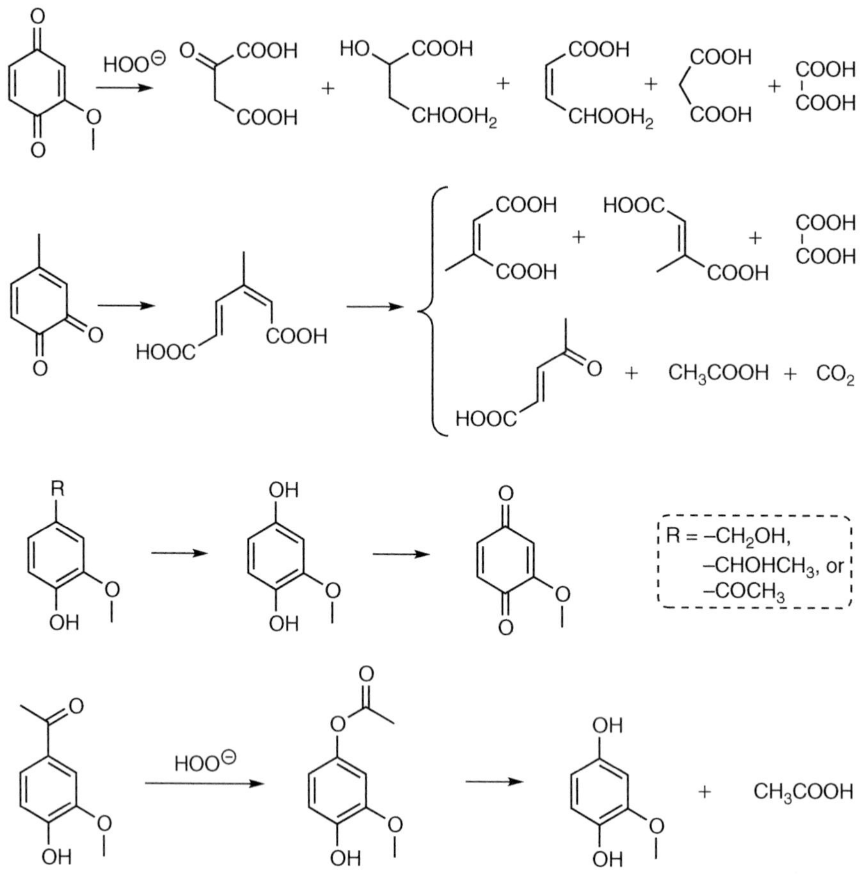

Figure 7.22 Oxidation meditate by H_2O_2 of several lignin model compounds [27]

Lignins were oxidized at 100 °C using aqueous solutions of H_2O_2 and HCl catalyst [288]. Treatment of rye straw with 2% H_2O_2 at pH = 11.5 for 12 h at 20, 30, 40, 50, 60, and 70 °C resulted in a dissolution of 52.7, 75.7, 81.8, 83.1, 85.8, and 87.8% of the original lignin [289], respectively.

7.3.6.3 Organometal Catalyzed Oxidation

Several reviews cover the oxidative valorization of lignins [278, 285, 290].

MnO_2, which is found in decayed wood, combined with an oxalate mixture can oxidize the aromatic components in wheat straw [291], and this represents a possible route for fungi to begin

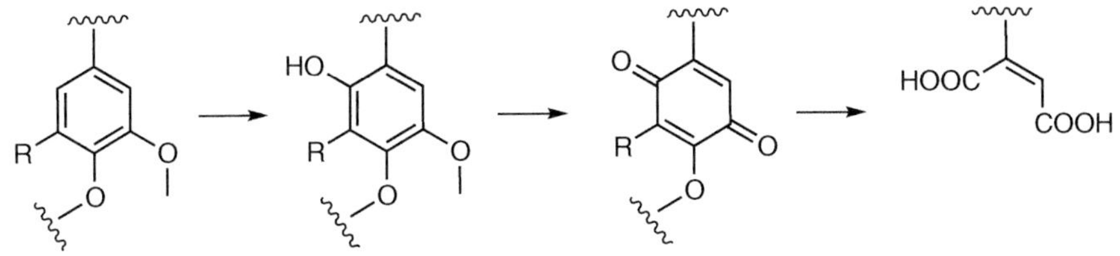

Figure 7.23 *Oxidative rupture of the aromatic ring of lignin with peracids [27]*

the natural degradation of wood. A mixture of $Co(CH_3CO_2)_2/Mn(CH_3CO_2)_2/HBr$ in acetic acid can oxidize the lignin model compound 3,4-dimethoxytoluene to its corresponding benzaldehyde [292]. Five different lignin samples, from wood and bagasse, were oxidized in air with a $Co(CH_3CO_2)_2/Mn(CH_3CO_2)_2/Zr(CH_3CO_2)_4/HBr$ mixture in acetic acid [293]. The oxidation of lignin model compounds to its corresponding aldehyde has been developed by Korpi et al. [294], with Cu^{2+} catalysts and O_2 as the oxidant under alkaline conditions, as a process for the O_2 bleaching of pulp. The iron-TAML catalysts of Collins [295] activate H_2O_2 in the bleaching of wood pulp. Two Mn^{4+} catalysts were studied in the bleaching of pine Kraft-anthraquinone pulp with H_2O_2 under alkaline conditions at 80 °C for 2 h [296]. The oxidation by H_2O_2 of lignin with several water-soluble anionic and cationic iron and manganese porphyrins has been explored by Crestini et al. [297]. Methyltrioxorhenium was found to be a powerful and promising catalyst for the oxidation of technical lignins by using H_2O_2 as a primary oxidant [298]. Oxidation of lignin using aqueous polyoxometalates (POMs) in the presence of alcohols [299], and the use of salen complexes in the presence of H_2O_2 [300] have also been reported.

Four metal organic frameworks (MOFs) are being evaluated as possible catalysts for alkaline oxidation of lignin. A complex MOF of Cu(II) and Fe(III) converted the lignin into monomers in 50% yield. The S/G ratio obtained was similar to those of nitrobenzene oxidation [301].

7.3.6.4 Others Oxidants

Peracid and its salts have proved to be effective oxidants for the bleaching of pulps [27]. This oxidant causes oxidative rupture of the aromatic nuclei to dicarboxylic acids (see Figure 7.23).

Different hydroxyl groups in lignin can be determined using a oxidative process: free phenolic hydroxyl groups with metaperiodate (IO_4^-) [302] oxidation (see Figure 7.24) or with nitrosodisulfonate (Fremy's salt) [303] (see Figure 7.25); benzyl alcohols are converted into aryl ketones by DDQ [304] (see Figure 7.26).

Advanced oxidation processes (AOPs) are based on the photodegradation of soluble lignin using TiO_2/UV photocatalytic technique. The photodegradation of the lignin black liquor has shown products of great interest such as vanillin and vanillic acid [305].

An industrial Kraft lignin was subjected to oxo-ammoniation in ammonia solution and under oxygen atmosphere at 150 °C. The resulting ammoniated samples were isolated from the onset of reaction (OAR). The carboxylic groups in OARs probably occur as ammonium salts or as amide [306].

7.3.7 Other Modifications of Lignins

Lignins present a large number of reactive sites: ether linkages, hydroxyl groups, carbonyl groups, carboxyl and ester functions, ethylenic moieties, sulfur-containing groups. Modifications of side chains and reactive sites on aromatic rings are also possible [307].

Ungureanu subjected lignin to the hydroxymethylation reaction, in order to introduce hydroxyl groups into its structure, thus assuring a more complete exploitation of this natural aromatic

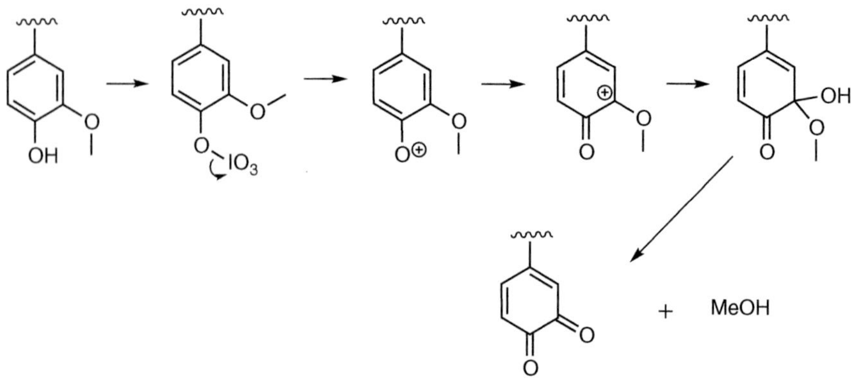

Figure 7.24 Lignin oxidation mediated by IO_4^- [27, p. 472]

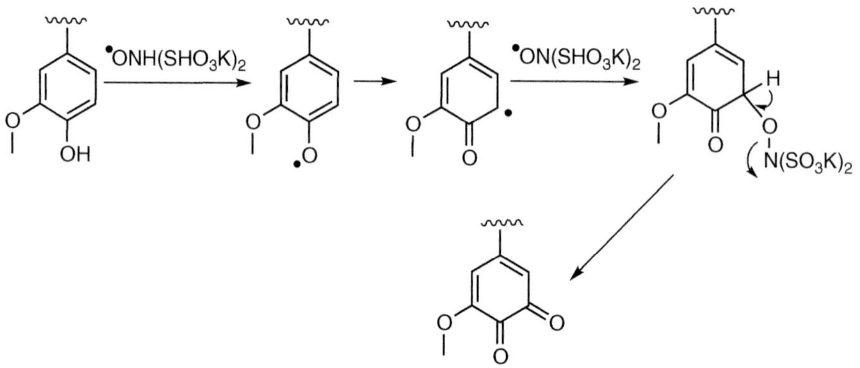

Figure 7.25 Lignin oxidation mediated by Fremy's salt [27, p. 473]

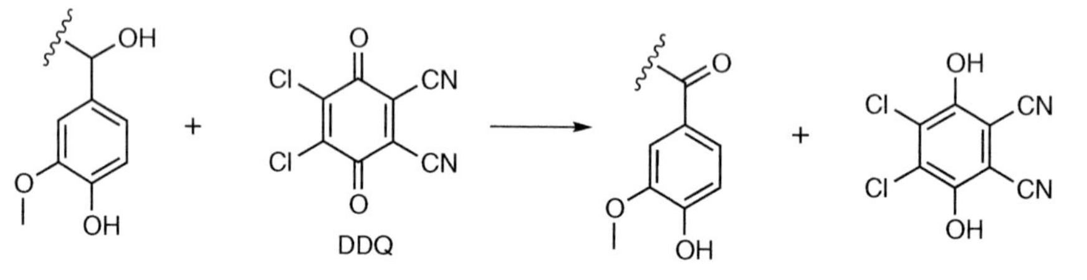

Figure 7.26 Lignin oxidation mediated by DDQ [27, p. 473]

polymer [308]. The hydroxymethylation of alkali lignin with formaldehyde in alkaline solution was also studied by Malutan et al. [309]. The researchers also studied the epoxidation by reaction with epichlorohydrin [310]. El Mansouri et al. [311] studied the glyoxalation reaction (Mannich reaction) of lignins in alkaline medium. The amount of introduced hydroxyl groups augmented with increasing reaction time. These lignins may be good raw material for lignin-based phenol–formaldehyde resins.

Reaction of the lignin with formaldehyde has also been reported for the application to phenolic resins, epoxy resins, adhesives, and polyolefins [312] (see Chapter 9).

Depolymerization of lignin with molecular HI in a nonpolar solvent is a selective, high-yield reaction that releases a 1,3-diiodo-1-(4-hydroxyaryl)propane of potential synthetic value into the solution [313].

7.4 Thermolysis (Pyrolysis) of Lignins

Thermal decomposition of lignin has been investigated mainly for three different purposes:

- The pyrolytic method is appropriate to investigate lignin structure due to rapid analysis using small amounts of samples.
- The reaction process of thermal degradation is required in order to produce carbon from lignin or charcoal from wood.
- Characterization is necessary in order to utilize decomposed materials as source of chemicals.

The topic has been recently reviewed [314–319]. The thermal decomposition of lignins produces four fraction: a carbonaceous residue (coke, about 55%), an aqueous distillate (20%, water, methanol, acetone, and acetic acid), tar (15%, phenolic compounds), and gaseous products (12%, CO, CH_4, CO_2, and ethane) [320].

Various kinds of lignin have been investigated by pyrolysis measurement, for example, KL [321], solvolysis lignin [322], DL [323], MWL [324], MWL [325], Alcell lignin [326], KL [327], KL [328, p. 144], KL [329], and KL [330].

In 2013, Azadi et al. [318] reviewed thermochemical processes that can be used to isolate lignin from the lignocellulosic biomass and subsequently convert it into liquid fuels, hydrogen, and aromatic monomers. They indicated four main processes:

- Pyrolysis of isolated lignins.
- Catalytic hydropyrolysis.
- Sub- and supercritical water treatment.
- Supercritical solvents.

Various bioresources for further conversion to produce biofuels have attracted attention [330, 331]. de Wild et al. [332] have developed a pyrolysis based on the lignin biorefinery approach, called LIBRA, to transform lignin into phenolic bio-oil and biochar using bubbling fluidized bed reactor technology. Bio-oil is a potential source for value-added products that can replace petrochemical phenol in wood adhesives, resins, and polymer applications. Biochar can be used, for example, as a fuel, a soil improver, a solid bitumen additive, and a precursor for activated carbon. Results indicate that ≈80 wt% of the dry lignin can be converted into bio-oil (with a yield of 40–60%) and biochar (30–40%).

From alkali fusion of lignin protocatechuic acid (3,4-dihydroxybenzoic acid), catechol (1,2-dihydroxybenzene), and oxalic acid are the most abundant products isolated [320].

7.5 Biochemical Transformations of Lignins

None of the structural components of wood is easily biodegradable. Cellulose and hemicellulose are polysaccharides that form plant cell walls. The third wood polymer, lignin, is a disordered aromatic polymer that protects the cellulose and hemicellulose against microbial attack. The combination of these materials in wood is even less biodegradable than the individual components. There is, however, a group of microorganisms called white-rot fungi that can degrade all three structural components of wood [333]. Biodegradation chemistry takes advantage of existing reactivity differences between lignin and carbohydrates. The employed enzymes often have high specificity for phenolic structures [334].

Lignin is highly resistant to biodegradation and only higher fungi are capable of degrading polymers *via* an oxidative process. This process has been studied extensively in the past 20 years, but the actual mechanism has not yet been fully elucidated. Lignin is found to be degraded by enzyme lignin peroxidases produced by some fungi.

In recent years, a research line has been opened in order to develop genetically modified plant species with several purposes. Because plants can tolerate large variations in lignin composition, often without apparent adverse effects, substitution of some fraction of the traditional monolignols by alternative monomers through genetic engineering is a promising strategy to tailor lignin in bioenergy crops [335].

7.5.1 Biodegradation of Lignin

Research on lignin biodegradation has accelerated greatly in recent years, mainly because of the broad potential applications of bioligninolytic systems in pulping, bleaching, converting lignins into useful products, and treating wastes.

Despite numerous studies, it is not entirely clear which microbes, other than certain fungi, degrade the lignin polymer. Fungi are the real lignin degraders. The major degraders of fully lignified plant tissues (> 15 % lignin) are filamentous fungi. Although there are over 2000 species of wood-rotting fungi, the substantial majority of these (over 90%) are white-rot fungi. Many review articles discuss the mechanism of lignin degradation by these organisms [336–338].

The first studies of cellulose and lignin degradation by fungi were performed in the 1920s by Falck and Haag [339]. As summarized by Lyr [340], it became clear in those early studies that fungi were able to attack all the structural components of wood. In all cases, the polysaccharides were degraded simultaneously with lignin.

The degradation of lignin by white-rot fungi has some special features [333]: Firstly, lignin is not degraded during fungal growth but only after nutrient depletion triggers secondary metabolism. Secondly, despite that complete oxidation of lignin is highly exothermic, fungal degradation of lignin actually needs an energy source. Finally, fungi use the same kinds of enzymes (peroxidases and laccases) to initiate lignin degradation that plants use to make lignin. Leisola *et al.* [333] call these curious features "the lignin enigma." Figure 7.27 shows a proposed mechanism for the degradation of veratryl alcohol (a model compound) by lignin peroxidase [341].

Bacterial lignin degradation has been most extensively studied in actinomycetes; particularly *Streptomyces* spp., *Streptomyces viridosporus*, and *Streptomyces setonii* have caused losses of 32–44% of the lignin in spruce, maple, and Agropyron lignocelluloses [342]. What can bacteria not do? They cannot metabolize lignin, because "the molecule is too large for most bacteria to handle, and its activation energy is too high" [343].

Plant litter is composed of complex mixtures of organic components, mainly polysaccharides and lignin. Kogel-Knabner [344] investigated the changes during decay and formation of humic substances by solid-state ^{13}C NMR spectroscopy.

The effect of several white-rot fungi on the lignin in birch wood [42] (see Table 7.5) reveal that the syringylpropane lignin units were preferentially degraded.

Quantification of the protocol for DFRC monomer analysis of products of lignins extracted from biotreated wood samples with *Ceriporiopsis subvermispora*, indicated that β-*O*-aryl cleavage was a significant route for lignin biodegradation but that β-β, β-5, β-1, and 4-*O*-5 linkages were more resistant to the biological attack. The amount of aromatic hydroxyls did not increase with the split of β-*O*-4 linkages, suggesting that the β-*O*-4 cleavage products remain quinone-type structures as detected

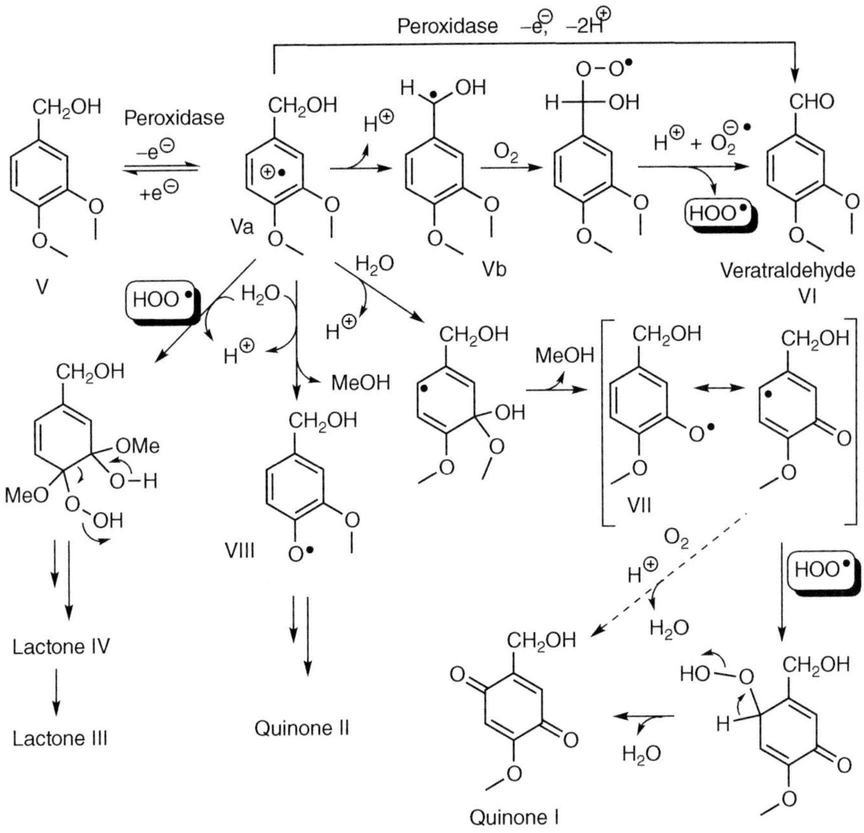

Figure 7.27 Proposed mechanism of the role of activated oxygen in degradation of a monomeric lignin model compound, veratryl alcohol (V), by lignin peroxidase. In the absence of oxygen, veratraldehyde (VI) is the only product formed from the cation radical. In the presence of oxygen, seven other products resulting from radical chemical reactions were detected [341]

by UV/vis spectroscopy. NMR techniques have also indicated the formation of new substructures containing nonoxygenated, saturated aliphatic carbons (CH_2 and CH_3) in the side chains [345].

Wood exposed outdoors undergoes photochemical degradation caused by UV radiation. This degradation takes place primarily in the lignin component, which is responsible for the characteristic color changes [346].

7.5.2 Enzyme-Based Oxidation of Lignin

In nature, lignin is selectively oxidized by white-rot fungi that utilize a series of enzymes to selectively oxidize lignin in the presence of cellulose and hemicellulose, which are enzymatically decomposed by soft-rot fungi and/or red-rot fungi species [347]. The most active lignin degrading enzymes are laccases, manganese peroxidases, and lignin peroxidases. The use of lignolytic enzymes in industrial processes has recently been discussed [348].

Laccase is a multicopper oxidase that oxidizes lignin under concomitant reduction of O_2 to H_2O [349, 350]. The enzyme contains four copper atoms [351]. Blocked phenolic substrates can be indirectly oxidized by laccases, when these are used in combination with a radical mediator species.

Manganese peroxidases represent the second class of enzymes that oxidatively depolymerize lignin under concomitant reduction of O_2 or H_2O_2 to H_2O [352].

Large-scale applications of the enzyme-based valorization of lignin seem to be prohibitive because the need to recycle the costly, active ingredient is compromised by a rapid loss of activity of the enzymatic catalyst [353].

Lignin is hydrolyzed by ligninases, thus releasing cellulose, which can be used to produce EtOH.

Recently, Sethi *et al.* [354] used four marine microorganisms to degrade 18 different lignocellulosic biomasses for a period of 2 months. Maximum degradation was observed in about 10 substrates commonly degraded by all the four organisms in the first 3 weeks. *Mesorhizobium* spp. showed better degradation in all the substrates compared to the other organisms [355].

References

[1]   Lapierre C, Pollet B, Rolando C. New insights into the molecular architecture of hardwood lignins by chemical degradative methods. Res Chem Intermed. 1995;21(3-5):397–412.

[2]   Tanahashi M, Higuchi T. Chemical degradation methods for characterization of lignins. Methods Enzymol. 1988;161(Biomass, Pt. B):101–109.

[3]   Adler E. Lignin chemistry-past, present and future. Wood Sci Technol. 1977;11:169–218.

[4]   Lundquist K. Low-molecular weight lignin hydrolysis products. Appl Polym Symp. 1976;3(28):1393–1407.

[5]   Nimz H. Beech lignin-proposal of a constitutional scheme. Angew Chem Int Ed. 1974;13(5):313–321.

[6]   Sakakibara A. Hydrogenolysis. In: Lin SY, Dence CW, editors. Methods in Lignin Chemistry, Springer Series in Wood Science. Springer Berlin Heidelberg; 1992. p. 350–368.

[7]   Freudenberg K. Lignin. Angew Chem. 1939;52:362–363.

[8]   Freudenberg K, Lautsch W, Engler K. Lignin. XXXIV. Formation of vanillin from spruce lignin. Naturwissenschaften. 1940;73:167–171.

[9]   Leopold B. Aromatic keto- and hydroxy-polyethers as lignin models. III. Acta Chem Scand. 1950;4:1523–1537.

[10] Leopold B, Lignin X. Nitrobenzene oxidation of the products formed by the condensation of resorcinol with lignin models. Acta Chem Scand. 1951;5:1393–13934.

[11] Leopold B, Malmström IL, Motzfeldt K, Finsnes E, Sörensen JS, Sörensen NA. Studies on lignin. III. Oxidation of wood from *Picea abies* (L.) karst. (Norway spruce) with nitrobenzene and alkali. Acta Chem Scand. 1952;6:38–48.

[12] Erickson M, Larsson S, Miksche GE, Kaipainen K, Aaltonen R, Swahn CG. Gaschromatographische analyse von ligninoxydationsprodukten. VII. Ein verbessertes verfahren zur charakterisierung von ligninen durch methylierung und oxydativen abbau. Acta Chem Scand. 1973;27:127–140.

[13] Gellerstedt G. Chemical degradation methods: permanganate oxidation. In: Lin SY, Dence CW, editors. Methods in Lignin Chemistry, Springer Series in Wood Science. Springer Berlin Heidelberg; 1992. p. 322–333.

[14] Nimz H. Über ein neues abbauverfahren des lignins. Chem Ber. 1969;102(3):799–810.

[15] Sakakibara A. A structural model of softwood lignin. Wood Sci Technol. 1980;14(2):89–100.

[16] Lapierre C, Rolando C, Monties B. Characterization of poplar lignins acidolysis products: capillary gas-liquid and liquid-liquid chromatography of monomeric compounds. Holzforschung. 1983;37(4):189–198.

[17] Lapierre C, Monties B, Rolando C. Thioacidolysis of lignin: Comparison with acidolysis. J Wood Chem Technol. 1985;5(2):277–292.

[18] Rolando C, Monties B, Lapierre C. Thioacidolysis. In: Lin SY, Dence CW, editors. Methods in Lignin Chemistry, Springer Series in Wood Science. Springer Berlin Heidelberg; 1992. p. 334–349.

[19] Lu F, Ralph J. Derivatization followed by reductive cleavage (DFRC method), a new method for lignin analysis: protocol for analysis of DFRC monomers. J Agric Food Chem. 1997;45(7):2590–2592.

[20] Lu F, Ralph J. The DFRC method for lignin analysis. 2. Monomers from isolated lignins. J Agric Food Chem. 1998;46(2):547–552.

[21] Gellerstedt G, Henriksson G. Lignins: major sources, structure and properties. In: Belgacem MN, Gandini A, editors. Monomers, Polymers and Composites from Renewable Resources. Amsterdam: Elsevier; 2011. p. 201–224.

[22] Freudenberg K, Lautsch W. The constitution of pine lignin. Naturwissenschaften. 1939;27:227–228.

[23] Leopold B, Malmstrom IL. Lignin. IV. Investigation on the nitrobenzene oxidation products of lignin from different woods by paper partition chromatography. Acta Chem Scand. 1952;6:49–54.

[24] Creighton RHJ, Hibbert H. Studies on lignin and related compounds. LXXVI. Alkaline nitrobenzene oxidation of corn stalks. isolation of p-hydroxybenzaldehyde. J Am Chem Soc. 1944;66(1):37–38.

[25] Creighton RHJ, Gibbs RD, Hibbert H. Studies on lignin and related compounds. LXXV. Alkaline nitrobenzene oxidation of plant materials and application to taxonomic classification. J Am Chem Soc. 1944;66(1):32–37.

[26] Higuchi T, Ito Y, Shimada M, Kawamura I. Chemical properties of milled wood lignin of grasses. Phytochemistry. 1967;6(11):1551–1556.

[27] Chang HM, Allan GG. Oxidation. In: Sarkanen KV, Ludwig CH, editors. Lignins: Occurrence, Formation, Structure and Reactions. New York: Wiley-Interscience; 1971. p. 433–485.

[28] Chen CL. Nitrobenzene and cupric oxide oxidations. In: Lin SY, Dence CW, editors. Methods in Lignin Chemistry, Springer Series in Wood Science. Springer Berlin Heidelberg; 1992. p. 301–321.

[29] Pew JC. Nitrobenzene oxidation of lignin model compounds, spruce wood and spruce 'native lignin'. J Am Chem Soc. 1955;77(10):2831–2833.

[30] Bagby MO, Nelson GH, Helman EG, , Clark TF. Determination of lignin in non-wood plant fiber sources. Tappi. 1971;54:1876–1878.

[31] Brauns FE, Brauns DA. The Chemistry of Lignin: Supplement. New York: Academic Press; 1960.

[32] Pearl IA. The Chemistry of Lignin. New York: M. Dekker; 1967.

[33] Meshitsuka G, Nakano J. Structural characteristics of compounds middle lamella lignin. J Wood Chem Technol. 1985;5(3):391–404.

[34] Obst JR, Sachs IB, Kuster TA. The quantity and type of lignin in tyloses of bur oak (*Quercus macrocarpa*). Holzforschung. 1988;42(4):229–231.

[35] Leopold B. Lignin. XI. The alkaline nitrobenzene oxidation of lignin and lignin models. Sven Kem Tidskr. 1952;64:18–26.

[36] Abbott TP, James C, Bagby MO. Oxidation and quantification of ^{14}C-lignin at different ages in wheat, pine, oak, and kenaf. J Wood Chem Technol. 1986;6(4):473–486.

[37] Kratzl K, Silbernagel H. Über das verhalten von fichtenholz und lignin bei thermischer behandlung mit wasser. Monatsh Chem. 1952;83(4):1022–1037.

[38] Jayne JE. A study of the alkaline nitrobenzene oxidation of chlorite lignin. Tappi. 1953;36:571–576.

[39] Cymbaluk NF, Neudoerffer TS. A quantitative gas-liquid chromatographic determination of aromatic aldehydes and acids from nitrobenzene oxidation of lignin. J Chromatogr A. 1970;51(C):167–174.

[40] Van Buren JB, Dence CW. Chlorination behavior of pine Kraft lignin. Tappi. 1970;53(12):2246–2253.

[41] Blanchette RA, Obst JR, Hedges JI, Weliky K. Resistance of hardwood vessels to degradation by white rot *Basidiomycetes*. Can J Bot. 1988;66(9):1841–1847.

[42] Nilsson T, Daniel G, Kirk TK, Obst JR. Chemistry and microscopy of wood decay by some higher ascomycetes. Holzforschung. 1989;43(1):11–17.

[43] Winston MH, Chen CL, Gratzl JS, Goldstein IS. Characterization of the lignin residue from hydrolysis of sweetgum wood with superconcentrated hydrochloric acid. Holzforschung. 1986;40(Suppl.):45–50.

[44] Pearl IA. Vanillin from lignin materials. J Am Chem Soc. 1942;64(6):1429–1431.

[45] Pearl IA, Beyer DL. Lignin and related products. VI. The oxidation of fermented sulfite spent liquor with cupric oxide under pressure. Tappi. 1950;33:544–548.

[46] Pepper JM, Casselman BW, Karapally JC. Lignin oxidation. Preferential use of cupric oxide. Can J Chem. 1967;45(23):3009–3012.

[47] Freudenberg K, Janson A, Knopf E, Haag A, Meister M. Lignin. XV. Ber Dtsch Chem Ges B. 1936;69:1415–1425.

[48] Freudenberg K. The constitution and biosynthesis of lignin. In: Freudenberg K, Neish AC, editors. Constitution and Biosynthesis of Lignin, Molecular Biology, Biochemistry and Biophysics. vol. 2. Berlin: Springer-Verlag; 1968. p. 45–122.

[49] Larsson S, Miksche GE. Gas chromatographic analysis of lignin oxidation products. Diphenyl ether linkage in lignin. Acta Chem Scand. 1967;21(7):1970–1971.

[50] Larsson S, Miksche GE, Larsen C, Nielsen PH, Werner PE, Junggren U, et al. Gaschromatographische analyse von ligninoxydationsprodukten. II. Nachweis eines neuen verknüpfungsprinzips von phenylpropaneinheiten. Acta Chem Scand. 1969;23:917–923.

[51] Larsson S, Miksche GE, Cowling EB, Lin SY, Chan RPK, Craig JC. Gaschromatographische analyse von ligninoxydationsprodukten. III. Oxydativer abbau von methyliertem Björkman–lignin (fichte). Acta Chem Scand. 1969;23:3337–3351.

[52] Larsson S, Miksche GE, Sood MS, Nielsen BE, Ljunggren H, Ehrenberg L. Gaschromatographische analyse von ligninoxydationsprodukten. IV. Zur struktur des lignins der birke. Acta Chem Scand. 1971;25:647–662.

[53] Larsson S, Miksche GE, Sood MS, Nielsen BE, Ljunggren H, Ehrenberg L. Gaschromatographische analyse von ligninoxydationsprodukten. V. Zwei trimere abbauprodukte aus fichtenlignin. Acta Chem Scand. 1971;25:673–679.

[54] Larsson S, Miksche GE, Danielsen J, Haaland A, Svensson S. Gaschromatographische analyse von ligninoxydationsprodukten. VI. 4,4-Bis-aryloxy-cyclohexa-2,5-dienonstrukturen im lignin. Acta Chem Scand. 1972;26:2031–2038.

[55] Miksche GE, Yasuda S. Structure of beech (*Fagus silvatica* L.) lignin. Justus Liebigs Ann Chem. 1976;1976(7–8):1323–1332.

[56] Erickson M, Miksche GE, Enzell CR, Liaaen-Jensen S, Enzell CR, Mannervik B. Two dibenzofurans obtained on oxidative degradation of the moss *Polytrichum commune* Hedw. Acta Chem Scand, Ser B. 1974;28:109–113.

[57] Erickson M, Miksche GE. Charakterisierung der lignine von Gymnospermen durch oxidativen abbau. Holzforschung. 1974;28(4):135–138.

[58] Erickson M, Miksche GE. Charakterisierung der lignine von pteridophyten durch oxidativen abbau. Holzforschung. 1974;28(5):157–159.

[59] Erickson M, Miksche GE. On the occurrence of lignin or polyphenols in some mosses and liverworts. Phytochemistry. 1974;13(10):2295–2299.

[60] Freudenberg K, Chen CL, Cardinale G. Die oxydation des methylierten natürlichen und künstlichen lignins. Chem Ber. 1962;95(11):2814–2828.
[61] Andersson A, Erickson M, Fridh H, Miksche GE. Zur struktur des lignins der rinde von laub- und nadelhölzern. Holzforschung. 1973;27(6):189–193.
[62] Glasser WG, Barnett CA. Structure of lignins in pulps - 3. The association of isolated lignins with carbohydrates. Tappi. 1979;62(8):101–105.
[63] Morohoshi N, Glasser WG. The structure of lignins in pulps - Part 4: comparative evaluation of five lignin depolymerization techniques. Wood Sci Technol. 1979;13(3):165–178.
[64] Morohoshi N, Glasser WG. The structure of lignins in pulps - Part 5: gas and gel permeation chromatography of permanganate oxidation products. Wood Sci Technol. 1979;13(4):249–264.
[65] Gellerstedt G, Lindfors EL. Structural changes in lignin during Kraft pulping. J Wood Chem Technol. 1984;38(3):151–158.
[66] Gellerstedt G, Gustafsson K. Structural changes in lignin during Kraft cooking. Part 5. Analysis of dissolved lignin by oxidative degradation. J Wood Chem Technol. 1987;7(1):65–80.
[67] Gellerstedt G, Gustafsson K, Lindfors EL. Structural changes in lignin during oxygen bleaching. Nord Pulp Pap Res J. 1986;1(3):14–17.
[68] Gellerstedt G, Gustafsson K, Northey RA. Structural changes in lignin during kraft cooking. Part 8. Birch lignins. Nord Pulp Pap Res J. 1988;3(2):87–94.
[69] Gellerstedt G, Lindfors EL, Lapierre C, Monties B. Structural changes in lignin during Kraft cooking. Part 2. Characterization by acidolysis. Sven Papperstidn. 1984;87(9):R61–R67.
[70] Erickson M, Dence CW. Phenolic and chlorophenolic oligomers in chlorinated pine Kraft pulp and in bleach plant effluents. Sven Papperstidn. 1976;79(10):316–322.
[71] Lindström K, Österberg F. Characterization of the high molecular mass chlorinated matter in spent bleach liquors (SBL) - Part 1. Alkaline SBL. Holzforschung. 1984;38(4):201–212.
[72] Österberg F, Lindström K. Characterization of the high molecular mass chlorinated matter in spent bleach liquors (SBL). Part II. Acidic SBL. Holzforschung. 1985;39(3):149–158.
[73] Österberg F, Lindström K. Characterization of the high molecular mass chlorinated matter in spent bleach liquors (SBL): 3-mass spectrometric interpretation of aromatic degradation products in SBL. Org Mass Spectrom. 1985;20(8):515–524.
[74] Erickson M, Larsson S, Miksche GE, Wiehager AC, Lindgren BO, Swahn CG. Gaschromatographische analyse von ligninoxydationsprodukten. VIII. Zur struktur des lignins der fichte. Acta Chem Scand. 1973;27:903–914.
[75] Bose SK, Wilson KL, Hausch DL, Francis RC. Lignin analysis by permanganate oxidation. Part 2. Lignins in acidic organosolv pulps. Holzforschung. 1999;53(6):603–610.
[76] Bose SK, Wilson KL, Francis RC, Aoyama M. Lignin analysis by permanganate oxidation. Part 1. Native spruce lignin. Holzforschung. 1998;52(3):297–303.
[77] Parkås J, Brunow G, Lundquist K. Quantitative lignin analysis based on permanganate oxidation. BioResources. 2007;2(2):169–178.
[78] Javor T, Buchberger W, Faix O. Capillary electrophoretic determination of lignin degradation products obtained by permanganate oxidation. Anal Chim Acta. 2003;484(2):181–187.
[79] Nimz H. Isolation of guaiacylglycerol and its dimeric β-aryl ether from spruce lignin. Chem Ber. 1967;100(1):181–186.
[80] Nimz H. Isolation of guaiacylglycerol β-coniferyl ether from spruce wood. Chem Ber. 1965;98(2):533–537.
[81] Nimz H. Oligomeric degradation phenols of pine lignin. Chem Ber. 1966;99(8):2638–2651.
[82] Nimz H. Two aldehydic dilignols from spruce lignins. Chem Ber. 1967;100(8):2633–2639.
[83] Nimz H. Mild hydrolysis of beech lignin. III. Isolation of two additional degradation products with a 1,2-diarylpropane structure. Chem Ber. 1966;99(2):469–474.

[84] Sakakibara A, Sano Y. Chemistry of lignin. In: Hon DNS, editor. Wood and Cellulosic Chemistry. New York: Marcel Dekker; 2001. p. 109–174.

[85] Sakakibara A, Nakayama N. Hydrolysis of lignin with dioxane and water. I. Formation of cinnamic alcohols and aldehydes. Mokuzai Gakkaishi. 1961;7:13–18.

[86] Sakakibara A, Nakayama N. Hydrolysis of lignin with dioxane and water. II. Identification of hydrolysis products. Mokuzai Gakkaishi. 1962;8:153–156.

[87] Sakakibara A, Nakayama N. Hydrolysis of lignin with dioxane and water. III. Hydrolysis products of various woods and lignin preparations. Mokuzai Gakkaishi. 1962;8:157–162.

[88] Sano Y, Sakakibara A. Hydrolysis of lignin with dioxane and water. VII. Isolation of dimeric and trimeric compounds with 1,2-diarylpropane structure. Mokuzai Gakkaishi. 1970;16(3):121–125.

[89] Omori S, Sakakibara A. Hydrolysis of lignin with dioxane and water. X. Isolation of arylglycerols and diarylpropanediol from yachidamo (Fraxinus mandshurica). Mokuzai Gakkaishi. 1972;18(7):355–360.

[90] Sano Y, Sakakibara A. Hydrolysis of lignin with dioxane and water. XIV. Isolation of a new trimeric compound. Mokuzai Gakkaishi. 1975;21(8):461–465.

[91] Omori S, Sakakibara A. Hydrolysis of lignin with dioxane-water. XVI. Isolation of syringylglycerol-β-syringaresinol ether from hardwood lignin. Mokuzai Gakkaishi. 1979;25(2):145–148.

[92] Aoyama M, Sakakibara A. Hydrolysis of lignin with dioxane-water. XVIII. Isolation of a new lignol from hardwood lignin. Mokuzai Gakkaishi. 1979;25(10):644–646.

[93] Sano Y, Sakakibara A. Hydrolysis of lignin with dioxane and water. Hydrolysis of some model compounds. Mokuzai Gakkaishi. 1976;22(9):526–531.

[94] Sano Y, Endo S, Sakakibara A. Hydrolysis of lignin with dioxane and water. Hydrolysis of some model compounds. (3). Mokuzai Gakkaishi. 1977;23(4):193–198.

[95] Sarkanen KV, Ludwig CH. Lignins: Occurrence, Formation, Structure and Reactions. New York: Wiley-Interscience; 1971.

[96] Sano Y, Sakakibara A. Delignification of woods by solvolysis. III. Isolation, identification and mechanism of formation of low molecular-weight solvolysis lignins. Mokuzai Gakkaishi. 1985;31(2):109–118.

[97] Sano Y. Reactivity of β-O-4 linkages in lignin during solvolysis pulping. Degradation of β-O-4 lignin model compounds. Mokuzai Gakkaishi. 1989;35(9):813–819.

[98] Westermark U, Samuelsson B, Lundquist K. Homolytic cleavage of the β-ether bond in phenolic β-O-4 structures in wood lignin and in guaiacylglycerol-β-guaiacyl ether. Res Chem Intermed. 1995;21(3–5):343–352.

[99] Pepper JM, Baylis PET, Adler E. Isolation and properties of lignins obtained by the acidolysis of spruce and aspen woods in dioxane-water medium. Can J Chem. 1959;37:1241–1248.

[100] Lundquist K. Acidolysis. In: Lin SY, Dence CW, editors. Methods in Lignin Chemistry, Springer Series in Wood Science. Springer Berlin Heidelberg; 1992. p. 289–300.

[101] Pla F, Froment P, Mouttet B, Robert A. Étüde de la délignification des végétaux par acidolyse. Holzforschung. 1984;38:127–132.

[102] Stumpf W, Freudenberg K. Lösliches lignin aus fichten- und buchenholz. Angew Chem Int Ed. 1950;62(22):537–.

[103] Stumpf W, Weygand F, Grosskinsky OA. Synthesis of C14-labeled dioxane and its use as an extraction solvent for soluble lignin. Chem Ber. 1953;86:1391–1401.

[104] Yasuda S, Ota K. Chemical structures of sulfuric acid lignin. Part X. Reaction of syringylglycerol-β-syringyl ether and condensation of syringyl nucleus with guaiacyl lignin model compounds in sulfuric acid. Holzforschung. 1987;41(1):59–65.

[105] Karlsson O, Lundquist K, Meuller S, Westlid K, Lönnberg H, Berg JE, et al. On the acidolytic cleavage of arylglycerol β-aryl ethers. Acta Chem Scand, Ser B. 1988;42:48–51.

[106] Lundquist K, Kirk TK. Acid degradation of lignin. IV. Analysis of lignin acidolysis products by gas chromatography, using trimethylsilyl derivatives. Acta Chem Scand. 1971;25(3):889–894.

[107] Nakatsubo F, Tanahashi M, Higuchi T. Acidolysis of bamboo lignin. II. Isolation and identification of acidolysis products. Wood Res. 1972;53:9–18.

[108] Higuchi T, Tanahashi M, Sato A. Acidolysis of bamboo lignin. I. Gas-liquid chromatography and mass spectrometry of acidolysis monomers. Mokuzai Gakkaishi. 1972;18(4):183–189.

[109] Higuchi T, Tanahashi M, Nakatsubo F. Acidolysis of bamboo lignin. III. Estimation of arylglycerol-β-aryl ether groups in lignin. Wood Res. 1973;54:9–18.

[110] Wolter KE, Harkin JM, Kirk TK. Guaiacyl lignin associated with vessels in Aspen *Callus Cultures*. Physiol Plant. 1974;31(2):140–143.

[111] Kutsuki H, Higuchi T. The formation of lignin of *Erythrina crista-galli*. Mokuzai Gakkaishi. 1978;24(9):625–631.

[112] Westermark U. The occurrence of *p*-hydroxyphenylpropane units in the middle-lamella lignin of spruce (*Picea abies*). Wood Sci Technol. 1985;19(3):223–232.

[113] Lapierre C, Rolando C, Monties B. Chromatographic en phase gazeuse sur colonne couplee a la spectrometrie de masse (CGCC-SM) et chromatographic en phase liquide a haute performance (CLHP) des monomeres d'acidolyse des lignines. Bull Liaison - Groupe Polyphenols. 1982;11:381–387.

[114] Lapierre C, Guittet E. Enrichissement photosynthetique en carbone 13 de lignines de peuplier: caractérisation préliminaire par acidolyse et RMN ^{13}C. Holzforschung. 1983;37(5):217–224.

[115] Lapierre C, Monties B, Rolando C. Structural studies of lignins: estimation of arylglycerol aryl ether bonds by thioacidolysis. C R Acad Sci, Ser III. 1984;299(11):441–444.

[116] Lapierre C, Monties B, Rolando C. Thioacidolysis of poplar lignins: Identification of monomeric syringyl products and characterization of guaiacyl–syringyl lignin fractions. Holzforschung. 1986;40(2):113–118.

[117] Pometto ALI, Crawford DL. Simplified procedure for recovery of lignin acidolysis products for determining the lignin-degrading abilities of microorganisms. Appl Environ Microbiol. 1985;49(4):879–881.

[118] Lundquist K. On the occurrence of β–1 structures in lignins. J Wood Chem Technol. 1987;7(2):179–185.

[119] Govender M, Bush T, Spark A, Bose SK, Francis RC. An accurate and non-labor intensive method for the determination of syringyl to guaiacyl ratio in lignin. Bioresour Technol. 2009;100(23):5834–5839.

[120] Lewis NG, Yamamoto E. Lignin: occurrence, biogenesis and biodegradation. Annu Rev Plant Physiol Plant Mol Biol. 1990;41:455–496.

[121] Holmberg B. Thioglycolic acid lignin in spruce wood. Sven Papperstidn. 1930;33:679–686.

[122] Nimz H, Das K. Low molecular weight degradation products of lignin. II. Dimeric phenolic degradation products obtained by degradation of beech lignin with thioacetic acid. Chem Ber. 1971;104(8):2359–2380.

[123] Lapierre C, Polet B, Tollier MT, Chabbert B, Monties B. Molecular profiling of lignins by thioacidolysis: advantages and limitations through ten years of practice. In: International Chemical Recovery Conference. vol. 2. 7th International Symposium, Wood & Pulping Chemistry; (Beijing; China); 1991. p. 818–828.

[124] Node M, Hori H, Fujita E. Demethylation of aliphatic methyl ethers with a thiol and boron trifluoride. J Chem Soc [Perkin 1]. 1976:2237–2240.

[125] Fuji K, Ichikawa K, Node M, Fujita E. Hard acid and soft nucleophile system. New efficient method for removal of benzyl protecting group. J Org Chem. 1979;44(10):1661–1664.

[126] Lapierre C, Pollet B, Monties B, Rolando C. Thioacidolysis of spruce lignin: gas chromatography-mass spectroscopy analysis of the main dimers recovered after Raney nickel desulfurization. Holzforschung. 1991;45(1):61–68.

[127] Lapierre C. Hétérogénéité des lignines de peuplier: Mise an évidence systématique. Paris: Université Paris-Sud; 1986.
[128] Terashima N, Atalla RH, Ralph SA, Landucci LL, Lapierre C, Monties B. New preparations of lignin polymer models under conditions that approximate cell wall lignification. II. Structural characterization of the models by thioacidolysis. Holzforschung. 1996;50(1):9–14.
[129] Lin SY, Dence CW, editors. Methods in Lignin Chemistry. Berlin: Springer; 1992.
[130] Brauns FE. The Chemistry of Lignin. New York: Academic Press; 1952.
[131] Goheen DW. Low molecular weight chemicals. In: Sarkanen KV, Ludwig CH, editors. Lignins: Occurrence, Formation, Structure and Reactions. New York: Wiley-Interscience; 1971. p. 797–831.
[132] Hrutfiord BF. Reduction and hydrogenolysis. In: Sarkanen KV, Ludwig CH, editors. Lignins: Occurrence, Formation, Structure and Reactions. New York: Wiley-Interscience; 1971. p. 487–509.
[133] Lindblad AR. Preparation of oils from wood by hydrogenation. Ing Vetenskap Akad Handl. 1931;107:7–59.
[134] Kashima K, Osada T. The hydrocracking of lignin. Kogyo Kagaku Zasshi. 1961;64:916–919.
[135] Sakakibara A, Kubota M, Oda K. Hydrogenolysis of lignin V. Hydrogenolysis with the catalysts metal carbonyl and nickel carbonate. Ringyo Shikenjo Kenkyu Hokoku. 1964;166:159–171.
[136] Parkhurst HJJ, Huibers DTA, Jones MW. Production of phenol from lignin. Prepr - Am Chem Soc, Div Pet Chem. 1980;25(3):657–667.
[137] Harris EE, D'Ianni J, Adkins H. Reaction of hardwood lignin with hydrogen. J Am Chem Soc. 1938;60(6):1467–1470.
[138] Lautsch W. Oxidative and hydrogenating degradation of wood, lignin and sulfur-containing waste liquor from spruce. Cellul-Chem. 1941;19:69–87.
[139] Freudenberg K, Lautsch W, Piazolo G, Scheffer A. Lignin. XLII. The pressure hydrogenation of lignin and the lignin-containing waste liquors of the spruce. Ber Dtsch Chem Ges B. 1941;74:171–183.
[140] Coscia CJ, Schubert WJ, Nord FF. Investigations on lignins and lignification. XXIV. The application of hydrogenation, hydrogenolysis, and vapor phase chromatography in the study of lignin structure. J Org Chem. 1961;26(12):5085–5091.
[141] Olcay A. Investigations on lignins and lignification. XXIX. Hydrogenation products of spruce milled-wood lignin and of related model compounds. Holzforschung. 1963;17(4):105–110.
[142] Pepper JM, Steck W. The effect of time and temperature on the hydrogenation of aspen lignin. Can J Chem. 1963;41(11):2867–2875.
[143] Pepper JM, Lee YW. Lignin and related compounds. I. Comparative study of catalysts for lignin hydrogenolysis. Can J Chem. 1969;47(5):723–727.
[144] Pepper JM, Lee YW. Lignin and related compounds. II. Studies using ruthenium and Raney nickel as catalysts for lignin hydrogenolysis. Can J Chem. 1970;48(3):477–479.
[145] Pepper JM, Fleming RW. Lignin and related compounds. V. The hydrogenolysis of aspenwood lignin using rhodium-on-charcoal as catalyst. Can J Chem. 1978;56(7):896–898.
[146] Pepper JM, Supathna P. Lignin and related compounds. VI. A study of variables affecting the hydrogenolysis of sprucewood lignin using a rhodium-on-charcoal catalyst. Can J Chem. 1978;56(7):899–902.
[147] Pepper JM, Rahman MD. Lignin and related compounds. XI. Selective degradation of aspen poplar lignin by catalytic hydrogenolysis. Cellul Chem Technol. 1987;21(3):233–239.
[148] Rahman MD, Pepper JM. Lignin and related compounds XII. Catalytic degradation of proto and isolated aspen lignins under initially alkaline conditions. J Wood Chem Technol. 1988;8(3):313–322.

[149] Hwang BH, Sakakibara A. Hydrogenolysis of protolignin. XVIII. Isolation of a new dimeric compound with a heterocycle involving α-,β-diether. Holzforschung. 1981;35(6):297–300.

[150] Hwang BH, Sakakibara A. Hydrogenolysis of protolignin. XV. Further isolation of some compounds from hardwood lignin. Mokuzai Gakkaishi. 1979;25(10):647–652.

[151] Hwang BH, Sakakibara A, Miki K. Hydrogenolysis of protolignin. XVII. Isolation of three dimeric compounds with γ–O–4, β–1, and β–O–4 linkages from hardwood lignin. Holzforschung. 1981;35(5):229–232.

[152] Matsukura M, Sakakibara A. Hydrogenolysis of protolignin. IV. Isolation of a dimeric material of 'condensed type'. Mokuzai Gakkaishi. 1969;15(7):297–302.

[153] Matsukura M, Sakakibara A. Hydrogenolysis of protolignin. V. Isolation of some dimeric compounds with carbon to carbon linkage. Mokuzai Gakkaishi. 1973;19(3):131–135.

[154] Matsukura M, Sakakibara A. Hydrogenolysis of protolignin. VI. Isolation of a new trimeric compound with carbon to carbon linkage. Mokuzai Gakkaishi. 1973;19(3):137–140.

[155] Matsukura M, Sakakibara A. Hydrogenolysis of protolignin. VIII. Isolation of a dimer with Cβ–Cβ′ linkage and a biphenyl. Mokuzai Gakkaishi. 1973;19(4):171–176.

[156] Ohta M, Sakakibara A. Hydrogenolysis of protolignin. III. Isolation of diguaiacylpropanol and chromatographic detection of some monomers. Mokuzai Gakkaishi. 1969;15(6):247–250.

[157] Sakakibara A, Sudo K, Kishi M, Aoyama M, Hwang BH. Hydrogenolysis of protolignin. XVI. Isolation of β–O–4 and β–β type compounds. Mokuzai Gakkaishi. 1980;26(9):628–632.

[158] Sudo K, Hwang BH, Sakakibara A. Isolation of α–O–γ compound from the hydrogenolysis products of lignin. Mokuzai Gakkaishi. 1978;24(6):424–425.

[159] Sudo K, Sakakibara A. Hydrogenolysis of protolignin. VII. Isolation of DL-syringaresinol, biphenyl, and diarylpropane from hardwood lignin. Mokuzai Gakkaishi. 1973;19(4):165–169.

[160] Sudo K, Sakakibara A. Hydrogenolysis of protolignin. IX. Isolation of a trimeric compound linked with biphenyl- and Cβ–C5-linkages. Mokuzai Gakkaishi. 1974;20(7):327–330.

[161] Sudo K, Sakakibara A. Hydrogenolysis of protolignin. X. Isolation of two dimeric compounds with a β-aryl ether linkage. Mokuzai Gakkaishi. 1974;20(7):331–335.

[162] Sudo K, Sakakibara A. Hydrogenolysis of protolignin. XI. Isolation of a dimer with Cβ–C6 and a trimer with two Cβ–C5 linkages. Mokuzai Gakkaishi. 1974;20(8):396–401.

[163] Sudo K, Sakakibara A. Hydrogenolysis of protolignin. XII. Isolation of dimeric and trimeric compounds with carbon to carbon linkages. Mokuzai Gakkaishi. 1975;21(3):164–169.

[164] Sudo K, Sakakibara A. Hydrogenolysis of protolignin. XIII. Isolation of trimeric biphenyl and lignan type compounds. Mokuzai Gakkaishi. 1977;23(3):151–155.

[165] Sudo K, Hwang BH, Sakakibara A. Hydrogenolysis of protolignin. XIV. Isolation of a dimeric compound with α–O–γ linkage. Mokuzai Gakkaishi. 1979;25(1):61–66.

[166] Wada I, Sakakibara A. Hydrogenolysis of protolignin. I. Hydrogenolysis products under various reaction conditions. Mokuzai Gakkaishi. 1969;15(5):214–218.

[167] Yasuda S, Sakakibara A. Hydrogenolysis of protolignin in compressed wood materials. I. Isolation of two dimers with Cβ–C5 and Cβ–C3 composed of p-hydroxyphenyl and guaiacyl nuclei and two p-hydroxyphenyl nuclei, respectively. Mokuzai Gakkaishi. 1975;21(6):370–375.

[168] Yasuda S, Sakakibara A. Hydrogenolysis of protolignin in compression wood. II. Isolation of two dimers with β-aryl ether linkage and phenylisochroman structure. Mokuzai Gakkaishi. 1976;22(11):606–612.

[169] Yasuda S, Sakakibara A. Hydrogenolysis of protolignin in compression wood. III. Isolation of four dimeric compounds with carbon to carbon linkage. Mokuzai Gakkaishi. 1977;23(2):114–119.

[170] Yasuda S, Sakakibara A. Hydrogenolysis of protolignin in compression wood. IV. Isolation of a diphenyl ether and three dimeric compounds with carbon to carbon linkage. Mokuzai Gakkaishi. 1977;23(8):383–387.

[171] Yasuda T, Sakakibara A. Hydrogenolysis of protolignin in compression wood. V. Isolation of two trimeric compounds with γ-lactone ring. Holzforschung. 1981;35(4):183–187.
[172] Barta K, Matson TD, Fettig ML, Scott SL, Iretskii AV, Ford PC. Catalytic disassembly of an organosolv lignin *via* hydrogen transfer from supercritical methanol. Green Chem. 2010;12(9):1640–1647.
[173] Ralph J, Peng J, Lu F, Hatfield RD, Helm RF. Are lignins optically active? J Agric Food Chem. 1999;47(8):2991–2996.
[174] Lu F, Ralph J. Chapter 6 - lignin. In: Cereal Straw as a Resource for Sustainable Biomaterials and Biofuels. Amsterdam: Elsevier; 2010. p. 169–207.
[175] Lu F, Ralph J. DFRC method for lignin analysis. 1. New method for β-aryl ether cleavage: lignin model studies. J Agric Food Chem. 1997;45(12):4655–4660.
[176] Del Rio JC, Marques G, Rencoret J, Martinez AT, Gutierrez A. Occurrence of naturally acetylated lignin units. J Agric Food Chem. 2007;55(14):5461–5468.
[177] Funaoka M, Abe I. The reaction of lignin under the presence of phenol and boron trifluoride. I. On the formation of catechol from MWL, dioxane lignin and Kraft lignin. Mokuzai Gakkaishi. 1978;24(4):256–261.
[178] Funaoka M, Abe I, Chiang VL. Nucleus exchange reaction. In: Lin SY, Dence CW, editors. Methods in Lignin Chemistry, Springer Series in Wood Science. Springer Berlin Heidelberg; 1992. p. 301–321.
[179] Ishikawa H. Phenolation of lignin. Ehime Daigaku Kiyo, Dai-6-bu. 1958;4:1–94.
[180] Kratzl K, Buchtela K, Gratzl J, Zauner J, Ettingshausen O. Lignin and plastics. The reactions of lignin with phenol and isocyanates. Tappi. 1962;45:113–119.
[181] Tai S, Nakano J, Migita N. Lignin. LXIII. Utilization of lignin. 6. Activation of thiolignin with phenol. Mokuzai Gakkaishi. 1968;14(1):40–45.
[182] Funaoka M, Abe I, Yoshimura M. Reaction of hydrotropic lignin with phenol and formaldehyde. Mokuzai Gakkaishi. 1977;23(11):571–578.
[183] Kratzl K, Oburger M. Kondensationsreaktionen von lignin-modellsubstanzen mit phenol unter saurer katalyse. Holzforschung. 1980;34(1):11–16.
[184] Kratzl K, Oburger M. Kondensationsreaktionen von lignin-modellsubstanzen mit phenol unter saurer katalyse - 2. Mitteilung. Holzforschung. 1980;34(6):191–196.
[185] Chiang VL, Funaoka M. The formation and quantity of diphenylmethane type structures in residual lignin during Kraft delignification of Douglas-Fir. Holzforschung. 1988;42(6):385–391.
[186] Chiang VL, Stokke DD, Funaoka M. Lignin fragmentation and condensation reactions in middle lamella and secondary wall regions during Kraft pulping of Douglas fir. J Wood Chem Technol. 1989;9(1):61–83.
[187] Chiang VL, Funaoka M. The dissolution and condensation reactions of guaiacyl and syringyl units in residual lignin during kraft delignification of sweet gum. Holzforschung. 1990;44(2):147–156.
[188] Criegee R. Mechanism of ozonolysis. Angew Chem Int Ed. 1975;14(11):745–752.
[189] Geletneky C, Berger S. The mechanism of ozonolysis revisited by ^{17}O-NMR spectroscopy. Eur J Org Chem. 1998;(8):1625–1627.
[190] Dorée C, Cunningham M. LXXVI. The action of ozone on cellulose. Part III. Action on beech wood (lignocellulose). J Chem Soc, Trans. 1913;103:677–686.
[191] Sarkanen KV, Islam A, Anderson CD. Ozonation. In: Lin SY, Dence CW, editors. Methods in Lignin Chemistry, Springer Series in Wood Science. Springer Berlin Heidelberg; 1992. p. 387–406.
[192] Matsumoto Y, Ishizu A, Nakano J. Studies on chemical structure of lignin by ozonation. Holzforschung. 1986;40(Suppl.):81–85.

[193] Aulin-Erdtman G, Tomita Y, Forsen S. Degradation of lignin and model compounds. I. Configuration of dehydrodiisoeugenol. Acta Chem Scand. 1963;17:535–536.
[194] Tanahashi M, Nakatsubo F, Higuchi T. Structural elucidation of bamboo lignin by acidolysis and ozonolysis I. J Chem Res. 1975;58:1–11.
[195] Haluk JP, Metche M. Chemical and spectrographic characterization of poplar lignin by acidolysis and ozonolysis. Cellul Chem Technol. 1986;20(1):31–50.
[196] Kratzl K, Claus P, Reichel G. Reactions of lignin and lignin model compounds with ozone. Tappi. 1976;59(11):86–87.
[197] Eriksson T, Gierer J. Studies on the ozonation of structural elements in residual Kraft lignins. J Wood Chem Technol. 1985;5(1):53–84.
[198] Bonnet MC, De Laat J, Dore M. Identification of ozonation by-products of lignin and of carbohydrates in water. Environ Technol Lett. 1989;10(6):577–590.
[199] Matsumoto Y, Ishizu A, Nakano J. Determination of glyceraldehyde-2-aryl ether type structure in lignin by the use of ozonolysis. Mokuzai Gakkaishi. 1984;30(1):74–78.
[200] Matsumoto Y, Ishizu A, Nakano J, Terasawa K. Residual sugars in Klason lignin. J Wood Chem Technol. 1984;4(3):321–330.
[201] Habu N, Matsumoto Y, Ishizu A, Nakano J. The role of the diarylpropane structure as a minor constituent in spruce lignin. Holzforschung. 1990;44(1):67–71.
[202] Tsutsumi Y, Islam A, Anderson CD, Sarkanen KV. Acidic permanganate oxidations of lignin and model compounds: comparison with ozonolysis. Holzforschung. 1990;44(1):59–66.
[203] Akiyama T, Sugimoto T, Matsumoto Y, Meshitsuka G. Erythro/threo ratio of β-0-4 structures as an important structural characteristic of lignin. I: Improvement of ozonization method for the quantitative analysis of lignin side-chain structure. J Wood Sci. 2002;48(3):210–215.
[204] Phillips M, Goss MJ. Lignin. VIII. Oxidation of alkali lignin. J Am Chem Soc. 1933;55:3466–3470.
[205] Kolsaker P, Bailey PS. Ozonation of compounds of the type Ar-CH = CH-G; ozonation in methanol. Acta Chem. Scand. 1967; 21(2): 537--546, doi: 10.3891/acta.chem.scand.21-0537.
[206] Tomita B, Kurozumi K, Takemura A, Hosoya S. Ozonized lignin-epoxy resins. In: Glasser WG, Sarkanen S, editors. Lignin. Washington, DC: American Chemical Society; 1989. p. 496–505.
[207] Kondo T, Meshitsuka G, Ishizu A, Nakano J. Preparation and ozonation of completely allylated and methallylated lignins. Mokuzai Gakkaishi. 1987;33(9):724–727.
[208] Dobinson F. Ozonization of malonic acid in aqueous solution. Chem Ind. 1959;26:853–854.
[209] Soteland N. Attempts to characterize the oxidized lignin after ozone treatment of western hemlock groundwood. II. Nor Skogind. 1971;25(5):135–139.
[210] Balousek PJ, McDonough TJ, McKelvey RD, Johnson DC. The effects of ozone upon a lignin model containing the β-aryl ether linkage. Sven Papperstidn. 1981;84(9):R49–R54.
[211] Tishchenko DV. Ozonization of lignin. Zh Prikl Khim. 1959;32:686–690.
[212] Deslongchamps P, Atlani P, Frehel D, Malaval A, Moreau C. Oxidation of acetals by ozone. Can J Chem. 1974;52(21):3651–3664.
[213] Miksche GE, Kjekshus A, Mukherjee AD, Nicholson DG, Southern JT. Zum alkalischen abbau der p-alkoxy-arylglycerin-β-aryläthersturkturen des lignins. Versuche mit erythro-veratrylglycerin-β-guajacyläther. Acta Chem Scand. 1972;26:3275–3281.
[214] Meier D, Faix O. Pyrolysis-gas chromatography-mass spectrometry. In: Lin SY, Dence CW, editors. Methods in Lignin Chemistry, Springer Series in Wood Science. Springer Berlin Heidelberg; 1992. p. 177–199.
[215] Ralph J, Hatfield RD. Pyrolysis-GC-MS characterization of forage materials. J Agric Food Chem. 1991;39(8):1426–1437.

[216] Saake B, Lehnen R. Lignin. In: Arpe HJ, editor. Ullmann's Encyclopedia of Industrial Chemistry, Ullmann's Encyclopedia of Industrial Chemistry. vol. 21. Weinheim: Wiley-VCH Verlag GmbH; 2012. p. 21–36.
[217] Moodley B, Mulholland DA, Brookes HC. The chemical oxidation of lignin found in sappi saiccor dissolving pulp mill effluent. Water SA. 2012;38(1):1–8.
[218] Rowell RM. Acetylation of wood. For Prod J. 2006;56(9):4–12.
[219] Rowell RM. Chemical modification of wood: a short review. Wood Mater Sci Eng. 2006;1(1):29–33.
[220] Pringsheim H, Magnus H. The acetyl content of lignin. Z Physiol Chem. 1919;105:179–186.
[221] Heuser E, Ackermann W. Acetylation of lignin. Cellul-Chem. 1924;5:13–21.
[222] Fuchs W. Genuine lignin. I. Acetylation of pine wood. Ber Dtsch Chem Ges B. 1928;61:948–951.
[223] Suida H, Titsch H. Chemistry of beech wood; acetylation of beech wood and cleavage of the acetylbeech wood. Ber Dtsch Chem Ges B. 1928;61:1599–1604.
[224] Suida H, Titsch H. Acetylated wood, the combination of the incrustation and it method of separation of the constituents of wood. Monatsh Chem. 1929;53/54:687–706.
[225] Friese H. The water-soluble degradation products of lignin. II. Ber Dtsch Chem Ges B. 1930;63:1902–1910.
[226] Beckmann E, Liesche O, Lehmann F. Lignin aus winterroggenstroh. Angew Chem. 1921;34(50):285–288.
[227] Friedrich A, Diwald J. Lignin. I. Lignin of pine wood. Monatsh Chem. 1925;46:31–46.
[228] Freudenberg K, Hess H. Method for the recognition of different kinds of hydroxyl groups. Its application to lignin. Justus Liebigs Ann Chem. 1926;448:121–133.
[229] Klason P. Beiträge zur Kenntnis der chemischen Zusammensetzung des Fichtenholzes, Schriften des Vereins der Zellstoff- und Papierchemiker und -Ingenieure. Borntraeger; 1911.
[230] Dorée C, Hall L. The lignosulfonic acid obtained by the action of sulfurous acid on spruce wood. J Soc Chem Ind, Lond. 1924;43:257–263.
[231] Brauns FE, Lewis HF, Brookbank EB. Lignin ethers and esters. Preparation from Pb and other metallic derivatives of lignin. Ind Eng Chem. 1945;37:70–73.
[232] Thielemans W, Wool RP. Lignin esters for use in unsaturated thermosets: lignin modification and solubility modeling. Biomacromolecules. 2005;6(4):1895–1905.
[233] Ambily C, Sunny K, Tessymol M. Synthesis, characterization and thermal studies on natural polymers modified with 2-(5-(4-dimethylamino-benzylidin)-4-oxo-2-thioxo-thiazolidin-3-yl)acetic acid. Res J Chem Sci. 2012;2(12):37–45.
[234] Chandran A, Kuriakose S, Mathew T. Synthesis, characterization and photoresponsive studies of lignin functionalized with 2-(5-(4-dimethylaminobenzyliden)-4-oxo-2-thioxothiazolidin-3-yl)acetic acid. Open J Org Polym Mater. 2012;2(4):63–67.
[235] Chandran A, Kuriakose S, Mathew T. Synthesis, characterization and photoinduced cis-trans isomerization studies on lignin modified with 4-[(e)-2-(3-hydroxynaphthalen-2-yl)diazen-1-yl] benzoic acid. Open J Org Polym Mater. 2012;2(4):68–73.
[236] Heuser E, Schmitt R, Gunkel L. Methylation of lignin. Cellul-Chem. 1921;2:81–86.
[237] Klason P. Constitution of the lignin of pine wood. Ber Dtsch Chem Ges B. 1920;53:1864–1873.
[238] Heuser E, Samuelsen S. Oxidation of lignin- and lignosulfonic acid methyl ether. Cellul-Chem. 1922;3:78–83.
[239] Jones GM, Brauns FE. Ethers of certain lignin derivatives. Pap Trade J. 1944;119(11):108–1011.
[240] Adler E, Gierer J. The alkylation of lignin with alcoholic hydrochloric acid. Acta Chem Scand. 1955;9:84–93.

[241] Adler E, Brunow G, Lundquist K. Investigation of the acid-catalyzed alkylation of lignins by means of NMR spectroscopic methods. Holzforschung. 1987;41(4):199–207.
[242] Siochi EJ, Ward TC, Haney MA, Mahn B. The absolute molecular weight distribution of hydroxypropylated lignins. Macromolecules. 1990;23(5):1420–1429.
[243] Wu LCF, Glasser WG. Engineering plastics from lignin. I. Synthesis of hydroxypropyl lignin. J Appl Polym Sci. 1984;29(4):1111–1123.
[244] Lin L, Yoshioka M, Yao Y, Shiraishi N. Liquefaction of wood in the presence of phenol using phosphoric acid as a catalyst and the flow properties of the liquefied wood. J Appl Polym Sci. 1994;52(11):1629–1636.
[245] Lin L, Yoshioka M, Yao Y, Shiraishi N. Preparation and properties of phenolated wood/phenol/formaldehyde co-condensed resin. J Appl Polym Sci. 1995;58(8):1297–1304.
[246] Alma MH, Maldas D, Shiraishi N. Liquefaction of several biomass wastes into phenol in the presence of various alkalis and metallic salts as catalysts. J Polym Eng. 1998;18(3):161–177.
[247] Kobayashi M, Asano T, Kajiyama M, Tomita B. Analysis on residue formation during wood liquefaction with polyhydric alcohol. J Wood Sci. 2004;50(5):407–414.
[248] Kunaver M, Jasiukaityte E, Cuk N, Guthrie JT. Liquefaction of wood, synthesis and characterization of liquefied wood polyester derivatives. J Appl Polym Sci. 2010;115(3):1265–1271.
[249] Krzan A, Kunaver M, Tisler V. Wood liquefaction using dibasic organic acids and glycols. Acta Chim Slov. 2005;52(3):253–258.
[250] Zhang T, Zhou Y, Liu D, Petrus L. Qualitative analysis of products formed during the acid catalyzed liquefaction of bagasse in ethylene glycol. Bioresour Technol. 2007;98(7):1454–1459.
[251] Lin L, Nakagame S, Yao Y, Yoshioka M, Shiraishi N. Liquefaction mechanism of β-O-4 lignin model compound in the presence of phenol under acid catalysis. Part 2. Reaction behavior and pathways. Holzforschung. 2001;55(6):625–630.
[252] Lin L, Yao Y, Shiraishi N. Liquefaction mechanism of β-O-4 lignin model compound in the presence of phenol under acid catalysis. Part 1. Identification of the reaction products. Holzforschung. 2001;55(6):617–624.
[253] Pomilio U. Pomilio cellulose process by gas chlorination. Ind Eng Chem, News Ed. 1937;15:73–74.
[254] Hardman H, Cole EJ. Paper-Making Practice. Manchester University Press; 1960.
[255] Dence CW. Halogenatin and nitration. In: Sarkanen KV, Ludwig CH, editors. Lignins: Occurrence, Formation, Structure and Reactions. New York: Wiley-Interscience; 1971. p. 373–432.
[256] Cross CF, Bevan EJ. XVI.- The chemistry of bast fibres. J Chem Soc, Trans. 1882;38:90–110.
[257] Heuser E, Sieber R. Action of chlorine on spruce wood. Angew Chem. 1913;26(103):801–806.
[258] Hägglund E. Lignin. Ark Kemi, Mineral Geol. 1918;7:1–20.
[259] Jonas KG. Lignin and humin substances. Angew Chem. 1921;34(Aufsatzteil):289–291.
[260] Tropsch H. Action of antimony pentachloride on lignin. Gesammelte Abh Kennt Kohle. 1921;6:301–302.
[261] Polčín J. Study of alkali sulfite-lignin. I. Chlorination with hypochlorites. Chem Pap. 1954;8(4):227–234.
[262] Karrer P, Widmer K. Polysaccharides. IX. Cellulose and lignin. Helv Chim Acta. 1921;4:700–702.
[263] Brage C, Eriksson T, Gierer J. Reactions of chlorine dioxide with lignins in unbleached pulps. Part III. Reactions with model compounds representing olefinic structures in native and residual lignins. Holzforschung. 1995;49(2):127–138.
[264] Tropsch H, Schellenberg A. Effect of 5 N nitric acid on cellulose, dextrose, natural and artificial humic acids, and lignin. Gesammelte Abh Kennt Kohle. 1921;6:257–262.
[265] Fischer F, Tropsch H. Action of nitric acid on lignin. Gesammelte Abh Kennt Kohle. 1921;6:279–288.

[266] Powell WJ, Whittaker H. Chemistry of lignin. Part I. Flax lignin and some derivatives. J Chem Soc, Trans. 1924;125:357–364.
[267] Fuchs W. The relation between humic acids and lignin. Brennst-Chem. 1928;9:298–302.
[268] Kurschner K. Separation of wood into cellulose and nitrolignin. Cellul-Chem. 1931;12:281–286.
[269] Hibbert H, Marion L. Lignin and related compounds. IV. The nitration of glycol-lignin. Can J Res. 1930;3:130–139.
[270] Phelps EB. The pollution of streams by sulphite pulp waste. A study of possible remedies. water-supply paper 226. Washington, DC: Department of the interior. United States Geological Survey. Sanitary Research Laboratory and Sewage Experiment Station of the Massachusetts Institute of Technology; 1909. http://pubs.usgs.gov/wsp/0226/report.pdf (accessed March 13, 2015).
[271] Öman E. A method of preparing nitro compounds from lignine sulphonic acid or its salts; 1918.
[272] Lindsey JB, Tollens B. Wood sulphite-liquor and lignin. Justus Liebigs Ann Chem. 1892;267(2–3):341–366.
[273] Klason P. Chemical structure of the lignin of spruce wood. Ark Kemi, Mineral Geol. 1917;6(15):1–21.
[274] Gierer J, Lindeberg O. Reactions of lignin during sulfate pulping. Part XIX. Isolation and identification of new dimers from a spent sulfate liquor. Acta Chem Scand, Ser B. 1980;34(3):161–170.
[275] Calabria GMM, Goncalves AR. Obtainment of chelating agents through the enzymatic oxidation of lignins by phenol oxidase. Appl Biochem Biotechnol. 2006;129–132:320–325.
[276] Babu BR, Parande AK, Raghu S, Kumar TP. Cotton textile processing: waste generation and effluent treatment. J Cotton Sci. 2007;11(3):141–153.
[277] Brooks RE, Moore SB. Alkaline hydrogen peroxide bleaching of cellulose. Cellulose. 2000;7(3):263–286.
[278] Collinson SR, Thielemans W. The catalytic oxidation of biomass to new materials focusing on starch, cellulose and lignin. Coord Chem Rev. 2010;254(15–16):1854–1870.
[279] Fischer F, Schrader H, Friedrich A. Pressure oxidation of lignin. Gesammelte Abh Kennt Kohle. 1921;6:1–21.
[280] Marshall HB, Vincent DL. Production of syringealdehyde and/or vanillin from hardwood waste pulping liquors; 1978.
[281] Deng H, Lin L, Sun Y, Pang C, Zhuang J, Ouyang P, et al. Activity and stability of perovskite-type oxide lacoo$_3$ catalyst in lignin catalytic wet oxidation to aromatic aldehydes process. Energy Fuels. 2009;23(1):19–24.
[282] Wong Z, Chen K, Li J. Formation of vanillin and syringaldehyde in an oxygen delignification process. BioResources. 2010;5(3):1509–1516.
[283] Rodrigues Pinto PC, Borges da Silva EA, Rodrigues AE. Insights into oxidative conversion of lignin to high-added-value phenolic aldehydes. Ind Eng Chem Res. 2011;50(2):741–748.
[284] Kadla JF, Chang HM, Jameel H. The reactions of lignins with hydrogen peroxide at high temperature. Part 1. The oxidation of lignin model compounds. Holzforschung. 1997;51(5):428–434.
[285] Crestini C, Crucianelli M, Orlandi M, Saladino R. Oxidative strategies in lignin chemistry: a new environmental friendly approach for the functionalisation of lignin and lignocellulosic fibers. Catal Today. 2010;156(1–2):8–22.
[286] Phillips M. The chemistry of lignin. Chem Rev. 1934;14(1):103–170.
[287] Li M, Foster C, Kelkar S, Pu Y, Holmes D, Ragauskas A, et al. Structural characterization of alkaline hydrogen peroxide pretreated grasses exhibiting diverse lignin phenotypes. Biotechnol Biofuels. 2012;5:38.

[288] Quintana GC, Rocha GJM, Goncalves AR, Velasquez JA. Evaluation of heavy metal removal by oxidised lignins in acid media from various sources. BioResources. 2008;3(4):1092–1102.
[289] Sun RC, Fang JM, Tomkinson J. Delignification of rye straw using hydrogen peroxide. Ind Crops Prod. 2000;12(2):71–83.
[290] Lange H, Decina S, Crestini C. Oxidative upgrade of lignin - Recent routes reviewed. Eur Polym J. 2013;49(6):1151–1173.
[291] Lequart C, Kurek B, Debeire P, Monties B. MnO_2 and oxalate: an abiotic route for the oxidation of aromatic components in wheat straw. J Agric Food Chem. 1998;46(9):3868–3874.
[292] Partenheimer W. The unusual characteristics of the aerobic oxidation of 3,4-dimethoxytoluene with metal/bromide catalysts. Adv Synth Catal. 2004;346(12):1495–1500.
[293] Partenheimer W. The aerobic oxidative cleavage of lignin to produce hydroxyaromatic benzaldehydes and carboxylic acids *via* metal/bromide catalysts in acetic acid/water mixtures. Adv Synth Catal. 2009;351(3):456–466.
[294] Korpi H, Lahtinen P, Sippola V, Krause O, Leskela M, Repo T. An efficient method to investigate metal-ligand combinations for oxygen bleaching. Appl Catal, A. 2004;268(1–2):199–206.
[295] Collins TJ. Taml oxidant activators: a new approach to the activation of hydrogen peroxide for environmentally significant problems. Acc Chem Res. 2002;35(9):782–790.
[296] Alves V, Capanema E, Chen CL, Gratzl J. Comparative studies on oxidation of lignin model compounds with hydrogen peroxide using Mn(IV)-Me3TACN and Mn(IV)-Me4DTNE as catalyst. J Mol Catal A: Chem. 2003;206(1-2):37–51.
[297] Crestini C, Saladino R, Tagliatesta P, Boschi T. Biomimetic degradation of lignin and lignin model compounds by synthetic anionic and cationic water soluble manganese and iron porphyrins. Bioorg Med Chem. 1999;7(9):1897–1905.
[298] Crestini C, Pro P, Neri V, Saladino R. Methyltrioxorhenium: a new catalyst for the activation of hydrogen peroxide to the oxidation of lignin and lignin model compounds. Bioorg Med Chem. 2005;13(7):2569–2578.
[299] Voitl T, von Rohr PR. Oxidation of lignin using aqueous polyoxometalates in the presence of alcohols. ChemSusChem. 2008;1(8-9):763–769.
[300] Badamali SK, Luque R, Clark JH, Breeden SW. Microwave assisted oxidation of a lignin model phenolic monomer using Co(salen)/SBA-15. Catal Commun. 2009;10(6):1010–1013.
[301] Michael P. Masingale, Ericka F. Alves, Theresah N. Korbieh, Samar K. Bose, Raymond C. Francis. An oxidant to replace nitrobenzene in lignin analysis. BioResources 2009;4(3):1139–1146.
[302] Adler E, Falkehag I, Smith B. Periodate oxidation of phenols. VIII. ^{18}o-Studies of the mechanism of oxidative cleavage of the monoether of pyrocatechol and of hydroquinone. Acta Chem Scand. 1962;16:529–540.
[303] Adler E, Lundquist K. Estimation of uncondensed phenolic units in spruce lignin. Acta Chem Scand. 1961;15:223–224.
[304] Beck HD, Adle E. Oxidation with quinones. Acta Chem Scand. 1961;15:218–219.
[305] Ksibi M, Amor SB, Cherif S, Elaloui E, Houas A, Elaloui M. Photodegradation of lignin from black liquor using a UV/TiO_2 system. J Photochem Photobiol, A. 2003;154(2-3):211–218.
[306] Lapierre C, Monties B, Meier D, Faix O. Structural investigation of Kraft lignins transformed *via* oxo-ammoniation to potential nitrogenous fertilizers. Holzforschung. 1994;48(Suppl.):63–68.
[307] Allan GG. Modification reactions. In: Sarkanen KV, Ludwig CH, editors. Lignins: Occurrence, Formation, Structure and Reactions. New York: Wiley-Interscience; 1971. p. 511–573.
[308] Ungureanu E, Ungureanu O, Capraru AM, Popa VI. Chemical modification and characterization of straw lignin. Cellul Chem Technol. 2009;43(7-8):263–269.
[309] Malutan T, Nicu R, Popa VI. Contribution to the study of hydroxymethylation reaction of alkali lignin. BioResources. 2008;3(1):13–20.

[310] Malutan T, Nicu R, Popa VI. Lignin modification by epoxidation. BioResources. 2008;3(4):1371–1376.

[311] El Mansouri NE, Yuan Q, Huang F. Study of chemical modification of alkaline lignin by the glyoxalation reaction. BioResources. 2011;6(4):4523–4536.

[312] Stewart D. Lignin as a base material for materials applications: chemistry, application and economics. Ind Crops Prod. 2008;27(2):202–207.

[313] Shevchenko SM. Depolymerization of lignin in wood with molecular hydrogen iodide. Croat Chem Acta. 2000;73(3):831–841.

[314] Brebu M, Vasile C. Thermal degradation of lignin - a review. Cellul Chem Technol. 2010;44(9):353–363.

[315] Pandey MP, Kim CS. Lignin depolymerization and conversion: a review of thermochemical methods. Chem Eng Technol. 2011;34(1):29–41.

[316] Mohan D, Pittman J, Charles U, Steele PH. Pyrolysis of wood/biomass for bio-oil: a critical review. Energy Fuels. 2006;20(3):848–889.

[317] Acharya B, Sule I, Dutta A. A review on advances of torrefaction technologies for biomass processing. Biomass Convers Biorefin. 2012;2(4):349–369.

[318] Azadi P, Inderwildi OR, Farnood R, King DA. Liquid fuels, hydrogen and chemicals from lignin: a critical review. Renewable Sustainable Energy Rev. 2013;21:506–523.

[319] Hatakeyama H, Hatakeyama T. Lignin structure, properties, and applications. Adv Polym Sci. 2010;232(1):1–63.

[320] Allan GG, Mattila T. High energy degradation. In: Sarkanen KV, Ludwig CH, editors. Lignins: Occurrence, Formation, Structure and Reactions. New York: Wiley-Interscience; 1971. p. 575–596.

[321] Hatakeyama T, Nakamura K, Hatakeyama H. Differential thermal analysis of styrene derivatives related to lignin. Polymer. 1978;19(5):593–594.

[322] Hatakeyama T, Hatakeyama H. Thermal Properties of Green Polymers and Biocomposites. Dordrecht: Kluwer Academic; 2004.

[323] Hirose S, Hatakeyama H. A kinetic study on lignin pyrolysis using the integral method. Mokuzai Gakkaishi. 1986;32(8):621–625.

[324] Jakab E, Faix O, Till F, Szekely T. Thermogravimetry/mass spectrometry study of six lignins within the scope of an international round robin test. J Anal Appl Pyrolysis. 1995;35(2):167–179.

[325] Baumberger S, Dole P, Lapierre C. Using transgenic poplars to elucidate the relationship between the structure and the thermal properties of lignins. J Agric Food Chem. 2002;50(8):2450–2453.

[326] Serio MA, Charpenay S, Bassilakis R, Solomon PR. Measurement and modeling of lignin pyrolysis. Biomass Bioenergy. 1994;7(1-6):107–124.

[327] Hirose S, Kobashigawa K, Izuta Y, Hatakeyama H. Thermal degradation of polyurethanes containing lignin studied by TG-FTIR. Polym Int. 1998;47(3):247–256.

[328] Hatakeyama T, Liu Z. Handbook of Thermal Analysis. Chichester: John Wiley & Sons, Ltd; 1998.

[329] Caballero JA, Font R, Marcilla A. Pyrolysis of kraft lignin: yields and correlations. J Anal Appl Pyrolysis. 1997;39(2):161–183.

[330] Fahmi R, Bridgwater AV, Thain SC, Donnison IS, Morris PM, Yates N. Prediction of Klason lignin and lignin thermal degradation products by Py-GC/MS in a collection of lolium and festuca grasses. J Anal Appl Pyrolysis. 2007;80(1):16–23.

[331] Yang H, Yan R, Chen H, Zheng C, Lee DH, Liang DT. In-depth investigation of biomass pyrolysis based on three major components: Hemicellulose, cellulose and lignin. Energy Fuels. 2006;20(1):388–393.

[332] de Wild PJ, Huijgen WJJ, Heeres HJ. Pyrolysis of wheat straw-derived organosolv lignin. J Anal Appl Pyrolysis. 2012;93:95–103.
[333] Leisola M, Pastinen O, Axe DD. Lignin-designed randomness. J Med Chem. 2012;2012(3):1.
[334] Cho DW, Parthasarathi R, Pimentel AS, Maestas GD, Park HJ, Yoon UC, et al. Nature and kinetic analysis of carbon-carbon bond fragmentation reactions of cation radicals derived from SET-oxidation of lignin model compounds. J Org Chem. 2010;75(19):6549–6562.
[335] Vanholme R, Morreel K, Darrah C, Oyarce P, Grabber JH, Ralph J, et al. Metabolic engineering of novel lignin in biomass crops. New Phytol. 2012;196(4):978–1000.
[336] Reid ID. Biodegradation of lignin. Can J Bot. 1995;73(Suppl. 1, Sect. E-H, Fifth International Mycological Congress, Sect. E-H, 1994):S1011–S1018.
[337] Wong DWS. Structure and action mechanism of ligninolytic enzymes. Appl Biochem Biotechnol. 2009;157(2):174–209.
[338] Hatakka A, Hammel KE. Fungal biodegradation of lignocelluloses. In: Hofrichter M, editor. Industrial Applications, The Mycota. vol. 10. Heidelberg: Springer-Verlag; 2011. p. 319–340.
[339] Falck R, Haag W. Decomposition of lignin and of cellulose: two different processes by wood-destroying fungi. Ber Dtsch Chem Ges B. 1927;60:225–232.
[340] Lyr H. Der holzabbau durch pilze. Arch Forstwes. 1961;10:615–626.
[341] Haemmerli SD, Schoemaker HE, Schmidt HWH, Leisola MSA. Oxidation of veratryl alcohol by the lignin peroxidase of phanerochaete chrysosporium. Involvement of activated oxygen. FEBS Lett. 1987;220(1):149–154.
[342] Kirk TK, Farrell RL. Enzymic 'combustion': the microbial degradation of lignin. Annu Rev Microbiol. 1987;41:465–505.
[343] Lane N. Microbiology: batteries not included. Nature. 2006;441(7091):274–277.
[344] Kogel-Knabner I. The macromolecular organic composition of plant and microbial residues as inputs to soil organic matter. Soil Biol Biochem. 2002;34(2):139–162.
[345] Guerra A, Mendonca R, Ferraz A, Lu F, Ralph J. Structural characterization of lignin during *Pinus taeda* wood treatment with *Ceriporiopsis subvermispora*. Appl Environ Microbiol. 2004;70(7):4073–4078.
[346] Rowell RM. Chemical modification. Enclypedia of Forest Sciences. vol. 3. Oxford: Elsevier Academic Press; 2004.
[347] Ferm R, Kringstad KP, Cowling EB. Formation of free radicals in milled wood lignin and syringaldehyde by phenol-oxidizing enzymes. Sven Papperstidn. 1972;75(21):859–865.
[348] Maciel MJM, Castro e Silva A, Ribeiro HCT. Industrial and biotechnological applications of ligninolytic enzymes of the basidiomycota: a review. Electron J Biotechnol. 2010;13(6):14–15.
[349] Cole JL, Clark PA, Solomon EI. Spectroscopic and chemical studies of the laccase trinuclear copper active site: geometric and electronic structure. J Am Chem Soc. 1990;112(26):9534–9548.
[350] Solomon EI, Lowery MD. Electronic structure contributions to function in bioinorganic chemistry. Science. 1993;259(5101):1575–1581.
[351] Reinhammar B, Malmström BG. 'Blue' copper-containing oxidases. In: Spiro TG, editor. Copper Proteins. New York: John Wiley & Sons, Inc.; 1981. p. 109–149.
[352] Paice MG, Bourbonnais R, Reid ID, Archibald FS, Jurasek L. Oxidative bleaching enzymes: a review. J Pulp Pap Sci. 1995;21(8):J280–J284.
[353] Rodriguez Couto S, Osma JF, Saravia V, Guebitz GM, Toca Herrera JL. Coating of immobilized laccase for stability enhancement: a novel approach. Appl Catal, A. 2007;329:156–160.
[354] Sethi R, Padmavathi T, Sullia SB. Lignocellulose biomass degradation by marine microorganisms. Eur J Exp Biol. 2013;3(2):129–138.
[355] Sweeney MD, Xu F. Biomass converting enzymes as industrial biocatalysts for fuels and chemicals: recent developments. Catalysts. 2012;2(2):244–263.

Part IV
Lignins Applications

8
Applications of Modified and Unmodified Lignins

8.1 Introduction

As has been mentioned earlier, lignin is the second most abundant natural biopolymer after cellulose. Natural sources of lignin are annual plants, hardwood, and softwood, each presenting different contents of lignin. In fact, lignin represents approximately 24% on average of the total components of a plant, as shown in Figure 8.1. This may be mainly a result of direct biomass extraction or as a by-product of paper industry, although it might be found in other industrial processes as well. Despite the different sources and types of lignin, summarized in Figure 8.2, it is clear that the main source of lignin by far is the cooking liquor generated in wood-pulping processes. Between 40 and 50 million cubic meter per annum are produced worldwide as a mostly noncommercialized waste product.[1] Such a remarkable quantity means that a huge mass of this very sophisticated and complex polymer is available every year for use in many applications, either directly or chemically modified. Consequently, different methods have been proposed for obtaining lignin from different sources, leading to a wide range of structures, molecular weights, and chemical functionalization that make all these applications possible. For decades, researchers have tried to find practical applications of lignin obtained from either biomass or pulping liquors.

In the case of the paper industry, lignin is usually burnt at the same factory due to its high caloric content, which makes it an excellent fuel. Any other high-quality product requires further processing of lignin in order to separate it from cellulose. Once processed and separated, lignin is a complex biopolymer with remarkable properties, such as biodegradability and low toxicity or chemical functionalization. Pulping lignins are highly altered, while lignins extracted directly from biomass are closer to native lignin. Both present a challenge toward commercially valuable end uses. In either case, the range of useful chemical functionalities and structures in lignins is wide. There are lignins

---

[1] http://www.ili-lignin.com/aboutlignin.php (accessed September 17, 2014).

Lignin and Lignans as Renewable Raw Materials: Chemistry, Technology and Applications, First Edition.
Francisco G. Calvo-Flores, José A. Dobado, Joaquín Isac-García and Francisco J. Martín-Martínez.
© 2015 John Wiley & Sons, Ltd. Published 2015 by John Wiley & Sons, Ltd.

250 *Lignin and Lignans as Renewable Raw Materials*

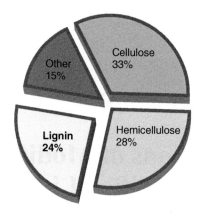

Figure 8.1 Percentages of different plant components

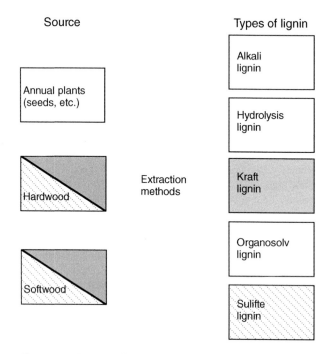

Figure 8.2 Sources of lignin and main types produced in industry

with sulfonic groups in the polymer, lignins that are sulfur-free, lignins with several molecular mass average, and many others with different chemical structures. As an example, Table 8.1 lists some characteristics of lignosulfonate and Kraft lignins, which are the most popular among the pulping-derived ones (see Figure 8.2).

From an economic standpoint, lignin isolated either from biomass or the paper industry is cheaper than many other manufactured materials, but it plays the same role that many other chemicals do. This makes lignin very competitive in any target market where it might be introduced. However, even though enormous amounts of lignin are available on the market every year worldwide, and despite

Table 8.1 Physical and chemical properties of the two main types of lignin[a]

	Lignosulfonate	Kraft lignin
Elemental composition		
C (%)	53.0	66.0
H (%)	5.4	5.8
S (%)	6.5	1.6
Contaminants	Different products from carbohydrate degradation	None
Solubility	Water	
Molecular weight (Da)	4000–150 000	700
Main interunit bond types	β-O-4	C–C polystyrene type with lat

[a]Data obtained from ref. [1].

its low-price advantage, both biomass and paper industry by-product lignins are underutilized. There is a relatively wide range of lignin types that can be found on the market with different physical and chemical properties [1], but still most of their applications remain to be exploited. A remarkable fact, already mentioned throughout this textbook, is that the structure of lignin depends heavily on the source of biomass and the method used for isolation. Therefore, specific applications depend on the type of lignin.

In any case, lignin can be chemically modified in order to improve its reactivity, that is, polymerization reactions, or in order to increase its efficacy for any other application. The chemical modification of lignin is performed by many different reactions, namely oxidation or reduction reactions; alkylation to form ether bonds; Friedel–Crafts acylation; transformation of functional groups by reactions of elimination, substitution, or addition; introduction of heteroatoms by nitration, amidation, sulfonation, or isocyanate synthesis; organometallic reactions; and copolymerization.

From all these different possibilities, probably the main field of application of modified lignin is polymer chemistry, where it is used for the preparation of polymeric materials. In this field, lignin can be used as an additive in a preformed polymer, or as macromonomer incorporated into the polymer structure during the polymerization process. As a component of new polymeric materials, it improves different properties such as thermal stability, tensile strength, and biodegradability. As a direct application with polymers, it has, however, serious limitations because the lignin matrix presents relatively inaccessible reactive centers for polymerization reactions, due mainly to the steric hindrance that prevents phenol and other groups such as –OH ones, from reacting properly, as it occurs in other simpler molecules.

Despite these minor issues concerning the use of lignin in polymerization reactions, several strategies are being developed in order to improve its reactivity. The first approach consists of increasing the number of phenol units available by the so-called phenolation procedure. In this case, lignin, native or not, reacts with phenol, generally at the α-position of the side chain, and new phenol moieties are incorporated into the structure of the biopolymer, making it a more reactive material due to the increased number of phenol rings. The second strategy is to selectively hydrolyze methoxy groups of aromatic rings. More free hydroxyl groups are generated by this procedure, so that new more reactive positions in the aromatic rings are available for reaction, that is, with formaldehyde. The third strategy is the use of spacers attached to appropriate functional groups of lignin in order to bring about some possible reactive centers away from the lignin core. Other methods include chemical transformations, such as oxidation, reduction, or controlled hydrolysis of lignin. Hu et al. [2] provided a review in this regard that describes the main transformations performed on lignin toward producing more reactive polymers useful in the construction of new polymeric materials.

In this chapter, some of the more striking direct applications of modified and unmodified lignins are described. The information has been compiled from diverse sources such as scientific papers, commercial companies, government agencies, and academic institutions. Such dispersed information available from so many sources is combined in an overall discussion in this chapter.

8.2 Lignin as Fuel

Lignin is a material with a remarkable heating value that can be used directly as a fuel, or be incorporated as additive to several types of biomass-derived fuels for industrial and domestic uses. Some examples of these applications are described in the following sections.

8.2.1 Combustion in the Paper Industry

Lignin is a carbonaceous material with a combustion heat (heating value) of 26.7 MJ/kg [3]. This value implies holding the highest energy content among all natural carbon polymeric compounds. Bulk lignin produced in the paper industry can be burnt to recover this energy, and since the 1930s many paper factories have used black liquor as an energy source.

About 7 m^3 of black liquor are produced for every cubic meter of pulp. Most Kraft pulp mills use recovery boilers to recover this black liquor in order to burn much of it afterward, thus, generating steam and recovering chemicals from the cooking process. These chemicals include sodium hydroxide (NaOH) and sodium sulfide (Na_2S), which have been used for separating lignin from cellulose during the papermaking process. Black liquor is the most remarkable biomass fuel in a Kraft pulp mill, because it still contains around 50% of wood substance. Thus, reusing black liquor is an efficient technique that has helped paper mills to reduce environmental problems with emissions, such as carbon dioxide (CO_2), which is not counted in the greenhouse gas (GHG) inventory [4]. It has also helped to reduce the use of chemicals by recovering and reusing them. Burning lignin can make manufacturers nearly energy self-sufficient, producing on-site 66% on average of their own electric power needs. For example, since the 1990s in the United States, paper factories have consumed nearly all the black liquor produced. As a result, the forest product industry has become one of the leading generators of carbon-neutral renewable energy in the United States, producing some 28.5 million megawatt hours of electricity per year. Only about 1–2% of the lignin produced is used to make other products [5].

8.2.2 Heating and Power

In recent years, some fuels have been designed for powering production and residential heating, in order to produce added-value goods from lignin, and to take advantage of its combustion heat. This is related to the increased cost of petroleum-based energy production and the renewed interest in the use of wood for producing energy for residential heating.

Many efforts have been devoted by fuel manufacturers to produce high-quality and high fuel value wood derivatives. Lignin itself has arisen as a good candidate for the production of fuel or as a valuable additive for fuel value wood derivatives. Two main preparations of wood derivatives have been developed for power production and heating: briquettes and pellets. Biomass is dried and compressed under high pressure to convert it into high-density solid blocks for briquettes or cylinders roughly 6–25 mm in diameter and 5–50 mm long for pellets. They are easy to handle and cleaner than other untreated biomass wastes. Briquettes and pellets present higher energy density and smaller volume than do uncompressed solid biofuels, and they are easier to transport and store.

Lignin is an excellent additive for biomass heating fuels because of its properties. It is a natural binder and its heating value helps improve fuel properties. A representative example on this application area can be found in a patent of Weyerhaeuser NR Company for the manufacture of artificial fire logs [6]. Flame properties can be improved by the addition of lignin to cellulosic materials together with propane-1,3-diol, obtained from renewable resources, and some noncrude-based waxes. The performance of these artificial fire logs is basically equivalent to that of existing commercial fire logs.

In the case of pellets, it is quite easy to find on the market lignin products for both industrial and domestic applications for generating power or residential heating. For example, the IBICOM biorefinery plant at Kalundborg (Denmark) converts the woody part of the straw into sugars, EtOH, and lignin. The plant produces lignin that requires no further treatment or purification. The lignin is pelletized and used to replace some of the coal burned by an adjacent power plant, generating the so-called green electricity[2] [7]. This factory produces 11 400 m^3 of lignin pellets per year for fuel.

Northern European countries are the best primary producers and consumers of wood pellets, and Sweden has emerged as the leading producer and consumer in the world.[3] However, many companies worldwide sell wood pellets that are made from biomass waste, especially wood that is compressed under high pressure. The final product has a strong constitution resulting from the wood lignin and resin of the own material, which acts as a natural binding agent [8]. In addition, it bears considering the diversity of lignin, which makes pellet-burning properties dependent on the type of wood they are made from, the moisture content, and the average and origin of the lignin.

In this sense, some academic articles have been published to study the fuel value of pellets according to composition. For example, Stevens and Gardner [9] have studied the addition of lignin to pellets in order to enhance their fuel value. Two types of lignin were examined in the production of wood pellets: Kraft black liquor and indulin AT (IAT). Lignin was added to a softwood furnish and pellets were prepared by a commercial manufacturer. The pellets were analyzed, demonstrating that those prepared with IAT produced better-quality pellets, yielding a higher fuel value than with Kraft black liquor. The Kraft black liquor pellets were soft and spongy and easily fell apart. A cost analysis indicated that lignin preparation can have a major impact on the feasibility of adding lignin to wood pellets to enhance fuel value.

8.3 Lignin as a Binder

The application of lignin as a binder in fuel pellets has been discussed so far. But, there are many other examples of pelletizing and related techniques associated with lignin. One of the most significant examples of lignin as a binder is based on lignosulfonates, which are effective and economical adhesives, acting as binding agents for many industrial goods. This property of lignin has been applied for the preparation of several compressed materials. A common binding agent for this purpose is a system called LignoBond, based on calcium lignosulfonates. This technology was invented in the late 1980s [10].

8.3.1 Coal Briquettes

Malhotra, in Denver, CO, United States, developed a US patent [11] in 2009, which describes a procedure for the preparation of coal briquettes from coal particles without including binders other than lignin. The method includes grinding, drying, mixing, and briquetting, and according to the authors, it can be located either proximally to a power plant or delivered by any transport means (truck, train, ship, or airfreight) to the power plant.

---

[2] http://www.inbicon.com/en/key-advantages/end-products (accessed March 24, 2014).
[3] http://webberenergyblog.wordpress.com/2013/03/03/the-hitchhikers-guide-to-wood-pellets (accessed December 10, 2014).

Recently, Lumadue et al. [12] made a notable study on briquetted anthracite fines[4] for a foundry coke substitute made with 86–92% anthracite fines, 2.3–8.6% lignin, 4.5% silicon metal powder, and 0.9% hydrolyzed collagen (denatured collagen) by mass. Lignin extracted from Kraft black liquor has been shown to act as a fusing binder for anthracite-fine briquettes, which can serve as a substitute for coke in foundries when 3–8% lignin is included. The bindered briquettes burn as fast as or faster than coke, contain no more net sulfur than does coke, and contain 38% higher energy per volume than does coke.

8.3.2 Packing

Lignin has been used for the manufacture of packing material, in a similar way as in pellets with *LignoBond* technology, this being a procedure reinforcing several packaging materials. It was invented in the late 1980s by Forss et al. [10] in 1988. VTT[5] has developed sheets used for packing with a remarkable increase in air and water resistance and with high paper strength. These packing sheets have been manufactured from recycled paper fibers (50% magazine and 50% newsprint) and a mixture of Kraft and organosolv lignins (20% calculated as dry/dry fibers). The addition of lignin creates a barrier on the sheet, producing a higher quality material than conventional recycled paper.

8.3.3 Pelleted Feeds

Pelletizing of livestock feed with lignosulfonates is an effective response to several problems. Lignosulfonates are nontoxic for livestock and compatible with many feeds used in this activity. For example, calcium lignosulfonates can be mixed with different feed compositions with a great diversity of ratios of fat, grain, food fibers, and supplements. Livestock health and pellet properties are actually improved by the addition of lignin because of its natural antibiotic and antioxidant activity. Lignosulfonates are determinant in physical properties, such as the hardness and durability of pellets, stability with respect to humidity[6] and facilitates transport, storage, and general handling of these feeds.

8.4 Lignin as Chelating Agent

A molecular entity in which there is chelation (and the corresponding chemical species) is called a "chelate".[7,8] This term suggests the way that structures such as lignin, clamps metallic ions.

Lignosulfonates are nonhazardous and low-cost materials that have many polar groups able to form chelates with metals. Chelation can be a reversible process, making lignins an excellent candidate to sequester or liberate metal ions in aquatic media or soils. For example, lignin can be used as a carrier to provide essential ionic nutrients to plants and as a material to remove dangerous pollutants, such as heavy metals, in soil or water.

Furthermore, as biodegradable carriers, calcium lignosulfonates may be used in crop production as a chelating agent for micro- and macronutrients. By a chelation mechanism, low-charge micronutrient ions, such as boron, manganese, and iron, are incorporated into a soluble form, which

---

[4] Small pieces of material from an anthracite coal preparation plant, usually below 3 mm diameter.
[5] http://www.vtt.fi (accessed December 14, 2014).
[6] http://www.lignotechfeed.com/Pelleting-Aids/LignoBond-DD (accessed December 13, 2014).
[7] http://goldbook.iupac.org/C01012.html (accessed December 13, 2014).
[8] The word chelate was coined by Morton and Drew in 1920, derived from the Greek term "chela," meaning "great claw" of the lobster or other crustaceans.

is capable of supplying nutrients to plants. In this way, nutrients are slowly released into the soil in a bioavailable form.

On the other hand, transition metals such as copper, cadmium, nickel, zinc, lead, and mercury, pose an environmental problem for both human health and the sustainability of ecosystems. Removing these ions from wastewater or accidental spills is a primary problem to address in order to provide clean and nontoxic water. Heavy metals have been traditionally removed from polluted waters by several processes such as chemical precipitation, solvent extraction, ion-exchange resins, reverse osmosis, and adsorption, but all these methods have serious disadvantages, namely high cost, large input of chemicals, and incomplete removal of waste pollutants from water [13]. The adsorption process with lignin is a cheaper, simpler, and an effective technique for the removal of heavy metals from wastewater [14], and it offers an alternative to more expensive methods. Some academic work in this regard will be discussed in the following text.

Guo et al. [15] investigated in 2008 the sorption of the heavy-metal ions, such as Pb(II), Cu(II), Zn(II), and Ni(II) on lignin isolated from black liquor (Kraft lignin), and they found that lignin had affinity with metal ions in the following order: Pb(II) > Cu(II) > Cd(II) > Zn(II) > Ni(II). Adsorption kinetics of metal ions on lignin with an initial metal concentration of 0.8 mM showed a pseudosecond-order equation.

Šćiban and Klasnja [16] have investigated the removal of copper, zinc, cadmium, and chromium ions by Kraft lignin in several water solutions of $CuSO_4$, $Cd(NO_3)_2$, $ZnSO_4$, and $K_2Cr_2O_7$. The results of this investigation showed that adsorbed heavy-metal ions followed the following order: Cr(VI) Cd(II) > Cu(II) > Zn(II). The adsorption of cations (Cu, Zn, and Cd) is somewhat better described by Freundlich adsorption isotherm and adsorption of Cr(VI) by Langmuir adsorption isotherm.

The influence of other ions, such as Ni(II), Cd(II), and Pb(II), on Cu(II) adsorption was evaluated too for the adsorption capacities of lignin to the former four heavy-metal ions and were determined by measuring equilibrium isotherms. The influence of investigated interfering ions in decreasing adsorption efficiency of Cu(II) followed the order Ni(II) > Cd(II) > Pb(II). These results support the idea that the adsorption behavior of heavy-metal ions in this biopolymer should be perceived from the aspect of possible influence of interfering ion species.

Another example of the use of Kraft lignin as a sequestrant of heavy metals has been provided by Mohan et al. [17], studying the adsorption of Cd and Cu with lignin extracted from black liquor waste at different temperatures, lignin particle sizes, pHs, and solid-to-liquid ratios by a batch method to determine equilibrium and kinetic parameters. The maximum lignin adsorption capacities at 25 °C were 87.05 mg/g (1.37 mmol/g) and 137.14 mg/g (1.22 mmol/g) for Cu(II) and Cd(II), respectively. Adsorption of Cu(II) (68.63 mg/g at 10 °C and 94.68 mg/g at 40 °C) and Cd(II) (59.58 mg/g at 10 °C and 175.36 mg/g at 40 °C) increased at higher temperatures, in all cases following a pseudosecond-order rate kinetics.

On the other hand, Acemioğlu et al. [18] have studied the sorption of Cu(II) by EtOH-based organosolv lignin, concluding that the amount of Cu(II) ions adsorbed onto the lignin increased with higher concentrations and pH. However, it decreased with increasing temperatures. Even though the abundance and very low cost of lignocellulosic wastes from agricultural operations are advantages that render them suitable alternatives for heavy-metal remediation, further successful studies on these materials are essential to demonstrate the efficacy of this technology [19].

The capacity of forming chelates of lignin formulations is evidenced in other commercial applications. Metals in their ionic forms can be tied up in water-soluble lignin derivatives as lignosulfonates, so that the ionic metals can remain in water, avoiding the formation of insoluble precipitates.

Some important micronutrient ions for plants, such as cobalt, copper, manganese, zinc, boron, and molybdenum, remain available in water, and they are key micronutrients for the development of many plants. They can be easily supplied in stable commercial preparations for this procedure.[9]

---

[9] http://www.tnn.com.au/lignin-chelated-micronutrients (accessed December 13, 2014)

Other applications related to lignin chelates are water treatments for boilers and cooling systems for cleaning and preventing solid salt deposits.

8.5 Lignin in Biosciences and Medicine

Native lignin and some of its preparations and derivatives have been shown to have promising results in fields such as biosciences and medicine because of their interactions with living organisms. Antioxidant, antiviral, antitumor, and anticarcinogenic activities are some of the properties that have been attributed to lignin.

Lignin is a known free-radical scavenger, and the radical-scavenging properties of lignins imply an efficient antioxidant effect. Dizhbite *et al.* [20] have studied the radical-scavenging efficiency of a series of lignins isolated from several deciduous and coniferous wood species and 10 lignin fragments against the well-known scavenger power of 1,1-diphenyl-2-picrylhydrazyl (DPPH) radical by instrumental methods such as ESR and spectrophotometry. These researchers have pointed out the importance of the moieties constituted by nonetherified OH phenolic groups, *ortho*-methoxy groups, hydroxyl groups, and the double bonds between the outermost carbon atoms in the side chain of lignin for increasing scavenger activity.

The relationship between lignin structure, radical scavenging, and antioxidant activity has been demonstrated by Pan *et al.* [21] from organosolv lignin samples. Thus, lignins that present more phenolic and less aliphatic hydroxyl groups in their structure and with a low molecular weight showed high antioxidant activity. The dispersity index of lignin is another factor to be considered in order to explain the antioxidant activity. The narrowest dispersity index showed high antioxidant activity.

The antioxidant activity of some industrial lignins derived from different sources has been studied by Vinardell *et al.* [22] and its potential for cosmetic formulations, such as topical applications, has been evaluated to test the irritation of eyes and skin. The high antioxidant capacity of the lignins, studied together with their high safety when applied on sensible parts such as eyes and skin, has opened new perspectives for their potential use in cosmetic or pharmaceutical topical formulations. The same researchers have comparatively studied the antioxidant and cytotoxic effects of several lignins from different sources. In all cases, samples showed a level similar to that of the antioxidant activity exhibited by the known antioxidant epicatechin [22, 23].

The antiviral activity of lignins is also known in early 1990s. Sakagami *et al.* [24] reviewed the antiviral and immunostrengthening activity of lignin-related materials produced from pine cone extracts obtained by NaOH treatment and acid–EtOH precipitation. In this case, the extracts were complex mixtures of lignin moieties and carbohydrates. The extracts were tested on mice and showed a potent activity against herpes simplex and influenza virus, and they induced antimicrobial activity against *Staphylococcus aureus*, *Escherichia coli*, *Pseudomonas aeruginosa*, *Klebsiella pneumoniae*, and *Candida albicans*, and induced antiparasite activity against *Hymenolepis nana*.

In lignin samples and derivatives, there is a parallel between antiviral and antitumor activities. Most of these studies demonstrate that lignins with antiviral activity inhibit the growth of tumor cells.

Many other studies have been conducted in this field using native lignins or plant extracts containing lignin moieties as antivirals or anticarcinogens. Thus, Yamamoto *et al.* [25] have tested the immunostrengthening activity of water-soluble fractions formed by a combination of polysaccharides and lignin-rich mixtures, prepared from the solid culture medium of a fungus called *Lentinus edodes* mycelia (LEM). This is one of the most popular foods in several Far Eastern countries. There are some clinical reports indicating that the conditions of patients with viral diseases, such as B hepatitis and AIDS, have been improved by oral administration of LEM. This effect might result from the immunostrengthening activity of LEM. These lignin-rich extracts have also shown a hepatoprotective effect after induced liver injury in mice [26]. González López *et al.* [27] have proposed an alternative treatment against *Herpes simplex* virus, with LCC tablets from pine cone extracts. Lignin and lignin fragments have been shown to inhibit the growth of bacteria. Organosolv, sulfite, and Kraft lignins inhibit the growth of *Sporobolomyces roseus*, *Candida tropicalis*, *Trichosporon cutaneum*, and *Candida albicans*. These tests conducted on bacteria have demonstrated that antimicrobial activity depends strongly on the lignin sample origin and on the isolation methods [28].

Some of the aforementioned properties of lignins have led these biopolymers to be tested in order to diminish the genotoxicity of some drugs in order to reduce the side effects in patients. Naik *et al.* [29] have recently studied the genoprotective effects of lignin isolated from oil palm black liquor waste on mice against cyclophosphamide (CP, 50 mg/kg b.w.), showing remarkable efficacy as an antioxidant and a genoprotective agent.

As mentioned earlier, lignin is commonly used as an additive for livestock feed. Two aspects of such applications can be noted: one concerning the benefits related to livestock health and the second one associated with the physical properties of the feed (see Section 8.3). First, lignin is considered a prebiotic substance [30], and especially sulfur-free lignins, such as Alcell, have shown these prebiotic effects on monogastric animals. Thus, their antioxidant, antibiotic, and antibacterial properties are extremely useful as a feed additive in livestock breeding. These properties are generally beneficial to gastrointestinal tract health, because lignin may help to control intestinal pathogens, increasing the weight gain of livestock. However, much research is still required to determine the optimum lignin dosages for livestock and the potential effects on the environment [31].

Finally, natural lignin, together with some carbohydrates, is considered to be part of dietary fiber, resistant to digestion, but has been shown to play a key role in the absorption of bile acid and thereby influence lipid metabolism [32]. Therefore, a diet rich in vegetables contributes both quantitatively and qualitatively to essential fiber compounds. The effects of fiber consumption have been analyzed by Sánchez-Muñiz [33]. They have demonstrated the influence on cholesterol and lipoprotein levels, systolic and diastolic blood pressures, and antioxidant availability and profile when fiber components, such as cellulose, hemicellulose, gums, mucilages, pectins, oligosaccharides, and lignins, are incorporated into the diet.

8.6 Lignin in Agriculture

Some applications of unmodified lignin have been developed in the field of agriculture or feedstock. Lignins have been used as binders for the pelletization of feeds (see Section 8.3.3). Feed pellets with lignin for feedstock are prepared not only for handling and preservation of feeds. The use of lignin presents advantages other than the feeding properties of pellets. Lignin can be considered to be part of insoluble dietary fibers, but for livestock, native lignin is regarded mostly as a barrier to nutrient digestibility. However, lignins such as Alcell and Kraft can offer health benefits in animals in the absence of antibiotics. For example, the addition of 12.5 g/kg of dry matter Alcell lignin improves body weight gain in Holstein calves; the addition of 40 and 80 g/kg of dry matter of Kraft lignin improves weight gain and feed efficiency in broiler chickens. These lignins have prebiotic effects such as antimicrobial activity in monogastric animals by favoring the development of beneficial bacteria and improving morphological structures in the intestines [31].

On the market, several commercial preparations of lignin, especially lignosulfonates, are available as additives for the fertilizer fabrication. The addition of lignin to fertilizers can improve granulation and reduce caking (formation of lumps) during storage, for example, of formulations of phosphates or ureas, increasing hardness of granules and long-term durability.[10] Some academic work has evaluated the influence of lignin as an additive. For example, Kraft lignins have been tested for improving fertilizers as a coating for controlled release of water-soluble fertilizers or as a pelletizer. These fertilizers

---

[10] http://legnochem.ca (accessed September 17, 2014).

have been tested and some physical-chemical properties have been evaluated. The most efficiency was found for those in which the coating had a mixture of urea, lignin, and several rosins as adhesives [34].

Another effect of lignosulfonate fertilizer formulations is related to pollution prevention. The massive use of fertilizers has the problem of groundwater pollution. For example, it has been demonstrated on calcareous soil that the release rate of pesticides can be controlled by a lignin-based controlled release formulations mixing the pesticide with activated carbon and Kraft lignin [35].

Lignosulfonates are used also for the controlled application of agrochemicals. Pesticides and other chemicals employed in agriculture contain moderate amounts of commercial lignosulfonates as dispersant agents, or binders, that improved handling, transport, and storage of agrochemical formulations of suspension concentrates, water-dispersible granules, and wettable powders. Lignin is very efficient in improving dispersion and suspension concentrates as well as particles in solution and can bind extrusion and suspension for extruded granules. In many cases, lignosulfonates are able to protect UV photosensitive agrochemical molecules, making them more light stable.

Because of the chelating effect, lignin is the ideal vehicle to administer micronutrients to soil lacking this essential component for chlorophyll synthesis as well as leaf and fruit development and plant growth in general. Cobalt, zinc, iron, manganese, copper, and boron deficiencies of soils can be corrected by the addition of lignin-chelated products. These products are usually compatible with liquid fertilizers, such as ammonium nitrate and urea.

8.7 Polymers with Unmodified Lignin

For decades, the incorporation of diverse formulations of lignin into other polymers or copolymers has been investigated in order to improve the mechanical properties of such copolymers and the environmental behavior of formulations. However, practical results of this research remain modest. In this field, two approaches have been developed. Firstly, lignin formulations have been used without chemical modification within the polymer, and the second approach involves the chemical transformation of lignin before incorporation into the polymer.

Lignin, in any of its forms, can be incorporated into the polymer as a macromonomer, or as an additive. In the first case, lignin must show similar functional groups such as the monomers used onto the polymers or copolymers. In addition, they must undergo the same type of reaction as the monomers (i.e., esterification, amide formation, and condensation), when they are incorporated into polymer chains. In the second case, there is not reaction of lignin with monomers, but it is mixed only with the polymer as an additive, being part of its formulation. In the latter case, the antioxidant properties of lignin are an added value, and as a free-radical scavenger, lignin is used in several polymers to protect them from oxidation, light, or temperature in rubbers, polyalkenes, polyesters, and other synthetic polymers.

8.7.1 Phenol–Formaldehyde Binders

Urea formaldehyde and phenol formaldehyde (PF) adhesives (see Figure 8.3) are widely used to manufacture plywood, chipboards, medium-density fiberboards (MDFs), and oriented strand boards (OSBs). Modern plywood was invented in the 19th century by Immanuel Nobel, father of Alfred Nobel [36, p. 67] , from wood and adhesives. For indoor applications, plywood is generally manufactured with urea formaldehyde as a binder. This is an inexpensive glue but is strongly limited for outdoor uses because of the very low water resistance, while outdoor and marine-grade plywood are made with phenol–formaldehyde glue to prevent delamination and to retain strength in high humidity. However, these adhesives are relatively hazardous due the toxic nature of phenol and formaldehyde. Moreover, formaldehyde is suspected of being carcinogenic in very high concentrations. As a

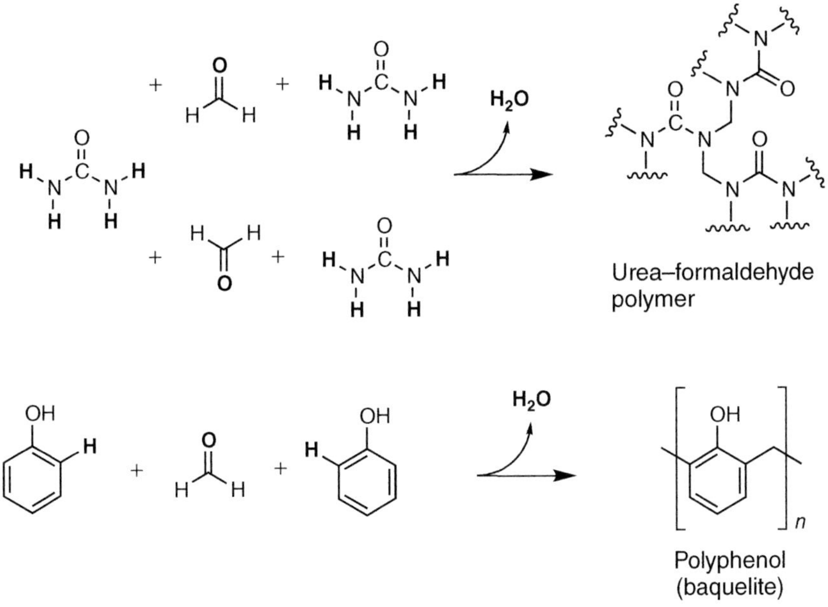

Figure 8.3 Syntheses of urea–formaldehyde and phenol–formaldehyde polymers

result, many manufacturers are turning to lower formaldehyde-emitting adhesives, denoted by an "E" rating.[11] A way to improve environmental consequences in the fabrication of these wood derivatives is the use of lignin as an environmentally friendly choice to avoid or diminish toxicity and hazards of the starting materials. Figure 8.3 shows a polymerization reaction for both adhesives.

Because of its high availability, low cost, and nontoxic nature, lignin may be an excellent alternative to phenol to prepare safer adhesives. The so-called *Karatex* process, either sulfite lignin or Kraft lignin, is copolymerized with phenol–formaldehyde resins to make adhesives for particle boards [37]. Lignin is able to bind the release of harmful formaldehyde and thus improve the environmental performance of an adhesive capable of binding wood derivatives such as plywood, water boards, OSBs, and particleboards.

The use of unmodified lignin in the preparation of such adhesives has increased since the 1970s [38]. Many assays have been conducted concerning the replacement of phenol by lignin, especially lignosulfonate, in the manufacture of plywood or fiberboards, because of its price and availability. This replacement of phenols by lignins is not only environmentally friendly but also less expensive than other binders used in the wood composite industry [39, pp. 219–241].[12]

Acetosol, organosolv, or hydrolyzed lignins have been used instead of phenol in phenolic resins to produce modified polymers, which exhibit properties with good curing characteristics [40]. In binders, about 50% of Kraft lignin or lignosulfonate lignin can be added to phenol–formaldehyde resins instead of phenol, without substantially altering the final product properties [41–43]. One of the most widely used methods for lignin methyolation is the Wooten method [44], in which lignin reacts with formaldehyde for 5 h at 553 °C in the presence of NaOH.

Some physical and mechanical properties of resins are altered; for example, in novolac-type resins, the addition of lignin leads to polymeric systems with stronger rigidity [45].

---

[11] "E" possessing the lowest formaldehyde emissions.
[12] The reaction between lignin and formaldehyde is called methyolation.

260 Lignin and Lignans as Renewable Raw Materials

The final outcome of applying this technology in a massive way has not yet been very successful in many cases. A serious problem is due to the low reactivity of unmodified lignin to formaldehyde because of the relatively small number of free phenolic –OH groups available to react with formaldehyde and the steric hindrance, which prevents the attack of formaldehyde [46].

Currently, efforts to increase the use of lignin for the production of adhesives in this field is related to the chemical transformation of lignin to increase active phenolic –OH group ratio, or the possibility of cross-linking Kraft lignin or sulfur-free organosolv lignin moieties with other cross-linking agents, such as glutaraldehyde, furfuryl alcohol, or polyethylene glycol diglycidyl ether instead of formaldehyde as the main cross-linking agent in these adhesives [47].

Brake friction material can be prepared from novolac-type phenolic resin with hexamethylenetetramine (HMTA) as a curing agent. Lignin-modified phenolic (LPF) resin prepared from MeOH-soluble soda lignin by solvent blend and *in situ* polymerization, leads to an environmentally friendly brake friction material. Friction tests have shown that the replacement of PF with lignin improved fade resistance but reduced reactivity with HMTA in the curing process and flexural strength [48]. This can be considered an eco-friendly method for this kind of material.

8.7.2 Polyolefin–Lignin Polymers

The so-called polyolefins are macromolecules synthesized by polymerization of alkenes (see Figure 8.4). In fact, there are no double bonds in polymer chains, but the name retains the word "olefin." Because of the nature of the monomer, and the mechanism of the polymerization, there is no chemical parallel between the lignin and alkenes, and only in case of functionalized side chain in the monomer, could lignin be incorporated into the polymer structure. For this reason, in most polyolefins, lignin plays the role only of additive that can modify some physical properties of the former polymer.

Polyolefin polymers such as polypropylene (PP), polyethylene (PE), and polystyrene (PS) are widely used materials for the manufacture of many objects, including food and drink packages. Most of these are highly resistant to biodegradation. Therefore, when such polymers are blended with naturally occurring macromolecules, such as lignin, some physical properties can be altered. For example, the addition of 30% of dry lignin powder to high-density polyethylene (HDPE), low-density polyethylene (LDPE), or PP can modify the mechanical and technical properties of polymeric materials. On the other hand, lignin has been shown to be a stabilizer against the photooxidative degradation of polyolefins [49].

8.7.3 Polyester–Lignin Polymers

Polyesters are polymers with ester functional groups in the chain (see Figure 8.5). There are many families of polyesters (aliphatic, aromatic, linear, ramified, etc.). Even though most of them are produced by chemical synthesis, there are some naturally occurring polyesters. Polyesters have many practical applications and are relatively easily biodegraded. In these polymers, lignin can act as a macromonomer or as an additive. Because of the presence of free phenolic –OH groups in the biopolymer, lignins can be inserted into polyester chains. Another effect is that lignin presents a complex three-dimensional structure in which a few units of lignin can substantially change the shape of the final polymer. Some examples of these materials are as follows.

Figure 8.4 General scheme for the polymerization of alkene into polyethylene

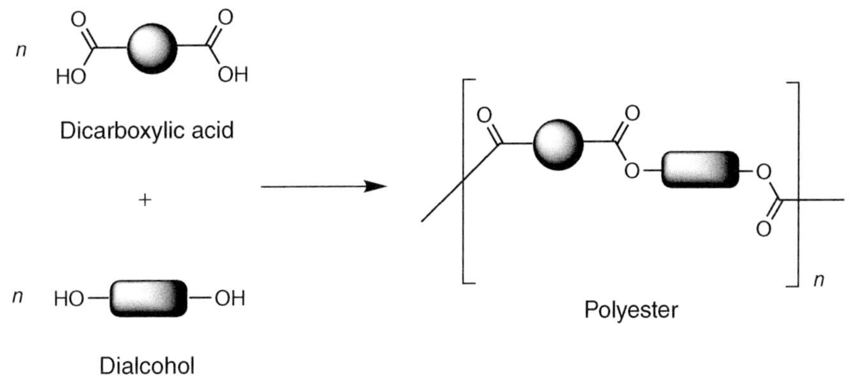

Figure 8.5 General scheme for the formation of polyester from dicarboxylic acid and dialcohol

With Kraft lignin as a macromonomer, a thermoplastic copolyester can be synthesized. When Kraft lignin is polymerized with sebacoyl chloride (decanedioyl dichloride) in the presence of triethylamine in N,N-dimethylacetamide (DMAc), a high-M_w polymer is formed that exhibits good thermal stability up to 200 °C. In addition, this is an eco-friendly material because it is biodegraded very easily [50].

Another lignin derivative, calcium lignosulfonate, can be added as filler (an inert substance added to polymers to fill up space) to a polyester such as polybutylene succinate. The addition of this lignin increases the hydrophilicity of the blends, shows higher values of water uptake at equilibrium, and facilitates the biodegradation of this hydrophobic polyesters, enhancing rigidity and lowering synthesis costs [51].

Pucciariello et al. [52] have developed a biodegradable polyester from poly(ε-caprolactone) blended with lignin prepared from straw using the steam explosion method. They employed high-energy ball milling, an innovative technique for the preparation of blends. The composites showed that lignin strongly stabilized the polymer against UV radiation. The modulus of the blends increased with the addition of lignin but decreased the tensile strength and the elongation at breakage.

Some patents related to these kinds of polymers have been claimed. For example, different polyesters have been synthesized from several diacids, and diacid derivatives with diols have been used as a polyester backbone. The polyester backbone together with lignin or lignin derivatives has been condensed during polymerization in order to produce colored, shapable polyesters suitable for use in food, beverage, pharmaceutical, and cosmetic container applications [53].

8.7.4 Acrylamide–Lignin Polymers

Acrylamide–lignin polymers can be prepared by copolymerization of alkali lignin from paper sludge with acrylamide in basic media initiated by $K_2S_2O_8$ (see Figure 8.6). A polymeric material is made that has been tested for industrial sludge water treatment as a flocculant together with aluminum salts.

8.7.5 Polyurethane–Lignin Polymers

Polyurethanes (PUs) are polymers composed of chains of organic moieties joined by carbamate link (urethane). Conventional PUs are synthesized by the reaction of an isocyanate and a polyol (see Figure 8.7). They are used to make foams, durable elastomeric wheels and tires, and other consumer goods.

Lignin can be employed as a macromonomer for the synthesis of PU because of its great amounts of hydroxyl groups available in the matrix polymer, capable of forming urethane bonds between lignin

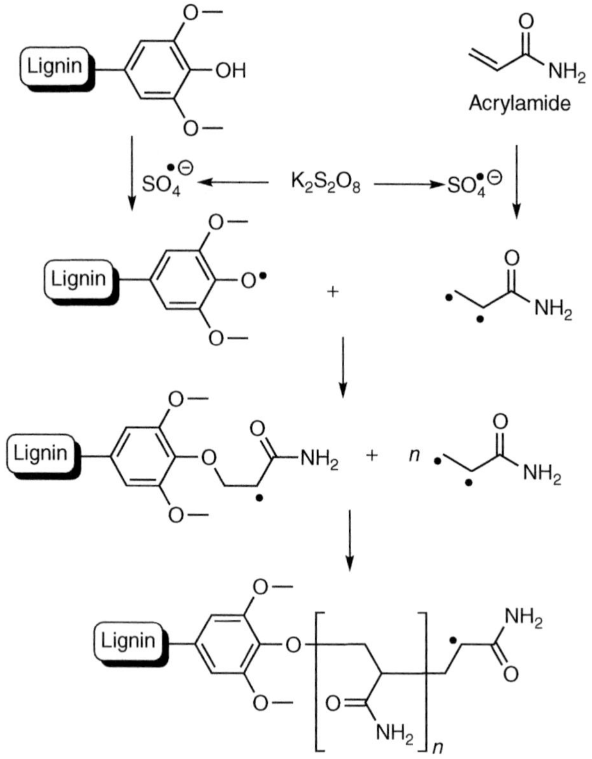

Figure 8.6 Lignin polymerization with acrylamide [54]

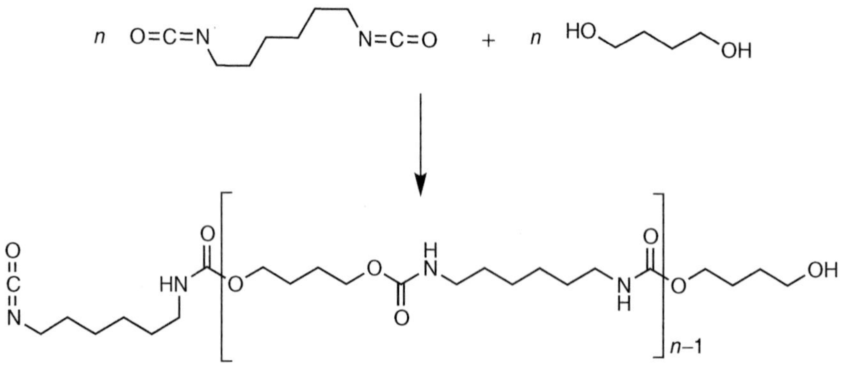

Figure 8.7 Synthesis of polyurethane from 1,6-diisocyanatohexane and butane-1,4-diol

units and other copolymers. Precured PU can be synthesized mainly by the reaction of mixtures of organosolv lignins and polyols such as polyethylene glycol (PEG), polypropylene glycol (PPG), or adipic acid [55], and an isocyanate such as methylene diphenyl isocyanate (MDI) together with a plasticizer [56]. In this case, some lignin–PU blends have been prepared, dissolving PU elastomers and adding flax/soda pulping lignin. An example of the structures that result from this procedure is summarized in Figure 8.8.

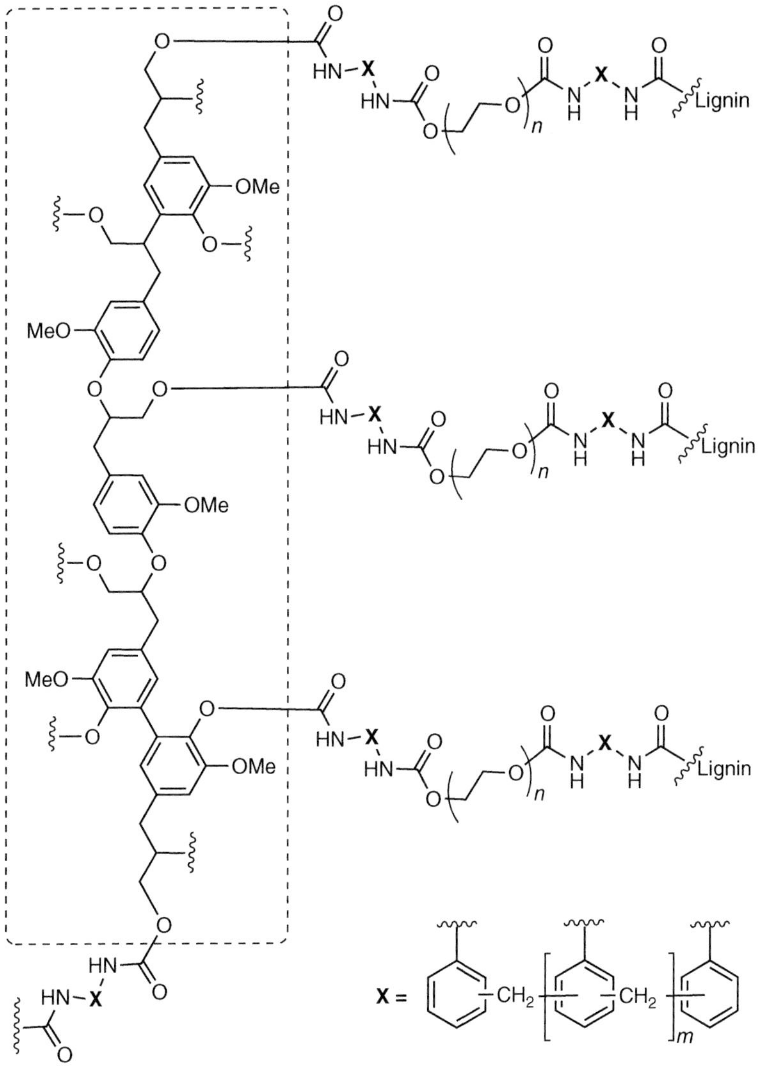

Figure 8.8 Schematic representation of the network structure of the polyurethane–lignin polymer

Cateto et al. [57] have tested the synthesis of PUs obtained by the reaction of lignin of several origins. A wide range of PU elastomers and rigid PU foams has been prepared. As expected, properties of polymers depend on the lignin type and the average of lignin used in every sample. Alcell and Indulin AT proved to be the most viable raw materials to be used as macromonomers. Rigid PU sheets and foams can be prepared by this procedure.

8.7.6 Bioplastics (Liquid Wood)

The term "liquid wood" refers to a bioplastic composed of three natural components: lignin, cellulose fibers, and some natural additives (plasticizers, dyes, antioxidants,, etc.), the trade name of which

Table 8.2 Comparison of physical properties of Arboform® with wood and plastics [58]

Properties	Arboform®	PE (LD, HD)	PP[a]	PS	PA 66[a]	Beech (cross)
Tension at break (N/mm^2)	14–22	8–30	30–40	45–65	65	7
Modulus of elasticity in tension (N/mm^2)	2000–7000	50–500	600–1700	1200–3300	2000	1500
Linear expansion coefficient (1/K) 10^{-6}	10–50	170–200	100–200	70	80	45
Decrease in size (%)	0.1–0.3	2–3	2–3	1–3	1–3	–
Vicat temperature (°C %)	80	40–70	90	80–90	250	–

[a] Not reinforced.

is Arboform [58]. This material was developed in 1996 by Pfitzer and Nägele at the Fraunhofer Institute for Chemical Technology.[13] Composed basically of low-content sulfur lignins (approximately 30%) and cellulose (approximately 60%), this material is mixed with natural additives such as wax. This material is a plastic granulate that can be melted and injection-molded, so that it can be applied with conventional plastic-processing machines in just the same way as a petrochemical thermoplastic material at temperatures below 160 °C and it is thermally stable up to at least 95 °C and has mechanical properties similar to those of a polyamide.

Table 8.2 shows a comparison of some physical properties of Arboform® with wood and some common plastics. Arboform is considered as "sustainable" plastic that is sold as granules of various colors. It can be molded into desired shapes, with the aid of additives. The bioplastic can be modified in such a way that it survives undamaged after contact with water or saliva, and therefore is the basis to manufacture any item. Speakers, musical instruments, helmets, watches, simple household products, furniture, car interior bits, toys, and so on, can be produced using this material. The final product resembles highly polished wood, or may have a more matte finish and look like the plastic used in most household items. Furthermore, it is highly recyclable, and is one of the most widely used polymers produced from lignin.

8.7.7 Hydrogels

Hydrogels are materials constituted by a polymeric cross-linked network able to absorb large amounts of water. In most cases, absorbing several times their own weight, they remain insoluble in water. Many practical applications of hydrogels are reported in the field of biomedical sciences for the manufacture of implants, soft lenses, surgical materials, or in pharmaceutical formulations, health-care products, and water reservoirs for dry soils. Most of these hydrogels must be biocompatible, and they are prepared using cellulose as the main component. Ciolacu et al. [59] have published a method for preparing new cellulose–lignin-based hydrogels, and their use in controlled release of polyphenols has been tested. Moreover, hydrogel products have also been developed as water absorbents for specific applications (personal hygiene products, underwater devices, etc.). The low toxicity of lignin and its abundance offer a promising future for this application.

---

[13] In 2010, both scientist received the European Inventor Award for their invention.

8.7.8 Foams and Composites

Lignin has been used as an additive in the preparation of PU foam and composites. As an example of this possibility for using lignin in this field, is the mechanical and thermal stabilities of a hybrid material based on PU foam, which has been synthesized from polyether polyols, with a percentage of alkaline lignin (0.5% of total mass). Because of lignin addition used as part of the composite, new properties of the foam can be developed [60].

Other attempts to prepare foams from lignin have been described by Stevens *et al.* [61], who prepared biodegradable foams from starch and Kraft pine lignin foam as panels approximately 0.2 cm thick. The replacement of 20% of the starch with lignin does not prevent foam formation and has no harmful effect on foam density or morphology. Molasses has been used, too, for the preparation of PU foams. Hatakeyama *et al.* [62] have mixed sodium lignosulfonate and molasses as raw materials for PU polymerization. Hydroxyl groups of both reacted with several isocyanates to prepare diverse foams. Physical-chemical properties of these foams derived from biopolymers showed that they can be used as insulation materials.

Another possibility is the preparation of inorganic–organic hybrid composites. The so-called geocomposites are new materials specifically designed for a single propose, that is, separation and filtration, reinforcement of structures in construction or soil stabilization, drainage, and containment. They must be produced at low cost from affordable raw materials, in many cases forming biodegradable structures. Lignin has also been used for fabricating geocomposites derived from PU and two kinds of lignin: Kraft lignin and lignosulfonate and molasses [63]. The hydroxyl groups of the aforementioned lignins or molasses and the isocyanate solutions, the base of PU polymerization, were injected into sand, and as a result, a mixed organic–inorganic material was formed with a high degree of cross-linking and foaming.

Another example of this organic–inorganic hybrid material is described by Lippach *et al.* [64] Kraft lignin, alkoxysilanes, and organic linkers, such as 3-glycidyloxypropyltrimethoxysilane, 3-(triethoxysilyl)propylisocyanate (IPTES), and Bis(trimethoxysilyl)hexane have been investigated, as have the mechanical properties of the composite.

8.7.9 Conducting Polymers

Conducting polymers have a great potentiality on some technical applications. There are several conductivity levels for conducting polymers, depending on their structure, and this diversity can be applied to the fabrication, for example, of electrostatic dissipative materials on coatings of equipment for explosives, fuel tanks, or radar-invisible aircraft.

The use of the so-called dopants is very common to improve handling and stability, as well as for modulating the electrical properties of conducting polymers. Dopants are usually soluble ions, or soluble polymeric acids that act as a counter on which the polymer is formed during the polymerization process. Soluble lignosulfonic acids are highly soluble and can be used as acid templates in the polymerization of aniline and pyrrole to give the corresponding conducting polymer (see Figure 8.9). The conducting polymer resulting from this procedure is a trademark Ligno-Pani™. The polymer has redox activity, is highly dispersive, can be cross-linked, and can be incorporated into a wide range of binders and coatings. One of the most remarkable properties is that it inhibits corrosion.

Milczarek and Inganäs [66] have prepared polymeric cathodes made of a polypyrrole/lignin composite. This cathode is prepared from pyrrole by its electrochemical oxidation in solutions to polypyrrole of lignin derivatives (see Figure 8.10). The quinone group in lignin is used for electron and proton storage and exchange during redox cycling. Lignosulfonate and polypyrrole form a renewable interpenetrating polymeric network that can be used as a cathode in batteries for charge and energy storage.

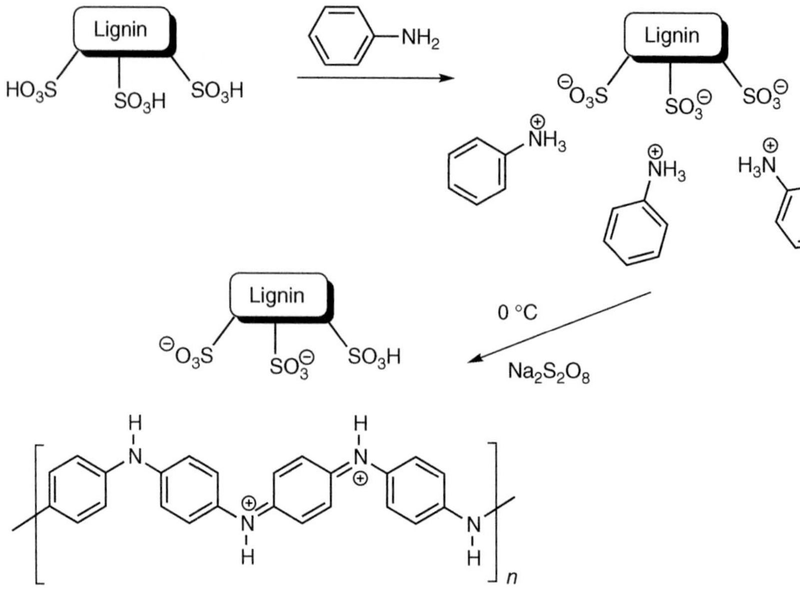

Figure 8.9 Template-guided polymerization of aniline with lignosulfonic acid as the dopant [65]

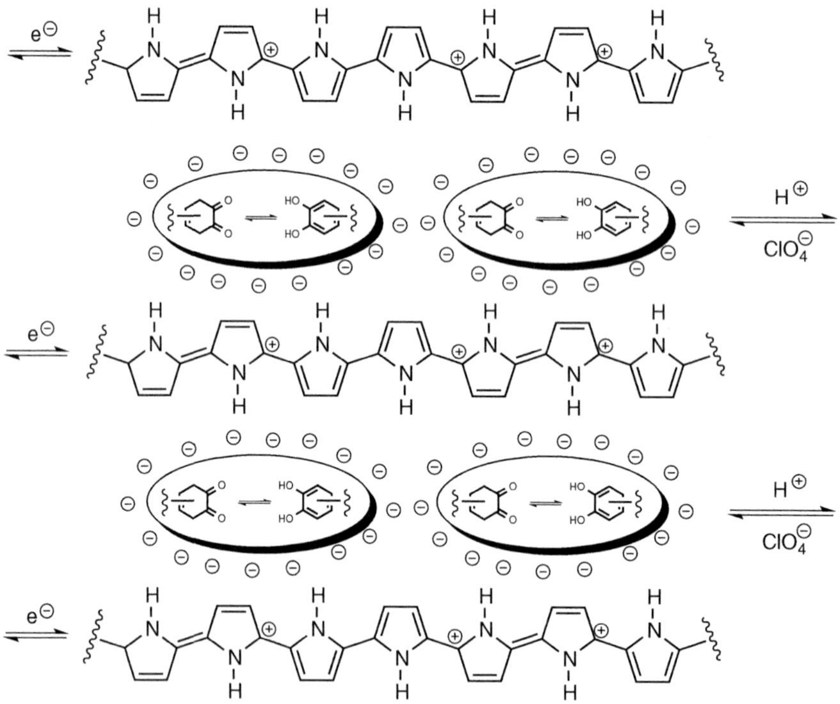

Figure 8.10 Schematic representation of the oxidative reactions of quinone groups in a lignosulfonate biopolymer with a polypyrrole matrix [66]

8.8 Other Applications of Unmodified Lignins

In the following sections, other remarkable applications for unmodified lignins, according to the literature, are described.

8.8.1 Lignin in Lead-Acid Batteries

Lead-acid batteries are the oldest rechargeable ones, and they are massively used in the automotive industry. These batteries for improved performance have many additives, such as inorganic salts and carbon black. These additives are called expanders, and without them, the sponge lead in the negative plates of the battery would lose performance rapidly when cycled [67].

Water-soluble lignosulfonates are the main components of battery expanders, and they improve performance of batteries in low temperatures at high rates of discharge. Different lignosulfonates have been employed as expanders, and these exert widely different effects on the performance of lead-acid batteries.

8.8.2 Lignin-Based Nanoparticles and Thin Films

Nanostructured materials are organic, inorganic, or miscellaneous ones with sizes ranging between 1 and 100 nm. Because of the so-called "size effect," new properties arise with the possibility of new roles in many scientific fields such as surface and new materials sciences, organic and inorganic chemistry, molecular biology, medicine, and electronics. A notable group of nanostructures includes metal nanoparticles (NPs). For these kinds of NPs, the potentiality of their applications can sometimes be severely limited by the toxicity of these nanostructures and the environmental impact of the synthesis methods. For this reason, it is critical to provide researchers with biocompatible NPs and sustainable procedures for the preparation of metal NPs. Recently, Luque and Varma [68] have summarized some topics related to this area in a comprehensive monograph.

For biological systems, NP applications could, in principle, be carried out with lower impact using biocompatible materials. Thus, research focusing on the preparation of such nanomaterials is a relevant goal to be developed in the coming years. For example, in biological fluids and organs, new functions and applications could be found in the near future because of the size and structure of NPs, which make them easier to integrate into a large number of biomedical devices [69], such as for the delivery of genes, vaccines or drugs for tumor targeting, oral delivery, or treatment of the central nervous system. However, there are other many fields in which environmentally innocuous NPs can be potentially employed, such as foam and emulsion stabilizers or matrices for environmental remediation systems. Lignin-based NPs are very good candidates for all these purposes because they are biodegradable and environmentally benign, so that an efficient preparation of lignin NPs can produce simple, inexpensive, biocompatible, and nontoxic NPs.

There are two main ways to use lignin as a starting material for the preparation of NPs. The first consists of the synthesis of pure lignin NPs, and the second one is the preparation of combined systems in which lignin is employed as a stabilizer or in the core of an NP. Using the first strategy, Frangville *et al.* [70] have developed two methods for synthesizing NPs from lignin. The first method was based on precipitation of low-sulfonated lignin from an ethylene glycol solution, using diluted acidic aqueous solutions, and the second method was based on the acidic precipitation of lignin from a high-pH aqueous solution. The former method reveals that lignin NPs contain highly porous and densely packed lignin domains which have good stability of the NPs even at high pH values. The latter one allows the preparation of stable NPs only at low pH values. Both methods are able to produce biodegradable NPs compatible with the growth of microalgae, such as *Chlamydomonas reinhardtii* and yeast. Such biodegradable and environmentally compatible NPs have potential applications as drug-delivery vehicles, stabilizers of cosmetic and pharmaceutical formulations, and in other areas to replace or substitute more expensive and potentially toxic nanomaterials.

Some examples of lignin as a stabilizer of metal NPs have been described by Milczarek et al. [71], who reported the synthesis of silver NPs in which technical, ultrafiltrated sodium lignosulfonate from softwood plays a double role as a reducing agent of silver ions and as a stabilizer of the silver metal NPs. A similar procedure has been employed for palladium NPs and has been used as catalyst in carbon–carbon coupling reactions [72]. The same authors have described a one-pot synthesis of palladium and platinum NPs with ammonium and calcium derivatives of the lignin samples in water and under aerobic conditions [73].

Wurm and Weiss [74] have recently reviewed the literature on the topic of using lignin in NP formulations, showing, for example, how lignin was doped with multiwalled carbon nanotubes (MWCNTs) or how lignin is used as a macromonomer in a step-growth polymerization with toluene-2,4-diisocyanate (TDI) terminated poly(propylene glycol) for chemical sensing applications [75].

Another aspect of lignin related to nanotechnology is the preparation of thin films of lignin on inert supports. Two main domains are involved in this field: the study of surface forces of lignin thin layer with the support material; and with external systems and the energy related to these processes, together with investigations into the role of these materials in biocatalysis.

The surface energy of lignin has many implications, especially for the pulp and paper industry, because the wettability of wood fibers depends on average of lignin, and a dramatic effect can be appreciated in the development of capillary forces during the drying and consolidation phases of papermaking [76, p. 191]. Ultrathin films of lignocellulose can facilitate a better understanding of the complex events that occur during the bioconversion of cellulosic biomass. Therefore, researchers are inspired to produce structures that resemble the cell wall of fibers to monitor enzyme binding and cellulolytic reactions [77].

8.8.3 Lignin in Dust Control

Dirt roads can provide a good and economical solution for rural transportation with a low traffic volume, but unpaved surfaces are the largest source of particulate air pollution in many countries.[14] Other problems associated with dust on these kinds of roads are the slowing of plant growth around the road, which damages the surface for the circulation vehicles and poses a serious problem of safety for drivers due the limited visibility on many occasions. Such surfaces can be found also at other sites such as open-air sports areas, surroundings of racing circuits or airports, and public works. In all these cases, it is necessary to have an effective system to control dust particles.

Lignosulfonates have been used as a treatment for dirt roads in Europe and the United States since the 1920s. In unpaved dirt roads, it is necessary to hold down dust and stabilize the surface for proper road maintenance, in order to avoid dust clouds that disturb visibility. Lignosulfonates are cheaper alternatives to chloride salts, such as calcium chloride, which can hold down dust and stabilize unpaved road surfaces because they attract water and retain moisture from the atmosphere for prolonged periods.

The method of applying lignins to road surfaces for dust control consists of spraying lignin solutions onto a dirt surface, (1.0–2.0 lb/yd^2), and, over time, road surfaces begin to show improved stabilization.[15] Water evaporates from the lignosulfonates as they dries, and the dust particles are trapped by the highly viscous, naturally adhesive material. In addition, some lignosulfonates become completely insoluble because of solar heating.[16] During rain, clay disperses and in turn swells and plugs pores, thereby reducing water penetration.[17]

---

[14] http://www.epa.gov/pm/basic.html (accessed December 13, 2014).

[15] "Lignins: A Safe Solution for Roads." Dialogue/Newsletters Vol.1 No. 3. Lignin Institute. July 1992. Retrieved 2007-10-16. http://saferoadservices.com/wp-content/uploads/2012/09/LigninSafeSolution.pdf

[16] http://www.pacificdustcontrol.com/lignin-sulfonate (accessed December 13, 2014).

[17] Wisconsin Transportation Bulletin No. 13, 1997. http://epdfiles.engr.wisc.edu/pdf_web_files/tic/bulletins/Bltn_013_DustControl.pdf

The application of lignosulfonates to dirt roads has significant advantages, such as a dramatic improvement in driving comfort and safety and a denser and firmer road cap, preventing the loosening of dirt and gravel, thus reducing road repairs.

This lignosulfonate treatment tends to be more effective than inorganic chloride salts on gravel roads containing higher levels of sand. On the other hand, lignosulfonates are environmentally safe products and nontoxic when properly applied, making them safe for plants and surface water surrounding the roadways. Being noncorrosive, they can be applied without special equipment or clothing.

The binding property of lignin-based products is used in many aspects of dust control and prevention, not only for dirt roads, but also for dust suppression in roads, parking lots, racing tracks, quarries, paddocks, and construction sites.

8.8.4 Lignin in Concrete Admixtures

Concrete is manufactured from three major components: portland cement, water, and aggregate. The world uses more than 7 billion m^3 of concrete each year. Admixtures are ingredients in concrete other than the aforementioned components that are added to the mixture immediately before or during mixing. The American Concrete Institute describes the reasons for using admixtures: to improve workability without increasing water content or to decrease the water content at the same level of workability, to retard or accelerate the initial setting time, to reduce or prevent shrinkage, to create slight expansion, to alter the rate or capacity for bleeding, to reduce segregation, to improve pumpability, to reduce rate of slump loss, to retard or reduce heat evolution during early hardening, to accelerate the rate of strength development at early ages, to augment strength (compressive, tensile, or flexural), and to increase durability or resistance to severe exposure conditions. In addition, the application of deicing salts and other chemicals reduces the permeability of concrete, controls expansion caused by the reaction of alkalies with potentially reactive aggregate constituents, strengthens the bond of concrete to steel reinforcement, improves the bond between existing and new concrete, bolsters impact and abrasion resistance, inhibits corrosion of embedded metal, and produces colored concrete or mortar.

Sodium and calcium lignosulfonates can be directly added in the concrete as a water-reducing agent, that is, approximately 10% of water can be reduced with a dosage from 0.2 to 0.3% of lignin. They provide good strength characteristics (15% of strength in 3–28 days), because they tend to entrain air. The final concrete is sticky and hard to finish. Lignin improves the mix capacity and the life of concrete, and the durability in freeze/thaw cycles. Workability is better, too, because the setting time is retarded and the concrete shows a lower heat of hydration. On the other hand, lignin-treated concrete contains no chloride and no basic active matter, and hence it is harmless to steel bars embedded in the works.

8.8.5 Lignin as a Dispersant, Emulsifier, and Surfactant

Lignin is capable of diminishing surface tension on liquids when added to a suspension or a colloid. Therefore, the separation of insoluble particles is improved in this media, avoiding their settling or clumping. This is the main reason why lignin is used as a dispersant [78, p. 361] or surfactant. Most of lignin-based dispersants are manufactured from lignosulfonates, which are considered to be anionic surfactants because of the presence of ionized sulfonic groups in the polymer matrix as salts. Some applications of this use of lignin are the following:

In agriculture, many pesticides are applied as wettable powders in which the active ingredients of the pesticides are used in a finely ground state combined with wetting agents in order to prepare diluted suspensions. The formulations for pesticide application are normally formed by a complex mixture, which includes wetting agents, dispersants, spreaders, and carriers. Lignosulfonates are commonly used dispersants for this purpose. Other formulations for dispensing pesticides in agriculture are the so-called water-dispersible granules (WG). These have become a relatively new type of pesticide formulation, being developed as a safer alternative for storing and handling pesticide formulations than are wettable powders [79].

A family of surfactants has been developed by corporations such as Texaco from lignin as a renewable resource, seeking chemically enhanced oil recovery by combining lignosulfonates and fatty amines, derived from other renewable resources. The lignosulfonate–amine combination is surface active and replaces the oil-soluble component of conventional IOR surfactant systems. A variety of lignosulfonates can be combined with many fatty amines to produce these novel materials [80].

Tanning is the industrial process of converting the animal skin into leather. Proteins such as collagen and other related proteins are chemically transformed into a more resistant and durable material. There are two main procedures for tanning: chrome tanning and vegetable tanning. The former is one or the most widespread procedures because of its efficiency, but it has a serious environmental drawback due to the nature of chrome salts[18] utilized in the process. Vegetable tanning, which uses natural ingredients to tan leather, is environmentally friendly, and thus, vegetable-tanned leather products can be recycled. But, this process is more expensive and a color application in final products is limited. In this industrial branch, lignin has been used too as a sulfite lignin extract. The leather is tanned with lignin adjusted to pH within the 3.0–5.0 range. This mixture has a tanning action resembling that of the vegetable tannins. When lignin is used as the only tanning agent, it produces a brown leather that tends to be thin and hard, with low tan fixation. For this reason, it is used in combination with other vegetable tanning agents in order to improve the fullness and firmness of the leather.[19]

Lignosulfonates can stabilize many types of emulsions of heterogeneous mixtures and immiscible liquids. Magnesium lignosulfonate is an additive sold[20] for the manufacture of refractory ceramics or silicon carbide ceramics. It is mixed to the raw ceramic mass or the slurry as a bonding agent, because the greater strength of the blanks, when pressed, keeps the emulsions homogeneous, and avoids flocculation of the mixture components. When lignosulfonates are added to asphalt emulsions, they are capable of reinforcing and stabilizing them, and can retard the aging of the material because of their antioxidant capacity. Some experimental studies [81, 82] have analyzed key properties of asphalt according to the average and the type of lignin added. Another example of the application of lignosulfonates in this field is their use as an emulsifying agent in the fabrication of liquid ink, constituted by a solid pigment and a water carrier [83].

Other applications of lignosulfonates are their use as additives in the fabrication of drilling fluids. In the process of drilling boreholes, mainly for gas or oil extraction, the drilling fluids are used as lubricants, because enormous friction is produced that could damage the drill working at very high pressures. The drilling fluid is made of water, oil, and gas-based fluids, and the composition and viscosity of the material is determined by a drilling fluids engineer, depending on the circumstances. Emulsifiers such as lignosulfonates are key components of drilling fluids to maintain homogeneity and stability during drilling with an adequate viscosity. Among the most commonly used drilling fluids have been ferrochrome lignosulfonates, but the environmental impact of these salts is well known [84]. Another problem associated with drilling fluids involves the thermal stability of the formulation. Lignosulfonates improve thermal resistance of these drilling fluids.

Wax emulsions are other industrial products widely used as lubricants in the fabrication of polymer objects, for example, polyvinyl chloride (PVC), dispersing agents in additives for plastic, coloring plastics (masterbatches), or as additives in inks or coatings. In the paper industry, wax emulsions are used for the manufacture of common paper, paperboard, boxboard, and insulating board. Wax emulsions must be stabilized by emulsifiers. The addition of sodium lignosulfonate within an average range of 2–6% can suffice. Moreover, the wax emulsion prepared by this method is acid stable.[21]

---

[18] Chromium sulfate and other salts of chromium.

[19] http://www.cbi.nl (accessed December 13, 2014).

[20] http://www.zellwerke.de (accessed December 13, 2014).

[21] http://www.spx.com (accessed December 14, 2014).

8.8.6 Lignin as Floating Agent

The addition of lignin sulfonates to water changes its surface tension and increases its density. This effect is employed for the selective flotation of minerals, such as fluorite or barite, or the opposite effect, that is, the selective depression of molybdenite or some rare earth minerals [85, p. 180].

A further application in this field is that when added to water, lignin sulfonates are able to improve floatability of fruits, such as pears and apples, in postharvesting procedures, such as classification to avoid damaging the fruit.[22]

8.9 New Polymeric Materials Derived from Modified Lignins and Related Biomass Derivatives

Another remarkable field of lignin applications in polymer chemistry comes from the chemical modification of different lignins before incorporation of lignin into a polymer. Several approaches can be followed for this; for example, increasing the number of reactive groups of the native lignin. Another possibility is to introduce new phenolic moieties to improve the reactivity of lignin and the third one is to link a spacer with a functional group to avoid steric hindrance of most of the reactive sites in the polymer due the high degree of cross-linking of the lignin structure that hides reactive functional groups.

8.9.1 Modified Lignin in Phenol–Formaldehyde Wood Adhesives

Lignin has been extensively used in the preparation adhesives for plywood and other industrial wood derivatives as has been pointed out in this Chapter, but lignin can replace phenol molecules in phenol–formaldehyde resins only at a relatively low proportion [86, p. 84], because the properties of the resin do not have adequate quality standards to be extensively used as a phenol substitute. In addition to the steric hindrance effects, it has to be considering the decreasing reactivity of aromatic rings of lignin compared with unsubstituted phenol.

It is remarkably difficult to formulate a very effective glue due to almost nonaccessibility of phenolic moieties in the matrix of the polymer, making the cross-linked net of the phenol–formaldehyde resin less effective (see Figure 8.11). An estimation of the reactive aromatic sites available for the formaldehyde condensation yielded approximately 0.3 positions for every nine carbon units in Kraft lignin. This average assumes that only 10% of phenol moieties are reactive (see Figure 8.12). Thus, it is critical to improve this average by nonexpensive methods in order to increase active positions for the reaction between lignin polymer and formaldehyde. Two main chemical transformations can be performed for this purpose: phenolation (or phenolysis), which consists of the treatment of lignin with phenol in the presence of polar protic organic solvents, such as MeOH and EtOH [87]; and the demethylation method. Other procedures are also available to improve reactivity, as described in subsequent sections.

8.9.1.1 Phenolation

The phenolation process involves the treatment of lignosulfonates with phenol in an acidic medium with heating at approximately 70 °C, the boiling point of the common organic solvents (see Figure 8.13). Under these conditions, phenol reacts with lignin aromatic rings and side chains. During the process, some ether bonds are cleaved so that a new sort of lignin is formed, decreasing the M_w of the molecule. The viscosity of this modified lignin is variable, depending on the phenol/lignin ratio. The lignins produced by this procedure can be used for polymerization with formaldehyde as

---

[22] http://postharvest.tfrec.wsu.edu/pages/J4I1D (accessed September 17, 2014).

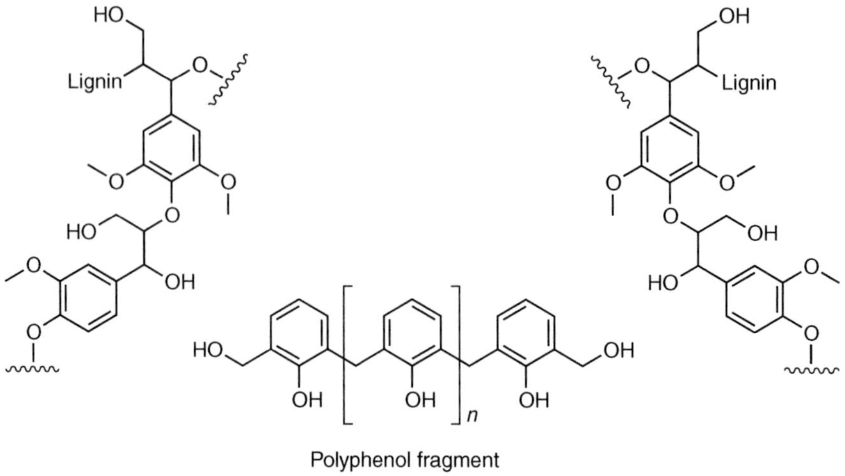

Figure 8.11 Schematic representation of the lignin resin composed by a polyphenol fragment [86, p. 84]

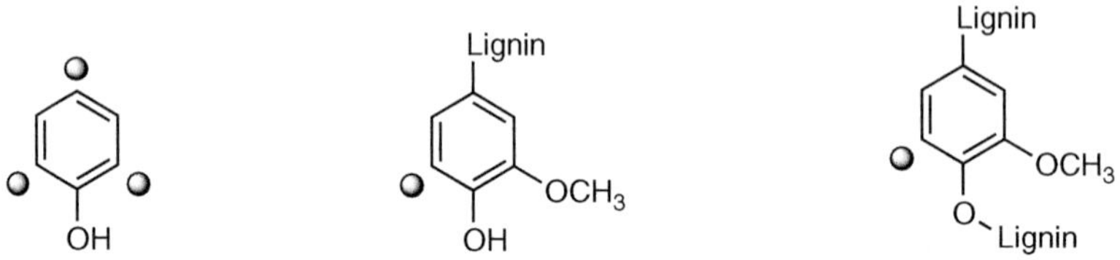

Figure 8.12 Steric hindrance effects that hamper cross-inking in lignin polymers. Steric hindrance sites are marked with solid spheres

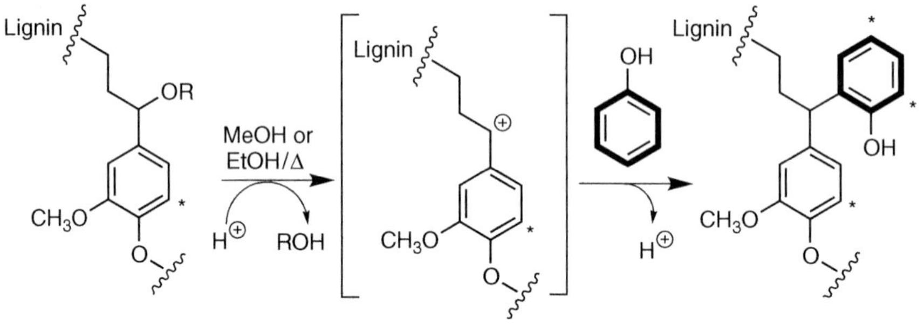

Figure 8.13 Phenolization reaction in lignin. Active sites are indicated with an asterisk (*) [88]

a binder in plywood panels. This method suffers from a drawback that phenol cannot be replaced 100% in the glue, and hence the material does not meet quality standards.

8.9.1.2 Demethylation

The second most widespread method for chemically modifying lignin for this purpose is the demethylation procedure (see Figure 8.14). For example, in Kraft lignin, about 50% of hydroxyl

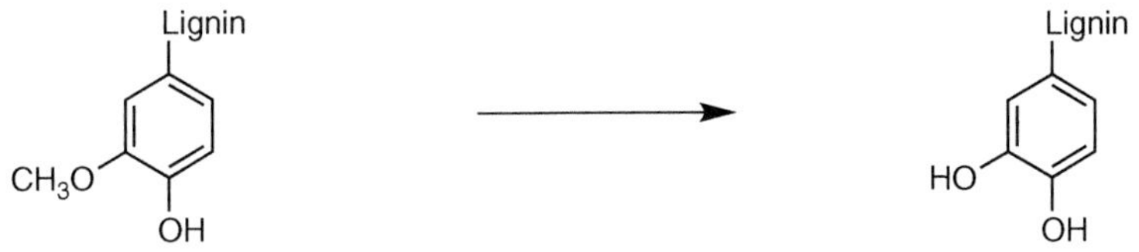

Figure 8.14 Demethylation reaction in lignin

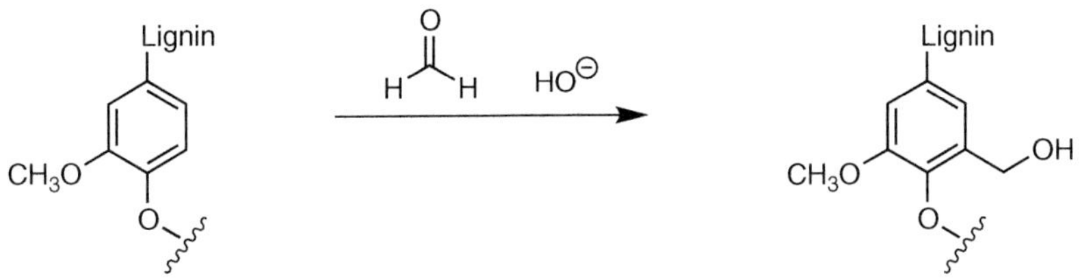

Figure 8.15 Lederer–Manasse reaction of lignin with formaldehyde in dilute basic medium [89]

groups on aromatic rings in the polymer are methylated, meaning that the reactivity of this lignin is quite diminished for reaction with formaldehyde. The demethylation of Kraft lignin allows a considerable number of additional free hydroxyl groups and a higher reactivity to formaldehyde. Several methods of demethylation can be followed, but reaction with sulfur is an inexpensive and efficient procedure.

8.9.1.3 Hydroxymethylation

Another way to improve lignin reactivity on wood phenol–lignin adhesives is by the hydroxymethylation reaction. There is no uniform performance of the modified lignins by this procedure related to their behavior as macromonomers. When lignins are treated with formaldehyde and diluted alkali, mainly the Lederer–Manasse reaction (see Figure 8.15) takes place and hydroxymethyl groups are introduced into aromatic rings of the lignin [89]. Some competitive reactions may occur, such as Cannizzaro reaction, in which formaldehyde undergoes a deproportionation reaction to give MeOH and formic acid. Hydroxymethylated lignin can replace phenol in several proportions around 40% to produce resins. In some cases, these have very low amounts of free formaldehyde and the final products usually show similar characteristics of commercial formulations. However, other studies have yielded opposite results concerning residual free formaldehyde and have pointed out some difficulties regarding storage stability.

8.9.1.4 Glyoxalation

Calcium lignosulfonate lignin has been treated with glyoxal water solution in basic medium to give glyoxalated lignin. Depending on the percentage composition of the reagents, several glyoxalated lignins can be produced. Formulations of adhesives for exterior-grade particleboards have been tested. The nonvolatile and nontoxic aldehydes, such as glyoxal, comfortably pass the international standard specifications for these materials, including press times comparable to those of formaldehyde-based commercial adhesives used in particleboard panels. The main structural differences between lignosulfonate lignins and the modified one were discerned by ^{13}C NMR spectroscopy. ^{13}C signals corresponding to benzylic alcohol and quinone methide from methoxylated guayacyl-type lignin appear after the reaction with glyoxal (see Figure 8.16) [90].

Figure 8.16 Fragments of modified lignin with carbon atoms (in bold marked with arrows) identified by ^{13}C NMR spectroscopy [90]

8.9.1.5 Pyrolytic Lignins and Hydrolysis Products

A strategy that has stirred growing interest is the use of pyrolytic lignin as a renewable resin from the pyrolysis of biomass. Pyrolysis of lignocellulosic biomass produces bio-oil, a material which has a high potentiality as a raw material for the production of liquid fuels and chemicals [91], because of the high yield of pyrolytic lignin and its easy assimilation into phenol formaldehyde formulations. Most of these pyrolytic lignins appear to be oligomeric alkylated phenolic units probably linked by C–C bonds [92]. These oligomers can be incorporated into the phenol–formaldehyde polymer.

A further option is to carry out a partial hydrolysis of lignin with HCl, or other acids, before polymerization with formaldehyde. The amount of lignin incorporated into the polymer can be increased by this procedure [93, p. 158]. In both cases, smaller lignin fragments can be synthesized, diminishing steric hindrance of reactive groups for polymerization.

8.9.2 Modified Lignins for Epoxy Resin Synthesis

Epoxy resins are well-known thermostable polymers used for adhesives, binders, and coatings, among many other applications. They are prepared *in situ*, by a mixture of a preexisting polyepoxy polymer and other kinds of polymers with a wide range of functionalization as amines, acids, anhydrides, phenols, alcohols, or thiols that react when a catalyst is added. The catalyst produces a curing effect by cross-linking reactions. Epoxy resins are usually very hard materials because of the high degree of cross-linking. On the market, there are many kinds of epoxy resins, depending upon the nature of the starting components. One of the most widely used procedures to prepare epoxy resins involves epichlorohydrin as a major raw material to induce polymerization together with bisphenol A.

Epichlorohydrin can be synthesized from a petrochemical starting product, such as propene, or glycerol, a renewable starting product [94] (Figure 8.17). In both cases, epichlorohydrin is synthesized with good yields, from the same mixture of chlorinated alcohols as are used without separation.

Epichlorohydrin reacts with bisphenol A to give the corresponding epoxy resin, that is, bisphenol A diglycidyl ether (BADGE) (see Figure 8.18). This polymer is technically very efficient, but presents serious environmental risk because of the hazards associated with bisphenol A [95].[23]

Ligninosulfonates can be added as a cross-linking material in this epoxy resin. This is a method to improve its environmental characteristics because of the decrease in the amount of bisphenol A in the polymer. The problem of this technology is the same as has been discussed earlier, that is, the need to increase reactivity of lignin. To improve the lignin reactivity, some additional phenol moieties may be introduced into the lignin structure by the reaction of lignosulfonate with phenol and sulfuric acid. It has been demonstrated that two reactions take place: Some new phenolic units are incorporated into the lignin matrix; and the lignin is partially hydrolyzed to give diphenolic moieties (see Figure 8.19). Both are able to react with epichlorohydrin and to polymerize together with bisphenol A to give a

---

[23] http://www.epa.gov/oppt/existingchemicals/pubs/actionplans/bpa.html (accessed September 17, 2014).

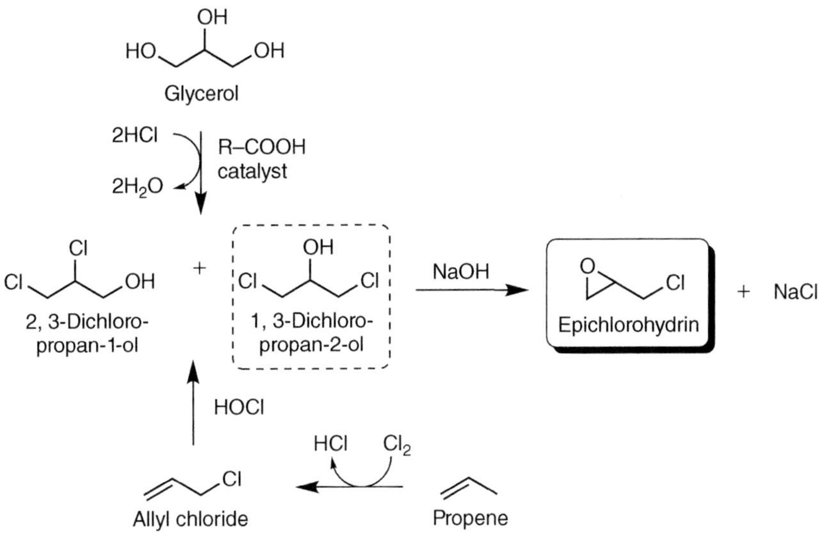

Figure 8.17 Synthesis of epichlorohydrin from glycerol and propene [94]

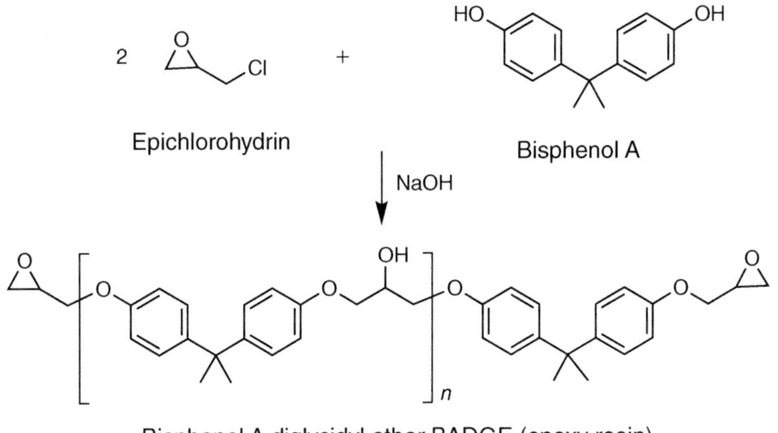

Figure 8.18 Synthesis of epoxy resin (bisphenol A diglycidyl ether) from epichlorohydrin and bisphenol A

hybrid epoxy resin formed with a renewable nontoxic raw material, such as lignin, and a molecule from the petrochemical industry, such as bisphenol A.

8.9.3 Polyurethanes

Modified lignins can be used as polyol precursors for the synthesis of PUs (Figures 8.20 and 8.21). This method has been known since the 1980s. For example, Kelley et al. [96] have made PU films from chain-extended hydroxypropyl lignins (CEHPLs) which were prepared from organosolv and Kraft lignins, by reaction with propylene oxide.

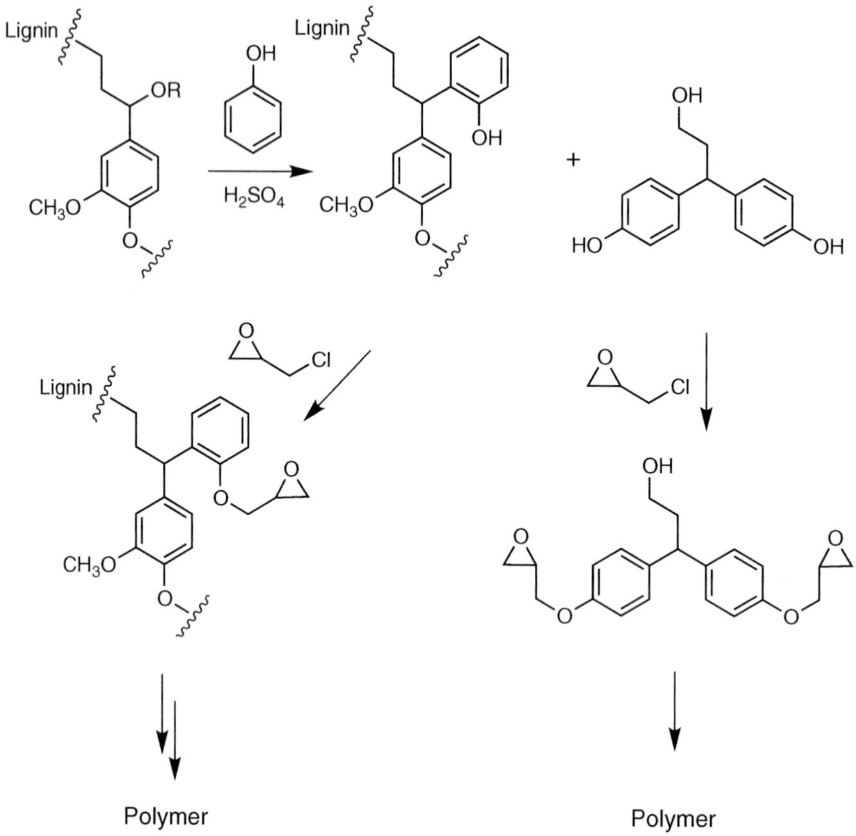

Figure 8.19 *Polymerization reactions mediated by epichlorohydrin [88]*

Modified lignin can be used as a starting material for the synthesis of PU by reaction with isocyanates, for example, by Lewis acid treatment. This leads to a partially demethylated lignin, with a higher average of free hydroxyl group [97], or polyurethanes from steam-exploded lignin previously reacted with (4,4)-methylenebis(phenylisocyanate) and polymerized together with ethylene glycol [98]. Hydroxypropylated lignin has been used to synthesize PU [99].

8.9.3.1 Polyesters

Some examples or preparation of polyesters from modified lignin can be found in the literature. For example, alkylated and acylated Kraft lignins yield polymeric materials with mechanical properties very similar to those of conventional PS. These thermoplastic materials can also be plasticized by commercially available aliphatic polyesters or PEG derivatives [101]. In a study of this procedure, Li and Sarkanen [102] in 2002 prepared some promising thermoplastic blends from simple alkylated Kraft lignin derivatives and aliphatic polyesters with low T_g. Such blends are miscible when the value of CH_2/COO ratio of the polyester lies between 2.0 and 4.0. Polyesters interact sufficiently but not too strongly with the alkylated Kraft lignin species by intermolecular attraction, which should compete for the Kraft lignin components in the peripheral regions of the supramolecular complexes, improving mechanical properties of the blend.

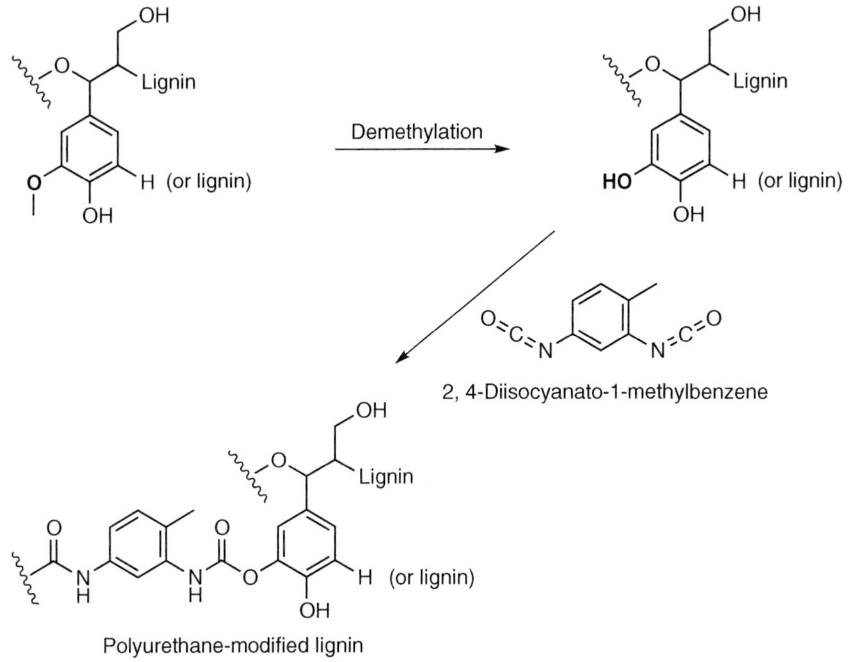

Figure 8.20 Lignin modification with polyurethane [100]

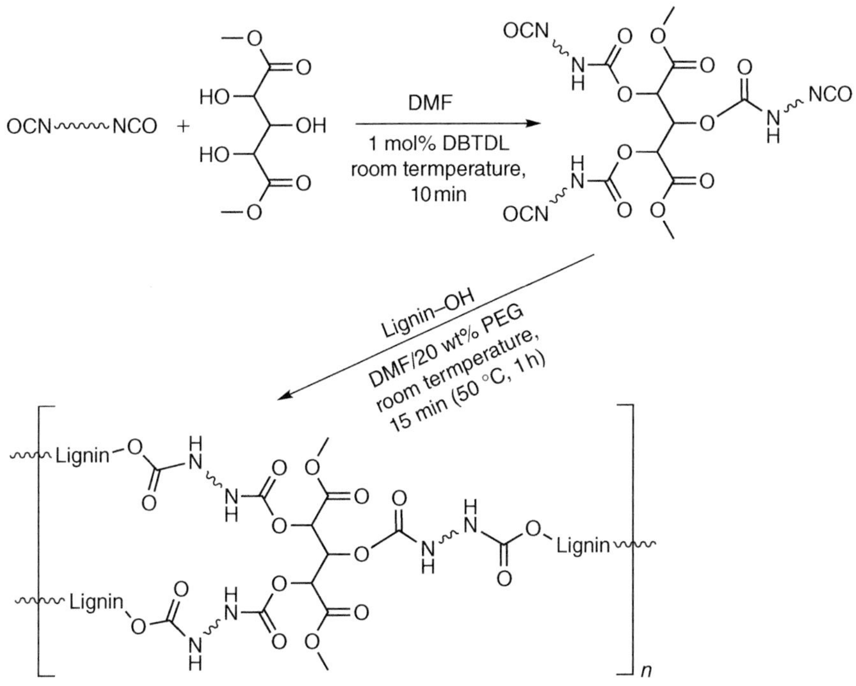

Figure 8.21 Scheme of the lignin–PU preparation [100]

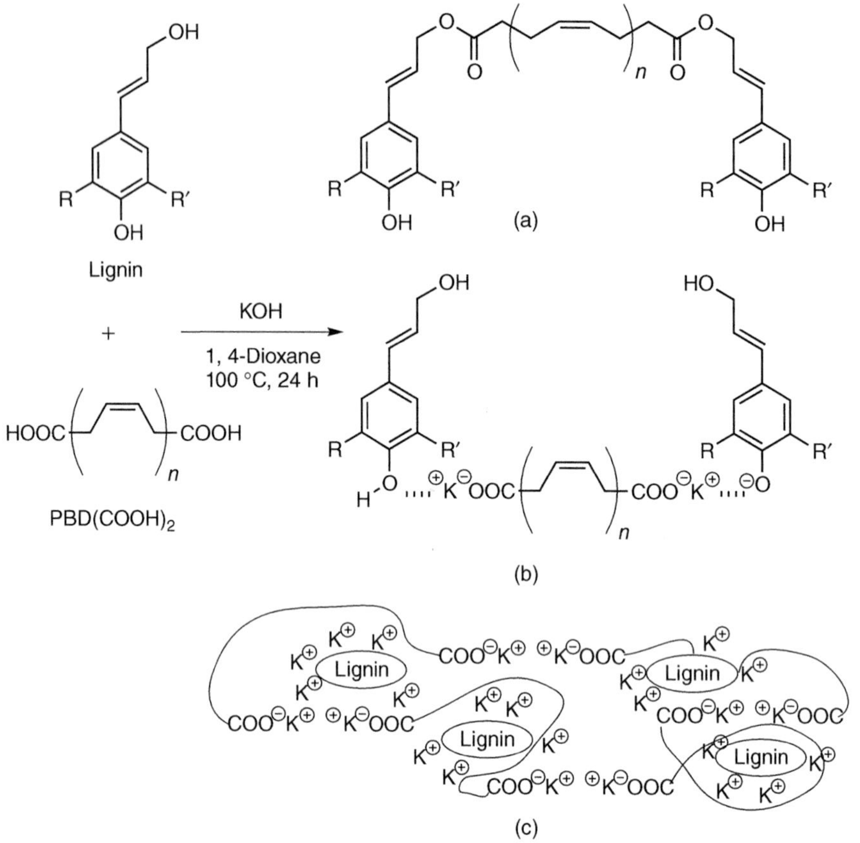

Figure 8.22 Scheme for the lignin–polybutadiene linkage: ester linkage (a), ionic linkage (b), and interpenetrating network (c) [103]

8.9.4 Lignin–Polybutadiene Copolymers

Saito et al. [103] have developed a thermoplastic copolymer using several lignins that reacted with dicarboxy-terminated polybutadiene PBD–(COOH)$_2$ (see Figure 8.22). Thermoplastic copolymers were made using unmodified lignins, which were solvent extracted from hardwood in the polymerization reaction. The same lignin was washed with MeOH and then was treated with PBD–(COOH)$_2$ and finally it was modified by reaction with formaldehyde before polymerization. In this case, a higher M_w lignin resulted because this reaction induces a higher degree of cross-linking. This modified lignin facilitated the preparation of freestanding films of lignin-based thermoplastic material.

8.10 Polymers Derived from Chemicals Obtainable from Lignin Decomposition

de Wild et al. [104] have developed a pyrolysis-based lignin biorefinery approach, called LIBRA, to transform lignin into phenolic bio-oil and biochar using a bubbling fluidized bed reactor technology. Bio-oil is a potential source for value-added products that can replace petrochemical phenol in wood adhesives, resins, and polymer applications. The biochar, for example, can be used as a fuel, as a soil amendment, as a solid bitumen additive, and as a precursor for activated carbon.

8.11 Other Applications of Modified Lignins

Kraft lignin has been modified using different methods, such as oxidation with oxygen, and polyoxometalate catalysis, solely or in the presence of laccase. The resulting Kraft lignin has been employed as adsorbent of heavy metals and pesticides. This chemical modification of lignin by these procedures avoids a high grade of depolymerization and increases the amounts of carboxyl (up to 15%) and carbonyl (up to 500%) groups. Therefore, it is possible to improve its sorption capacity against metals. The resulting material is a sorbent mimicking humic matter. This technique shows promising results for bioremediation applications [105], and it has been tested, for example, for biosorption of nickel using unmodified and modified lignins extracted from agricultural waste [106]. Modified lignin has shown better results than unmodified lignin, providing a way to valorize residual biomass.

Amine functionalization by the Mannich reaction of industrial softwood Kraft lignin is a method to modify and improve lignin reactivity. The potential value-added applications of these modified lignins with high amine contents include the use as surfactant chemicals, polycationic materials, slow-release fertilizers, and, among others, as a cationic flocculant in wastewater treatment for removing diazo dyes [107].

Urea-modified lignin has been prepared according to the Mannich reaction (see Figure 8.23). Ammonium polyphosphate (APP) and urea-modified lignin have been tested as intumescent flame retardant (IFR) system to improve flame retardancy in poly(lactic acid) (PLA) polymer. The flammability of IFR–PLA composites has been studied using several standard tests, showing that the urea-modified lignin combined with APP exhibited much better flame retardancy and thermal stability than the combination of native lignin and APP [108].

Hydroxymethylated lignin-based or copper–epoxylignin NPs have been tested for birch veneer treatment, ensuring strong biological wood stability and high hydrophobicity. Veneer samples were analyzed after burial in soil for 6 months, showing low mass loss values [109].

An increase in the carboxylic content of the lignin preparation resulted in greater hydrolysis yield. These results suggest that the carboxylic acids within the lignin partially alleviate nonproductive binding of cellulases to lignin [110].

8.11.1 Nanoparticles (NPs)

The hydroxyl groups of lignin films have been modified with poly(N-isopropylacrylamide), maleimide terminated through atom-transfer radical polymerization under aqueous conditions to prepare ion-responsive nanofibers [111]. Recently, a straightforward strategy has been developed to

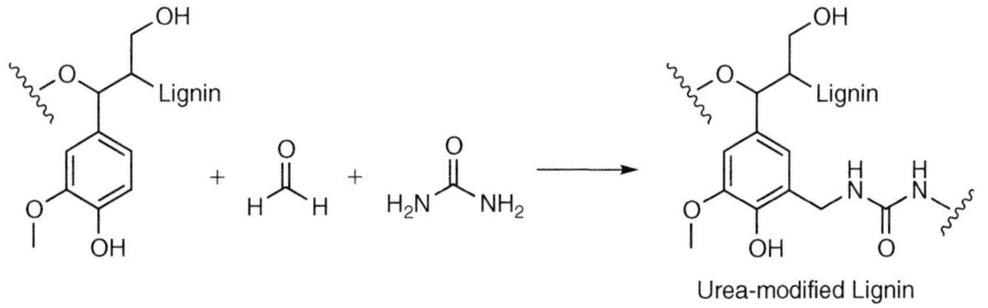

Figure 8.23 Lignin modification by means of the Mannich reaction

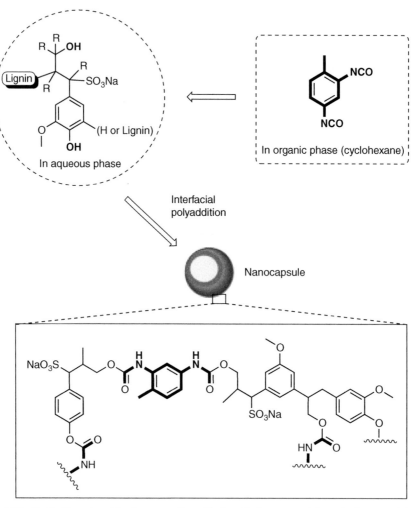

Figure 8.24 Synthetic protocol for the generation of hollow lignin nanocontainers by inverse miniemulsion [112]

generate hollow nanocapsules by interfacial polyaddition of lignin with toluene-2,4-diisocyanate in an inverse miniemulsion. Lignin derivatives were dissolved in water and dispersed in organic solvent, by means of a polyaddition reaction within the hydroxyl groups of lignin at the droplet surface. By this approach, hydrophilic substances can be incorporated into biodegradable lignin nanocontainers (see Figure 8.24) [112].

Gold NP composites containing lignin or oxidized lignin, adsorbed onto graphite, have been used to prepare carbon-paste-modified electrodes (CPMEs). Gold NPs were produced on both composites using the reducing properties of the lignin, which was able to reduce Au(III), from $HAuCl_4$ at pH 4.7. Au/graphite/lignin(oxi) composite has shown better characteristics for electrochemical purposes. The best composite to prepare CPME was the one containing 0.2% gold and 2.5% oxidized lignin, with the percentages expressed in relation to the graphite mass. The system has been catalytically tested to the oxidation of dopamine and ascorbic acid and reduction of nitrite and iodate in the positive potential range [113].

Copper and zinc NPs have been used together with lignin in which the biopolymer had been previously hydroxymethylated. Solutions of this material have been tested as biocides against soil microorganisms in order to preserve wood. In this way, wood biodegrades more slowly than wood subjected to conventional treatments [109, 114].

8.11.2 Cationic Amphiphilic Lignin Derivatives

A novel cationic amphiphilic lignin derivative with high surface activity was prepared from Kraft lignin *via* the introduction of dehydroabietyl groups as lipophilic groups, and diethylenetriamine groups as hydrophilic groups by the Mannich and ketone–amine condensation reactions. Solubility, surface tension, hydrophilic–lipophilic balance (HLB) values, foamability, and zeta potential were used to evaluate the basic physicochemical properties of the cationic amphiphilic lignin derivative [115].

8.11.3 Soil Preservation

One serious environmental problem is soil recovery after forest fires caused accidentally or related to wood production in tropical areas. Reforestation becomes a difficult task in both cases, because of the high acidity of the soils after the forest fire. The low pH of these soils has been attributed primarily as a consequence of fire on biomass. At acidic pH values, the aluminum ion concentration increases in the soil water, inhibiting the growth of plant roots [116]. Katsumata and Meshitsuka [117] have developed some tests to evaluate the influence of modified lignins on plant root growth under significant amounts of Al(III). Modified lignins have been prepared by alkaline treatment of commercial Kraft lignins from softwood. These alkaline lignins have been submitted to two reactions: treatment with oxygen under pressure and the sulfonation by treatment with sodium sulfite. In the first case, the formation of carboxylic acids and phenolic hydroxyl groups take place, and in the second one, sulfonic groups are introduced into the polymer by radical sulfonation. These modified lignins are able to remove Al(III) toxicity for plant root growth at pH $\sim$ 4.5, because of the formation of complexes between the modified lignins and the aluminum ions, as confirmed by ^{27}Al NMR spectra. Roots of several plants are able to grow faster and stronger in the presence of modified lignins. This is a promising result for soil recovery.

8.11.4 Fertilizers

The modification of lignin leads to new fertilizers with a higher performance than native lignin. Ammoxidized Kraft lignin has been tested as a slow-release fertilizer for the cultivation of such species as sorghum. The experimental plots showed that modified lignin is a satisfactory source of nitrogen, with the advantage that percolated water after application had essentially lower nitrate levels than after conventional fertilization with inorganic salts, with a similar grain and plant development [118].

References

[1] Glasser WG. Potential role of lignin in tommorrow's wood utilization technologies. For Prod J. 1981;31(3):24–29.

[2] Hu L, Pan H, Zhou Y, Zhang M. Methods to improve lignin's reactivity as a phenol substitute and as replacement for other phenolic compounds: a brief review. BioResources. 2011;6(3):3515–3525.

[3] Jenkins BM, Baxter LL, Miles TR Jr, Miles TR. Combustion properties of biomass. Fuel Process Technol. 1998;54(1–3):17–46.

[4] Gavrilescu D. Energy from biomass in pulp and paper mills. Environ Eng Manage J. 2008;7(5):537–546.

[5] Glasser WG, Northey RA, Schultz TP, editors. Lignin: Historical, Biological, and Materials Perspectives. vol. 742. Washington, DC: American Chemical Society; 1999.

[6] Park DW, Neogi AN, Furtner L. Artificial fire log; 2011.

[7] Ehrenfeld J, Gertler N. Industrial ecology in practice: the evolution of interdependence at Kalundborg. J Ind Ecol. 1997;1(1):67–79.

[8] Global Wood Pellets Markets and Industry. *Policy drivers, market status, and raw material potential. N.p.: IEA Bioenergy Task 40*; 2007. http://www.bioenergytrade.org/downloads/t40-global-wood-pellet-market-study_final.pdf (accessed 2 March 2013).

[9] Stevens J, Gardner DJ. Enhancing the fuel value of wood pellets with the addition of lignin. Wood Fiber Sci. 2010;42(4):439–443.

[10] Forss KG, Fuhrmann AGM, Toroi M. Procedure for manufacturing lignocellulosic material products [Patent]; 1988. WO Patent App. PCT/FI1988/000,033.

[11] Malhotra R. Coal particles briquette where the binder is lignin and methods and systems of preparing the same [Patent]; 2010. US20100154296A1.

[12] Lumadue MR, Cannon FS, Brown NR. Lignin as both fuel and fusing binder in briquetted anthracite fines for foundry coke substitute. Fuel. 2012;97:869–875.

[13] Laus R, Costa TG, Szpoganicz B, Fávere VT. Adsorption and desorption of Cu(II), Cd(II) and Pb(II) ions using chitosan crosslinked with epichlorohydrin-triphosphate as the adsorbent. J Hazard Mater. 2010;183(1-3):233–241.

[14] Ahmaruzzaman M. Industrial wastes as low-cost potential adsorbents for the treatment of wastewater laden with heavy metals. Adv Colloid Interface Sci. 2011;166(1-2):36–59.

[15] Guo X, Zhang S, Shan XQ. Adsorption of metal ions on lignin. J Hazard Mater. 2008;151(1):134–142.

[16] Šćiban M, Klasnja M. Study of the adsorption of copper(II) ions from water onto wood sawdust, pulp and lignin. Adsorpt Sci Technol. 2004;22(3):195–206.

[17] Mohan D, Pittman CU Jr, Steele PH. Single, binary and multi-component adsorption of copper and cadmium from aqueous solutions on Kraft lignin-a biosorbent. J Colloid Interface Sci. 2006;297(2):489–504.

[18] Acemioğlu B, Samil A, Alma MH, Gundogan R. Copper(II) removal from aqueous solution by organosolv lignin and its recovery. J Appl Polym Sci. 2003;89(6):1537–1541.

[19] Krishnani KK, Ayyappan S. Heavy metals remediation of water using plants and lignocellulosic agrowastes. Rev Environ Contam Toxicol. 2006;188:59–84.

[20] Dizhbite T, Telysheva G, Jurkjane V, Viesturs U. Characterization of the radical scavenging activity of lignins - natural antioxidants. Bioresour Technol. 2004;95(3):309–317.

[21] Pan X, Kadla JF, Ehara K, Gilkes N, Saddler JN. Organosolv ethanol lignin from hybrid poplar as a radical scavenger: relationship between lignin structure, extraction conditions, and antioxidant activity. J Agric Food Chem. 2006;54(16):5806–5813.

[22] Vinardell MP, Ugartondo V, Mitjans M. Potential applications of antioxidant lignins from different sources. Ind Crops Prod. 2008;27(2):220–223.

[23] Ugartondo V, Mitjans M, Vinardell MP. Comparative antioxidant and cytotoxic effects of lignins from different sources. Bioresour Technol. 2008;99(14):6683–6687.

[24] Sakagami H, Kawazoe Y, Komatsu N, Simpson A, Nonoyama M, Konno K, *et al.* Antitumor, antiviral and immunopotentiating activities of pine cone extracts: potential medicinal efficacy of natural and synthetic lignin-related materials. Anticancer Res. 1991;11(2):881–888.

[25] Yamamoto Y, Shirono H, Kono K, Ohashi Y. Immunopotentiating activity of the water-soluble lignin rich fraction prepared from LEM - the extract of the solid culture medium of *Lentinus edodes* Mycelia. Biosci Biotechnol Biochem. 1997;61(11):1909–1912.

[26] Chen MF, Chung HH, Lu HL. Protection of the extracts of *Lentinus edodes* mycelia against carbon-tetrachloride-induced hepatic injury in rats. Sci World J. 2012;2012:231586.
[27] González López BS, Yamamoto M, Sakagami H. Treatment of Herpes Simplex Virus with Lignin-Carbohydrate Complex Tablet, an Alternative Therapeutic Formula. InTech; 2012.
[28] Cazacu G, Capraru M, Popa VI. Advances concerning lignin utilization in new materials. Adv Struct Mater. 2013;18:255–312.
[29] Naik P, Rozman HD, Bhat R. Genoprotective effects of lignin isolated from oil palm black liquor waste. Environ Toxicol Pharmacol. 2013;36(1):135–141.
[30] Roberfroid M. Prebiotics: the concept revisited. J Nutr. 2007;137(3):830S–837S.
[31] Baurhoo B, Ruiz-Feria CA, Zhao X. Purified lignin: nutritional and health impacts on farm animals - a review. Anim Feed Sci Technol. 2008;144(3):175–184.
[32] Eastwood M, Kritchevsky D. Dietary fiber: how did we get where we are? Annu Rev Nutr. 2005;25(1):1–8.
[33] Sánchez-Muñiz FJ. Dietary fibre and cardiovascular health. Nutr Hosp. 2012;27(1):31–45.
[34] García MC, Díez JA, Vallejo A, García L, Cartagena MC. Use of Kraft pine lignin in controlled-release fertilizer formulations. Ind Eng Chem Res. 1996;35(1):245–249.
[35] Fernandez-perez M, Garrido-Herrera FJ, Gonzalez-Pradas E. Alginate and lignin-based formulations to control pesticides leaching in a calcareous soil. J Hazard Mater. 2011;190(1-3):794–801.
[36] Fant K, Ruuth M. Alfred Nobel: A Biography. Arcade Publishing; 2006.
[37] Forss KG, Fuhrmann AGM, Toroi M. Manufacture of lignocellulosic products with improved mechanical strength and water resistance [Patent]; 1988. WO 8807104A1.
[38] Falkehag SI. Lignin in materials. Appl Polym Symp. 1975;28(Proc. Cellul. Conf., 8th, 1975, Vol. 1):247–257.
[39] Pizzi A. Advanced Wood Adhesives Technology. Marcel Dekker; 1994.
[40] Stewart D. Lignin as a base material for materials applications: chemistry, application and economics. Ind Crops Prod. 2008;27(2):202–207.
[41] Donmez Cavdar A, Kalaycioglu H, Hiziroglu S. Some of the properties of oriented strandboard manufactured using Kraft lignin phenolic resin. J Mater Process Technol. 2008;202(1–3):559–563.
[42] Park Y, Doherty WOS, Halley PJ. Developing lignin-based resin coatings and composites. Ind Crops Prod. 2008;27(2):163–167.
[43] Tejado A, Kortaberria G, Peña C, Labidi J, Echeverría JM, Mondragon I. Lignins for phenol replacement in novolac-type phenolic formulations. Part I: lignophenolic resins synthesis and characterization. J Appl Polym Sci. 2007;106(4):2313–2319.
[44] Wooten AL, Sellers TJ, Tahir PM. Reaction of formaldehyde with lignin. For Prod J. 1988;38(6):45–46.
[45] Tejado A, Kortaberria G, Peña C, Blanco M, Labidi J, Echeverría JM, *et al.* Lignins for phenol replacement in novolac-type phenolic formulations. II. Flexural and compressive mechanical properties. J Appl Polym Sci. 2008;107(1):159–165.
[46] Kuo M, Hse CY, Huang DH. Alkali treated Kraft lignin as a component in flakeboard resins. Holzforschung. 1991;45(1):47–54.
[47] Poppius-Levlin K. Lignin – new openings for applications. VTT Res Highlights. 2013;5:35–45.
[48] Kuroe M, Tsunoda T, Kawano Y, Takahashi A. Application of lignin-modified phenolic resins to brake friction material. J Appl Polym Sci. 2013;129(1):310–315.
[49] Gandini A. Polymers from renewable resources: a challenge for the future of macromolecular materials. Macromolecules. 2008;41(24):9491–9504.
[50] Thanh Binh NT, Luong ND, Kim DO, Lee SH, Kim BJ, Lee YS, *et al.* Synthesis of lignin-based thermoplastic copolyester using Kraft lignin as a macromonomer. Compos Interfaces. 2009;16(7–9):923–935.

[51] Lin N, Fan D, Chang PR, Yu J, Cheng X, Huang J. Structure and properties of poly(butylene succinate) filled with lignin: a case of lignosulfonate. J Appl Polym Sci. 2011;121(3):1717–1724.

[52] Pucciariello R, Bonini C, D'Auria M, Villani V, Giammarino G, Gorrasi G. Polymer blends of steam-explosion lignin and poly(ε-caprolactone) by high-energy ball milling. J Appl Polym Sci. 2008;109(1):309–313.

[53] Pruett WP, Hyatt JA, Hilbert SD. Lignin and lignin derivatives as copolymerizable colorants for polyesters; 1989.

[54] Rong H, Gao B, Zhao Y, Sun S, Yang Z, Wang Y, et al. Advanced lignin-acrylamide water treatment agent by pulp and paper industrial sludge: synthesis, properties and application. J Environ Sci. 2013;25(12):2367–2377.

[55] Ciobanu C, Ungureanu M, Ignat L, Ungureanu D, Popa VI. Properties of lignin–polyurethane films prepared by casting method. Ind Crops Prod. 2004;20(2):231–241.

[56] Chahar S, Dastidar MG, Choudhary V, Sharma DK. Synthesis and characterization of polyurethanes derived from waste black liquor lignin. J Adhes Sci Technol. 2004;18(2):169–179.

[57] Cateto CA, Barreiro MF, Ottati C, Lopretti M, Rodrigues AE, Belgacem MN. Lignin-based rigid polyurethane foams with improved biodegradation. J Cell Plast. 2014;50(1):81–95.

[58] Nägele H, Pfitzer J, Nägele E, Inone ER, Eisenreich N, Eckl W, et al. Arboform® - a thermoplastic, processable material from lignin and natural fibers. In: Hu TQ, editor. Chemical Modification, Properties, and Usage of Lignin. New York: Springer-Verlag GmbH; 2002. p. 101–119.

[59] Ciolacu D, Oprea AM, Anghel N, Cazacu G, Cazacu M. New cellulose–lignin hydrogels and their application in controlled release of polyphenols. Mater Sci Eng, C. 2012;32(3):452–463.

[60] Li CY, Liu LL. Preparation and test of polyurethane foam composites with alkali lignin/renewable PUF. Adv Mater Res. 2011;150–151(Pt. 2, Advances in Composites):1167–1170.

[61] Stevens ES, Klamczynski A, Glenn GM. Starch-lignin foams. eXPRESS Polym Lett. 2010;4(5):311–320.

[62] Hatakeyama H, Kosugi R, Hatakeyama T. Thermal properties of lignin- and molasses-based polyurethane foams. J Therm Anal Calorim. 2008;92(2):419–424.

[63] Hatakeyama H, Nakayachi A, Hatakeyama T. Thermal and mechanical properties of polyurethane-based geocomposites derived from lignin and molasses. Composites, Part A. 2005;36(5):698–704.

[64] Lippach AKW, Krämer R, Hansen MR, Roos S, Stöwe K, Stommel M, et al. Synthesis and mechanical properties of organic-inorganic hybrid materials from lignin and polysiloxanes. ChemSusChem. 2012;5(9):1778–1786.

[65] Berry BC, Viswanathan T. Lignosulfonic acid-doped polyaniline (ligno-pani™)- a versatile conducting polymer. In: Hu TQ, editor. Chemical Modification, Properties, and Usage of Lignin. New York: Springer-Verlag GmbH; 2002. p. 21–40.

[66] Milczarek G, Inganäs O. Renewable cathode materials from biopolymer/conjugated polymer interpenetrating networks. Science. 2012;335(6075):1468–1471.

[67] Boden DP. Selection of pre-blended expanders for optimum lead/acid battery performance. J Power Sources. 1998;73(1):89–92.

[68] Luque R, Varma RS, editors. Sustainable Preparation of Metal Nanoparticles. RSC Green Chemistry. The Royal Society of Chemistry; 2013.

[69] Xu T, Zhang N, Nichols HL, Shi D, Wen X. Modification of nanostructured materials for biomedical applications. Mater Sci Eng, C. 2007;27(3):579–594.

[70] Frangville C, Rutkevi?ius M, Richter AP, Velev OD, Stoyanov SD, Paunov VN. Fabrication of environmentally biodegradable lignin nanoparticles. ChemPhysChem. 2012;13(18):4235–4243.

[71] Milczarek G, Rebis T, Fabianska J. One-step synthesis of lignosulfonate-stabilized silver nanoparticles. Colloids Surf, B. 2013;105:335–341.

[72] Coccia F, Tonucci L, D'Alessandro N, D'Ambrosio P, Bressan M. Palladium nanoparticles, stabilized by lignin, as catalyst for cross-coupling reactions in water. Inorg Chim Acta. 2013;399:12–18.

[73] Coccia F, Tonucci L, Bosco D, Bressan M, D'Alessandro N. One-pot synthesis of lignin-stabilised platinum and palladium nanoparticles and their catalytic behaviour in oxidation and reduction reactions. Green Chem. 2012;14(4):1073–1078.

[74] Wurm FR, Weiss CK. Nanoparticles from renewable polymers. Front Chem. 2014;2:49.

[75] Faria FAC, Evtuguin DV, Rudnitskaya A, Gomes MTSR, Oliveira JABP, Graca MPF, et al. Lignin-based polyurethane doped with carbon nanotubes for sensor applications. Polym Int. 2012;61(5):788–794.

[76] Notley SM, Norgren M. In: Lucia LA, Rojas OJ, editors. Lignin: Functional Biomaterial with Potential in Surface Chemistry and Nanoscience. John Wiley & Sons, Ltd; 2009. p. 173–205.

[77] Hoeger IC, Filpponen I, Martin-Sampedro R, Johansson LS, Österberg M, Laine J, et al. Bicomponent lignocellulose thin films to study the role of surface lignin in cellulolytic reactions. Biomacromolecules. 2012;13(10):3228–3240.

[78] Sjoblom J. Encyclopedic Handbook of Emulsion Technology. Taylor & Francis; 2010.

[79] Li Z, Pang Y, Lou H, Qiu X. Influence of lignosulfonates on the properties of dimethomorph water-dispersible granules. BioResources. 2009;4(2):589–601.

[80] DeBons FE, Whittington LE. Improved oil recovery surfactants based on lignin. J Pet Sci Eng. 1992;7(1-2):131–138.

[81] Chen HX, Xu QW. Experimental study of fibers in stabilizing and reinforcing asphalt binder. Fuel. 2010;89(7):1616–1622.

[82] Hesp SAM, Iliuta S, Shirokoff JW. Reversible aging in asphalt binders. Energy Fuels. 2007;21(2):1112–1121.

[83] Hale NS, Xu M. A printing method of applying a polymer surface preparation to a substrate; 1995.

[84] Chaffee C, Spies RB. The effects of used ferrochrome lignosulphonate drilling muds from a Santa Barbara channel oil well on the development of starfish embryos. Mar Environ Res. 1982;7(4):265–277.

[85] Somasundaran P, Wang D. Solution Chemistry: Minerals and Reagents, Developments in Mineral Processing. Elsevier Science; 2006.

[86] Feldman D. Lignin and its polyblends – a review. In: Hu TQ, editor. Chemical Modification, Properties, and Usage of Lignin. New York: Springer-Verlag GmbH; 2002. p. 81–99.

[87] Effendi A, Gerhauser H, Bridgwater AV. Production of renewable phenolic resins by thermochemical conversion of biomass: a review. Renewable Sustainable Energy Rev. 2008;12(8):2092–2116.

[88] Calvo-Flores FG, Dobado JA. Lignin as renewable raw material. ChemSusChem. 2010;3(11):1227–1235.

[89] Malutan T, Nicu R, Popa VI. Contribution to the study of hydroxymetylation reaction of alkali lignin. BioResources. 2008;3(1):13–20.

[90] El Mansouri NE, Pizzi A, Salvado J. Lignin-based polycondensation resins for wood adhesives. J Appl Polym Sci. 2007;103(3):1690–1699.

[91] Gayubo AG, Valle B, Aguayo AT, Olazar M, Bilbao J. Pyrolytic lignin removal for the valorization of biomass pyrolysis crude bio-oil by catalytic transformation. J Chem Technol Biotechnol. 2010;85(1):132–144.

[92] of Consultants & Engineers NB. Synthetic Resins Technology Handbook. NIIR Project Consultancy Services; 2005.

[93] Fink JK. Reactive Polymers Fundamentals and Applications: A Concise Guide to Industrial Polymers, Plastics Design Library. Elsevier Science; 2013.

[94] Bell BM, Briggs JR, Campbell RM, Chambers SM, Gaarenstroom PD, Hippler JG, et al. Glycerin as a renewable feedstock for epichlorohydrin production. The GTE process. Clean: Soil, Air, Water. 2008;36(8):657–661.

[95] Kang JH, Aasi D, Katayama Y. Bisphenol A in the aquatic environment and its endocrine-disruptive effects on aquatic organisms. Crit Rev Toxicol. 2007;37(7):607–625.

[96] Kelley SS, Glasser WG, Ward TC. Engineering plastics from lignin XIV. Characterization of chain-extended hydroxypropyl lignins. J Wood Chem Technol. 1988;8(3):341–359.

[97] Chung H, Washburn NR. Improved lignin polyurethane properties with Lewis acid treatment. ACS Appl Mater Interfaces. 2012;4(6):2840–2846.

[98] Bonini C, D'Auria M, Emanuele L, Ferri R, Pucciariello R, Sabia AR. Polyurethanes and polyesters from lignin. J Appl Polym Sci. 2005;98(3):1451–1456.

[99] Hofmann K, Glasser W. Engineering plastics from lignin. 23. Network formation of lignin-based epoxy resins. Macromol Chem Phys. 1994;195(1):65–80.

[100] Pohjanlehto H, Setälä HM, Kiely DE, McDonald AG. Lignin-xylaric acid-polyurethane-based polymer network systems: preparation and characterization. J Appl Polym Sci. 2014;131(1):39714/1–39714/7.

[101] Sarkanen S, Li Y. Alkylation, acylation in added plasticizer; measurable cohesive strength; 2001.

[102] Li Y, Sarkanen S. Alkylated Kraft lignin-based thermoplastic blends with aliphatic polyesters. Macromolecules. 2002;35(26):9707–9715.

[103] Saito T, Brown RH, Hunt MA, Pickel DL, Pickel JM, Messman JM, et al. Turning renewable resources into value-added polymer: development of lignin-based thermoplastic. Green Chem. 2012;14(12):3295–3303.

[104] de Wild PJ, Huijgen WJJ, Heeres HJ. Pyrolysis of wheat straw-derived organosolv lignin. J Anal Appl Pyrolysis. 2012;93:95–103.

[105] Simões Dos Santos DA, Rudnitskaya A, Evtuguin DV. Modified kraft lignin for bioremediation applications. J Environ Sci Health, Part A: Toxic/Hazard Subst Environ Eng. 2012;47(2):298–307.

[106] Okoronkwo AE, Olusegun SJ. Biosorption of nickel using unmodified and modified lignin extracted from agricultural waste. Desalin Water Treat. 2013;51(7–9):1989–1997.

[107] Fang R, Cheng X, Xu X. Synthesis of lignin-base cationic flocculant and its application in removing anionic azo-dyes from simulated wastewater. Bioresour Technol. 2010;101(19):7323–7329.

[108] Zhang R, Xiao X, Tai Q, Huang H, Hu Y. Modification of lignin and its application as char agent in intumescent flame-retardant poly(lactic acid). Polym Eng Sci. 2012;52(12):2620–2626.

[109] Popa VI, Capraru AM, Grama S, Malutan T. Nanoparticles based on modified lignins with biocide properties. Cellul Chem Technol. 2011;45(3–4):221–226.

[110] Nakagame S, Chandra RP, Kadla JF, Saddler JN. Enhancing the enzymatic hydrolysis of lignocellulosic biomass by increasing the carboxylic acid content of the associated lignin. Biotechnol Bioeng. 2011;108(3):538–548.

[111] Gao G, Dallmeyer JI, Kadla JF. Synthesis of lignin nanofibers with ionic-responsive shells: Water-expandable lignin-based nanofibrous mats. Biomacromolecules. 2012;13(11):3602–3610.

[112] Yiamsawas D, Baier G, Thines E, Landfester K, Wurm FR. Biodegradable lignin nanocontainers. RSC Adv. 2014;4(23):11661–11663.

[113] Buoro RM, Bacil RP, da Silva RP, da Silva LCC, Lima AWO, Cosentino IC, et al. Lignin-AuNp modified carbon paste electrodes - preparation, characterization, and applications. Electrochim Acta. 2013;96:191–198.

[114] Gilca IA, Capraru AM, Grama S, Popa VI. Agents for wood bioprotection based on natural aromatic compounds and their complexes with copper and zinc. Cellul Chem Technol. 2011;45(3-4):227-231.

[115] Liu Z, Zhao L, Cao S, Wang S, Li P. Preparation and evaluation of a novel cationic amphiphilic lignin derivative with high surface activity. BioResources. 2013;8(4):6111-6120.

[116] Aimi R, Murakami T. Cell-physiological studies on the effect of aluminum on the growth of crop plants. Nogyo Gijutsu Kenkyusho Hokoku D: Seiri Iden. 1964;11:331-392, 393-396.

[117] Katsumata K, Meshitsuka G. Modified Kraft lignin and its use for soil preservation. In: Hu TQ, editor. Chemical Modification, Properties, and Usage of Lignin. New York: Springer-Verlag GmbH; 2002. p. 51-165.

[118] Ramírez F, González V, Crespo M, Meier D, Faix O, Zúñiga V. Ammoxidized kraft lignin as a slow-release fertilizer tested on *Sorghum vulgare*. Bioresour Technol. 1997;61(1):43-46.

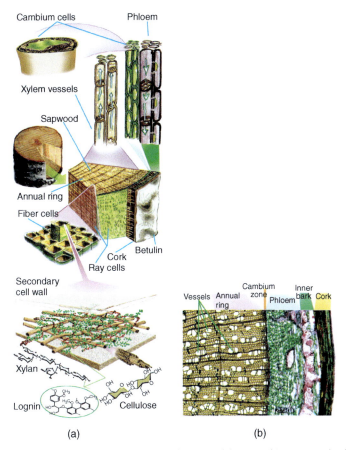

Figure 2.6 Structural features of wood. (a) General structural features; (b) micrograph of birch surface structures. Copyright from ref. [15].

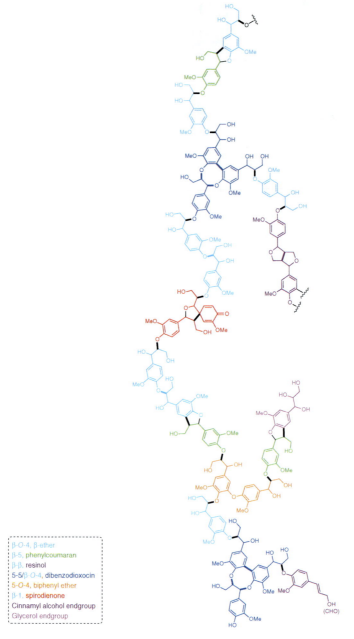

Figure 2.18 *Spruce lignin model proposed by Brunow [31].*

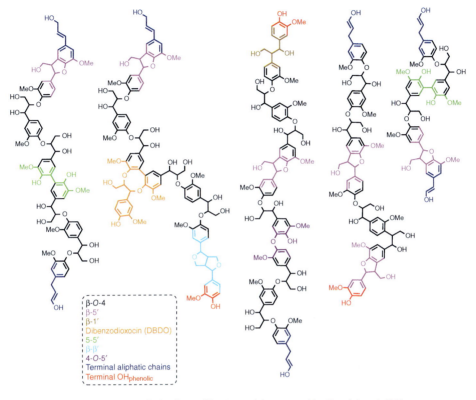

Figure 2.19 Milled softwood-lignin model proposed by Crestini et al. [33].

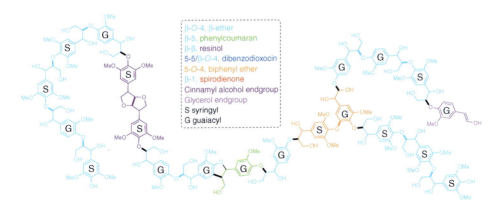

Figure 2.21 Hardwood lignin model proposed by Boerjan et al. [36].

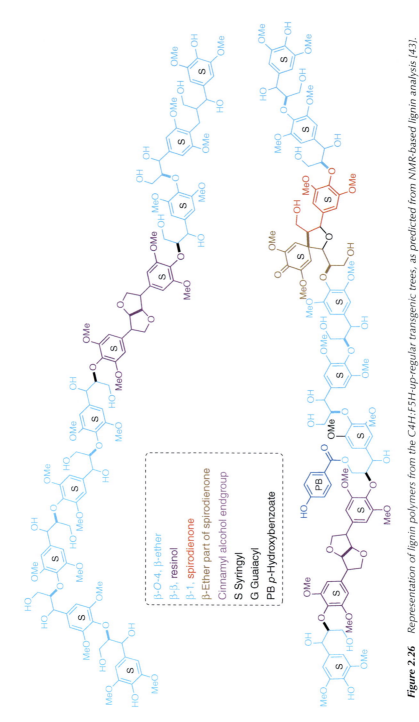

Figure 2.26 Representation of lignin polymers from the C4H:F5H-up-regular transgenic trees, as predicted from NMR-based lignin analysis [43].

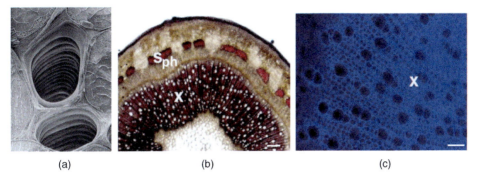

Figure 3.1 Lignification pattern in Populus tissues. (a) Scanning electron micrograph of xylem elements in a Zinnia stem. Courtesy of Kim Findlay and Copyright from ref. [7]. (b) Transverse section of stem segment. Lignin deposition, visualized under the light microscope after phloroglucinol–HCl staining (red color) (x-xylem, ph-phloem, s-sclerenchyma. Bars = 100 μm). Copyright from ref. [8]. (c) Secondary xylem from stem. Lignin distribution by fluorescent microscopy (autofluorescence). Copyright from ref. [8].

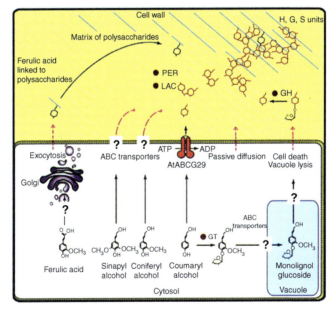

Figure 4.9 Monolignol biosynthesis and transport in the plant cell. Three main lignin building blocks (p-coumaryl alcohol, coniferyl alcohol, and sinapyl alcohol) are synthesized in the cytosol. Monolignols are transported from the cytosol to different locations: to the cell wall for oxidative cross-linking by apoplastic peroxidases (PER) and laccases (LAC) into lignins; into the vacuole for storage as glucoconjugates and, for ferulic acid, into the Golgi apparatus for incorporation into polysaccharides (pectins or arabinoxylans). Most of these transport routes still remain to be discovered. In principle, these hydrophobic molecules may passively diffuse through membranes, undergo active transport through membrane transporters or through Golgi-derived vesicles or simply be released from dying cells. p-Coumaryl alcohol is exported across the plasma membrane by the AtABCG29 transporter, whereas free coniferyl alcohol and sinapyl alcohol are exported by other, so far unknown ABC transporters. Monolignols are selectively imported into the vacuole as glucoconjugates by unknown ABC transporters. Monolignol glucoconjugates are generated by cytosolic glucosyl transferases (GT) and need to be deglucosylated by the cell wall-associated glucosylhydrolases (GH) before incorporation into lignin polymers. A hypothetical transport route of ferulic acid into Golgi vesicles is also represented. Copyright from ref. [67].

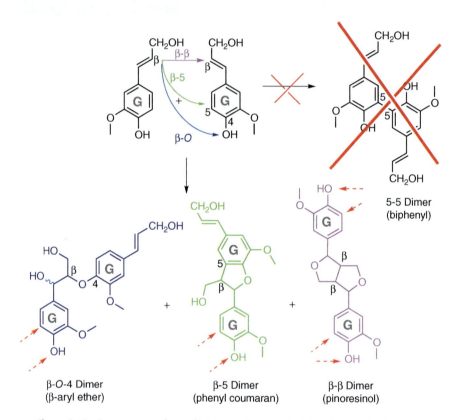

Figure 4.14 Dimerization of monolignols: coniferyl alcohol dehydrodimerization [100].

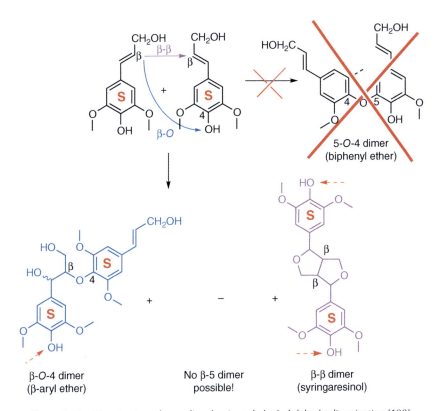

Figure 4.15 Dimerization of monolignols: sinapyl alcohol dehydrodimerization [100].

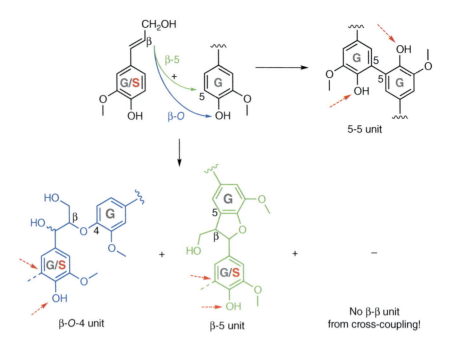

Figure 4.17 Lignification of monolignols: monolignol cross-coupling with a G-end unit [100].

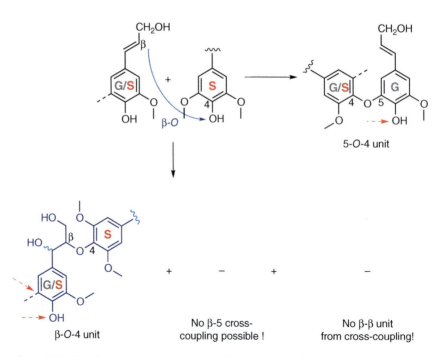

Figure 4.18 Lignification of monolignols: monolignol cross-coupling with an S-end unit [100].

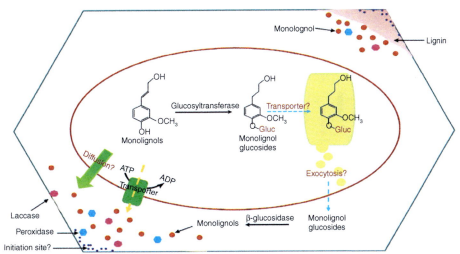

Figure 4.26 The scheme for monolignol deposition and the subsequent initiation of lignin polymerization within the cell wall. Symplastic transport of monolignols may export them to the cell wall through active transport or by passive diffusion. Alternatively, they may be sequestrated and stored as glucoconjugates into the vacuoles in gymnosperms, before their subsequent transport to the cell wall and hydrolysis to free monolignols for polymerization. The deposited monolignols in the cell wall diffuse to initiation sites where the polymerization process begins. The polymerization to form different bond-linkages of lignin is known to be a random chemical process. However, the nature of initiating sites and the way in which the amount and type of lignin formation is controlled across the cell wall are poorly understood. Copyright from ref. [2].

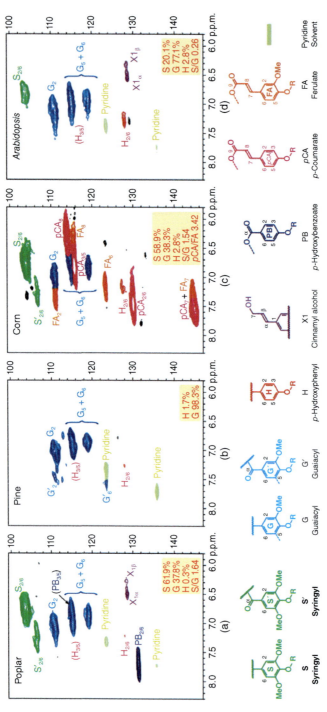

Figure 6.14 2D NMR spectra revealing lignin unit compositions. Partial short-range ^{13}C-^{1}H (HSQC) correlation spectra (aromatic regions only) of cell wall gels in DMSO-d_6/pyridine-D_5 (4:1, v/v) from (a) two-year-old greenhouse-grown poplar wood, (b) mature pine wood, (c) senesced corn stalks and (d) senesced Arabidopsis inflorescence stems. Contours in this region are used to measure S/G/H ratios, as well as relative p-hydroxybenzoate (PB, in poplar), p-coumarate (pCA in corn) and ferulate (FA in corn) levels [238].

Figure 11.22 *Illustration of* Podophyllum peltatum *(American Podophyllum). Courtesy of David Nesbitt, 2014.*

9
High-Value Chemical Products

9.1 Introduction

The unique chemical composition of lignin makes it a very versatile polymer as well as a natural source of many high-value chemical products beyond the straightforward and bulk applications already described in this book. Native lignins and pulping-industry lignins are complex polymers with a matrix of phenolic moieties that have no parallel in nature. In fact, lignin has been highlighted as the only renewable source for producing industrial aromatics by the market research company Frost and Sullivan in its report of 2012. Theoretically, a depolymerization process of lignin may break the lignin matrix into a family of phenolic structures, such as cresols, catechols, resorcinols, quinones, vanillin, or guaiacols, all of them high-value chemicals that are difficult to obtain even from classical petrochemical industry (see Figure 9.1). Most of these aromatic products are high-value molecules because of their properties, pharmacological applications, and use as chemical intermediates for the synthesis of other products [1]. Even more important from the economic standpoint is that this source of aromatics is decorrelated from the fluctuating price of oil.[1]

Despite that lignins are such a promising source of chemicals, pathways for getting these high-value products out of lignin are still in a very early stage of development, and only few applications are sufficiently developed to be commercially available. The selective transformation of lignin still remains elusive [2], and these cases where conversion into aromatic compounds takes places are more basic research and laboratory-scale methods than authentic industrial procedures. Furthermore, although these chemicals were produced from a renewable raw material (see Figure 9.1), which is a great advantage itself, the process also has to be economically competitive with petrochemical processes, and this is still far from being the case. As *someone* involved in the industrial production and development of chemicals said: "Our aim is to obtain useful substances from lignin and demonstrate in turn that you can indeed make money out of them".[2]

---

[1] http://www.frost.com/prod/servlet/press-release-print.pag?docid=269974856 (accessed September 4, 2014).
[2] https://www.ecn.nl/news/newsletter-en/2010/june-2010/chemicals-from-biomass/ (accessed September 4, 2014).

Lignin and Lignans as Renewable Raw Materials: Chemistry, Technology and Applications, First Edition.
Francisco G. Calvo-Flores, José A. Dobado, Joaquín Isac-García and Francisco J. Martín-Martínez.
© 2015 John Wiley & Sons, Ltd. Published 2015 by John Wiley & Sons, Ltd.

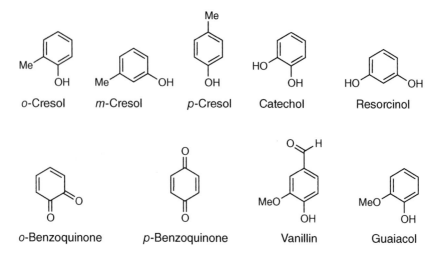

Figure 9.1 Phenolic parent compounds obtained from depolymerization of lignin

Nonetheless, several potential strategies have been put forward for breaking lignin into simple chemical structures [3]. Thus, these strategies might be real alternatives as the state of the art of current technology evolves. Most of these approaches fit under the concept of biorefinery [4], which is the "facility that integrates biomass conversion processes and equipment to produce fuels, power, and chemicals from biomass",[3] although some of them can be also developed from paper industry technology. Either under biorefinery concept or considering lignin *per se* from the paper industry, more time is still needed to show promising results over the short and middle terms.

In some cases, biomass is either thermally transformed by gasification to produce synthesis gas (syngas) or degraded by pyrolysis to produce mixtures of smaller molecules. Once these smaller aromatic compounds are produced and isolated, other high-value chemical products can be synthesized by means of the conventional chemical industry [5]. Other approaches consist of obtaining very simple aromatic molecules by removing many of the oxygenated functional groups of lignin monomers, producing benzene, toluene, xylene (the so-called BTX), and phenol [6]. Nowadays, most BTX production is based on the catalytic transformation of naphtha at petroleum refineries, while phenol is synthesized from isopropylbenzene (cumene) by oxidation. It is also an alternative to perform a direct conversion of lignin into the desired value-added chemical by using selective catalysts in a one-pot transformation. This latter strategy is the most challenging and the greatest breakthrough of all, since it would enable the direct production of chemicals that are not easily produced even at the present status of existing petrochemical industry [7]. The key factor might be related to the development of new catalysts for breaking down the polymeric lignin structure with high selectivity.

Figure 9.2 shows schematically the major nonselective thermochemical and chemical transformations of lignin, by depolymerization according to Pandey and Kim [8].

Several main fields need to be considered in order to produce aromatics from a renewable raw material such as lignin by the catalytic valorization to it. In the following sections, some of these methods will be reviewed, showing the state of the art in lignin valorization, but the great challenge is to develop new catalysts to dramatically improve the selectivity.

---

[3] http://www.nrel.gov/biomass/biorefinery.html (accessed September 4, 2014).

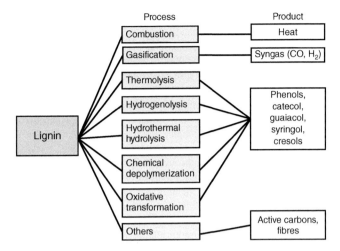

Figure 9.2 Thermochemical and chemical lignin transformation processes [8]

9.2 Gasification: Syngas from Lignin

Gasification is a well-known process in which carbon-based materials are converted into a mixture of H_2 and CO gases, usually known as synthesis gas or syngas. This is a versatile reactive mixture used as fuel in combined cycle power plants or in the preparation of many basic organic compounds for chemical industry, depending on conditions of pressure, temperature, and catalyst. For example, MeOH is produced by a reaction of syngas in the presence of zinc/copper oxide catalyst at high temperatures. As fuel, CO is converted into CO_2 and H_2 into water vapor. Many impurities, such as SO_2, mercury, and other particles, are formed during gasification, and therefore it is essential to purify H_2 and CO before burning them. The present technology allows the conversion of about a fifth of the reacting mixture of syngas gases in every catalytic cycle, so that it is a challenge for the coming years to develop new catalysts to improve yield, and to carry out the procedure at lower temperatures.

In the specific case of biomass, the conversion takes place at high temperature (700–1000 °C), producing syngas and some residual tar, a black mixture of hydrocarbons and free carbon. The advantage of syngas from lignin is that this kind is a sulfur-free and metal-free material, which makes lignin a promising raw material for this purpose. Nevertheless, biomass gasifiers based on lignin are still at an early stage of development, and producing syngas efficiently from lignin is still a great challenge for researchers as well as for companies. To mention just one of the cases in more advanced state of development, we cite the University of North Dakota Energy and Environmental Research Center (Grand Forks, North Dakota, USA) is, for instance, one of the organizations pursuing improved gasification technology by thermocatalytic conversion of lignin into synthetic fuels.[4]

9.3 Thermolysis of Lignin

Thermolysis is a pyrolytic decomposition of organic matter at elevated temperatures by a rapid heating in the absence of oxygen. This pyrolysis process degrades biomass to bio-oil, gas, and char (or ash). The anaerobic conditions avoid combustion of lower M_w fragments obtained from biomass.

---

[4] http://biomassmagazine.com/articles/2928/cellulosic-ethanol-what-to-do-with-the-lignin (accessed September 4, 2014).

In case of lignin, it is a quite complex mechanism where many parallel radical coupling and rearrangement reactions take place simultaneously. A handicap for developing the technology needed for the thermolysis of lignin might be that, according to several studies, laboratory-scale results found with multigram-scale experiments between 150 and 900 °C cannot be extrapolated to large-scale experiments [9]. However, some of the results of these experiments are considered erratic, and research continues in this regard.

Furthermore, some technology is well developed. Indeed, not only performing the pyrolysis of lignin but also modifying the composition of the products is possible when pyrolysis vapors are passed over a heterogeneous catalyst. For example, zeolite HZSM-5 (Zeolite Socony Mobil-5), an aluminum silicate zeolite patented by the company Mobil, gave a 45 wt% yield of liquid bio-oil, with 78% of BTX mixture, when Alcell lignin was treated at 550 °C. This is a promising result that is somehow the consequence of solubility properties of organosolv lignin. This type of lignin is soluble in most organic solvents because of its lower M_w compared to pulp lignins or native lignins, making it suitable for *feeding in solution* in a reactor. In addition, when larger pore size zeolitic heterogeneous catalyst is employed, the procedure can be extended to Kraft lignin with a 75 wt% conversion into liquid, a minimum formation of nonuseful char, and approximately 40% composition of BTX. In addition, native lignin from wheat straw has been converted into ethylbenzene in a two-step transformation by pyrolytic depolymerization over a composite prepared from Re and Y/HZSM-5(25) zeolite and then with EtOH [10] (see Figure 9.3). The first step has a yield of approximately 15 wt% of organic liquid, containing 90% benzene. This yield can go up to 72% in the presence of the same HZSM-5(25) zeolite.

Some other pyrolytic transformations have been described in literature. Pyrolysis of Kraft lignin on TiO_2 at 550 °C produces phenols, cresols, and xylenols with a global yield of 7.5 wt% *via* catalytic defunctionalization of complex moieties. The treatment of commercially available lignins with zeolite HZSM-5 impregnated of 6 wt% of lanthanum at 600 °C give 6 wt% of a mixture of aromatics and approximately 9% of light alkenes.

In addition, the pyrolysis of lignin under microwave irradiation, combined with an extraction process with switchable hydrophilicity solvents (SHS), has been described. Fu *et al.* [11] developed this method using SHS, such as *N,N*-dimethylcyclohexylamine, to extract phenolic compounds from bio-oil, resulting in microwave pyrolysis of lignin. An amount of 10 g of bio-oil is used in each case, and even at that small scale, material losses are small. Around 96% of bio-oil is recovered, as well as 72% of guaiacol and 70% of 4-methylguaiacol. The SHS solvent can be also partially recovered after extraction (approximately 91%). The starting material (microwave pyrolysis lignin oil) and the fractions resulted from SHS extraction have been characterized by gas chromatography-mass spectrometry (GC-MS) and quantitative $^{13}C\{^1H\}$ and $^{31}P\{^1H\}$ NMR spectroscopy. The final amount of pure phenolic compounds is about 500 mg (5%).

Despite the different attempts of developing an advanced technology for depolymerization of lignin, controlled selectivity is still challenging. In an attempt to improve selectivity in lignin

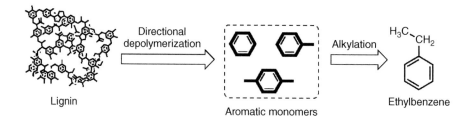

Figure 9.3 *Directional synthesis of ethylbenzene through catalytic transformation of lignin [10]*

depolymerization to obtain high-value products, many metallic catalysts have been tested on several lignins with this pyrolitic procedure. Different catalysts and many different conditions have been described in literature, some of them recently summarized by Wang et al. [12]. Overall, metallic oxides and metals supported on carbon, zeolite, or mesoporous silica gel have been employed. Remarkable lower temperatures and higher yields have been reached in many cases, and this metallic catalytic thermolysis has been employed on Kraft lignin, Kraft spruce lignin, lignosulfonate, several organosolv lignins, birch sawdust lignin, and lignins partially hydrolyzed previously even by enzymatic hydrolysis. Table 9.1 presents some remarkable results, showing catalysts or mixture of catalysts, reaction conditions, the main aromatic compounds produced, and yields.

In general, conventional pyrolysis of lignin can be used to produce mixtures of char, gases, and bio-oil, which are composed of valuable phenolic compounds, such as phenol, guaiacol, and catechol.

Table 9.1 Metallic catalyzed lignin depolymerization [12]

Lignin	Catalyst	Reaction conditions		Main products (yield, %)
		T (°C)	P (MPa)	
Kraft [13]	(1) Si–Al cat./H_2O/BuOH	200–350	1.1–23	Phenols (6.5)
	(2) ZrO_2–Al_2O_3–FeO_x	300		
Organosolv switchgrass [14]	Formic acid; Pt/C; EtOH	350		4-Propylguaiacol (7.8) 4-methylguaiacol (5.0)
Acidic hydrolysis spruce [15]	Formic acid; Pd cat., Nafion SAC-13	300	9.6	Guaiacol (2.0) Pyrocatechol (1.8) Resorcinol (0.5)
Enzymatic hydrolysis spruce [15]	Formic acid; Pd cat., Nafion SAC-13	300	9.6	Guaiacol (1.7) Pyrocatechol (1.3) Resorcinol (1.0)
Kraft spruce [15]	Formic acid; Pd cat., Nafion SAC-13	300	9.6	Guaiacol (4.7) Pyrocatechol (4.9) Resorcinol (1.2)
Lignosulfonate [16]	Ni/C, NiLa/C, NiPt/C, NiCu/C, NiPd/C, and NiCe/C	200	5	Guaiacol (10)
Birch sawdust [17]	Ni/C	200	?	Propenylguaiacol (12) Propenylsyringol (36)
Enzymatic hydrolysis corn stalk [18]	Pt/C, Pd/C, Ru/C	200–250	1–6	4-Ethylphenol (0.1–3.1) 4-Ethylguaicol (0.1–1.4)
Organosolv olive tree pruning [19]	Ni, Pd, Pt, or Ru supported by mesoporous Al-SBA-15	140		Diethyl phthalate (1.1)
Guaiacyl dehydrogenation oligomers [20]	K10 montmorillonite clay (Al_2O_3–$4SiO_2$–H_2O)/HCl	100		Low-M_w products (–)

9.4 Hydrodeoxygenation (Hydrogenolysis)

Hydrodeoxygenation (HDO) of lignin is a pyrolytic transformation under reducing conditions in the presence of a catalyst and a reducing agent such as H_2. The process is often called hydrogenolysis, and thus appears as such in the literature as well. In a similar way as the petrochemical industry prepares conventional fuels such as gas or diesel, other hydrogen donors such as formic acid or tetralin can be employed, although in these cases the results after enzymatic and mild acidic conditions are not very relevant due to the complexity of the resulting mixtures [21].

HDO is one of the most promising methods for lignin depolymerization, because it achieves high yields and produces simpler product mixtures. Because of the presence of a reducing agent, the oxygen content of the fragments and simpler molecules is lower than that of starting lignin. However, the procedure requires relatively high pressures and specific catalysts that might be expensive in some cases.

Several examples of HDO for isolated lignin are described in the literature. The Noguchi–Crown–Zellerbach process is one of the earliest known methods, developed in the 1950s [7]. The reaction is performed with lignosulfonated lignin in phenolic liquid media at 100–200 atm and temperatures in the range of 370–430 °C. As a catalyst, an iron/copper/zinc sulfide mixture is used, which leads to a 21% yield of phenol production.

Kraft lignin can also be converted into phenol-rich mixtures by using oxides of earth-abundant metals as catalysts. Temperatures between 650 and 850 °C and hydrogen pressures between 3.4 and 17.2 MPa are employed. The total yield is around 37.5 wt% of the starting lignin, and the main components of the resulting mixture are p-ethylphenol (33%), p-propylphenol (20%), and m-cresol (12%). This method is also called "Hydrocarbon Research Institute Lignol Process," following the developer's name [22]. Depending on pyrolysis conditions, up to 74% of the oil shows mixtures with a M_w below 200, 22% having a M_w ranging from 200 to 500, and a small amount ranging as high as 2000. The lighter fraction includes phenols (8.7%) cyclohexanes (5%), naphthalenes (4%), and phenanthrenes (1.2%) [23]. When Kraft pine lignin is treated with hydrogen and ammonium heptamolybdate[5] at 430 °C a 61 wt% low-M_w oil is produced. Similar results can be achieved when this method is applied to hardwood Kraft lignin or organosolv lignin by using a mixture of NiMo or Cr_2O_3 metal oxides as catalyst.

Moreover, HDO of lignin can be carried out as a part of a combined several-stage process that involves pyrolytic treatment in a first stage followed by a single or double HDO. Examples of this methodology are described in the following.

Pine organosolv lignin is first pyrolyzed and later subjected to a combined two-stage HDO with Ru/C and hydrogen (14 MPa) at 300 °C and 250 °C. Carbon content of the resulting material in each step is 33% and 35%, respectively. The results by applying these means suggest a possible method for employing lignin as raw material in the preparation of gasoline [24]. The method has been applied to organosolv and Kraft lignins as well with Pt/γ-Al_2O_3 catalyst on inert atmosphere (Ar, 5.8 MPa) at 225 °C, producing an extractable oil composed of monomeric species with a 12% yield. The bio-oil was later treated in a subsequent step with H_2 (5.5 MPa) and CoMo/Al_2O_3 or Mo_2/CNF as catalyst at 300 °C in dodecane solvent. The deoxygenation of the monomeric species and a mixture of arenes and partially oxygenated arenes take place with an overall yield of 9%, of which 24% is oxygen-free atmosphere [25]. Some degradation of the catalyst has been noted with a crystalline nucleus formation that diminished the active surface, and therefore the efficiency of the process.

The two-stage process has also been used on organosolv lignins produced from herbaceous and deciduous plants. The pyrolysis is performed in a continuous fluidized process, followed by hydrogenolysis (10 MPa, H_2 in on Ru/C) in dodecane, which gives small amounts of cycloalkanes and mostly low-M_w aromatic compounds [26]. The reaction mechanism is still unclear because of

---

[5] A water-soluble catalyst.

the complexity of the process and different options of bond breaking, but some studies have been performed on simpler models [27].

9.5 Hydrothermal Hydrolysis

Lignins can be depolymerized and converted into a liquid mixture of hydrolytically transformed products by treatment with water, and even supercritical water, at high temperatures and pressures. Hydrothermal hydrolysis, compared to other lignin-degradation methods, shows four main advantages as described by Kang et al. [28]. First of all, it is not necessary to carry out a pre-drying procedure, directly transforming papermaking wastewater lignins. Second, nitrogen and sulfur oxides formed in other degradation processes,[6] can be dissolved directly in water, and further treatment is not needed. Third, under hydrothermal conditions, water acts as a hydrogen source, which is necessary for gasification of lignin. Finally, the required hydrothermal temperature is usually lower than other biomass thermochemical methods, including pyrolysis and steam gasification.

As an example, the liquefaction of organosolv lignin in compressed hot water has been performed in the presence and absence of catalysts. A reaction time of 1 h at 250 °C leads to a 97% conversion. Some 53% of this material is a bio-oil that contains 74% of 20 phenolic compounds on average [29]. If the process is catalyzed by $Ba(OH)_2$ and $RbCO_3$, the amount of oil production is reduced by 15% and 23%, respectively. If Kraft lignin is employed, a series of 100 components spread over six fractions are identified by GC-MS [30]. In this case, similar conditions, and later purification with a combined solvent extraction followed by a pH shift, are used.

Both organosolv and Kraft lignins have also been treated at 374 °C for 10 min to give bio-oil. In the case of Kraft lignin, 58–72% of bio-oil was produced and for organosolv lignin, it was 79%. These remarkable differences show the strong dependency of the results on the type of lignin used. In all cases, complex mixtures of aromatic compounds are resulted [31].

Furthermore, when oxidizing conditions are used with Kraft lignin (0.1% H_2O_2) in compressed hot water at a temperature range of 150–200 °C, a mixture of carboxylic acids (formic, acetic, and succinic acids, 45%), CO_2 (19%), and a high-M_w residue were formed. In the case of organosolv lignin, an oligomer of lignin with a M_w of approximately 300 and a 20% wt of organic acids were produced [31].

The alkaline hydrolysis of Kraft lignin at 300 °C in the presence of EtOH and phenol (as radical capture agent) leads to a depolymerization process to give fragments of 440–480 and 900–1200 and negligible formation of char.

Finally, Kraft, soda, Alcell, and sugarcane bagasse lignins have been submitted to an aqueous phase reforming process at low temperatures ($T \leq 498$ K) and pressures ($P \leq 29$ bar) for the production of aromatic chemicals and H_2, using water or EtOH/water mixtures and catalysts. The composition of the isolated yields of monomeric aromatic compounds and overall lignin conversion based on these isolated yields varied from 10% to 15%, depending on the lignin sample, with the balance consisting of gaseous products and residual solid material [32]. The method has been more fully studied and extended to other catalysts that are able to give reduction products [33] (see Figure 9.4).

9.6 Chemical Depolymerization

Chemical depolymerization is probably the most direct and logical method to produce high-value chemicals from lignin, but it is also probably the most challenging. Theoretically, conventional

---

[6] N and S are common elements of lignin.

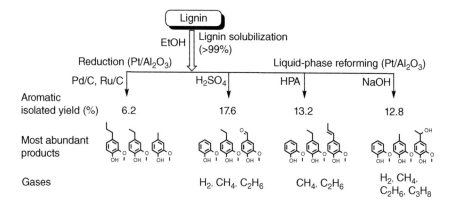

Figure 9.4 *Valorization of lignin via reduction or liquid-phase reforming [33]*

hydrolysis of lignin by chemical agents gives simple monolignol structures or related compounds, but in reality is quite different. In most cases, there is little selectivity in depolymerization processes, and yields are modest. An additional problem is the difficulty of isolation of the products from the degradation of the polymer. The key steps to obtain chemicals from lignin are the selectivity of the process and the reaction conditions. Carrying out the process under milder reaction conditions is important to avoid by-products and complex mixtures.

In the following subsections, some of common methods for chemically induced depolymerization of lignin are described. The main results of this approach have recently been reported [12].

9.6.1 Acid Media Depolymerization

Acid treatment of lignin to obtain low-M_w aromatic compounds is one of most direct methods for depolymerization of lignin. One of the first attempts to hydrolyze lignin in an acidic medium was made by Hewson and Hibbert in the 1940s [34].

They used mixtures of inorganic acids, such as HCl, and organic acids, such as formic acid, in alcohol solvents, such as EtOH and ethylene glycol, respectively. The reaction conditions were in a relatively low-temperature range (78–200 °C), not high enough, in fact, to break the lignin polymer into monomeric compounds for further use. The main products from this procedure were methoxyphenols, catechols, and phenols. In all cases, methoxyphenols were the main product when temperatures rise from 360 to 400 °C.

Better results are achieved with higher temperatures and using several proportions of formic acid and EtOH under pressure. Methoxyphenol, catechol, and phenol are produced when temperatures rise from 360 to 400 °C. In the case of wheat straw lignin mixtures treated with formic acid and EtOH at temperatures of the range of 360–380 °C and pressures of 25 MPa, it is possible to produce methoxyphenols, catechols, and phenols with yields in wt% in the ranges of 2.9–1.3, 1.5–0.5, and 2.0–0.3, respectively [35, 36].

Several attempts to improve yield and selectivity of this process have been made by employing cocatalysts such as Pd or Pt, but no better results have been achieved yet, since applying these catalysts does not decrease the activation energy for the reaction [12].

9.6.2 Base Media Depolymerization

Treating lignin with aqueous NaOH solution at high temperatures leads to phenols and derivatives. The process is relatively simple, but its selectivity remains unsatisfactory. Lavoie *et al.* [37] have

Table 9.2 Base-catalyzed lignin depolymerization [12]

Lignins	Base catalyst aqueous solution	Reaction conditions		Main products	Yield (wt.%)
		T (°C)	P (MPa)		
Steam explosion hemp [37]	5 wt.% NaOH	300–330	3.5	Guaiacol, Catechol, Vanillin	0.9–2.8, 0.8–3.0, 0.5–0.8
Steam explosion softwood [37]	5 wt.% NaOH	300–330	3.5	Guaiacol, Catechol, Vanillin	1.2–2.1, 0.1–3.2, 0.3–0.5
Organosolv [39]	2 wt.% NaOH	300	25	Syringol, Hydroxy-acetophenone, Guaiacol	4.1, 1.6, 1.1
Kraft [40]	5 wt.% NaOH	270–315	13	Pyrocatechol	0.5–4.9
Organosolv olive tree pruning [38]	4 wt.% NaOH, KOH, Ca(OH)$_2$, LiOH, or K$_2$CO$_3$	300	90	Catechol	0.1–2.4

employed steam-exploited lignin from softwood and hemp with a 5 wt% solution of NaOH at a temperature between 300 and 330 °C under 9–13 MPa. Guaiacol, catechol, and vanillin, among a mixture of 26 products (identified by GC-MS), were the main low-M_w products under these conditions. Depending on the origin of lignin, either from softwood or hemp, the average content of these three main phenolic molecules in the final mixture varies. When commercial organosolv lignin is treated under similar conditions, syringol, hydroxyacetophenone, and catechol are the main products. Contrarily, from Kraft lignin, pyrocatechol was the main product.

In a more complete study, Toledano *et al.* [38] carried out a depolymerization of organosolv lignin with several bases such as KOH, NaOH, Ca(OH)$_2$, LiOH, and K$_2$CO$_3$ in water solutions, catechol being the main product with a yield ranging from 0.1 to 2.4 wt%. The results are consistent with the breaking of aryl–alkyl bonds in the polymer when temperatures above 270 °C are reached, β-*O*-4 bonds being the most common to break.

In general, most of depolymerization processes described here have been carried out at about 300 °C, which is expected to provide enough energy for this purpose. Results are summarized in Table 9.2.

9.6.3 Ionic Liquid-Assisted Depolymerization

Ionic liquids are salts in which ions are poorly coordinated. It results in these solvents being liquid below 100 °C, or even at rt where cations, anions, or both have a delocalized charge distribution and one component is organic, preventing the formation of a stable crystal lattice.[7] Ionic liquids have been extensively applied as solvents and some of them dissolve wood, which makes it possible to extract and separate lignin from biomass as an alternative to traditional methods, although conditioned by its cost [41].

---

[7] http://www.organic-chemistry.org/topics/ionic-liquids.shtm (accessed September 4, 2014).

Table 9.3 Ionic liquid-assisted lignin depolymerization [12]

Lignins	Ionic liquid	Reaction conditions		Main products	Yield (wt.%)
		T (°C)	P (MPa)		
Organosolv beech [42]	[EMIM][CF$_3$SO$_3$][a] + Mn(NO$_3$)$_2$	100	8.4	2,6-Dimethoxy-1,4-benzoquinone	11.5
Eugenol [43]	[EMIM][CF$_3$SO$_3$][a] + Brønsted acid	200		Guaiacol	7.9
Oak wood [44]	[HMIM]Cl[b]	110–150		Alkyl-aryl ether linkages cleavage	–
Guaiacylglycerol-β-guaiacyl ether [45]	[PMIM]Cl[c]	150		Guaiacol	71.5
Veratrylglycerol-β-guaiacyl ether [45]	[PMIM]Cl[c]	150		Guaiacol	70
Guaiacylglycerol-β-guaiacyl ether [46]	[HMIM]Cl[b]	150		β-O-4 ether bond cleavage	–

[a] 1-Ethyl-3-methylimidazolium trifluoromethanesulfonate [EMIM][CF$_3$SO$_3$].
[b] 1-H-3-methylimidazolium chloride [HMIM]Cl.
[c] 1-Propyl-3-methylimidazolium chloride [PMIM]Cl.

The depolymerization of lignin in ionic liquids takes place under more gentle conditions than thermal procedures (100–200 °C range). The method is similar to deoxygenation and requires a hydrogen source that is usually provided by a Brønsted acid. However, it has two serious drawbacks. First, the high cost of ionic liquids makes this procedure limited to low and moderate quantities of lignin. Second, recovering the solvent is necessary, and this process of separation and isolation of the products from the solvent remains very difficult because of the π–π interactions between the solvent and the resulting aromatic molecules. Therefore, the isolation process of the final products is sometimes difficult. The main results of this technique are summarized in Table 9.3.

9.6.4 Supercritical Fluids-Assisted Depolymerization

Supercritical fluids can be used for lignin depolymerization. Lignin has shown good solubility in this media and the depolymerization process proceeds with high selectivity, and the aromatic products from this procedure are easily separated from reaction mixtures. Several methods have been described in the literature and the reaction can be developed in alcohols, MeOH or EtOH, H$_2$O, or mixtures of CO$_2$, acetone, and H$_2$O. In some cases, a base such as KOH is added to the solvent, while in others no additional reactant other than solvent is added. As a starting material, Kraft, organosolv, or alkaline lignins have been used. The temperature mentioned for these processes is in the range of 290–420 °C. The aromatic compounds isolated from this procedure together with the yields are summarized in Table 9.4.

The use of supercritical fluids for lignin depolymerization presents a similar problem as ionic liquids, related to the high cost or to the equipment employed to perform the reaction. Thus, the application remains restricted to a quite specific procedure for special occasions.

Table 9.4 Supercritical fluid-assisted depolymerization [12]

Lignins	Supercritical fluids	T (°C)	P (MPa)	Main products	Yield (wt.%)
Kraft [47] or Organosolv [47]	MeOH/KOH EtOH/KOH	290		Catechol Phenol	–
Alkaline [48]	H_2O	300	25	Catechol Phenol Cresol	28.37 7.53 11.67
Organosolv [49]	H_2O/ p-cresol	350–420		2-(Hydroxy-benzyl)-4-methyl-phenol	< 75
Organosolv hardwood [50]	CO_2/acetone/ H_2O	300–370	10	Syringol Guaiacol 2-Methoxy-4-methyl-phenol	3.6 1.6 1.6
Organosolv wheat straw [50]	CO_2/acetone/ H_2O	300–370	10	Syringic acid Guaiacol	2.2 1.6

9.7 Oxidative Transformation of Lignin

Another approach to lignin depolymerization consists of oxidative treatment. This can be achieved in two different ways: first, as a procedure for pulp bleaching[8] during high-quality paper manufacture, where cellulose is separated from lignin; and second, as a method for chemical preparation from side-product lignin in paper chemistry or as part of an integrated biorefinery process. In the latter case, certain selectivity and yields are needed to produce chemicals, especially aromatic compounds, for a process to be considered economically viable.

The state of the art of lignin oxidation is similar to the blenching process, but producing high-value molecules at this moment is in fact in a very early step, and only a few examples can be found in the literature in this regard [51]. The different oxidative treatments of lignin will be covered in the following sections.

9.7.1 Oxidation with Chlorinated Reagents

Molecular chlorine is capable of oxidizing lignin with high reactivity to produce mixtures of both polychlorinated and nonchlorinated molecules [52, pp. 125–160] (see Figure 9.5). It is assumed that degradation begins by the attack of aromatic rings by chloronium ions to form cations at those positions with higher electron density.

In this process, many chlorinated and polychlorinated molecules are formed, which, due to environmental implications, cause chlorine to be replaced by other less-pollutant reagents such as chlorine dioxide. With chlorine dioxide, the formation of chlorinated derivatives is much lower than with chlorine, and therefore about 95% of the pulp blenching of Kraft processes is carried out by this method. The reaction takes place by a radical mechanism preferentially with phenolic units of lignin and olefinic side chains. Figure 9.6 shows the proposed reaction mechanism. The final products of

---

[8] See Glossary for a definition.

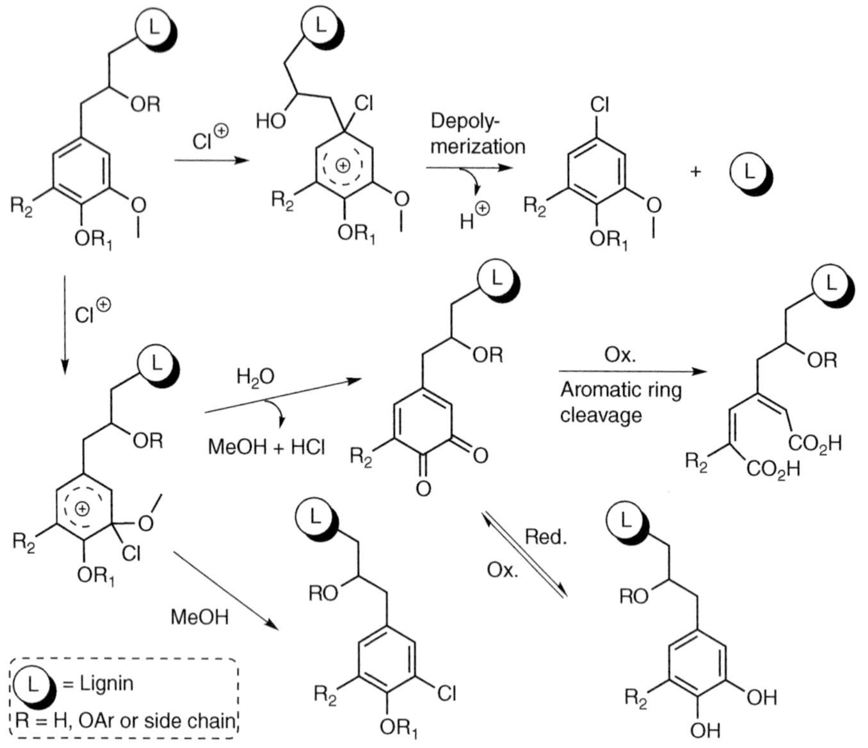

Figure 9.5 Proposed lignin chlorination mechanism [52, pp. 125–160]

this oxidation are simple dicarboxylic acids, while quinones are formed as intermediate oxidation products [53]. Nonetheless, as a method for preparing chemicals, it remains under development.

9.7.2 Oxidation with Ozone

Ozone is a powerful oxidant capable not only of breaking lignin matrix but also of yielding simpler phenolic compounds. When lignin is treated with ozone, the oxidation produces very simple molecules such as carboxylic acids (acetic or formic ones), as well as low-M_w alcohols such as MeOH and CO_2, because aromatic compounds formed in an initial step are easily degraded by ozone. As a result of this treatment, a nonselective degradation process takes place, limiting the application as a conventional method for preparing high-value organic compounds [54].

9.7.3 Oxidation with Hydrogen Peroxide

Hydrogen peroxide has been also used for bleaching processing in alkaline media. This oxidation reaction leads to the oxidation of unsaturated side-chain moieties of some aromatic units of lignin, such as p-hydroxyphenyl, guaiacyl, and syringyl units. Some of these reactive products can be intermediates to be converted into quinones, and, due to the multiple reaction sites for hydroperoxide anions, this leads to different ring-opening products (see Figure 9.7).

In fact, p-hydroquinone, p-quinone, and o-quinone have been detected in the effluent lines from hydrogen peroxide bleaching of mechanical pulp [55].

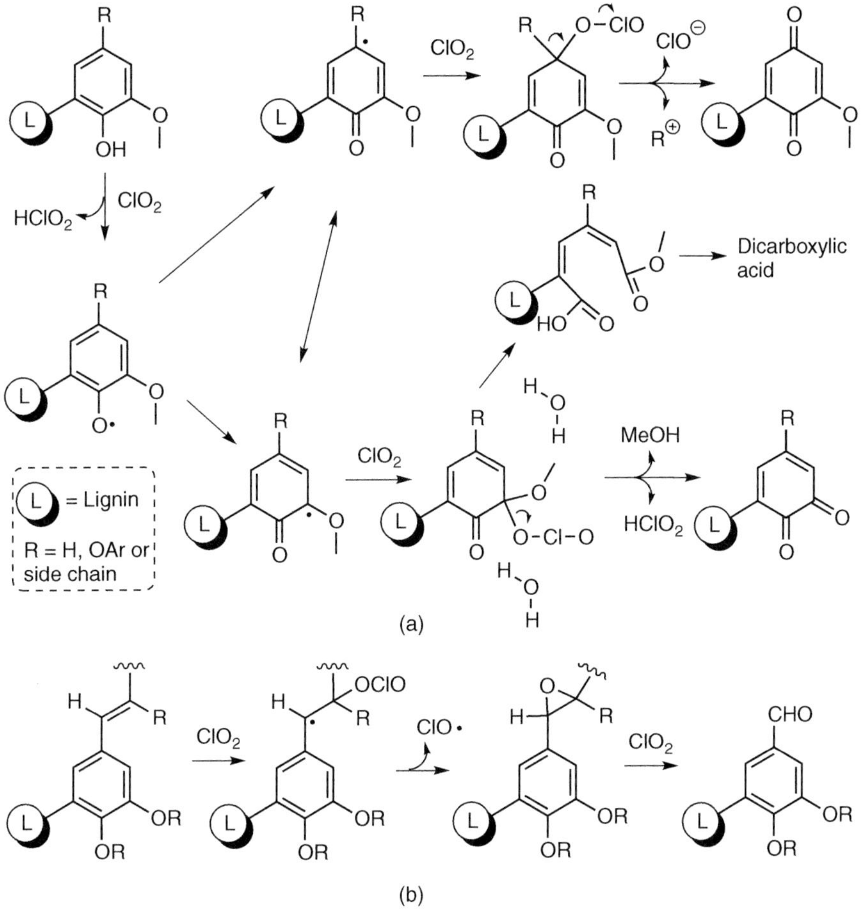

Figure 9.6 *Proposed mechanism for chlorinated reagents at the (a) phenolic units of lignin and (b) the corresponding conjugate double bonds [52, pp. 125–160]*

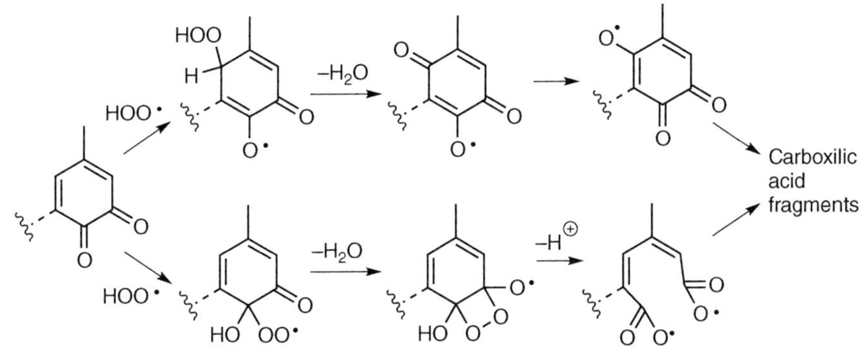

Figure 9.7 *Proposed mechanism for the oxidation of lignin by hydrogen peroxide [55]*

The degradation of lignin with hydrogen peroxide is in essence focused of pulp bleaching, but due to its reactivity and ecofriendly status as an oxidant reagent, it presents a high potentiality for the preparation of low-M_w aromatic aldehyde and carboxylic/dicarboxylic acids as well as simple acids, especially when combined with adequate catalyst.

9.7.4 Oxidation with Peroxy Acids

The heterolytic cleavage of the peroxy bond in a peroxyacid compound produces the hydroxonium ion, HO^+. This species is very reactive and readily reacts with electron-rich sites, such as aromatic rings and double bonds on side-chain structures of lignin. In many cases, an oxygen insertion takes place between lignin and peroxy acids.

9.7.5 Catalytic Oxidation

Catalytic oxidation is a middle-term technology to be improved in order to achieve high-value products. Conventional oxidations used for pulp blenching shows poor selectivity without high yields, so the processes described below are not competitive with the petrochemical industry. The main products that could come from these reactions include quinones, phenols, aldehydes, ketones, acids, and/or dicarboxylic acids. Many reaction conditions have been tested using oxidants as air, molecular oxygen, hydrogen peroxide, and others in the presence of such catalysts as metallic salts, metal oxides and coordination complexes, composites of metal oxides, polyoxometalates, organometallics and metalloporphirines, metal-organic frameworks (MOFs),, and organocatalysts. Much work has been done in this field, including lignin models (recently summarized), but with very poor results to be applied on a large scale [51]. Only in the case of vanillin, does lignin oxidations have a commercial application.

9.8 High-Value Chemicals from Lignin

Lignin has an enormous potential as a source of high-value products. Controlled depolymerization of lignin is one of the most challenging and promising technologies over the middle-long term to produce small aromatic chemicals that now are obtained only from petrochemical industry. On the other hand, a fuller transformation of lignin can produce other more sophisticated materials attractive to the market. In the following, different procedures to make high-value products from lignins as well as the state of the art of these processes are described.

Table 9.5 lists the main chemicals that are potentially available from lignin. This is a summary of the state of the art of current technology related to extensive chemicals produced by the PNNL in 2007. This report offers an idea of the most remarkable areas of research to be developed. However, only some modest results have been forthcoming so far in producing high-value chemicals from this raw material.

9.8.1 Vanillin

Vanillin is a phenolic derivative, used as a flavoring agent in foods, beverages, pharmaceuticals, and in the fragrance industry. The best known source of vanillin is the vanillin plant (*Vanilla planifolia*), a member of the orchid family. In this plant, vanillin occurs as a glycoside, which hydrolyzes to vanillin and its carbohydrate. But, it has also been identified as a component of many oils, balsams, resins, and woods. The demand for vanilla flavoring far exceeds the supply of vanilla beans, so that for many years synthetic methods have been developed.

Table 9.5 Potential lignin-derived products [56]

Product	Technology status[a]	Market risk[b]	Challenges	Market volume[b]
Hydrocarbon and aromatic chemicals				
BTX	PD	L	Catalytic challenges: selective dehydroxylation, demethoxylation and dealkylation	H
Phenol/substituted phenols	PD	L	-do- and secondary derivatization of BTX chemicals	H
Aromatic polyols	E	?	-do-	?
Biphenyls	E	?	-do-	M
Cyclohexane	E	L	-do-	H
Aromatic monomers	E	?	Selective hydrogenolysis	?
Oxidized products (vanillin/DMSO)	D	H	Biocatalytic route for selective oxidation	L
$C_1 - C_7$ gases and mixed liquid fuels	E	L	Catalyst life, reduce process steps, process scale-up	H
Macromolecules and their derivatives				
Carbon fiber	PD	M	Economic challenges: isolation of lignin, spinning rate, carbon yield, varied lignin sources	H
Polymer extender	PD	M	Modification of lignin to compatibilize with polymer matrices, color of lignin-extended product	M
Thermosets	E	M	M_w and viscosity control of the product, varied lignin sources	?
Formaldehyde-free adhesives	PD	M	Consistency of lignin to ensure constant cure rate	H
Syngas products				
MeOH/dimethyl ether	D	L	Technological challenges: economic syngas purification, process scale-up	H
EtOH/mixed alcohols	E	M	Economic syngas purification, catalyst and process improvement to produce 2-C alcohols, process scale-up	H

[a] D = developed, PD = partially developed, E = emerging.
[b] L = low, M = moderate, H = high.

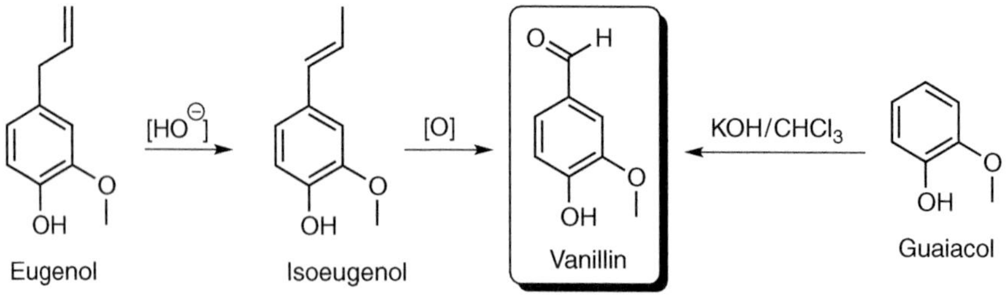

304 Lignin and Lignans as Renewable Raw Materials

Figure 9.8 Syntheses of vanillin

One of the first syntheses of vanillin was made by Reimer [57] in 1876, from guaiacol. Guaiacol is converted into vanillin, refluxing the starting product with chloroform in basic media (see Figure 9.8).

Toward the end of the 19th century, a commercial synthesis of vanillin was developed using eugenol isolated from clove oil, which is an essential oil from the clove plant (*Syzygium aromaticum*). Eugenol is transformed into isoeugenol by treatment with a base, and isoeugenol oxidized to vanillin with nitrobenzene (see Figure 9.8).

Other starting materials have also been used, including coniferyl alcohol and curcumin [58, pp. 294–295] (see Figure 9.9).

In the case of curcumin, the reaction takes place in one step under microwave irradiation catalyzed by a bismuth salt [60] (see Figure 9.10).

A subsequent method uses guaiacol as a starting product, which can be produced from catechol. This procedure was developed by Rhoda in 1970, and most of vanillin produced nowadays is made by this method (see Figure 9.11). Guaiacol is treated with glyoxylic acid, by electrophilic aromatic substitution to give vanillylmandelic acid, which is then transformed into 4-hydroxy-3-methoxyphenylglyoxylic acid by oxidation and finally converted into vanillin by decarboxylation.

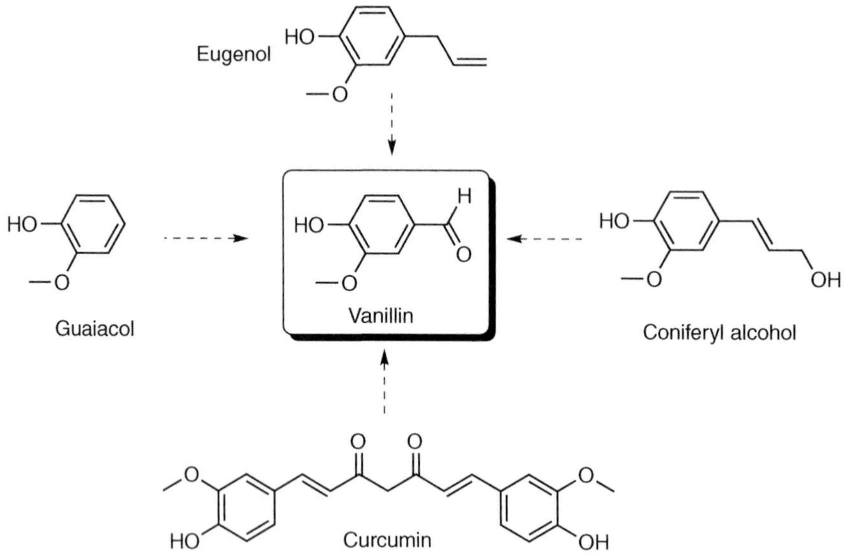

Figure 9.9 Four classical syntheses of vanillin from guaiacol, eugenol, coniferyl alcohol, and curcumin [59]

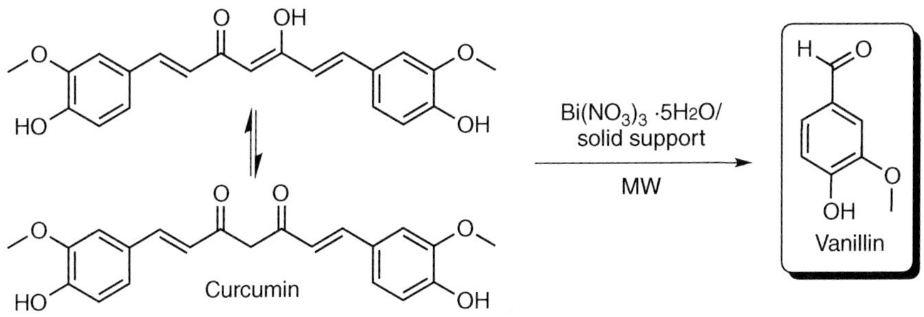

Figure 9.10 Synthesis of vanillin from curcumin induced by bismuth salts under MW irradiation [60]

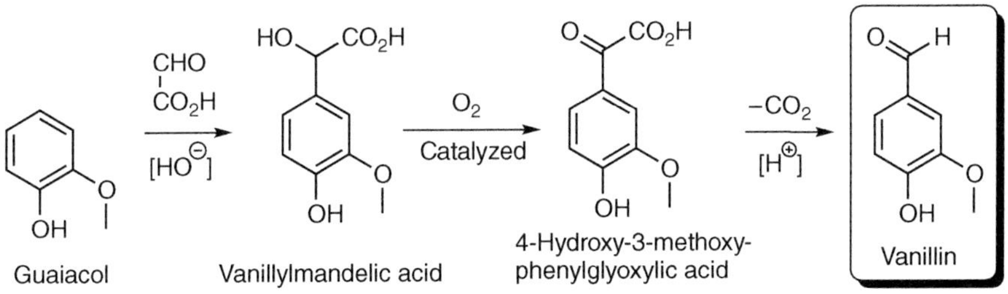

Figure 9.11 Rhoda's synthesis of vanillin

In the early 20th century, it was discovered that vanillin could be obtained as a by-product of the sulfite process in the pulping procedure of paper industry. Vanillin production from lignosulfonates started in 1937, and remained the dominant method for many years. By 1981, a single pulp and paper mill in Ontario supplied 60% of the world market for synthetic vanillin [61]. This method produces huge amounts of waste effluents that, combined with the growing public awareness on environmental issues, led to unsustainable effluent-treatment costs, causing these mills to close.

Despite that most vanillin is produced nowadays by the Rhoda technique, the Norwegian company Borregaard continues manufacturing vanillin from lignin, and since 1993 has been the only vanillin producer from lignosulfonate (see Table 9.6). The method can be considered an integrated procedure for the preparation of high-value wood products, because 1000 kg of wood is converted into 400 kg specialty cellulose, 400 kg lignin, 3 kg vanillin, 20 kg yeast, 50 kg EtOH, and 45 kg CO_2, together with bioenergy recovery. The key step is an oxidation with a copper catalyst, which is recycled due the strict limitations of copper in the effluent. Figure 9.12 schematically shows the process.

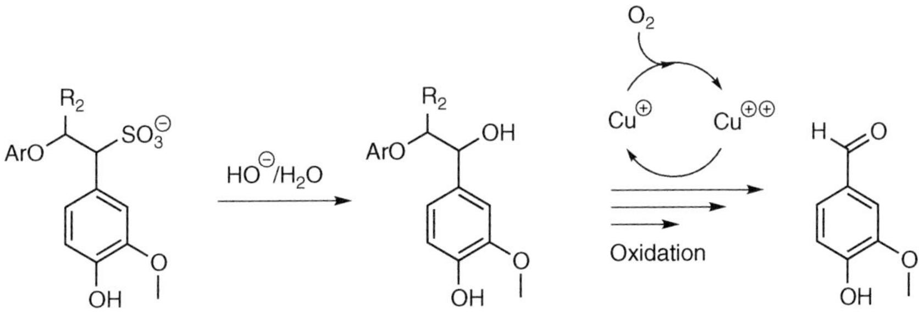

Figure 9.12 Borregaard's synthesis of vanillin [59]

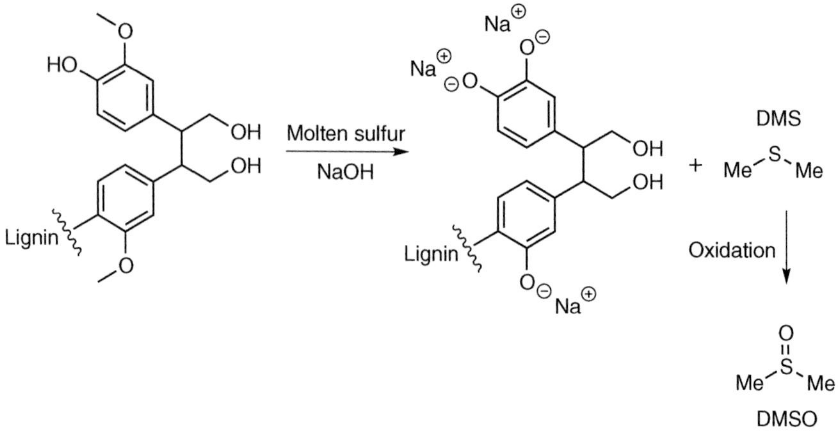

Figure 9.13 Production of DMSO from lignin [59]

The Borregaard method is economically viable and can be considered an ecofriendly industrial activity, because CO_2 emissions linked to vanillin from lignin in timber have been proven to be 90% lower than the emissions resulting from vanillin based on the petrochemical industry. In 2009, a procedure was reported for producing vanillin and lignin-based polyurethanes from Kraft lignin [63, 64].

9.8.2 Dimethyl Sulfide and Dimethylsulfoxide

Dimethyl sulfoxide (DMSO) is a colorless polar aprotic solvent that can dissolve polar and nonpolar molecules, including many salts, so that it is used as a solvent in many reactions and shows very low toxicity[9]. DMSO was produced from 1961 to 2010 by oxidation of dimethyl sulfide (DMS), from Kraft lignin by Gaylord Chemical. When lignin is treated with molten sulfur in alkaline media, two methyl groups are transferred from lignin to sulfur, yielding DMS. Then, DMS can be converted into DMSO by oxidation (see Figure 9.13).

The lignin method becomes an alternative to the petrochemical pathway, which manufactures DMS and DMSO from MeOH and carbon disulfide or hydrogen sulfide. However, the lignin process has been replaced by the petrochemical procedure despite that it generates highly odorous sulfur-containing compounds that are contaminants in their process for making DMSO.

9.8.3 Active Carbon

Kraft lignin is a good precursor for the preparation of char and active carbon [65]. Carbonization is the first stage and involves the formation of char by pyrolysis of lignin with a temperature range of 600–900 °C. In this process, a nonporous material is formed that must be activated. Activation is the second stage and it can be performed physically or chemically. Physical activation can be carried out by treating char with an oxidant gas, such as steam or CO_2 in a range of 600–1200 °C. Chemical activation takes place on char by impregnation with chemicals such as H_3PO_4, KOH, or NaOH, followed by heating under a nitrogen flow in a temperature range of 450–900 °C. In either case, the chosen method depends on its chemical use.

Table 9.7 summarizes some procedures for preparing active carbon from lignin and some of the properties of these active carbons.

---

[9] http://www.dmso.org (accessed September 4, 2014).

Table 9.6 Lignin producers of vanillin [62]

Producer	Capacity (TPA)[a]	Comment
Ontario Paper	3000	Closed in 1988
Monsanto	2000	Closed in 1991
ITT Rayonier	1500	Closed in 1993
Borregaard	1500	Still in production

[a] TPA = Tons per annum

Table 9.7 Physical and chemical activation of lignins [65][a]

Lignin used	Activation	Conditions	Surface area[d]	Micropore volume[e]
Indulin C[b] [66]	PA: carbonization – N_2	C(300 °C, 2 h)	< 10	< 0.01
Kraft lignin [67]	PA: carbonization – N_2	C(350 °C, 2 h) + A(800 °C + 40 h)	1613	0.47
	PA: carbonization – CO_2	C(350 °C, 2 h) + A(850 °C + 20 h)	1853	0.57
Lignin[c] [68]	PA: carbonization – N_2	C(500–900 °C)	10-50	–
Hydrolytic lignin [69]	PA: carbonization – Ar activation – steam	C(600 °C, 2 h) + steam(800 °C)	865	0.365
Hydrolytic lignin [70]	PA: pyrolysis (fluidized bed) – air (with Al–Cu–Cr catalyst) activation-steam	Temp: pyrolysis – 700 °C, steam activation – 780 °C	769	–
Hydrolytic lignin [71]	PA: steam activation	700 °C, 2 h	–	0.33
Indulin [66]	C[b] CA: carbonized then activated with KOH	Lignin:KOH/4 : 1(700 °C, 1 h)	514	0.214
Lignin[c] [68]	CA: $ZnCl_2$, H_3PO_4, K_2CO_3, Na_2CO_3, KOH, NaOH	Impregnation ratio 1 for all	~ 800–2000	–
Kraft lignin [72]	CA: $ZnCl_2$	Lignin:$ZnCl_2$/1 : 2.3(500 °C, 1 h)	~ 1800	1.039
Kraft lignin [73]	CA: H_3PO_4	Lignin:H_3PO_4/1 : 2(427 °C, 2 h)	1459	0.82
Hydrolysis lignin [74]	A: carbonized then activated with KOH	Lignin:KOH/4 : 1(850 °C, 15 min)	2753	1.37
Kraft lignin [75]	CA: H_3PO_4	Lignin:H_3PO_4/1 : 1.4(600 °C, 1 h)	1370	0.78

[a] C = carbonization, A = activation, PA = physical activation, CA = chemical activation.
[b] From black liquors of kraft pulping.
[c] From strong black liquor of kraft pulping, acidulated with CO_2 to obtain the lignin.
[d] $m^2\ g^{-1}$.
[e] $cm^3\ g^{-1}$

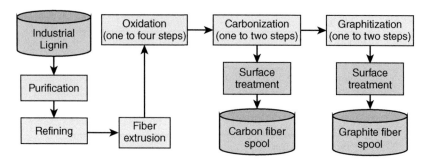

Figure 9.14 Schematic production of carbon fiber from industrial lignin [77]

Active carbon from lignin shows remarkable adsorption activity compared with the one produced from other carbonaceous source materials, and they can be used as a low-cost alternative to graphite and petroleum-based carbons, enabling the production of sustainable, functional carbon materials for various applications [76].

9.8.4 Carbon Fiber

Carbon fibers are remarkable materials manufactured mainly from polyacrylonitrile (PAN) precursors. Other procedures use pitches, notably mesophase pitch (MPP), although, the amount of carbon fibers prepared in this way is minor. Because of the high cost of the necessary petroleum-based precursors and their high production costs, carbon fiber remains a specialty product and has been limited to use in a few specific fields, such as aerospace technology, sporting goods, high-end automotive, and other very specialized industrial applications. It is estimated that approximately 51% of production costs of carbon fibers depend on precursor prices. In this sense, lignin becomes a low-cost raw precursor material for carbon-fiber manufacturing [77].

More common methods for preparing carbon fiber from lignin involves the production of a suitable lignin capable of being melt-spun into fiber in an inert atmosphere. The lignin fiber is submitted to two processes, namely oxidative thermostabilization, which depends on the ability to cross-link, followed by carbonization. The method is very complex and requires careful control of the lignin, spinning conditions, treatment temperatures, and ramping profiles to produce carbon fibers of superior strength. This is a serious drawback when lignin from paper mill streams is used. Unfortunately, the current standard of recovering lignin from paper mill streams is not able to yield lignin with the purity levels and physical properties for fast and economically viable melt-spinning and eventual conversion into carbon fiber with the quality that industry demands.

Several synthetic methods exist for preparing carbon fiber from lignin by carbonization of dry-spun lignin fibers dissolved, for example, in a basic solution and plasticized with polyhydroxylate molecules, such as polyvinyl alcohol or glycerol [78]. Figure 9.14, schematically depicts the process of producing of carbon fibers from lignin.

References

[1] Bjørsvik HR, Liguori L. Organic processes to pharmaceutical chemicals based on fine chemicals from lignosulfonates. Org Process Res Dev. 2002;6(3):279–290.

[2] Bozell JJ. Approaches to the selective catalytic conversion of lignin: a grand challenge for biorefinery development. Top Curr Chem. 2014;353:229–255.

[3] Gallezot P. Catalytic routes from renewables to fine chemicals. Catal Today. 2007;121(1–2):76–91.
[4] Kamm B, Gruber PR, Kamm M, editors. Biorefineries-Industrial Processes and Products: Status Quo and Future Directions. Wiley-VCH Verlag GmbH; 2008.
[5] Chen GQ, Patel MK. Plastics derived from biological sources: present and future: a technical and environmental review. Chem Rev. 2012;112(4):2082–2099.
[6] Parkhurst HJJ, Huibers DTA, Jones MW. Production of phenol from lignin. Prepr - Am Chem Soc, Div Pet Chem. 1980;25(3):657–667.
[7] Zakzeski J, Bruijnincx PCA, Jongerius AL, Weckhuysen BM. The catalytic valorization of lignin for the production of renewable chemicals. Chem Rev. 2010;110(6):3552–3599.
[8] Pandey MP, Kim CS. Lignin depolymerization and conversion: a review of thermochemical methods. Chem Eng Technol. 2011;34(1):29–41.
[9] Nowakowski DJ, Bridgwater AV, Elliott DC, Meier D, de Wild P. Lignin fast pyrolysis: results from an international collaboration. J Anal Appl Pyrolysis. 2010;88(1):53–72.
[10] Fan M, Jiang P, Bi P, Deng S, Yan L, Zhai Q, et al. Directional synthesis of ethylbenzene through catalytic transformation of lignin. Bioresour Technol. 2013;143:59–67.
[11] Fu D, Farag S, Chaouki J, Jessop PG. Extraction of phenols from lignin microwave-pyrolysis oil using a switchable hydrophilicity solvent. Bioresour Technol. 2014;154:101–108.
[12] Wang H, Tucker M, Ji Y. Recent development in chemical depolymerization of lignin: a review. Adv Appl Microbiol. 2013;2013:838645.
[13] Yoshikawa T, Yagi T, Shinohara S, Fukunaga T, Nakasaka Y, Tago T, et al. Production of phenols from lignin via depolymerization and catalytic cracking. Fuel Process Technol. 2013;108:69–75.
[14] Xu W, Miller SJ, Agrawal PK, Jones CW. Depolymerization and hydrodeoxygenation of switchgrass lignin with formic acid. ChemSusChem. 2012;5(4):667–675.
[15] Liguori L, Barth T. Palladium-nafion *SAC*-13 catalysed depolymerisation of lignin to phenols in formic acid and water. J Anal Appl Pyrolysis. 2011;92(2):477–484.
[16] Song Q, Wang F, Xu J. Hydrogenolysis of lignosulfonate into phenols over heterogeneous nickel catalysts. Chem Commun. 2012;48(56):7019–7021.
[17] Song Q, Wang F, Cai J, Wang Y, Zhang J, Yu W, et al. Lignin depolymerization (*LDP*) in alcohol over nickel-based catalysts via a fragmentation-hydrogenolysis process. Energy Environ Sci. 2013;6(3):994–1007.
[18] Ye Y, Zhang Y, Fan J, Chang J. Selective production of 4-ethylphenolics from lignin via mild hydrogenolysis. Bioresour Technol. 2012;118:648–651.
[19] Toledano A, Serrano L, Pineda A, Romero AA, Luque R, Labidi J. Microwave-assisted depolymerisation of organosolv lignin via mild hydrogen-free hydrogenolysis: catalyst screening. Appl Catal, B. 2014;145:43–55.
[20] Bouxin F, Baumberger S, Pollet B, Haudrechy A, Renault JH, Dole P. Acidolysis of a lignin model: investigation of heterogeneous catalysis using montmorillonite clay. Bioresour Technol. 2009;101(2):736–744.
[21] Kleinert M, Gasson JR, Barth T. Optimizing solvolysis conditions for integrated depolymerisation and hydrodeoxygenation of lignin to produce liquid biofuel. J Anal Appl Pyrolysis. 2009;85(1+2):108–117.
[22] Huibers DTA, Parkhurst HJ. Lignin hydrocracking process to produce phenol and benzene; 1983.
[23] Oasmaa A, Johansson A. Catalytic hydrotreating of lignin with water-soluble molybdenum catalyst. Energy Fuels. 1993;7(3):426–429.
[24] Ben H, Mu W, Deng Y, Ragauskas AJ. Production of renewable gasoline from aqueous phase hydrogenation of lignin pyrolysis oil. Fuel. 2013;103:1148–1153.

[25] Jongerius AL, Bruijnincx PCA, Weckhuysen BM. Liquid-phase reforming and hydrodeoxygenation as a two-step route to aromatics from lignin. Green Chem. 2013;15(11):3049–3056.
[26] de Wild P, Van der Laan R, Kloekhorst A, Heeres E. Lignin valorisation for chemicals and (transportation) fuels *via* (catalytic) pyrolysis and hydrodeoxygenation. Environ Prog Sustainable Energy. 2009;28(3):461–469.
[27] Jongerius AL, Jastrzebski R, Bruijnincx PCA, Weckhuysen BM. CoMo sulfide-catalyzed hydrodeoxygenation of lignin model compounds- an extended reaction network for the conversion of monomeric and dimeric substrates. J Catal. 2012;285(1):315–323.
[28] Kang S, Li X, Fan J, Chang J. Hydrothermal conversion of lignin: a review. Renewable Sustainable Energy Rev. 2013;27:546–558.
[29] Holmelid B, Kleinert M, Barth T. Reactivity and reaction pathways in thermochemical treatment of selected lignin-like model compounds under hydrogen rich conditions. J Anal Appl Pyrolysis. 2012;98:37–44.
[30] Kang S, Li X, Fan J, Chang J. Classified separation of lignin hydrothermal liquefied products. Ind Eng Chem Res. 2011;50(19):11288–11296.
[31] Zhang B, Huang HJ, Ramaswamy S. Reaction kinetics of the hydrothermal treatment of lignin. Appl Biochem Biotechnol. 2008;147(1-3):119–131.
[32] Zakzeski J, Weckhuysen BM. Lignin solubilization and aqueous phase reforming for the production of aromatic chemicals and hydrogen. ChemSusChem. 2011;4(3):369–378.
[33] Zakzeski J, Jongerius AL, Bruijnincx PCA, Weckhuysen BM. Catalytic lignin valorization process for the production of aromatic chemicals and hydrogen. ChemSusChem. 2012;5(8):1602–1609, S1602/1–S1602/24.
[34] Hewson WB, Hibbert H. Lignin and related compounds. LXV. Re-ethanolysis of isolated lignins. J Am Chem Soc. 1943;65:1173–1176.
[35] Gasson JR, Forchheim D, Sutter T, Hornung U, Kruse A, Barth T. Modeling the lignin degradation kinetics in an ethanol/formic acid solvolysis approach. Part 1. Kinetic model development. Ind Eng Chem Res. 2012;51(32):10595–10606.
[36] Forchheim D, Gasson JR, Hornung U, Kruse A, Barth T. Modeling the lignin degradation kinetics in an ethanol/formic acid solvolysis approach. Part 2. Validation and transfer to variable conditions. Ind Eng Chem Res. 2012;51(46):15053–15063.
[37] Lavoie JM, Bare W, Bilodeau M. Depolymerization of steam-treated lignin for the production of green chemicals. Bioresour Technol. 2011;102(7):4917–4920.
[38] Toledano A, Serrano L, Labidi J. Organosolv lignin depolymerization with different base catalysts. J Chem Technol Biotechnol. 2012;87(11):1593–1599.
[39] Roberts VM, Stein V, Reiner T, Lemonidou A, Li X, Lercher JA. Towards quantitative catalytic lignin depolymerization. Chem-Eur J. 2011;17(21):5939–5948.
[40] Beauchet R, Monteil-Rivera F, Lavoie JM. Conversion of lignin to aromatic-based chemicals (L-chems) and biofuels (l-fuels). Bioresour Technol. 2012;121:328–334.
[41] Yinghuai Z, Yuanting KT, Hosmane NS. Applications of ionic liquids in lignin chemistry. In: Kadokawa JI, editor. Ionic Liquids - New Aspects for the Future. New York: InTech; 2013. p. 315–346.
[42] Stäerk K, Taccardi N, Boesmann A, Wasserscheid P. Oxidative depolymerization of lignin in ionic liquids. ChemSusChem. 2010;3(6):719–723.
[43] Binder JB, Gray MJ, White JF, Zhang ZC, Holladay JE. Reactions of lignin model compounds in ionic liquids. Biomass Bioenergy. 2009;33(9):1122–1130.
[44] Cox BJ, Ekerdt JG. Depolymerization of oak wood lignin under mild conditions using the acidic ionic liquid 1-H-3-methylimidazolium chloride as both solvent and catalyst. Bioresour Technol. 2012;118:584–588.
[45] Jia S, Cox BJ, Guo X, Zhang ZC, Ekerdt JG. Decomposition of a phenolic lignin model compound over organic *N*-bases in an ionic liquid. Holzforschung. 2010;64(5):577–580.

[46] Jia S, Cox BJ, Guo X, Zhang ZC, Ekerdt JG. Cleaving the β-*O*-4 bonds of lignin model compounds in an acidic ionic liquid, 1-*H*-3-methylimidazolium chloride: an optional strategy for the degradation of lignin. ChemSusChem. 2010;3(9):1078–1084.

[47] Miller JE, Evans L, Littlewolf A, Trudell DE. Batch microreactor studies of lignin and lignin model compound depolymerization by bases in alcohol solvents. Fuel. 1999; 78(11):1363–1366.

[48] Diono W, Sasaki M, Goto M. Recovery of phenolic compounds through the decomposition of lignin in near and supercritical water. Chem Eng Process. 2008;47(9-10):1609–1619.

[49] Takami S, Okuda K, Man X, Umetsu M, Ohara S, Adschiri T. Kinetic study on the selective production of 2-(hydroxybenzyl)-4-methylphenol from organosolv lignin in a mixture of supercritical water and *p*-cresol. Ind Eng Chem Res. 2012;51(13):4804–4808.

[50] Gosselink RJA, Teunissen W, van Dam JEG, de Jong E, Gellerstedt G, Scott EL, *et al*. Lignin depolymerisation in supercritical carbon dioxide/acetone/water fluid for the production of aromatic chemicals. Bioresour Technol. 2012;106:173–177.

[51] Ma R, Xu Y, Zhang X. Catalytic oxidation of biorefinery lignin to value-added chemicals to support sustainable biofuel production. ChemSusChem. 2015;8(1):24–51.

[52] Dence CW, Reeve DW. Pulp Bleaching: Principles and Practice. 4th ed. Tappi Atlanta, GA; 1996.

[53] Ni Y, Shen X, Heiningen ARP. Studies on the reactions of phenolic and non-phenolic lignin model compounds with chlorine dioxide. J Wood Chem Technol. 1994;14(2):243–262.

[54] Akiyama T, Magara K, Matsumoto Y, Meshitsuka G, Ishizu A, Lundqvist K. Proof of the presence of racemic forms of arylglycerol-β-aryl ether structure in lignin: studies on the stereo structure of lignin by ozonation. J Wood Sci. 2000;46(5):414–415.

[55] Gellerstedt G, Pettersson I. Chemical aspects of hydrogen peroxide bleaching. Part II. The bleaching of Kraft pulps. J Wood Chem Technol. 1982;2(3):231–250.

[56] Holladay J, White JF, Bozell JJ, Johnson D. Top Value-Added Chemicals from Biomass - Volume II-Results of Screening for Potential Candidates from Biorefinery Lignin. PNNL-16983, Richland, WA: Pacific Northwest National Laboratory; 2007. http://www.pnl.gov/publications/abstracts.asp?report=230923.

[57] Reimer K. Über eine neue bildungsweise aromatischer aldehyde. Ber Dtsch Chem Ges. 1876; 9(1):423–424.

[58] Berger RG. Flavours and Fragrances: Chemistry, Bioprocessing and Sustainability. Springer-Verlag; 2007.

[59] Calvo-Flores FG, Dobado JA. Lignin as renewable raw material. ChemSusChem. 2010; 3(11):1227–1235.

[60] Bandyopadhyay D, Banik BK. Bismuth nitrate-induced microwave-assisted expeditious synthesis of vanillin from curcumin. Org Med Chem Lett. 2012;2(1):15–18.

[61] Hocking MB. Vanillin: synthetic flavoring from spent sulfite liquor. J Chem Educ. 1997; 74(9):1055–1059.

[62] Thiel L, Hendricksr F. Study into the establishment of an aroma and fragrance fine chemicals value chain in South Africa (tender number t79/07/03). Part 3: Aroma chemicals derived from petrochemical feedstocks. Triumph Venture Capital (Pty) Limited; 2004.

[63] Voitl T, Rohr PRv. Demonstration of a process for the conversion of kraft lignin into vanillin and methyl vanillate by acidic oxidation in aqueous methanol. Ind Eng Chem Res. 2010;49(2):520–525.

[64] Borges da Silva EA, Zabkova M, Araujo JD, Cateto CA, Barreiro MF, Belgacem MN, *et al*. An integrated process to produce vanillin and lignin-based polyurethanes from Kraft lignin. Chem Eng Res Des. 2009;87(9):1276–1292.

[65] Carrott-Suhas PJM, Ribeiro-Carrott MML. Lignin - from natural adsorbent to activated carbon: a review. Bioresour Technol. 2007;98(12):2301–2312.

[66] Khezami L, Chetouani A, Taouk B, Capart R. Production and characterisation of activated carbon from wood components in powder: Cellulose, lignin, xylan. Powder Technol. 2005; 157(1–3):48–56.
[67] Rodriguez-Mirasol J, Cordero T, Rodriguez JJ. Activated carbons from carbon dioxide partial gasification of eucalyptus kraft lignin. Energy Fuels. 1993;7(1):133–138.
[68] Hayashi J, Kazehaya A, Muroyama K, Watkinson AP. Preparation of activated carbon from lignin by chemical activation. Carbon. 2000;38(13):1873–1878.
[69] Baklanova ON, Plaksin GV, Drozdov VA, Duplyakin VK, Chesnokov NV, Kuznetsov BN. Preparation of microporous sorbents from cedar nutshells and hydrolytic lignin. Carbon. 2003;41(9):1793–1800.
[70] Kuznetsov BN, Shchipko ML. The conversion of wood lignin to char materials in a fluidized bed of Al-Cu-Cr oxide catalysts. Bioresour Technol. 1995;52(1):13–19.
[71] Gergova K, Petrov N, Eser S. Adsorption properties and microstructure of activated carbons produced from agricultural by-products by steam pyrolysis. Carbon. 1994;32(4):693–702.
[72] Gonzalez-Serrano E, Cordero T, Rodríguez-Mirasol J, Rodríguez JJ. Development of porosity upon chemical activation of kraft lignin with $ZnCl_2$. Ind Eng Chem Res. 1997;36(11):4832–4838.
[73] Gonzalez-Serrano E, Cordero T, Rodriguez-Mirasol J, Cotoruelo L, Rodriguez JJ. Removal of water pollutants with activated carbons prepared from h3po4 activation of lignin from kraft black liquors. Water Res. 2004;38(13):3043–3050.
[74] Zou Y, Han BX. Preparation of activated carbons from Chinese coal and hydrolysis lignin. Adsorpt Sci Technol. 2001;19(1):59–72.
[75] Montané D, Torné-Fernández V, Fierro V. Activated carbons from lignin: kinetic modeling of the pyrolysis of kraft lignin activated with phosphoric acid. Chem Eng J. 2005;106(1):1–12.
[76] Chatterjee S, Clingenpeel A, McKenna A, Rios O, Johs A. Synthesis and characterization of lignin-based carbon materials with tunable microstructure. RSC Adv. 2014;4(9):4743–4753.
[77] Baker DA, Rials TG. Recent advances in low-cost carbon fiber manufacture from lignin. J Appl Polym Sci. 2013;130(2):713–728.
[78] Kadla JF, Kubo S, Gilbert RD, Venditti RA. Lignin-based carbon fibers. In: Hu TQ, editor. Chemical Modification, Properties, and Usage of Lignin. New York: Springer-Verlag GmbH; 2002. p. 121–137.

Part V
Lignans

Part 4

10
Structure and Chemical Properties of Lignans

10.1 Introduction

The first evidence toward the recognition of lignans was in 1928, when Robinson [1] proposed that a common feature of many natural products was a C_6–C_3 unit (i.e., a propylbenzene skeleton) perhaps derived from cinnamyl units. Later, in 1936, in a review of natural resins, Haworth [2, 3] defined lignans as a group of phenols found in plants, the structure of which was determined by the union of two cinnamic acid residues (two C_6–C_3 units linked by a β–β' carbon bond, see Figure 10.1) or their biogenetic equivalents. From those early stages, the science of lignans underwent extraordinary progress.

Lignans and the so-called neolignans are secondary metabolites commonly included in the human diet, and they are widespread in the plant kingdom [4–6]. Their biological role in plants is unclear, but they are commonly related to defensive mechanisms against external agents [7, 8]. In addition, they possess outstanding physiological properties as antioxidants, phytoestrogens, and anticancer compounds, just to mention a few (see Chapter 11).

In this chapter, the most relevant aspects on these issues will be outlined, although the chemistry and biology of lignans is extensive and complex. For further reading, specific books [9], certain book chapters [10–14], several reviews [15–24], and PhD theses [25–31] concerning chemical and biological properties of lignans are available.

10.2 Structure and Classification of Lignans

According to the structure, lignans can be classified into five families of compounds as follows: proper lignans, neolignans, sesquilignans and dineolignans, norlignans, and hybrid lignans. Lignans and neolignans are natural products derived from the same phenylpropane (C_6–C_3) unit that

---

Lignin and Lignans as Renewable Raw Materials: Chemistry, Technology and Applications, First Edition.
Francisco G. Calvo-Flores, José A. Dobado, Joaquín Isac-García and Francisco J. Martín-Martínez.
© 2015 John Wiley & Sons, Ltd. Published 2015 by John Wiley & Sons, Ltd.

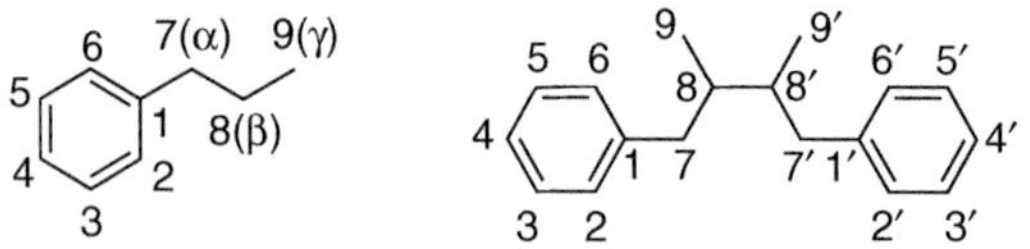

Figure 10.1 Atom numbering in monolignols and lignans

makes up the lignin polymer. Lignans (a group of about 900 compounds) have two units linked by a bond between the β-carbons of the side chains. Sesquilignans and dineolignans have three and four units of this phenylpropane monomer, respectively, while neolignans have two units linked by a carbon–carbon bond aside from the one between the two β-carbons or by a carbon–oxygen bond. About 500 compounds constitute the former group.

A class of natural phenolic compounds with diphenylpentane carbon skeletons ($C_6-C_5-C_6$) is composed of the so-called norlignans, found mainly in conifers and Monocotyledons.

10.2.1 Lignans

According to the "Type of Compound Index," lignans are classified essentially into eight subgroups (see Figure 10.2): simple dibenzylbutane lignans, dibenzylbutyrolactol lignans, dibenzylbutyrolactone lignans, furanoid lignans, furofuranoid lignans, simple aryltetralin lignans, arylnaphthalene lignans, and dibenzocyclooctadiene lignans.

These eight subgroups can also be further classified, now according to structural criteria, into lignans with and without $C_{9(9')}$-oxygen [17, 32].

Regarding their chemical structure, lignans can present a wide range of chemical substitutions along the aromatic rings. The main classes of these aromatic substitutions are depicted in Figure 10.3.

In addition, lignans display a substantial variation in their enantiomeric composition. Therefore, they can be found as both pure enantiomers and enantiomeric compositions, including racemates [33].

10.2.1.1 Neolignans

Neolignans are dimers of cinnamyl units. However, they are different from classical lignans in that their structure is formed by two units coupled by other than the β–β' link typical of lignans (see Figure 10.4) [4, 34]. Even though more diverse in structural type, neolignans are less numerous than classical lignans, and they have been isolated only from the botanical orders *Magnoliales* and *Piperales* [9].

10.2.1.2 Sesquilignans and Dineolignans

Sesquilignans are formed by three units of either cinnamic acid or cinnamyl alcohol or any combination of these (e.g., saucerneol D (**1**) [35–38], a lignan constituent of *Saururus chinensis*) and dineolignans by four units of phenylpropanoid (e.g., manassantin (**2**) [39–42], a dineolignan that has recently been isolated from active root extracts of *Saururus cernuus* L., also referred to as "lizard's tail."), respectively [4] (see Figure 10.5).

Moss [43][1] recommends the use of the names of sesquineolignans and dineolignans, because these tri- and tetramers must always present neolignan-type linkages.

---

[1] IUPAC recommendations.

Structure and Chemical Properties of Lignans 317

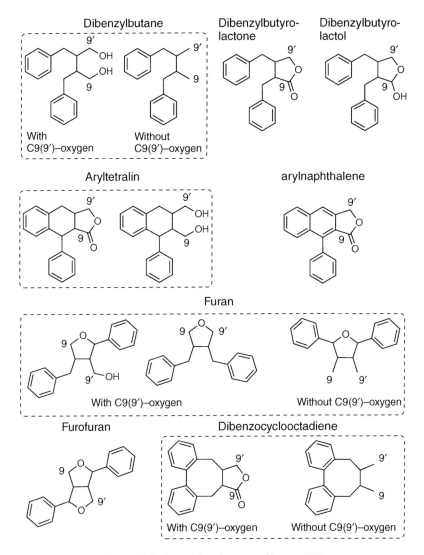

Figure 10.2 The eight subgroups of lignans [32]

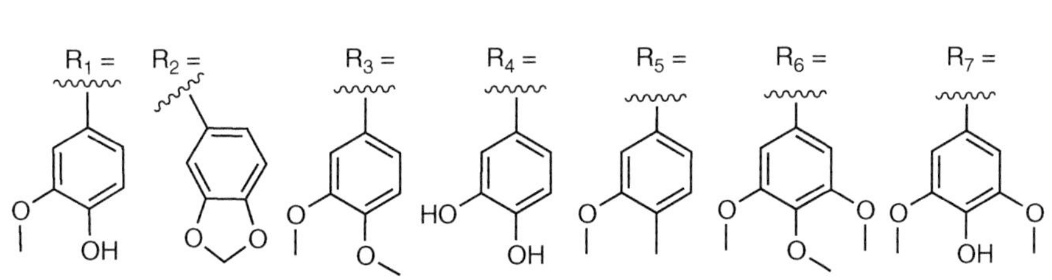

Figure 10.3 Main aromatic substituents of lignans

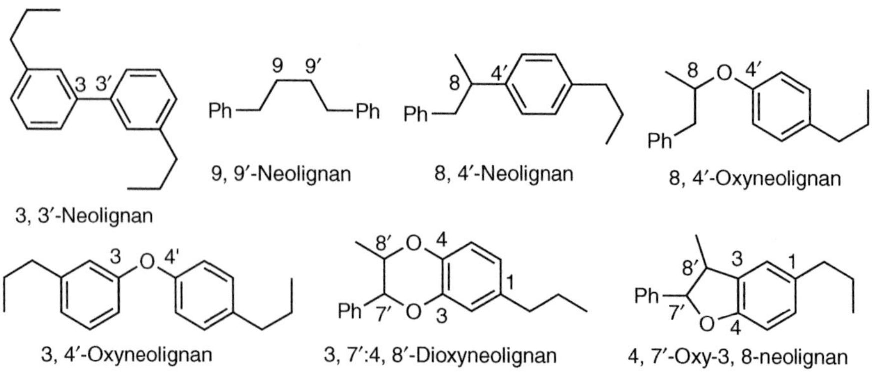

Figure 10.4 Neolignan structures resulted by the different bonding of two cinnamyl units

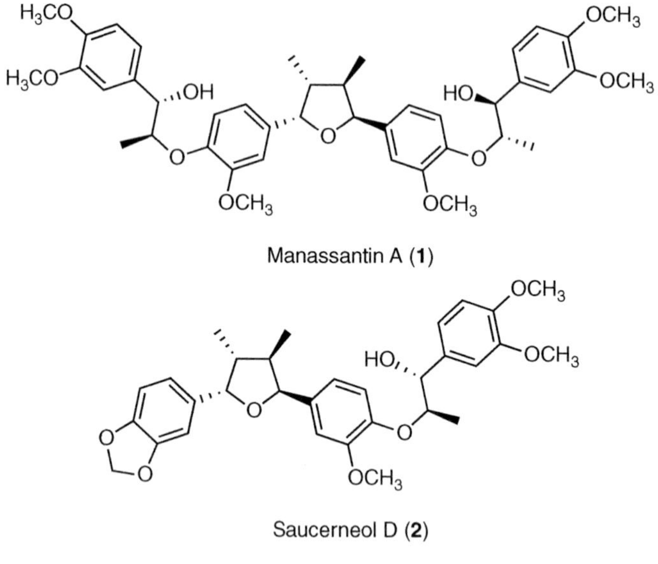

Manassantin A (**1**)

Saucerneol D (**2**)

Figure 10.5 Sesquilignans and dineolignans

10.2.1.3 Norlignans

Norlignans can be seen as lignans that lost one carbon atom, or as another different class of secondary metabolites with a different composition. Some authors, such as Moss [43] consider them to be a subclass of lignans, and thus we have included them in this section because their metabolism is closely related to that of lignans. The prefix "Nor-" is therefore included when the structure is derived from lignane, neolignane, or oxyneolignane and one or more carbon atoms are lost [43].

According to Suzuki and Umezawa [32], norlignans are apparently composed of phenylpropane (C_6–C_3) and phenylethane (C_6–C_2) units. On the basis of the linkage position between the two units,

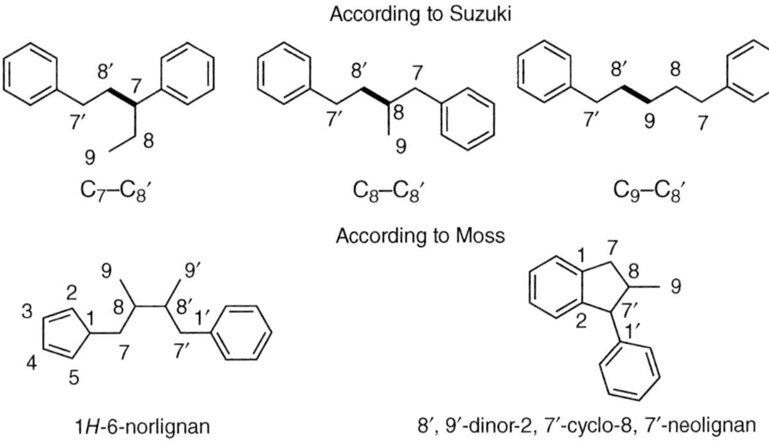

Figure 10.6 Chemical structures of norlignans

their chemical structure is classified into three groups: C_7–$C_{8'}$ linkage type, C_8–$C_{8'}$ linkage type, and C_9–$C_{8'}$ linkage type (see Figure 10.6).

10.2.2 Hybrid Lignans

Lignans also include the structure of another type of natural product: flavonolignans, coumarinolignans, stilbenolignans, xantholignans, and so on. [8, 44–50]. These natural products are commonly classified as "hybrid lignans."

The structures of some examples of hybrid lignans are depicted in Figures 10.7 and 10.8. Silibinin (**3**), a flavonolignan that in 1968 became the first known hybrid to be described, is found in milk thistle extract [51]; pseudotsuganol (**4**) is the first isolated true flavonolignan [52]; cleomiscosin D (**5**) is a coumarino-lignan from seeds of *Cleome viscosa* [53–55]; daphnecin (**6**) is a coumarino-lignan from *Daphne mucronata* [56]; gnetucleistol F (**7**) is a stilbenolignans from *Gnetum cleistostachyum* [57]; maackolin (**9**) is isolated from *Maackia amurensis* and was identified as a stilbenolignan. The stilbenolignan aiphanol (**8**) can be found in the seeds of *Aiphanes aculeata* [58, 59]; cadensin G (**10**) is a xantholignan; examples of sesquiterpene-neolignans are eudesmagnolol (**11**), piperitylmagnolol (**13**), clovanemagnolol, caryolanemagnolol, eudeshonokiol A, and eudesobovatol A [60, 61]; and palliation (**12**) is a macrocyclic lignan derivative from *Pellia epiphylla* [62, 63].

10.3 Nomenclature of Lignans

The nomenclature of the diverse range of structures classified as lignans depends largely on trivial names and, if necessary, on the appropriate numbering derived from the systematic name. The structural diversification of such biogenetically related group of compounds limits the utility of a systematic nomenclature, since it disguises structural similarities. The system followed for naming the structures is based on the original system introduced by Freudenberg and Weinges [64], later extended by Moss [43], and now accepted by IUPAC [43]. It follows the trivial or semisystematic

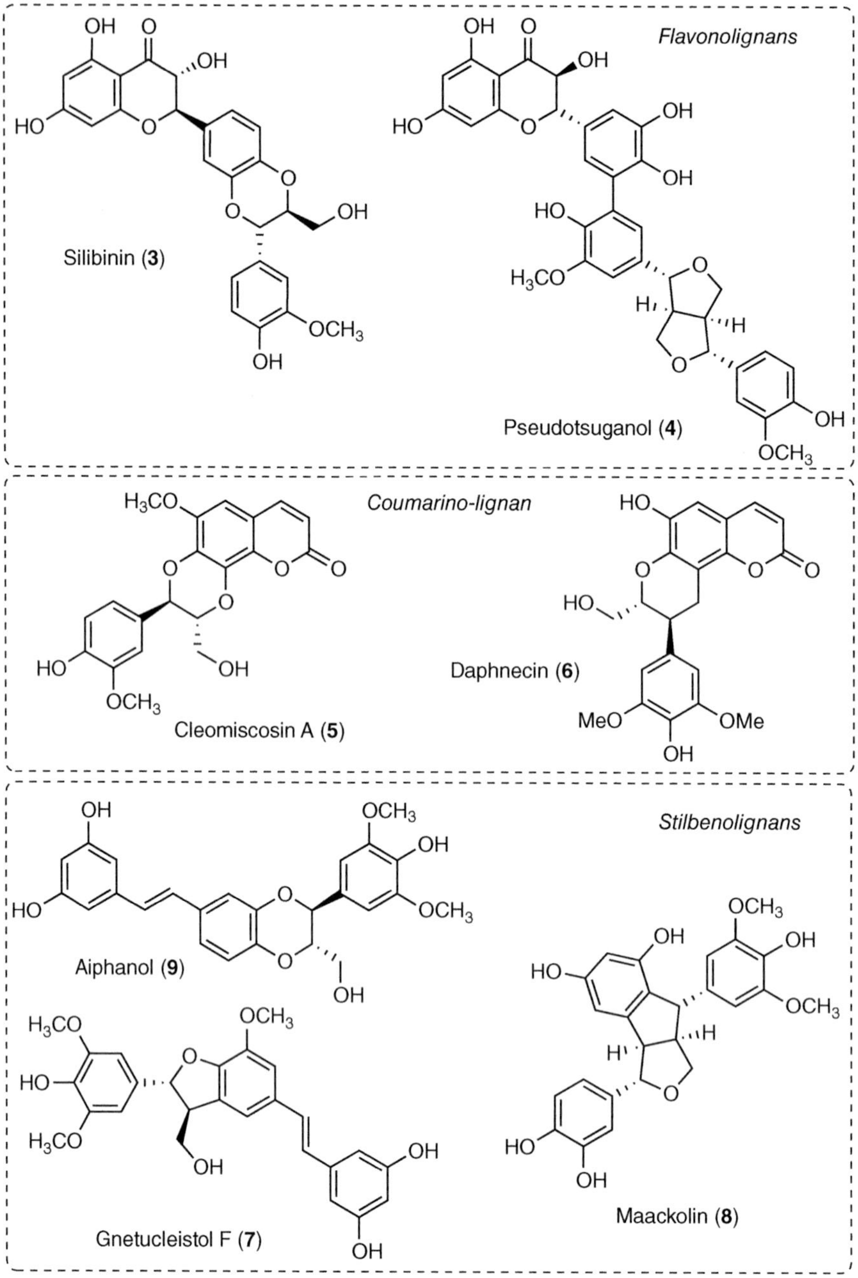

Figure 10.7 Hybrid lignans (Flavono-, coumarino-, and stilbenolignans)

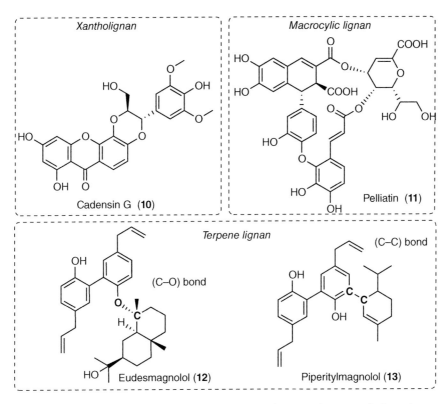

Figure 10.8 Hybrid lignans (xantholignan, terpene lignan, and macrocyclic lignan)

names, which are commonly given as synonyms within the entries to provide the chemical abstract service (CAS) name. Some examples of these trivial names are collected in Table 10.1.

According to Moss, the existing trivial names for lignans and neolignans may still be used, but the semisystematic name is recommended, and also be quoted when first encountered in the corresponding text. This systematic name is assigned according to the following steps:

1. We choose a name for the fundamental parent structure: for example, lignane, neolignane, and oxyneolignane (see Figure 10.9).
2. A prefix that indicates any modification to the fundamental parent structure is added. For example, we use the following prefixes:

   cyclo- when the carbon skeleton presents an additional ring,
   seco- when there is a cleavage in a ring bond,
   nor- when there is a loss of one or more carbon atoms,
   homo- when there is an addition of one or more carbon atoms,
   oxa- or aza- when a carbon atom is replaced by a heteroatom,
   epoxy- when additional rings are generated by bridging two positions by an oxygen atom, and finally
   abeo- when a parent structure does not possess an accepted skeleton but may be considered as such by migration of one or more bonds (see Figure 10.10).

Table 10.1 Synonyms of most common lignan names used in this textbook[a]

Name used in this textbook	Synonyms
γ-Schizandrin	(±)-γ-Schisandrin; (±)-γ-Schizandrin; Schisandrin B; Schizandrin B; Wuweizisu B; gamma-Schisandrin; γ-Schisandrin; Deoxygomisin A
4-Demethyl-PPT	4′-Demethylpodophyllotoxin; 4′-O-demethylpodophyllotoxin
(−)-Asarinin	(−)-Episesamin; Desgeranyloxyarmatumin; L-Asarinin; Xanthoxylin S
cis-Hinokiresinol	(−)-Nyasol; Nyasol
Deangeloylgomisin B	Schisandrol A; Schizandrol A
Deoxy-PPT	Desoxypodophyllotoxin; 4-Deoxypodophyllotoxin; AS 2-3; Anthricin; Silicicolin
Schizandrin A	Schisandrin A; (+)-Deoxyschizandrin; Deoxyschisandrin; Wuweizisu A
(+)-Diayangambin	(+)-Syringaresinol dimethyl ether; Lirioresinol C dimethyl ether
Enterolactone	(±)-enterolactone; HPMF
(+)-Epimagnolin	(+)-Epimagnoline A; (+)-epi-Magnolin; Epimagnolin; epimagnolin
(+)-Epimagnolin A	Epimagnolin A
Etoposide	Celltop; EPE; Epipodophyllotoxin VP 16213; Eposin; Eto-Gry; Etosid; Fytosid; Lastet; NSC 141540; Sintopozid; Toposar; VP 16; VP 16-123; VePesid; Vepesid J; Zuyeyidal; trans-Etoposide
Eucommin A	(+)-Medioresinol 4-O-β-D-glucopyranoside
(+)-Gomisin A	Besigomsin; Gomisin A; Schisandrol B; Schisantherinol B; Schizandrol B; TJN 101; Wuweizi alcohol B; Wuweizichun B
Gomisin B	Schisantherin B; Schizantherin B; Wuweizi ester B
Gomisin C	Schisantherin A; Schizantherin A; Wuweizi ester A
(−)-Gomisin N	Gomisin N; Isokaduranin; Deoxygomisin O
(+)-Gomisin N	Kaduranin; Gomisin M methyl ether
Heteroclitin G	Kadsulignan K
Hinokinin	(−)-Hinokinin; Cubebinolide; Dihydroisohibalactone; Hinoquinin
Justicidin C	Neojusticin B
Justicidin D	Justicidine D; Lignan J1; Neojusticin; Neojusticin A
Justicidin E	Justicidine E; NSC 309692
Justiciresinol	(+)-Justiciresinol; (+)-Jusglaucinol; Jusglaucinol
Kobusin	(+)-Demethoxyaschantin; (+)-Kobusin; (+)-Methylpiperitol; (+)-Spinescin; Kubosin; Methylpiperitol; O-Methylpiperitol; Spinescin
(−)-Lirioresinol A	(−)-Epi-syringaresinol
Liriodendrin	(+)-Syringaresinol di-O-β-D-glucopyranoside; (+)-Syringaresinol di-O-β-D-glucoside; (+)-Syringaresinol-di-β-D-glucopyranoside; Acanthoside D
Longipedunin A	1-Demethylkadsuphilin A
(−)-Syringaresinol	(−)-Lirioresinol B
Machilin A	erythro-Austrobailignan-5; meso-Austrobailignan-5
Matairesinol	(−)-Matairesinol; (8R,8′R)-(−)-Matairesinol
Nordihydroguaiaretic acid	Dihydronorguaiaretic acid; Dihydro-dinorguaiaretic acid; NSC 4291

Table 10.1 (continued)

Name used in this textbook	Synonyms
o-Demethyl-SECO	3-Demethyl-(−)-secoisolariciresinol
(+)-Phillygenin	(+)-Phylligenin; Epipinoresinol methyl ether; Forsythigenol; Phillygenin; Phillygenol; Phillyrin aglycone; Sylvatesmin
Phyllyrin	Phillyrin; Phillygenin β-D-glucopyranoside; Phillyroside
Podophyllotoxin	Condyline; Condylox; NSC 24818; Podofilox
Savinin	Hibalactone; (−)-Hibalactone; (−)-Savinin; NSC 150442; Savinine
Schisantherin D	Schizantherin D; Shisantherin D
Schizandrin	(+)-Schizandrin; Schisandrin; Schisandrine; Schizandrine; Schizandrol; Wuweizi alcohol A; Wuweizichun A
Schizandrin C	(S)-(−)-Schisandrin C; Schisandrin C; Wuweizisu C; (−)-Wuweizisu C
Secoisolarisiresinol	(−)-Secoisolariciresinol; (8R,8′R)-(−)-Secoisolariciresinol; Knotolan
Sesamin	(+)-Sesamin; Fagarol; NSC 36403; Sezamin; Zengxiaomin; d-Sesamin
Sesaminol	Justisolin
Teniposide	EPT; NSC 122819; S 122819; Tenoposide; VM 26; Vehem; Vehem-Sandoz; Vumon
(+)-Yangambin	(+)-Yangabin; Lirioresinol B dimethyl ether

^aData from SciFinderScifinder.®
^bScifinder, version 2014; Chemical Abstracts Service: Columbus, OH, 2014; http://scifinder.cas.org (accessed December 13, 2014).

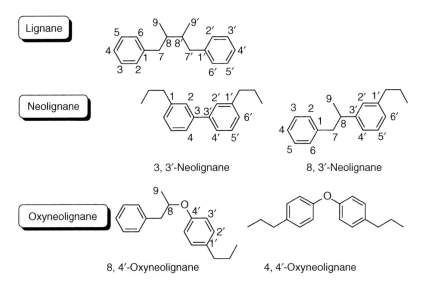

Figure 10.9 Nomenclature of lignans. Step 1: Fundamental parent structure according to Moss [43]

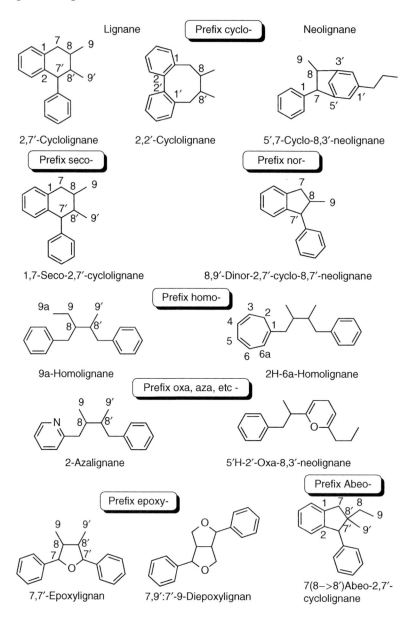

Figure 10.10 Nomenclature of lignans. Step 2: Modifications of the fundamental parent structures

3. Then, we must indicate, if applicable, the changes introduced at the hydrogenation level (see Figure 10.11).
4. We add the name of derivative functional groups: carboxylic acids, lactones, and so on (see Figure 10.12).
5. We must address the absolute configuration of each of the stereocenters, and, if appropriate, the atropoisomery of the structure, for example, absolute configuration (see Figure 10.13).

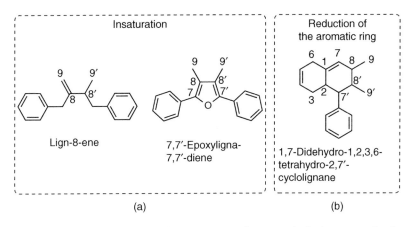

Figure 10.11 Nomenclature of lignans. Step 3: Changes in the hydrogenation level

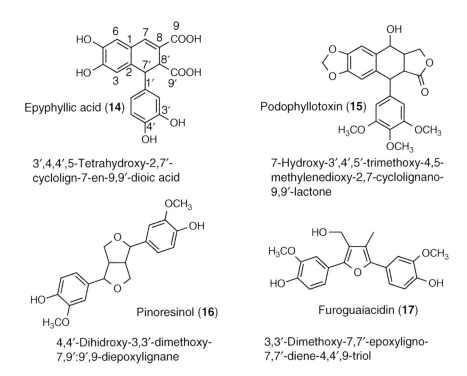

Figure 10.12 Nomenclature of lignans. Step 4: Derivatives

10.4 Lignan Occurrence in Plants

Lignans are present in several different sources. According to Gordaliza et al. [65], lignans can be found in more than 60 families of vascular plants and have been isolated from different plant parts: roots and rhizomes, woody parts, stems, leaves, fruits and seeds, as well as, in other cases, from

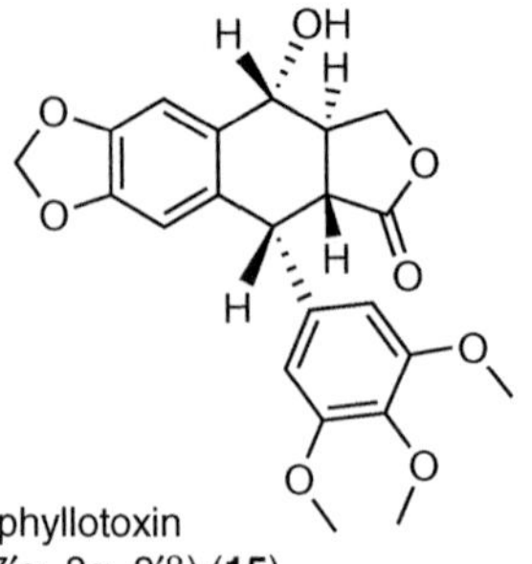

Podophyllotoxin
(7α, 7′α, 8α, 8′β) **(15)**

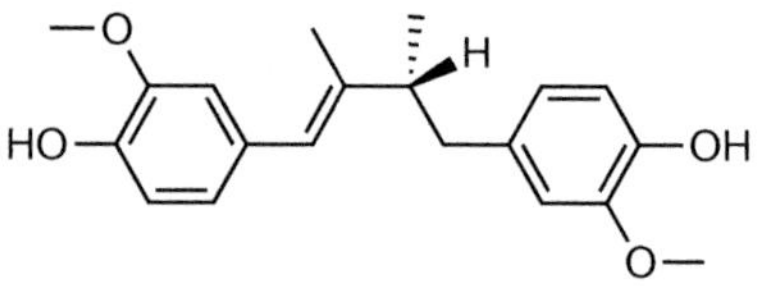

Guaiaretic acid (18)
7(E)-(8′R)-3, 3-dimethoxylign-7-ene-4, 4′-diol

Figure 10.13 Nomenclature of lignans. Step 5: Stereochemistry

Table 10.2 Resin and heartwood sources[a,b,c]

Lignan	Source	Yield (%)	[Ref.]
Guaiaretic acid (18)	Guaiacun officinale	12	[71]
Hinokinin (20)	Chameacyparis obtusa	30	[72]
Matairesinol (21)	Podocarpus spicatus	50	[73, 74]
Galbacin (24)	Himantandra baccata	b	[75]
Galgravin (25)	Himantandra baccata	0.6	[75]
Lariciresinol (22)	Larix decidua	11	[76, 77]
Olivil (23)	Olea europa	45	[78]
Eudesmin (30)	Eucalyptus hemiphloia	10	[79]
Gmelinol (34)	Gmelina leichardtii	2.3	[80, 81]
Pinoresinol (16)	Pinus, Picea sp.	35	[82]
Sylvatesmin (31)	Symplocos lucida	0.3	[83, 84]
Conidendrin (29)	Tsuga, Picea sp.	b	[85]
Cyclotaxiresinol (26)	Taxus baccata	1.1	[86]
Isolariciresinol (27)	Fitzroyia cuppresssoides	0.3	[87]
Cycloolivil (28)	Olea cunninghamii	b	[88]

[a]Data from ref. [9, p. 139].
[b]Appreciable amounts obtained.
[c]See Figure 10.3 for the meaning of R groups.

exudates and resins [12, 66–68]. Apart from being found in plants, lignans have also been detected in the urine of humans and other mammals, and several *Streptomyces* species are lignan producers [69].

The fact that more than 250 new lignans were isolated in just five years (2004–2008) from different sources could serve as an example of the broad occurrence of lignans and the recent development of this field [70].

In 2003, a remarkable study by Umezawa [4] on the phylogenetic distribution of lignans, listed by Ayres and Loike [9], provided valuable results on the families of lignans. Despite the lack of quantitative data, the study concludes that 108 families within the Magnoliophyta division, 8 within Gymnospermophyta, and 2 within Pteridophyta have been found. In addition, in the division of Bryophyta, only *p*-hydroxycinnamate dimers have been isolated.

Even though lignans occur typically in vascular plants, these sources do not provide commercially useful quantities [9], whereas wound resins of trees constitute the major source of lignans (see Table 10.2 and Figure 10.14).

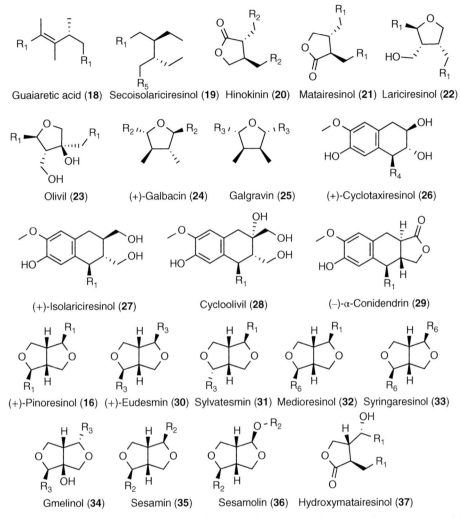

Figure 10.14 Lignans most commonly distributed in plants [9]. See Figure 10.3 for the meaning of R groups

Holmbom et al. [89] have revealed that knots, that is, the branch bases inside tree stems, commonly contain 5–10% (w/w) of lignans. Norway spruce (*Picea abies*) knots contain as much as 6–24% lignans, with 7-hydroxymatairesinol (HMR, (**37**)) as the predominant (70–85%) lignan. Some other spruce species also contain HMR as the main lignan, but some spruce species also have other dominating lignans. Most fir (*Abies*) species contain secoisolariciresinol and lariciresinol as the main lignans. Knots are detrimental in the manufacture of pulp and paper and should preferably be removed before pulping, which is possible using a recently developed industrial process called ChipSep [90].

In plants, lignans usually occur freely or bound to sugars [16, 91–105].

The most commonly distributed lignans in food plants are: secoisolariciresinol (**19**), found in cashews, chickpeas, coffee, cranberry, flax, peas, sunflower seeds, and wine; matairesinol (**21**), found in flax, oats, pineapple, rye, and wine; lariciresinol (**22**), found in buckwheat, aubergine, oats, pineapple, and rye; medioresinol (**32**), found in lemon, rye, and sesame seeds; pinoresinol (**16**), found in asparagus, flax, lemon, and rye; syringaresinol (**33**), found asparagus, barley, buckwheat, millet, oranges, rye, and wheat; and sesamin (**35**) and sesamolin (**36**), found in sesame seeds. The structures of these compounds are depicted in Figure 10.14).

Lignans are found in most fiber-rich plants, including grains such as wheat, barley, and oats; legumes such as beans, lentils, and soybeans; and vegetables such as garlic, asparagus, broccoli, and carrots [106, 107]. Despite exceptional cases, such as flax and sesame seeds (> 300 mg/100 g), the quantity of lignans is generally low (< 2 mg/100 g). Quantitative data of total amount of lignans in common food are presented in Table 10.3 (see ref. [91] and references therein) [108].

10.5 Methods of Determination and Isolation of Lignans from Plants

Mass spectrometers coupled or not with a chromatographic separation have become the technique of choice to analyze lignans. Reverse phase high-performance liquid chromatography (RP-HPLC) is the most commonly used analytical technique for detection and quantification of lignans [109]. However, gas chromatography (GC) coupled to a mass spectrometer has also been used to detect and quantify lignans [110, 111].

The choice of method for the extraction of lignans varies widely depending on the sample and the molecular structure of the product. A wide range of methods is available as isolation procedures for lignans [112].

For extracting lignans, there are two main steps: first, the application of polar solvents to different parts of the plants containing the lignans; and second, the dissolution of the resulting extracts in water and reextraction with nonpolar solvents [4, 113–115].

However, new and appropriate methods for the analysis of lignans from plant sources and body fluids have also been applied during the last decade. In this regard, conventional chromatographic methods, including RP-HPLC, remain the most useful and commonly applied techniques [116]. Among the novel techniques, supercritical CO_2 provides an effective pathway for the extraction of lignans from leaves, seeds, and fruits of *Schisandra chinensis* [117], while micellar electrokinetic chromatography (MEKC) separates 12 lignans from *Phyllanthus* plants [118].

Even though conventional methods, including HPLC, have been employed in general to isolate *Taxus* lignans, more advanced techniques have also been applied. In fact, stopped-flow HPLC-NMR spectroscopy [119], for isolating lignans from the plant *Torreya jackii* (family *Taxaceae*), and GC-MS, for analyzing trimethylsilyl derivatives of the lignans present in galls and shoots of *Picea glauca* [120], are some remarkable examples.

Nowadays, the qualitative and quantitative analytic methodology, most commonly employed for isolating lignans, is HPLC with photodiode array detector (DAD, diode array detector) and multiple-stage mass spectrometry (MSn). The stereoisomeric lignans that cannot be separated by these achiral columns are separated by various chiral chromatography methods [109]. Nevertheless, other methodologies such as high-speed countercurrent chromatography (HSCCC) and MEKC have also been applied for the separation of some plant lignans.

10.5.1 Lipid Extraction

Lipid substances are present in all plants and since they are soluble in organic solvents, it is advisable to separate them at an early stage by extraction with hexane or petroleum ether. Phenolic natural products have only limited solubility in hexane, especially when a free hydroxyl group is present. Nevertheless, in a search for lignans, it is necessary to monitor a concentrated extract by thin layer chromatography (TLC) and/or, if possible, by HPLC. The tendency to extract lignan, including free phenols, is accentuated if a mixture of hexane and diethyl ether is used to remove lipids [121].

Table 10.3 Total (fresh weight) content of lignans (as aglycones) in common foods and their botanical origin[a]

Common food	Genus and species	Total[b] (µg per 100 g)
Beverages		
Coffee, arabica nescafe	Coffea arabica	694
Coffee, maxwell house	C. arabica	485
Wine, Cabernet Sauvignon, France, red	Vitis vinifera	760
Wine, Chardonnay, France, white	V. vinifera	196
Wine, Chardonnay, Italy, white	V. vinifera	153
Wine, Chianti, reserve, Italy, red	V. vinifera	1378
Cereals		
Barley, whole grain	Hordeum vulgare	370
Buckwheat, whole grain	Fagopyrum esculentum	867
Millet, common whole grain	Panicum miliaceum	245
Oat, whole grain	Avena sativa	859
Rye, whole grain	Secale cereale	1891
Rye, whole meal	S. cereale	100
Wheat whole grain	Triticum aestivum	539
Fruits		
Cranberry	Vaccinium macrocarpon	136
Grape	V. vinifera	126
Guava	Psidium guajava	134
Kiwi	Actinidia deliciosa	147
Lemon	Citrus limon	335
Orange	Citrus sinensis	122
Pineapple	Ananas comosus	172
Strawberry	Fragaria anamassa	143
Nuts, seeds and spices		
Cashew	Anacardium occidentale	247
Hazelnut, European hazel	Corylus avellana	116
Walnuts	Juglans nigra	160
Caraway seed	Carum carvi	204
Cumin	Cuminum cymicum	208
Flax seed	Linum usitatissimum	335 002
Sesame seed	Sesamum indicum	132 275
Sunflower seed	Helianthus annuus	581
Vegetables and legumes		
Asparagus	Asparagus officinalis	344
Chickpea	Cicer arietinum	35 067
Chives	Allium schoenoprasum	117
Aubergine	Solanum melongena	107
Garlic	Allium sativum	158
Pea	Pisum sativum	8355
Peanut	Arachis hypogaea	279
Soybean	Glycine max	131

[a] Data from ref. [91].
[b] Only foods with the total amount higher than 100 µg in 100 g.

Fonseca et al. [122] removed lipids from the evaporated benzene extract with hexane. Another approach is to attempt the complete removal of soluble organic compounds from the raw material using a hot polar solvent, such as acetone, EtOH, or MeOH, and then to extract lipids by treating the initial evaporate with hexane [9]. Steam distillation has also been used to remove volatile compounds from nonvolatile lignans.

10.5.2 Solvent Extraction

Several lignans were obtained by an initial extraction with EtOH and fractionally extracting the residue successively with hexane, chloroform, n-butanol, and water [123]. It is to be expected that hot alcoholic solvents will remove all lignans and their glycosides from plant material. Glycosides may be concentrated by partitioning them into the aqueous phase of a two-phase system [9].

In 2007, a microwave-assisted extraction (MAE) method was applied for the first time to extract lignans. The values were compared with those resulting with conventional extraction methods and the results demonstrated that MAE was more effective in terms of both yield and time consumption [124]. Recently, an optimized MAE method was evaluated through repeatability, recovery (97.5%), and efficiency testing by Nemes and Orsat [125]. The MAE method is efficient for extracting lignans from the plant matrix, and it achieves significantly higher extraction yields than the established reference methods.

Today, organic solvents have to be replaced for safety and environmental issues, and therefore different strategies have been proposed in this regard. Supercritical fluids are greener alternatives to nonpolar solvents, for example, supercritical CO_2 [126]. Another alternative in order to obtain polar lignans [127] is to use a pressurized low-polarity water (PLPW) extraction.

10.5.3 Separation by Precipitation

Those lignans with free phenolic hydroxyl groups may be obtained from alcoholic extracts by the addition of concentrated KOH solution to precipitate their potassium salts. The more-soluble sodium salts can sometimes be leached directly from the plant material into NaOH, followed by acidification and extraction into an organic solvent [9].

10.5.4 Chromatographic Methods

Recently, Willför et al. have reviewed the methods and procedures for analyzing lignans in trees and other plants [109].

TLC has now come into routine use for the initial examination of plant material and for monitoring the composition of extracts and fractions obtained from chromatographic columns at the various stages of purification. Kiesegel G or alumina plates are generally satisfactory. Since all lignans absorb ultraviolet (UV) light in the region of 254 nm, they are clearly shown on dye-marked plates (GF_{254}) as dark spots on a green fluorescent background. Other detection methods are exposure to iodine vapor, to spray with a 5% solution of concentrated H_2SO_4 in EtOH (followed by heating for a few minutes at approximately 100 °C), and heating after a spray with phosphotungstic acid in EtOH. The sensitivity of these methods is on the order of 0.1 µg. General surveys of reagents suitable for color development on TLC plates have been published and include detection methods and choice of eluents for TLC of lignans [128, 129]. The choice of conditions for the column chromatography is best made by preliminary trials with TLC and the most useful media are alumina and silica gel. Eluents and solid phases used depend on the nature of the lignans to be separated.

HPLC is superior to the methods discussed so far in terms of resolution, recovery, detection, and speed of operation. It may be applied analytically, in a semipreparative way (100 mg scale), or on a larger scale.

Ayres and Chater, in 1969, [130] determined suitable conditions for the GC of the main classes of lignans.

Droplet countercurrent chromatography (DCCC) is applicable to the isolation of polar lignans [131].

10.5.5 Extraction of Polar Lignans from Biological Materials

The first evidence of extraction of polar lignans from biological materials [9] such as urine was reported by Bradlow [132], who developed a method for removing lignans and/or their glycosides from dilute aqueous solutions or urine samples. The method consists of passing the sample through a column of neutral PS resin and the successive elution with a small volume of ethanol or methanol solvent. For the concentration of large volumes of water, another method developed in 1982 by Fotsis *et al.* [133] passes the sample through a column of octadecylsilane-bonded silica, followed by graded elution using mixed solvents. Since then, in 2003, a method consisting in HPLC equipped with a coulometric electrode array detector was developed to measure plant and mammalian lignans in human urine [134], and a HPLC-MS-MS method was also described for the detection of HMR and its potential metabolites in human plasma [135]. In 2006, Knust *et al.* [136] applied a method involving the coupling of HPLC with electrospray ionization mass spectrometry (HPLC-ESI-MS) for the quantitative determination of the mammalian lignans enterolactone and enterodiol in human blood and urine. In contrast to the techniques previously published, the method allows direct measurement of free enterolignans[2] as well as their monoglucuronide conjugates in human biofluids with minimal sample preparation.

10.6 Structure Determination of Lignans

The structures of extracted and synthesized lignans have been confirmed using NMR techniques, together with the classical melting point determination, IR and UV spectroscopy, for more than 60 years [44, 137, 138].

Single-crystal X-ray analysis remains the most reliable method for determining the absolute configurations of lignans [70]. Nevertheless, many other techniques are involved in the structural elucidation of lignans. Among them, 1D and 2D NMR analyses as well as high-resolution mass spectrometry (HRMS), UV and infrared (IR) spectra, are the most important. In addition, the analysis of their circular dichroism (CD) and nuclear Overhauser effect spectroscopy (NOESY) is useful when more than one chiral center is present, and the lignans have optical properties. Chemical methods, such as Mosher's ester modification, may also help [139].

Ayres and Loike [9, pp. 175–195] extensively discussed structural determination of the different classes of lignans. Beyond NMR (1D and 2D), UV, IR, and HRMS, an excellent review of CD and optical rotatory dispersion, together with chemical correlation for absolute configurations assignation, is discussed. Some of the most remarkable characteristics are described as follows.

Lignans, since they contain aromatic rings, present the typical aromatic three bands in the regions 210, 230, and 280 nm of the UV spectra [140]. In the case of free phenolic lignans, the UV spectra is quite similar to the one shown by etherified compounds, although in the case of the free phenolic lignans the phenols are shifted to longer wavelength regions of the spectra, the so-called bathochromic effect in their anions [141].

Differently, in dibenzocyclooctadiene lignans, the conjugation of aromatic rings enhances the UV absorption [142]. In lignans with rotation, the spectra are not very different from those of normal acyclic lignans, due to steric restrictions. On the other hand, arylnaphthalene lignans show an unmistakable intense absorption and strong UV fluorescence [9, p. 170].

---

[2] Any lignan formed from another by metabolism in the gut.

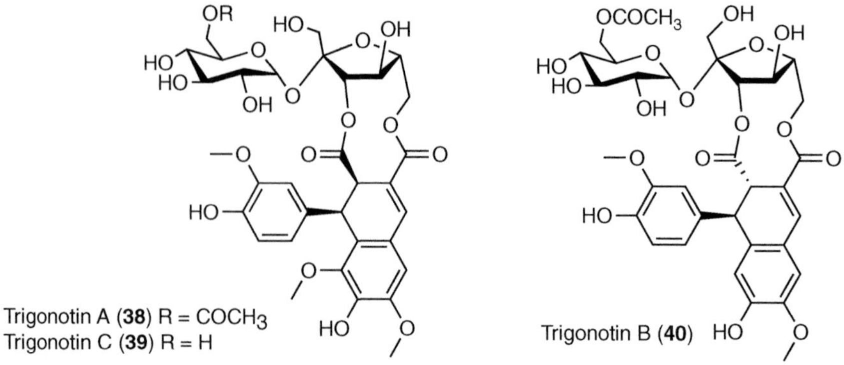

Figure 10.15 Trigonotins A–C [143]

Trigonotin A (**38**) R = COCH₃
Trigonotin C (**39**) R = H
Trigonotin B (**40**)

Three similar 7′,8′-dihydroarylnaphthalene oligosaccharides, trigonotins A–C (**38–40**) [143] (Figure 10.15), showed strong yellow-green fluorescence under basic conditions. This is the first time that such a property has been reported for ayldihydronaphthalenes.

Copious relevant data on the IR spectra of lignans have been published [144]. The technique employed for the mass spectra determination depends basically on the volatility and functionalization of the sample. An extensive discussion on mass spectrometry of different types of lignans is provided in ref. [9, pp. 175–195].

10.7 The Chemical Synthesis of Lignans

Covering all the information available at present concerning the synthesis of lignans is not the purpose of this chapter. There are many structures, including stereochemical variations, and huge numbers of synthetic possibilities.[3] Therefore, we summarize only some of them as examples, showing the most general synthetic pathways to obtain the different structural subclasses, or emphasize those that due to pharmacological applications have attracted major interest.

Today, two main trends can be highlighted with respect to the synthesis of lignans. The first is based on the traditional organic chemistry pathways including asymmetric synthesis, while the second and more recent one implies chemical pathways coming from biosynthesis strategies [44]. By way of example, some ligand syntheses of the different types of structures are collected; in each and every case, the most synthesized molecules or the most original synthesis will be included: dibenzylbutane lignans (dihydroguaiaretic acid (**41**) and secoisolariciresinol (**19**)), dibenzylbutyrolactone lignans (matairesinol (**21**)), simple aryltetralin lignans (podophyllotoxin (PPT) (**15**)), and dibenzocyclooctadiene lignans (steganone (**42**)).

10.7.1 Generalities on the Asymmetric Total Synthesis of Lignans

According to Del Signore and Berner [145], the majority of the approaches used in the asymmetric total synthesis of chiral lignans can be divided into four general groups regardless of the type or class of lignan.

---

[3] For example, almost 2000 articles can be found under the topic "lignan synthesis" in SciFinder.

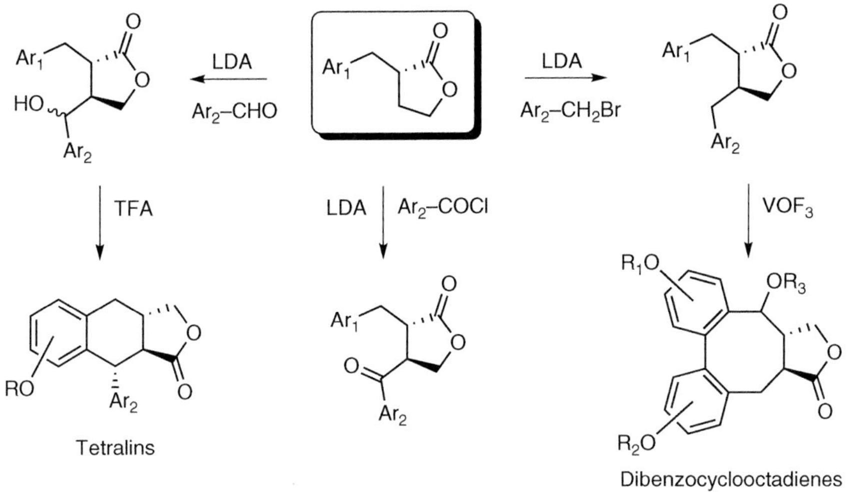

Figure 10.16 *Monobenzylbutyrolactone transformations [145]*

10.7.1.1 Diastereoselective Alkylation of Chiral Butyrolactones

This is one of the first and most widely used approaches for synthesizing different classes of lignans (see Figure 10.16): the different diarylbutyrolactone lignans are obtained through deprotonation of lactone and subsequent alkylation or aldol reaction. This synthesis gives dibenzocyclooctadiene lignans *via* biaryl coupling, or tetraline lignans *via* deshydrative ring closure. The main problem arising from such strategy is obtaining the chiral benzylbutyrolactone itself with high yield and enantioselectivity. In this regard, many synthetic pathways can be found in the literature [146–159], some examples being the strategies described below:

The synthesis of (−)-enterolactone (**43**) by Yoda *et al.* [160], Chenevert *et al.* [161], and Sibi *et al.* [159]; the synthesis of (+)-neoisostegane (a dibenzocyclooctadiene lignan (**44**)) and (−)-deoxypodophyllotoxin (**45**) by Morimoto *et al.* [150–152], or obtaining several dibenzocyclooctadiene lignans by Tanaka *et al.* [162–172].

10.7.1.2 The Diastereoselective Conjugate Addition to 2(5H)-Furanones

Tomioka *et al.* [173–175] developed this approach in the early 1980s. It is based in the diastereoselective Michael addition to a chiral butenolide, which was used to obtain different lignans such as burseran (**46**) (a diphenylbutyrolactone) and steganacin (**47**) (a dibenzocyclooctadiene lignan) (see Figure 10.17). Similar strategies have been followed by different authors for obtaining lignans: Jansen and Feringa [176–180], Pelter *et al.* [137, 181, 182], and Enders *et al.* [183–185].

10.7.1.3 The Use of Chiral Oxazolidinones

Chiral oxazolidinones have been employed as chiral auxiliaries due to their high diastereoselectivities in aldol and other reactions. Kise *et al.* [186, 187] applied this strategy to the synthesis of dibenzylbutyrolactones and dibenzylbutanediols (see Figure 10.18). Also, Maioli *et al.* [188] synthesized (−)-sesaminone **49** using *N*-4-pentenoyloxazolidinone as a chiral starting material, while Sibi *et al.* obtained (−)-enterolactone (**43**) and various dibenzylbutyrolactones with a similar approach [189].

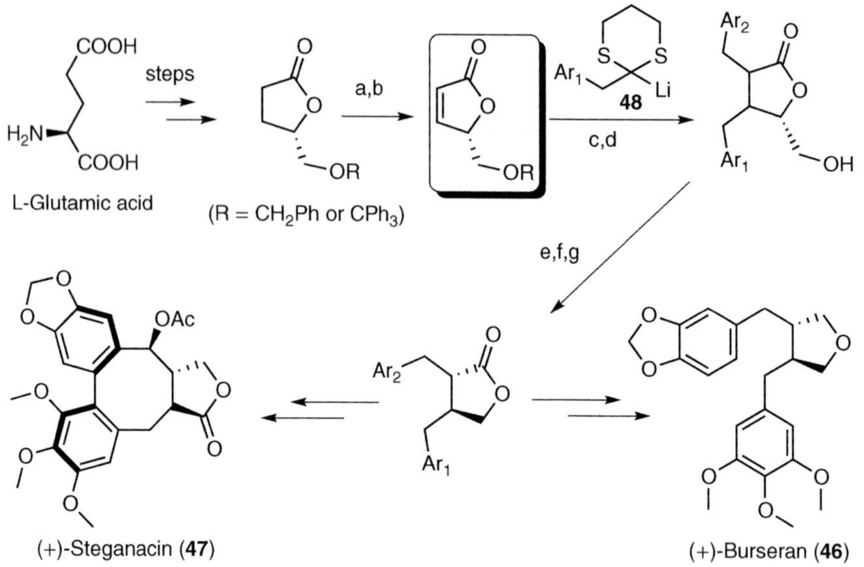

Figure 10.17 *Asymmetric synthesis of lignans using the Koga approach [145]. (a) i. LDA, HMPA, −78 °C ii. PhSeBr at −20 °C → 19 h rt; (b) NaIO$_4$, MeOH/H$_2$O, 30 min, rt; (c) i. (**48**), THF ii. substituted benzyl bromide; (d) Ra-Ni, EtOH, reflux; (e) THF, LiAlH$_4$, rt; (f) NaIO$_4$, t-BuOH, rt; (g) CrO$_3$, pyridine, CH$_2$Cl$_2$, rt*

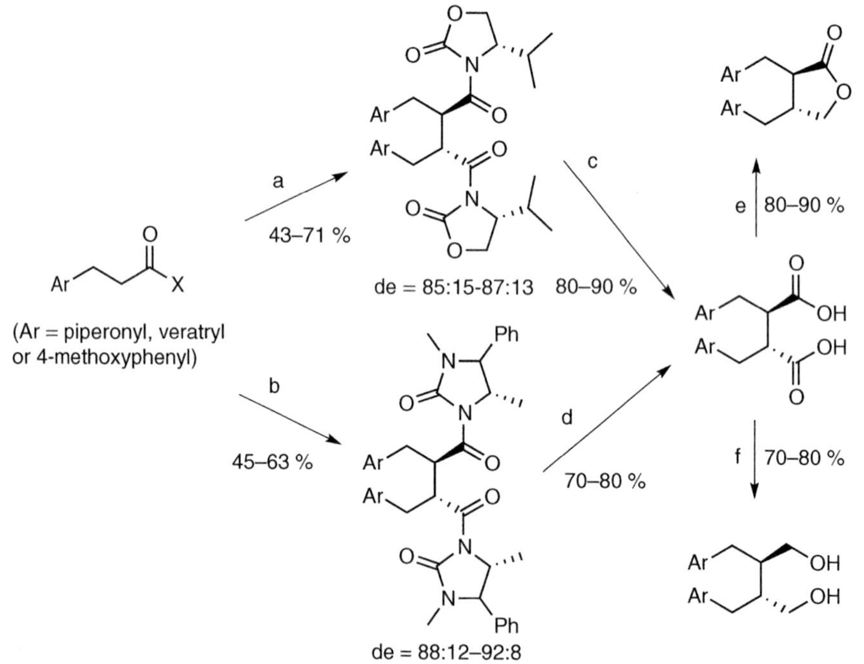

Figure 10.18 *Asymmetric synthesis of dibenzylbutyrolactones and dibenzylbutanediols [186, 187]. (a) LDA, THF, TiCl$_4$, −78 °C → rt, 24 h; (b) LDA, hexane/THF, CuCl$_2$, DMPU, −78 °C → rt, 12 h; (c) THF, LiOH, H$_2$O$_2$, rt 12–24 h; (d) LiOH, THF/H$_2$O$_2$, reflux 24–48 h; (e) i. Ac$_2$O, MeOH −78 °C ii. NaBH$_4$, 1 h; (f) LiAlH$_4$*

10.7.1.4 The Use of Asymmetric Cycloaddition Reactions (Asymmetric Diels–Alder Reaction)

Despite this being a widely used reaction in the synthesis of many natural products, only few examples concerning its use in the synthesis of lignans can be found. Charlton et al. used the asymmetric Diels–Alder reaction for the synthesis of several aryltetralines (see Figure 10.19) [190–192], Pelter et al. used the same strategy for the synthesis of (−)-isopodophyllotoxin (**50**) [193], and Jones et al. [194, 195] for an efficient synthesis of (−)-podophyllotoxin (**15**).

10.7.2 Dibenzylbutane Lignans

Within the dibenzylbutane type of lignans, two different classes of compounds have been selected to exemplify the synthesis, namely dihydroguaiaretic acid and secoisolariciresinol [44, 196, 197].

10.7.2.1 Dihydroguaiaretic Acid

In 1918, many years before, dihydroguaiaretic acid (**41**) was finally isolated as a naturally occurring compound by demethylation of hydrogenated dimethyl ether of (−)-guaiaretic acid, the synthetic preparation of this acid had been firstly pursued [198] (Figure 10.20). The already isolated compound was finally confirmed through synthesis in 1934 [71, 199, 200].

In 1947, Lieberman et al. [201] reported a more systematic synthesis [202]. In this pathway, 1-piperonyl-1-bromoethane and its Grignard derivative reacted in the key step, although the initial method has been modified for decades. Despite these changes, it failed to increase yield and/or selectivity toward desired stereoisomers [203].

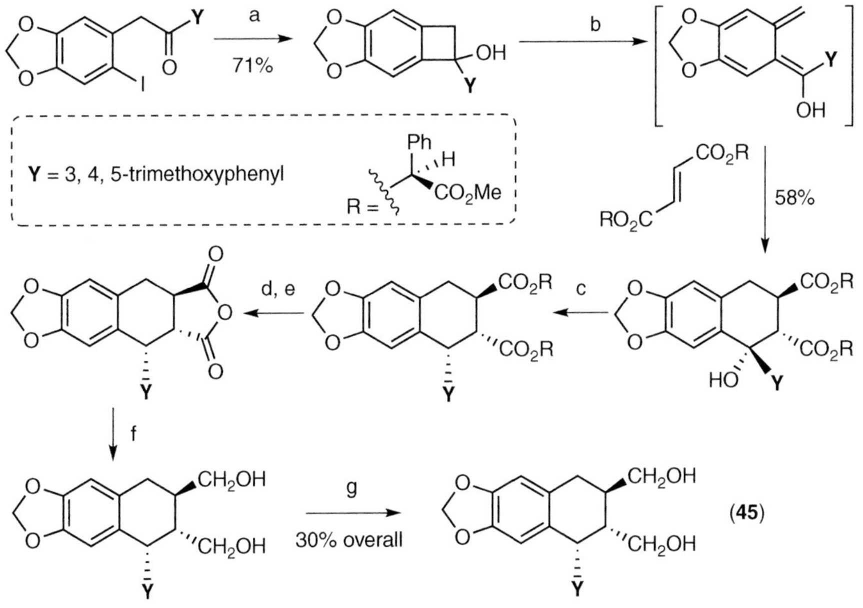

Figure 10.19 Asymmetric synthesis of (−)-deoxypodophyllotoxin (**45**) of Bogucki and Charlton [191]. (a) n-BuLi, THF, −78 °C; (b) toluene, reflux, 44 h; (c) BF_3Et_2O, CH_2Cl_2, −20 °C then $LiAlH_4$, −55 °C → rt; (d) Pd/C, H_2, MeOH/AcOH, rt, 89 h; (e) $(CF_3CO)_2O$, reflux, 2 h; (f) $NaBH_4$, i-PrOH, 15 h, rt; (g) benzene, p-TsOH, reflux, 17.5 h

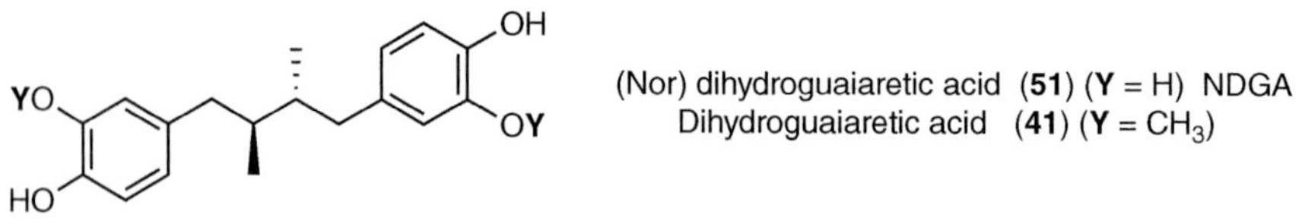

Figure 10.20 Dihydroguaiaretic acid structure

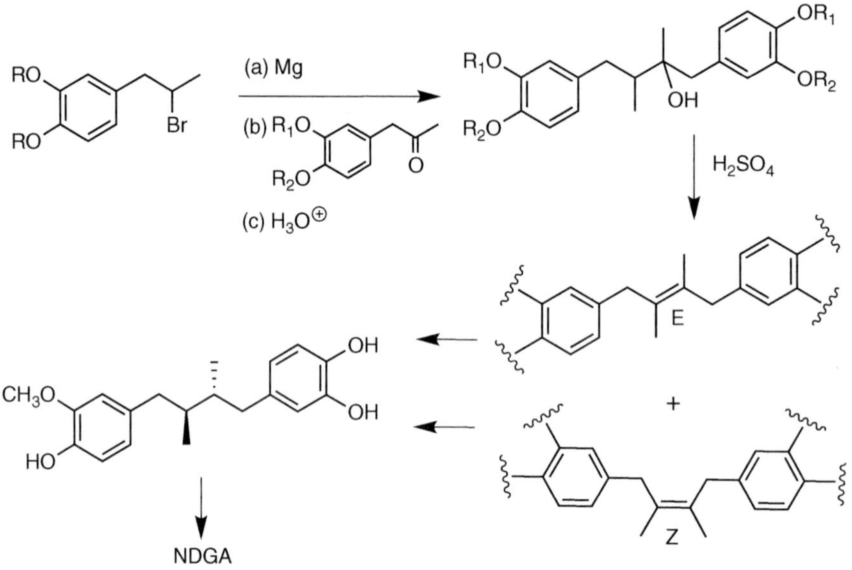

(Nor) dihydroguaiaretic acid (**51**) (Y = H) NDGA
Dihydroguaiaretic acid (**41**) (Y = CH$_3$)

Figure 10.21 Synthesis of NDGA from 1,4-di-(substituted-phenyl)-2,3-dimethyl-butan-2-ol [197]

However, not the expected McMurry-type butanes but 1,4-disubstituted butane-2,3-diols were yielded by Ti-induced carbonyl coupling reactions of the substituted phenylacetones [204]. Also, other strategies have been pursued, such as oxidative coupling of β-keto esters [205], double condensation of piperonal with diethyl succinate followed by a couple of reductive steps [206], and transition metalphosphine complex-catalyzed Grignard coupling reaction of halothiophenes followed by catalytic reduction [207].

Even though both stereoselectivity and yield have been improved by these methods, the cost of the reactions, their yield, and the existence of lengthy reaction sequences in some cases remain drawbacks to be overcome in order to achieve a general application of especially unsymmetrically substituted NDGAs (Nordihydroguaiaretic acids).

Son et al. [197] described a procedure for the preparation of nordihydroguaiaretic acid (**51**) via 1,4-di-(substituted-phenyl)-2,3-dimethyl-butan-2-ol as a key intermediate (see Figure 10.21).

10.7.2.2 Secoisolariciresinol

In 2006, Wang et al. [196] investigated the chemical transformation of coniferyl alcohol in secoisolariciresinol (**19**) via an oxidative coupling. This synthesis has three stages: in the first, the commercial aldehyde (**52**) is converted to 5-*tert*-butylferulate (**53**); in the second, two molecules of (**53**) are linked by oxidative coupling (this being the critical step) to yield (**54**); and in the third, reduction of the

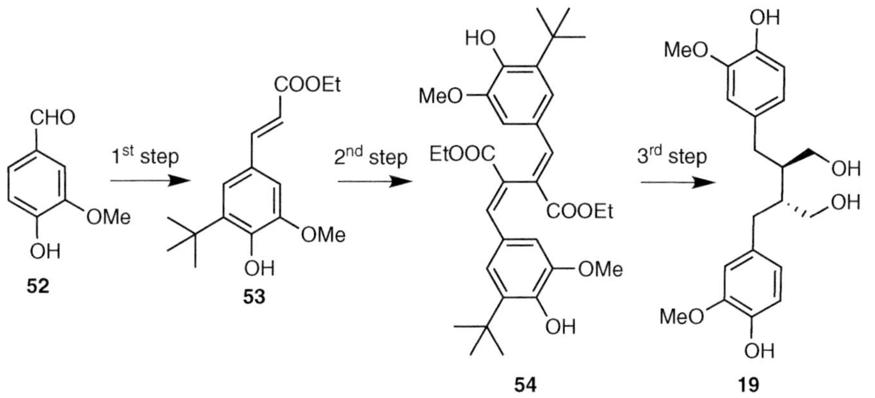

Figure 10.22 Synthesis of secoisolariciresinol of Wang et al. [44, 196]

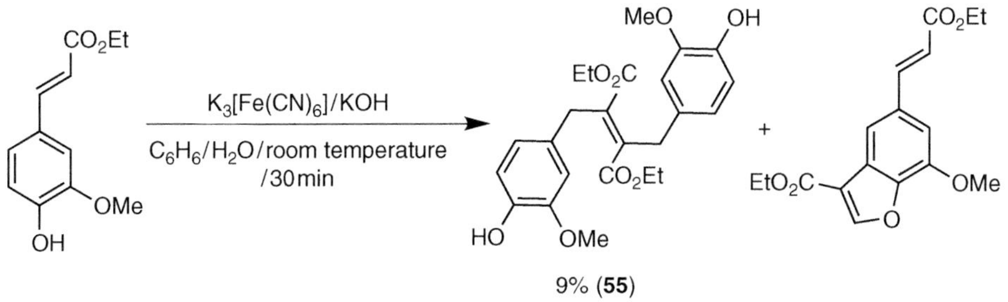

Figure 10.23 Direct oxidative coupling of ethyl ferulate [196]

unsaturated link, removal of the *tert*-butyl groups, and final reduction of the esters give secoisolariciresinol (**19**) (see Figure 10.22).

The direct oxidative coupling of ethyl ferulate gives only a 9% yield of the desired secoisolariciresinol precursor (**55**), the main product, benzofurane, being indicated in Figure 10.23.

Recently, Morreel *et al.* [208] described the synthesis of secoisolariciresinol starting from feluric acid,[4] the radical coupling gives an 8,8′-dilactone (**56**), which is easily transformed into secoisolariciresinol (see Figure 10.24).

Secoisolariciresinol can also be produced by chemical modification of other lignans such as matairesinol (**21**), a dibenzylbutyrolactone lignan. Reduction of **19** with LiAlH$_4$ gives secoisolariciresinol. Similarly, hydroxysecoisolariciresinol can be obtained from hydroxymatairesinol (see Figures 10.27 and 10.28 in Section 10.7.3.1) [138, 209–211].

Enzymes such as pinoresinol/lariciresinol reductase (PLR) have also been used to catalyze the conversion of lignans into secoisolariciresinol (see Figure 10.25) [212, 213].

10.7.3 Dibenzylbutyrolactone Lignans

Trazzi *et al.* [214] showed the relevant role of the butyrolactones as key intermediates in the synthesis of other lignans such as aryltetralin lignan or dibenzylcyclooctadiene lignan (see Figure 10.26).

---
[4] The most abundant hydroxycinnamic acid found in plants.

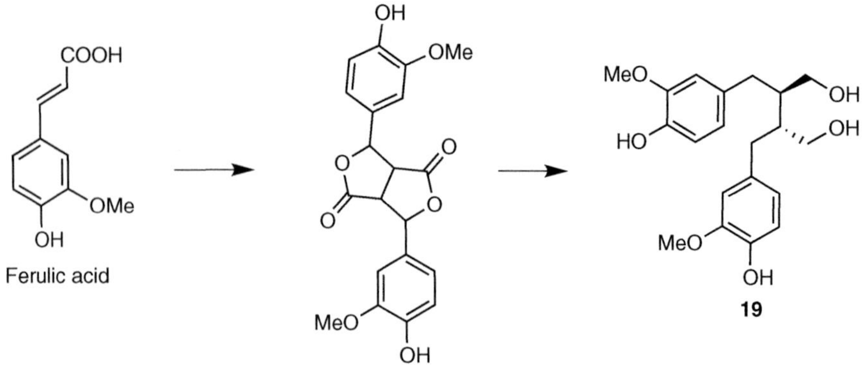

Figure 10.24 Synthesis of secoisolariciresinol of Morreel et al. from ferulic acid [208]

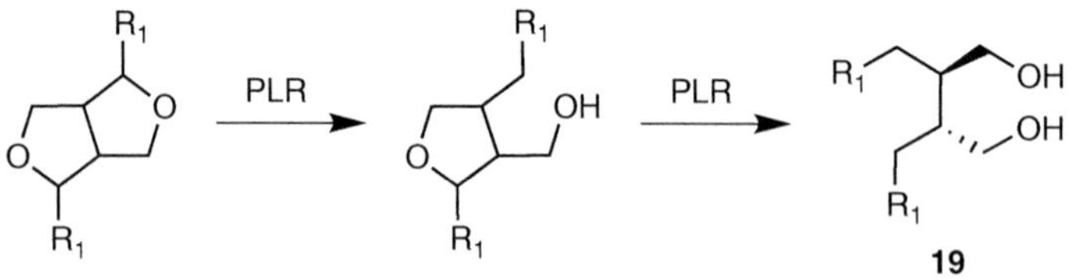

Figure 10.25 Synthesis of secoisolariciresinol from pinoresinol mediated by PLR enzyme. See Figure 10.3 for the meaning of R groups

10.7.3.1 Matairesinol

Mäkelä et al. [215] described the synthesis of matairesinol (**21**) *via* the classical pathway using a Michael conjugated addition with butenolide (see Figure 10.27).

Fischer et al. [216] synthesized hydroxymatairesinol (**57**) by following a radical carboxyarylation approach (see Figure 10.28).

10.7.4 Cyclolignans (Aryltetralin Lignans)

The most prominent member of the aryltetralin lignans is PPT, isolated from plant species of the genus *Podophyllum* and which presents remarkable medicinal properties.

10.7.4.1 Podophyllotoxin

Since its isolation by Hartwell et al. [217] in 1953, PPT (**15**) [65, 218] and its isomers have been the subject of numerous synthetic endeavors [19, 49, 219–223]. Traditionally, it has been isolated from podophyllin (resin of *Podophyllum rhizome*), reaching a content in podophyllotoxin of 4–5% (dry weight) in *Podophyllum emodi* [224, 225]. It has also been found in other types of plants [226], but its natural resources are insufficient and limited to meet pharmaceutical demand [65].

The key challenge for the synthesis of PPT is to establish the 1,2-*cis*-stereochemistry together with a *trans*-lactone ring fusion [223]. Another challenge is to deal with its easy epimerization, in basic media, to picro-PPT. Because of this epimerization, only 3% of the PPT exists when equilibrium is

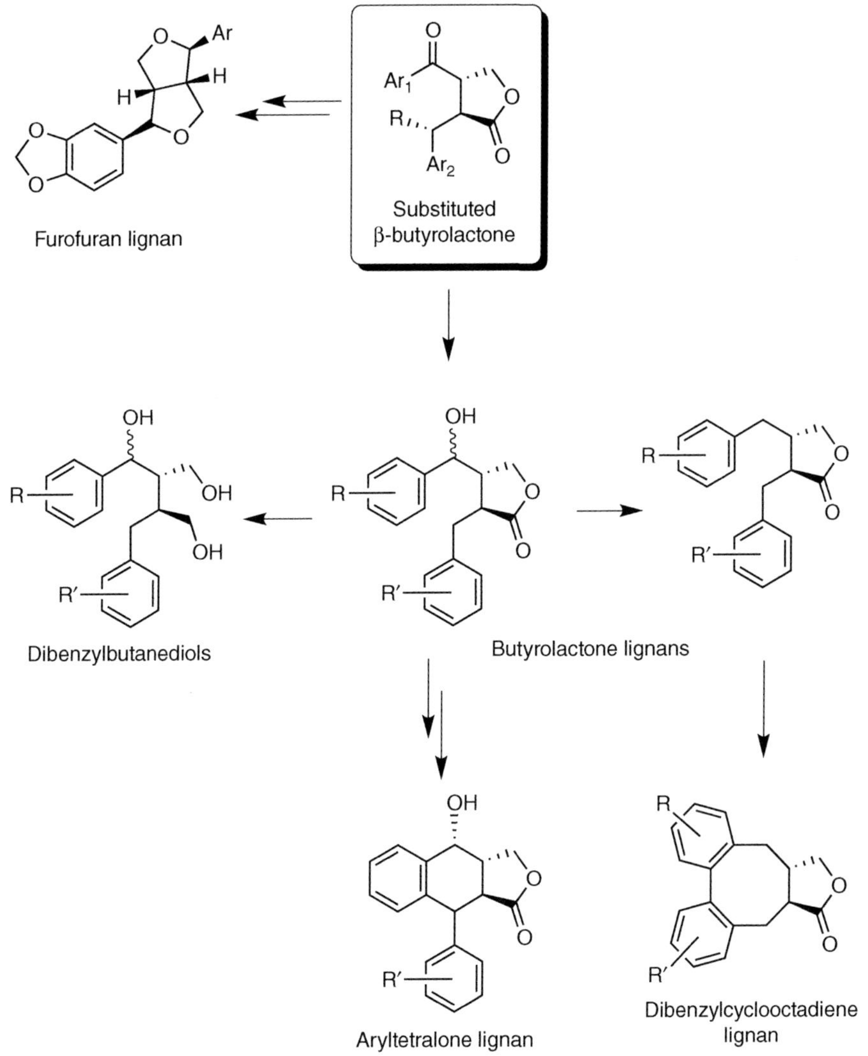

Structure and Chemical Properties of Lignans 339

Figure 10.26 *Butyrolactones as key intermediates in the synthesis of lignans [214]*

reached. In 1966, Gensler and Gatsonis [227] designed a procedure to isolate PPT in a reasonable yield using several crystallizations (see Figure 10.29).

For the construction of the aryltetralin ring in the chemical synthesis of PPT derivatives, four general approaches have been developed (see Figure 10.30): (a) The oxo-ester pathway [228]; (b) the dihydroxy acid pathway [229]; (c) the pathway based in a Diels–Alder reaction [230]; and (d) the tandem conjugate addition pathway [231]. Full chemical synthesis is not an option from the commercial standpoint, due to the presence of four chiral positions and the *trans*-γ-lactonic ring.

The most common approach used for the synthesis of aryltetralin lactone skeleton is an electrophilic aromatic substitution [220, 232, 233] as shown in Figure 10.31.

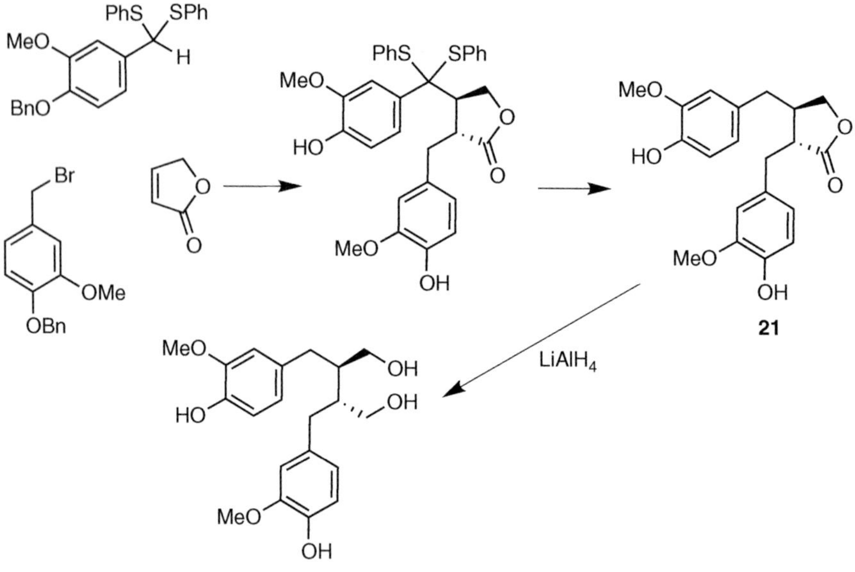

Figure 10.27 Synthesis of matairesinol (**21**) and secoisolariciresinol (**19**) [215]

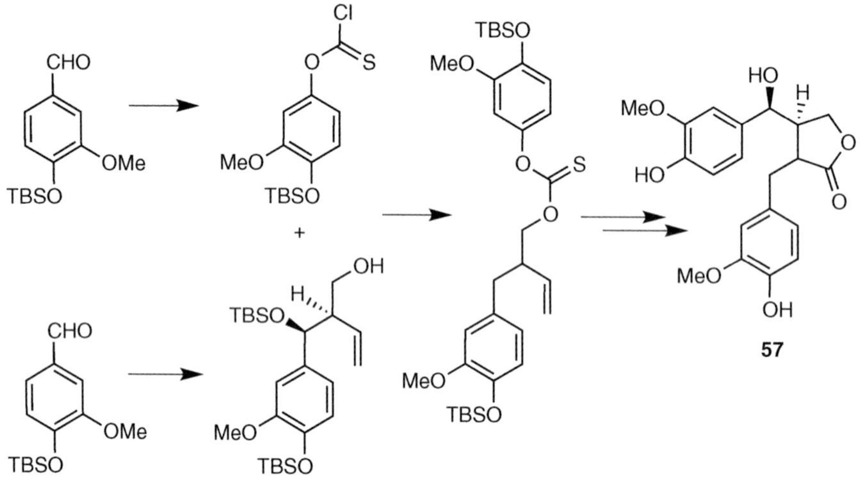

Figure 10.28 Synthesis of hydroxymatairesinol (**57**) [216]

The other principal strategy used is the Diels–Alder reaction [234–238] (Figure 10.32).

Alternatively, PPT can be synthesized by the use of a Michael-induced ring closure (i.e., in the reaction of a tungsten–carbene with a chiral epoxide) [228, 239] (see Figure 10.33) or an intramolecular Heck reaction from an iodine derivative [240] (see Figure 10.34).

Using a chiral auxiliary is another synthetic pathway that has been performed to obtain PPT with high stereoselectivity [190, 191, 194, 241–245]. Yet another strategy is the so-called biomimetic

Structure and Chemical Properties of Lignans 341

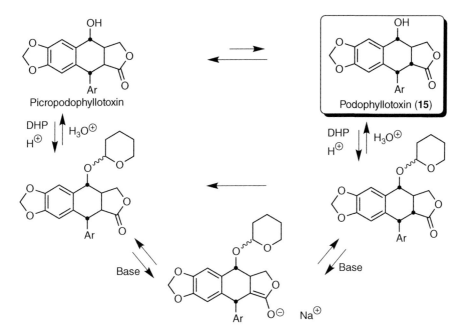

Figure 10.29 Isomerization reactions for podophyllotoxin [227]

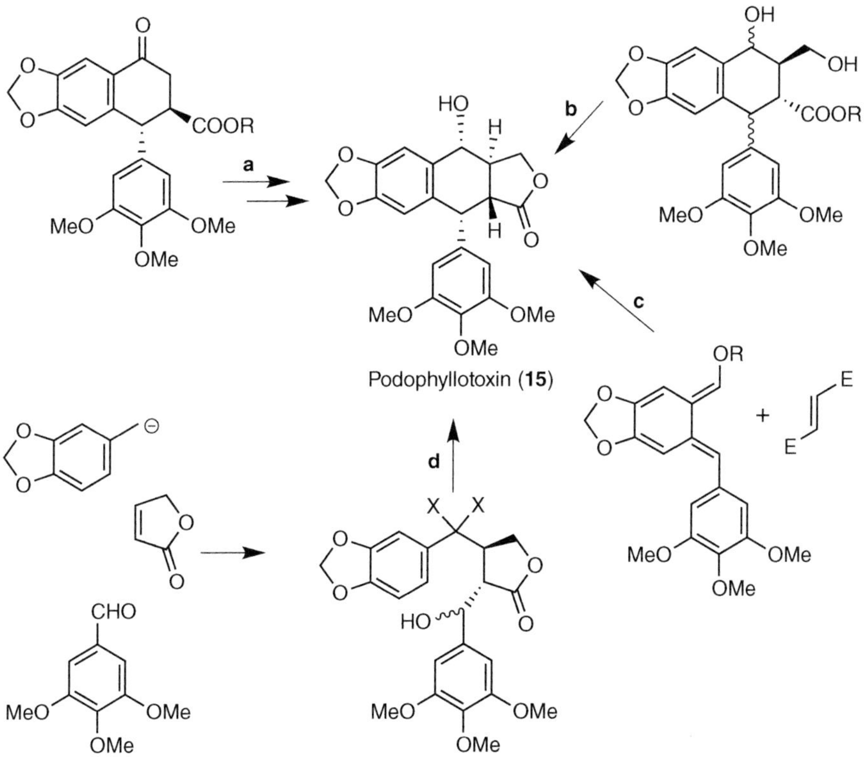

Figure 10.30 Four main pathways to the synthesis of podophyllotoxin. (a) The oxo-ester pathway; (b) the dihydroxy acid pathway; (c) Diels–Alder pathway, and (d) the tandem conjugate addition pathway [49]

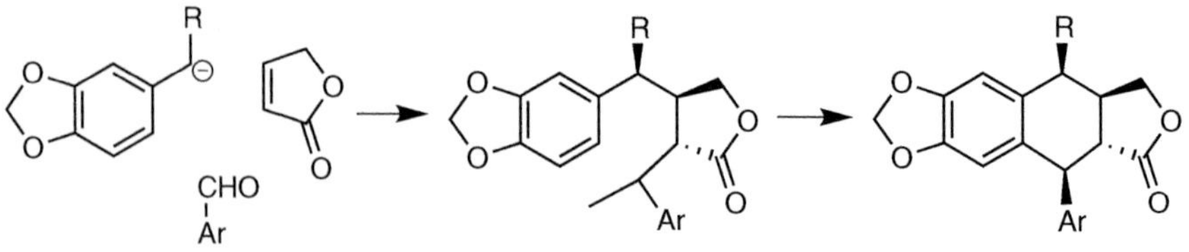

Figure 10.31 Synthesis of aryltetralin lactone

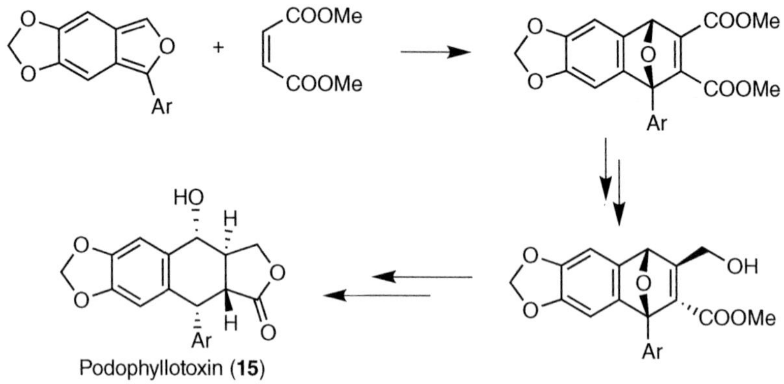

Figure 10.32 Synthesis of podophyllotoxin via Diels–Alder reaction

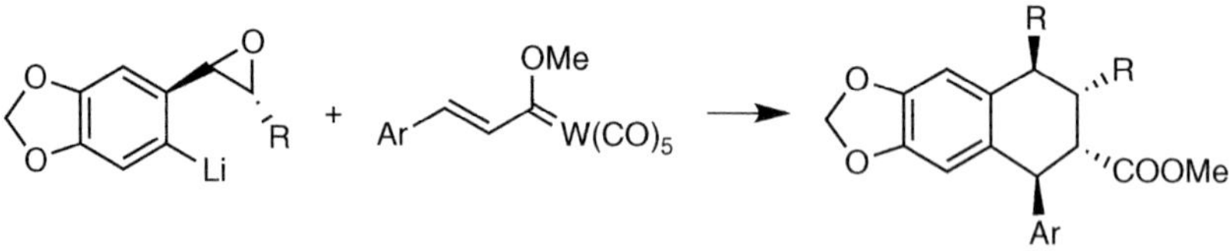

Podophyllotoxin (**15**)

Figure 10.33 Synthesis of podophyllotoxin by using a Michael-induced ring closure [228, 239]

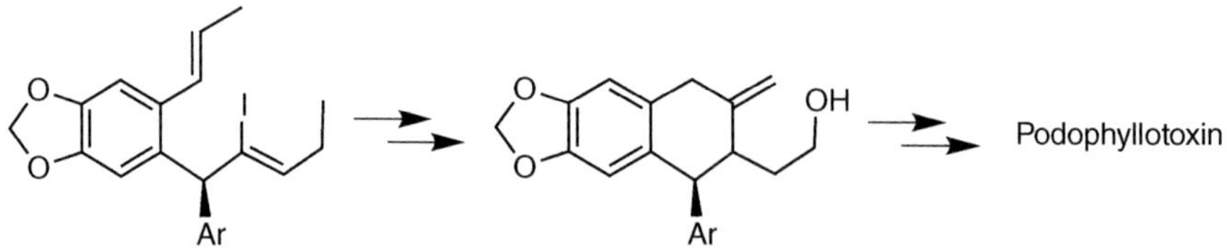

Podophyllotoxin

Figure 10.34 Synthesis of podophyllotoxin using an intramolecular Heck reaction [240]

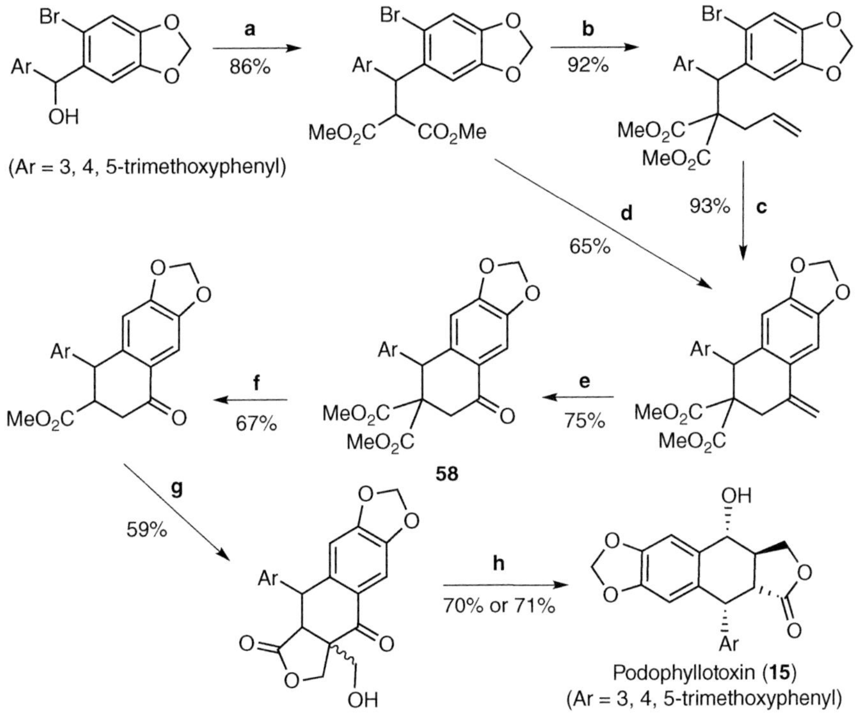

Figure 10.35 Synthesis of podophyllotoxin in nine steps [252]. a: (1) Ac_2O, Et_3N, DMAP cat., CH_2Cl_2; (2) dimethylmalonate, $TiCl_3$, toluene; b: $Pd(OAc)_2$, dppe, NaH, allyl acetate, DMF; c: same as b and emphn-Bu_4NOAc; d: same as c and NaH; e: (1) NMO, $OsCl_3$, THF/H_2O; (2) $NaIO_4$, acetone/H_2O; f: (1) NaOH, MeOH; (2) descarboxylation; g: NaOH; h: Zinc borohydride, then Gensler enolate quenching procedure or Jones reagent, then acid workup

strategy, an oxidative phenol coupling reaction that is mimetic to the reaction used by cells for the biosynthesis of lignans [218]. This reaction has been performed in two ways:

- Metal-mediated oxidative phenol coupling.
- Enzyme-catalyzed oxidative phenol coupling.

The oxidative phenol coupling has been catalyzed with different metal salts: silver oxide [246, 247], MnO_2 [248], and $FeCl_3$ [249], yielding different results. The oxidative phenol coupling catalyzed with different peroxidases has also been employed to obtain aryltetralin lignans with different stereochemistry [250, 251].

Mingoia et al. [252] successfully synthesized podophyllotoxin in nine steps. A pseudodomino sequence is employed, involving a successive intermolecular Pt-catalyzed allylic alkylation and an intramolecular Mizoroki–Heck coupling (see Figure 10.35) toward the intermediate ketone **58**. This ketone **58** could be converted to (±)-PPT **15** in three steps [253, 254].

10.7.5 Dibenzocyclooctadiene Lignans

Approximately 100 lignan derivatives possessing this skeleton have been isolated from the Schizandraceae family of flowering plants [9, 255].

Among these, the most complex structure is the analogous steganicin one, since it contains both a cyclooctadiene and a lactonic ring.

The core structure of dibenzocyclooctadiene lignan derivatives (($\pm$)-steganone) can be obtained by two major pathways: cyclization of biphenyl compound and 1,4-diaryl compound (see Figure 10.36) [17].

These strategies differ in the ordering of the biaryl coupling and eight-membered ring closure steps. Pathway A uses an intermolecular biaryl coupling before eight-membered ring closure, whereas in pathway B the two aryl groups are coupled in an intramolecular reaction to produce the eight-membered ring.

10.7.5.1 Pathway A: Cyclization of Biphenyl Compounds

Pathway A uses an intermolecular biaryl coupling before eight-membered ring closure.

Intermolecular Biphenyl Coupling Reactions. The Ullmann and Bielecki [256] reaction is a classical method to synthesize biphenyl derivatives. The coupling reaction of aryl halides in the presence of active copper provides the symmetrical biphenyl derivative [257] (see Figure 10.37).

During the 1970s and 1980s, some new methods for aryl–aryl bond formation were discovered, as the Kharasch et al. [258, 259], Negishi et al. [260, 261], Stille [262], and Suzuki [263] reactions. These coupling reactions are used to synthesize various nonsymmetrical biphenyl derivatives in the presence of different metal complexes as catalysts.

In the Kharasch coupling reaction [258], a Grignard reagent Ar_1MgX is reacted with an aryl halide Ar_2X catalyzed by Ni or Pd complexes to yield a biphenyl compound. Aryl halides substituted with electron-withdrawing groups, such as RCdO, COOR, or NO_2, fail to react with the Grignard reagent.

In the related Negishi reaction, aryl zinc reagents are coupled with an aryl halide or aryl triflate also catalyzed by Ni(0) or Pd(0). Because the aryl zinc complex is a milder reagent, many functional groups, such as RCdO, COOR, NO_2, and CN, on the substrates are not deleterious to the coupling reaction.

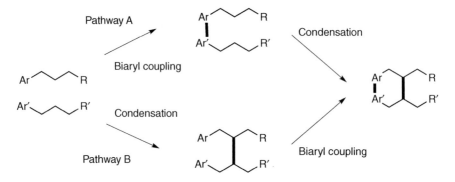

Figure 10.36 The two major strategies for the synthesis of dibenzocyclooctadiene lignan derivatives

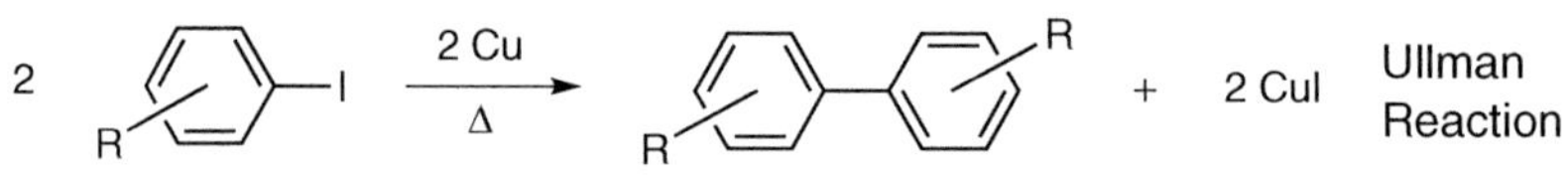

Figure 10.37 Ullmann reaction of biphenyl coupling

Structure and Chemical Properties of Lignans 345

In the Stille reaction, an aryl tin reagent is used as the aryl metal for the coupling reaction. The neutral conditions in the Stille reaction can be applied to a wide range of substrates having a variety of functional groups. However, the organotin reagents and tin by-products are quite toxic.

Similar to the Stille reaction, the Suzuki one has been widely utilized for the synthesis of natural products. In this coupling reaction, aryl-boronic acid reacts with an aryl halide or aryl triflate in the presence of a Pd complex (i.e., $Pd(PPh_3)_4$). Typically, an aqueous solution of a weak base, such as Na_2CO_3, K_3PO_4, or $Ba(OH)_2$, is used as the reaction solvent. The Suzuki reaction gives high yields of biphenyl derivatives, even with highly substituted substrates.

Figure 10.38 shows four different synthetic pathways (a–d) to the precursors of steganone (**42**) using the pathway A.

In pathway "a," developed by Larson and Raphael [264], the key biphenyl intermediate was obtained through a Negishi reaction. In pathway "b," developed by Monovich et al. [265] and Uemura et al. [266], the key biphenyl intermediate was obtained through a Suzuki reaction. In pathway "c," developed by Robin et al. [267, 268], the key biphenyl intermediate was obtained through an aldol condensation. In pathway "d," developed by Ziegler et al. [269] and by Robin et al. [267], the key biphenyl intermediate was obtained through a standard Ullman and Bielecki [256] coupling reaction.

Cyclization of Biphenyl Derivatives. These compounds can be cyclized by various intramolecular coupling strategies to close the eight-membered ring of lignan structures. Therefore, Carroll et al. [270] have utilized $TiCl_4$ and Mg–Hg for the intramolecular coupling of a biphenyl ketone as a viable synthesis of the benzocyclooctadiene ring system of lignans. Molander et al. [271–273], Curran et al. [274], and Kunishima et al. [275] have recently used the Sm(II) diyodide as a reagent for the reductive ketyl-olefin cyclization of a wide variety of carbocyclic skeletons. The cyclation reactions depicted in Figure 10.38 are discussed in the synthesis of steganone (see section 10.7.5.3).

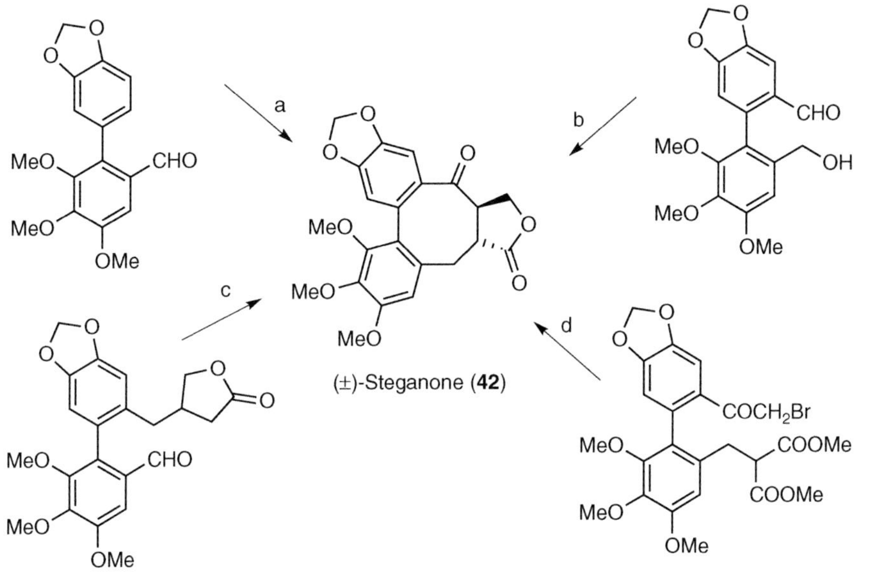

Figure 10.38 Four different synthetic pathways to the synthesis of (±)-steganone (**42**) by pathway A strategies

10.7.5.2 Pathway B: Intramolecular Biphenyl Coupling Reactions

Pathway B uses an intermolecular condensation before biaryl coupling to the eight-membered ring. It is believed that in nature the enzyme that catalyzes the formation of lignans proceeds through an oxidative coupling process from the phenyl groups of acyclic lignans [276–281]. This pathway has stimulated researchers to consider a biomimetic strategy for the synthesis of these type of lignans (see Figure 10.39). The first step in the pathway entails the synthesis of acyclic dibenzylbutane lignans.

Synthesis of Acyclic Lignans. Despite the different synthetic approaches considered so far, it is important to stress the different available options depending upon the nature of the acyclic lignan (symmetric or asymmetric one).

The dibenzylbutane lignans can be synthesized from two arylpropane units by different strategies:

By an oxidative coupling reaction [282] or a Grignard coupling [168, 283], the stereochemistry of the product can be controlled depending on the reduction conditions [282] (see Figure 10.40).

The McMurry [284] reaction is a reductive coupling of carbonyl groups mediated by Ti, which has also been used for the synthesis of acyclic lignans [285, 286] (see Figure 10.41).

A more general synthetic strategy, developed by Minato et al. [207], for the synthesis of nonsymmetrical lignans is depicted in Figure 10.42.

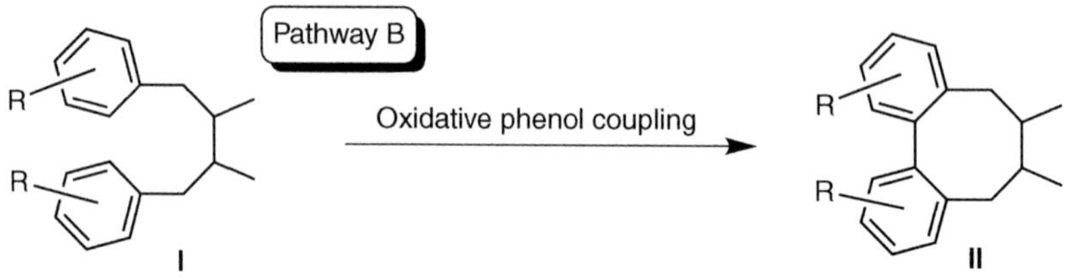

Figure 10.39 Intramolecular coupling of eight-membered ring

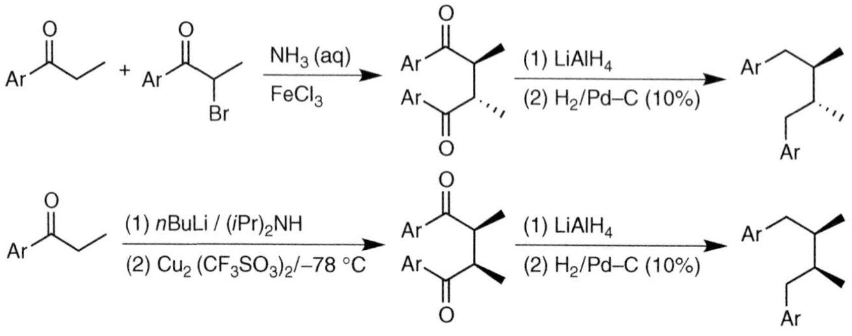

Figure 10.40 Synthesis of dibenzylbutane lignans [168, 282, 283]

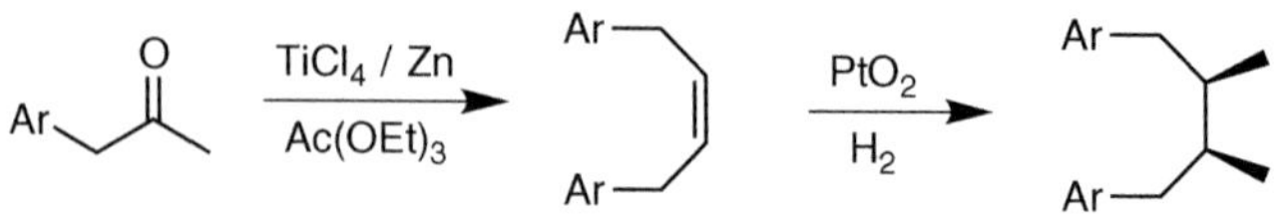

Figure 10.41 McMurry reaction for the coupling of two carbonyl groups mediated by Ti [285, 286]

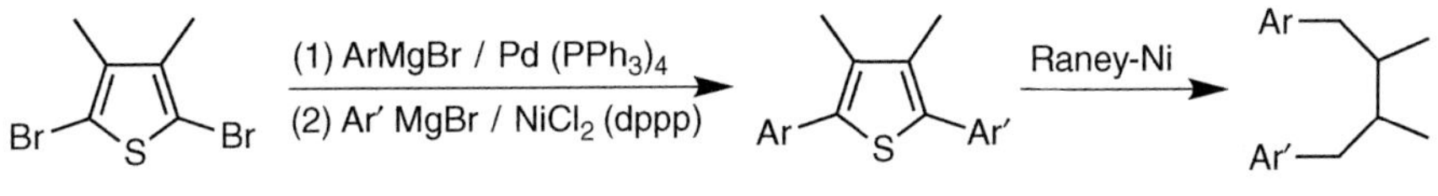

Figure 10.42 Synthesis of nonsymmetrical lignans using the strategy of Minato et al. [207]

Nonsymmetrical acyclic lignans have also been obtained by Krauss *et al.* [287–291] using an acyclic anion equivalent, or by means of the well-developed samarium Grignard reaction of Takeya *et al.* [292], as illustrated in Figures 10.43 and 10.44.

As commented earlier, the dibenzylbutyrolactone lignans have been used to synthesize many types of lignans. In particular, the dibenzylbutyrolactone series of lignans, as **62** and **63**, has been employed as precursors for the synthesis of the stegacin series of lignans. The key methodology for the synthesis of unsymmetrical dibenzyl-γ-lactone lignans is shown in Figure 10.45. As also discussed earlier, the most commonly used synthetic pathways to the benzylbutyrolactones as **61** are by benzylation [175, 221, 267] of the lactone **59**, or by hydrogenation [150, 167–169] of **60** and subsequent lactonization. A new benzylation of **61** gives **62**, or by condensation gives **63**.

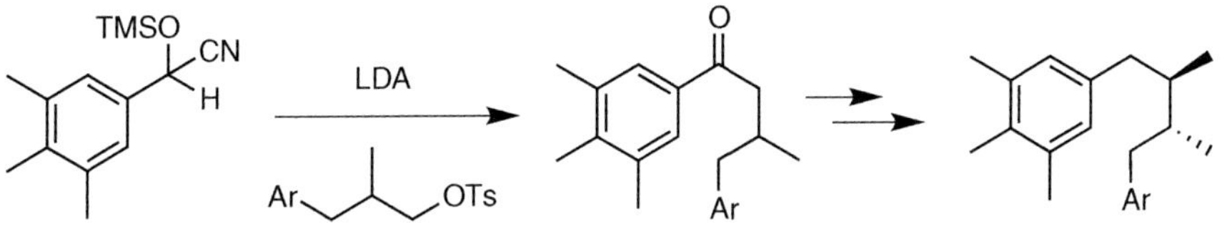

Figure 10.43 Synthesis of nonsymmetrical diaryl lignans using the strategy of Krauss et al. (acyl anion equivalent) [287–291]

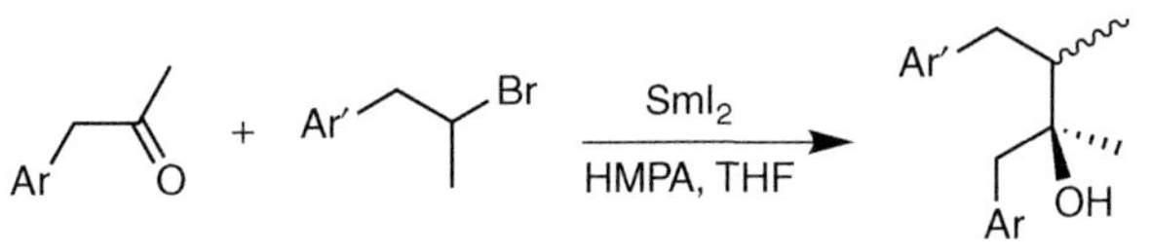

Figure 10.44 Synthesis of nonsymmetrical diaryl lignans using the strategy of Takeya et al. [292] (SmI_2)

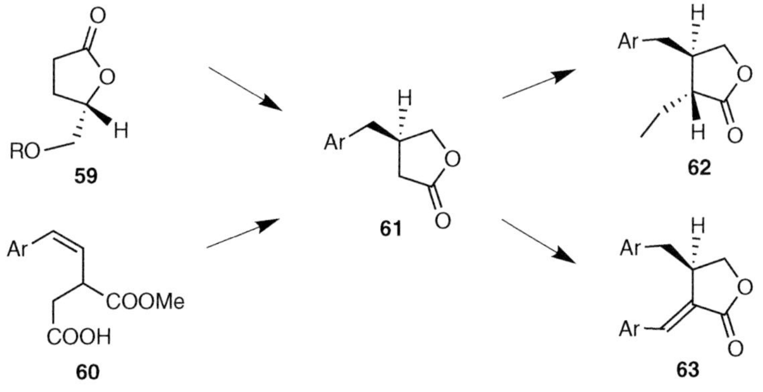

Figure 10.45 Synthesis of dibenzylbutyrolactone lignans

Another strategy is the coupling of a dianion with two aryl-substituted electrophiles [293, 294] (see Figure 10.46).

Intramolecular biaryl oxidative coupling. The intramolecular aryl oxidative coupling reaction is the key step in the synthesis of lignans following a biomimetic strategy. The reagents used are critical for the success of the reaction. Most of the older coupling reagents are complicated by side reactions. Magnus *et al.* [295] employed Tl(III) trifluoroacetate (TTFA), an efficient nonphenol oxidative coupling reagent [296]) and used a Simmons–Smith ring expansion reaction for the total synthesis of steganone (see Figure 10.47).

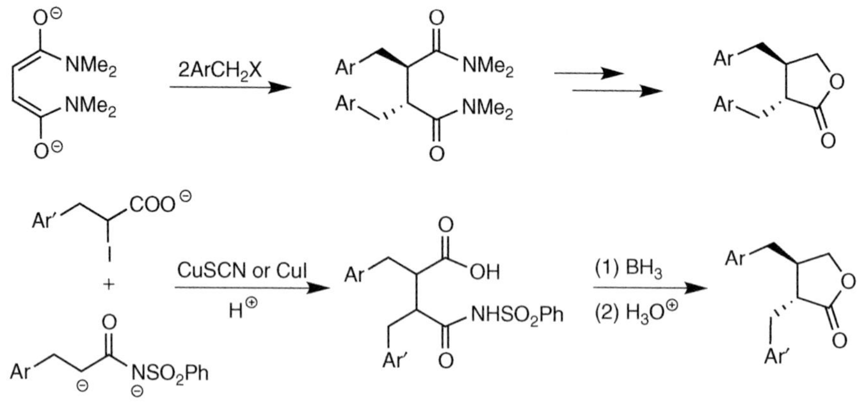

Figure 10.46 Synthesis of dibenzylbutyrolactone lignans using the strategy of Belletire et al. [293]

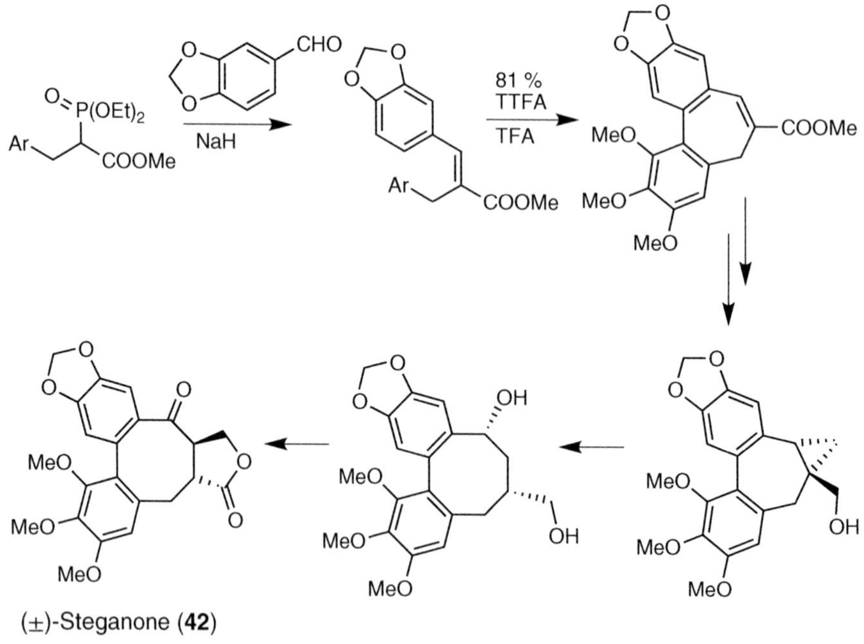

Figure 10.47 Total synthesis of steganone by Magnus et al. [295]

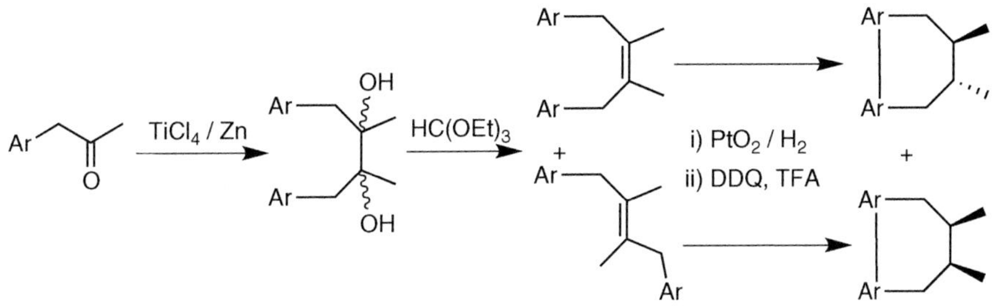

Figure 10.48 Strategy for the synthesis of a series of lignans type wuweizi by Chang et al. [298, 299].

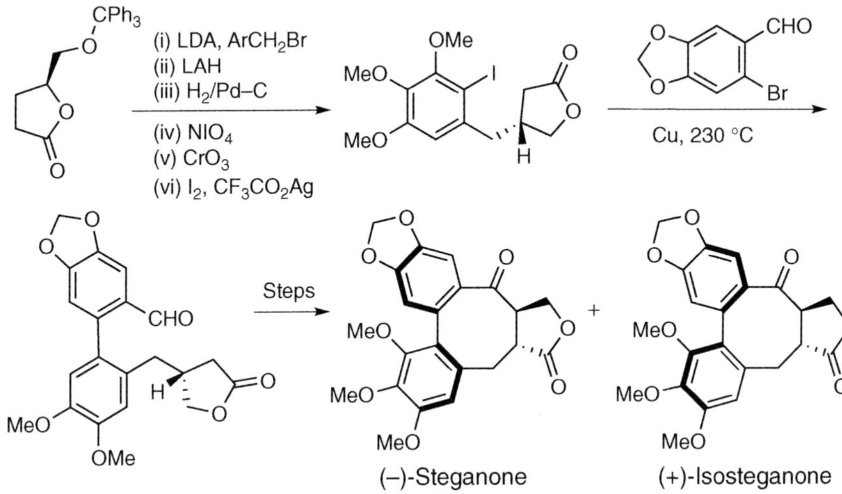

Figure 10.49 Synthesis of (−)-steganone and (+)-isosteganone by Robin et al. [267]

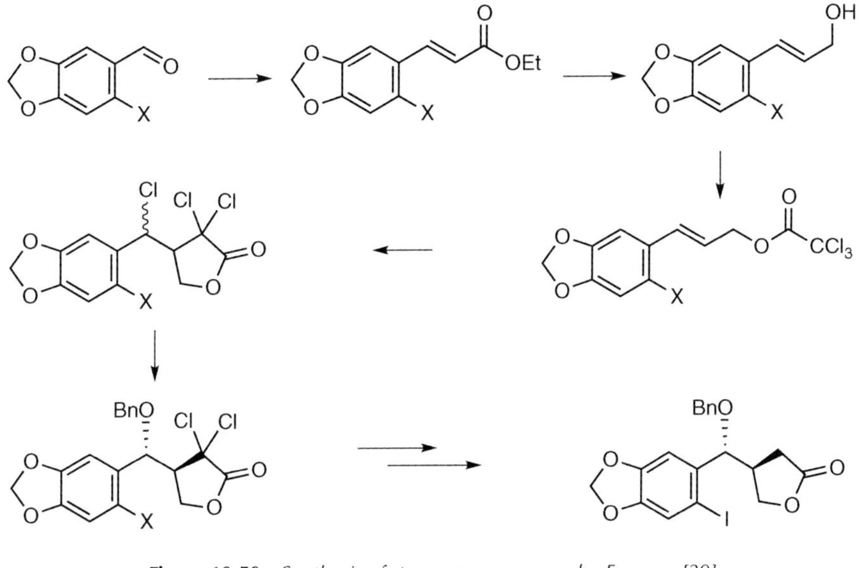

Figure 10.50 Synthesis of steganone precursor by Faruque [29]

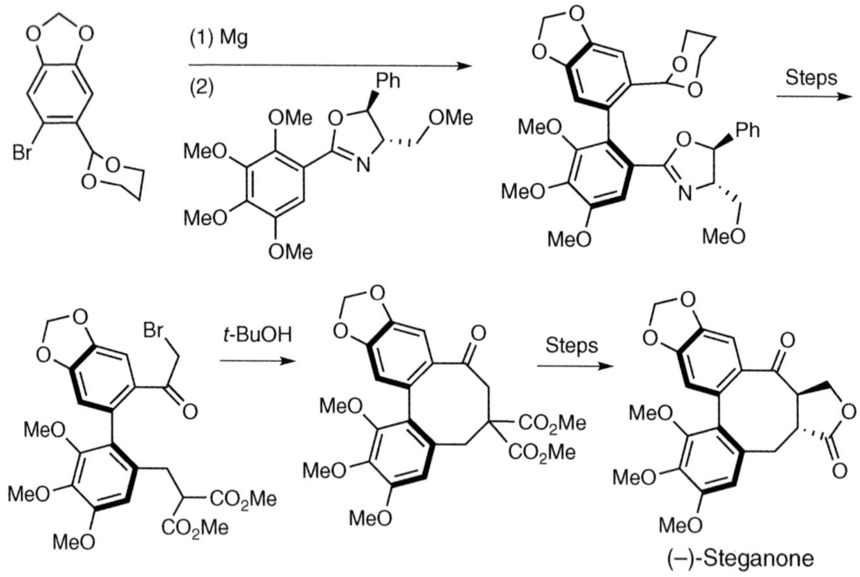

Figure 10.51 Asymmetric synthesis of (−)-steganone by Meyers et al. [311]

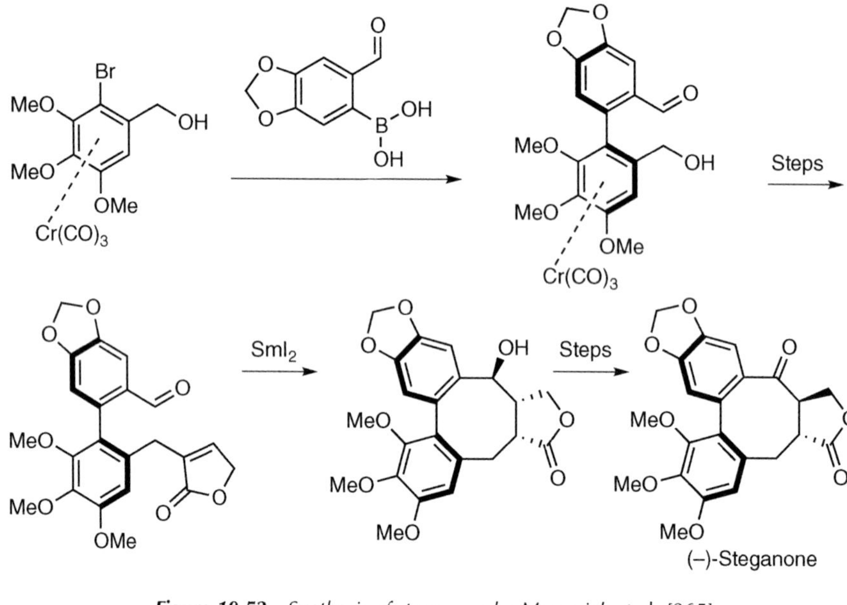

Figure 10.52 Synthesis of steganone by Monovich et al. [265]

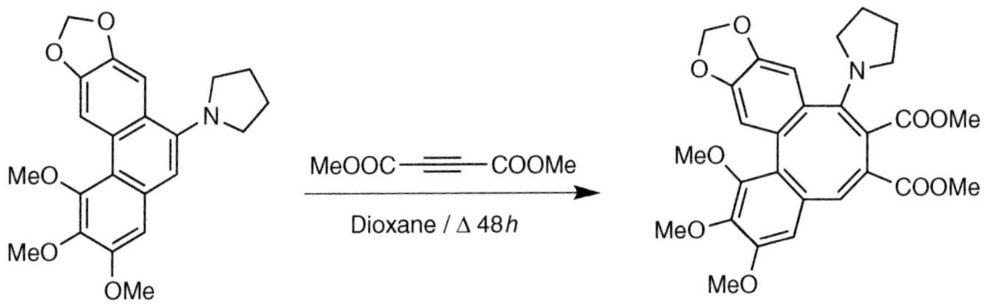

Figure 10.53 Synthesis of steganone precursor by Hughes and Raphael [309]

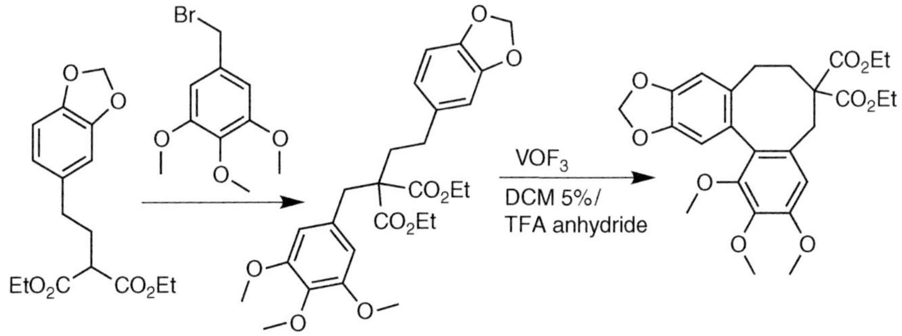

Figure 10.54 Synthesis of steganone precursor by Kende and Liebeskind [312]

The first nonphenol oxidative coupling reagent that was not a transition metal was DDQ, and it was used in the synthesis of a series of lignans type wuweizi by Chen *et al.* [297], and Chang *et al.* [298–302] (see Figure 10.48).

10.7.5.3 Asymmetric Total Synthesis of Steganone and Analogs

Dibenzocyclooctadiene lignans presented in nature contain several asymmetric centers and an axial chiral biaryl moiety, which is configurationally stable. The absolute configuration of the biphenyl units and the conformation of the lactone ring in the steganacin and its analogs have been found to be related to antitumor activity by Kupchan *et al.* [303, 304], Hicks and Sneden [305], and Tomioka *et al.* [306].

To date, nine successful total syntheses of steganone have been reported. Of these, six afforded racemic material with three generating (−)-steganone [29, 266, 307–315].

The different methods reported for the asymmetric synthesis of steganone can be classified into four general approaches:

1. Diastereoselective alkylation of chiral butyrolactones.
   Tomioka *et al.* [174, 316] and Tanaka *et al.* used this approach to synthesize a variety of steganacin and schizandrin [165–170] analogs. Robin *et al.* [267] described the total synthesis of (−)-steganone and (+)-isosteganone from a chiral butyrolactone (see Figure 10.49). This synthesis differs from other pathways by the moment in which the two aryl groups are introduced onto the lactone core structure. Recently, Faruque [29] published a new synthesis of the same iodine precursor (Figure 10.50).

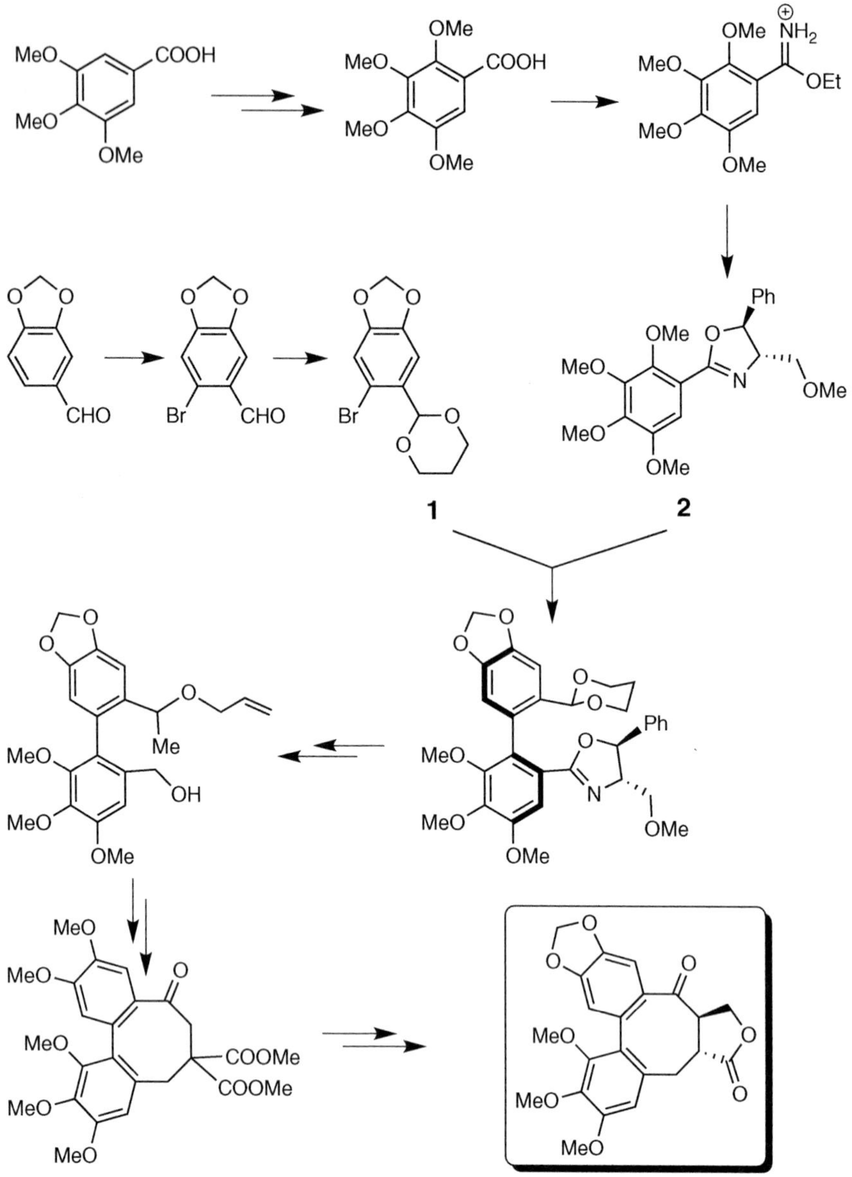

Figure 10.55 *Total synthesis of steganone by Meyers et al. [311]*

2. Diastereoselective addition to chiral butenolactones.
   Tomioka *et al.* [175] and Pelter *et al.* [181] utilized this approach to synthesize a series of isostegane analogs.
3. Use of chiral oxazolines.
   Meyers *et al.* [311] reported the total synthesis of (−)-steganone using an oxazoline chiral auxiliary in the first step, a Kharasch coupling reaction (see Figure 10.51).

4. Use of chromium tricarbonyl complexes.

   The aromatic chromium tricarbonyl complexes are organometallic compounds that present a planar chirality based on a geometry known as "piano stool," due to the planar arrangement of the aryl group and the three CO ligands as "legs" on the chromium-bond axis [317]. Monovich *et al.* [265] completed an enantioselective synthesis of (−)-steganone using a chiral chromium tricarbonyl complex. The key step of this synthesis is based on a Suzuki coupling with an aryl-boronic acid that yields a chiral biaryl complex. The complex obtained in such a way is dissociated by a photooxidation reaction before the cyclation step with SmI_2 (see Figure 10.52).

   Others synthesis of steganone precursor has been performed by Hughes and Raphael [309] (see Figure 10.53), Kende and Liebeskind [312] (see Figure 10.54) and a total synthesis of steganone by Meyers *et al.* [311] (see Figure 10.55).

References

[1] Robinson R. The relationship of some complex natural products to the simple and amino acids. Proc Univ Durham Philos Soc. 1928;8(Part 1):14–59.

[2] Haworth RD. Constituents of natural phenolic resins. Nature. 1941;147:255–257.

[3] Haworth RD. The chemistry of the lignan group of natural products. J Chem Soc. 1942: 448–456.

[4] Umezawa T. Diversity in lignan biosynthesis. Phytochem Rev. 2003;2(3):371–390.

[5] Milder IEJ, Feskens EJM, Arts ICW, Bueno-de-Mesquita HB, Hollman PCH, Kromhout D. Intake of the plant lignans secoisolariciresinol, matairesinol, lariciresinol, and pinoresinol in Dutch men and women. J Nutr. 2005;135(5):1202–1207.

[6] Penalvo JL, Adlercreutz H, Uehara M, Ristimaki A, Watanabe S. Lignan content of selected foods from Japan. J Agric Food Chem. 2008;56(2):401–409.

[7] Harmatha J, Dinan L. Biological activities of lignans and stilbenoids associated with plant-insect chemical interactions. Phytochem Rev. 2003;2(3):321–330.

[8] Schroeder FC, Del Campo ML, Grant JB, Weibel DB, Smedley SR, Bolton KL, et al. Pinoresinol: a lignol of plant origin serving for defense in a caterpillar. Proc Natl Acad Sci USA. 2006;103(42):15497–15501.

[9] Ayres DC, Loike JD. Lignans: Chemical, Biological and Clinical Properties, Chemistry and Pharmacology of Natural Products. Cambridge: Cambridge University Press; 1990.

[10] Deyama T, Nishibe S. Pharmacological properties of lignans. In: Heitner C, Dimmel D, Schmidt J, editors. Lignin and Lignans: Advances in Chemistry. CRC Press; 2010. p. 585–629.

[11] Begum S, Sahai M, Ray A. Non-conventional lignans: coumarinolignans, flavonolignans, and stilbenolignans. In: Kinghorn AD, Falk H, Kobayashi J, editors. Progress in the Chemistry of Organic Natural Products, Fortschritte der Chemie organischer Naturstoffe/Progress in the Chemistry of Organic Natural Products. vol. 93. Vienna: Springer-Verlag GmbH; 2010. p. 1–70.

[12] Cunha WR, Andrade e Silva ML, Sola Veneziani RC, Ambrósio SR, Kenupp Bastos J. Lignans: chemical and biological properties. In: Rao V, editor. Phytochemicals - A Global Perspective of their Role in Nutrition and Health. vol. 33. InTech; 2012. p. 213–234.

[13] Hwang DY. Therapeutic effects of lignans and blend isolated from Schisandra chinesis on hepatic carcinoma. In: Kuang H, editor. Recent Advances in Theories and Practice of Chinese Medicine. InTech; 2012. p. 390–406.

[14] Lewis NG, Davin LB, Sarkanen S. Lignin and lignan biosynthesis: distinctions and reconciliations. In: Lewis NG, Sarkanen S, editors. Lignin and Lignan Biosynthesis, ACS Symposium Series. vol. 697. American Chemical Society; 1998. p. 1–27.

[15] Del Signore G, Berner OM, ur Rahman A. In: Recent Developments in the Asymmetric Synthesis of Lignans. vol. 33, Part M. Elsevier; 2006. p. 541–600.
[16] Saleem M, Hyoung JK, Ali MS, Yong SL. An update on bioactive plant lignans. Nat Prod Rep. 2005;22(6):696–716.
[17] Chang J, Reiner J, Xie J. Progress on the chemistry of dibenzocyclooctadiene lignans. Chem Rev. 2005;105(12):4581–4609.
[18] Apers S, Vlietinck A, Pieters L. Lignans and neolignans as lead compounds. Phytochem Rev. 2004;2(3):201–207.
[19] Ward RS. Different strategies for the chemical synthesis of lignans. Phytochem Rev. 2003;2(3):391–400.
[20] Ward RS. Lignans, neolignans, and related compounds. Nat Prod Rep. 1999;16(1):75–96.
[21] Ward RS. Lignans, neolignans, and related compounds. Nat Prod Rep. 1997;14(1):43–74.
[22] Ward RS. Lignans, neolignans, and related compounds. Nat Prod Rep. 1995;12(2):183–205.
[23] Ward RS. Lignans, neolignans, and related compounds. Nat Prod Rep. 1993;10(1):1–28.
[24] MacRae WD, Towers GHN. Biological activities of lignans. Phytochemistry. 1984;23(6):1207–1220.
[25] Raffaelli B. Synthesis of Lignano-9,9'-Lactones and Rearrangement Studies. Helsinki: University of Helsinki; 2012.
[26] Nemes SM. Practical Methods for Lignans Quantification. Montreal: McGill University; 2012.
[27] Sullivan R. Régulation transcriptionnelle de la biosynthèse des lignanes du lin (*Linum usitatissimum* et *Linum flavum*) et amélioration de l'extraction des lignanes. Orleans: Université d'Orléans; 2011.
[28] Sedlaák É. Separation, Identification and Quantification of Lignans in Native Plant Samples and Enhancement of Lignan Production in *Forsythia in vitro* Cell Cultures. Budapest: Semmelweis University; 2011.
[29] Faruque J. Towards Total Synthesis of Steganone *via* Copper Mediated Atom Transfer Radical Cyclisation. Manchester: The University of Manchester; 2010.
[30] Mahendra KC. Bioactivity and Bioavailability of Lignans from Sesame (*Sesamum indicum* L.). Mysore: Mysore University; 2009.
[31] Fuß E. Biosynthesis of Lignans. Düsseldorf: University of Düsseldorf; 2007.
[32] Suzuki S, Umezawa T. Biosynthesis of lignans and norlignans. J Wood Sci. 2007;53:273–284.
[33] Macías FA, López A, Varela RM, Torres A, Molinillo JMG. Bioactive lignans from a cultivar of *Helianthus annuus*. J Agric Food Chem. 2004;52(21):6443–6447.
[34] Gottlieb OR. Neolignans. In: Herz W, Grisebach H, Kirby GW, editors. Progress in the Chemistry of Organic Natural Products. vol. 35. Vienna: Springer-Verlag GmbH; 1978. p. 1–72.
[35] Moon TC, Kim JC, Song SE, Suh SJ, Seo CS, Kim YK, et al. Saucerneol D, a naturally occurring sesquilignan, inhibits LPS-induced iNOS expression in RAW264.7 cells by blocking NF-κB and MAPK activation. Int Immunopharmacol. 2008 10;8(10):1395–1400.
[36] Yun JY, Roh E, Son JK, Lee SH, Seo CS, Hwang BY, et al. Effect of saucerneol D on melanin production in cAMP-elevated melanocytes. Arch Pharmacal Res. 2011;34(8):1339–1345.
[37] Kim SN, Kim MH, Kim YS, Ryu SY, Min YK, Kim SH. Inhibitory effect of (−)-saucerneol on osteoclast differentiation and bone pit formation. Phytother Res. 2009;23(2):185–191.
[38] Lu Y, Li Y, Seo CS, Murakami M, Son JK, Chang HW. Saucerneol D inhibits eicosanoid generation and degranulation through suppression of Syk kinase in mast cells. Food Chem Toxicol. 2012;50(12):4382–4388.
[39] Chang JS, Lee SW, Kim MS, Yun BR, Park MH, Lee SG, et al. Manassantin A and B from *Saururus chinensis* inhibit interleukin-6-induced signal transducer and activator of transcription 3 activation in Hep3B cells. J Pharmacol Sci. 2011;115(1):84–88.
[40] Lee J, Huh MS, Kim YC, Hattori M, Otake T. Lignan, sesquilignans and dilignans, novel HIV-1 protease and cytopathic effect inhibitors purified from the rhizomes of *Saururus chinensis*. Antiviral Res. 2010;85(2):425–428.

[41] Seo CS, Lee WH, Chung HW, Chang EJ, Lee SH, Jahng Y, et al. Manassantin A and B from *Saururus chinensis* inhibiting cellular melanin production. Phytother Res. 2009; 23(11):1531–1536.
[42] Hanessian S, Reddy GJ, Chahal N. Total synthesis and stereochemical confirmation of manassantin A, B, and B1. Org Lett. 2006;8(24):5477–5480.
[43] Moss GP. Nomenclature of lignans and neolignans (IUPAC recommendations 2000). Pure Appl Chem. 2000;72(8):1493–1523.
[44] Sainvitu P, Nott K, Richard G, Blecker C, Jérôme C, Wathelet JP, et al. Structure, properties and obtention routes of flaxseed lignan secoisolariciresinol: a review. Biotechnol, Agron, Soc Environ. 2012;16(1):115–124.
[45] Onocha PA, Ali MS. Pycnanolide A and B: new lignan lactones from the leaves of *Pycnanthus angolensis* (myristicaceae). Res J Phytochem. 2011;5(3):136–145.
[46] Wróbel A, Eklund P, Bobrowska-Hägerstrand M, Hägerstrand H. Lignans and norlignans inhibit multidrug resistance protein 1 (MRP1/ABCC1)-mediated transport. Anticancer Res. 2010;30(11):4423–4428.
[47] Ono E, Kim HJ, Murata J, Morimoto K, Okazawa A, Kobayashi A, et al. Molecular and functional characterization of novel furofuranclass lignan glucosyltransferases from *Forsythia*. Plant Biotechnol. 2010;27(4):317–324.
[48] Pereira DM, Valentão P, Pereira JA, Andrade PB. Phenolics: from chemistry to biology. Molecules. 2009;14(6):2202–2211.
[49] Canel C, Moraes RM, Dayan FE, Ferreira D. Molecules of interest: podophyllotoxin. Phytochemistry. 2000;54(2):115–120.
[50] St[ä]helin H, Von Wartburg A. The chemical and biological route from podophyllotoxin glucoside to etoposide: ninth cain memorial award lecture. Cancer Res. 1991;51(1):5–15.
[51] Cheung CWY, Gibbons N, Johnson DW, Nicol DL. Silibinin - a promising new treatment for cancer. Anti-Cancer Agents Med Chem. 2010;10(3):186–195.
[52] Foo LY, Karchesy J. Pseudotsuganol, a biphenyl-linked pinoresinol-dihydroquercetin from Douglas-fir bark: isolation of the first true flavonolignan. Chem Commun. 1989;(4):217–219.
[53] Kumar S, Ray AB, Konno C, Oshima Y, Hikino H. Cleomiscosin D, a coumarino-lignan from seeds of *Cleome viscosa*. Phytochemistry. 1988;27(2):636–638.
[54] Ray AB, Chattopadhyay SK, Kumar S, Konno C, Kiso Y, Hikino H. Structures of cleomiscosins, coumarinolignoids of *Cleome viscosa* seeds. Tetrahedron. 1985;41(1):209–214.
[55] Chen JJ, Wang TY, Hwang TL. Neolignans, a coumarinolignan, lignan derivatives, and a chromene: anti-inflammatory constituents from *Zanthoxylum avicennae*. J Nat Prod. 2008;71(2):212–217.
[56] Rasool MA, Khan R, Malik A, Bibi N, Kazmi SU. Structural determination of daphnecin, a new coumarinolignan from *Daphne mucronata*. J Asian Nat Prod Res. 2010;12(4):324–327.
[57] Yao CS, Lin M, Wang L. Isolation and biomimetic synthesis of anti-inflammatory stilbenolignans from *Gnetum cleistostachyum*. Chem Pharm Bull. 2006;54(7):1053–1057.
[58] Banwell MG, Bezos A, Chand S, Dannhardt G, Kiefer W, Nowe U, et al. Convergent synthesis and preliminary biological evaluations of the stilbenolignan (±)-aiphanol and various congeners. Org Biomol Chem. 2003;1(14):2427–2429.
[59] Lee D, Cuendet M, Vigo JS, Graham JG, Cabieses F, Fong HHS, et al. A novel cyclooxygenase-inhibitory stilbenolignan from the seeds of *Aiphanes aculeata*. Org Lett. 2001;3(14):2169–2171.
[60] Syu WJ, Shen CC, Lu JJ, Lee GH, Sun CM. Antimicrobial and cytotoxic activities of neolignans from *Magnolia officinalis*. Chem Biodivers. 2004;1(3):530–537.
[61] Matsuda H, Kageura T, Oda M, Morikawa T, Sakamoto Y, Yoshikawa M. Effects of constituents from the bark of magnolia obovata on nitric oxide production in lipopolysaccharide-activated macrophages. Chem Pharm Bull. 2001;49(6):716–720.

[62] Cullmann F, Adam KP, Zapp J, Becker H. Pelliatin, a macrocyclic lignan derivative from *Pellia epiphylla*. Phytochemistry. 1996;41(2):611–615.

[63] Scher JM, Zapp J, Becker H. Lignan derivatives from the liverwort *Bazzania trilobata*. Phytochemistry. 2003;62(5):769–777.

[64] Freudenberg K, Weinges K. Systematik und nomenklatur der lignane. Tetrahedron. 1961;15(1-4):115–128.

[65] Gordaliza M, Garcia PA, Miguel del Corral JM, Castro MA, Gomez-Zurita MA. Podophyllotoxin: distribution, sources, applications and new cytotoxic derivatives. Toxicon. 2004;44(4):441–459.

[66] Rao CBS. Chemistry of Lignans, Andhra University Series. Andhra University Press; 1978.

[67] Landete JM. Plant and mammalian lignans: a review of source, intake, metabolism, intestinal bacteria and health. Food Res Int. 2012;46(1):410–424.

[68] Castro MA, Gordaliza M, Miguel Del Corral JM, Feliciano AS. The distribution of lignanoids in the order coniferae. Phytochemistry. 1996;41(4):995–1011.

[69] Chiung YM, Hayashi H, Matsumoto H, Otani T, Yoshida K, Huang MY, et al. New metabolites, tetrahydrofuran lignans, produced by *Streptomyces* sp. IT-44. J Antibiot. 1994;47(4):487–491.

[70] Pan JY, Chen SL, Yang MH, Wu J, Sinkkonen J, Zou K. An update on lignans: natural products and synthesis. Nat Prod Rep. 2009;26(10):1251–1292.

[71] Haworth RD, Mavin CR, Sheldrick G. 311. The constituents of guaiacum resin. Part II. Synthesis of DL-guaiaretic acid dimethyl ether. J Chem Soc. 1934:1423–1429.

[72] Yoshiki Y, Ishiguro T. Crystalline constituents of Hinoki oil. Yakugaku Zasshi. 1933;53:73–151.

[73] Haworth RD, Richardson T. 141. The constituents of natural phenolic resins. Part I. Matairesinol. J Chem Soc. 1935:633–636.

[74] Easterfield TH, Bee J. LXXXII. The resin acids of the coniferae. Part II. Matairesinol. J Chem Soc, Trans. 1910;97:1028–1032.

[75] Hughes GK, Ritchie E. The chemical constituents of *Himantandra* species. I. The lignans of *Himantandra baccata* and *H. belgraveana*. Aust J Chem. 1954;7:104–112.

[76] Haworth RD, Kelly W. The constituents of natural phenolic resins. Part VIII. Lariciresinol, cubebin, and some stereochemical relationships. J Chem Soc. 1937:384–391.

[77] Bamberger M, Landsiedl A. Zur kenntniss der überwallungsharze. III. Monatsh Chem. 1897;18(1):481–509.

[78] Korner G, Vanzetti L. Olivil, its composition and constitution. Atti Accad Naz Lincei, Cl Sci Fis, Mat Nat, Rend. 1903;12(I(V)):122–125.

[79] Smith HG. On aromadendrin or aromadendric acid from the turbid group of eucalyptus kinos. Am J Pharm. 1896;68:679–687.

[80] Smith HG. Crystalline deposit occurring in the timber of the 'colonial beech'. Chem News J Ind Sci. 1913;108:169–172.

[81] Birch AJ, Lions F. The constitution of gmelinol. I. J Proc R Soc N S W. 1938;71:391–405.

[82] Bamberger M. Zur kenntniss der überwallungsharze. II. Monatsh Chem. 1894;15(1):505–518.

[83] Nishida K, Sumimoto M, Kondo T. Studies on the chemical constituents of bark of *Symplocos lucida*. II. On the hydrolysis of the glucoside. Nippon Rin Gakkaishi. 1951;33:235–239.

[84] Nishida K, Sumimoto M, Kondo T. Studies on the chemical constituents of bark of *Symplocos lucida*. III. The examination of active groups on the aglucon. Nippon Rin Gakkaishi. 1951;33:269–272.

[85] Pearl IA. Conidendrin from western hemlock sulfite waste liquor. J Org Chem. 1945;10(3):219–221.

[86] King FE, Jurd L, King TJ. Isotaxiresinol (3'-demethylisolariciresinol), a new lignan extracted from the heartwood of the English yew, *Taxus baccata*. J Chem Soc (Resumed). 1952:17–24.

[87] Erdtman H, Tsuno K. Chemistry of the order cupressales. LVI. Heartwood constituents of *Fitzroya cupressoides*. Acta Chem Scand. 1969;23(6):2021–2024.
[88] Briggs LH, Frieberg AG. The resinol of Olea, Cunninghamii (Maire). J Chem Soc (Resumed). 1937:271–273.
[89] Holmbom B, Eckerman C, Eklund P, Hemming J, Nisula L, Reunanen M, *et al*. Knots in trees - a new rich source of lignans. Phytochem Rev. 2003;2(3):331–340.
[90] Eckerman C, Holmbom B. Method for recovery of compression wood and/or normal wood from oversize chips [US 6,739,533]; 2004.
[91] Peterson J, Dwyer J, Adlercreutz H, Scalbert A, Jacques P, McCullough ML. Dietary lignans: physiology and potential for cardiovascular disease risk reduction. Nutr Rev. 2010;68(10):571–603.
[92] Raffaelli B, Hoikkala A, Leppälä E, Wähälä K. Enterolignans. J Chromatogr, B: Anal Technol Biomed Life Sci. 2002;777(1-2):29–43.
[93] Heinonen S, Nurmi T, Liukkonen K, Poutanen K, Wähälä K, Deyama T, *et al*. In Vitro metabolism of plant lignans: new precursors of mammalian lignans enterolactone and enterodiol. J Agric Food Chem. 2001;49(7):3178–3186.
[94] Mazur WM, Duke JA, Wähälä K, Rasku S, Adlercreutz H. Isoflavonoids and lignans in legumes: nutritional and health aspects in humans. J Nutr Biochem. 1998;9(4):193–200.
[95] Milder IEJ, Arts ICW, Venema DP, Lasaroms JJP, Wähälä K, Hollman PCH. Optimization of a liquid chromatography-tandem mass spectrometry method for quantification of the plant lignans secoisolariciresinol, matairesinol, lariciresinol, and pinoresinol in foods. J Agric Food Chem. 2004;52(15):4643–4651.
[96] Smeds AI, Eklund PC, Sjöholm RE, Willför SM, Nishibe S, Deyama T, *et al*. Quantification of a broad spectrum of lignans in cereals, oilseeds, and nuts. J Agric Food Chem. 2007;55(4):1337–1346.
[97] Katsuzaki H, Kawakishi S, Osawa T. Sesaminol glucosides in sesame seeds. Phytochemistry. 1994;35(3):773–776.
[98] Kim KS, Park SH, Choung MG. Nondestructive determination of lignans and lignan glycosides in sesame seeds by near infrared reflectance spectroscopy. J Agric Food Chem. 2006;54(13):4544–4550.
[99] Ryu SN, Ho CT, Osawa T. High performance liquid chromatographic determination of antioxidant lignan glycosides in some varieties of sesame. J Food Lipids. 1998;5(1):17–28.
[100] Li X, Yuan JP, Xu SP, Wang JH, Liu X. Separation and determination of secoisolariciresinol diglucoside oligomers and their hydrolysates in the flaxseed extract by high-performance liquid chromatography. J Chromatogr, A. 2008;1185(2):223–232.
[101] Muir AD. Flax lignans-analytical methods and how they influence our understanding of biological activity. J AOAC Int. 2006;89(4):1147–1157.
[102] Strandås C, Kamal-Eldin A, Andersson R, Åman P. Composition and properties of flaxseed phenolic oligomers. Food Chem. 2008;110(1):106–112.
[103] Struijs K, Vincken JP, Verhoef R, van Oostveen-van Casteren WHM, Voragen AGJ, Gruppen H. The flavonoid herbacetin diglucoside as a constituent of the lignan macromolecule from flaxseed hulls. Phytochemistry. 2007;68(8):1227–1235.
[104] Struijs K, Vincken JP, Verhoef R, Voragen AGJ, Gruppen H. Hydroxycinnamic acids are ester-linked directly to glucosyl moieties within the lignan macromolecule from flaxseed hulls. Phytochemistry. 2008;69(5):1250–1260.
[105] Struijs K, Vincken JP, Doeswijk TG, Voragen AGJ, Gruppen H. The chain length of lignan macromolecule from flaxseed hulls is determined by the incorporation of coumaric acid glucosides and ferulic acid glucosides. Phytochemistry. 2009;70(2):262–269.
[106] Tham DM, Gardner CD, Haskell WL. Potential health benefits of dietary phytoestrogens: a review of the clinical, epidemiological, and mechanistic evidence. J Clin Endocrinol Metab. 1998;83(7):2223–2235.

[107] Murphy PA, Hendrich S. Phytoestrogens in foods. In: Advances in Food and Nutrition Research. vol. 44. Academic Press; 2002. p. 195–246, IN1-IN4.
[108] Toure A, Xu X. Flaxseed lignans: source, biosynthesis, metabolism, antioxidant activity, bioactive components, and health benefits. Compr Rev Food Sci Food Saf. 2010;9(3):261–269.
[109] Willför SM, Smeds AI, Holmbom BR. Chromatographic analysis of lignans. J Chromatogr, A. 2006;1112(1-2):64–77.
[110] Penalvo JL, Heinonen SM, Nurmi T, Deyama T, Nishibe S, Adlercreutz H. Plant lignans in soy-based health supplements. J Agric Food Chem. 2004;52(13):4133–4138.
[111] Penalvo JL, Haajanen KM, Botting N, Adlercreutz H. Quantification of lignans in food using isotope dilution gas chromatography/mass spectrometry. J Agric Food Chem. 2005; 53(24):9342–9347.
[112] Marston A, Hostettmann K. Modern separation methods. Nat Prod Rep. 1991;8(4):391–414.
[113] Topcu G, Demirkiran O. Lignans from *Taxus* species. Top Heterocycl Chem. 2007;11(Bioactive Heterocycles V):103–144.
[114] Lee KH, Xiao Z. Lignans in treatment of cancer and other diseases. Phytochem Rev. 2003;2(3):341–362.
[115] Das B, Kashinath A. Part VII. Review on the chemical constituents of medicinal plants and bioactive natural products' new phytoconstituents of yew plants. J Sci Ind Res. 1996;55(4):246–258.
[116] Wu MD, Huang RL, Kuo LMY, Hung CC, Ong CW, Kuo YH. The anti-HBsAg (human type B hepatitis, surface antigen) and anti-HBeAg (human type B hepatitis, e antigen) C_{18} dibenzocyclooctadiene lignans from *Kadsura matsudai* and *Schizandra arisanensis*. Chem Pharm Bull. 2003;51(11):1233–1236.
[117] Bártlová M, Opletal L, Chobot V, Sovová H. Liquid chromatographic analysis of supercritical carbon dioxide extracts of *Schizandra chinensis*. J Chromatogr, B: Anal Technol Biomed Life Sci. 2002;770(1–2):283–289.
[118] Kuo CH, Lee SS, Chang HY, Sun SW. Analysis of lignans using micellar electrokinetic chromatography. Electrophoresis. 2003;24(6):1047–1053.
[119] Cavin A, Potterat O, Wolfender JL, Hostettmann K, Dyatmyko W. Use of on-flow LC/^{1}H NMR for the study of an antioxidant fraction from *Orophea enneandra* and isolation of a polyacetylene, lignans, and a tocopherol derivative. J Nat Prod. 1998;61(12):1497–1501.
[120] Kraus C, Spiteller G. Comparison of phenolic compounds from galls and shoots of *Picea glauca*. Phytochemistry. 1997;44(1):59–67.
[121] Greger H, Hofer O. New unsymmetrically substituted tetrahydrofurofuran lignans from *Artemisia absinthium*: assignment of the relative stereochemistry by lanthanide induced chemical shifts. Tetrahedron. 1980;36(24):3551–3558.
[122] Fonseca SF, de Paiva Campello J, Barata LES, Rúveda EA. ^{13}C NMR spectral analysis of lignans from *Araucaria angustifolia*. Phytochemistry. 1978;17(3):499–502.
[123] Agrawal PK, Rastogi RP. Two lignans from *Cedrus deodara*. Phytochemistry. 1982;21(6): 1459–1461.
[124] Beejmohun V, Fliniaux O, Grand E, Lamblin F, Bensaddek L, Christen P, et al. Microwave-assisted extraction of the main phenolic compounds in flaxseed. Phytochem Anal. 2007;18(4):275–282.
[125] Nemes SM, Orsat V. Evaluation of a microwave-assisted extraction method for lignan quantification in flaxseed cultivars and selected oil seeds. Food Anal Methods. 2012;5(3):551–563.
[126] Bozan B, Temelli F. Supercritical CO_2 extraction of flaxseed. J Am Oil Chem Soc. 2002;79(3):231–235.
[127] Cacace JE, Mazza G. Pressurized low polarity water extraction of lignans from whole flaxseed. J Food Eng. 2006;77(4):1087–1095.
[128] Stahl E. Thin-Layer Chromatography: A Laboratory Handbook. Springer-Verlag; 1969.

[129] Kirchner JG. Thin-Layer Chromatography, Techniques of Chemistry. John Wiley & Sons, Inc.; 1978.
[130] Ayres DC, Chater RB. Lignans and related phenols-X: assignment of structure to the principal classes of lignans by gas chromatography. Tetrahedron. 1969;25(17):4093–4098.
[131] Hostettmann K. Droplet counter-current chromatography and its application to the preparative scale separation of natural products. Planta Med. 1980;39(1):1–18.
[132] Bradlow HL. Extraction of steroid conjugates with a neutral resin. Steroids. 1968;11(3): 265–272.
[133] Fotsis T, Heikkinen R, Adlercreutz H, Axelson M, Setchell KDR. Capillary gas chromatographic method for the analysis of lignans in human urine. Clin Chim Acta. 1982;121(3):361–371.
[134] Nurmi T, Voutilainen S, Nyyssonen K, Adlercreutz H, Salonen JT. Liquid chromatography method for plant and mammalian lignans in human urine. J Chromatogr, B: Anal Technol Biomed Life Sci. 2003;798(1):101–110.
[135] Smeds A, Hakala K. Liquid chromatographic-tandem mass spectrometric method for the plant lignan 7-hydroxymatairesinol and its potential metabolites in human plasma. J Chromatogr, B: Anal Technol Biomed Life Sci. 2003;793(2):297–308.
[136] Knust U, Hull WE, Spiegelhalder B, Bartsch H, Strowitzki T, Owen RW. Analysis of enterolignan glucuronides in serum and urine by HPLC-ESI-MS. Food Chem Toxicol. 2006;44(7):1038–1049.
[137] Pelter A, Ward RS, Jones DM, Maddocks P. Asymmetric syntheses of lignans of the dibenzylbutyrolactone, dibenzylbutanediol, aryltetralin and dibenzocyclooctadiene series. Tetrahedron: Asymmetry. 1992;3(2):239–242.
[138] Eklund P, Sillanpää R, Sjöholm R. Synthetic transformation of hydroxymatairesinol from Norway spruce (*Picea abies*) to 7-hydroxysecoisolariciresinol, (+)-lariciresinol and (+)-cyclolariciresinol. J Chem Soc [Perkin 1]. 2002;(16):1906–1910.
[139] Dale JA, Dull DL, Mosher HS. α-methoxy-α-trifluoromethylphenylacetic acid, a versatile reagent for the determination of enantiomeric composition of alcohols and amines. J Org Chem. 1969;34(9):2543–2549.
[140] Platt JR. Classification of spectra of catacondensed hydrocarbons. J Chem Phys. 1949; 17(5):484–495.
[141] Evcim U, Gozler B, Freyer AJ, Shamma M. Haplomyrtin and (−)-haplomyrfolin: two lignans from *Haplophyllum myrtifolium*. Phytochemistry. 1986;25(8):1949–1951.
[142] Laboratories SR, Simons. The Sadtler Handbook of Ultraviolet Spectra, The Sadtler Handbooks Series. Sadtler Research Laboratories; 1979.
[143] Otsuka H, Kuwabara H, Hoshiyama H. Identification of sucrose diesters of aryldihydronaphthalene-type lignans from *Trigonotis peduncularis* and the nature of their fluorescence. J Nat Prod. 2008;71(7):1178–1181.
[144] Yamaguchi K. Spectral Data of Natural Products, Spectral Data of Natural Products. Elsevier Publishing Company; 1970.
[145] Del Signore G, Berner OM. Recent developments in the asymmetric synthesis of lignans. In: ur Rahman A, editor. Studies in Natural Products Chemistry. vol. 33, Part M. Elsevier; 2006. p. 541–600.
[146] Tomioka K, Mizuguchi H, Koga K. Studies directed towards the asymmetric total synthesis of antileukemic lignan lactones. Synthesis of (−)-podorhizon. Tetrahedron Lett. 1978;19(47):4687–4690.
[147] Tomioka K, Koga K. Studies directed towards the asymmetric total synthesis of antileukemic lignan lactones. Synthesis of optically pure key intermediate and its utility. Tetrahedron Lett. 1979;20(35):3315–3318.

[148] Yoda H, Naito S, Takabe K, Tanaka N, Hosoya K. Conjugate addition to chiral *gamma*-heterosubstituted δ-lactones as pivotal synthons from *L*-glutamic acid. Synthesis of an optically active lignan lactone; (−)-hinokinin. Tetrahedron Lett. 1990;31(52):7623–7626.

[149] Morimoto T, Chiba M, Achiwa K. Highly efficient asymmetric hydrogenation of itaconic acid derivatives catalyzed by a modified DIOP-rhodium complex. Tetrahedron Lett. 1989;30(6):735–738.

[150] Morimoto T, Chiba M, Achiwa K. Asymmetric reactions catalyzed by chiral metal complexes. XXXVI. An efficient synthesis of natural (+)-neoisostegane using asymmetric hydrogenation catalyzed by a chiral bisphosphine-rhodium(i) complex. Heterocycles. 1990;30(1, Spec. Issue):363–366.

[151] Morimoto T, Chiba M, Achiwa K. Asymmetric total synthesis of (−)-deoxypodophyllotoxin. Tetrahedron Lett. 1990;31(2):261–264.

[152] Morimoto T, Chiba M, Achiwa K. Asymmetric reactions catalyzed by chiral metal complexes. LV. Efficient asymmetric syntheses of naturally occurring lignan lactones using catalytic asymmetric hydrogenation as a key reaction. Tetrahedron. 1993;49(9):1793–1806.

[153] Brown E, Daugan A. An easy preparation of (−) and (+)-β-piperonyl-γ-butyrolactones, key-intermediates for the synthesis of optically active lignans. Tetrahedron Lett. 1985;26(33):3997–3998.

[154] Doyle MP, Protopopova MN, Zhou QL, Bode JW, Simonsen SH, Lynch V. Optimization of enantiocontrol for carbon-hydrogen insertion with chiral dirhodium(II) carboxamidates. synthesis of natural dibenzylbutyrolactone lignans from 3-aryl-1-propyl diazoacetates in high optical purity. J Org Chem. 1995;60(21):6654–6655.

[155] Honda T, Kimura N, Sato S, Kato D, Tominaga H. Chiral synthesis of lignan lactones, (−)-hinokinin, (−)-deoxypodorhizone, (−)-isohibalactone and (−)-savinin by means of enantioselective deprotonation strategy. J Chem Soc [Perkin 1]. 1994;(8):1043–1046.

[156] Itoh T, Chika J, Takagi Y, Nishiyama S. An efficient enantioselective total synthesis of antitumor lignans: synthesis of enantiomerically pure 4-hydroxyalkanenitriles *via* an enzymic reaction. J Org Chem. 1993;58(21):5717–5723.

[157] Filho HCA, Filho UFL, Pinheiro S, Vasconcellos MLAA, Costa PRR. Enantioselective synthesis of (*R*)-(+)-β-piperonyl-γ-butyrolactone. Tetrahedron: Asymmetry. 1994;5(7):1219–1220.

[158] Charlton JL, Chee GL. Asymmetric synthesis of lignans using oxazolidinones as chiral auxiliaries. Can J Chem. 1997;75(8):1076–1083.

[159] Sibi MP, Liu P, Johnson MD. A short synthesis of both enantiomers of enterolactone. Can J Chem. 2000;78(1):133–138.

[160] Yoda H, Kitayama H, Katagiri T, Takabe K. Novel stereoselective synthesis of (−)-enterolactone employing chiral unsaturated lactam. Tetrahedron. 1992;48(16):3313–3322.

[161] Chenevert R, Mohammadi-Ziarani G, Caron D, Dasser M. Chemoenzymic enantioselective synthesis of (−)-enterolactone. Can J Chem. 1999;77(2):223–226.

[162] Tanaka M, Mukaiyama C, Mitsuhashi H, Wakamatsu T. Synthesis of optically pure gomisin A and schizandrin: the first total synthesis of gomisin a and schizandrin having naturally occurring configurations. Tetrahedron Lett. 1992;33(29):4165–4168.

[163] Tanaka M, Itoh H, Mitsuhashi H, Maruno M, Wakamatsu T. The stereoselective first total synthesis of isoschizandrin having the natural configuration. Tetrahedron: Asymmetry. 1993;4(4):605–608.

[164] Tanaka M, Mitsuhashi H, Maruno M, Wakamatsu T. First total synthesis of optically pure metabolites of gomisin A. Synlett. 1994;(8):604–606.

[165] Tanaka M, Mitsuhashi H, Maruno M, Wakamatsu T. First total synthesis of optically pure deoxyschizandrin and wuweizisu C. The thermal stability of biaryl configuration. Tetrahedron Lett. 1994;35(22):3733–3736.

[166] Tanaka M, Ohshima T, Mitsuhashi H, Maruno M, Wakamatsu T. The first total synthesis of γ-schizandrin and gomisin N having natural configuration. Heterocycles. 1994;37(2):739–742.

[167] Tanaka M, Mukaiyama C, Mitsuhashi H, Maruno M, Wakamatsu T. Synthesis of optically pure gomisi lignans: the total synthesis of (+)-schizandrin, (+)-gomisin A, and (+)-isoschizandrin in naturally occurring forms. J Org Chem. 1995;60(14):4339–4352.

[168] Tanaka M, Ohshima T, Mitsuhashi H, Maruno M, Wakamatsu T. Total syntheses of the lignans isolated from *Schisandra Chinensis*. Tetrahedron. 1995 10;51(43):11693–11702.

[169] Ohshima T, Tanaka M, Mitsuhashi H, Maruno M, Wakamatsu T. Total synthesis of homochiral kadsurin having the natural configuration. Tetrahedron: Asymmetry. 1995;6(1):139–146.

[170] Tanaka M, Ikeya Y, Mitsuhashi H, Maruno M, Wakamatsu T. Total syntheses of the metabolites of schizandrin. Tetrahedron. 1995;51(43):11703–11724.

[171] Tanaka M, Mitsuhashi H, Maruno M, Wakamatsu T. Total synthesis of the major metabolites of gomisin A. Synthesis of homochiral met aA-II, met A-III, met D, and met F. Heterocycles. 1996;42(1):359–374.

[172] Yamauchi S, Tanaka T, Kinoshita Y. First highly stereoselective synthesis of (+)-dihydrosesamin, a trisubstituted tetrahydrofuran-type of lignan, by using highly erythro-selective aldol condensation. J Chem Soc [Perkin 1]. 2001;(18):2158–2160.

[173] Tomioka K, Ishiguro T, Koga K. Asymmetric total synthesis of the antileukemic lignans (+)-trans-burseran and (−)-isostegane. J Chem Soc, Chem Commun. 1979;(15):652–653.

[174] Tomioka K, Ishiguro T, Koga K. First asymmetric total synthesis of (+)-steganacin. Determination of absolute stereochemistry. Tetrahedron Lett. 1980;21(31):2973–2976.

[175] Tomioka K, Ishiguro T, Iitaka Y, Koga K. Stereoselective reactions. VI. Asymmetric total synthesis of natural (−)- and unnatural (+)-steganacin. Determination of the absolute configuration of natural antitumor steganacin. Tetrahedron. 1984;40(8):1303–1312.

[176] Jansen JFGA, Feringa BL. Asymmetric 1,4-additions to γ-(menthyloxy)butenolides. Enantiospecific synthesis of chiral 1,4-butanediols. Tetrahedron Lett. 1989;30(40):5481–5484.

[177] Jansen JFGA, Feringa BL. Asymmetric 1,4-additions to γ-menthyloxybutenolides. (Part III). Enantioselective synthesis of (−)-eudesmin. Tetrahedron Lett. 1991;32(27):3239–3242.

[178] van Oeveren A, Jansen JFGA, Feringa BL. Enantioselective synthesis of natural dibenzylbutyrolactone lignans (−)-enterolactone, (−)-hinokinin, (−)-pluviatolide, (−)-enterodiol, and furofuran lignan (−)-eudesmin *via* tandem conjugate addition to γ-alkoxybutenolides. J Org Chem. 1994;59(20):5999–6007.

[179] van der Deen H, Cuiper AD, Hof RP, van Oeveren A, Feringa BL, Kellogg RM. Lipase-catalyzed second-order asymmetric transformations as resolution and synthesis strategies for chiral 5-(acyloxy)-2(5*H*)-furanone and pyrrolinone synthons. J Am Chem Soc. 1996;118(16):3801–3803.

[180] Brinksma J, van der Deen H, van Oeveren A, Feringa BL. Enantioselective synthesis of benzylbutyrolactones from 5-hydroxyfuran-2(5*H*)-one. new chiral synthons for dibenzylbutyrolactone lignans by a chemoenzymic route. J Chem Soc [Perkin 1]. 1998;(24):4159–4164.

[181] Pelter A, Ward RS, Jones DM, Maddocks P. Asymmetric synthesis of lignans of the dibenzylbutanediol and tetrahydrodibenzocyclooctene series. J Chem Soc [Perkin 1]. 1993;(21):2631–2637.

[182] Pelter A, Ward RS, Jones DM, Maddocks P. Asymmetric synthesis of dibenzylbutyrolactones and aryltetralin lignan lactones by tandem conjugate addition to a chiral butenolide. J Chem Soc [Perkin 1]. 1993;(21):2621–2629.

[183] Enders D, Kirchhoff J, Lausberg V. Diastereo- and enantioselective synthesis of lignan building blocks by tandem michael addition/electrophilic substitution of lithiated α-amino nitriles to furan-2(5*H*)-one. Justus Liebigs Ann Chem. 1996;1996(9):1361–1366.

[184] Enders D, Lausberg V, Del Signore G, Berner OM. A general approach to the asymmetric synthesis of lignans: (−)-methyl piperitol, (−)-sesamin, (−)-aschantin, (+)-yatein, (+)-dihydroclusin, (+)-burseran, and (−)-isostegane. Synthesis. 2002;(4):515–522.

[185] Enders D, Del Signore G, Berner OM. First asymmetric synthesis of (−)-lintetralin *via* intramolecular Friedel-Crafts-type cyclization. Chirality. 2003;15(6):510–513.

[186] Kise N, Tokioka K, Aoyama Y, Matsumura Y. Enantioselective synthesis of 2,3-disubstituted succinic acids by oxidative homocoupling of optically active 3-acyl-2-oxazolidones. J Org Chem. 1995;60(5):1100–1101.

[187] Kise N, Ueda T, Kumada K, Terao Y, Ueda N. Oxidative homocoupling of chiral 3-arylpropanoic acid derivatives. Application to asymmetric synthesis of lignans. J Org Chem. 2000;65(2):464–468.

[188] Maioli AT, Civiello RL, Foxman BM, Gordon DM. Asymmetric synthesis of sesaminone: Confirmation of its structure and determination of its absolute configuration. J Org Chem. 1997;62(21):7413–7417.

[189] Sibi MP, Liu P, Ji J, Hajra S, Chen JX. Free-radical-mediated conjugate additions. enantioselective synthesis of butyrolactone natural products: (−)-enterolactone, (−)-arctigenin, (−)-isoarctigenin, (−)-nephrosteranic acid, and (−)-roccellaric acid. J Org Chem. 2002; 67(6):1738–1745.

[190] Maddaford SP, Charlton JL. A general asymmetric synthesis of (−)-α-dimethylretrodendrin and its diastereomers. J Org Chem. 1993;58(15):4132–4138.

[191] Bogucki DE, Charlton JL. An asymmetric synthesis of (−)-deoxypodophyllotoxin. J Org Chem. 1995;60(3):588–593.

[192] Coltart DM, Charlton JL. The asymmetric synthesis of aryltetralin lignans: (−)-isolariciresinol dimethyl ether and (−)-deoxysikkimotoxin. Can J Chem. 1996;74(1):88–94.

[193] Pelter A, Ward RS, Li Q, Pis J. An asymmetric synthesis of isopodophyllotoxin. Tetrahedron: Asymmetry. 1994;5(5):909–910.

[194] Bush EJ, Jones DW. Asymmetric total synthesis of (−)-podophyllotoxin. J Chem Soc [Perkin 1]. 1996;(2):151–155.

[195] Jones DW. Diels-Alder additions to 1-phenyl-2-benzopyran-3-one and transformations of the adducts: model experiments for podophyllotoxin synthesis. J Chem Soc [Perkin 1]. 1994;(4):399–405.

[196] Wang Q, Yang Y, Li Y, Yu W, Hou ZJ. An efficient method for the synthesis of lignans. Tetrahedron. 2006;62(25):6107–6112.

[197] Son JK, Seung HL, Nagarapu L, Jahng Y. A simple synthesis of nordihydroguaiaretic acid and its analogues. Bull Korean Chem Soc. 2005;26(7):1117–1120.

[198] Schroeter G, Lichtenstadt L, Irineu D. Über die konstitution der guajacharz-subztanzen. (i). Ber Dtsch Chem Ges. 1918;51(2):1587–1613.

[199] Haworth RD, Richardson T. Constituents of guaiacum resin. III. Synthesis of DL-guaiaretic diethyl ether. J Chem Soc. 1935:120–122.

[200] Perry CW, Kalnins MV, Deitcher KH. Synthesis of lignans. I. Nordihydroguaiaretc acid. J Org Chem. 1972;37(26):4371–4376.

[201] Lieberman SV, Mueller GP, Stiller ET. A synthesis of nordihydroguaiaretic acid. J Am Chem Soc. 1947;69(6):1540–1541.

[202] Sakakibara Y. Synthesis of 2,3-dipiperonylbutane. Nippon Kagaku Zasshi. 1952;73:235–236.

[203] Hearon WM, MacGregor WS. The naturally occurring lignans. Chem Rev. 1955;55:957–1068.

[204] Gezginci MH, Timmermann BN. A short synthetic route to nordihydroguaiaretic acid (NDGA) and its stereoisomer using Ti-induced carbonyl-coupling reaction. Tetrahedron Lett. 2001;42(35):6083–6085.

[205] Gu WX, Wu AX, Gao Q, Pan XF. Total synthesis of machilin A. Chin Chem Lett. 2000;11(1):15–16.

[206] Xia YM, Cao XP, Peng K, Ren XF, Pan XF. An efficient synthetic method of nordihydroguaiaretic acid (NDGA). Chin Chem Lett. 2003;14(4):359–360.

[207] Minato A, Tamao K, Suzuki K, Kumada M. Synthesis of a lignan skeleton *via* nickel- and palladium-phosphine complex catalyzed Grignard coupling reaction of halothiophenes. Tetrahedron Lett. 1980;21(41):4017–4020.

[208] Morreel K, Kim H, Lu F, Dima O, Akiyama T, Vanholme R, et al. Mass spectrometry-based fragmentation as an identification tool in lignomics. Anal Chem. 2010;82(19):8095–8105.

[209] Eich E, Pertz H, Kaloga M, Schulz J, Fesen MR, Mazumder A, et al. (−)-arctigenin as a lead structure for inhibitors of human immunodeficiency virus type-1 integrase. J Med Chem. 1996;39(1):86–95.

[210] Ward RS, Hughes DD. Oxidative cyclisation of 3,4-dibenzyltetrahydrofurans using ruthenium tetra(trifluoroacetate). Tetrahedron. 2001;57(10):2057–2064.

[211] Fryatt T, Botting NP. The synthesis of multiply ^{13}C-labeled plant and mammalian lignans as internal standards for *LC-MS* and *GC-MS* analysis. J Labelled Comp Radiopharm. 2005;48(13):951–969.

[212] Chu A, Dinkova A, Davin LB, Bedgar DL, Lewis NG. Stereospecificity of (+)-pinoresinol and (+)-lariciresinol reductases from *Forsythia intermedia*. J Biol Chem. 1993;268(36):27026–27033.

[213] Dinkova-Kostova AT, Gang DR, Davin LB, Bedgar DL, Chu A, Lewis NG. (+)-pinoresinol/(+)-lariciresinol reductase from *Forsythia intermedia*. Protein purification, cDNA cloning, heterologous expression and comparison to isoflavone reductase. J Biol Chem. 1996;271(46):29473–29482.

[214] Trazzi G, André MF, Coelho F. Diastereoselective synthesis of β-piperonyl-γ-butyrolactones from Morita-Baylis-Hillman adducts. Highly efficient synthesis of (±)-yatein, (±)-podorhizol and (±)-*epi*-podorhizol. J Braz Chem Soc. 2010;21(12):2327–2339.

[215] Makela TH, Wahala KT, Hase TA. Synthesis of enterolactone and enterodiol precursors as potential inhibitors of human estrogen synthetase (aromatase). Steroids. 2000;65(8):437–441.

[216] Fischer J, Reynolds AJ, Sharp LA, Sherburn MS. Radical carboxyarylation approach to lignans. Total synthesis of (−)-arctigenin, (−)-matairesinol, and related natural products. Org Lett. 2004;6(9):1345–1348.

[217] Hartwell JL, Johnson JM, Fitzgerald DB, Belkin M. Podophyllotoxin from *Juniperus* species; savinin. J Am Chem Soc. 1953;75:235–236.

[218] Bruschi M, Orlandi M, Rindone M, Rindone B, Saliu F, Suarez-Bertoa R, et al. Podophyllotoxin and antitumor synthetic aryltetralines. Toward a biomimetic preparation. In: Mukherjee A, editor. Biomimetics Learning from Nature. Advances in Biomimetics. InTech; 2010. p. 305–324.

[219] Pelter A, Ward RS, Pritchard MC, Kay IT. Synthesis of lignans related to the podophyllotoxin series. J Chem Soc [Perkin 1]. 1988;(6):1603–1613.

[220] Pelter A, Ward RS, Jones DM, Maddocks P. Asymmetric synthesis of homochiral dibenzyl-butyrolactone lignans by conjugate addition to a chiral butenolide. Tetrahedron: Asymmetry. 1990;1(12):857–860.

[221] Ward RS. Asymmetric synthesis of lignans. Tetrahedron. 1990;46(15):5029–5041.

[222] Ward RS. Synthesis of podophyllotoxin and related compounds. Synthesis. 1992;(8):719–730.

[223] Sellars JD, Steel PG. Advances in the synthesis of aryltetralin lignan lactones. Eur J Org Chem. 2007;(23):3815–3828.

[224] Jackson DE, Dewick PM. Aryltetralin lignans from *Podophyllum hexandrum* and *Podophyllum peltatum*. Phytochemistry. 1984;23(5):1147–1152.

[225] Canel C, Dayan FE, Ganzera M, Khan IA, Rimando A, Burandt J Charles L, et al. High yield of podophyllotoxin from leaves of *Podophyllum peltatum* by *in situ* conversion of podophyllotoxin 4-*O*-β-D-glucopyranoside. Planta Med. 2001;67(1):97–99.

[226] Bedir E, Khan I, Moraes RM. Bioprospecting for podophyllotoxin. In: Janick J, Whipkey A, editors. Trends in New Crops and New Uses. ASHS Press; 2002. p. 545–549.

[227] Gensler WJ, Gatsonis CD. Synthesis of podophyllotoxin. J Org Chem. 1966;31(12):4004–4008.

[228] Kende AS, King ML, Curran DP. Total synthesis of (±)-4′-demethyl-4-epipodophyllotoxin by insertion-cyclization. J Org Chem. 1981;46(13):2826–2828.
[229] Macdonald DI, Durst T. A highly stereoselective synthesis of podophyllotoxin and analogues based on an intramolecular Diels-Alder reaction. J Org Chem. 1988;53(16):3663–3669.
[230] Klemm LH, Olson DR, White DV. Intramolecular Diels-Alder reactions. VII. Electroreduction of α,β-unsaturated esters. I. Synthesis of rac-deoxypicropodophyllin by intramolecular Diels-Alder reaction plus trans addition of hydrogen. J Org Chem. 1971;36(24):3740–3743.
[231] Ziegler FE, Schwartz JA. Synthetic studies on lignan lactones: aryl dithiane route to (±)-podorhizol and (±)-isopodophyllotoxone and approaches to the stegane skeleton. J Org Chem. 1978;43(5):985–991.
[232] Gonzalez AG, Perez JP, Trujillo JM. Constituents of the *Umbelliferae*. 18. Synthesis of two arylnaphthalene lignans. Tetrahedron. 1978;34(7):1011–1013.
[233] Van Speybroeck R, Guo H, Van der Eycken J, Vandewalle M. Enantioselective total synthesis of (−)-epipodophyllotoxin and (−)-podophyllotoxin. Tetrahedron. 1991;47(26):4675–4682.
[234] Rodrigo R. A stereo- and regiocontrolled synthesis of podophyllum lignans. J Org Chem. 1980;45(22):4538–4540.
[235] Rajapaksa D, Rodrigo R. A stereocontrolled synthesis of antineoplastic podophyllum lignans. J Am Chem Soc. 1981;103(20):6208–6209.
[236] Forsey SP, Rajapaksa D, Taylor NJ, Rodrigo R. Comprehensive synthetic route to eight diastereomeric podophyllum lignans. J Org Chem. 1989;54(18):4280–4290.
[237] Jones DW, Thompson AM. Synthesis of podophyllum lignans *via* an isolable *o*-quinonoid pyrone. J Chem Soc, Chem Commun. 1987;(23):1797–1798.
[238] Kuroda T, Takahashi M, Kondo K, Iwasaki T. Efficient synthesis of 1-aryl-3,4-dihydro-4-hydroxynaphthalene: Application to the stereocontrolled synthesis of (±)-isopicropodophyllin and (±)-isopodophyllotoxin. J Org Chem. 1996;61(26):9560–9563.
[239] Capriati V, Florio S, Luisi R, Perna FM, Salomone A, Gasparrini F. An efficient route to tetrahydronaphthols *via* addition of *ortho*-lithiated stilbene oxides to α,β-unsaturated Fischer carbene complexes. Org Lett. 2005;7(22):4895–4898.
[240] Kennedy-Smith JJ, Young LA, Toste FD. Rhenium-catalyzed aromatic propargylation. Org Lett. 2004;6(8):1325–1327.
[241] Andrews RC, Teague SJ, Meyers AI. Asymmetric total synthesis of (−)-podophyllotoxin. J Am Chem Soc. 1988;110(23):7854–7858.
[242] Engelhardt U, Sarkar A, Linker T. Efficient enantioselective total synthesis of (−)-epipodophyllotoxin. Angew Chem, Int Ed. 2003;42(22):2487–2489.
[243] Ward RS, Pelter A, Brizzi A, Sega A, Paoli P. Synthesis of diversely functionalized dibenzylbutyrolactones and aryltetralins from silylated cyanohydrin anions. J Chem Res. 1998;(5):226–227.
[244] Charlton JL, Plourde GL, Koh K, Secco S. Asymmetric synthesis of podophyllotoxin analogs. Can J Chem. 1990;68(11):2022–2027.
[245] Charlton JL, Koh K. Asymmetric synthesis of (−)-neopodophyllotoxin. J Org Chem. 1992;57(5):1514–1516.
[246] Maeda S, Masuda H, Tokoroyama T. Studies on the preparation of bioactive lignans by oxidative coupling reaction. II. Oxidative coupling reaction of methyl (e)-3-(4,5-dihydroxy-2-methoxyphenyl)propenoate and lipid peroxidation inhibitory effects of the produced lignans. Chem Pharm Bull. 1994;42(12):2506–2513.
[247] Maeda S, Masuda H, Tokoroyama T. Studies on the preparation of bioactive lignans by oxidative coupling reaction. V. Oxidative coupling reaction of methyl (*E*)-3-(2-hydroxyphenyl)propenoate derivatives and lipid peroxidation inhibitory effects of the produced lignans. Chem Pharm Bull. 1995;43(6):935–940.

[248] Daquino C, Rescifina A, Spatafora C, Tringali C. Biomimetic synthesis of natural and unnatural lignans by oxidative coupling of caffeic esters. Eur J Org Chem. 2009;2009(36):6289–6300.

[249] Bogucki DE, Charlton JL. A non-enzymic synthesis of (S)-(−)-rosmarinic acid and a study of a biomimetic route to (+)-rabdosiin. Can J Chem. 1997;75(12):1783–1794.

[250] Lajide L, Escoubas P, Mizutani J. Termite antifeedant activity from tropical plants. Part 1. Termite antifeedant activity in *Xylopia aethiopica*. Phytochemistry. 1995;40(4):1105–1112.

[251] Zoia L, Bruschi M, Orlandi M, Tolppa EL, Rindone B. Asymmetric biomimetic oxidations of phenols: the mechanism of the diastereo- and enantioselective synthesis of thomasidioic acid. Molecules. 2008;13(1):129–148.

[252] Mingoia F, Vitale M, Madec D, Prestat G, Poli G. Pseudo-domino palladium-catalyzed allylic alkylation/mizoroki-heck coupling reaction: a key sequence toward (±)-podophyllotoxin. Tetrahedron Lett. 2008;49(5):760–763.

[253] Kende AS, Liebeskind LS, Mills JE, Rutledge PS, Curran DP. Oxidative aryl-benzyl coupling. A biomimetic entry to podophyllin lignan lactones. J Am Chem Soc. 1977;99(21):7082–7083.

[254] Bertz SH, Adams WO, Silverton JV. 2,5-bis(methoxycarbonyl)-4-hydroxycyclopent-2-en-1-one as an intermediate in Weiss' glyoxal reaction. Analogous chemistry of malondialdehyde. J Org Chem. 1981;46(13):2828–2830.

[255] Herz W, Gordon W, Falk H, Hunek S. Progress in the Chemistry of Organic Natural Products 81, Fortschritte der Chemie Organischer Naturstoffe / Progress in the Chemistry of Organic Natural Products Series. Springer-Verlag; 2001.

[256] Ullmann F, Bielecki J. Über synthesen in der biphenylreihe. Ber Dtsch Chem Ges. 1901;34(2):2174–2185.

[257] Fanta PE. Ullmann synthesis of biaryls. Synthesis. 1974;(1):9–21.

[258] Kharasch MS, Jensen EV, Urry WH. Addition of carbon tetrachloride and chloroform to olefins. Science. 1945;102:128.

[259] Stanforth SP. Catalytic cross-coupling reactions in biaryl synthesis. Tetrahedron. 1998;54(3–4):263–303.

[260] Negishi E, King AO, Okukado N. Selective carbon-carbon bond formation *via* transition metal catalysis. 3. A highly selective synthesis of unsymmetrical biaryls and diarylmethanes by the nickel- or palladium-catalyzed reaction of aryl- and benzylzinc derivatives with aryl halides. J Org Chem. 1977;42(10):1821–1823.

[261] Negishi E, Takahashi T, King AO. Synthesis of biaryls *via* palladium-catalyzed cross coupling: 2-methyl-4′-nitrobiphenyl. Org Synth. 1988;66:67–74.

[262] Stille JK. Palladium catalyzed coupling of organotin reagents with organic electrophiles. Pure Appl Chem. 1985;57(12):1771–1780.

[263] Suzuki A. New synthetic transformations *via* organoboron compounds. Pure Appl Chem. 1994;66(2):213–222.

[264] Larson ER, Raphael RA. Synthesis of (−)-steganone. J Chem Soc [Perkin 1]. 1982;(2):521–525.

[265] Monovich LG, Le Huérou Y, Rönn M, Molander GA. Total synthesis of (−)-steganone utilizing a samarium(II) iodide promoted 8-endo ketylolefin cyclization. J Am Chem Soc. 2000;122(1):52–57.

[266] Uemura M, Daimon A, Hayashi Y. An asymmetric synthesis of an axially chiral biaryl *via* an (arene)chromium complex: formal synthesis of (−)-steganone. J Chem Soc, Chem Commun. 1995;(19):1943–1944.

[267] Robin JP, Gringore O, Brown E. Asymmetric total synthesis of the antileukemic lignan precursor (−)-steganone and revision of its absolute configuration. Tetrahedron Lett. 1980;21(28):2709–2712.

[268] Robin JP, Dhal R, Brown E. Syntheses totales et etudes de lignanes biologiquement actifs–III: application de l'α-hydroxyalkylation de β-benzyl γ-butyrolactones a la creation des squelettes phenyl tetraline et bisbenzocyclooctadiene. Premiere synthese des picrosteganes, synthese formelle de la (±)-steganacine. Tetrahedron. 1984;40(18):3509–3520.

[269] Ziegler FE, Fowler KW, Sinha ND. A total synthesis of (±) steganacin *via* the modified Ullmann reaction. Tetrahedron Lett. 1978;19(31):2767–2770.

[270] Carroll AR, Read RW, Taylor WC. Intramolecular oxidative coupling of aromatic compounds. VII. a convenient synthesis of (±)-deoxyschizandrin. Aust J Chem. 1994;47(8):1579–1589.

[271] Molander GA, Harris CR. Sequencing reactions with samarium(II) iodide. Chem Rev. 1996;96(1):307–338.

[272] Molander GA, McKie JA. Synthesis of substituted cyclooctanols by a samarium(II) iodide promoted 8-endo radical cyclization process. J Org Chem. 1994;59(11):3186–3192.

[273] Molander GA, George KM, Monovich LG. Total synthesis of (+)-isoschizandrin utilizing a samarium(II) iodide-promoted 8-endo ketylolefin cyclization. J Org Chem. 2003;68(25):9533–9540.

[274] Curran DP, Fevig TL, Jasperse CP, Totleben MJ. New mechanistic insights into reductions of halides and radicals with samarium(II) iodide. Synlett. 1992;(12):943–961.

[275] Kunishima M, Hioki K, Kono K, Sakuma T, Tani S. Barbier-type reactions of aryl halides with ketones mediated by samarium diiodide. Chem Pharm Bull. 1994;42(10):2190–2192.

[276] Huneck S. New results on the chemistry of lichen substances. In: Herz W, Falk H, Kirby GW, Moore RE, editors. Progress in the Chemistry of Organic Natural Products, Progress in the Chemistry of Organic Natural Products. vol. 81. Vienna: Springer-Verlag GmbH; 2001. p. 1–276.

[277] Ikeya Y, Taguchi H, Yosioka I, Kobayashi H. The constituents of *Schizandra chinensis* Baill. IV. The structures of two new lignans, pre-gomisin and gomisin J. Chem Pharm Bull. 1979;27(7):1583–1588.

[278] Robin JP, Landais Y. Ruthenium(IV) dioxide in fluoro acid medium. An efficient biaryl phenol coupling process, exemplified with a biomimetic access to the skeleton of steganacin from presteganes. J Org Chem. 1988;53(1):224–226.

[279] Cambie RC, Clark GR, Craw PA, Rutledge PS, Woodgate PD. Synthesis and structure of a stegane from dimethylmatairesinol. Aust J Chem. 1984;37(8):1775–1784.

[280] Burden JK, Cambie RC, Craw PA, Rutledge PS, Woodgate PD. Oxidative coupling of lignans. IV. Monophenolic oxidative coupling. Aust J Chem. 1988;41(6):919–933.

[281] Pelter A, Ward RS, Venkateswarlu R, Kamakshi C. Oxidative transformations of lignans. Reactions of dihydrocubebin and a derivative with DDQ. Tetrahedron. 1991;47(7):1275–1284.

[282] Biftu T, Hazra BG, Stevenson R. Synthesis of (±)-deoxyschizandrin. J Chem Soc [Perkin 1]. 1979:2276–2281.

[283] Xie L, Chen L, Xie J. Progress on the synthesis of dibenzocyclooctadiene lignans. Youji Huaxue. 1991;11(4):371–381.

[284] McMurry JE. Carbonyl-coupling reactions using low-valent titanium. Chem Rev. 1989;89(7):1513–1524.

[285] Qi XX, Chang JB, Guo RY, Chen RF. The synthesis of a novel lignan. Chin Chem Lett. 2000;11(11):971–974.

[286] Chang JB, Xie JX. Synthesis of (±)-deoxyschisandrin and the corresponding *trans*-isomer. Chin Chem Lett. 1996;7(9):801–802.

[287] Krauss AS, Taylor WC. Intramolecular oxidative coupling of aromatic compounds. I. Oxidation of diphenolic substrates. Aust J Chem. 1991;44(9):1307–1333.

[288] Krauss AS, Taylor WC. Intramolecular oxidative coupling of aromatic compounds. II. The synthesis of (2*RS*,3*SR*)- and (2*RS*,3*RS*)-2,3-dimethyl-1,4-bis(3,4,5-trimethoxyphenyl)butan-1-one and the determination of stereochemistry. Aust J Chem. 1991;44(9):1335–1340.

[289] Krauss AS, Taylor WC. Intramolecular oxidative coupling of aromatic compounds. III. Monophenolic substrates. Aust J Chem. 1992;45(5):925–933.

[290] Krauss AS, Taylor WC. Intramolecular oxidative coupling of aromatic compounds. IV. Oxidation of non-phenolic substrates. Aust J Chem. 1992;45(5):935–939.

[291] Carroll AR, Krauss AS, Taylor WC. Intramolecular oxidative coupling of aromatic compounds. V. *para-para* Diphenolic oxidative coupling as a possible route to the eupodienone skeleton. Aust J Chem. 1993;46(3):277–292.

[292] Takeya T, Ohguchi A, Ara Y, Tobinaga S. Synthesis of (±)-dibenzocyclooctadiene lignans, (±)-schizandrin, (±)-gomisin a and their stereoisomers, utilizing the samarium-Grignard reaction. Chem Pharm Bull. 1994;42(3):430–437.

[293] Belletire JL, Fry DF, Fremont SL. The role of dianion coupling in the synthesis of dibenzylbutane lignans. J Nat Prod. 1992;55(2):184–193.

[294] Mahalanabis KK, Mumtax M, Snieckuz V. Dimetalated tertiary succinamides. Synthesis of several classes of lignans including the mammalian urinary lignans enterolactone and enterodiol. Tetrahedron Lett. 1982;23(39):3975–3978.

[295] Magnus P, Schultz J, Gallagher T. A short synthesis of (±)-steganone and (±)-steganacin. J Chem Soc, Chem Commun. 1984;(17):1179–1180.

[296] Taylor EC, Andrade JG, Rall GJH, McKillop A. Thallium in organic synthesis. 59. Alkaloid synthesis *via* intramolecular nonphenolic oxidative coupling. Preparation of (±)-ocoteine, (±)-acetoxyocoxylonine, (±)-3-methoxy-n-acetylnornantenine, (±)-neolitsine, (±)-kreysigine, (±)-*O*-methylkreysigine, and (±)-multifloramine. J Am Chem Soc. 1980;102(21):6513–6519.

[297] Chen LH, Xie L, Xie JX. Total synthesis of an analogue of schizandrin. Yaoxue Xuebao. 1991;26(1):20–24.

[298] Chang JB, McPhail AT, Kong M, Xie JX. Intramolecular oxidative transformations of 1,4-diaryl-2,3-dimethyl-2,3-epoxybutane with DDQ. Chin Chem Lett. 1996;7(8):691–692.

[299] Chang J, Xie J. Total synthesis of schizandrin, the main active ingredient isolated from the Chinese herbal medicine fructus schizandrae. Yaoxue Xuebao. 1998;33(6):424–428.

[300] Chang J, Xie J, Chen R, Liu P. Total synthesis of (±)-deoxyschisandrin and its analogs. Yaoxue Xuebao. 1999;34(12):913–917.

[301] Chang J, Xie J. Synthesis of some new lignans and the mechanism of intramolecular nonphenolic oxidative coupling of aromatic compounds. Sci China, Ser B: Chem. 2000;43(3):323–330.

[302] Chang JB, Xie JX. Synthesis of new schizandrin analogues. Chin Chem Lett. 2001;12(8):667–670.

[303] Kupchan SM, Britton RW, Ziegler MF, Gilmore CJ, Restivo RJ, Bryan RF. Tumor inhibitors. LXXXX. Steganacin and steganangin, novel antileukemic lignan lactones from *Steganotaenia araliacea*. J Am Chem Soc. 1973;95(4):1335–1336.

[304] Wang RWJ, Rebhun LI, Kupchan SM. Antimitotic and antitubulin activity of the tumor inhibitor steganacin. Cancer Res. 1977;37(9):3071–3079.

[305] Hicks RP, Sneden AT. Neoisostegane, a new bisbenzocyclooctadiene lignan lactone from *Steganotaenia araliacea* Hochst. Tetrahedron Lett. 1983;24(29):2987–2990.

[306] Tomioka K, Ishiguro T, Mizuguchi H, Komeshima N, Koga K, Tsukagoshi S, *et al*. Stereoselective reactions. XVII. Absolute structure-cytotoxic activity relationships of steganacin congeners and analogs. J Med Chem. 1991;34(1):54–57.

[307] Becker D, Hughes LR, Raphael RA. Total synthesis of the antileukaemic lignan (±)-steganacin. J Chem Soc [Perkin 1]. 1977;14:1674–1681.

[308] Becker D, Hughes LR, Raphael RA. Synthesis of a steganone analog. J Chem Soc, Chem Commun. 1974;(11):430–431.

[309] Hughes LR, Raphael RA. Synthesis of the antileukemic lignan precursor (±)-steganone. Tetrahedron Lett. 1976;(18):1543–1546.

[310] Magnus P, Schultz J, Gallagher T. Synthesis of the antileukemic agent (±)-steganone using a stereoconvergent biaryl coupling reaction. J Am Chem Soc. 1985;107(17):4984–4988.
[311] Meyers AI, Flisak JR, Aitken RA. Asymmetric synthesis of (−)-steganone. Further application of chiral biaryl syntheses. J Am Chem Soc. 1987;109(18):5446–5452.
[312] Kende AS, Liebeskind LS. Total synthesis of (±)-steganacin. J Am Chem Soc. 1976;98(1): 267–268.
[313] Ziegler FE, Chliwner I, Fowler KW, Kanfer SJ, Kuo SJ, Sinha ND. The ambient temperature Ullmann reaction and its application to the total synthesis of (±)-steganacin. J Am Chem Soc. 1980;102(2):790–798.
[314] Monovich LG, Le Huérou Y, Rönn M, Molander GA. Total synthesis of (−)-steganone utilizing a samarium(II) iodide promoted 8-endo ketylolefin cyclization. J Am Chem Soc. 1999;122(1):52–57.
[315] Narasimhan NS, Aidhen IS. Radical mediated intramolecular arylation using tributyltinhydride/AIBN: a formal synthesis of steganone. Tetrahedron Lett. 1988;29(24):2987–2988.
[316] Tomioka K, Mizuguchi H, Ishiguro T, Koga K. Stereoselective reactions. VII. Synthesis of racemic and optically pure stegane, isostegane, picrostegane, and isopicrostegane *via* highly selective isomerization. Chem Pharm Bull. 1985;33(1):121–126.
[317] Uemura M, Kamikawa K. Stereoselective induction of an axial chirality by Suzuki cross coupling of tricarbonyl(arene)chromium complexes with arylboronic acids. J Chem Soc, Chem Commun. 1994;(23):2697–2698.

11
Biological Properties of Lignans

11.1 Introduction

Lignans are known to exhibit a rich structural diversity and a varied biological activity. Both have always attracted considerable attention among phytochemists, botanists, pharmacologists, environmentalists, and recently even experts in food production. This growing interest in bioactive lignans is motivated mainly by their potential use as either phytopharmaceuticals or nutraceuticals [1].

In fact, among the several families of secondary metabolites synthesized by plants, lignans are recognized as a class of natural products with a wide range of remarkable biological activities (see Table 11.1).

According to such interest, much effort has been devoted to understanding the biosynthesis of lignans and their biological activity [15–17] during the last decade. In this chapter, the most outstanding results on both issues will be highlighted.

Concerning biological activity, two different approaches are reported in the literature when discussing lignans. One approach focuses on biological activity itself, listing the different lignans that present such activity. This trend is followed, for instance, by Deyama and Nishibe [18] or Saleem *et al.* [8]. By contrast, the other approach describes the different types of lignans, while discussing the biological activity of each of them. For example, MacRae and Towers [5], and also Ward [19–22] with respect to the synthesis, structure, and isolation of lignans, followed in this direction. Using one approach or another depends only on the viewpoint and the preferences of the author. In this chapter, even though developing both approaches together might lead the study into certain repetitions, we will thoughtfully mix the best of each approach in an effort to provide the most detailed information to the reader with the most flexible searching criteria.

When a new lignan is isolated from a natural source, generally a "trivial" name is given. These trivial names are related to their biological origin. Although trivial names are considered to be ephemeral and replaced by names describing the skeleton and the characteristic groups, when the full structure is known, an IUPAC "systematic name" may be generated. However, this name is too cumbersome to be continually inserted into the text of the present work, and therefore, we use trivial names (see Table 10.1 on page 322 for synonyms).

Lignin and Lignans as Renewable Raw Materials: Chemistry, Technology and Applications, First Edition.
Francisco G. Calvo-Flores, José A. Dobado, Joaquín Isac-García and Francisco J. Martín-Martínez.
© 2015 John Wiley & Sons, Ltd. Published 2015 by John Wiley & Sons, Ltd.

Table 11.1 Main reported biological activities of lignans [2]

Biological activity	References
Antiviral	[3–6]
Anticancer	[5–8]
Cancer prevention	[9, 10]
Anti-inflammatory	[8]
Antimicrobial	[8]
Antioxidant	[7, 8, 11]
Immunosuppressive	[8]
Hepatoprotective	[12]
Osteoporosis prevention	[13, 14]

11.2 Biosynthesis of Lignans

The biosynthesis of lignans is closely related to the biosynthesis of many other phenylpropanoid compounds such as lignin, neolignans, and norlignans [17]. In fact, lignins as well as lignans are biosynthesized through the same pathway in the earlier steps, starting from hydroxycinnamyl alcohols (see Chapter 4).

The synthesis of the hydroxycinnamyl alcohol monomers, precursors of lignans, according to Dewick [23], takes place from the reaction of L-phenylalanine (L-Phe) and L-tyrosine (L-Tyr), mediated by a series of cinnamic acid derivatives. More specifically, the chemical reduction of these acids forms three different alcohols, namely *p*-coumaryl alcohol (**1**), coniferyl alcohol (**2**), and sinapyl alcohol (**3**), which are the main precursors of lignins and lignans (see Figure 11.1).

According to the proposed mechanism, the peroxidase enzyme induces a one-electron oxidation of the phenol group allowing the delocalization of the unpaired electron by resonant forms (see Figure 4.10 on p. 85).

11.2.1 Pinoresinol Synthase (Dirigent Protein)

The radical pairing of the monomer resonant structures described above gives rise to reactive dimeric species vulnerable of nucleophilic attack. Consequently, a wide range of lignins and lignans can be produced. Among these products, the dimerization of two coniferyl alcohol monomeric units (see resonance form IV depicted in Figure 4.10) into pinoresinol (PIN) (**4**) *via* intermolecular 8, 8′ oxidative coupling with the aid of dirigent protein (DIR) is the most well-known process to date (see Figure 11.2) [2].

Despite being biosynthesized from the same immediate monomeric precursor, that is, coniferyl alcohol (**2**), lignans and lignins differ fundamentally in their optical activity. Lignans are frequently found in optically active form, where the particular enantiomer observed can vary according to the plant species [24]. This implies that lignan biosynthesis is under strict enantioselective control of monomer coupling, whereas lignin biosynthesis is not. For example, (+)-pinoresinol is present in *Forsythia* species [25], whereas the (−)-enantiomer occurs in *Daphne tangutica* [26]. It is commonly assumed that the coupling mode depends on the plant species.

The first example of dirigent protein (DIR) activity was in (+)-pinoresinol synthase activity, which was found in *Forsythia intermedia* [27] and was then purified and called DIR [28]. Notably, the DIR that controls this transformation lacks oxidative catalytic capacity by itself, but in the presence of an appropriate oxidase, it can confer absolute specificity to the coupling reaction [24].

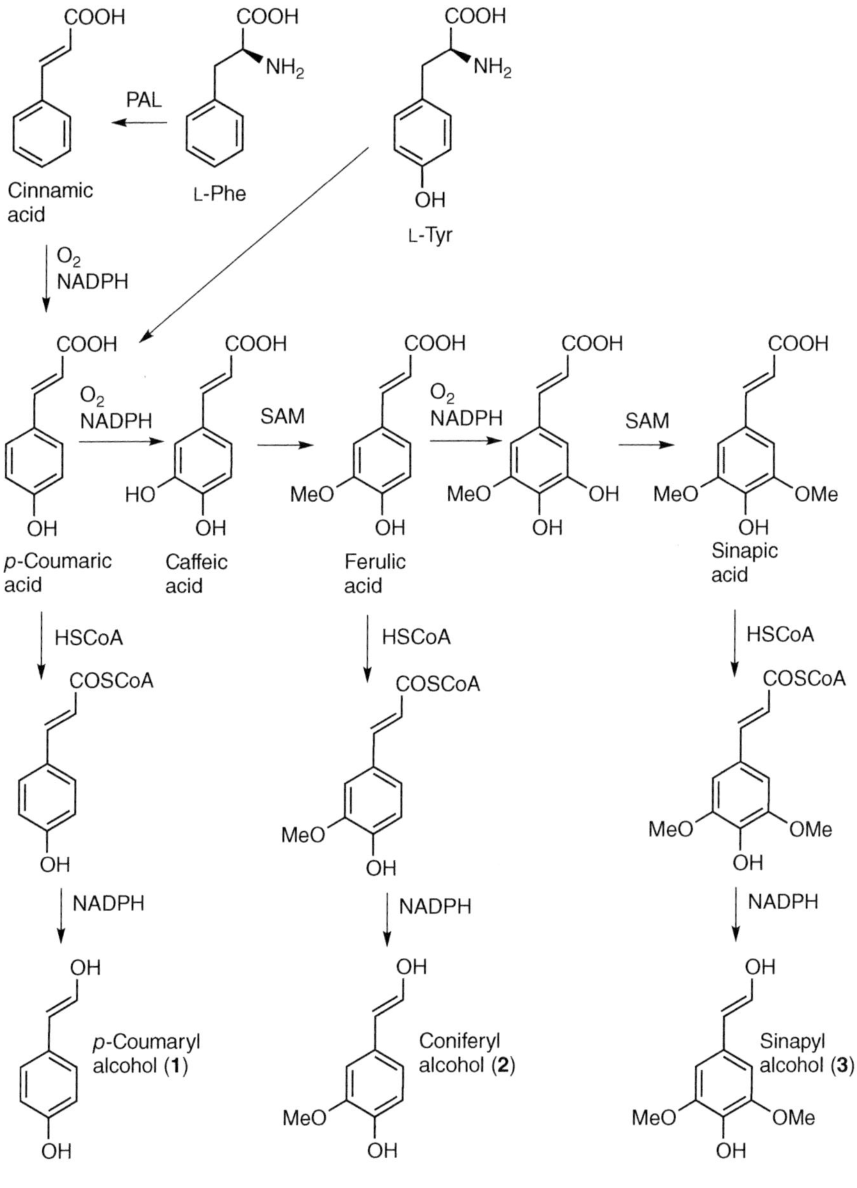

Figure 11.1 Biosynthesis of the precursors of lignans (hydroxycinnamyl alcohol monomers) [23]

11.2.2 The General Biosynthetic Pathway of Lignans

Biosynthesis of lignans with and without $C_9(C_{9'})$ oxygen diverges mainly in the pathways of the phenylpropanoid monomers, and it is after phenylpropanoid dimerization step when the diversity of the aromatic substitution is introduced [16].

Biosynthesis of lignans with $C_9(C_{9'})$ oxygen is the most studied in depth [17]. In this synthesis, an enantioselective dimerization of two coniferyl alcohol units with the aid of DIR protein is used to form

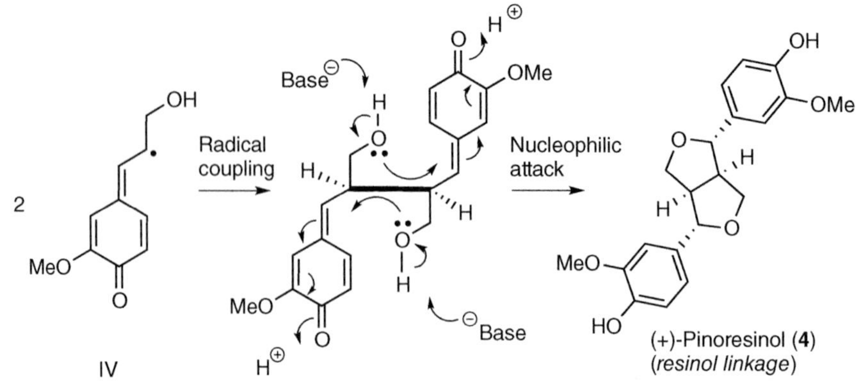

Figure 11.2 Enantioselective formation of (+)-pinoresinol with the aid of dirigent protein (DIR) [23]

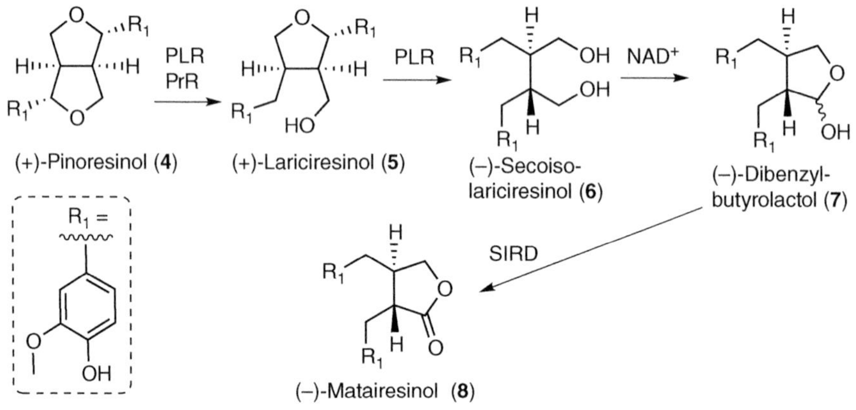

Figure 11.3 Conversion of pinoresinol into matairesinol [16]

the lignan pinoresinol (**4**) (see Figure 11.2). The general lignan biosynthetic pathway suggests that the conversion occurs from coniferyl alcohol (**2**) to matairesinol (MAT) (**8**), as has been demonstrated in various plant species. "*It is worth emphasizing that all biochemical processes involving coupling and post-coupling modifications in lignans are under explicit control*" [24].

The conversion of pinoresinol (**4**) into secoisolariciresinol (SECO) (**6**) is mediated by pinoresinol/lariciresinol reductases (PLR) in the presence of NADPH and hydrogen peroxide [29]. Later, secoisolariciresinol dehydrogenase (SIRD) catalyzed the selective oxidation of secoisolariciresinol (**6**) into matairesinol (**8**) in the presence of NADH, *via* dibenzyl butyrolactol (**7**) (see Figure 11.3) [30, 31].

11.2.3 Other Biosynthetic Pathways of Lignans

Biosynthetic pathways for many other lignans can be regarded as starting from these four lignans: pinoresinol (**4**), lariciresinol (LARI) (**5**), secoisolariciresinol (**6**), and matairesinol (**8**). The remaining structural variations are reached from different synthetic pathways described in the literature (see Figure 11.4) [2, 15, 16].

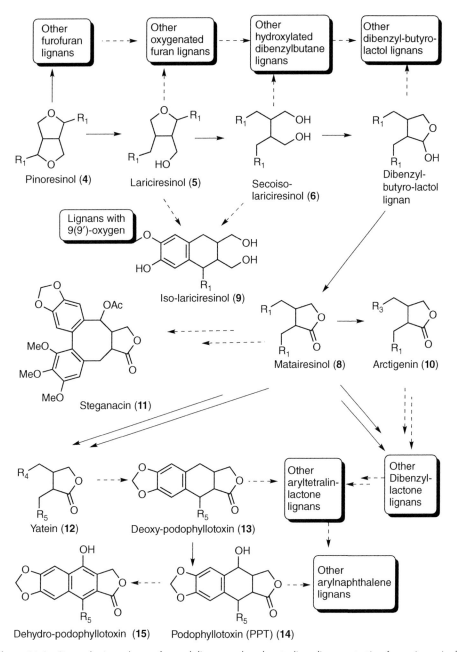

Figure 11.4 Biosynthetic pathways for cyclolignan and cyclooctadiene lignans starting from pinoresinol [16]. See Figure 11.7 for the meaning of R groups

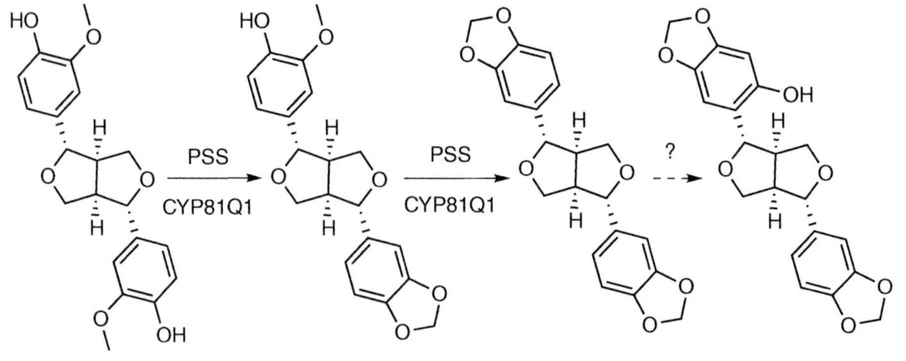

Figure 11.5 Biosynthetic pathway for (+)-sesaminol from (+)-pinoresinol. Undetermined pathways are indicated with a dashed arrow and "?" symbol [32]

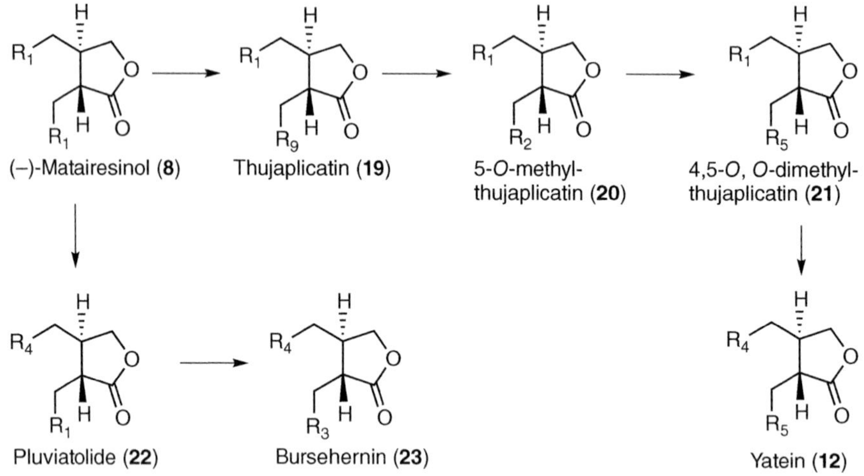

Figure 11.6 Biosynthetic pathway for yatein and bursehemin in Anthriscus sylvestris [17]. See Figure 11.7 for the meaning of R groups

Sesamin (**17**), the most abundant lignan in sesame seeds, and other furofuran lignans are synthesized from pinoresinol (**4**) by the (+)-piperitol/(+)-sesamin synthase (PSS) enzyme CYP81Q1 (see Figure 11.5) [32].

Sakakibara *et al.* [33] demonstrated a direct pathway for the transformation of matairesinol (**8**) to another dibenzylbutirolactone lignan called yatein (**12**), *via* thujaplicatin (**19**) (see Figure 11.6) [16].

Toward the end of the 1980s, Dewick [34] revealed the conversion of yatein (**12**) into podophyllotoxin (PPT) (**14**) *via* deoxy-PPT (**13**) [16, 17, 35], although the enzyme that converts yatein (**12**) into deoxy-PPT (**13**) is still unknown (see Figure 11.8).

Despite the enzymatic data reported on the late steps of the biosynthesis of PPT (**14**), the early steps are not yet well understood, and therefore such biosynthesis is not completely known. Different synthetic pathways have been proposed. Sakakibara *et al.* [33] suggest a sequence that starts with matairesinol and forms yatein in four steps (see Figure 11.6). It implies a biosynthetic pathway in

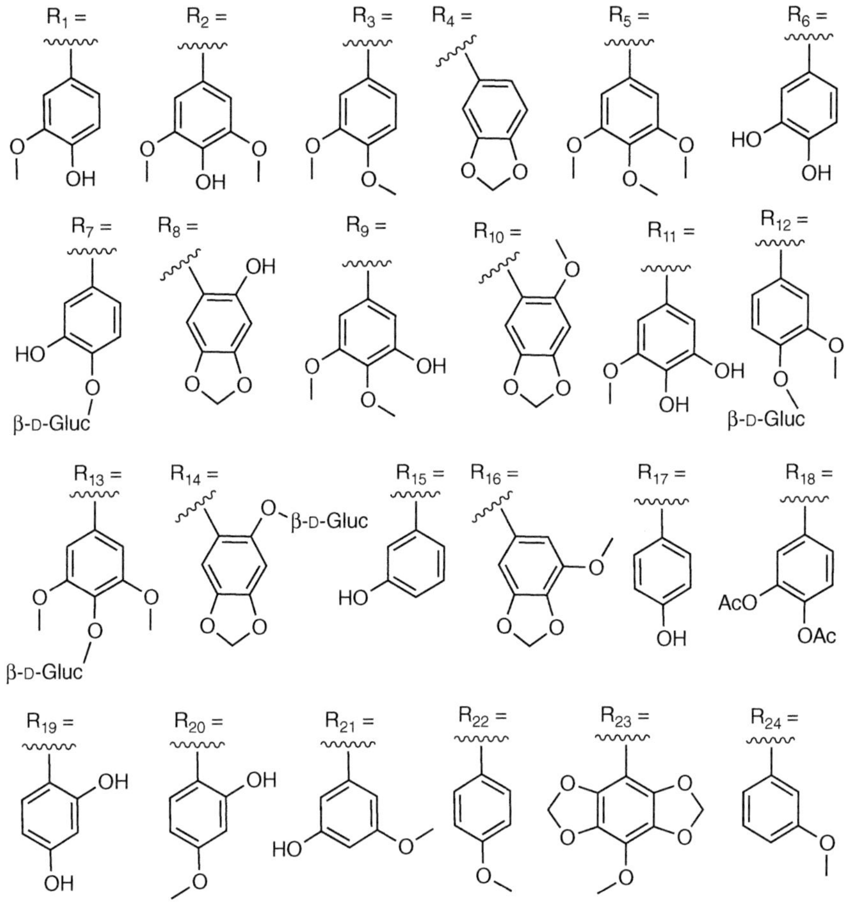

Figure 11.7 Common lignan substituents $R_1 - R_{24}$ used in figures for clarity

which the lignan yatein is an intermediate toward PPT (**14**) and 6-methoxy-PPT (**26**). Unfortunately, the catalyzing enzymes remain uncharacterized [33], although the presumed intermediates in plant (e.g., see ref. [36]) have been identified. In this way, (−)-matairesinol (**8**) has been established as a direct precursor for PPT (**14**), while yatein (**12**) has been to be effectively converted into PPT (**14**) [35]. The process would imply the stereospecific coupling of coniferyl alcohol (**2**) to form (−)-matairesinol (**8**) through a series of reactions, where ferulic acid and methylenedioxy-substituted cinnamic acid could also be incorporated into lignans [37].

Concerning the enzymes, deoxy-PPT 6-hydroxylase (DOP6H) is the first one established to be involved in the late steps of the formation of PPT (**14**) and derivatives. According to the studies, the formation of 6-methoxy-PPT (**26**) does not proceed via PPT (**14**) [38], but through a 6-hydroxylation of deoxy-PPT (**13**) to β-peltatin (**24**) carried out by the DOP6H enzyme, which is a P450-dependent monooxygenase. It also requires a 7-hydroxylase to be formed from deoxy-PPT (**13**) [39]. The drawback to this synthetic pathway is that the expected cytochrome P450 protein remains to be isolated and characterized.

The second enzyme established as being involved in the biosynthetic pathway is β-peltatin 6-*O*-methyltransferase (β-peltatin 6OMT). This enzyme has been partially purified from cell-suspension

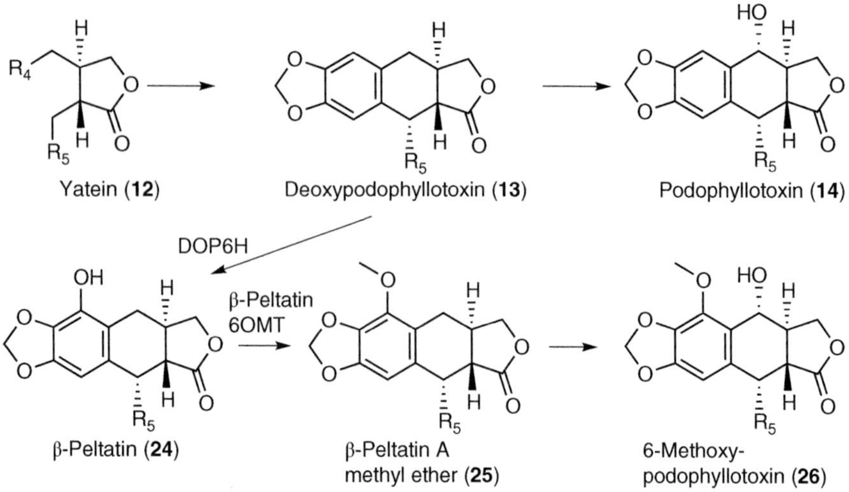

Figure 11.8 Biosynthetic pathway for podophyllotoxin (PPT) (**14**) and 6-methoxy-PPT [36]. See Figure 11.7 for the meaning of R groups

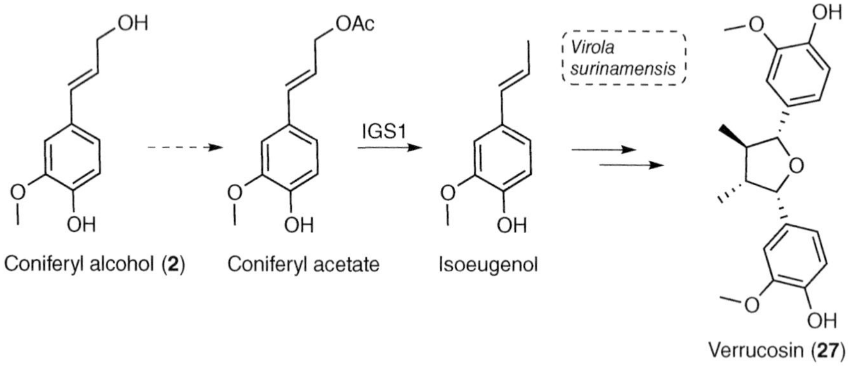

Figure 11.9 Isoeugenol formation from coniferyl alcohol by isoeugenol synthase and its conversion into verrucosin lignan in *Virola surinamensis* [16]

cultures of *Linum nodiflorum* [40]. In this pathway, 6-methoxy-PPT (**26**) is formed by hydroxylation at position 6 of β-peltatin A methyl ether (**25**), although the catalyzing enzyme still remains to be identified (see Figure 11.8).

The same problem (identification of the actual enzymes involved in the pathway) happens also with another alternative pathway that involves the formation of 7′-hydroxymatairesinol (**49**), which is efficiently metabolized into 5-methoxy-PPT [41].

Given the structural similarity, it seems likely that arylnaphthalene lignans are formed from aryltetralin lignans by dehydrogenation or by hydroxylation followed by a dehydration step [17] (see Figure 11.4).

The biosynthesis of lignans without $C_9(C_{9'})$ oxygen diverges mainly in the pathways of the phenylpropanoid monomers, since they are formed from the related phenylpropanoid monomers with and without $C_9(C_{9'})$ oxygen, respectively. For example, the conversion of coniferyl alcohol (**2**) into verrucosin (**27**) (see Figure 11.9) [16, 42–44].

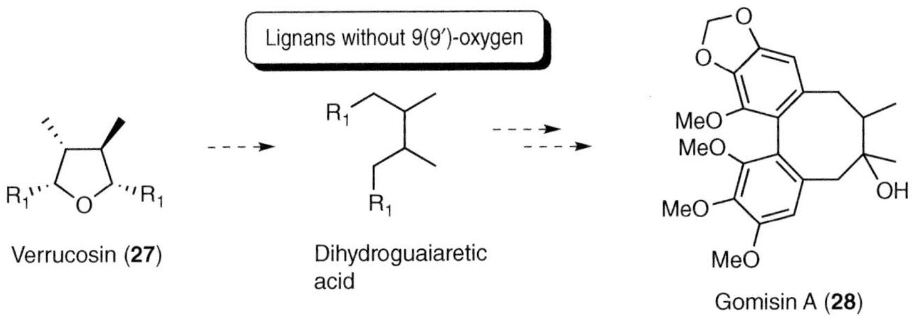

Figure 11.10 *Biosynthetic pathways for cyclooctadiene lignans, starting from coniferyl alcohols [16]. See Figure 11.7 for the meaning of R groups*

Dibenzocyclooctadiene lignans with oxygen at $C_9(C_{9'})$ (e.g., steganacin, **11**) might be formed from the "seco-" analog without the biphenyl bond, dibenzocyclooctadiene lignans without $C_9(C_{9'})$ oxygen (e.g., wuweizisu and gomisin) may be formed by the coupling of propenylphenols (see Figure 11.10) [42, 43].

11.2.4 Biosynthetic Pathways for Neolignans

Neolignans are formed by the coupling of two propenylphenol units, such as in the case of lignans, but attached differently than $\beta - \beta(C_8(C_{8'}))$ bonds. The fact that DIR proteins participate in this coupling catalyzed by oxidases is supported by the chirality of the majority of neolignans. For example, an abundant lignin type is the one containing the $C_8(C_{5'})$ linkage motif. These substances are formed by reduction of the allylic side chain of dehydrodiconiferyl alcohol. A tentative biosynthetic pathway is shown in Figure 11.11.

11.2.5 Biosynthetic Pathways for Norlignans

Similar to lignans, biosynthesis of norlignans has also been suggested to be related to lignin biosynthesis, given that norlignans have two aromatic rings and a side chain with five carbons [45].

In this way, norlignans, such as hinokiresinol (**32**), the most simple of norlignans, were demonstrated by Suzuki *et al.* [46–48] to be formed from phenylpropanoid monomers *via* p-coumaryl p-coumarate (**31**) as a dimeric intermediate (see Figures 11.12 and 11.13).

Sequirin D (**33**) may be formed *via* a dienonephenol rearrangement (see Figure 11.14) [49].

11.3 Metabolism of Lignans

Components of orally administered herbal medicines are often converted into pharmacologically active compounds by intestinal flora. Some lignans in plants and natural medicines are called phytoestrogens, because they are transformed by intestinal microflora into molecules as enterodiol (END) (**36**) and enterolactone (ENL) (**38**) which show estrogen-like biological activity [50–53].

Mazur *et al.* [54] confirmed that enterolignans[1] are produced and absorbed in the colon mainly from lignan intake in the diet. Later, the metabolism of plant lignans by human faecal microflora was investigated by Heinonen *et al.* [53]. END has been found, as glucuronides, in the urine of humans, baboons, vervet monkeys, and rats [55–57].

---
[1] Any lignan formed from another by metabolism in the gut.

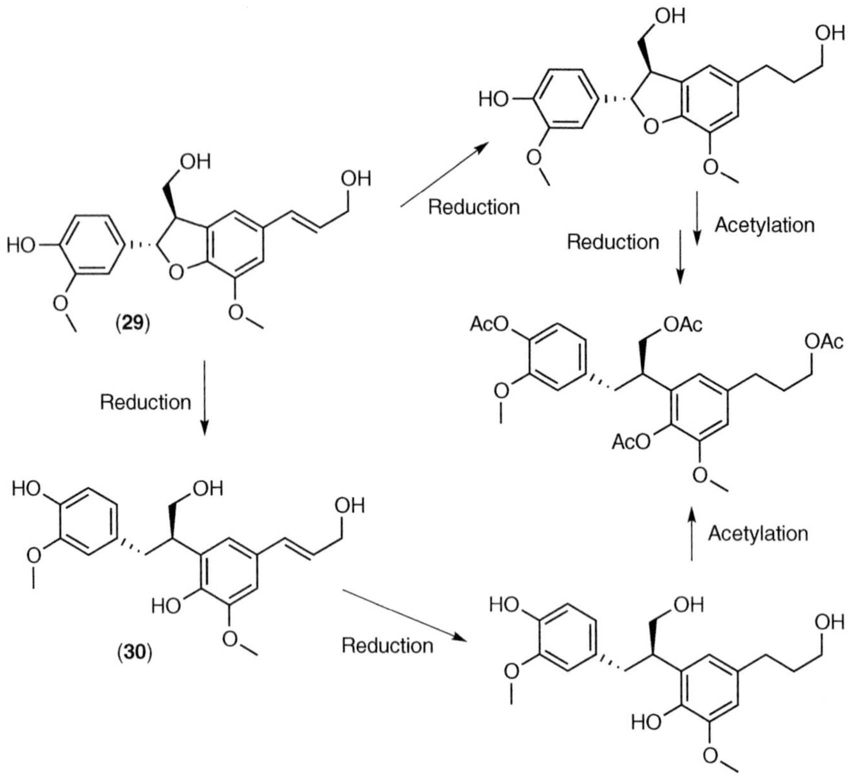

Figure 11.11 *Proposed biosynthetic pathway to neolignans containing 8–5′ linkage present in Cryptomeria japonica [24]*

Mammalian lignans are formed in the human body in the gastrointestinal tract, where gastrointestinal bacteria hydrolyze the sugar moiety of secoisolariciresinol diglucoside (SDG) (**34**) to release SECO [58, 59]. This is followed by dehydroxylation and demethylation by the colonic microflora to give the mammalian lignan END (see Figure 11.15). END is presumed to be oxidized by the gastrointestinal microbiota to produce ENL.

Figure 11.15 depicts the bacterial metabolites formed from SDG. Intestinal bacteria in human gut catalyze the *O*-deglycosylation of SDG and yield SECO, and then, by means of a *O*-demethylation and dehydroxylation of SECO, the skeleton is transformed to END, and later END is oxidized to ENL.

One or two additional reduction steps are involved in enterolignan production from lariciresinol (**5**) and pinoresinol (**4**), respectively [66].

The structure of syringaresinol (SYR) (**47**) differs from that of pinoresinol (**4**) in two additional methoxy groups at meta positions in both phenolic rings, and this relatively complex substitution pattern can yield several metabolites during incubation. Most of the metabolites that were tentatively identified for SYR share an oxydiarylbutane structure, suggesting that SYR breakdown occurs similarly to that of pinoresinol (**4**), involving the corresponding lariciresinol and SECO intermediates (see Figure 11.15).

Two reactions of demethylation and dehydroxylation are needed to transform matairesinol (**8**) into ENL, with 4,4′-dihydroxy ENL (**39**) being found after demethylation reactions. However, this is

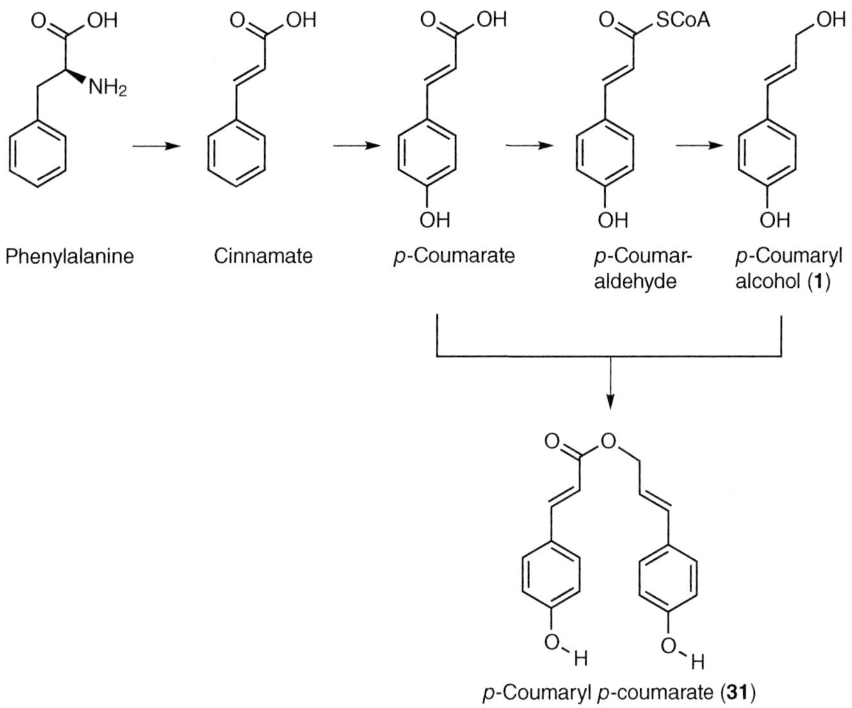

Biological Properties of Lignans 379

Phenylalanine → Cinnamate → p-Coumarate → p-Coumaraldehyde → p-Coumaryl alcohol (1)

p-Coumaryl p-coumarate (31)

Figure 11.12 The cinnamate/monolignol biosynthetic pathway [48]

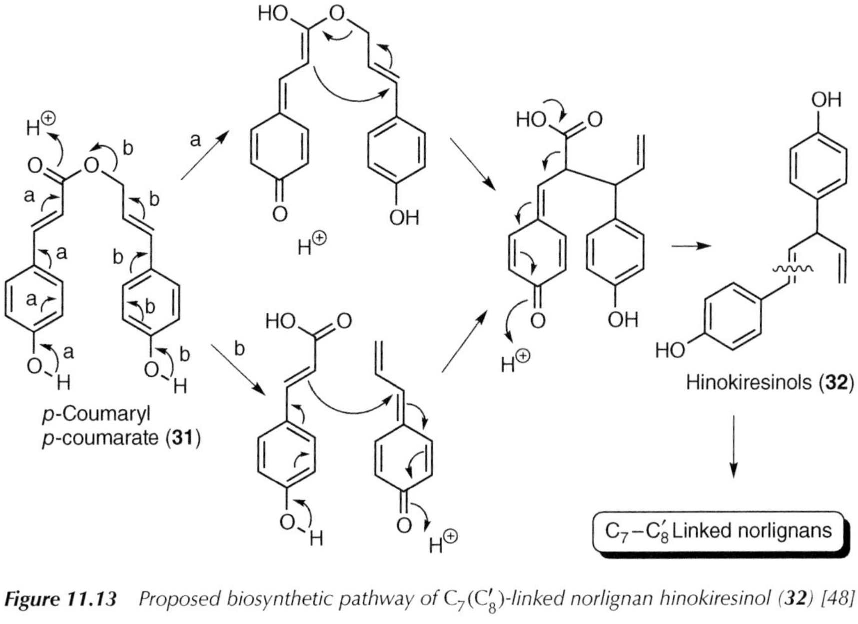

p-Coumaryl p-coumarate (31) → Hinokiresinols (32) → C_7–C_8' Linked norlignans

Figure 11.13 Proposed biosynthetic pathway of $C_7(C_8')$-linked norlignan hinokiresinol (32) [48]

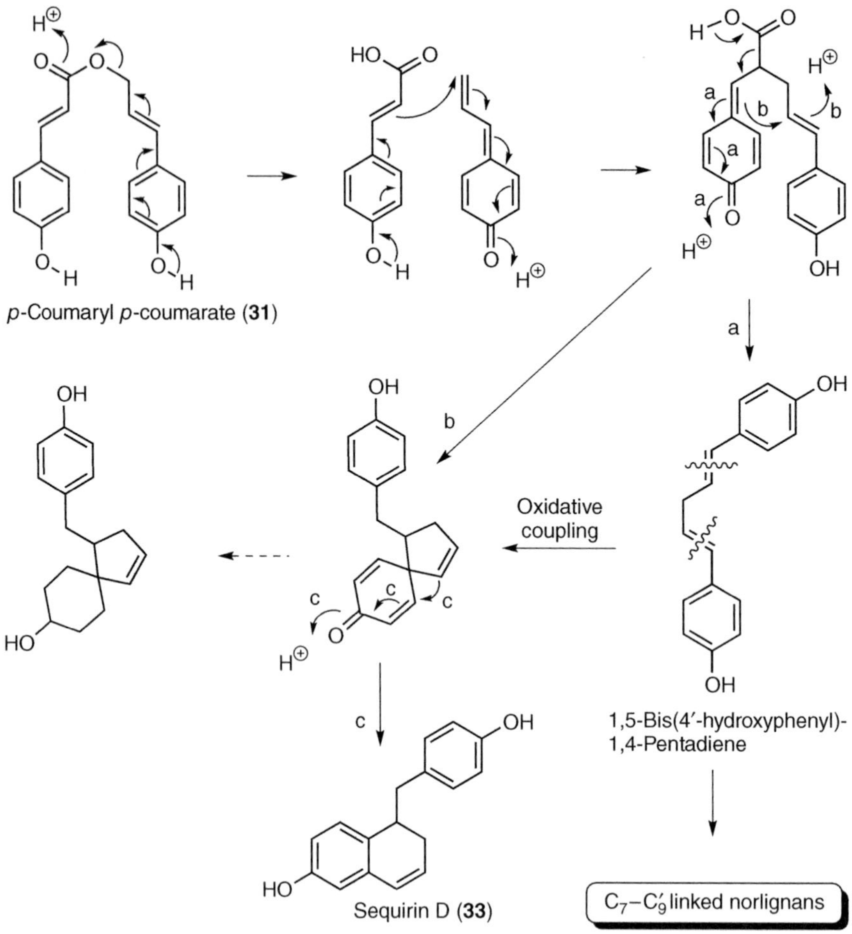

Figure 11.14 Proposed biosynthetic pathway of $C_7(C_9')$-linked norlignan Sequirin D (**33**) [49]

likely to be a minor metabolic pathway if other lignans are present in the diet [59] (see Figure 11.16). Compounds such as 4,4′-dihydroxy ENL (**39**) and 4,4′-dideoxy ENL are theoretical metabolites of (+)-matairesinol (**8**), which is most likely a plant precursor of ENL (**38**) [67].

The transformation of arctiin (**40**), already bearing a butyrolactone structure, consists of the first three types of reactions mentioned for SDG; the hydrolysis of lignan glucosides to their aglycones is the first transformation step; second, demethylation of a methoxy group adjacent to a hydroxy group occurs; third, dehydroxylation; fourth and fifth demethylation; and the last dehydroxylation reactions produce ENL (**38**) (see Figure 11.17).

Also, (+)-sesamin (**17**) is converted into enterolactone (**38**) via sesaminol triglucoside (STG) (**35**), showing that the methylene dioxy ring was cleaved by intestinal bacteria (see Figure 11.18).

Tracheloside (**42**) was converted into trachelogenin (**44**) and then metabolized to deoxytrachelogenin (**43**) [70, 71] (see Figure 11.19).

(−)-Olivil 4′,4″-di-*O*-β-D-glucoside (**45**) and (+)-1-hydroxypinoresinol 4′,4″-di-*O*-β-D-glucoside (**46**) were transformed to (−)-enterolactone (**38**) by human intestinal bacteria via 2-hydroxyenterodiol

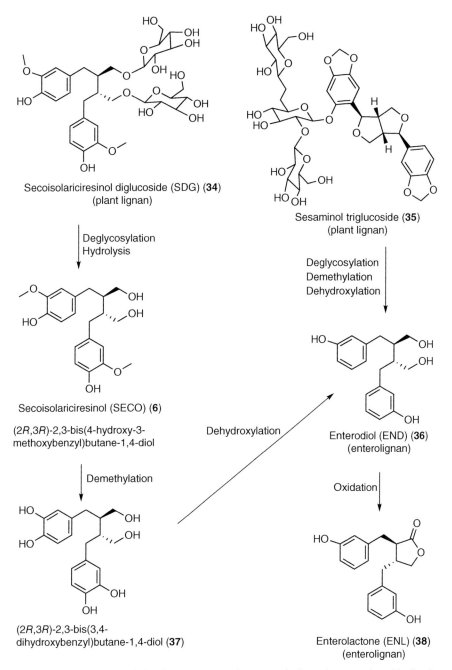

Figure 11.15 Bioconversion of plant lignans to enterolignans in the human gut, mediated by facultative aerobes [60–65]

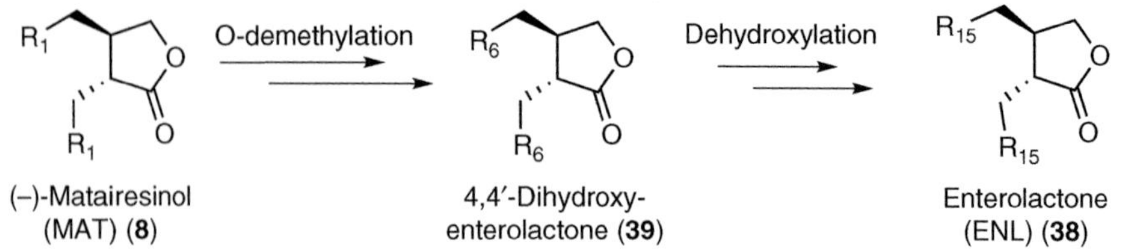

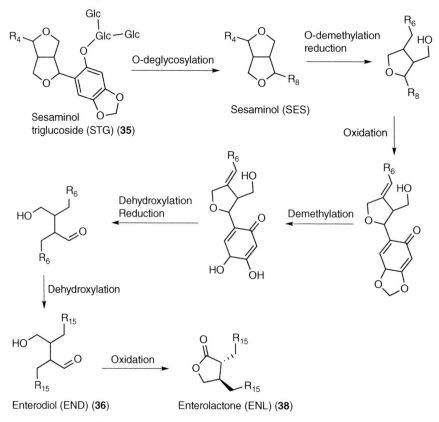

Figure 11.16 Matairesinol (MAT) metabolism by intestinal bacteria [68]. See Figure 11.7 for the meaning of R groups

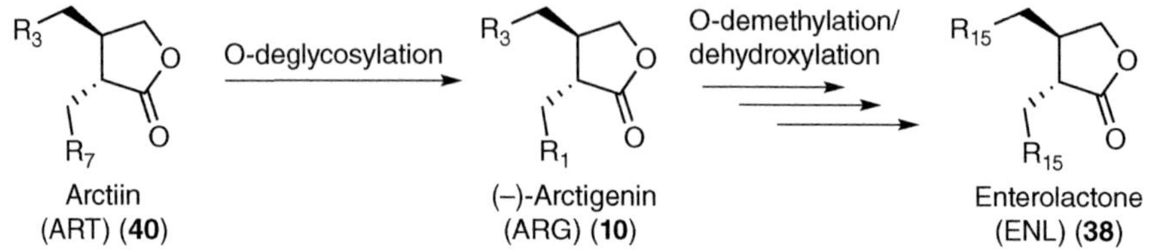

Figure 11.17 Arctiin (ART) metabolism by intestinal bacteria [68]. See Figure 11.7 for the meaning of R groups

Figure 11.18 Sesaminol triglucoside (STG) metabolism by intestinal bacteria [68, 69]. See Figure 11.7 for the meaning of R groups

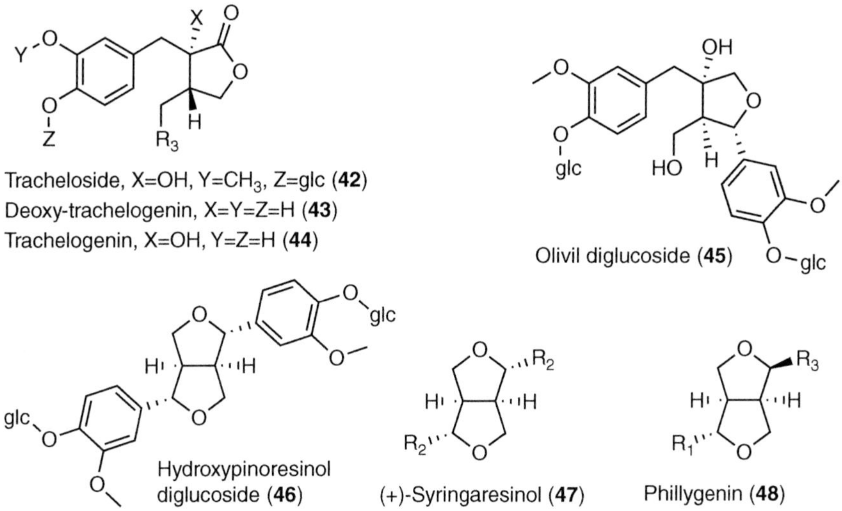

Tracheloside, X=OH, Y=CH$_3$, Z=glc (**42**)
Deoxy-trachelogenin, X=Y=Z=H (**43**)
Trachelogenin, X=OH, Y=Z=H (**44**)

Olivil diglucoside (**45**)

Hydroxypinoresinol diglucoside (**46**)

(+)-Syringaresinol (**47**)

Phillygenin (**48**)

Figure 11.19 Metabolism of lignans: tracheloside and olivil glucoside. See Figure 11.7 for the meaning of R groups

(see Figure 11.19). The structure of this newly isolated compound was elucidated by spectral analysis.

The metabolism by intestinal bacteria of the following lignin compounds and their glycosides to enterodiol (**36**) and enterolactone (**38**) has been confirmed by GC/MS and HPLC/MS analyses: pinoresinol (**4**), syringaresinol (**47**), lariciresinol (**5**), secoisolariciresinol (**6**), matairesinol (**8**), arctigenin (**10**), and phillygenin (**48**) [66, 72–74].

Table 11.2 lists the bacteria involved in lignan metabolism.

11.4 Plant Physiology and Plant Defense

In spite of the extensive distribution of lignans, their biological functions in plants remain unclear. Because some lignans have potent antimicrobial, antifungal, antiviral, antioxidant, insecticidal, and antifeeding properties, they probably play a notable role in plant defense against various biological pathogens and pests [75–78]. Furthermore, they may participate in plant growth and development [79]. Lignans are also responsible, at least partially, for properties such as durability, quality, color, and texture of certain species of noble wood [80].

No more than two decades ago, the current wave of interest was still focused on chemical ecology, mainly on protection of plants against harmful organisms and adverse environmental effects. Later, it turned toward their insect feeding regulatory activity and insect antihormonal effect. Recently, Harmatha et al. [1] have attempted to learn more on the immunomodulatory properties of lignans.

There is evidence that many lignans are produced by plants as response to a fungal attack (see Figure 11.20). For instance, *Picea abies* infection by *Fomes annosus* fungi induces the accumulation of (−)-matairesinol (**8**), 7′-hydroxymatairesinol (**49**), α-conidendrin (**50**), and liovil (**51**) to high levels [81]. The antifungi activity of lignans is believed to be an inhibition effect over fungal enzymes, such as celulase, polygalacturonase, and laccase [82].

Reportedly, 7′-hydroxymatairesinol (**49**) inhibits fungal growth [81]. Shain and Hillis [81] proposed that (**49**) alkalinity contributes to the *in vivo* resistance of the sapwood to infections by

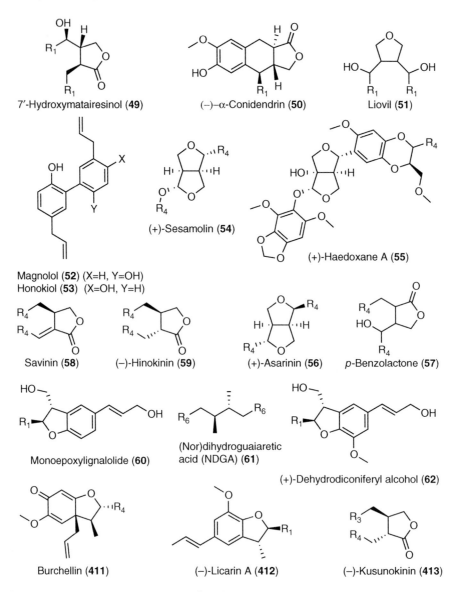

Figure 11.20 Lignans and neolignans with fungal activity. See Figure 11.7 for the meaning of R groups

F. annosus. Erdemoglu et al. [83] have reported that isolariciresinol (**9**) also has antifungal activity. Podophyllotoxin (**14**) and some of its isomers have been tested for such aspects as insecticidal, phytogrowth inhibitory, ichthytoxic, and antiparasitic activities [84–87]. Bang et al. [88] reported the antifungal activity of the neolignans magnolol (**52**) and honokiol (**53**).

Lignans show antibacterial properties, as in the case of the neolignans magnolol (**52**) and honokiol (**53**), isolated from *Magnolia grandiflora*, which inhibit the growth and progression of *Staphylococcus aureus*, *Bacillus subtilis*, and *Mycobacterium smegmatic* [89].

Table 11.2 Bacteria implicated in the metabolism of lignans[a,b]

Microorganisms metabolism	Transformation
In vitro human fecal inoculum	SES to END and ENL
Bacteroides distasonis, B. fragilis, and *B. ovatus*; and *Clostridium cocleatum* and sp. SDG-Mt85-3Db	Deglycosylation of SDG
Bifidobacterium pseudocatenulatum WC 401	SDG to SECO
Butyribacterium methylotrophicum, Eubacterium callanderi and *E. limosum*; and *Peptostreptococcus productus*	Demethylation of SECO
Clostridium saccharogumia, Eggerthella lenta, Blautia producta, and *Lactonifactor longoviformis*	SDG to END and ENL
Clostridium scindens and *Eggerthella lenta*	Dehydroxylation of SECO
Consortium END-49	Flaxseeds to END
Clostridium coccoides–Eubacterium rectale	ENL production
ED-Mt61/PYG-s6 Dehydrogenation of END to ENL	
E. lenta	PIN and LARI to SECO
E. lenta SDG-2	(+) DHEND to (+) END, but not for (−) DHEND to (−) END
Eggerthella sp. SDG-2, *Ruminococcus* (R.) sp. END-1	ARC and SDG to (−) ENL and (+) END
Enterococcus fecalis strain PDG-1	(+) PIN to (+) LARI
Eubacterium sp. ARC-2	Demethylating arctigenin to 4, 4′-diOH-ENL (**39**)
Human fecal microbiota	ART to ENL and MAT to ENL
Peptostreptococcus productus SECO-Mt75m3 and *E. lenta* SECO-Mt75m2	Demethylation and dehydroxylation of SECO
Peptostreptococcus sp. SDG-1 and *Eubacterium* sp. SDG-2	SECO to END and ENL
Rat intestinal microflora	STG to END and ENL
Ruminal microbiota, Fecal microbiota	ENL and END production
Ruminococcus productus	Demethylation of MAT
Ruminococcus sp. END-1 and strain END-2	(−) END to (−) ENL, and (+) END to (+) ENL
Strain ARC-1	DHEND to (−) END, but also (+) DHENL to (+) ENL
Strain END-2	END to ENL
Strain END-2 and ARC-1	(+) and (−) SECO to (+) ENL and (−) END

[a] See text for the meaning of acronyms.
[b] Data from ref. [68].

Some of the lignans isolated from *Piper* species possess antifeedant[2] activity against stored pests [90].

Deoxypodophyllotoxin (**13**), isolated from *Juniperus sabina* L. possess insecticidal activity against the larvae of *Pieris rapae* L. [91]. (−)-Kusunokinin (**413**) showed insecticidal activity

---

[2] A naturally occurring substance in certain plants which adversely affects insects or other animals which eat them.

against *Anticarsia gemmatalis* and its activity was dose dependent [92]. The lignans, sesamin (**17**) and (+)-sesamolin (**54**) combined with certain synthetic compounds, synergize their insecticidal activities [93]. In one report, the inflorescence material of *Piper mullesua* was examined for sesamin (**17**) content, where the insecticidal and growth inhibitory effects of purified sesamin were quantified against the larvae of *Spilarctia oblique* Walker [94]. Sintim *et al.* [95] showed that the sesame leaves were either antifeedants or reduced insect development.

Pinoresinol (**4**) is found to occur as a minor component in the defensive secretion produced by glandular hairs of caterpillars of the cabbage butterfly, *P. rapae* [96].

Podophyllotoxin (**14**), pinoresinol (**4**), sesamin (**17**), nordihydroguaiaretic acid (NDGA) (**61**), and the neolignans burchellin (**411**), and licarin A (**412**) added to the diet of *Rhodnius prolixus* larvae induce antifeedant effects. In addition, pinoresinol and NDGA significantly inhibit ecdysis [97].

Several studies have evaluated the larvicidal activity of the furan lignan grandisin (**179**), a leaf extract from *Piper solmsianum*, against *Aedes aegypti*[3]. The toxicity of grandisin was also evaluated against the larvae of *Chrysomya megacephala* F. (Diptera: Calliphoridae) [100].

NDGA (**61**) showed larvicidal activity against *Culex pipiens* [101]. Two furofuran lignans identified in *Phryma leptostachya* var. Asiatica roots showed larvicidal activity against three mosquito species [102, 103] and housefly (*Musca domestica* L.) [104].

(+)-Haedoxane A (**55**), which is isolated from *Phyrma leptostachya* roots, is probably the best known sesquilignan with insecticide properties against *Lepidoptera larvae* and housefly [105–107].

Extracts from lignan-rich plants of the Schisandraceae family exhibited insecticidal [107–110] and antifeedant activities [111]. For example, Gomisin B (**370**) and N (**400**) showed a high activity against *Drosophila melanogaster* [112]. Sesamin (**17**) and (−)-asarinin (**56**) are effective in heightening the toxicity of a wide variety of insecticides [113, 114], and *p*-benzolactone (**57**) is an insect feeding inhibitor [115]. Other methylenedioxy-containing lignans, such as savinin (**58**) and hinokinin (**59**), are also known to be insecticide synergists [116].

Certain lignans such as (−)-matairesinol (**8**) and (−)-bursehernin (**23**) provide protection against the nematodes *Globodera pallida* and *Globodera rostochiensis* [117, 118]. Germination inhibition has also been detected for a monoepoxylignanolide (**60**), isolated from potato root [119]. This compound is remarkable because it can be synthesized in response to infection by a nematode. The monoepoxylignanolide (**60**) isolated from *Aegdops ouata*, is a germination inhibitor with activity that depends heavily upon radiation of specific wavelengths [119, 121, 122].

The allelopathic activity[4] of NDGA (**61**) is manifested by dramatically reducing the seedling root growth of barnyard grass, green foxtail, perennial ryegrass, annual ryegrass, red millet, lambsquarters, lettuce, and alfalfa, and also reduces the hypocotyl growth of lettuce and green foxtail [120].

Cytokinins (CK) are a class of plant growth compounds produced in plant roots and shoots that promote cell division or cytokinesis. Only adenine-type cytokinins such as kinetin, zeatin, and 6-benzylaminopurine are found in plants. It is remarkable that some neolignans, such as the 4-*O*-glucoside derived from (+)-deshydrodiconiferilic alcohol (**62**), yield a similar activity as CK, although at higher concentrations [123].

11.5 Podophyllotoxin

Most of the current growing interest in lignans is focused on the antitumoral activity shown by podophyllotoxin-related cyclolignans. Podophyllotoxin (PPT, (**14**)) is probably the most widely

---

[3] Dengue is a tropical disease caused by an arbovirus transmitted by the mosquito *Aedes aegypti*. Because no effective vaccine is available for the disease, the strategy for its prevention has focused on vector control by the use of natural insecticides [98, 99].
[4] Allelopathy is the inhibition of growth in one species of plants by chemicals produced by another species.

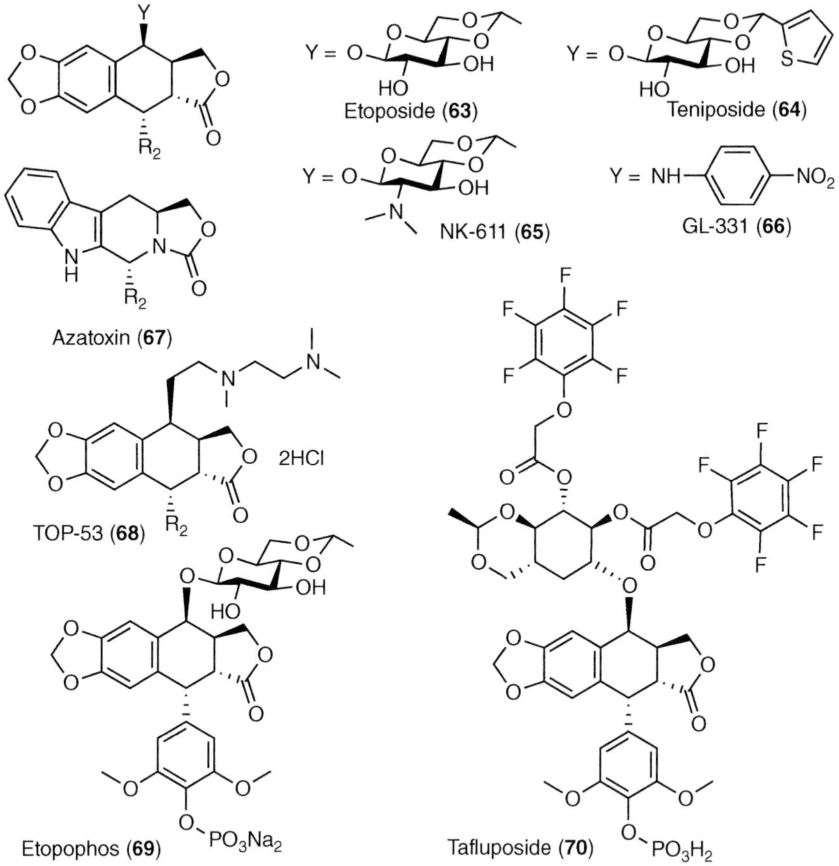

Figure 11.21 Antitumor PPT-glycosyl derivatives. See Figure 11.7 for the meaning of R groups

known and studied lignan so far. This compound alone and its properties could constitute a complete book chapter (e.g. see ref. [79]), and therefore we will indeed dedicate a complete section to the topic. However, whether dedicating a chapter or a section, it is impossible to summarize in a short space the huge amount of available data concerning PPT, and therefore we will highlight only some of the most relevant reviews that have appeared in literature since the year 2000. This information will serve not only as an example but also as an index of the vast data already published. These reviews cover all different aspects of PPT and its derivatives: distribution and isolation in plants, synthesis (see Section 10.7.4), biosynthesis (see Section 11.2.3), and biological activity [6, 124–134]. It is also notable to remark that this topic has also been covered by many other books, general reviews on lignans, and other types of publications that cannot be covered in this textbook.

Among all the known natural plant lignans, which number in the hundreds, the best known is PPT, a cytotoxic aryltetralin lactone originally isolated from *Podophyllum peltatum* L.[5] and related species [126]. Podophyllin, an alcoholic extract of the plant was first cited, in 1942, as a topical treatment for

---

[5] Podophyllum means foot-shaped leaf and peltatum means shield-like.

Table 11.3 Physicochemical properties of podophyllotoxin

Property[a]	Value
Structure[b]	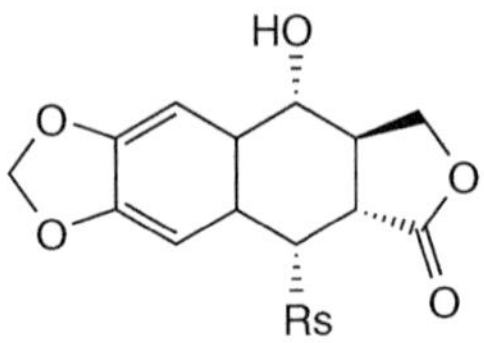 (chemical structure shown with HO, O, Rs, and O groups)
IUPAC systematic name	(5R,5aR,8aR,9R)-9-hydroxy-5-(3,4,5-trimethoxyphenyl)-5,5a,8a,9-tetrahydrofuro[3′,4′ : 6,7]naphtho[2,3-d][1,3]dioxol-6(8H)-one
Molecular formula	$C_{22}H_{22}O_8$
Molecular weight	414.41
Melting point	183–184 °C
Solubility in water	120 mg/L (at rt)
Log P	0.591
Optical rotation $[\alpha]_D^{20}$	135.5 (C 0.2 $CHCl_3$)
Refractive index	1.362(22.5 °C)

[a] Data from refs [127, 138–141].
[b] See Figure 11.7 for the meaning of R groups.

venereal warts (*Condyloma acuminatum*), an ailment caused by a papilloma virus [135]. This would appear to be one of the first reported examples of the antiviral activity of a lignan natural product [3].

PPT is the first and perhaps the best known example of the use of a lignan as a lead compound, since its glucopyranoside derivative was recognized as a potent antitumor factor. This invention entails a particularly fascinating account, involving a multitude of investigations conducted over a period of more than a century [124]. The studies culminated in the structure elucidation of PPT, the assessment of its biological activity, and the invention of its mode of action. Moreover, PPT is used as a precursor for the synthesis of relevant antitumor drugs, such as etoposide (**63**) and teniposide (**64**), which are used in the treatment of lung cancer, testicular cancer, a variety of leukemias, and other solid tumors [124, 132, 136, 137] (see Figure 11.21).

The main physical and chemical characteristics of PPT are summarized in Table 11.3.

11.5.1 Extraction, Synthesis, and Biotechnological Approaches

PPT has traditionally been isolated from podophyllin, resin of *Podophyllum rhizome*. *Podophyllum emodi* (Indian Podophyllum) is preferred to *P. peltatum* (American Podophyllum, see Figure 11.22) because the content of PPT is about 4.3% of dry weight in *P. emodi* against 0.3% in *P. peltatum* [127, 130, 142, 143]. In 2001, a new extraction process was described increasing the yield of PPT of rhizomes and leaves to about 5.2% of dry weight [144]. The finding that leaves of *P. peltatum* are a rich source of PPT is noteworthy since leaves are renewable organs that store lignans as glucopyranosides [77]. The new extraction protocol was applied to other genera resulting in alternative source of PPT. However, the collection of known plants that are natural sources of PPT is limited and insufficient to supply the increasing demand of this compound [130]. The possibility of *in vitro* propagation and optimization of the cultivation of *Podophyllum* species have also been studied [77, 145–149].

Even though PPT has traditionally been isolated from *P. peltatum* and *P. emodi*, it has also been found in around 20 genera such as *Diphylleia, Dysosma, Catharanthus, Polygala, Anthriscus, Linum,*

Figure 11.22 Illustration of Podophyllum peltatum *(American Podophyllum). Courtesy of David Nesbitt, 2014. (See insert for color representation of this figure.)*

Hyptis, Teucrium, Nepeta, Thuja, Juniperus, Cassia, Haplophyllum, Commiphora, and *Hernandia* [130, 151][6] (see Figure 11.23).

11.5.2 Biological Activities of PPT

Extracts of *Podophyllum* species have been used by diverse cultures since ancient times as antidotes against poisons, or as cathartic, purgative, antihelminthic, vesicant, and suicidal agents [79][and references therein]. Podophyllin was included in the US Pharmacopoeia in 1820 and the use of this resin was prescribed for the treatment of venereal warts, attributing this action to PPT [130].

---

[6] *Linum album* Ky. ex Boiss. (Linaceae), known as "Katan-e-Golsefid" in Persian, is an endemic herbaceous perennial plant widely distributed to mountainous areas, sandy slopes, and sandy clay soils in fields at an altitude of 1200–3200 m in Irano-Turanian regions. Its flowering time is restricted from April to June.

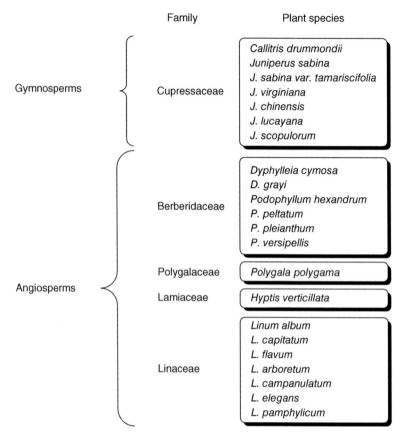

Figure 11.23 Plant sources of podophyllotoxin [132, 150]

Initial expectations regarding the clinical utility of PPT were tempered largely due to its unacceptable gastrointestinal toxicity [126], and thus too toxic for therapeutic purposes [132, 152]. This led chemists in the Sandoz pharmaceutical research department to investigate the possibility that the podophyllum lignans might occur naturally as glycosides [124]. Using special procedures, these researchers indeed produced several glucosides. Being less hydrophobic, the glucosides displayed lower toxicity than the aglycons, but their cytostatic activity was reduced to the same degree [126]. The research efforts were then focused on a programme for chemical modification of both the glucosides and aglycones of a wide range of PPT derivatives. Nearly 600 derivatives were prepared and tested over a period of about 20 years [124] (Table 11.4).

The "antiviral activity" of an aqueous extract of *P. peltatum* was investigated [154]. From this extract, PPT was found to be the most active component in inhibiting the replication of Measles and Herpes simplex type I viruses [155, 156]. In fact, PPT is included in many pharmacopoeias and used as an antiviral agent in the treatment of *C. acuminatum* caused by human papilloma virus (HPV) [157] and other venereal and perianal warts [79, 158–161]. The application of PPT cured almost all the warts completely in lesser time than other treatments and with fewer side effects. PPT and analogous compounds are also active against cytomegalovirus and Sindbis virus [162]. Moreover, PPT is also effective in the treatment of anogenital warts in children and against *Molluscum contagiosum*, that is generally a self-limiting benign skin disease that affects mostly children, young adults, and

Table 11.4 *Short chronology of discovery and development of podophyllum drugs [153]*

Year	Event
1753	Linnaeus first described *Podophyllum peltatum* and first gave its modern botanical name
1820	Podophyllin was included in the US Pharmacopoeia
1861	Bently mentioned local antitumor effects of podophyllin
1880	Podwyssotzki isolated and chemically investigated podophyllotoxin
1942	Kaplan described effects of podophyllin in benign tumors
1946	King and Sullivan reported the mechanism of action of podophyllotoxin
1966	Synthesis and biological evaluation of etoposide
1967	Start of clinical trials of teniposide (**64**)
1971	Start of clinical trials of etoposide (**63**)
1978	Sandoz reported further development of teniposide (**64**) and etoposide (**63**) to Bristol-Myers
1983	Approval by the FDA of etoposide (**63**) as VePesid for testicular cancer
1984	Studies on mechanism of action: inhibition of topoisomerase II by stabilization of cleavable complex
1986	The institute of Microbial Chemistry and Nippon-Kayaku identified NK 611 (**65**)
1990	Kuo-Hisiung Lee synthesized GL-331 (**66**) and Genelabs Technologies Inc. had patented this technology and proceeded with phase II clinical trials in Taiwan
1992	Pomier *et al.* identified azatoxin (**67**) using molecular modeling of the pharmacophore defined by topoisomerase II inhibitors
1993	TOP53 (**68**) was synthesized by scientists at the Taiho company and had been progressed to phase II clinical trials
1996	US launch of the prodrug etopophos (**69**)
2000	Tafluposide (**70**), a novel catalytic inhibitor of topoisomerases I and II had been isolated and developed in preclinical trials

See Figure 11.21 for structures.

human immunodeficiency virus (HIV) patients [163]. This compound inhibits these viruses at an essential early step in the replication cycle after entry of the virus into the cells or reduces the capacity of infected cells to release viruses. There are several reports regarding the formulations of PPT [159, 164–166]. PPT has other uses in dermatology: it is also a useful agent in *Psoriasis vulgaris* [167–169].

Antitumor activity is another outstanding property of PPT. It is effective in treating Wilms tumors, different types of genital tumors (e.g., *Carcinoma verrucosus*) and in non-Hodgkin and other lymphomas [79], and lung cancer [170, 171].

It has also proved effective in the treatment of *Rheumatoid arthritis* as a result of the reduction it brings about in the activation of complement system [79]. Immunostimulatory activities of PPT have also been described [172].

Furthermore, aqueous extract of rhizome of *P. emodi* protects against radiation-induced damage to spermatogenesis [173], and the influence of *P. emodi* on endogenous antioxidant defense system in mice was correlated to its radioprotective effect [174].

PPT and some of its isomers have been tested for other activities, such as insecticidal, phytogrowth inhibitory, ichthytoxic, and antiparasitic activities [84–87].

Studies on the penetration of PPT into human bioengineered skin have demonstrated that the lignan induces acantholysis and cytolysis in the skin-equivalent model used for a wide variety of

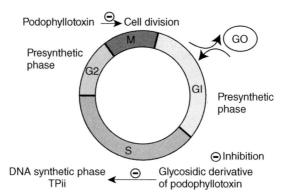

Figure 11.24 *Podophyllotoxin and its glycosidic derivative mode of action [133]*

pharmacotoxicological trials. This might apply to claims of efficacy for cosmetic compounds [175]. In combination with vinblastine, it was used as a mitotic agent for preparing embryonic chromosomes for trials [176].

Etoposide (**63**) also showed a strong cytoprotective effect against human immunodeficiency virus type-1 (HIV-1) *in vitro* [177, 178].

11.5.3 The Action Mechanism of PPT

PPT interrupts the cell cycle by inhibiting microtubule assembly. The antimitotic activity of PPT was reported at least half a century ago, and is due to reversible binding to tubulin. Thus, PPT was found to interact with tubulin at the colchicine binding site [179]. Compounds that inhibit tubulin polymerization, such as the alkaloids vinblastine, and vincristine, are commonly used as chemotherapeutic agents against cancer, and PPT has also been explored as a lead compound for new antitumoral agents [180]. The action mechanism of PPT is based on inhibiting the polymerization of tubulin and arresting the cell cycle in the metaphase [79, 181, 182] and references cited therein. Studies on cyclolignans by Gordaliza *et al.* [183] proposed that cyclolignanolides of the PPT group might work as alkylating agents, rather than as acylating agents (Figure 11.24).

Several articles have been published related to the action mechanism of this cyclolignan. Schonbrunn *et al.* [184] crystallized PPT linked to a tubulin fragment. Effects of microtubule-damaging agents, such as PPT or colchicine, on DNA and the cell cycle have been described [185, 186]. Chaudhuri *et al.* [187] and Pal *et al.* [188] have studied the interactions of B-ring of colchicine with α-tubulin and López-Pérez *et al.* have discussed the role of the dipole moment in the activity of cyclolignans [189]. Moreover, PPT acts by inhibiting the assembly of microtubules. It binds to α/β-tubulin dimer, giving rise to a PPT–tubulin complex and this stops the polymerization of microtubules at one end, but does not stop the disassembly at the other end leading to the degradation of microtubules. Thus, cells are arrested in the metaphase stage of mitosis and ultimately cell growth stops [150, 190].

When rat tumors injected with PPT were analyzed, a decrease in cytochrome oxidase, cytochrome C, and succinoxidase activities was found, as well as a reduction in respiration [191–193].

In addition to its cytotoxic activity, PPT is also known for its antiviral properties [126]. The antiviral activity (including HIV) of naturally occurring lignans and synthetic analogs was reviewed by Charlton [3], who concluded that there were several modes of antiviral activity associated with lignans: tubulin binding (inhibition of tubulin polymerization interferes with the formation of the cellular cytoskeleton and with some critical viral processes), reverse transcriptase (RT) inhibition, integrase inhibition, and topoisomerase inhibition (although the latter association appeared to be weaker) [180].

11.5.4 Congeners and Derivatives (Other Aryltetralin Lactones)

Among the plethora of physiological activities and potential medicinal and agricultural applications, the antineoplastic and antiviral properties of PPT congeners and their derivatives are arguably the most eminent from a pharmacological perspective. Numerous new PPT derivatives are currently under development and evaluation as topoisomerase inhibitors and potential anticancer drugs [126].

PPT derivatives as potential new antitumoral agents have been the subject of review, which will not be repeated here [181, 194–198]. PPT analogs are still being explored nowadays by different research groups [199–202] (Figure 11.25).

It has been observed that treatment of cancer patients with cytotoxic agents also diminished their immune responses. Therefore, cytotoxic agents, such as PPT derivatives (cyclolignans), have been investigated as immunosuppressive drugs to prevent rejection of transplanted organs [203, 204].

11.5.4.1 Isolation of Other PPT-Related Compounds

Using special procedures to inhibit enzymatic degradation, Stahelin and Von Wartburg [124] produced PPT-β-D-glucopyranoside (**71**) as the main component and its 4′-demethyl derivative (**72**) from the Indian *Podophyllum* species. Both of these glucosides (**73**) and (**74**) and the glucosides of α- and β-peltatin (PT) (**75**) and (**24**), respectively, were also isolated from the American *P. peltatum*. Being less hydrophobic, the glucosides displayed lower toxicity than the aglycones, but their cytostatic activity was reduced to the same degree.

Deoxy-PPT (**13**), which has been found in a great variety of plants (such as *Libocedrus* [205], *Linum* [206], *Bursera* [207], *Podophyllum* [142, 208, 209] *Anthriscus* [210], *Diphylleia* [209] *Dysosma* [211], and *Hernandia* [212]), is the most rife aryltetralin lignan, with well-known cytotoxic activity [130].

β-PT A methyl ether (**25**) has been studied due to its activity against different cancer cell lines [213], as well as its antitumor activity in the WA16 tumor system [214]. This lignan is present in a great variety of plants (*Juniperus phoenicea* [215], *Bursera permollis* [213], *B. fagaroides* [214], *B. simaruba* [216], *Anthriscus sylvestris* [210], *Libocedrus plumose* [205], and some *Linum* species and cultures [40, 206, 217, 218]). 5′-Desmethoxy-β-PT A methyl ether (**76**) has been reported only from *B. fagaroides* with activity against the WA16 tumor system [214]. Burseranin (**77**) has been only described as a constituent of *B. graveolens*, and showed activity against HT1080 cell line [219].

5′-Demethoxy 5-methoxy-PPT and other PPT-related lignans have been isolated, and their structure elucited, from plants and cell cultures of *Linum flavum* [220].

11.5.4.2 PPT analogs activity

The evaluation of the cytotoxic activity of the aforementioned natural compounds (PPT (**14**), desoxy-PPT (**13**), β-PT A methyl ether (**25**), 5′-desmethoxy-β-PT A methyl ether (**76**), burseranin (**77**), and acetyl-PPT (**78**)) against the human cancer cell lines KB, PC-3, MCF-7, and HF-6 showed that, except for burseranin (**77**) and acetyl-PPT (**78**) (it is not a constituent of *Podophyllum*, and has been isolated from PPT for structure–activity relationship (SAR) studies [221]), all the isolated compounds displayed high activity [222]. The cytotoxic activity of PPT (**14**), desoxy-PPT (**13**), and their congeners is well known [223–226]. Some SAR studies, using several PPT analogs, showed that the core structure of deoxy-PPT (**13**) is responsible for this cytotoxicity. The extra methoxy group on the 6-position in 5′-desmethoxy-β-PT A methyl ether (**76**) significantly changed the *in vitro* cytotoxicity when compared to desoxy-PPT (**13**).

A conservative estimate suggests that there are more than 200 known natural and semisynthetic PPT derivatives on their role in cancer chemotherapy [125].

In 2007, Liu *et al.* [153] reported the structure of 353 chemically modified PPT derivatives. The modifications were made in one or several rings of the structure, and the activity of these modified

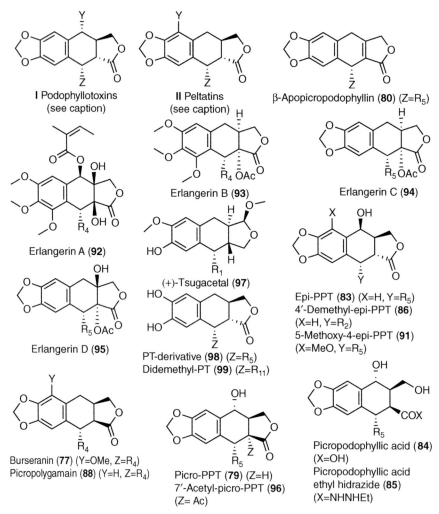

Figure 11.25 Main aryltetralin lactone lignans with biological activity. I: Podophyllotoxin (PPT) (**14**) (Y = OH, Z = R$_5$); deoxy-PPT (**13**) (Y = H, Z = R$_5$); 6-methoxy-PPT (**26**) (Y = = O, Z = R$_5$); PPT glucoside (**71**) (Y = Ogluc, Z = R$_5$); 4′-demethyl-PPT (**72**) (Y = OH, Z = R$_2$); acetyl-PPT (**78**) (Y = OAc, Z = R$_5$); 3′-demethyl-PPT (**81**) (Y = OH, Z = R$_2$); 5′-desmethoxy-PPT (**82**) (morelsin) (Y = OH, Z = R$_2$); 4′-demethyldeoxy-PPT (**87**) (Y = OH, Z = R$_2$). II: α-Peltatin (PT) (**75**) (Y = OH, Z = R$_2$); β-PT (**24**) (Y = OH, Z = R$_5$); β-PT A methyl ether (**25**) (Y = OMe, Z = R$_5$); α-PT-glucoside (**73**) (Y = Ogluc, Z = R$_2$); β-PT-glucoside (**74**) (Y = Ogluc, Z = R$_5$); 5′-desmethoxy-β-PT -glucoside (**76**) (Y = Ogluc, Z = R$_5$). See Figure 11.7 for the meaning of R groups

derivatives was also evaluated. Castro et al. [129] had previously made a similar evaluation for congeners and derivatives.

Markkanen et al. [227] published a detailed study in 1981 of the antiviral effects of 21 different lignans, including many PPT derivatives, on *Herpes simplex* virus type 1 (HSV-1). Many of these compounds showed strong antiviral activity. A series of 4β-triazole-linked glucose PPT conjugates have been synthesized by employing a "click-Chemistry" approach. Most of these triazole derivatives have good anticancer activity against a panel of five human cancer cell lines (HL-60, SMMC-7721, A-549, MCF-7, SW480) [228].

Many cyclolignans similar to PPT are active against viruses such as measles virus, types I and II herpes, vaccinia virus, and cytomegalovirus [154, 162, 229]. Meanwhile, peltatins (**24**), (**75**) and 4′-demethyl PPT (**72**) have a cathartic effect over the gastrointestinal tract [5].

In 1982, Bedows and Hatfield [154] studied the effect of four of the podophyllin lignans (PPT (**14**), picro-PPT (**79**) (otherwise picropodophyllin), α-PT (**75**), β-PT (**24**), and deoxy-PPT (**13**)) on both HSV-1 (a DNA virus) and the measles virus (an RNA virus).

PPT (**14**), 4′-demethyldeoxy-PPT (**72**), and β-apopicro-podophyllin (**80**) strongly inhibited P-388 (murine lymphocytic leukemia) [230]. Picro-PPT (**79** is inhibitor of the insulin-like growth factor-1 receptor and malignant cell growth [231].

Other PPT-related lignans have been shown to possess immunosuppressive properties and are seen as candidates for use in organ transplanting [203, 204].

PPT (**14**) [232–235], deoxy-PPT (**13**) [232, 235, 236], 3′-demethyl-PPT (**81**) [232], 4′-demethyl-PPT (**72**) [232–235], 5′-desmethoxy-PPT (morelsin, (**82**)) [237], PPT glucoside (**71**) [234, 235], picro-PPT (**79**) [234, 235], epi-PPT (**83**) [234, 235], picropodophyllic acid (**84**) [234], podophylic acid ethylhydrazide (SP-1, (**85**)) [238], 4′-demethyl epi-PPT ethylidene-β-D-glucoside (etoposide, (**63**)) [239], and teniposide (**64**) [239, 240] show antitumor activity.

Most of PPT semisynthetic derivatives used in chemotherapy are glycosides at the C_7 hydroxy position. Lopez-Perez et al. [241] studied the cytotoxicity of several PPT esters, concluding that cytotoxicity is not due to the lipophilicity at the C_7 position, but to the spacial arrangement of such a voluminous group. Thus, the configuration at C_8 and the presence of a lactonic group could play a role [240].

From Kelleher's [240] SAR study of PPT analogs, it is possible to conclude that the configuration of the hydroxyl group at C_4 of the B-ring is of some relevancy to antitumor activity, but the situation is complex. Epi-PPT (**83**), which differs from PPT (**14**) only in the stereochemistry at C_4, is an order of magnitude less effective as a cytotoxic agent. Positioning the hydroxyl group elsewhere on the molecule, however, does not have such a marked effect. β-Peltatin (PT) (**24**), with a hydroxyl group at C_5 of the A-ring, is an even more potent antitumor agent, and deoxy-PPT (**13**), with no hydroxyl group, is almost equally effective. Replacing the hydroxyl group of β-PT with the more bulky methoxyl group gives β-PT A methyl ether (**25**), or with glucose gives β-PT β-D-glucopyranoside (**74**). However, the results indicate a significant reduction in cytotoxic activity [240]. The polarity of the C_4 substituent rather than its size is the key factor in antitumor activity. The configuration at C_2 seems to play a significant role in determining antitumor action. Picro-PPT (**79**) differs only with respect to stereochemistry at this carbon and shows a markedly attenuated potency as a cytostatic agent.

Several SAR studies have been carried out on the antimitotic activity of PPT-type lignans, and a number of conclusions can be drawn (Figure 11.26). There is apparently no difference in the tubulin-binding activity of PPT (**14**) and deoxy-PPT (**13**) [242, 243] suggesting that the C_4 hydroxyl is of little consequence. Epi-PPT (**83**), the C_4 stereoisomer of PPT [242, 243] and 4′-demethyl-epi-PPT (**86**) [244] are more than an order of magnitude less effective at binding tubulin than PPT. This illustrates the importance of the stereochemical configuration of the C_4 proton in contributing to tubulin binding. As in the case of antitumor activity, a glucose moiety at the C_4 position of PPT markedly reduces the tubulin-binding capacity [243]. It is especially remarkable that the synthetic epi-PPT derivative contains a bulky substituent at C_4, etoposide (**63**), and has no detectable effect upon microtubule assembly [245]. This contrasts with its antitumor activity, which is quite appreciable [238]. The stereochemistry at C_2/C_3 appears also to be relevant. Picro-PPT (**79**) displays a much reduced ability to bind tubulin [242, 243, 246], and picropodophyllic acid (**84**) has no detectable activity [243]. The relevance of the lactone ring of PPT to tubulin assembly has been well documented. Replacing it with a furan ring results in a compound with severalfold diminished activity

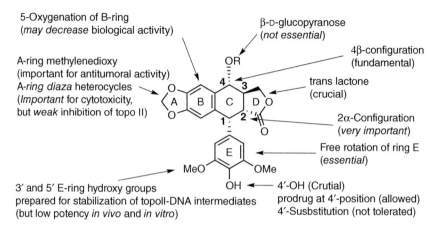

Figure 11.26 SAR keypoints for podophyllotoxin derivatives [133]

[242, 243, 245, 247]. Replacing the oxygen of the cyclic ether with a carbon, sulfur, or sulfone group reduces the activity 10-fold, 20-fold, and completely, respectively [242]. As in the case of antitumor activity, the hydroxyl group of PPT to C_5 (for β-PT (**24**)) results in a moderate increase in the inhibition of tubulin aggregation [243]. Kelleher [240] has emphasized the correlation between antitumor activity and antimitotic activity of derivatives of PPT. Even though correlations exist between these two activities for some PPT derivatives, suggesting a common mode of action, this is true for only a limited number of compounds.

Picropolygamain (**88**) inhibited A-549, MCF-7, and HT-29 cells [248]. Deoxy-PPT (**13**), β-PT A methyl ether (**25**), picro-β-PT methyl ether (**89**), and dehydro-β-PT methyl ether (**90**) showed cytotoxic activities against 12 cell lines[7] [213]. Burseranin (**77**) showed inhibitory activity against HT1080 human fibrosarcoma cell line [219] and 5-methoxy-4-epi-PPT (**91**) exhibited cytotoxicity against human epidermoid carcinoma KB and murine lymphocytic leukemia L1210 cell lines [249]. It also inhibited the assembly of tubulin into microtubules.

Four major lignans, Erlangerins A–D (**92–95**), analogous to the structure of PPT, were isolated from the resin of *Commiphora erlangerian*[8] [250, 251]. Compounds (**92**) and (**93**) suppressed cell viability only at higher concentrations, in contrast to PPT. The observed biological activities of (**94**) and (**95**) are closely related to PPT (**14**) and 4-deoxy-PPT (**13**), and hence may indicate a similar mechanism of action.

7′-Acetyl-picro-PPT (**96**) exhibited significant inhibition of the transformation of murine epidermal JB6 cells [252], and (+)-tsugacetal (**97**) showed weak activity in the BST [253].

Ito *et al*. [254] isolated antitumor lignans similar to PPT (**13**) and (**98**) from the seeds of *Hernandia ovigera*. These compounds exhibited significant inhibitory effects at high concentration and were found to be slightly weaker than that of β-carotene, a vitamin A precursor that has been used commonly in cancer prevention studies.

On the other hand, lignans (**13**) and (**87**) derived from PPT inhibited lipid peroxidation in rat brain and kidney homogenates and rat erythrocyte hemolysis [255].

---

[7] A431 (human epidermoid carcinoma), BC1 (human breast cancer), Col2 (human colon cancer), HT (human fibrocarcinoma), KB (human nasopharyngeal carcinoma), KB-V1 (vinblastine-resistant KB), LNCaP (hormone-dependent prostate cancer), Lu1 (human lung cancer), Mel2 (human melanoma), U373 (human glioblastoma), ZR75-1 (human breast cancer), and ASK (astrocytoma).

[8] Poisonous to humans and animals, and has traditionally been used as an arrow poison.

Finally, modified aryltetralins (**24, 25, 75**) and PPT derivative (**99**) display strong activity as potent anti-HIV agents [256–260].

11.5.4.3 Mode of action of PPT congeners and derivatives

While PPT inhibits microtubule assembly during mitosis, PPT congeners and derivatives act due to their capacity to form topoisomerase and DNA complexes. Such activity leads to a double-helix breaking, which interrupts the cellular cycle in the G2 phase, leading to a cell death.

Whereas PPT is the most prominent representative of the tubulin-binding lignans, and its derivatives are inhibitors of DNA topoisomerase II[9], inhibition of RT has been observed for various other classes of lignans, such as dibenzylbutyrolactones, dibenzylbutanes, dibenzocyclooctadienes, and aryltetralins [180].

PPT derivatives can be divided into two groups: (1) inhibitors of tubulin polymerization, such as PPT itself, and (2) inhibitors of DNA topoisomerase II, such as etoposide (**63**) and teniposide (**64**). The main structural difference between these two groups is the presence of a small equatorial substituent at position 4 (C-ring) for inhibitors of tubulin polymerization, and of a bulky substituent (such as a glucose moiety) in axial position for topoisomerase II inhibitors [261]. The latter group of compounds are 4β-congeners of PPT, and are called the epi-PPT (**83**) [180]. Semisynthetic derivatives of epi-PPT (**83**), for example, etoposide (**63**) [262], etopophos (**69**), and teniposide (**64**), induce a premitotic block in late S or early G2 stage [263].

MacRae and Towers [5] reviewed the antiviral activity of lignans in 1984, and analogous to Bedows and Hatfield [154], concluded that the antiviral effects of lignans were related to inhibition of microtubule formation.

The ability of PPT and several of its derivatives to influence nucleic acid metabolism has been studied. Both PPT (**14**) and the derivative, etoposide (**63**), inhibit the synthesis of DNA, RNA, and protein in HeLa cells at a concentration of 100 pM [247]. The compounds, 4'-demethyl-PPT (**72**), 4'-demethyldeoxy-PPT (**87**), 4'-demethyl-epi-PPT (**86**), α-PT (**75**), etoposide (**63**), and teniposide (**64**) all cause DNA fragmentation while PPT (**14**), deoxy-PPT (**13**), epi-PPT (**83**), and β-PT (**24**) were all inactive [264].[10]

A decrease in cytochrome oxidase, cytochrome C, and succinoxidase activities, as well as a reduction in respiration, was found with the derivatives β-PT (**24**) [265] and a synthetic derivative, acetyl-PPT (**78**) [266].

11.6 Biological Activity of Different Lignan Structures

As discussed in Chapter 10, lignans can be classified into eight different groups according to their chemical structure. Within this classification, this chapter describes in detail the main biological properties of some of the most representative derivatives from each group.

Despite the impressive progress made in organic chemistry synthetic procedures, about 25% of all prescription medicines to date are still of plant origin. Also, approximately 60% of all drugs under clinical trial for multiple cancer treatments are either natural products, derived from natural products, or contain pharmacophores derived from natural products [267, 268]. These facts make lignan biological activity a key issue in developing final applications, but unfortunately, some of the most interesting products are only in very small amounts in plants. Metabolic engineering offers new perspectives for improving the production of compounds of interest. The production of plant secondary metabolites by means of large-scale culture of plant cells in bioreactors is technically feasible [269].

---

[9] DNA topoisomerases are involved in conformational and topological changes in DNA during replication.

[10] Note that all of the active compounds possess a 4'-hydroxyl group, while none of the inactive ones do.

Among these pharmacological activities of lignans, their antitumor, anti-inflammatory, immunosuppressive, cardiovascular, antioxidant, and antiviral actions deserve special mention [3, 270–275].

In the case of cancer treatments, the search for natural products as potential anticancer agents dates back to the *Ebers papyrus* in 1550 BC, but the scientific period of this search is much more recent, beginning in the 1950s with the discovery and development of the vinca alkaloids, vinblastine, and vincristine, and the isolation of the cytotoxic PPT [276].

In other areas of medicine, extracts from roots and rhizomes of *Podophyllum* present cathartic and poisonous properties [79], as well as purgative, anthelmintic, and vesicant [234, 277]. *Olea europaea* bark has been employed as antipyretic, antirheumatic, tonic, and also anti-scrofulosis. All these properties are thought to be related to isolated lignans of *O. europaea* [278]. Only lignans obtained from *Podophyllum* and NDGA seem to be involved in human and animal intoxications, with some exceptions such as plicatic acid from red cedar (*Thuja plicata*), which produces asthma and hay fever in some people [279–281].

11.6.1 Dibenzylbutane Lignans

There are many dibenzylbutane lignans with biological activity, and only the most representative among them are discussed below. Figure 11.27 depicts a selected group of dibenzylbutane lignans with biological activity.

Anolignan A (**100**) and anolignan B (**101**) showed moderate cytotoxicity against BC1, HT1080 (human fibro sarcoma), Lu1, Mel2, Col1 (human colon cancer), LNCaP, and U373 cell lines. In addition, anolignan A (**100**) also showed significant cytotoxicity against P-388, while anolignan C (a furanoid lignan **102**) yielded specific, but only moderate, cytotoxicity against ZR-75-1 [282].

Quite differently, termilignan **103** is a monomethyl ether of anolignan A and has been isolated together with thannilignan (**104**) and anolignan B (**101**) from *Terminalia bellerica*. All three compounds possess demonstrable anti-HIV-1, antimalarial, and antifungal activities *in vitro* [283].

Hattalin (2,3-dibenzylbutane-1,4-diol) (**105**) showed strong inhibitory activity against ZR-75-1 [284] and cinnamophilin (**106**), while its *O,O*-dimethyl ether (**107**) and *O,O*-diacetyl derivatives (**108**) showed significant antiplatelet aggregation activity [285].

In addition, Lee *et al.* [286] evaluated machilin A (**109**) for its ability to inhibit phospholipase C γ1 (PLCG1) *in vitro*. This compound exhibited inhibitory activities on PLCG1, and showed good inhibitory activity on three human cancer cell lines: A549 (lung), MCF-7 (breast), and HCT-15 (colon).

In addition, Li *et al.* [287] reported that *erytro*-austrobailignan-6 (**110**) showed inhibition of topoisomerases I and II.

Lignan (**111**) was isolated from *Larrea tridentata*, and it showed antioxidative activity by the DCFH method without cytotoxicity [288].

Rhinacanthins A and F (**112** and **113**) show antiviral activity against type A influenza virus (flu) [289].

Pregomisin (**114**) isolated from *Schisandra chinensis* has been reported to show PAF antagonist activity [290].

Niranthin (**115**) exhibits a wide spectrum of pharmacological activities. In 2012, Chowdhury *et al.* [291] have shown for the first time that niranthin (**115**) is a potent antileishmanial agent. Niranthin (**115**) poisons *Leishmania donovani* topoisomerase IB and favors a Th1 immune response in mice.

In addition, *cis*-Hinokiresinol (**116**) has been included as an example of one of the active compounds with c-AMP phosphodiesterase inhibition[11] [293].

---

[11] Cyclic adenosine monophosphate (c-AMP) is a secondary messenger within cells. The c-AMP phosphodiesterase inhibition test is a useful means for screening biologically active compounds [292].

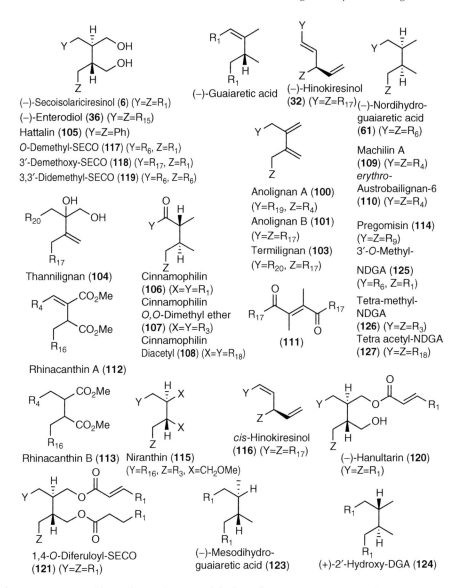

Figure 11.27 Main dibenzylbutane lignans with biological activity. See Figure 11.7 for the meaning of R groups

11.6.1.1 Secoisolariciresinol (SECO)

Flax, *Linum usitatissimum*, is particularly rich in the lignan SDG (**34**) [294], and it also contains small amounts of the lignans MAT (**8**), pinoresinol (PIN) (**4**), and isolariciresinol (**9**) and ω-3 fatty acid [53, 295–297].

Secoisolariciresinol (SECO, (**6**)) is converted into enterolignans enterodiol (END) (**36**), and enterolactone (ENL) (**38**)] by the anaerobic intestinal microflora [298–300]. Four compounds (SECO, (**6**)),

MAT (**8**), lariciresinol (LARI) (**5**), and PIN (**4**)) are mammalian estrogen precursors, also called phytoestrogens [301, 302]. The phytoestrogens could inhibit aromatase activity, and thus decrease intracellular estrogen production [303]. Because flaxseed affects the reproductive development of offspring, caution is suggested when consuming flaxseed during pregnancy and lactation [304].

SECO and its metabolite END induced a significant decrease in cell invasion of a breast cancer cell line (MDAMB-231) [305]. However, this occurred only at high concentration [306]. A smaller decrease in aromatase activity was observed with END (**36**) and its theoretical precursors, O-demethyl-SECO (ODSI) (**117**), 3′-demethoxy-SECO (DMSI) (**118**), and 3, 3′-didemethyl-SECO (DDSI) (**119**) [307].

The mammalian lignans END and ENL have been shown to inhibit breast and colon carcinomas. In addition, Lin *et al.* [308] were the first to report the effect of lignans on prostate carcinoma.

Rickard *et al.* [309] studied the effects of flaxseed and its constituent lignan SDG, which inhibits mammary tumor development in rats. SDG from flaxseed has been shown to have a beneficial antitumor effect throughout the early promotional phase of carcinogenesis [310, 311]. SDG has also been reported to have potent lung radioprotective properties without abrogating the efficacy of radiotherapy [312]. Beneficial effects of SDG in cancer and lupus nephritis showed that these beneficial effects could be due to the ability of SDG to scavenge. OH radicals [274]. The antioxidant activity of SDG was lower than that of END and ENL [274]. Prasad [313] investigated the ability of SDG to scavenge OH using HPLC method.

A study revealed that SECO has a high antioxidative power for the inhibition of lipid peroxidation *in vitro* [314]. Both (−)-SECO and its (+)-enatiomer showed antioxidative activity [294].

While (−)-SECO, isolated from *Taxus* species, exhibited potent cytotoxicity against KB-16, A-549, and HT-29 tumor cell lines [315], it did not show any cytotoxicity against ZR-75-1 [282]. The highest activity against the HT-1080 fibrosarcoma cell line was registered for SECO [316]. It has also been shown to possess potent DPPH radical scavenging activity and significant inhibitory activity against nitric oxide (NO) production in lipopolysaccharide-activated murine macrophage-like J774.1 cells [317]. (−)-SECO prevent D-galactosamine/lipopolysaccharide-induced hepatic injury by inhibiting hepatocyte apoptosis [318], and also possess significant anti-inflammatory and antinociceptive *in vivo* [319]. SECO was also found to be an antiallergic compound [320].

ODSI (**117**) was found to possess antifungal activity [321]. On the other hand, from the seeds of *Trichosanthes kirilowii* are isolated (−)-1-O-feruloyl-SECO (**120**), called hanultarin, and a homolog, 1,4-O-diferuloyl-SECO (**121**) [322]. Both compounds exhibited strong cytotoxic effects against the cell lines of human lung carcinoma A549, melanoma SK-Mel-2, and mouse skin melanoma B16F1. Hanultarin (**120**) also showed an inhibitory effect against the polymerization of actin cytoskeleton in normal epidermal keratinocytes (HaCaT cells).

11.6.1.2 Nordihydroguaiaretic Acid (NDGA)

NDGA (**61**), which makes up 5–10% of the dry weight of the leaves of creosote bush (*L. tridentata*) [323], possesses numerous biological activities [324]. It has been reported to be active against bacteria [324, 325], amoebae [326], viruses [327], fungi [328], and cancer cells [329, 330], and to possess antiherbivore properties [331]. NDGA is also a well-known antioxidant [324] for fats and oils and has been used in human foods for over 30 years (until 1972) for this purpose. Renal and hepatotoxicity are also reported for chronic use of creosote bush and NDGA [324, 332, 333].

NDGA dramatically reduces the seedling root growth of barnyard grass, green foxtail, perennial ryegrass, annual ryegrass, red millet, lambsquarters, lettuce, and alfalfa, and reduces the hypocotyl growth of lettuce and green foxtail [120]. Moreover, NDGA (**61**) was found to suppress HIV-1 replication in infected cells by preventing proviral transcription and HIV Tat-transactivated transcription [289].

NDGA (**61**) has been reported to be inhibitory to the growth of *Streptococcus* species, *S. aureus*, *B. subtilis* and, in the case of norisoguaiacin, to *Pseudomonas aeruginosa* as well [334].

NDGA (**61**) interacts with steroid-binding proteins (SBPs). It induces conformational changes in them and causes a reduction or loss of immunorecognition of SBP by anti-SBP antibodies [303]. Also, NDGA (**61**), commonly used for the inhibition of lipoxygenase isoenzymes, showed a strong growth inhibition [305].

Yu et al. [335] reported the potent antioxidative properties of *meso*-DGA (**123**) isolated from *Machilus thunbergii*. In another study on antioxidant lignans, Filleur et al. [336] reported *meso*-DGA, nectandrin B, and erytro-austrobailignan to exert an antiproliferative effect on MCF-7 cells, as well as antioxidative activity on DPPH radical.

The lignan (+)-2′-hydroxy-DGA (**124**), isolated from *Saururus chinensis*, exhibited antioxidative activity in various assay systems and DPPH radical scavenging activity [337].

Also, 3′-*O*-methyl NDGA (**125**) inhibits *in vivo* the HIV life cycle [289]. In addition, NDGA is a PAF inhibitor, and consequently, it is a potential antiasthmatic, antiallergic, and anti-inflammatory agent [79].

A series of methylated derivatives of NDGA was prepared, and tetramethyl NDGA (**126**) was found to be more active than the original lead [338]. The same derivatives, as well as tetraacetyl NDGA (**127**), were found to be potentially useful in the treatment of papilloma virus infections and their associated induced human cancers [339]. In addition, it was shown that (**127**) inhibited melanoma *in vivo* [340].

Five analogs of NDGA synthesized by Xia et al. [341, 342] exhibited anti-HSV activity.

11.6.2 Dibenzylbutyrolactone Lignans

While (−)-yatein (**12**) showed antimitotic potency [230], desmethoxy-yatein (**128**) [207][12] showed activity against P388 lymphocytic leukemia cell line [212], and wikstromol (**129**) [345, 346] showed inhibitory activity against P-388 (see Figure 11.28 for structures).

Trachelogenin (**44**) showed the most potent activity as Ca_2^+ antagonists [347], while prestegances A (**131**) and B (**132**) showed inhibitory activity against PAF-induced rabbit platelet aggregation [79, 348].

Actaealactone (**133**) from *Actaea racemosa* [349] showed antioxidant activity. It also exhibited a small stimulating effect on the growth of MCF-7 breast cancer cells, and americanin (**134**) showed toxicity in the brine shrimp lethality bioassay [350].

In rabbits, (+)-nortrachelogenin (**135**) has been proved to be a central nervous system depressor [351], whereas savinin (**58**) isolated from *Acanthopanax chiisanensis* roots inhibits the TPA-induced production of prostaglandin E2, a notable inflammatory response mediator [352].

Bursehernin (**23**) and podorhizol (**136**) showed inhibitory effects on Epstein-Barr virus (EBV) activation [254].

The lignan 5′-methoxyyatein (**137**) has been shown to be a valuable inhibitor of both COX-1 and COX-2 [353][13].

Styraxlignolides (**138–140**) exhibit weak antioxidative activity against DPPH radicals [354], and nemerosin (**141**) and isochaihulactone (**142**) from *Bupleurum scorzonerifolium* were also found to possess immunosuppressive activity [355].

On the other hand, the dibenzylbutyrolactones (**143–145**) display strong activity as potent anti-HIV agents. One lignane, bis-2,4,6-trimethylbenzyllactone lignan, was active (IC_{50} = 35 µM) for the inhibition of the proliferation of HT29 colon cancer cells [356].

Cathartic activity is not restricted to PPT analogs. The compound, 2-hydroxyarctiin (**146**), has been found to be responsible for the cathartic activity of safflower meal [357].

---

[12] (Isolated from *H. ovigera* [254, 343], *Hernandia nymphaeifolia* [212], *Bursera schlechtendalii* [344], and *B. fagaroides*.
[13] Prostaglandins are a related family of chemicals that are produced within the cells of the body by the cyclooxygenases COX-1 and COX-2. They have several important functions, including the promotion of inflammation, pain, and fever [8].

Reportedly, (−)-*trans*-2-(3″,4″,5″-tri|methoxy|benzyl)-3-(3′,4-methyl|ene|dioxy|benzyl) butyrol-lactone (**147**) and (−)-*trans*-2-(3″,4″-di|methoxy|benzyl)-3-(3′,4′-methyl|ene|dioxy|benzyl) butyrollactone (**148**) posses cytostatic or antitumor activity [344].

Dihydroanhydropodorhizol (**149**) and its glucoside (**150**) show a much reduced activity when compared to the corresponding cyclic analog, deoxy-PPT (**5**) [227].

11.6.2.1 Matairesinol (MAT) and Enterolactone (ENL) (Phytoestrogens)

ENL (**38**) and its theoretical precursors, 3′-demethoxy-3-*O*-demethyl-MAT (**151**) and 4,4′-dideoxy ENL (**152**), decreased aromatase activity [307]. ENL has also been shown to inhibit breast and colon carcinomas [308] and preventing postmenopausal osteoporosis [358]. Moreover, several mammalian lignans: ENL (**38**), 3-*O*-methyl-ENL (**153**), and prestegane B (**132**) inhibited Na^+ and K^+ pump activity in human red blood cells [359].

It has been found that 4,4′-dihydroxy ENL (**145**) and 4,4′-dideoxy ENL (**152**) inhibit human placental aromatase[14]. Compounds (**145**) and (**152**) are theoretical metabolites of MAT (**8**), which is most likely a plant precursor of ENL (**38**) [67].

MAT (**8**) shows high activity as c-AMP phosphodiesterase inhibitor, being a potential antiasthmatic agent [289], and (+)-MAT (**8**) showed potent cytotoxic activity against L929 (murine) cells [360].

Kim *et al.* [361] reported the potent cytotoxic effect of MAT on human promyleocytic leukemia HL-60 cells.

7,12-dimethylbenz[a]anthracene (DMBA) induced rat mammary cancer. 7′-Hydroxy-MAT (**49**) had a statistically significant inhibitory effect on tumor growth [271] and prostate cancer [362, 363]. In addition, the antioxidative properties of 7′-hydroxy-MAT (**49**) were studied *in vitro* in lipid peroxidation, superoxide and peroxyl radical scavenging, and LDL-oxidation models in comparison with synthetic antioxidants [271]. Also, 7′-Hydroxy-MAT (**49**) inhibited the growth of the fungus [81].

Willför *et al.* [314] revealed that MAT (**8**), 7′-hydroxy-MAT (**49**), and nortrachelogenin (**135**) have high antioxidative potency. These compounds were also able to scavenge superoxide and peroxyl radicals *in vitro*. Also, mammalian lignans (**156**) showed antioxidative activity higher than that of ascorbic acid, and comparable to that of the known antioxidant NDGA (**61**) [364].

The common wood constituent MAT (**8**) showed a high degree of inhibitory activity. The guaiacyl ring seems to be the optimum substituent in eliciting the response. Arctigenin (**10**), with one veratryl ring, shows slightly reduced activity, and a compound with both aromatic substituents of the veratryl type (NC) is only one-third as active as MAT. The importance of the 4′-substituent is unclear; however, since glucosylation of one position eliminates inhibitory activity (**157**–**159**), glucosylation of two positions (**160**–**161**) results in a level of activity equivalent to that of the aglycone. The stereochemistry of MAT at C_2 appears to have very little effect on c-AMP phosphodiesterase inhibition; that is, (**10**), (**162**), and (**163**) have similar activities. Furthermore, hydroxylation of C_2 has no appreciable effect on enzyme inhibition, as indicated by the activity of derivatives (**49, 165**) and (**166**). This contrasts with the case of the C_5 position. Hydroxylation of MAT at C_5 produces an inactive compound [293].

11.6.2.2 Arctigenin

Dibenzylbutyrolactones derived from arctigenin (**10**) have been found to be active as inhibitors of viral integrase (see Figure 11.28) [180]. Arctiin (**40**) showed potent inhibitory activity against L1210 (mouse leukemia) [365]. Huang *et al.* [366] examined the effect of arctiin on growth regulation in prostate cancer PC-3 cells.

It has been reported that (−)-arctigenin (**10**) offers the strongest inducing activity toward mouse meyloid leukemia (Ml) cells of the 16 lignans isolated from *Arctium lappa* L. [365, 367]. Arctigenin

---

[14] Aromatase, human estrogen synthetase, catalyzes the conversion of androgens into estrogens in many tissues.

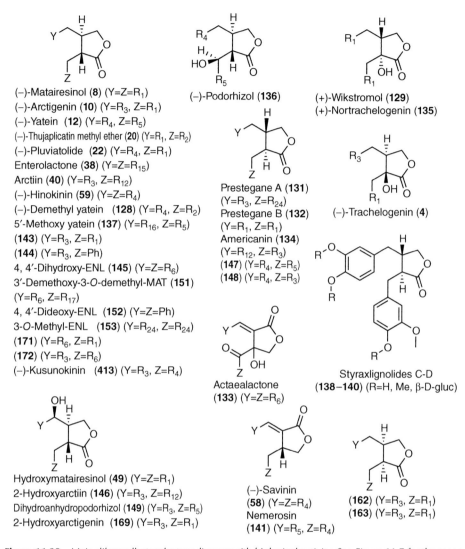

Figure 11.28 Main dibenzylbutyrolactone lignans with biological activity. See Figure 11.7 for the meaning of R groups

shows high activity, as c-AMP phosphodiesterase inhibitor being a potential antiasthmatic agent [289]. Arctigenin (**10**) inhibited lipooxygenase isoenzymes [305]. The effects of arctigenin (**10**) on mitogen-activated protein (MAP) kinase activation in Raw264.7 cells and MAP kinase kinase (MKK) activity were examined [368]. Lignan (**10**) also inhibited LPS-inducible nuclear NF-jB activation and nuclear translocation of p65, which was accompanied by inhibition of I-jB-a phosphorylation [369].

Three dibenzylbutyrolactone lignans (arctigenin (**10**, **167**, and **168**) showed significant hepatoprotective activity against CCl4-induced injury in primary cultures of rat hepatocytes [370].

In addition, (−)-arctigenin (**10**), (−)-trachelogenin (**44**), and their 2,3-benzylbutyramide derivatives showed remarkable cytotoxic activity against L5718Y (mouse lymphoma) cells *in vitro* [177, 178, 371]. Two butanolide lignans, arctigenin (**10**) and trachelogenin (**44**), inhibited the PAF receptor

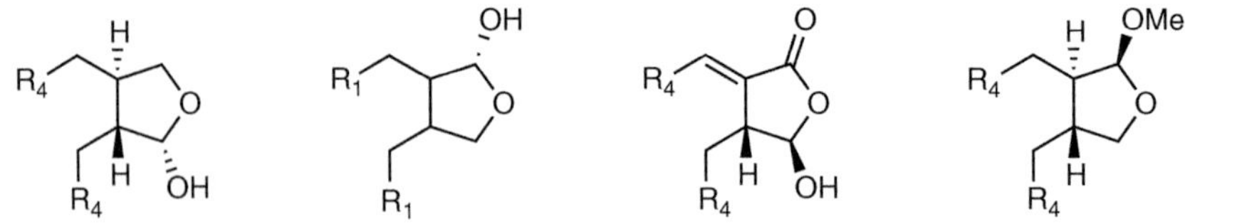

(−)-Cubebin (**175**)   Dibenzyl-butyrolactol (**7**)   (+)-Calocedrin (**173**)   (−)-Methylcubebin (**176**)

Figure 11.29 *Main dibenzylbutyrolactol lignans with biological activity [373]. See Figure 11.7 for the meaning of R groups*

[3, 372]. Arctigenin and trachelogenin reduced HIV viral protein production and inhibited viral RT activity [177]. The compounds were also found to inhibit topoisomerase II activity [178]. Eich *et al.* [257] reinvestigated the action of arctigenin (**10**) and trachelogenin (**44**) on HIV-1 in 1996 and again found them to be *in vivo* inhibitors of HIV-1 integrase as well as being topoisomerase II inhibitors. (−)-Arctigenin (**10**) and (−)-nortrachelogenin (**135**) inhibited superoxide production [372].

In vitro (−)-arctigenin (**10**) and (−)-trachelogenin (**44**) showed a strong cytoprotective effect against HIV-1, and efficiently inhibited both cellular topoisomerase II activity and the HIV-1 integrase reaction [177, 178, 371]. However, it is inactive against purified HIV-1 integrase [257].

Kim *et al.* [361] reported that 2-hydroxyarctigenin (**169**) showed potent cytotoxic effect on human promyleocytic leukemia HL-60 cells.

Three dibenzylbutyrolactones (**145**, **171**, and **172**) were active against HIV-1 integrase *in vitro*. Notably, all of these three compounds contain at least one catechol (*o*-dihydroxyphenyl) ring in their structure [256].

11.6.3 Dibenzylbutyrolactol Lignans

Calocedrin (**173**) significantly inhibited TNF-α production in LPS-stimulated RAW264.7 cells, and T-cell proliferation, without displaying cytotoxicity [369] (Figure 11.29).

Wood of the tree *Cedrus deodara* is composed of three lignans, namely wikstromol (**129**), MAT, and dibenzylbutyrolactol (**7**) [374], yielding its standardized herbal mixture (CD-3), a composition of 78% wikstromol, 11% MAT, and 11% dibenzylbutyrolactol (w/w), which has cytotoxic activity against several cancer cell lines.

Cubebin (**175**) [375–379] and methylcubebin (**176**) exerted a significant analgesic activity in the acetic acid-induced writhing in mice [380]. The results also showed that the analgesic activity of (**79**) was slightly more pronounced than that observed for (**81**).

De Souza *et al.* [381], in 2005, reported the trypanocidal activity of cubebin (**175**) and its semi-synthetic derivatives against free amastigote forms of *Trypanosoma cruzi*.[15]

Some derivatives of cubebin were prepared and evaluated for their antimycobacterial activity against *Mycobacterium tuberculosis* and *Mycobacterium avium*, suggesting that this class of compounds may lead to a new generation of antituberculosis agents [384].

11.6.4 Furanoid Lignans

Justiciresinol (**177**) exhibited cytotoxicity against A-549 (human lung carcinoma), MCF-7 (human breast carcinoma), and HT-29 (human colon adenocarcinoma) cell lines [346]. Anolignan C (**102**) showed specific, but only moderate, cytotoxicity against ZR-75-1 [282]. Magnosalicin (**178**) showed potent inhibitory activity on histamin release from rat mast cells [385] (see Figures 11.30 and 11.31).

---

[15] Chagas' disease, or American trypanosomiasis, is endemic in Central and South America and it is estimated that 16–18 million people are currently infected with the protozoan flagellate *Trypanosoma cruzi* [382] and more than 100 million worldwide are exposed to the risk of infection [383].

The lignan (−)-grandisin (**179**) has shown important pharmacological activities, such as citotoxicity and antiangiogenic, antibacterial, and trypanocidal activities. Hence, it has been considered as a promising anticancer agent [386].

PAF[16] receptor antagonists are expected to be useful as antiallergic, antiasthma, and anti-inflammatory drugs. Kadsurenone (a neolignan, **414**), veraguensin (**180**), galbelgin (**181**), galgravin (**182**), nectandrin A (**183**), nectandrin B (**184**), burseran (**185**), prestegane A (**131**), and prestegane B (**132**) showed inhibitory activity against PAF-induced platelet aggregation in rabbits [79, 348].

The synthetic bisphenyltetrahydro furan lignan, *trans*-2,5-bis(3,4,5-trimethoxyphenyl) tetrahydrofuran (L-652,731, (**186**)), is a potent and orally active PAF-specific and competitive receptor antagonist [387]. Also, a dihydrofuran lignan (**187**), isolated from *Stauntonia chinensis* (*Lardizabalaceae*), showed an analgesic effect in mice [388, 389].

On the other hand, two lignan glucosides (**188–189**) derived from germinated sesame seeds showed a less-potent antioxidant effect than α-tocopherol [390] and scavenged hydroxyl radicals [391]. Beilschmins A–C (**190–192**) [392] exhibit significant *in vitro* cytotoxicity against P-388 and HT-29 cell lines. Futokadsurins A–C (**193–195**) weakly inhibits activity against the generation of nitric oxide [393]. Moreover, olivil 9-*O*-β-D-xyloside (**196**) exhibited moderate antioxidant activity in a DPPH assay [394]. Dihydrosesamin (**197**) showed potent neurite outgrowth-promoting activity in PC12 cells [395]. Seselinone (**198**) shows moderate cytotoxic activity against the C6 rat glioma cell line [396]. Forsythialans A (**199**) and B (**200**) [397] both showed protective effects against renal epithelial cell injury induced by a peroxynitrite generator, 3-morpholinosydnonimine (SIN-1).

Santolinol (**201**) was a potent inhibitor of lipoxygenase [398] and (+)-lariciresinol (**5**) scavenged 50% diphenylpicrylhydrazyl (DPPH) radical and showed activity similar to that of superoxide dismutase (SOD) [399].

Fractionation of the extract of *Tripterygium wilfordii* led to the isolation of tripterygiol (**202**) and eight analogs, which were found to suppress pro-inflammatory gene expression [400]. The lignan compounds agastinol (**203**) and agastenol (**204**) inhibited etoposide-induced apoptosis in U937 cells. From these results, (**203**) and (**204**) seem to be worthy candidates for further research as potential antiapoptotic agents [401].

Lee *et al.* [286] evaluated machilin G (**205**) and (+)-galbacin (**206**) for their abilities to inhibit PLCG1 *in vitro*. The SAR results of these suggested that the benzene ring with the methylene dioxy group is responsible for the expression of inhibitory activities against PLCG1.

Li *et al.* [287] reported that nectandrin (**184**) inhibited topoisomerase I and II.

Lee *et al.* [402] isolated saucerneol B (a sesquineolignan (**207**)) and manassantins A and B (two dineolignans (**208–209**)), both having a furanoid ring, and evaluated them for their inhibitory activities (as anti-inflammatory) against hACAT-1 and hACAT-2. Saucerneol B (**207**) inhibited hACAT-1 threefold more strongly than hACAT-2, whereas (**208**) was more potent against hACAT-2 than hACAT-1. However, (**209**) inhibited mainly hACAT-1, but not hACAT-2. Manassantin and other saucerneol derivatives were studied using HeLa cells transfected with NFjB reporter construct [403]. Compounds (**207–209**) showed the most potent inhibitory activity.

It has recently been reported that magnones A and B (**210–211**) displayed anti-PAF activity [404], and (**210**) had TNF-α-suppressing activity in LPS-stimulated macrophages [405]. In view of these results, it is likely that these bioactive lignans are compounds worth developing as novel anti-inflammatory drugs.

Larreatricin derivatives (**212–216**) were isolated from *L. tridentata*. Among them, the lignans possessing a tetrahydrofuran moiety (**212–215**) showed strong antioxidative activity without cytotoxicity [288].

---

[16] platelet activating factor (PAF) is a bioactive phospholipid that causes hypertensive, inflammatory, and allergic responses and increases vascular permeability.

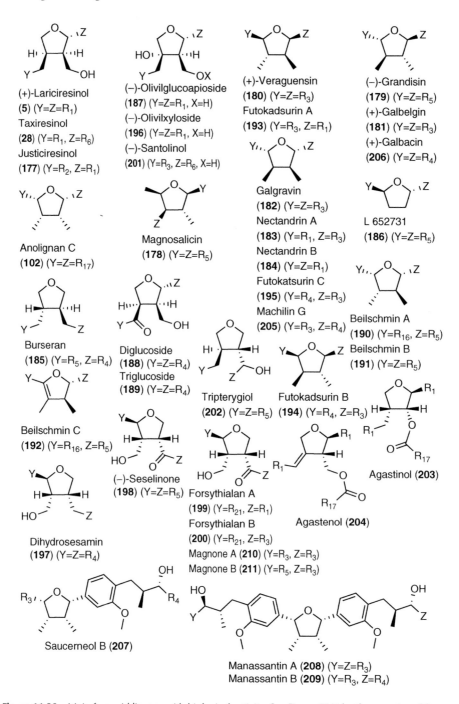

Figure 11.30 Main furanoid lignans with biological activity. See Figure 11.7 for the meaning of R groups

Larreatricin deriv. (**212**) (Y=Z=R$_5$)   Larreatricin deriv. (**213**) (Y=Z=R$_5$)   4-epi-Larreatricin (**214**) (Y=Z=R$_{17}$) 3′-Hydroxy-epi-larreatricin (**215**) (Y=R$_{17}$, Z=R$_6$)   3,4-Dehydro-larreatricin (**216**) (Y=Z=R$_5$)

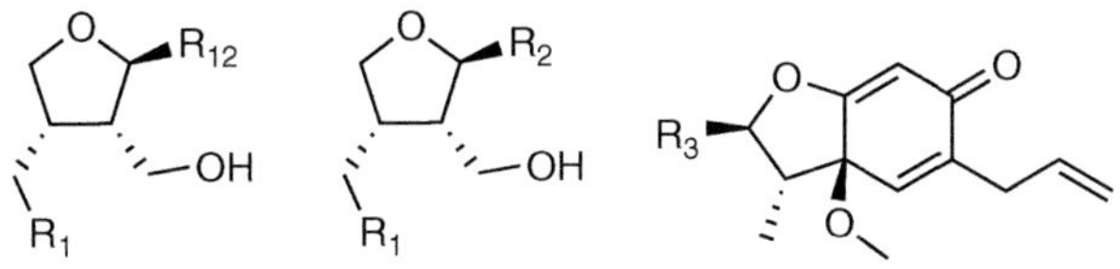

Lariciresinol glycoside (**217**)   5′-Methoxy-lariciresinol (**218**)   Kadsurenone (**414**)

Figure 11.30 Continuation

Filleur *et al.* [336] reported nectandrin (**183**) to exert an antiproliferative effect on MCF-7 cells, as well as antioxidative activity on DPPH radical. Burseran reportedly has antitumor activity [406].

(+)-Taxiresinol (**28**) was explored for anticancer properties against colon, liver, breast, and ovarian cancers [407]. Erdemoglu *et al.* [83] have reported that (−)-taxiresinol (**28**) has antifungal activity.

It has been revealed that (−)-lariciresinol (**5**) has a high antioxidative potency [314]. The lariciresinol glycoside (**217**) was tested to evaluate its *in vitro* anti-inflammatory effects [408]. It significantly inhibits TNF-α production [409, 410]. (±) − 5′-Methoxylariciresinol (**218**) and (±)-lariciresinol (**5**) showed inhibitory activity against P-388 [411].

11.6.5 Furofuranoid Lignans

(+)-Epi-yangambin (**225**) [412] and yangambin (SYR dimethyl ether (**226**)) [413] were potent selective antagonists of PAF, and (+)-diayangambin (**238**), magnolin (**227**), and fargesin (**228**) showed lower activity as Ca$_2^+$ antagonists[17] [414] (Figure 11.31).

Graminone B (**229**) inhibited the contractile response of isolated aorta in rabbits [415], while petaslignolide A (**230**) exhibited weak antioxidant activity [416].

5″-Methoxyhedyotisol A (**231**) exhibited potent antioxidant activity in the DPPH radical scavenging assay [417]. Also, the rearranged furolignan (+)-commiphorin (**232**) showed activity against a number of bacteria [418].

Both (+)-fargesin (**228**) and (+)-eudesmin (**233**), isolated from *Magnolia biondii* flowers[18], behave as PAF inhibitors [79].

Hausott *et al.* [305] showed the strong growth inhibition of lipoxygenase isoenzymes of (+)-epi-aschantin (**234**). epi-Yangambin (**225**) exhibited significant inhibition of the transformation of murine epidermal JB6 cells [252]. Ito *et al.* [254] isolated antitumor compounds from the seeds of *Hernandia ovigera* and tested them for their inhibitory effects on the EBV. epi-Magnolin (**235**) exhibited significant inhibitory effects at high concentrations. These results suggest that lignans might be valuable as antitumor compounds in chemical carcinogenesis [419].

The phillygenol (**236**) derivatives (phillygenin **48** and the disaccharide **237**) and their antitumor activities on human hepatoma cell SMMC-7721, human uterine cervix carcinoma cell HeLa, hamster lung fibroblast cell V79, and mouse melanoma cell B16 *in vitro* were studied [420]. The results

---
[17] They are widely used as therapeutic agents for coronary heart disease and hypertension.
[18] Used for treating headache and sinusitis.

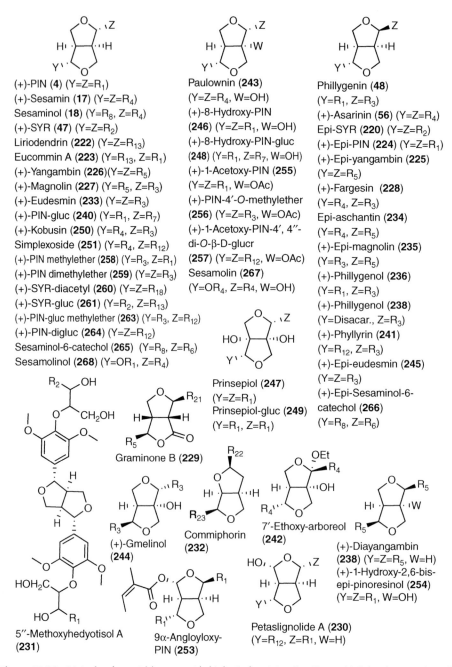

Figure 11.31 Main furofuranoid lignans with biological activity. See Figure 11.7 for the meaning of R groups

showed that phillygenin (**48**) had strong cytotoxicities on the four tested cell lines when compared to that of vincristine, whereas the glucosydes (**236**) and (**237**) had little effect on the proliferation of the cell lines tested. The loss of antitumor activity of (**236**) and (**237**) may be due to the substitution of the hydrogen of the phenolic hydroxyl group.

Evaluations have been made of (+)-diayangambin (**238**) *in vitro* and *in vivo* for its immunomodulatory and anti-inflammatory efficacy [421]. Human mononuclear cell proliferation was inhibited by (**238**). These findings indicate the potential of (**238**) for treating the immune and inflammatory responses.

The inhibitory activity of magnolin (**227**), lirioresinol-B dimethyl ether (yangambin, SYR dimethyl ether (**226**)), and epi-magnolin (**235**) [422] displayed relatively strong inhibitory activities on ICAM-1 expression induced by TNF-α.

Phyllyrin (**241**) was explored for interactions with the cyclooxygenase and 5-lipoxygenase pathways of arachidonate metabolism, and for its effects on cell viability [423].

Four lignans (**242–245**) were evaluated for their antifungal potential against basidiomycetes [380]. After comparisons of antifungal activity of different structures, it was concluded that the piperonyl nucleus present in gmelinol (**244**) and epi-eudesmin (**245**) contributed to the activity of these lignans. Of the lignans isolated, (+)-epi-eudesmin (**245**) appeared also to be an important antifungal constituent [380].

Compounds (**246–249**) showed antioxidant properties in trolox-equivalent antioxidant activity (TEAC) and chemiluminescence (CL) assays [424]. The aglycones 8-hydroxy-PIN (**246**) and prinsepiol (**247**) also displayed powerful antioxidant activity. It was found that the tertiary hydroxy group on the furofuran ring affected the degree of the antioxidant activity [275].

Encouraged by the noteworthy biological profile of lignans and neolignans, Prasad *et al.* [425] evaluated antiviral activities of representative examples of each structural type of lignans and neolignans: (−)-machilin G (**205**), belonging to 2,5-bisaryl-3,4-dimethyl-THF structural type; (+)-asarinin (**56**) and (+)-sesamin (**17**). The antiviral activity of these compounds was tested against Herpes simplex virus type 1 (HSV-1), coxsakie B2 (Cox B2), measles edmondston A (MEA), poliomyelitis virus type 1 (Polio 1), semliki forest L10 (SF L10), and vesicular stomatitis virus (VSV) at five different concentrations. While (+)-asarinin (**56**) was found to be highly active against Cox B2, MEA, and Polio 1 viruses, it was weakly active against SF L10 and VSV viruses.

The bis-epoxylignans, kobusin (**250**) and sesamin (**17**), were shown to inhibit the growth of silkworm (*Bombyx mori*) larvae [426]; and the bis-epoxylignan glycoside, simplexoside (**251**) has CNS-depressant activity in mice and rats while the aglycone is a stimulant [427, 428].

A monoepoxylignanolide (**60**) has been isolated from *A. ouata*, which is an inhibitor of germination [121, 122].

11.6.5.1 Pinoresinol (Phytoestrogens)

It has been found that (+)-PIN (**4**) significantly inhibits TNF-α (tumor necrosis factor) production [409, 410, 429] and showed *in vitro* anti-inflammatory effects [408].

Several PIN glucosides (di- and trisaccharides) were isolated from sesame seeds. The disaccharide derivatives showed almost the same antioxidant activity as α-tocopherol. These compounds can produce the antioxidant PIN (**4**) *via* hydrolysis by β-glucosidase from intestinal bacteria [430]. The diglucosid derivative of PIN (**264**) is also an antihypertension agent [431] and four complex glucosyl PIN derivatives isolated of *Rhus javanica* showed moderate inhibitory activity against the multiplication of the tobacco mosaic virus.

The lignan glucoside (+)-PIN mono-β-D-glucopyranoside (**240**) showed effective antioxidant potentials on the basis of oxygen radical absorbance capacity (ORAC) [432].

The lignans PIN (**4**) and the glucosides (liriodendrin, SYR glycoside **222**) and (**240**), isolated from the rhizomes of *Coptis japonica*, were tested to evaluate their *in vitro* anti-inflammatory effects

[408]. PIN (**4**) showed higher inhibitory effects on TNF-α production, whereas SYR glycoside (**222**) strongly suppressed lymphocyte proliferation [429].

PIN (**4**) and SYR (**47**) affected cytokine production from LPS, or phytohemagglutinin-stimulated human peripheralmononuclear cells [433]. The effects as inmunosuppressive were compared with the reference compound (prednisolone), and the results revealed that both compounds exhibited a significant inhibitory effect on all the cytokines tested [434].

Reportedly, (+)-PIN **4** and (+)-epi-PIN (**224**) relaxed norepinepherine-induced contraction in rat aortic strips without endothelium [435]. Also, 9α-angloyloxy-PIN (**253**), with HIV-1, reverses transcriptase inhibitory activity [436].

The lignan (+)-1-hydroxy-2,6-bis-epi-PIN (**254**) has antimycobacterium tuberculosis potential [437].

On the other hand, PIN is a very effective inhibitor of c-AMP phosphodiesterase, as well as its derivatives (+)-1-acetoxy- (**255**) and (+)-1-acetoxy-4″-O-methyl ether (**256**). The guayacyl group seems to be ideal for such activity [5]. Also, (+)-1-acetoxy-PIN (**255**) and (+)-1-acetoxy-PIN 4′,4″-di-O-β-D-glucoside (**257**) showed the strongest inhibitory activity toward c-AMP phosphodiesterase [438, 439].

The stereochemistry of the furan–phenyl bond is evidently relevant to the inhibitory activity of c-AMP phosphodiesterase. It was found that (+)-PIN (**4**), (+)-PIN monomethyl ether (**258**), (+)-PIN dimethyl ether (**259**), and (+)-PIN β-D-glucoside (**240**), in all of which both C_2 and C_6 are in the S configuration, are more effective in inhibiting c-AMP phosphodiesterase than their corresponding (+)-epi-PIN analogs ((**224**), respectively) in which C_2 is in the S and C_6 is in the R-configuration. Also, (−)-PIN (**4**) and (−)-PIN mono-β-D-glucoside (**240**), in which both C_2 and C_6 are in the R-configuration, display c-AMP phosphodiesterase inhibitory activities more than an order of magnitude less than (+)-PIN (**4**) or its β-D-glucoside (**240**) [5].

A wide variety of bis-epoxylignans are effective inhibitors of c-AMP phosphodiesterase [438]. As in the case of the lignanolides tested, the guaiacyl group appears to be most effective in contributing to the inhibitory activity. Compound (**258**) is more active than (**259**) which, in turn, is more active than (**4**). The syringyl group, however, is more effective than the veratryl group in conferring inhibitory activity, as seen from the activity of SYR (**47**). Remarkably, the peculiar relationship of the glucosides is evident with the bis-epoxylignans. A glucose moiety at the 4′-position of either aromatic group causes a marked reduction in inhibitory activity. There is a twofold decrease in the case of PIN β-D-glucoside (**240**) and (**263**). The presence of a glucose substituent at the 4′-positions of each ring, however, results in activity equal to that of the aglycone (e.g., (**264**) and (**222**) [5].

11.6.5.2 Syringaresinol

(+)-SYR O-β-D-glucoside (**261**), eucommin A ((+)-medioresinol O-β-D-glucoside **223**), and (+)-epi-pinoresinol (**224**) showed moderate anticomplementary activity [440].[19]

(−)-Epi-SYR (**220**) and (−)-SYR (**47**) showed potent cytotoxic activity against L-929 (murine) cells [360]. It was found that (±)-SYR (**47**) showed inhibitory activity against P-388 [411], while (−)-SYR (**47**) and (−)-SYR diacetate (**260**) showed significant inhibitory activity against P-388 [441]. On the other hand, (−)-SYR O-β-D-glucopyranoside (**261**) was inactive against 11 cancer cell lines: A431, BC1, Col2, HT, KB, KB-V1, LNCaP, Lu1, Mel2, ZR-75-1, and P-388 [442].

Liriodendrin (**222**) exhibited potent cytotoxic activity against P-388, KB, and lung cancer cells [443].

Similar to PIN, (+)-SYR di-D-glucoside (**222**) is much more active than its enantiomer (−)-syringaresinol di-D-glucoside (**222**) [444].

---

[19] The complementary system is a major immunity and is activated by a cascade mechanism *via* an antigen–antibody-mediated process.

Both leutheroside E (**262**) and liriodendrin (**222**) showed antistress activity [445].

Oral administration of liriodendrin (**222**) [446] prolonged the exercise time to exhaustion in chronic swimming stress tests in rats, increased the β-endorphin content in plasma [447], prevented stress-induced decrease in movement, and led to accelerated recoveries [446]. Liriodendrin (**222**), isolated from Siberian ginseng (*Acanthomax senticosus*), is used widely in Asia to help the cardiovascular system during prolongated exercise [447].

11.6.5.3 Sesamin

Even though (+)-sesamin (**17**), a main component of sesame lignans, showed high activity against SF L10 virus, it was inactive against other viruses, such as HSV-1, Cox B2, MEA, and Polio 1 [425]. (+)-Sesamin (**17**) showed cytotoxic potency against HeLa (human Hela cernix uteri tumor) cell [448]. (+)-Sesamin (**17**) reduced the concentration of lipoperoxide in plasma, liver, and tumors [449]. Sesamin (**17**) is also effective in enhancing the toxicity of a wide variety of insecticides [113, 114]. Sesamin has been shown to increase vitamin E levels by inhibiting its metabolism [450].

Kiso *et al.* [451] reported (−)-sesamin (**17**) to exhibit an antioxidative effect on lipid and alcohol metabolism in the rat liver. In another experiment, (**17**) and sesaminol (**18**) elevated α-tocopherol concentration and decreased thiobarbituric acid-reactive substance (TBARS) concentration in the blood plasma and liver of rats [451]. Sesaminol (**18**) showed strong antioxidant activity [452]. Compound (**18**) exhibited strong antioxidant effects in the autoxidation of linoleic acid [453] and also significantly inhibited peroxidation. Sesaminol glucosides, isolated from sesame seeds, can be hydrolyzed to the strong antioxidant (**18**) by intestinal β-glucosidase or bacteria [454].

Sesaminol 6-catechol (**265**) and epi-sesaminol 6-catechol (**266**), which were transformed from sesaminol triglucoside by *Aspergillus usamii*, exhibited higher antioxidant activities [455] than their precursor (**137**).

Other lignans present in sesame (*Sesamum indicum*), such as sesamolin (**54**), sesamin (**17**), and sesamolinol (**268**) yielded high antioxidant activity [452, 456].

Lee *et al.* [286] evaluated (−)-sesamin (**17**) for its ability to inhibit PLCG1 *in vitro*. The SAR results of several lignans (machilin A, sesamin, machilin G, and galbacin) suggest that the benzene ring with the methylene dioxy group is responsible for the expression of inhibitory activities against PLCG1.

Cui *et al.* [457] isolated four furofuran lignans [(−)-sesamin (**17**), yangambin (**226**), PIN (**4**), and SYR (**47**)] from *Chinese propolis*. Their antioxidative activity was evaluated by measuring the inhibition of lipid peroxidation in rat liver microsomes.

11.6.6 Aryltetralin Lignans (No Lactones)

Hara *et al.* [258] investigated the anti-HIV-1 activity of a large number of aryltetralin (aryltetrahydronaphthalene) lignan analogs, both *in vivo* and *in vitro*.

Many cyclolignans similar to PPT are active against viruses such as measles virus, types I and II herpes virus, influenza type A virus, vaccinia virus, and cytomegalovirus [154, 162, 229] (Figure 11.32).

San Feliciano *et al.* [458–460] reported on the antiviral effects of several lignans against HSV-1 and VSV infecting monkey kidney fibroblasts (CV-1) and hamster kidney fibroblasts (BHK), respectively. In general, tetrahydro- and dihydronaphthalenic lignans were more active than their naphthalenic counterparts. In addition, the lactonic lignans appeared to be more active than nonlactonic lignans, and trans-lactone compounds (e.g., PPT (**14**)) were more active than the corresponding *cis*-lactones (e.g., picro-PPT (**79**)). There was a rough correlation between antineoplastic and antiviral effects, suggesting a similar mechanism of action for the compounds tested [458–460].

Isolariciresinol (**9**) showed binding affinity to sex hormone–binding globulin (SHBG) [461]. Erdemoglu *et al.* [83] have reported that isolariciresinol (**9**) also possess antifungal activity.

Several lignans related to isolariciresinol (**269–271**) were investigated for their anti-inflammatory activities [462].

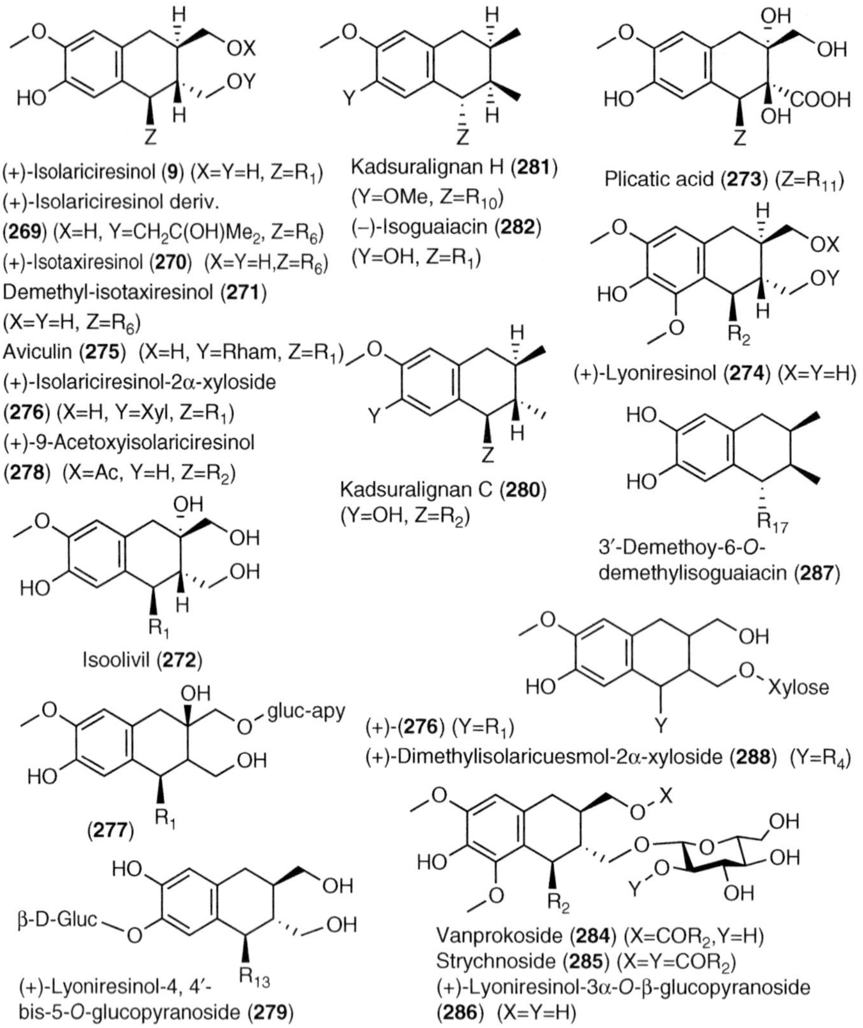

(+)-Isolariciresinol (9) (X=Y=H, Z=R$_1$)
(+)-Isolariciresinol deriv.
(269) (X=H, Y=CH$_2$C(OH)Me$_2$, Z=R$_6$)
(+)-Isotaxiresinol (270) (X=Y=H, Z=R$_6$)
Demethyl-isotaxiresinol (271)
(X=Y=H, Z=R$_6$)
Aviculin (275) (X=H, Y=Rham, Z=R$_1$)
(+)-Isolariciresinol-2α-xyloside
(276) (X=H, Y=Xyl, Z=R$_1$)
(+)-9-Acetoxyisolariciresinol
(278) (X=Ac, Y=H, Z=R$_2$)

Kadsuralignan H (281)
(Y=OMe, Z=R$_{10}$)
(−)-Isoguaiacin (282)
(Y=OH, Z=R$_1$)

Kadsuralignan C (280)
(Y=OH, Z=R$_2$)

Plicatic acid (273) (Z=R$_{11}$)

(+)-Lyoniresinol (274) (X=Y=H)

3′-Demethoxy-6-O-demethylisoguaiacin (287)

Isoolivil (272)

(277) R$_1$

β-D-Gluc

(+)-Lyoniresinol-4,4′-bis-5-O-glucopyranoside (279) R$_{13}$

(+)-(276) (Y=R$_1$)
(+)-Dimethylisolaricuesmol-2α-xyloside (288) (Y=R$_4$)

Vanprokoside (284) (X=COR$_2$, Y=H)
Strychnoside (285) (X=Y=COR$_2$)
(+)-Lyoniresinol-3α-O-β-glucopyranoside (286) (X=Y=H)

Figure 11.32 Main 9,9′-dihydroxyaryltetralin lignans with biological activity. See Figure 11.7 for the meaning of R groups

Isotaxiresinol (**270**) was equal to or even better than standard reference compounds, such as taxol and doxorubicin against colon adenocarcinoma (Caco-2) in assay systems [407].

Antioxidant activity of (+)-9-acetoxyisolariciresinol (**278**) [463] has been shown by the DCFH method [464].

Isoolivil ((+)-cycloolivil (**272**)) accumulates in the wood of *Prunus* sp. after fungal attack [465].

The sawdust of red cedar (*T. plicata*) causes asthma and rhinitis in certain exposed individuals. The compound responsible has been found to be the lignan, plicatic acid (**273**) [279].

Cyclolignans lyoniresinol (**274**) and aviculin (**275**) demonstrated potent antioxidant activity and only weak anti-NO production activity [466]. On the other hand, modified aryltetralin (**276**) displayed strong activity as potent anti-HIV agents [467].

An aryl tetrahydronaphthalene lignan (**277**) isolated from *Stauntonia chinensis* showed an analgesic effect in mice [388, 389].

Reportedly, (+)-lyoniresinol 4,4′-bis-*O*-β-D-glucopyranoside (**279**) was identified as a lipoxygenase inhibitor [468].

Kadsuralignan C (**280**) and kadsuralignan H (**281**) exhibited moderate nitric oxide production inhibitory activity against murine macrophage cell line RAW 264.7 [469, 470].

Aviculin (**275**) is a lignan glycoside, which exhibited a potent inhibitory effect on cancer cell invasion in an *in vitro* study [471].

Three lignan glucosides, vanprokoside, strychnoside, and a lyoniresinol glucoside (**284–286**), exhibited stronger radical scavenging activity against DPPH than ascorbic acid [472]. Compound (**287**) showed potent antioxidative activity [473].

It was found that (+)-dimethylisolaricuesmol-2α-xyloside (**288**) [237] has antitumor activity.

Yu *et al.* [335] reported the potent antioxidative properties of isoguaiacin (**282**).

11.6.7 Arylnaphthalene Lignans

A large variety of arylnaphthalene lignans has been isolated from species of *Justicia* (see Table 11.5) and are found in relatively high proportions [474]. These lignans may serve as lead compounds for the development of new therapeutic agents with cytotoxic activity [475, 476]. For example, lignans isolated from *Justicia pectoralis* are cytotoxic to leukemia and solid tumor cell lines [476]. Lignans also show antiangiogenic, antileishmanial, antifungal, hypolipidemic, antiasthmatic [477], antiviral [478], antineoplastic [181], antifeedant [479], insecticidal, cardiotonic, antidepressant [480], analgesic, antiplatelet [481], and anti-inflammatory [482] properties as well as acting as lipid peroxidation inhibitors. Potent anti-inflammatory activities were described for lignan glycosides isolated from *Justicia ciliata* [483] and phenolic compounds isolated from *Justicia prostrata* [484].

An extensive study of the antiviral activity of a series of 1-arylnaphthalene and 1-aryl-1,2-dihydronaphthalene lignans and their analogs was made by Cow *et al.* [510]. The compounds were tested against the human cytomegalovirus, but the antiviral activity did not extend into the nanomolar range and was often paired with high cytotoxicity (Figure 11.33).

Phyllanthostatin A (**289**) [511] showed inhibitory activity against P-388. Dehydro-β-PT methyl ether (**290**) showed cytotoxic activities against 11 cell lines [213].

The arylnaphthalene lignan TA-7552 (**291**) lowered the level of serum cholesterol and elevated HDL-C in rat [512]. The 2-pyridylmethyl derivatives of (**291**) exhibited more hypolipidemic activity than did (**291**) [513].

Diphyllin acetyl apioside (**292**) and diphyllin apioside (tuberculatin) (**293**) also showed an anti-inflammatory effect in rabbit [514]. These arylnaphthalene lignans are active against inflammation induced by tetradecailforbol acetate (TPA) [514]. Diphyllin acetylapioside (**292**) is a main inhibitor of 5-LOX [515].

Sacidumlignan A (**294**) showed moderate *in vitro* antimicrobial activity against two Gram-(+) bacteria [516].

Phyllanthusmins A–C (**295–297**) were isolated from *Phyllanthus oligospermus*. Of the three phyllanthusmins, phyllanthusmin A (**295**) showed significant cytotoxicity against KB and P-388 cancer cell lines [517].

Taxodiifoloside (**298**) exhibited moderate cytotoxic activities against five mammalian cancer cell lines [518].

Vitecannasides A (**299**) and B (**300**) exhibited greater radical-scavenging ability toward 1,1-diphenyl-2-picrylhydrazyl than that of L-cysteine [519].

Arylnaphthalene lignans such as restrojusticidin B (**301**) and phyllamycin B (**327**) (isolated from *Phyllantus myrtifolius*) act as no competitive HIV-1 inhibitors for the inverse transcriptase [520].

Prostalidins A, B, and C (**303–305**) appear to activate the central nervous system [521] in humans, but also cause a mild depression in rats and mice [521].

Table 11.5 Lignan isolates from the species of Justicia [485]

Compound	Biological activity	Species (extracts and [Refs])
Helioxanthin (**307**)	Inhibition of human hepatitis B viral replication antitumor	*Justicia flava* (EtOH [486, 487])
Taiwanin E (**336**)	Antiplatelet aggregation antitumor	*Justicia procumbens* (EtOH [481, 486])
Taiwanin E methyl ether (**337**)	Antiplatelet aggregation cytotoxicity against human cervical carcinoma	*Justicia purpurea* (MeOH [488]); *Justicia betonica* (MeOH [489]); *J. procumbens* (EtOH [481])
Justicidin E (**332**)	Inhibition of leukotriene biosynthesis by human leukocytes	*J. procumbens* (MeOH [475, 490])
Justicidin D (**331**)	Antiplatelet aggregation	*J. procumbens* (EtOH/MeOH [475, 481, 491]; MeOH [475, 478])
Justicidin B (**329**)	Anti-inflammatory, antiplatelet aggregation, cytotoxycity, antiviral, fungicidal, antiprotozoal against *T. cruzi*, antimalarial, and antirheumatic	*J. purpurea* (MeOH [488, 491–495]); *J. procumbens* (EtOH [481] and MeOH [478])
Diphyllin (**323**)	Cytotoxycity antiviral	*Justicia extensa* (EtOH [496]); *J. procumbens* (MeOH [475, 478, 481]); *J. ciliata* (MeOH [489])
Justicidin A (**309**)	Cytotoxycity, antiviral, "fish-killing" properties, induced apoptosis in human hepatoma cells	*J. extensa* (EtOH [496]); *J. betonica* (EtOH [489, 497]); *J. procumbens* (MeOH [475, 478, 481, 498]); *Justicia rhodoptera* (EtOH [499])
Cleistanthin B (**339**)	Antitumor	*J. purpurea* (MeOH [488, 500])
Patentiflorin A (**340**) and B (**341**)	Cytotoxicity against human carcinoma cells	*J. patentiflora* (EtOAc [501])
Tuberculatin (**293**)	Antitumor	*J.ciliata*; *J. betonica* (MeOH [489, 502])
Chinensinaphthol methyl ether (**335**)	Antiplatelet aggregation	*J. ciliata* (CH$_2$Cl$_2$ [489])
4'-Dimethyl chinensinaphthol methyl ether (**342**)	Antiplatelet aggregation	*J. ciliata* (CH$_2$Cl$_2$ [489]); *J. procumbens* (EtOH [481])
Elenoside (**311**)	Sedative, muscle relaxant, cytotoxic, antiviral, insecticidal, cardiotonic, analgesic, inhibition of lipid peroxidation, anti-inflammatory, stimulant	*Justicia hyssopifolia* (EtOAc [503–505])

Table 11.5 (continued)

Compound	Biological activity	Species (extracts and [Refs])
Justicidin C (**330**)	Antiplatelet aggregation	*J. ciliata* (CH$_2$Cl$_2$/Me$_2$CO [489]); *J. procumbens* (EtOH [475, 478, 481])
Justicidinoside A (**343**), C (**344**) and B (**345**)	Antiviral	*J. procumbens* (MeOH [478])
Justicinol (**346**)	Mild effect on the CNS	*J. patentiflora* (EtOAc [501])
Ciliatoside A (**347**)	Anti-inflammatory	*J. ciliata* (MeOH [483, 491])
Procumbenoside A (**310**)	Antitumor	*J. procumbens* (MeOH [502, 506])
Cilinaphthalide A (**348**)	Antitumor	*J. betonica* ( ); *J. ciliata* (CH$_2$Cl$_2$ [489])
Cilinaphthalide B (**349**)	Antiplatelet aggregation induced by adrenaline	*J. betonica* (MeOH); *J. ciliata* (CH$_2$Cl$_2$ [489]); *J. procumbens* (MeOH [507])
Diphyllin apioside-5-acetate (**392**)	Cytotoxycity and antiviral	*J. procumbens* (MeOH [478])
(+)-Isolariciresinol (**9**)	Anti-inflammatory	*J. flava* (EtOH [319])
Sesamin (**17**)	Angiogenic	*J. purpurea* (MeOH [488, 508])
PPT (**14**)	Cancer chemotherapy	*J. flava* (EtOH [126, 509])

Magnosinin (**306**) showed an anti-inflammatory effect comparable to the one of hydrocortisone acetate [522, 523].

Helioxanthin (**307**) and taiwanin C (**308**), isolated from *A. chiisanensis*, inhibit production, induced by TPA, of prostaglandin E2 (PGE2), a remarkable mediator of the inflammatory response, taiwanin C being the most potent inhibitor [352], whereas **308** showed no inhibitory effect on the release of radioactivity from [3*H*]arachidonic acid-labeled macrophages and the expression of COX-2 protein induced by TPA [352].

Tuberculatin (**293**) showed almost the same cytotoxic potencies against these cancer cell lines as justicidin A (**309**), with stronger cytotoxic activities against Hep3B, SiHa, HepG2, HT-29, HCT 116, MCF-7, and MCF-7-ras than its aglycone (**323**) (diphyllin). Compound procumbenoside (**310**) showed significant cytotoxic activity against the Hep3B and HepG2 cell lines [506].

Elenoside (**311**) showed moderate cytotoxicity and central depressive properties [505].

Phenylnaphthalene-type lignans, detetrahydroconidendrin and vitrofolal C–D (**312**–**314**) were found to afford antimicrobial activity against MRSA [524].

Compounds (**315**–**320**) showed stronger antioxidative activity than α-tocopherol. Moreover, vitedoamine A (**316**), α-conidendrinaldehyde (**317**), detetrahydroconidendrin (**318**), and vitrofolal F (**320**) were more potent antioxidants than *tert*-butylhydroxyanisole (BHA) [525].

Diesters of arylnaphthalene lignans and their heteroaromatic analogs have been shown to display hypolipidemic activity [513].

The naphthalene lactone L-702539 (**321**) is a potent and selective 5-lipoxygenase inhibitor [526].

Another series of arylnaphthalenes, including (**322**), have been evaluated as potential antiasthmatic agents [527].

Diphyllin (**323**), diphyllin monoacetate (**324**), diphyllin crotonate (**325**), and dehydroanhydro-PPT (**326**) have antitumor activity [237, 528].

Chang *et al.* [520, 529] studied the inhibition of HIV-RT by several naphthalene lignans and compared their activity of inhibiting human DNA polymerase-R (HDNAP-R). The two most potent lignans found in their study were phyllamycin B (**327**) and retrojusticidin B (**301**).

Rabdosiin (**328**) is a caffeic acid tetramer, and the sodium and potassium salts of this compound, and one of its diastereomers, are also strong inhibitors of HIV-1 [530, 531].

11.6.7.1 Justicidins

Justicidin A (**309**) has shown significant cytotoxic activities against Hep3B, HepG2, MCF-7, and MCF-7-ras [506]. Justicidin A (**309**) and diphyllin (**323**) inhibited 9-KB (human nasopharyngeal carcinoma), while justicidins C–E (**330**–**332**) did not [475]. Justicidin B (**329**) showed inhibition

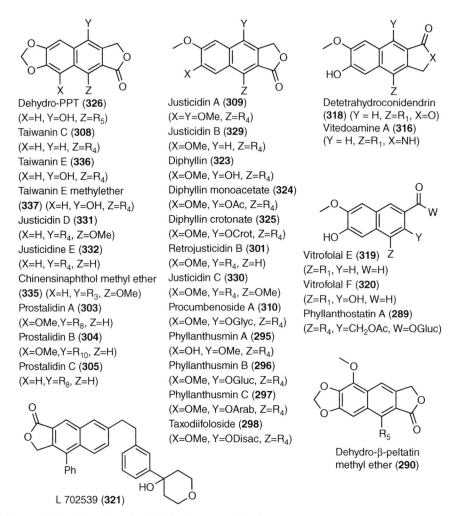

Figure 11.33 *The main arylnaphthalene lignans with biological activity. See Figure 11.7 for the meaning of R groups*

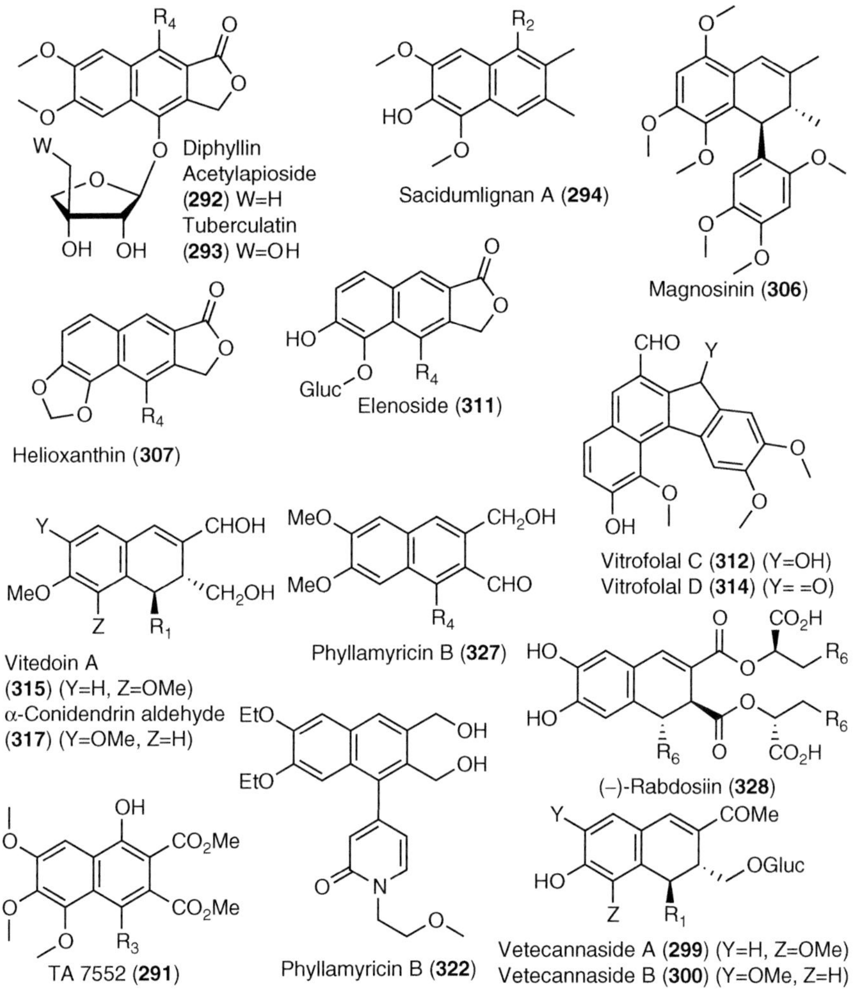

Figure 11.33 Continuation

against P-388 and NSCLCN6 (human bronchial epidermoid carcinoma) cells [532]. Justicidin B (**329**) inhibited the growth of the pathogenic fungi *Aspergillus fumigates*, *Aspergillus flavus*, and *Candida albicans*, but was not effective against other tested pathogens such as *Cryptococcus neoformans* or *Blastoschizomyces capitatus*. Justicidin B (**329**) also exhibited strong activity against the trypomastigote form of *Trypanosoma brucei* rhodesiense and moderate activity against *Trypanosoma cruzi*. In a test against *Plasmodium falciparum*, **329** showed only weak activity [493].

The arylnaphthalene derivatives justicidin A **309**) and B (**329**) and diphyllin (**323**) have been identified as pesticidal constituents of *Justicia hayatai* [533, 534].

Ten lignans, isolated from *Justicia procumbens var. leucantha*, have shown antiviral activity against VSV. The most active compounds were justicidin A (**309**), justicidin B (**329**), and diphyllin (**323**) [478].

Phenylnaphthalene-type lignans justicidin A (**309**), justicidin B (**329**), justicidin D (**331**), justicidin C (**330**), and chinensinaphthol methyl ether (**335**) exhibited *in vitro* effects on rat hepatic cytochrome

P450-catalyzed oxidation [535]. Among these lignans, justicidin C (**330**) had the strongest inhibitory effect on AHH activity, and caused a significant decrease of 7-methoxyresorufin O-demethylation activity, whereas other tested lignans showed no significant inhibitory potential. These compounds also decreased testosterone 6β-hydroxylation activity, while (**330**) had the least inhibitory effect.

Justicidin B (**329**), justicidin D (**331**), taiwanin E (**336**), and taiwanin E methyl ether (**337**) showed significant antiplatelet aggregation activity [481].

Retrojusticidin B (**301**) and phyllamyricin B (**338**) exhibit inhibitory activity against viral RT [529].

11.6.8 Dibenzocyclooctadiene Lignans

Approximately 100 lignan derivatives possessing the dibenzocyclooctadiene skeleton have been isolated from plants of the family Schisandraceae [79, 536], and a wide variety of biological activities exhibited by these lignans have been discovered [537]. Extracts from lignan-rich plants have been used in traditional Chinese medicine as antitussives and as tonics with antiviral activity [538] (Figures 11.34, 11.35, 11.36, and 11.37).

Some lignans isolated from *Schisandra hernyi* induce a DNA fragmentation, while they also possess relevant cytotoxicity against both leukemia and *in vitro* HeLa cells [539].

Some of the lignans of *Schisandra* fruit have been found to ameliorate the harmful effects of toxic drugs and to facilitate liver function and regeneration [540, 541].

A stegane series of dibenzocyclooctadiene analogs inhibited TPA [steganangin (**350**), episteganangin (**351**), steganacin (**352**), steganoate A (**353**), steganoate B (**354**), steganolide A (**355**), and (−)-steganone (**356**)] [542].

Steganacin (**352**), stegnanagin (**350**), steganol (**357**), and steganone (**356**) show antitumor activity [543].

It has been reported that steganacin (**352**) and steganangin (**350**) have significant antileukemic activity [543].

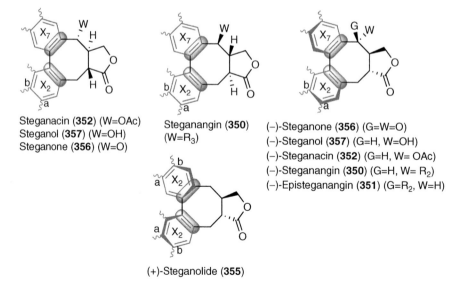

Figure 11.34 Main dibenzocyclooctadiene lignans (lactones) with biological activity. See Figure 11.35 for the meaning of fused rings X, Y, and R

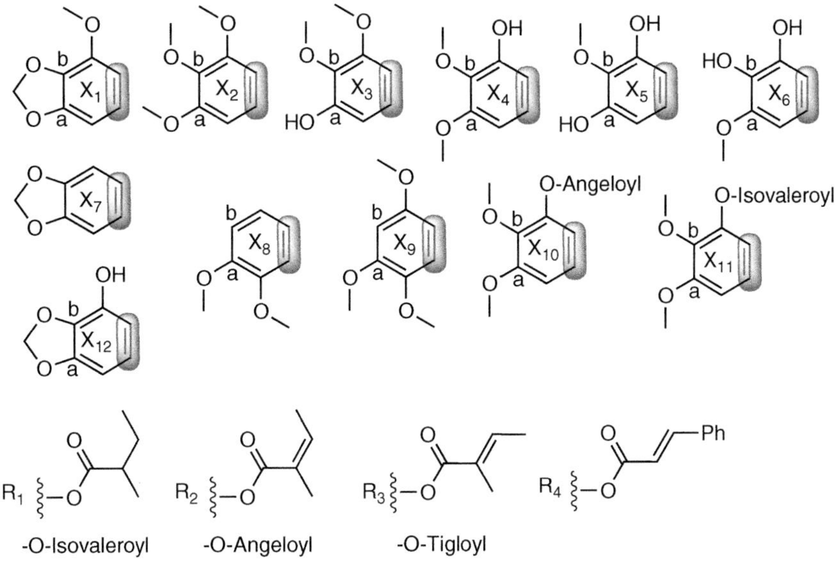

Figure 11.35 *Fused rings used in figures for clarity*

Found in traditional Chinese medicine formulations of wuweizi lignan containing plants, (−)-wuweizisu C (schisandrin C (**359**)) is considered a crucial component for the antihepatotoxic activity [544–546].

Longipedunins were obtained from the stems of *Kadsura longipedunculata*. Longipedunin A (**360**) showed moderate inhibitory activity against HIV-1 protease [547].

Propinquanins, isolated from *Schisandra propinqua* [548], were evaluated for their cytotoxic activities against tumor cell lines (using the MTT assay). Propinquanin B (**361**) exhibited significant cytotoxicity against HL-60 and Hep-G2 tumor cell lines [548]. Propinquanins E (**362**) and F (**363**) were found to exhibit moderate cytotoxic activity against the same two cell lines [549].

Kadsuphilins A (**364**) isolated from *Kadsura philippinensis* exhibited weak antioxidant activity [550].

Platelet activation factor (PAF) assay revealed that only kadsuphilin C (**365**) exhibited significant *in vitro* antiplatelet aggregation activity (IC_{50} 14 µM) [551].

Kadsuralignans were isolated from the roots of *Kadsura coccinea* [469, 470]. Bioassay of nitric oxide inhibitory activity showed that only kadsuralignan J (**366**) exhibited moderate activity [469, 470].

Schisandrin (**367**) showed its effect on hepatic mitochondrial glutathione antioxidant status in control and CCl4-intoxicated mice [552].

Schisandrin B (**368**) is effective in protecting the liver in such a way and increases resistance to the toxic effects of digitoxin and indomethacin [540]. Schisandrin B also attenuates cancer invasion and metastasis *via* inhibiting epithelial–mesenchymal transition [553].

The schizantherin compounds A, B, C, and D (**369–372**) are also effective in protecting the liver from injury and lowering serum GET levels [554]. This activity has been demonstrated in mice as well as human viral hepatitis patients [541]. Schisantherin E (**373**) and schisandrin A (**367**), however, are ineffective [554]. Even though it is by no means clear how these compounds act, it is remarkable to note that the four active compounds all contain at least one methylenedioxy group while neither of the inactive lignans possesses this moiety [5]. Since the methylenedioxy group is responsible for

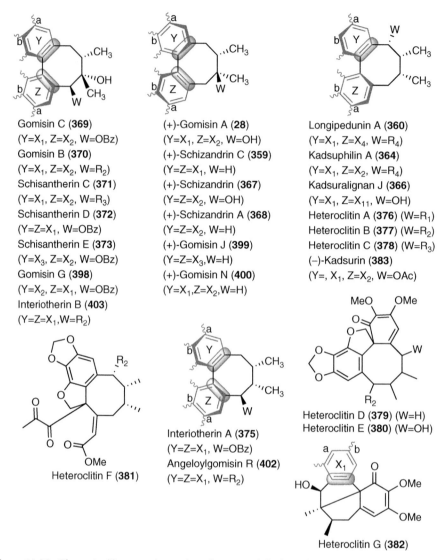

Figure 11.36 The main dibenzocyclooctadiene lignans with biological activity. See Figure 11.35 for the meaning of fused rings X, Y, and R.

mixed-function oxidase inhibition and since the liver is the primary location of this enzyme, it is reasonable to expect that this is the mechanism of action of these compounds.

Interiotherin A (**375**) and schisantherin D (**372**) showed powerful inhibitory activity against HIV replication in H9 lymphocytes. Heteroclitins A–G (**376–382**), kadsurin (**383**), and interiotherin (**375**) also inhibited lipid peroxidation in rat liver. Compounds heteroclitins A and D showed stronger inhibitory activity [555]. Preadministration of kadsurin (**383**) caused significant recovery of the SOD activity that is reduced by CCl4 intoxication [556].

In 2010, Yang *et al.* [557] reported anti-HIV-1 activity of 12 new dibenzocyclooctadiene lignans (marlignans (**384–395**)) isolated from *Schisandra wilsoniana* (see Figure 11.37).

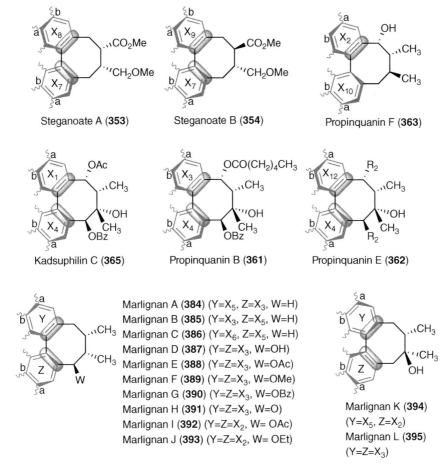

Figure 11.37 New dibenzocyclooctadiene lignans (marlignans A–L) with biological activity. See Figure 11.35 for the meaning of substituent [557]

11.6.8.1 Gomisins

In 1980s, Ikeya et al. [558–568] isolated a series of dibenzocyclooctadiene lignans ("Gomisins" in Japanese) from fruits of the *Schisandraceae*[20], which are then found to possess various biological activities, such as antihepatitis, antitumor, and antilipid peroxidation effects [569–573] (Figure 11.36).

In 1995, Fujihashi et al. [574] published on dibenzocyclooctadiene lignan anti-HIV activity that was related to inhibition of RT.

Gomisin A (**28**) inhibited early-stage hepatocarcinogenesis [575, 576], and suppressed tumor promotion [576–578]. Also, Gomisin A seems to act as liver protector against hepatotoxic compounds [579], and it boosts hepatic regeneration after a hepatectomy process [580, 581]. Gomisin A (**28**) inhibited early-stage hepatocarcinogenesis [575, 576], and suppressed tumor promotion [576–578].

---

[20] In traditional Chinese medicine, the fruits of *Schisandra* spp. are used as tonic, sedative, and astringent agents, while aerial parts are used for treatment of rheumatic lumbago and traumatic injury.

Gomisin A (**28**), gomisin J (**399**), and wuweizisu C (schisandrin C, **359**) inhibited inflammation induced by TPA in mice [578].

Gomisin A (**28**) and wuweizisu C (**359**) significantly inhibited lipid peroxidation due to ADP/Fe_3^+ or $ascorbate/Fe_2^+$ -induced CCl4 [546].

Gomisin C (schisantherin A (**369**)) reduced formation of superoxide by the peptide formyl-Met-Leu-Phe (FMLP) and phorbol myristate acetate (PMA). It can be attributed to inhibition of the activity of nicotinamide adenine dinucleotide phosphate (NADPH) oxidase [582].

Gomisin G (**398**) is a powerful anti-HIV agent [260, 583].

Gomisin J (**399**) and its derivatives were tested for anti-HIV activity. It appears that both the C_6 hydroxyl group and the C_7 benzoyloxy group in these compounds enhance activity [260, 584].

Gomisin J (**399**) from *Kadsura interior* exhibited good inhibitory influences on lipid peroxidation and the superoxide anion radical [585].

Gomisin N (**400**) has antioxidant activity [586].

Gomisin A (**28**), gomisin J (**399**), and wuweizisu C (**359**) inhibited inflammation induced by TPA in mice [578].

From a study on the anti-HIV activity of dibenzocyclooctadiene lignans, gomisin G (**398**) exhibited the most potent anti-HIV activity while schizantherin-D (**372**), kadsuranin (gomisin M methyl ether, **401**), and (−)-wuweizisu C (**359**) showed good activity. The results with these natural lignans suggested that 9-benzoyl and 8-hydroxy substituents might enhance the biological activity [260, 583].

Angeloylgomisin R (**402**) and interiotherin B (**403**) showed weak anti-HIV activity [584].

In the later 1970s, Ikeya *et al.* studied other dibenzocyclooctadiene lignans with different biological activities: (+)-gomisin K, (−)-gomisin L_2 [568], gomisin J [561], gomisin N [560, 587], gomisin O [560], gomisin S and T [588], angeloygomisin Q [589], angeloylisogomisin O [559], benzoylgomisin Q [590], epigomisin O [560], and schisandrin [591].

11.6.9 Neolignans

As an example, some biological properties of different types of neolignans such as benzofurans, biphenyls, benzodioxans, and alkyl phenyl ethers are discussed (see Figure 11.38).

- Benzofurans:

    The neolignans named tortoside E (**404**) and sakuraresinol (**405**) demonstrated potent antioxidant efficiency and only weak anti-NO production activity [466].

    The cisbenzofuran lignan (**406**) (isolated from *Tarenna attenuata*) showed strong antioxidant activity against H_2O_2-induced impairment in PC12 cells and exhibited DPPH radical scavenging activity [417].

    The (7S,8R)-benzofuran (**407**) was isolated as a 2:1 diastereomeric mixture from the leaves and small branches of *Cistus laurifolius*, which showed a potent inhibitory effect for prostaglandin E1 and a weak inhibitory effect for prostaglandin E2 [394].

    The (7S,8R)-benzofuran (**408**) was isolated from the leaves of *Neoalsomitra integrifoliola*, and found to exhibit weak anti-inflammatory activity [592].

- Benzodioxanes:

    The (7'S,8'S)-5-demethoxybilagrewin (**409**), isolated from other parts of the stem wood of *Zanthoxylum avicennae*, exhibited moderate inhibition activity against the generation of superoxide anion in human neutrophils [593].

- Alkyl aryl ethers:

    The alkyl aryl ether (**410**), isolated from *Tarenna attenuate*, showed potent antioxidant activity against H_2O_2-induced impairment in PC12 cells and exhibited DPPH radical scavenging activity [417].

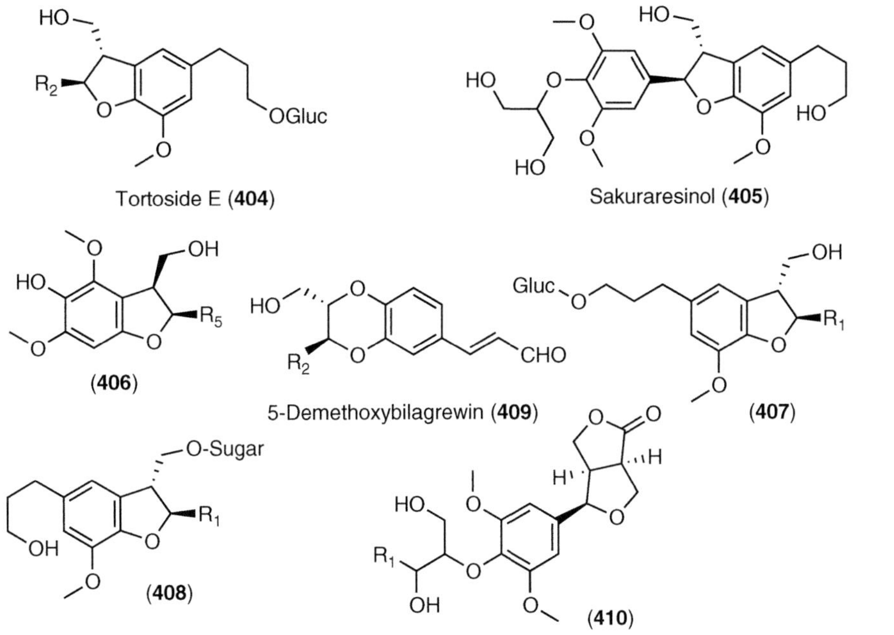

Figure 11.38 Neolignans with biological activity

- Biphenyls:

  Several reviews of the biological activity of magnolol and hinokiol (**52**) have been published [594–598].

11.6.10 Hybrid Lignans

Even though the hybrid lignans are beyond the scope of this textbook but may be of general interest to the reader, we briefly indicate in this section some recent reviews that cover the biological activity of the coumarino-, flavono-, and stilbeno-lignans.

While coumarinolignans have shown many noteworthy biological activities, the majority of these refer to their cytotoxic and hepatoprotective potential. The different activities shown by this group of compounds are summarized in ref. [599, pp. 25–27].

Several flavonolignans have proved to be excellent hepatoprotectants and are also potential cancer chemopreventive agents [599, pp. 48–49].

Stilbenolignans have been proposed as anti-inflammatory lead compounds [599, pp. 61–62].

References

[1] Harmatha J, Zídek Z, Kmoníkova E, Šmidrkal J. Immunobiological properties of selected natural and chemically modified phenylpropanoids. Interdiscip Toxicol. 2011;4(1):5–10.

[2] Cunha WR, Andrade e Silva ML, Sola Veneziani RC, Ambrósio SR, Kenupp Bastos J. Lignans: chemical and biological properties. In: Rao V, editor. Phytochemicals - A Global Perspective of their Role in Nutrition and Health. vol. 33. InTech; 2012. p. 213–234.

[3] Charlton JL. Antiviral activity of lignans. J Nat Prod. 1998;61(11):1447–1451.

[4] Cos P, Maes L, Vlietinck A, Pieters L. Plant-derived leading compounds for chemotherapy of human immunodeficiency virus (HIV) infection - an update (1998–2007). Planta Med. 2008;74(11):1323–1337.

[5] MacRae WD, Towers GHN. Biological activities of lignans. Phytochemistry. 1984;23(6):1207–1220.

[6] Yousefzadi M, Sharifi M, Behmanesh M, Moyano E, Bonfill M, Cusido RM, *et al*. Podophyllotoxin: current approaches to its biotechnological production and future challenges. Eng Life Sci. 2010;10(4):281–292.

[7] Pan JY, Chen SL, Yang MH, Wu J, Sinkkonen J, Zou K. An update on lignans: natural products and synthesis. Nat Prod Rep. 2009;26(10):1251–1292.

[8] Saleem M, Hyoung JK, Ali MS, Yong SL. An update on bioactive plant lignans. Nat Prod Rep. 2005;22(6):696–716.

[9] Huang WY, Cai YZ, Zhang Y. Natural phenolic compounds from medicinal herbs and dietary plants: potential use for cancer prevention. Nutr Cancer. 2010;62(1):1–20.

[10] Webb AL, McCullough ML. Dietary lignans: Potential role in cancer prevention. Nutr Cancer. 2005;51(2):117–131.

[11] Faure M, Lissi E, Torres R, Videla LA. Antioxidant activities of lignans and flavonoids. Phytochemistry. 1990;29(12):3773–3775.

[12] Negi AS, Kumar JK, Luqman S, Shanker K, Gupta MM, Khanuja SPS. Recent advances in plant hepatoprotectives: a chemical and biological profile of some important leads. Med Res Rev. 2008;28(5):746–772.

[13] Habauzit V, Horcajada MN. Phenolic phytochemicals and bone. Phytochem Rev. 2008;7(2):313–344.

[14] Rao LG, Kang N, Rao AV. Polyphenol antioxidants and bone health: a review. In: Rao V, editor. Phytochemicals - A Global Perspective of their Role in Nutrition and Health. vol. 33. InTech; 2012. p. 467–486.

[15] Kim HJ, Ono E, Morimoto K, Yamagaki T, Okazawa A, Kobayashi A, *et al*. Metabolic engineering of lignan biosynthesis in *Forsythia* cell culture. Plant Cell Physiol. 2009;50(12):2200–2209.

[16] Suzuki S, Umezawa T. Biosynthesis of lignans and norlignans. J Wood Sci. 2007;53:273–284.

[17] Umezawa T. Diversity in lignan biosynthesis. Phytochem Rev. 2003;2(3):371–390.

[18] Deyama T, Nishibe S. Pharmacological properties of lignans. In: Heitner C, Dimmel D, Schmidt J, editors. Lignin and Lignans: Advances in Chemistry. CRC Press; 2010. p. 585–629.

[19] Ward RS. Lignans, neolignans, and related compounds. Nat Prod Rep. 1999;16(1):75–96.

[20] Ward RS. Lignans, neolignans, and related compounds. Nat Prod Rep. 1997;14(1):43–74.

[21] Ward RS. Lignans, neolignans, and related compounds. Nat Prod Rep. 1995;12(2):183–205.

[22] Ward RS. Lignans, neolignans, and related compounds. Nat Prod Rep. 1993;10(1):1–28.

[23] Dewick PM. Medicinal Natural Products: A Biosynthetic Approach. John Wiley & Sons, Ltd; 2002.

[24] Lewis NG, Davin LB, Sarkanen S. Lignin and lignan biosynthesis: distinctions and reconciliations. ACS Symp Ser. 1998;697(Lignin and Lignan Biosynthesis):1–27.

[25] Kitagawa S, Nishibe S, Benecke R, Thieme H. Phenolic compounds from *Forsythia leaves*. II. Chem Pharm Bull. 1988;36(9):3667–3670.

[26] Zhuang L, Seligmann O, Jurcic K, Wagner H. Constituents of *Daphne tangutica*. Planta Med. 1982;45(7):172–176.

[27] Pare PW, Wang HB, Davin LB, Lewis NG. (+)-pinoresinol synthase: a stereoselective oxidase catalyzing 8,8′-lignan formation in *Forsythia intermedia*. Tetrahedron Lett. 1994;35(27):4731–4734.

[28] Davin LB, Wang HB, Crowell AL, Bedgar DL, Martin DM, Sarkanen S, et al. Stereoselective bimolecular phenoxy radical coupling by an auxiliary (dirigent) protein without an active center. Science. 1997;275(5298):362–367.

[29] Umezawa T, Davin LB, Lewis NG. Formation of the lignan, (−) secoisolariciresinol, by cell free extracts of *Forsythia intermedia*. Biochem Biophys Res Commun. 1990;171(3): 1008–1014.

[30] Umezawa T, Davin LB, Yamamoto E, Kingston DGI, Lewis NG. Lignan biosynthesis in *Forsythia* species. J Chem Soc, Chem Commun. 1990;(20):1405–1408.

[31] Umezawa T, Davin LB, Lewis NG. Formation of lignans (−)-secoisolariciresinol and (−)-matairesinol with *Forsythia intermedia* cell-free extracts. J Biol Chem. 1991;266(16): 10210–10217.

[32] Ono E, Nakai M, Fukui Y, Tomimori N, Fukuchi-Mizutani M, Saito M, et al. Formation of two methylenedioxy bridges by a *Sesamum* CYP81Q protein yielding a furofuran lignan, (+)-sesamin. Proc Natl Acad Sci U S A. 2006;103(26):10116–10121.

[33] Sakakibara N, Suzuki S, Umezawa T, Shimada M. Biosynthesis of yatein in *Anthriscus sylvestris*. Org Biomol Chem. 2003;1(14):2474–2485.

[34] Dewick PM. Biosynthesis of lignans. In: ur Rahman A, editor. Studies in Natural Product Chemistry. vol. 5. Elsevier; 1989. p. 459–503.

[35] Broomhead AJ, Rahman MMA, Dewick PM, Jackson DE, Lucas JA. Biosynthesis of *Podophyllum* lignans. Part 5. Matairesinol as precursor of *Podophyllum* lignans. Phytochemistry. 1991;30(5):1489–1492.

[36] Suzuki S, Sakakibara N, Umezawa T, Shimada M. Survey and enzymatic formation of lignans of *Anthriscus sylvestris*. J Wood Sci. 2002;48(6):536–541.

[37] Seidel V, Windhövel J, Eaton G, Alfermann WA, Arroo RR, Medarde M, et al. Biosynthesis of podophyllotoxin in *Linum album* cell cultures. Planta. 2002;215(6):1031–1039.

[38] Molog GA, Empt U, Kuhlmann S, van Uden W, Pras N, Alfermann AW, et al. Deoxypodophyllotoxin 6-hydroxylase, a cytochrome P450 monooxygenase from cell cultures of linum flavum involved in the biosynthesis of cytotoxic lignans. Planta. 2001;214(2):288–294.

[39] Van Uden W, Bouma AS, Waker JFB, Middel O, Wichers HJ, De Waard P, et al. The production of podophyllotoxin and its 5-methoxy derivative through bioconversion of cyclodextrin-complexed desoxypodophyllotoxin by plant cell cultures. Plant Cell, Tissue Organ Cult. 1995;42(1):73–79.

[40] Kranz K, Petersen M. β-peltatin 6-*O*-methyltransferase from suspension cultures of *Linum nodiflorum*. Phytochemistry. 2003;64(2):453–458.

[41] Xia ZQ, Costa MA, Proctor J, Davin LB, Lewis NG. Dirigent-mediated podophyllotoxin biosynthesis in *Linum flavum* and *Podophyllum peltatum*. Phytochemistry. 2000;55(6):537–549.

[42] Gottlieb OR. Plant chemosystematics and phylogeny. III. Chemosystematics of the *Lauraceae*. Phytochemistry. 1972;11(5):1537–1570.

[43] Moinuddin SGA, Hishiyama S, Cho MH, Davin LB, Lewis NG. Synthesis and chiral HPLC analysis of the diphenyltetrahydrofuran lignans, larreatricins, 8′-epi-larreatricins, 3,3′-didemethoxyverrucosins and meso-3,3′-didemethoxynectandrin B in the creosote bush (*Larrea tridentata*): evidence for regiospecific control of coupling. Org Biomol Chem. 2003;1(13):2307–2313.

[44] Lopes NP, Yoshida M, Kato MJ. Biosynthesis of tetrahydrofuran lignans in *Virola surinamensis*. Rev Bras Cienc Farm. 2004;40(1):53–57.

[45] Beracierta AP, Whiting DA. Stereoselective total syntheses of the (±)-di-*O*-methyl ethers of agatharesinol, sesquirin-A, and hinokiresinol, and of (±)-tri-*O*-methylsequirin-E, characteristic norlignans of coniferae. J Chem Soc [Perkin 1]. 1978;(10):1257–1263.

[46] Suzuki S, Umezawa T, Shimada M. Norlignan biosynthesis in *Asparagus officinalis* l.: the norlignan originates from two non-identical phenylpropane units. J Chem Soc [Perkin 1]. 2001;(24):3252–3257.

[47] Suzuki S, Nakatsubo T, Umezawa T, Shimada M. First *in vitro* norlignan formation with *Asparagus officinalis* enzyme preparation. Chem Commun. 2002;(10):1088–1089.

[48] Suzuki S, Yamamura M, Shimada M, Umezawa T. A heartwood norlignan, (e)-hinokiresinol, is formed from 4-coumaryl 4-coumarate by a *Cryptomeria japonica* enzyme preparation. Chem Commun. 2004;(24):2838–2839.

[49] Birch AJ, Liepa A. Biosynthesis. In: Rao CBS, editor. Chemistry of Lignans. Andhra University Press; 1978. p. 307–327.

[50] Fletcher RJ. Food sources of phyto-oestrogens and their precursors in Europe. Br J Nutr. 2003;89(Suppl. 1):S39–S43.

[51] Liggins J, Grimwood R, Bingham SA. Extraction and quantification of lignan phytoestrogens in food and human samples. Anal Biochem. 2000;287(1):102–109.

[52] Nurmi T, Heinonen S, Mazur W, Deyama T, Nishibe S, Adlercreutz H. Lignans in selected wines. Food Chem. 2003;83(2):303–309.

[53] Heinonen S, Nurmi T, Liukkonen K, Poutanen K, Wähälä K, Deyama T, et al. In Vitro metabolism of plant lignans: new precursors of mammalian lignans enterolactone and enterodiol. J Agric Food Chem. 2001;49(7):3178–3186.

[54] Mazur WM, Uehara M, Wahala K, Adlercreutz H. Phyto-oestrogen content of berries, and plasma concentrations and urinary excretion of enterolactone after a single strawberry-meal in human subjects. Br J Nutr. 2000;83(4):381–387.

[55] Stitch SR, Toumba JK, Groen MB, Funke CW, Leemhuis J, Vink J, et al. Excretion, isolation and structure of a new phenolic constituent of female urine. Nature. 1980;287(5784):738–740.

[56] Setchell KDR, Lawson AM, Mitchell FL, Adlercreutz H, Kirk DN, Axelson M. Lignans in man and in animal species. Nature. 1980;287(5784):740–742.

[57] Setchell KDR, Lawson AM, Conway E, Taylor NF, Kirk DN, Cooley G, et al. The definitive identification of the lignans trans-2,3-bis(3-hydroxybenzyl)-γ-butyrolactone and 2,3-bis(3-hydroxybenzyl)butane-1,4-diol in human and animal urine. Biochem J. 1981;197(2):447–458.

[58] Ford JD, Huang KS, Wang HB, Davin LB, Lewis NG. Biosynthetic pathway to the cancer chemopreventive secoisolariciresinol diglucoside?hydroxymethyl glutaryl ester-linked lignan oligomers in flax (*Linum usitatissimum*) seed. J Nat Prod. 2001;64(11):1388–1397.

[59] Toure A, Xu X. Flaxseed lignans: source, biosynthesis, metabolism, antioxidant activity, bioactive components, and health benefits. Compr Rev Food Sci Food Saf. 2010;9(3):261–269.

[60] Clavel T, Borrmann D, Braune A, Dore J, Blaut M. Occurrence and activity of human intestinal bacteria involved in the conversion of dietary lignans. Anaerobe. 2006;12(3):140–147.

[61] Kuijsten A, Arts ICW, Vree TB, Hollman PCH. Pharmacokinetics of enterolignans in healthy men and women consuming a single dose of secoisolariciresinol diglucoside. J Nutr. 2005;135(4):795–801.

[62] Lampe JW, Atkinson C, Hullar MAJ. Assessing exposure to lignans and their metabolites in humans. J AOAC Int. 2006;89(4):1174–1181.

[63] Jan KC, Hwang LS, Ho CT. Biotransformation of sesaminol triglucoside to mammalian lignans by intestinal microbiota. J Agric Food Chem. 2009;57(14):6101–6106.

[64] Liu Z, Saarinen NM, Thompson LU. Sesamin is one of the major precursors of mammalian lignans in sesame seed (*Sesamum indicum*) as observed *in vitro* and in rats. J Nutr. 2006;136(4):906–912.

[65] Peterson J, Dwyer J, Adlercreutz H, Scalbert A, Jacques P, McCullough ML. Dietary lignans: physiology and potential for cardiovascular disease risk reduction. Nutr Rev. 2010;68(10):571–603.

[66] Xie LH, Akao T, Hamasaki K, Deyama T, Hattori M. Biotransformation of pinoresinol diglucoside to mammalian lignans by human intestinal microflora, and isolation of *Enterococcus faecalis* strain PDG-1 responsible for the transformation of (+)-pinoresinol to (+)-lariciresinol. Chem Pharm Bull. 2003;51(5):508–515.

[67] Adlercreutz H, Bannwart C, Wahala K, Makela T, Brunow G, Hase T, et al. Inhibition of human aromatase by mammalian lignans and isoflavonoid phytoestrogens. J Steroid Biochem Mol Biol. 1993;44(2):147–153.

[68] Landete JM. Plant and mammalian lignans: a review of source, intake, metabolism, intestinal bacteria and health. Food Res Int. 2012;46(1):410–424.

[69] Yoder SC. Metabolism of Plant Lignans by Human Intestinal Bacteria. Washington, DC: University of Washington; 2013.

[70] Nose M, Fujimoto T, Takeda T, Nishibe S, Ogihara Y. Structural transformation of lignan compounds in rat gastrointestinal tract. Planta Med. 1992;58(6):520–523.

[71] Nose M, Fujimoto T, Nishibe S, Ogihara Y. Structural transformation of lignan compounds in rat gastrointestinal tract; II. Serum concentration of lignans and their metabolites. Planta Med. 1993;59(2):131–134.

[72] Saarinen NM, Smeds A, Makela SI, Ammala J, Hakala K, Pihlava JM, et al. Structural determinants of plant lignans for the formation of enterolactone *in vivo*. J Chromatogr B Analyt Technol Biomed Life Sci. 2002;777(1-2):311–319.

[73] Valentin-Blasini L, Blount BC, Rogers HS, Needham LL. HPLC-MS/MS method for the measurement of seven phytoestrogens in human serum and urine. J Expos Anal Environ Epidemiol. 2000;10(6, Pt. 2):799–807.

[74] Xie LH, Ahn EM, Akao T, Abdel-Hafez AAM, Nakamura N, Hattori M. Transformation of arctiin to estrogenic and antiestrogenic substances by human intestinal bacteria. Chem Pharm Bull. 2003;51(4):378–384.

[75] Fukuda Y, Osawa T, Namiki M, Ozaki T. Studies on antioxidative substances in sesame seed. Agric Biol Chem. 1985;49(2):301–306.

[76] Figgitt DP, Denyer SP, Dewick PM, Jackson DE, Williams P. Topoisomerase II: a potential target for novel antifungal agents. Biochem Biophys Res Commun. 1989;160(1):257–262.

[77] Moraes RM, Burandt CL Jr., Ganzera M, Li X, Khan I, Canel C. The American mayapple revisited: *Podophyllum peltatum* - still a potential cash crop? Econ Bot. 2000;54(4): 471–476.

[78] Harmatha J, Dinan L. Biological activities of lignans and stilbenoids associated with plant-insect chemical interactions. Phytochem Rev. 2003;2(3):321–330.

[79] Ayres DC, Loike JD. Lignans: Chemical, Biological and Clinical Properties, Chemistry and Pharmacology of Natural Products. Cambridge: Cambridge University Press; 1990.

[80] Carlos J, Boluda CJ, Duque B, Aragón Z. Lignanos (I): estructura y funciones en las plantas. Rev Fitoterapia. 2005;5(1):55–68.

[81] Shain L, Hillis WE. Phenolic extractives in Norway spruce and their effects on *Fomes annosus*. Phytopathology. 1971;61(7):841–845.

[82] Takemoto T, Miyase T, Kusano G. Boehmenan, a new lignan from the roots of *Boehmeria tricuspis*. Phytochemistry. 1975;14(8):1890–1891.

[83] Erdemoglu N, Sener B, Choudhary MI. Bioactivity of lignans from *Taxus baccata*. Z Naturforsch, C: J Biosci. 2004;59(7-8):494–498.

[84] Inamori Y, Kubo M, Tsujibo H, Ogawa M, Baba K, Kozawa M, et al. The biological activities of podophyllotoxin compounds. Chem Pharm Bull. 1986;34(9):3928–3932.

[85] Miyazawa M, Fukuyama M, Yoshio K, Kato T, Ishikawa Y. Biologically active components against *Drosophila melanogaster* from *Podophyllum hexandrum*. J Agric Food Chem. 1999;47(12):5108–5110.

[86] Garcia ES, Cabral MMO, Schaub GA, Gottlieb OR, Azambuja P. Effects of lignoids on a hematophagous bug, *Rhodnius prolixus*: feeding, ecdysis and diuresis. Phytochemistry. 2000;55(6):611–616.

[87] Oliva A, Moraes RM, Watson SB, Duke SO, Dayan FE. Aryltetralin lignans inhibit plant growth by affecting the formation of mitotic microtubular organizing centers. Pestic Biochem Physiol. 2002;72(1):45–54.

[88] Bang KH, Kim YK, Min BS, Na MK, Rhee YH, Lee JP, et al. Antifungal activity of magnolol and honokiol. Arch Pharmacal Res. 2000;23(1):46–49.

[89] Clark AM, El-Feraly FS, Li WS. Antimicrobial activity of phenolic constituents of *Magnolia grandiflora* L. J Pharm Sci. 1981;70(8):951–952.

[90] Nawrot J, Harmatha J. Natural products as antifeedants against stored products insects. Postharvest News Inf. 1994;5(2):17N–21N.

[91] Gao R, Gao C, Tian X, Yu X, Di X, Xiao H, et al. Insecticidal activity of deoxypodophyllotoxin, isolated from *Juniperus sabina* L, and related lignans against larvae of *Pieris rapae* L. Pest Manage Sci. 2004;60(11):1131–1136.

[92] Messiano GB, Vieira L, Machado MB, Lopes LMX, De Bortoli SA, Zukerman-Schpector J. Evaluation of insecticidal activity of diterpenes and lignans from *Aristolochia malmeana* against *Anticarsia gemmatalis*. J Agric Food Chem. 2008;56(8):2655–2659.

[93] Singh RP, Tomar SS, Attri BS, Parmar BS, Maheshwary ML, Mukerjee SK. Search for new *Pyrethrum synergists* in some botanicals. Pyrethrum Post. 1976;13(3):91–93.

[94] Srivastava S, Gupta MM, Prajapati V, Tripathi AK, Kumar S. Sesamin a potent antifeedant principle from *Piper mullesua*. Phytother Res. 2001;15(1):70–72.

[95] Sintim HO, Tashiro T, Motoyama N. Response of the cutworm *Spodoptera litura* to sesame leaves or crude extracts in diet. J Insect Sci. 2009;9:1–13.

[96] Schroeder FC, Del Campo ML, Grant JB, Weibel DB, Smedley SR, Bolton KL, et al. Pinoresinol: a lignol of plant origin serving for defense in a caterpillar. Proc Natl Acad Sci U S A. 2006;103(42):15497–15501.

[97] Garcia ES, Azambuja P. Lignoids in insects: chemical probes for the study of ecdysis, excretion and *Trypanosoma cruzi* - triatomine interactions. Toxicon. 2004;44(4):431–440.

[98] Cabral MMO, Alencar JA, Guimarães AE, Kato MJ. Larvicidal activity of grandisin against *Aedes Aegypti*. J Am Mosq Control Assoc. 2009;25(1):103–105.

[99] Leite ACCF, Kato MJ, Soares ROA, Guimarães AE, Santos-Mallet JR, Cabral MMO. Grandisin caused morphological changes larval and toxicity on *Aedes aegypti*. Braz J Pharmacog. 2012;22(3):517–521.

[100] Nogueira CDR, De Mello RP, Kato MJ, De Oliveira Cabral MM. Disruption of *Chrysomya megacephala* growth caused by lignan grandisin. J Med Entomol. 2009;46(2):281–283.

[101] Nishiwaki H, Hasebe A, Kawaguchi Y, Akamatsu M, Shuto Y, Yamauchi S. Larvicidal activity of (−)-dihydroguaiaretic acid derivatives against *Culex pipiens*. Biosci Biotechnol Biochem. 2011;75(9):1735–1739.

[102] Park IK, Shin SC, Kim CS, Lee HJ, Choi WS, Ahn YJ. Larvicidal activity of lignans identified in *Phryma leptostachya* var. asiatica roots against three mosquito species. J Agric Food Chem. 2005;53(4):969–972.

[103] Xiao XM, Hu ZN, Shi BJ, Wei SP, Wu WJ. Larvicidal activity of lignans from *Phryma leptostachya* L. against *Culex pipiens* pallens. Parasitol Res. 2012;110(3):1079–1084.

[104] Seo SM, Park IK. Larvicidal activity of medicinal plant extracts and lignan identified in *Phryma leptostachya* var. asiatica roots against housefly (*Musca domestica* l.). Parasitol Res. 2012;110(5):1849–1853.

[105] Taniguchi E, Imamura K, Ishibashi F, Matsui T, Nishio A. Structure of the novel insecticidal sesquilignan, haedoxan A. Agric Biol Chem. 1989;53(3):631–643.

[106] Yamauchi S, Ishibashi F, Taniguchi E. Synthesis and insecticidal activity of lignan analogs. Part 7. Insecticidal activity of sesquilignans with a 3-aryl-6-methoxy-2-methoxymethyl-1,4-benzodioxanyl group. Biosci Biotechnol Biochem. 1992;56(11):1760–1768.

[107] Yamauchi S, Taniguchi E. Synthesis and insecticidal activity of lignan analogs. Part 5. Influence on insecticidal activity of the 3-(3,4-methylenedioxyphenyl) group in the 1,4-benzodioxanyl moiety of haedoxan. Biosci Biotechnol Biochem. 1992;56(11):1744–1750.

[108] Yamauchi S, Taniguchi E. Synthesis and insecticidal activity of lignan analogs. (1). Agric Biol Chem. 1991;55(12):3075–3084.

[109] Yamauchi S, Taniguchi E. Synthesis and insecticidal activity of lignan analogs. II. Biosci Biotechnol Biochem. 1992;56(3):412–417.

[110] Yamauchi S, Nagata S, Taniguchi E. Synthesis and insecticidal activity of haedoxan analogs. (Part IV). Effect on insecticidal activity of substituents at the 1,4-benzodioxanyl moiety of haedoxan. Biosci Biotechnol Biochem. 1992;56(8):1193–1197.

[111] Nitao JK, Johnson KS, Scriber JM, Nair MG. Magnolia virginiana neolignan compounds as chemical barriers to swallowtail butterfly host use. J Chem Ecol. 1992;18(9):1661–1671.

[112] Miyazawa M, Hirota K, Fukuyama M, Ishikawa Y, Kameoka H. Insecticidal lignans against *Drosophila melanogaster* from fruits of *Schisandra chinensis*. Nat Prod Lett. 1998;12(3):175–180.

[113] Haller HL, McGovran ER, Goodhue LD, Sullivan WN. The synergistic action of sesamin with pyrethrum insecticides. J Org Chem. 1942;7:183–184.

[114] Haller HL, LaForge FB, Sullivan WN. Some compounds related to sesamin: their structures and their synergistic effect with pyrethrum insecticides. J Org Chem. 1942;7:185–188.

[115] Wada K, Munakata K. (−)-parabenzlactone, a new piperolignanolide isolated from parabenzoin *Trilobum nakai*. Tetrahedron Lett. 1970;11(23):2017–2019.

[116] Matsubara H. Synergist for insecticides. XXVII. Synergistic effect of several lignans on pyrethrins and allethrin. Bull Inst Chem Res, Kyoto Univ. 1972;50(3):197–205.

[117] Gonzalez JA, Estevez-Braun A, Estevez-Reyes R, Ravelo AG. Inhibition of potato cyst nematode hatch by lignans from bupleurum salicifolium (Umbelliferae). J Chem Ecol. 1994;20(3):517–524.

[118] Gonzalez JA, Estevez-Braun A, Estevez-Reyes R, Bazzocchi IL, Moujir L, Jimenez IA, et al. Biological activity of secondary metabolites from bupleurum salicifolium (Umbelliferae). Experientia. 1995;51(1):35–39.

[119] Yoshihara T, Yamaguchi K, Sakamura S. A lignan-type stress compound in potato infected with nematode (*Globodera rostochiensis*). Agric Biol Chem. 1982;46(3):853–854.

[120] Elakovich SD, Stevens KL. Phytotoxic properties of nordihydroguaiaretic acid, a lignan from *Larrea tridentata* (creosote bush). J Chem Ecol. 1985;11(1):27–33.

[121] Lavie D, Levy EC, Cohen A, Evenari M, Guttermann Y. New germination inhibitor from *Aegilops ovata*. Nature. 1974;249(5455):388–.

[122] Gutterman Y, Evenari M, Cooper R, Levy EC, Lavie D. Germination inhibition activity of a naturally occurring lignan from *Aegilops ovata* L. in green and infrared light. Experientia. 1980;36(6):662–663.

[123] Orr JD, Lynn DG. Biosynthesis of dehydrodiconiferyl alcohol glucosides: implications for the control of tobacco cell growth. Plant Physiol. 1992;98(1):343–352.

[124] St[ä]helin H, Von Wartburg A. The chemical and biological route from podophyllotoxin glucoside to etoposide: ninth cain memorial award lecture. Cancer Res. 1991;51(1):5–15.

[125] Bohlin L, Rosen B. Podophyllotoxin derivatives: drug discovery and development. Drug Discov Today. 1996;1(8):343–351.

[126] Canel C, Moraes RM, Dayan FE, Ferreira D. Molecules of interest: podophyllotoxin. Phytochemistry. 2000;54(2):115–120.

[127] Giri A, Narasu ML. Production of podophyllotoxin from *Podophyllum hexandrum*: a potential natural product for clinically useful anticancer drugs. Cytotechnology. 2000;34(1-2):17–26.

[128] Moraes RM, Dayan FE, Canel C. The lignans of *Podophyllum*. Stud Nat Prod Chem. 2002;26(Part G):149–182.

[129] Castro MA, Del Corral JMM, Gordaliza M, Gómez-Zurita MA, García PA, San Feliciano A. Chemoinduction of cytotoxic selectivity in podophyllotoxin-related lignans. Phytochem Rev. 2003;2(3):219–233.

[130] Gordaliza M, Garcia PA, del Corral JMM, Castro MA, Gomez-Zurita MA. Podophyllotoxin: distribution, sources, applications and new cytotoxic derivatives. Toxicon. 2004;44(4):441–459.

[131] Zheljazkov VD, Jones AM, Avula B, Maddox V, Rowe DE. Lignan and nutrient concentrations in American mayapple (*Podophyllum peltatum* L.) in the Eastern United States. HortScience. 2009;44(2):349–353.

[132] Majumder A, Jha S. Biotechnological approaches for the production of potential anticancer leads podophyllotoxin and paclitaxel: an overview. eJ Biol Sci. 2009;1(1):46–69.

[133] Nagar N, Jat RK, Saharan R, Verma S, Sharma D, Bansal K. Podophyllotoxin and their glycosidic derivatives. Pharmacophore. 2011;2(2):124–134.

[134] Chaurasia OP, Ballabh B, Tayade A, Kumar R, Kumar GP, Singh SB. Podophyllum l.: an endergered and anticancerous medicinal plant-an overview. Indian J Tradit Knowl. 2012;11(2):234–241.

[135] Kaplan IW. Condylomata acuminata. New Orleans Med Surg J. 1942;94:388–390.

[136] Holthuis JJM. Etoposide and teniposide. Bioanalysis, metabolism and clinical pharmacokinetics. Pharm Weekbl, Sci Ed. 1988;10(3):101–116.

[137] Imbert TF. Discovery of podophyllotoxins. Biochimie. 1998;80(3):207–222.

[138] Mellanoff IS, Schaeffer HJ. The resins of *Podophyllum peltatum* L. Am J Pharm. 1927;99:323–330.

[139] Jin Y, Chen SW, Tian X. Synthesis and biological evaluation of new spin-labeled derivatives of podophyllotoxin. Bioorg Med Chem. 2006;14(9):3062–3068.

[140] Kofod H, Jorgensen C. A note on the melting point of podophyllotoxin. Acta Chem Scand. 1955;9:347–348.

[141] Moss GP. Nomenclature of lignans and neolignans (IUPAC recommendations 2000). Pure Appl Chem. 2000;72(8):1493–1523.

[142] Jackson DE, Dewick PM. Aryltetralin lignans from *Podophyllum hexandrum* and *Podophyllum peltatum*. Phytochemistry. 1984;23(5):1147–1152.

[143] Purohit MC, Bahuguna R, Maithani UC, Purohit AN, Rawat MSM. Variation in podophylloresin and podophyllotoxin contents in different populations of *Podophyllum hexandrum*. Curr Sci. 1999;77(8):1078–1080.

[144] Canel C, Dayan FE, Ganzera M, Khan IA, Rimando A, Burandt CL Jr., et al. High yield of podophyllotoxin from leaves of *Podophyllum peltatum* by *in situ* conversion of podophyllotoxin 4-*O*-β-D-glucopyranoside. Planta Med. 2001;67(1):97–99.

[145] Moraes-Cerdeira RM, Burandt CL Jr., Bastos JK, Nanayakkara NP, McChesney JD. *In vitro* propagation of *Podophyllum peltatum*. Planta Med. 1998;64(1):42–45.

[146] Moraes RM, Bedir E, Barrett H, Burandt CL Jr., Canel C, Khan IA. Evaluation of *Podophyllum peltatum* accessions for podophyllotoxin production. Planta Med. 2002;68(4):341–344.

[147] Nadeem M, Palni LMS, Purohit AN, Pandey H, Nandi SK. Propagation and conservation of *Podophyllum hexandrum* Royle: an important medicinal herb. Biol Conserv. 2000;92(1):121–129.

[148] Watson MA, Scott K, Griffith J, Dieter S, Jones CS, Nanda S. The developmental ecology of mycorrhizal associations in mayapple, *Podophyllum peltatum*, *Berberidaceae*. Evol Ecol. 2001;15(4-6):425–442.

[149] Maqbool M. Mayapple: a review of the literature from a horticultural perspective. J Med Plants Res. 2011;5(7):1037–1045.
[150] Koulman A. Podophyllotoxin: A Study of the Biosynthesis, Evolution, Function and Use of Podophyllotoxin and Related Lignans. Groningen, Netherland: Rijksuniversiteit; 2003.
[151] Konuklugil B. The importance of aryltetralin (*Podophyllum*) lignans and their distribution in the plant kingdom. Ankara Univ Eczacilik Fak Derg. 1995;24(2):109–125.
[152] Boluda CJ, Duque B, Gulyas G, Aragón Z, Duque MA, Díez F. Lignanos (2): actividad farmacológica. Rev Fitoterapia. 2005;5(2):135–147.
[153] Liu YQ, Yang L, Tian X. Podophyllotoxin: current perspectives. Curr Bioact Compd. 2007;3(1):37–66.
[154] Bedows E, Hatfield GM. An investigation of the antiviral activity of *Podophyllum peltatum*. J Nat Prod. 1982;45(6):725–729.
[155] Hammonds TR, Denyer SP, Jackson DE, Irving WL. Studies to show that with podophyllotoxin the early replicative stages of herpes simplex virus type 1 depend upon functional cytoplasmic microtubules. J Med Microbiol. 1996;45(3):167–172.
[156] Sudo K, Konno K, Shigeta S, Yokota T. Inhibitory effects of podophyllotoxin derivatives on herpes simplex virus replication. Antiviral Chem Chemother. 1998;9(3):263–267.
[157] Syed TA, Cheema KM, Khayyami M, Ahmad SA, Ahmad SH, Ahmad S, *et al*. Human leukocyte interferon-α *versus* podophyllotoxin in cream for the treatment of genital warts in males. A placebo-controlled, double-blind, comparative study. Dermatology. 1995;191(2):129–132.
[158] Lassus A. Comparison of podophyllotoxin and podophyllin in treatment of genital warts. Lancet. 1987;2(8557):512–513.
[159] Beutner KR. Podophyllotoxin in the treatment of genital warts. Curr Probl Dermatol. 1996;24:227–232.
[160] Wantke F, Fleischl G, Gotz M, Jarisch R. Topical podophyllotoxin in psoriasis vulgaris. Dermatology. 1993;186(1):79.
[161] Wilson J. Treatment of genital warts - what's the evidence? Int J STD AIDS. 2002;13(4):216–220; quiz 221–222.
[162] MacRae WD, Hudson JB, Towers GHN. The antiviral action of lignans. Planta Med. 1989;55(6):531–535.
[163] Markos AR. The successful treatment of molluscum contagiosum with podophyllotoxin (0.5%) self-application. Int J STD AIDS. 2001;12(12):833.
[164] Syed TA, Lundin S, Ahmad M. Topical 0.3 % and 0.5 % podophyllotoxin cream for self-treatment of molluscum contagiosum in males. A placebo-controlled, double-blind study. Dermatology. 1994;189(1):65–68.
[165] Strand A, Brinkeborn RM, Siboulet A. Topical treatment of genital warts in men, an open study of podophyllotoxin cream compared with solution. Genitourin Med. 1995;71(6):387–390.
[166] Claesson U, Lassus A, Happonen H, Hogstrom L, Siboulet A. Topical treatment of venereal warts: a comparative open study of podophyllotoxin cream *versus* solution. Int J STD AIDS. 1996;7(6):429–434.
[167] Truedsson L, Sjoholm AG, Sturfelt G. Complement activating rheumatoid factors in rheumatoid arthritis studied by haemolysis in gel: relation to antibody class and response to treatment with podophyllotoxin derivatives. Int J Clin Rheumatol. 1985;3(1):29–37.
[168] Leander K, Rosen B. Medicinal uses for podophyllotoxins [patent]; 1988. US Patent 4788216.
[169] Schwartz J, Norton SA. Useful plants of dermatology. VI. The mayapple (*Podophyllum*). J Am Acad Dermatol. 2002;47(5):774–775.
[170] Utsugi T, Shibata H, Kumio S, Aoyagi K, Wierzba K, Kobunai T, *et al*. Antitumor activity of a novel podophyllotoxin derivative (TOP-53) against lung cancer and lung metastatic cancer. Cancer Res. 1996;56(12):2809–2814.

[171] Subrahmanyam D, Renuka B, Rao CVL, Sagar PS, Deevi DS, Babu JM, et al. Novel D-ring analogs of podophyllotoxin as potent anti-cancer agents. Bioorg Med Chem Lett. 1998;8(11):1391–1396.

[172] Pugh N, Khan IA, Moraes RM, Pasco DS. Podophyllotoxin lignans enhance IL-1βbut suppress TNF-αmRNA expression in LPS-treated monocytes. Immunopharmacol Immunotoxicol. 2001;23(1):83–95.

[173] Samanta N, Goel HC. Protection against radiation induced damage to spermatogenesis by *Podophyllum hexandrum*. J Ethnopharmacol. 2002;81(2):217–224.

[174] Mittal A, Pathania V, Agrawala PK, Prasad J, Singh S, Goel HC. Influence of *Podophyllum hexandrum* on endogenous antioxidant defence system in mice: possible role in radioprotection. J Ethnopharmacol. 2001;76(3):253–262.

[175] Hermanns-Lê T, Arrese JE, Goffin V, Pierard GE. Podophyllotoxin-induced acantholysis and cytolysis in a skin equivalent model. Eur J Morphol. 1998;36(3):183–187.

[176] Datt M, Sharma A. Vinblastine-podophylotoxin a good mitotic agent for preparing embryonic chromosomes. Indian J Anim Sci. 2000;70(9):912–913.

[177] Schröder HC, Merz H, Steffen R, Mueller WEG, Sarin PS, Trumm S, et al. Differential *in vitro* anti-HIV activity of natural lignans. Z Naturforsch, C: J Biosci. 1990;45(11-12):1215–1221.

[178] Pfeifer K, Merz H, Steffen R, Muller WEG, Trumm S, Schulz J, et al. In-vitro anti-hiv activity of lignans - differential inhibition of *HIV-1* integrase reaction, topoisomerase activity and cellular microtubules. J Pharm Med. 1992;2(2):75–97.

[179] Hamel E. Antimitotic natural products and their interactions with tubulin. Med Res Rev. 1996;16(2):207–231.

[180] Apers S, Vlietinck A, Pieters L. Lignans and neolignans as lead compounds. Phytochem Rev. 2004;2(3):201–207.

[181] Gordaliza M, Castro MA, Del Corral JMM, San Feliciano A. Antitumor properties of podophyllotoxin and related compounds. Curr Pharm Des. 2000;6(18):1811–1839.

[182] Buss AD, Cox B, Waigh RD, Abraham DJ. Natural products as leads for new pharmaceuticals. In: Abraham DJ, Rotella DP, editors. Burger's Medicinal Chemistry and Drug Discovery. New York: John Wiley & Sons, Inc.; 2003. p. 847–900.

[183] Gordaliza M, Del Corral JMM, Castro MA, Lopez-Vazquez ML, San Feliciano A, Garcia-Gravalos MD, et al. Synthesis and evaluation of pyrazolignans. A new class of cytotoxic agents. Bioorg Med Chem. 1995;3(9):1203–1210.

[184] Schonbrunn E, Phlippen W, Trinczek B, Sack S, Eschenburg S, Mandelkow EM, et al. Crystallization of a macromolecular ring assembly of tubulin liganded with the anti-mitotic drug podophyllotoxin. J Struct Biol. 1999;128(2):211–215.

[185] Cowan CR, Cande WZ. Meiotic telomere clustering is inhibited by colchicine but does not require cytoplasmic microtubules. J Cell Sci. 2002;115(19):3747–3756.

[186] Tseng CJ, Wang YJ, Liang YC, Jeng JH, Lee WS, Lin JK, et al. Microtubule-damaging agents induce apoptosis in HL 60 cells and G2/M cell cycle arrest in HT 29 cells. Toxicology. 2002;175(1-3):123–142.

[187] Chaudhuri AR, Seetharamalu P, Schwarz PM, Hausheer FH, Luduea RF. The interaction of the b-ring of colchicine with α-tubulin: a novel footprinting approach. J Mol Biol. 2000;303(5):679–692.

[188] Pal D, Mahapatra P, Manna T, Chakrabarti P, Bhattacharyya B, Banerjee A, et al. Conformational properties of α-tubulin tail peptide: implications for tail-body interaction. Biochemistry. 2001;40(51):15512–15519.

[189] Lopez-Perez JL, del Olmo E, de Pascual-Teresa B, Merino M, Barajas M, San Feliciano A. A role for dipole moment in the activity of cyclolignans. J Mol Struct: THEOCHEM. 2000;504:51–57.

[190] Petersen M, Alfermann AW. The production of cytotoxic lignans by plant cell cultures. Appl Microbiol Biotechnol. 2001;55(2):135–142.
[191] Waravdekar VS, Domingue A, Leiter J. Effect of podophyllotoxin, α-peltatin, and β-peltatin on the cytochrome oxidase activity of sarcoma 37. J Natl Cancer Inst. 1952;13:393–407.
[192] Waravdekar VS, Paradis AD, Leiter J. Enzyme changes induced in normal and malignant tissues with chemical agents. III. Effect of acetylpodophyllotoxin-ω-pyridinium chloride on cytochrome oxidase, cytochrome c, succinoxidase, succinic dehydrogenase, and respiration of sarcoma 37. J Natl Cancer Inst. 1953;14:585–592.
[193] Waravdekar VS, Powers O, Leiter J. Enzyme changes induced in normal and malignant tissues with chemical agents. VI. Effect of acetylpodophyllotoxin-ω-pyridinium chloride on malic oxidase and isocitric oxidase systems of sarcoma 37. J Natl Cancer Inst. 1956;16:1443–1452.
[194] Shi Q, Chen K, Morris-Natschke SL, Lee KH. Recent progress in the development of tubulin inhibitors as antimitotic antitumor agents. Curr Pharm Des. 1998;4(3):219–248.
[195] Lee KH. Novel antitumor agents from higher plants. Med Res Rev. 1999;19(6):569–596.
[196] Ramos AC, Pelaez-Lamamie De Clairac R, Medarde M. Heterolignans. Heterocycles. 1999;51(6):1443–1470.
[197] Xiao Z, Xiao YD, Feng J, Golbraikh A, Tropsha A, Lee KH. Antitumor agents. 213. Modeling of epipodophyllotoxin derivatives using variable selection k nearest neighbor QSAR method. J Med Chem. 2002;45(11):2294–2309.
[198] Demain AL, Vaishnav P. Natural products for cancer chemotherapy. Microb Biotechnol. 2011;4(6):687–699.
[199] Capilla AS, Sanchez I, Caignard DH, Renard P, Pujol MD. Antitumor agents. Synthesis and biological evaluation of new compounds related to podophyllotoxin, containing the 2,3-dihydro-1,4-benzodioxin system. Eur J Med Chem. 2001;36(4):389–393.
[200] Gordaliza M, Del Corral JMM, Castro MA, Garcia-Garcia PA, San Feliciano A. Cytotoxic cyclolignans related to podophyllotoxin. Farmaco. 2001;56(4):297–304.
[201] Madrigal B, Puebla P, Ramos A, Pelaez R, Gravalos D, Caballero E, et al. Synthesis and cytotoxic activities of analogues of thuriferic acid. Bioorg Med Chem. 2002;10(2):303–312.
[202] Roulland E, Magiatis P, Arimondo P, Bertounesque E, Monneret C. Hemi-synthesis and biological activity of new analogues of podophyllotoxin. Bioorg Med Chem. 2002;10(11):3463–3471.
[203] Gordaliza M, Faircloth GT, Castro MA, Del Corral JMM, Lopez-Vazquez ML, San Feliciano A. Immunosuppressive cyclolignans. J Med Chem. 1996;39(14):2865–2868.
[204] Gordaliza M, Castro MA, del Corral JMM, Lopez-Vazquez ML, San Feliciano A, Faircloth GT. In vivo immunosuppressive activity of some cyclolignans. Bioorg Med Chem Lett. 1997;7(21):2781–2786.
[205] Perry NB, Foster LM. Antitumor lignans and cytotoxic resin acids from a New Zealand gymnosperm, *Libocedrus plumosa*. Phytomedicine. 1994;1(3):233–237.
[206] Schmidt TJ, Hemmati S, Klaes M, Konuklugil B, Mohagheghzadeh A, Ionkova I, et al. Lignans in flowering aerial parts of linum species: chemodiversity in the light of systematics and phylogeny. Phytomedicine. 2010;71(14-15):1714–1728.
[207] Velazquez-Jimenez R, Torres-Valencia JM, Cerda-Garcia-Rojas CM, Hernandez-Hernandez JD, Roman-Marin LU, Manriquez-Torres JJ, et al. Absolute configuration of podophyllotoxin related lignans from *Bursera fagaroides* using vibrational circular dichroism. Phytochemistry. 2011;72(17):2237–2243.
[208] Jackson DE, Dewick PM. Tumor-inhibitory aryltetralin lignans from *Podophyllum pleianthum*. Phytochemistry. 1985;24(10):2407–2409.
[209] Broomhead AJ, Dewick PM. Tumor-inhibitory aryltetralin lignans in *Podophyllum versipelle*, *Diphylleia cymosa* and *Diphylleia grayi*. Phytochemistry. 1990;29(12):3831–3837.

[210] Hendrawati O, Woerdenbag HJ, Michiels PJA, Aantjes HG, van Dam A, Kayser O. Identification of lignans and related compounds in *Anthriscus sylvestris* by LC-ESI-MS/MS and LC-SPE-NMR. Phytochemistry. 2011;72(17):2172–2179.
[211] Jiang RW, Zhou JR, Hon PM, Li SL, Zhou Y, Li LL, et al. Lignans from dysosma versipellis with inhibitory effects on prostate cancer cell lines. J Nat Prod. 2007;70(2):283–286.
[212] Pettit GR, Meng Y, Gearing RP, Herald DL, Pettit RK, Doubek DL, et al. Antineoplastic agents. 522. *Hernandia peltata* (Malaysia) and *Hernandia nymphaeifolia* (Republic of Maldives). J Nat Prod. 2004;67(2):214–220.
[213] Wickramaratne DBM, Mar W, Chai H, Castillo JJ, Farnsworth NR, Soejarto DD, et al. Cytotoxic constituents of *Bursera permollis*. Planta Med. 1995;61(1):80–81.
[214] Bianchi E, Sheth K, Cole JR. Antitumor agents from *Bursera fagaroides* (*Burseraceae*) (β-peltatin A-methyl ether and 5'demethoxy-β-peltatin A-methyl ether). Tetrahedron Lett. 1969;(32):2759–2762.
[215] Cairnes DA, Ekundayo O, Kingston DGI. Plant anticancer agents. X. Lignans from *Juniperus phoenicea*. J Nat Prod. 1980;43(4):495–497.
[216] Noguera B, Diaz E, Garcia MV, Feliciano AS, Lopez-Perez JL, Israel A. Anti-inflammatory activity of leaf extract and fractions of *Bursera simaruba* (l.) Sarg (*Burseraceae*). J Ethnopharmacol. 2004;92(1):129–133.
[217] Berlin J, Wray V, Mollenschott C, Sasse F. Formation of β-peltatin-A methyl ether and coniferin by root cultures of *Linum flavum*. J Nat Prod. 1986;49(3):435–439.
[218] Federolf K, Alfermann AW, Fuss E. Aryltetralin-lignan formation in two different cell suspension cultures of linum album: deoxypodophyllotoxin 6-hydroxylase, a key enzyme for the formation of 6-methoxypodophyllotoxin. Phytomedicine. 2007;68(10):1397–1406.
[219] Nakanishi T, Inatomi Y, Murata H, Shigeta K, Iida N, Inada A, et al. A new and known cytotoxic aryltetralin-type lignans from stems of *Bursera graveolens*. Chem Pharm Bull. 2005;53(2):229–231.
[220] Wichers HJ, Versluis-De Haan GG, Marsman JW, Harkes MP. Podophyllotoxin related lignans in plants and cell cultures of *Linum flavum*. Phytochemistry. 1991;30(11):3601–3604.
[221] Xu H, Lv M, Tian X. A review on hemisynthesis, biosynthesis, biological activities, mode of action, and structure-activity relationship of podophyllotoxins: 2003-2007. Curr Med Chem. 2009;16(3):327–349.
[222] Rojas-Sepulveda AM, Mendieta-Serrano M, Mojica MYA, Salas-Vidal E, Marquina S, Villarreal ML, et al. Cytotoxic podophyllotoxin type-lignans from the steam bark of *Bursera fagaroides* var. fagaroides. Molecules. 2012;17:9506–9519.
[223] Hadimani SB, Tanpure RP, Bhat SV. Asymmetric total synthesis of (−)-podophyllotoxin. Tetrahedron Lett. 1996;37(27):4791–4794.
[224] Middel O, Woerdenbag HJ, van Uden W, van Oeveren A, Jansen JFGA, Feringa BL, et al. Synthesis and cytotoxicity of novel lignans. J Med Chem. 1995;38(12):2112–2118.
[225] Cho SJ, Tropsha A, Suffness M, Cheng YC, Lee KH. Antitumor agents. 163. Three-dimensional quantitative structure-activity relationship study of 4'-O-demethylepipodophyllotoxin analogs using the CoMFA/q2-GRS approach. J Med Chem. 1996;39(7):1383–1395.
[226] Thurston LS, Irie H, Tani S, Han FS, Liu ZC, Cheng YC, et al. Antitumor agents. 78. Inhibition of human DNA topoisomerase II by podophyllotoxin and α-peltatin analogs. J Med Chem. 1986;29(8):1547–1551.
[227] Markkanen T, Makinen ML, Maunuksela E, Himanen P. Podophyllotoxin lignans under experimental antiviral research. Drugs Exp Clin Res. 1981;7(6):711–718.
[228] Zi CT, Xu FQ, Li GT, Li Y, Ding ZT, Zhou J, et al. Synthesis and anticancer activity of glucosylated podophyllotoxin derivatives linked via 4β-triazole rings. Molecules. 2013;18(11):13992–4012.

[229] May G, Willuhn G. Antiviral effect of aqueous plant extracts in tissue culture. Arzneim Forsch. 1978;28(1):1–7.
[230] Novelo M, Cruz JG, Hernandez L, Pereda-Miranda R, Chai H, Mar W, et al. Chemical studies on Mexican *Hyptis* species. VI. Biologically active natural products from mexican medicinal plants. II. Cytotoxic constituents from *Hyptis verticillata*. J Nat Prod. 1993;56(10):1728–1736.
[231] Girnita A, Girnita L, Del Prete F, Bartolazzi A, Larsson O, Axelson M. Cyclolignans as inhibitors of the insulin-like growth factor-1 receptor and malignant cell growth. Cancer Res. 2004;64(1):236–242.
[232] Weiss SG, Tin-Wa M, Perdue JRE, Farnsworth NR. Potential anticancer agents. II. Antitumor and cytotoxic lignans from *Linum album* (*Linaceae*). J Pharm Sci. 1975;64(1):95–98.
[233] Hokanson GC. Podophyllotoxin and 4′-demethylpodophyllotoxin from *Polygala polygama* (*Polygalaceae*). Lloydia. 1978;41(5):497–498.
[234] Hartwell JL, Schrecker AW. Chemistry of podophyllum. Fortschr Chem Org Naturst. 1958;15:83–166.
[235] Stähelin H. Chemistry and mechanism of podophyllin derivatives. Planta Med. 1972;22(3): 336–347.
[236] Kupchan SM, Hemingway RJ, Hemingway NC. Tumor inhibitors. XIX. Deoxypodophyllotoxin, the cytotoxic principle of *Lebocedrus decurrens*. J Pharm Sci. 1967;56(3):408–409.
[237] Hartwell JL. Types of anticancer agents isolated from plants. Cancer Treat Rep. 1976;60(8):1031–1067.
[238] Stähelin H, Cerletti A. Experimental results with the podophyllum-cytostatics SP-1 and SP-G. Schweiz Med Wochenschr. 1964;94:1490–1502.
[239] Stähelin H. ′-Demethyl-epipodophyllotoxin thenylidene glucoside (VM 26), a podophyllum compound with a new mechanism of action. Eur J Cancer. 1970;6(4):303–311.
[240] Kelleher JK. Correlation of tubulin-binding and antitumor activities of podophyllotoxin analogs. Cancer Treat Rep. 1978;62(10):1443–1447.
[241] Lopez-Perez JL, del Olmo E, de Pascual-Teresa B, Abad A, San Feliciano A. Synthesis and cytotoxicity of hydrophobic esters of podophyllotoxins. Bioorg Med Chem Lett. 2004;14(5):1283–1286.
[242] Brewer CF, Loike JD, Horwitz SB, Sternlicht H, Gensler WJ. Conformational analysis of podophyllotoxin and its congeners. Structure-activity relationship in microtubule assembly. J Med Chem. 1979;22(3):215–221.
[243] Loike JD, Brewer CF, Sternlicht H, Gensler WJ, Horwitz SB. Structure-activity study of the inhibition of microtubule assembly *in vitro* by podophyllotoxin and its congeners. Cancer Res. 1978;38(9):2688–2693.
[244] Cortese F, Bhattacharyya B, Wolff J. Podophyllotoxin as a probe for the colchicine binding site of tubulin. J Biol Chem. 1977;252(4):1134–1140.
[245] Loike JD, Horwitz SB. Effects of podophyllotoxin and VP-16-213 on microtubule assembly *in vitro* and nucleoside transport in HeLa cells. Biochemistry. 1976;15(25):5435–5443.
[246] Kelleher JK. Tubulin binding affinities of podophyllotoxin and colchicine analogues. Mol Pharmacol. 1977;13(2):232–241.
[247] Gensler WJ, Murthy CD, Trammell MH. Nonenolizable podophyllotoxin derivatives. J Med Chem. 1977;20(5):635–644.
[248] Peraza-Sanchez SR, Pena-Rodriquez LM. Isolation of picropolygamain from the resin of *Bursera simaruba*. J Nat Prod. 1992;55(12):1768–1771.
[249] Zhang YJ, Litaudon M, Bousserouel H, Martin MT, Thoison O, Leonce S, et al. Sesquiterpenoids and cytotoxic lignans from the bark of *Libocedrus chevalieri*. J Nat Prod. 2007;70(8):1368–1370.

[250] Habtemariam S. Cytotoxic and cytostatic activity of erlangerins from *Commiphora erlangeriana*. Toxicon. 2003;41(6):723–727.

[251] Dekebo A, Lang M, Polborn K, Dagne E, Steglich W. Four lignans from *Commiphora erlangeriana*. J Nat Prod. 2002;65(9):1252–1257.

[252] Gu JQ, Park EJ, Totura S, Riswan S, Fong HHS, Pezzuto JM, *et al.* Constituents of the twigs of *Hernandia ovigera* that inhibit the transformation of JB6 murine epidermal cells. J Nat Prod. 2002;65(7):1065–1068.

[253] He K, Shi G, Zeng L, Ye Q, McLaughlin JL. Konishiol, a new sesquiterpene, and bioactive components from *Cunninghamia konishii*. Planta Med. 1997;63(2):158–160.

[254] Ito C, Itoigawa M, Ogata M, Mou XY, Tokuda H, Nishino H, *et al.* Lignans as anti-tumor-promoter from the seeds of *Hernandia ovigera*. Planta Med. 2001;67(2):166–168.

[255] Ng TB, Liu F, Wang ZT. Antioxidative activity of natural products from plants. Life Sci. 2000;66(8):709–723.

[256] Yang LM, Lin SJ, Yang TH, Lee KH. Synthesis and anti-HIV activity of dibenzylbutyrolactone lignans. Bioorg Med Chem Lett. 1996;6(8):941–944.

[257] Eich E, Pertz H, Kaloga M, Schulz J, Fesen MR, Mazumder A, *et al.* (−)-arctigenin as a lead structure for inhibitors of human immunodeficiency virus type-1 integrase. J Med Chem. 1996;39(1):86–95.

[258] Hara H, Fujihashi T, Sakata T, Kaji A, Kaji H. Tetrahydronaphthalene lignan compounds as potent anti-HIV type 1 agents. AIDS Res Hum Retroviruses. 1997;13(8):695–705.

[259] Lee CTL, Lin VCK, Zhang SX, Zhu XK, Vanvliet D, Hu H, *et al.* Anti-AIDS agents. 29. Anti-hiv activity of modified podophyllotoxin derivatives. Bioorg Med Chem Lett. 1997;7(22):2897–2902.

[260] Chen DF, Zhang SX, Xie L, Xie JX, Chen K, Kashiwada Y, *et al.* Anti-AIDS agents-XXVI. Structure-activity correlations of gomisin-G-related anti-HIV lignans from kadsura interior and of related synthetic analogs. Bioorg Med Chem. 1997;5(8):1715–1723.

[261] Ter-haar E, Rosenkranz HS, Hamel E, Day BW. Computational and molecular modeling evaluation of the structural basis for tubulin polymerization inhibition by colchicine site agents. Bioorg Med Chem. 1996;4(10):1659–1671.

[262] Allevi P, Anastasia M, Ciuffreda P, Bigatti E, Macdonald P. Stereoselective glucosidation of *Podophyllum* lignans. A new simple synthesis of etoposide. J Org Chem. 1993;58(15):4175–4178.

[263] Hainsworth JD, Greco FA. Etoposide: twenty years later. Ann Oncol. 1995;6(4):325–341.

[264] Loike JD, Horwitz SB. Effect of VP-16-213 on the intracellular degradation of DNA in hela cells. Biochemistry. 1976;15(25):5443–5448.

[265] Waravdekar VS, Paradis AD, Leiter J. Enzyme changes induced in normal and malignant tissues with chemical agents. IV. Effect of α-peltatin on glucose utilization by sarcoma 37 and on the adenosinetriphosphatase, hexokinase, aldolase, and pyridine nucleotide levels of sarcoma 37. J Natl Cancer Inst. 1955;16:31–39.

[266] Waravdekar VS, Paradis AD, Leiter J. Enzyme changes induced in normal and malignant tissues with chemical agents. V. Effect of acetylpodophyllotoxin-ω-pyridiniumchloride on uricase, adenosine deaminase, nucleoside phosphorylase, and glutamic dehydrogenase activities. J Natl Cancer Inst. 1955;16:99–105.

[267] Cragg GM, Newman DJ. Antineoplastic agents from natural sources: achievements and future directions. Expert Opin Investig Drugs. 2000;9(12):2783–2797.

[268] Cragg GM, Newman DJ, Snader KM. Natural products in drug discovery and development. J Nat Prod. 1997;60(1):52–60.

[269] Verpoorte R, Contin A, Memelink J. Biotechnology for the production of plant secondary metabolites. Phytochem Rev. 2002;1(1):13–25.

[270] Hirano T, Gotoh M, Oka K. Natural flavonoids and lignans are potent cytostatic agents against human leukemic HL-60 cells. Life Sci. 1994;55(13):1061–1069.

[271] Kangas L, Saarinen N, Mutanen M, Ahotupa M, Hirsinummi R, Unkila M, et al. Antioxidant and antitumor effects of hydroxymatairesinol (HM-3000, HMR), a lignan isolated from the knots of spruce. Eur J Cancer Prev. 2002;11(Suppl. 2):S48–S57.

[272] Lu H, Liu G. Antioxidant activity of dibenzocyclooctene lignans isolated from *Schisandraceae*. Planta Med. 1992;58(4):311–313.

[273] Ghisalberty EL. Cardiovascular activity of naturally occurring lignans. Phytomedicine. 1997;4(2):151–166.

[274] Kitts DD, Yuan YV, Wijewickreme AN, Thompson LU. Antioxidant activity of the flaxseed lignan secoisolariciresinol diglycoside and its mammalian lignan metabolites enterodiol and enterolactone. Mol Cell Biochem. 1999;202(1-2):91–100.

[275] Yamauchi S, Ina T, Kirikihira T, Masuda T. Synthesis and antioxidant activity of oxygenated furofuran lignans. Biosci Biotechnol Biochem. 2004;68(1):183–192.

[276] Srivastava V, Negi AS, Kumar JK, Gupta MM, Khanuja SPS. Plant-based anticancer molecules: a chemical and biological profile of some important leads. Bioorg Med Chem. 2005;13(21):5892–5908.

[277] Kelly M, Hartwell JL. The biological effects and the chemical composition of podophyllin: a review. J Natl Cancer Inst. 1954;14(4):967–1010.

[278] Tsukamoto H, Hisada S, Nishibe S. Lignans from bark of the olea plants. I. Chem Pharm Bull. 1984;32(7):2730–2735.

[279] Gandevia B, Milne J. Occupational asthma and rhinitis due to western red cedar (*Thuja plicata*), with special reference to bronchial reactivity. Br J Ind Med. 1970;27(3):235–244.

[280] Chan-Yeung M. Maximal expiratory flow and airway resistance during induced bronchoconstriction in patients with asthma due to western red cedar (thuja plicata). Am Rev Respir Dis. 1973;108(5):1103–1110.

[281] Chan-Yeung M, Barton GM, MacLean L, Grzybowski S. Occupational asthma and rhinitis due to western red cedar (thuja plicata). Am Rev Respir Dis. 1973;108(5):1094–102.

[282] Rimando AM, Pezzuto JM, Farnsworth NR, Santisuk T, Reutrakul V, Kawanishi K. New lignans from anogeissus acuminata with HIV-1 reverse transcriptase inhibitory activity. J Nat Prod. 1994;57(7):896–904.

[283] Valsaraj R, Pushpangadan P, Smitt UW, Adsersen A, Christensen SB, Sittie A, et al. New anti-HIV-1, antimalarial, and antifungal compounds from terminalia bellerica. J Nat Prod. 1997;60(7):739–742.

[284] Hirano T, Fukuoka K, Oka K, Naito T, Hosaka K, Mitsuhashi H, et al. Antiproliferative activity of mammalian lignan derivatives against the human breast carcinoma cell line, ZR-75-1. Cancer Invest. 1990;8(6):595–602.

[285] Wu TS, Leu YL, Chan YY, Yu SM, Teng CM, Su JD. Lignans and an aromatic acid from *Cinnamomum philippinense*. Phytochemistry. 1994;36(3):785–788.

[286] Lee JS, Kim J, Yu YU, Kim YC. Inhibition of phospholipase cγ1 and cancer cell proliferation by lignans and flavans from *Machilus thunbergii*. Arch Pharmacal Res. 2004;27(10):1043–1047.

[287] Li G, Lee CS, Woo MH, Lee SH, Chang HW, Son JK. Lignans from the bark of *Machilus thunbergii* and their DNA topoisomerases I and II inhibition and cytotoxicity. Biol Pharm Bull. 2004;27(7):1147–1150.

[288] Abou-Gazar H, Bedir E, Takamatsu S, Ferreira D, Khan IA. Antioxidant lignans from *Larrea tridentata*. Phytochemistry. 2004;65(17):2499–2505.

[289] Lewis NG, Davin LB. Lignans: biosynthesis and function. In: Barton SD, Nakanishi K, Meth-Cohn O, editors. Comprehensive Natural Products Chemistry. vol. 1. Oxford: Pergamon; 1999. p. 639–712.

[290] Lee IS, Jung KY, Oh SR, Kim DS, Kim JH, Lee JJ, et al. Platelet-activating factor antagonistic activity and ^{13}C NMR assignment of pregomisin and chamigrenal from schisandra chinensis. Arch Pharmacal Res. 1997;20(6):633–636.

[291] Chowdhury S, Mukherjee T, Mukhopadhyay R, Mukherjee B, Sengupta S, Chattopadhyay S, et al. The lignan niranthin poisons *Leishmania donovani* topoisomerase IB and favours a Th1 immune response in mice. EMBO Mol Med. 2012;4(10):1126–1143.

[292] Sutherland EW, Rall TW. Fractionation and characterization of a cyclic adenine ribonucleotide formed by tissue particles. J Biol Chem. 1958;232:1077–1091.

[293] Nikaido T, Ohmoto T, Noguchi H, Kinoshita T, Saitoh H, Sankawa U. Inhibitors of cyclic AMP phosphodiesterase in medicinal plants. Planta Med. 1981;43(1):18–23.

[294] Westcott ND, Muir AD. Flax seed lignan in disease prevention and health promotion. Phytochem Rev. 2004;2(3):401–417.

[295] Vanharanta M, Voutilainen S, Lakka TA, van der Lee M, Adlercreutz H, Salonen JT. Risk of acute coronary events according to serum concentrations of enterolactone: a prospective population-based case-control study. Lancet. 1999;354(9196):2112–2115.

[296] Halligudi N. Pharmacological properties of flax seeds: a review. Hygeia. 2012;4(2):70–77.

[297] Moree SS, Rajesha J. Secoisolariciresinol diglucoside: a potent multifarious bioactive phytoestrogen of flaxseed. Res Rev Biomed Biotechnol. 2011;2(3):1–24.

[298] Wang LQ, Meselhy MR, Li Y, Qin GW, Hattori M. Human intestinal bacteria capable of transforming secoisolariciresinol diglucoside to mammalian lignans, enterodiol and enterolactone. Chem Pharm Bull. 2000;48(11):1606–1610.

[299] Wang CZ, Ma XQ, Yang DH, Guo ZR, Liu GR, Zhao GX, et al. Production of enterodiol from defatted flaxseeds through biotransformation by human intestinal bacteria. BMC Microbiol. 2010;10:115.

[300] Sainvitu P, Nott K, Richard G, Blecker C, Jérôme C, Wathelet JP, et al. Structure, properties and obtention routes of flaxseed lignan secoisolariciresinol: a review. Biotechnol, Agron, Soc Environ. 2012;16(1):115–124.

[301] Raffaelli B, Hoikkala A, Leppälä E, Wähälä K. Enterolignans. J Chromatogr B Analyt Technol Biomed Life Sci. 2002;777(1-2):29–43.

[302] Bartkiene E, Juodeikiene G, Basinskiene L, Liukkonen KH, Adlercreutz H, Kluge H. Enterolignans enterolactone and enterodiol formation from their precursors by the action of intestinal microflora and their relationship with non-starch polysaccharides in various berries and vegetables. LWT–Food Sci Technol. 2011;44(1):48–53.

[303] Martin ME, Haourigui M, Pelissero C, Benassayag C, Nunez EA. Interactions between phytoestrogens and human sex steroid binding protein. Life Sci. 1995;58(5):429–436.

[304] Tou JCL, Chen J, Thompson LU. Flaxseed and its lignan precursor, secoisolariciresinol diglycoside, affect pregnancy outcome and reproductive development in rats. J Nutr. 1998;128(11):1861–1868.

[305] Hausott B, Greger H, Marian B. Naturally occurring lignans efficiently induce apoptosis in colorectal tumor cells. J Cancer Res Clin Oncol. 2003;129(10):569–576.

[306] Magee PJ, McGlynn H, Rowland IR. Differential effects of isoflavones and lignans on invasiveness of *MDA-MB*-231 breast cancer cells in vitro. Cancer Lett. 2004;208(1):35–41.

[307] Wang C, Makela T, Hase T, Adlercreutz H, Kurzer MS. Lignans and flavonoids inhibit aromatase enzyme in human preadipocytes. J Steroid Biochem Mol Biol. 1994;50(3-4):205–212.

[308] Lin X, Switzer BR, Demark-Wahnefried W. Effect of mammalian lignans on the growth of prostate cancer cell lines. Anticancer Res. 2001;21(6A):3995–3999.

[309] Rickard SE, Yuan YV, Thompson LU. Plasma insulin-like growth factor I levels in rats are reduced by dietary supplementation of flaxseed or its lignan secoisolariciresinol diglycoside. Cancer Lett. 2000;161(1):47–55.

[310] Thompson LU, Seidl MM, Rickard SE, Orcheson LJ, Fong HHS. Antitumorigenic effect of a mammalian lignan precursor from flaxseed. Nutr Cancer. 1996;26(2):159–165.
[311] Thompson LU, Rickard SE, Orcheson LJ, Seidl MM. Flaxseed and its lignan and oil components reduce mammary tumor growth at a late stage of carcinogenesis. Carcinogenesis. 1996;17(6):1373–1376.
[312] Pietrofesa R, Turowski J, Tyagi S, Dukes F, Arguiri E, Busch TM, et al. Radiation mitigating properties of the lignan component in flaxseed. BMC Cancer. 2013;13:179.
[313] Prasad K. Hydroxyl radical-scavenging property of secoisolariciresinol diglucoside (SDG) isolated from flax-seed. Mol Cell Biochem. 1997;168(1-2):117–123.
[314] Willför SM, Ahotupa MO, Hemming JE, Reunanen MHT, Eklund PC, Sjoeholm RE, et al. Antioxidant activity of knotwood extractives and phenolic compounds of selected tree species. J Agric Food Chem. 2003;51(26):7600–7606.
[315] Shen YC, Chen CY, Chen YJ, Kuo YH, Chien CT, Lin YM. Bioactive lignans and taxoids from the roots of Formosan *Taxus mairei*. Chin Pharm J. 1997;49(5-6):285–296.
[316] Banskota AH, Usia T, Tezuka Y, Kouda K, Nguyen NT, Kadota S. Three new C-14 oxygenated taxanes from the wood of *Taxus yunnanensis*. J Nat Prod. 2002;65(11):1700–1702.
[317] Banskota AH, Tezuka Y, Nguyen NT, Awale S, Nobukawa T, Kadota S. DPPH radical scavenging and nitric oxide inhibitory activities of the constituents from the wood of *Taxus yunnanensis*. Planta Med. 2003;69(6):500–505.
[318] Banskota AH, Nguyen NT, Tezuka Y, Le Tran Q, Nobukawa T, Kurashige Y, et al. Secoisolariciresinol and isotaxiresinol inhibit tumor necrosis factor-α-dependent hepatic apoptosis in mice. Life Sci. 2004;74(22):2781–2792.
[319] Küpeli E, Erdemoğlu N, Yeşilada E, Şener B. Anti-inflammatory and antinociceptive activity of taxoids and lignans from the heartwood of *Taxus baccata* L. J Ethnopharmacol. 2003;89(2-3):265–670.
[320] Koyama J, Morita I, Kobayashi N, Hirai K, Simamura E, Nobukawa T, et al. Antiallergic activity of aqueous extracts and constituents of *Taxus yunnanensis*. Biol Pharm Bull. 2006;29(11):2310–2312.
[321] Belletire JL, Fry DF. Total synthesis of (±)-wikstromol. J Org Chem. 1988;53(20):4724–4729.
[322] Moon SS, Rahman AA, Kim JY, Kee SH. Hanultarin, a cytotoxic lignan as an inhibitor of actin cytoskeleton polymerization from the seeds of *Trichosanthes kirilowii*. Bioorg Med Chem. 2008;16(15):7264–7269.
[323] Mabry TJ, DiFeo DR, Sakakibara M, Bohnstedt CF, Seigler D. The natural products: chemistry of larrea. In: Mabry TJ, Hunziker JH, DiFeo DR, editors. Creosote Bush: Biology and Chemistry of Larrea in New World Desserts, US/IBP Synthesis Series. vol. 6. Hutchinson Ross Inc.; 1977. p. 115–133.
[324] Oliveto EP. Nordihydroguaiaretic acid. A naturally occurring antioxidant. Chem Ind. 1972;(17):677–679.
[325] Shih AL, Harris ND. Antimicrobial activity of selected antioxidants. Cosmet Toiletries. 1980;95(2):75–76, 79–.
[326] Segura JJ. Effects of nordihydroguaiaretic acid and ethanol on the growth of *Entamoeba invadens*. Arch Invest Med. 1978;9(Suppl. 1):157–162.
[327] Lambert J, Dorr R, Timmermann B. Nordihydroguaiaretic acid: a review of its numerous and varied biological activities. Int J Pharmacogn (Lisse, Neth). 2004;42(2):149–158.
[328] Belmares H, Barrera A, Castillo E, Ramos LF, Hernandez F, Hernandez V. New rubber antioxidants and fungicides derived from *Larrea tridentata* (creosote bush). Ind Eng Chem Prod Res Dev. 1979;18(3):220–226.
[329] Burk D, Woods M. Hydrogen peroxide, catalase, glutathione peroxidase, quinones, nordihydroguaiaretic acid, and phosphopyridine nucleotides in relation to X-Ray action on cancer cells. Radiat Res Suppl. 1963;3:212–246.

[330] Smart CR, Hogle HH, Robins RK, Broom AD, Bartholomew D. An interesting observation on nordihydroguaiaretic acid (NSC-4291; NDGA) and a patient with malignant melanoma—a preliminary report. Cancer Chemother Rep, Part 1. 1969;53(2):147–151.
[331] Rhoades DF. Integrated antiherbivore, antidesiccant and ultraviolet screening properties of creosotebush resin. Biochem Syst Ecol. 1977;5(4):281–290.
[332] Lundberg WO, Halvorson HO, Burr GO. The antioxidant properties of nordihydroguaiaretic acid. Oil Soap. 1944;21:33–35.
[333] Yasumoto K, Yamamoto A, Mitsuda H. Soybean lipoxygenase. IV. Effect of phenolic antioxidants on lipoxygenase reaction. Agric Biol Chem. 1970;34(8):1162–1168.
[334] Gisvold O, Thaker E. Lignans from *Larrea divaricata*. J Pharm Sci. 1974;63(12):1905–1907.
[335] Yu YU, Kang SY, Park HY, Sung SH, Lee EJ, Kim SY, et al. Antioxidant lignans from *Machilus thunbergii* protect CCl4-injured primary cultures of rat hepatocytes. J Pharm Pharmacol. 2000;52(9):1163–1169.
[336] Filleur F, Le Bail JC, Duroux JL, Simon A, Chulia AJ. Antiproliferative, anti-aromatase, anti-17β-HSD and antioxidant activities of lignans isolated from *Myristica argentea*. Planta Med. 2001;67(8):700–704.
[337] Lee WS, Baek YI, Kim JR, Cho KH, Sok DE, Jeong TS. Antioxidant activities of a new lignan and a neolignan from *Saururus chinensis*. Bioorg Med Chem Lett. 2004;14(22):5623–5628.
[338] Hwu JR, Tseng WN, Gnabre J, Giza P, Huang RCC. Antiviral activities of methylated nordihydroguaiaretic acids. 1. Synthesis, structure identification, and inhibition of tat-regulated HIV transactivation. J Med Chem. 1998;41(16):2994–3000.
[339] Craigo J, Callahan M, Huang RCC, DeLucia AL. Inhibition of human papillomavirus type 16 gene expression by nordihydroguaiaretic acid plant lignan derivatives. Antiviral Res. 2000;47(1):19–28.
[340] Lambert JD, Meyers RO, Timmermann BN, Dorr RT. Tetra-*O*-methylnordihydroguaiaretic acid inhibits melanoma in vivo. Cancer Lett. 2001;171(1):47–56.
[341] Xia Y, Zhang Y, Wang W, Ding Y, He R. Synthesis and bioactivity of *erythro*-nordihydroguaiaretic acid, *threo*-(−)-saururenin and their analogues. J Serb Chem Soc. 2010;75(10):1325–1335.
[342] Xia YM, Bi WH, Zhang YY. Synthesis of dibenzylbutanediol lignans and their anti-HIV, anti-HSV, anti-tumor activities. J Chil Chem Soc. 2009;54(4):428–431.
[343] Ito C, Matsui T, Wu TS, Furukawa H. Isolation of 6,7-demethylenedesoxypodophyllotoxin from *Hernandia ovigera*. Chem Pharm Bull. 1992;40(5):1318–1321.
[344] McDoniel PB, Cole JR. Antitumor activity of *Bursera schlechtendalii* (Burseraceae). Isolation and structure determination of two new lignans. J Pharm Sci. 1972;61(12):1992–1994.
[345] Torrance SJ, Hoffmann JJ, Cole JR. Wikstromol, antitumor lignan from *Wikstroemia foetida* var. oahuensis gray and *Wikstroemia uva-ursi* gray (Thymelaeaceae). J Pharm Sci. 1979;68(5):664–665.
[346] Lee KH, Tagahara K, Suzuki H, Wu RY, Haruna M, Hall IH, et al. Antitumor agents. 49. Tricin, kaempferol-3-*O*-β-D-glucopyranoside and (+)-nortrachelogenin, antileukemic principles from *Wikstroemia indica*. J Nat Prod. 1981;44(5):530–535.
[347] Ichikawa K, Kinoshita T, Nishibe S, Sankawa U. The calcium-antagonist activity of lignans. Chem Pharm Bull. 1986;34(8):3514–3517.
[348] Braquet P, Godfroid JJ. PAF-acether specific binding sites: 2. Design of specific antagonists. Trends Pharmacol Sci. 1986;7(10):397–403.
[349] Nuntanakorn P, Jiang B, Einbond LS, Yang H, Kronenberg F, Weinstein IB, et al. Polyphenolic constituents of *Actaea racemosa*. J Nat Prod. 2006;69(3):314–318.
[350] Shoeb M, MacManus SM, Kumarasamy Y, Jaspars M, Nahar L, Thoo-Lin PK, et al. Americanin, a bioactive dibenzylbutyrolactone lignan, from the seeds of *Centaurea americana*. Phytochemistry. 2006;67(21):2370–2375.

[351] Kato A, Hashimoto Y, Kidokor M. (+)-Nortrachelogenin, a new pharmacologically active lignan from *Wikstroemia indica*. J Nat Prod. 1979;42(2):159–162.
[352] Ban HS, Lee S, Kim YP, Yamaki K, Shin KH, Ohuchi K. Inhibition of prostaglandin E2 production by taiwanin C isolated from the root of *Acanthopanax chiisanensis* and the mechanism of action. Biochem Pharmacol. 2002;64(9):1345–1354.
[353] Su BN, Jones WP, Cuendet M, Kardono LBS, Ismail R, Riswan S, et al. Constituents of the stems of *Macrococculus pomiferus* and their inhibitory activities against cyclooxygenases-1 and -2. Phytochemistry. 2004;65(21):2861–2866.
[354] Min BS, Na MK, Oh SR, Ahn KS, Jeong GS, Li G, et al. New furofuran and butyrolactone lignans with antioxidant activity from the stem bark of *Styrax japonica*. J Nat Prod. 2004;67(12):1980–1984.
[355] Chang WL, Chiu LW, Lai JH, Lin HC. Immunosuppressive flavones and lignans from *Bupleurum scorzonerifolium*. Phytochemistry. 2003;64(8):1375–1379.
[356] Sefkow M, Raschke M, Steiner C. Enantioselective synthesis and biological evaluation of α-hydroxylated lactone lignans. Pure Appl Chem. 2003;75(2-3):273–278.
[357] Palter R, Lundin RE, Haddon WF. Cathartic lignan glycoside isolated from *Carthamus tinctorus*. Phytochemistry. 1972;11(9):2871–2874.
[358] Chiechi LM, Micheli L. Utility of dietary phytoestrogens in preventing postmenopausal osteoporosis. Curr Top Nutraceutical Res. 2005;3(1):15–28.
[359] Braquet P, Senn N, Robin JP, Esanu A, Godfraind T, Garay R. Inhibition of the erythrocyte sodium-potassium pump by mammalian lignans. Pharmacol Res Commun. 1986;18(3):227–239.
[360] Lin RC, Skaltsounis AL, Seguin E, Tillequin F, Koch M. Phenolic constituents of *Selaginella doederleinii*. Planta Med. 1994;60(2):168–170.
[361] Kim JH, Park YH, Choi SW, Yang EK, Lee WJ. Lignan from safflower seeds induces apoptosis in human promyelocytic leukemia cells. Nutraceuticals Food. 2003;8(2):113–118.
[362] Bylund A, Saarinen N, Zhang JX, Bergh A, Widmark A, Johansson A, et al. Anticancer effects of a plant lignan 7-hydroxymatairesinol on a prostate cancer model *in vivo*. Exp Biol Med. 2005;230(3):217–223.
[363] McCann MJ, Gill CIR, McGlynn H, Rowland IR. Role of mammalian lignans in the prevention and treatment of Prostate cancer. Nutr Cancer. 2005;52(1):1–14.
[364] Niemeyer HB, Metzler M. Antioxidant activities of lignans in the FRAP assay (ferric reducing/antioxidant power assay). Spec Publ - R Soc Chem. 2001;269(Biologically-Active Phytochemicals in Food):394–395.
[365] Umehara K, Sugawa A, Kuroyanagi M, Ueno A, Taki T. Studies on differentiation-inducers from *Arctium Fructus*. Chem Pharm Bull. 1993;41(10):1774–1779.
[366] Huang DM, Guh JH, Chueh SC, Teng CM. Modulation of anti-adhesion molecule MUC-1 is associated with arctiin-induced growth inhibition in PC-3 cells. Prostate. 2004;59(3):260–267.
[367] Umehara K, Nakamura M, Miyase T, Kuroyanagi M, Uneo A. Studies on differentiation inducers. VI. Lignan derivatives from *Arctium Fructus*. (2). Chem Pharm Bull. 1996;44(12):2300–2304.
[368] Cho MK, Jang YP, Kim YC, Kim SG. Arctigenin, a phenylpropanoid dibenzylbutyrolactone lignan, inhibits MAP kinases and AP-1 activation *via* potent MKK inhibition: the role in TNF-α inhibition. Int Immunopharmacol. 2004;4(10-11):1419–1429.
[369] Cho JY, Park J, Kim PS, Yoo ES, Baik KU, Park MH. Savinin, a lignan from pterocarpus santalinus inhibits tumor necrosis factor-α–production and T cell proliferation. Biol Pharm Bull. 2001;24(2):167–171.

[370] Kim SH, Jang YP, Sung SH, Kim CJ, Kim JW, Kim YC. Hepatoprotective dibenzylbutyrolactone lignans of torreya nucifera against ccl4-induced toxicity in primary cultured rat hepatocytes. Biol Pharm Bull. 2003;26(8):1202–1205.

[371] Trumm S, Eich E. Cytostatic activities of lignanolides from *Ipomoea cairica*. Planta Med. 1989;55(7):658–659.

[372] Fujimoto T, Nose M, Takeda T, Ogihara Y, Nishibe S, Minami M. Studies on the Chinese crude drug Luoshiteng (II). On the biologically active components in the stem part of luoshiteng originating from *Trachelospermum jasminoides*. Shoyakugaku Zasshi. 1992;46(3): 224–229.

[373] Matsuda H, Kawaguchi Y, Yamazaki M, Hirata N, Naruto S, Asanuma Y, et al. Melanogenesis stimulation in murine B16 melanoma cells by *Piper nigrum* leaf extract and its lignan constituents. Biol Pharm Bull. 2004;27(10):1611–1616.

[374] Sachin BS, Koul M, Zutshi A, Singh SK, Tikoo AK, Tikoo MK, et al. Simultaneous high-performance liquid chromatographic determination of cedrus deodara active constituents and their pharmacokinetic profile in mice. J Chromatogr B Analyt Technol Biomed Life Sci. 2008;862(1-2):237–241.

[375] Pomeranz C. Über das cubebin. Monatsh Chem. 1887;8(1):466–470.

[376] Bogert MT, Powell G. The synthesis of simple and of substituted 2-alkylcinnamic alcohols, including a monomolecular cubebin. J Am Chem Soc. 1931;53(4):1605–1609.

[377] Haworth RD, Kelly W. The constituents of natural phenolic resins. Part VIII. Lariciresinol, cubebin, and some stereochemical relationships. J Chem Soc. 1937:384–391.

[378] Hänsel R, Schulz G, Rimpler H. Notice on the structure of cubebin. Arch Pharm Ber Dtsch Pharm Ges. 1967;300(6):559–560.

[379] Batterbee JE, Burden RS, Crombie L, Whiting DA. Chemistry and synthesis of the lignan (−)-cubebin. J Chem Soc C. 1969;(19):2470–2477.

[380] Kawamura F, Ohara S, Nishida A. Antifungal activity of constituents from the heartwood of *Gmelina arborea*: Part 1. Sensitive antifungal assay against Basidiomycetes. Holzforschung. 2004;58(2):189–192.

[381] de Souza VA, da Silva R, Pereira AC, Royo VdA, Saraiva J, Montanheiro M, et al. Trypanocidal activity of (−)-cubebin derivatives against free amastigote forms of *Trypanosoma cruzi*. Bioorg Med Chem Lett. 2005;15(2):303–307.

[382] Molfetta FA, Bruni AT, Honorio KM, da Silva ABF. A structure-activity relationship study of quinone compounds with trypanocidal activity. Eur J Med Chem. 2005;40(4):329–338.

[383] Takeara R, Albuquerque S, Lopes NP, Lopes JLC. Trypanocidal activity of *Lychnophora staavioides* mart. (vernonieae, asteraceae). Phytomedicine. 2003;10(6-7):490–493.

[384] Silva MLA, Coimbra HS, Pereira AC, Almeida VA, Lima TC, Costa ES, et al. Evaluation of *Piper cubeba* extract, (−)-cubebin and its semi-synthetic derivatives against oral pathogens. Phytother Res. 2007;21(5):420–422.

[385] Tsuruga T, Ebizuka Y, Nakajima J, Chun YT, Noguchi H, Iitaka Y, et al. Biologically active constituents of *Magnolia salicifolia*: inhibitors of induced histamine release from rat mast cells. Chem Pharm Bull. 1991;39(12):3265–3271.

[386] Messiano GB, Santos RAdS, Ferreira LDS, Simoes RA, Jabor VAP, Kato MJ, et al. In vitro metabolism study of the promising anticancer agent the lignan (−)-grandisin. J Pharm Biomed Anal. 2013;72:240–244.

[387] Hwang SB, Lam MH, Biftu T, Beattie TR, Shen TY. *trans*-2,5-Bis-(3,4,5-trimethoxyphenyl) tetrahydrofuran. An orally active specific and competitive receptor antagonist of platelet activating factor. J Biol Chem. 1985;260(29):15639–15645.

[388] Wan HB, Yuu DC, Riyan ST, Watanabe S, Tamai M, Okuyama S, Kitsukawa S, Omura S. et al. Lignane compound [patent]; 1989. JPH01242596.

[389] Wan HB, Yuu DC, Riyan ST, Watanabe S, Tamai M, Okuyama S, Kitsukawa S, Omura S. *et al.* Lignane compound [patent]; 1989. JPH01290693.
[390] Kuriyama K, Murai T. Inhibition effects of lignan glucosides on lipid peroxidation. Nippon Nogeikagaku Kaishi. 1996;70(2):161–167.
[391] Kuriyama K, Murui T. Scavenging of hydroxy radicals by lignan glucosides in germinated sesame seeds. Nippon Nogeikagaku Kaishi. 1995;69(6):703–705.
[392] Chen JJ, Chou ET, Duh CY, Yang SZ, Chen IS. New cytotoxic tetrahydrofuran- and dihydrofuran-type lignans from the stem of *Beilschmiedia tsangii*. Planta Med. 2006;72(4):351–357.
[393] Konishi T, Konoshima T, Daikonya A, Kitanaka S. Neolignans from *Piper futokadsura* and their inhibition of nitric oxide production. Chem Pharm Bull. 2005;53(1):121–124.
[394] Sadhu SK, Okuyama E, Fujimoto H, Ishibashi M, Yesilada E. Prostaglandin inhibitory and antioxidant components of *Cistus laurifolius*, a Turkish medicinal plant. J Ethnopharmacol. 2006;108(3):371–378.
[395] Kuroyanagi M, Ikeda R, Gao HY, Muto N, Otaki K, Sano T, *et al.* Neurite outgrowth-promoting active constituents of the japanese cypress (*Chamaecyparis obtusa*). Chem Pharm Bull. 2008;56(1):60–63.
[396] Vuckovic I, Trajkovic V, Macura S, Tesevic V, Janackovic P, Milosavljevic S. A novel cytotoxic lignan from *Seseli annuum* L. Phytother Res. 2007;21(8):790–792.
[397] Piao XL, Jang MH, Cui J, Piao X. Lignans from the fruits of *Forsythia suspensa*. Bioorg Med Chem Lett. 2008;18(6):1980–1984.
[398] Mehmood S, Riaz N, Ahmad Z, Afza N, Malik A. Lipoxygenase inhibitory lignans from *Salvia santolinifolia*. Pol J Chem. 2008;82(3):571–575.
[399] Hirano H, Tokuhira T, Yokoi T, Shingu T. Isolation of free radical scavenger from *Coptidis rhizoma*. Nat Med. 1997;51(6):539–540.
[400] Ma J, Dey M, Yang H, Poulev A, Pouleva R, Dorn R, *et al.* Anti-inflammatory and immunosuppressive compounds from *Tripterygium wilfordii*. Phytochemistry. 2007;68(8):1172–1178.
[401] Lee C, Kim H, Kho Y. Agastinol and agastenol, novel lignans from *Agastache rugosa* and their evaluation in an apoptosis inhibition assay. J Nat Prod. 2002;65(3):414–416.
[402] Lee WS, Lee DW, Baek YI, An SJ, Cho KH, Choi YK, *et al.* Human ACAT-1 and -2 inhibitory activities of saucerneol B, manassantin A and B isolated from *Saururus chinensis*. Bioorg Med Chem Lett. 2004;14(12):3109–3112.
[403] Hwang BY, Lee JH, Nam JB, Hong YS, Lee JJ. Lignans from *Saururus chinensis* inhibiting the transcription factor NF – κB. Phytochemistry. 2003;64(3):765–771.
[404] Jung KY, Kim DS, Oh SR, Park SH, Lee IS, Lee JJ, *et al.* Magnone A and B, novel anti-PAF tetrahydrofuran lignans from the flower buds of *Magnolia fargesii*. J Nat Prod. 1998;61(6):808–811.
[405] Chae SH, Kim PS, Cho JY, Park JS, Lee JH, Yoo ES, *et al.* Isolation and identification of inhibitory compounds on TNF-α production from *Magnolia fargesii*. Arch Pharmacal Res. 1998;21(1):67–69.
[406] Cole JR, Bianchi E, Trumbull ER. Antitumor agents from *Bursera microphylla*. II. Isolation of a new lignan: burseran. J Pharm Sci. 1969;58(2):175–176.
[407] Chattopadhyay SK, Kumar TRS, Maulik PR, Srivastava S, Garg A, Sharon A, *et al.* Absolute configuration and anticancer activity of taxiresinol and related lignans of *Taxus wallichiana*. Bioorg Med Chem. 2003;11(23):4945–4948.
[408] Cho JY, Kim AR, Park MH. Lignans from the rhizomes of *Coptis japonica* differentially act as anti-inflammatory principles. Planta Med. 2001;67(4):312–316.
[409] Yoshikawa K, Kinoshita H, Kan Y, Arihara S. Neolignans and phenylpropanoids from the rhizomes of *Coptis japonica* dissecta. Chem Pharm Bull. 1995;43(4):578–581.

[410] Cho JY, Park J, Yoo ES, Yoshikawa K, Baik KU, Lee J, et al. Inhibitory effect of lignans from the rhizomes of *Coptis japonica* var dissecta on tumor necrosis factor-α–production in lipopolysaccharide-stimulated RAW264.7 cells. Arch Pharmacal Res. 1998;21(1):12–16.

[411] Duh CY, Phoebe CHJ, Pezzuto JM, Kinghorn AD, Farnsworth NR. Plant anticancer agents, XLII. Cytotoxic constituents from *Wikstroemia elliptica*. J Nat Prod. 1986;49(4):706–709.

[412] Castro-Faria-Neto HC, Martins MA, Silva PMR, Bozza PT, Cruz HN, de Queiroz-Paulo M, et al. Pharmacological profile of epiyangambin: a furofuran lignan with PAF antagonist activity. J Lipid Mediat. 1993;7(1):1–9.

[413] Tibiriçá E. Cardiovascular properties of yangambin, a lignan isolated from brazilian plants. Cardiovasc Drug Rev. 2001;19(4):313–328.

[414] Chen CC, Huang YL, Chen HT, Chen YP, Hsu HY. On the calcium-antagonistic principles of the flower buds of *Magnolia fargesii*. Planta Med. 1988;54(5):438–440.

[415] Matsunaga K, Shibuya M, Ohizumi Y. Graminone B, a novel lignan with vasodilative activity from *Imperata cylindrica*. J Nat Prod. 1994;57(12):1734–1736.

[416] Min BS, Cui HS, Lee HK, Sok DE, Kim MR. A new furofuran lignan with antioxidant and antiseizure activities from the leaves of *Petasites japonicus*. Arch Pharmacal Res. 2005;28(9):1023–1026.

[417] Yang XW, Zhao PJ, Ma YL, Xiao HT, Zuo YQ, He HP, et al. Mixed lignan-neolignans from *Tarenna attenuata*. J Nat Prod. 2007;70(4):521–525.

[418] Sultana N, ur Rahman A, Jahan S. Studies on the constituents of *Commiphora mukul*. Z Naturforsch, B: J Chem Sci. 2005;60(11):1202–1206.

[419] Murakami A, Ohigashi H, Koshimizu K. Antitumor promotion with food phytochemicals: a strategy for cancer chemoprevention. Biosci Biotechnol Biochem. 1996;60(1):1–8.

[420] Zhao C, Qiu R, Zheng R. In vitro antitumor activities of furan lignans. Lanzhou Daxue Xuebao, Ziran Kexueban. 2000;36(4):66–68.

[421] De Leon EJ, Olmedo DA, Solis PN, Gupta MP, Terencio MC. Diayangambin exerts immunosuppressive and anti-inflammatory effects *in vitro* and *in vivo*. Planta Med. 2002;68(12):1128–1131.

[422] Ahn KS, Jung KY, Kim JH, Oh SR, Lee HK. Inhibitory activity of lignan components from the flower buds of *Magnolia fargesii* on the expression of cell adhesion molecules. Biol Pharm Bull. 2001;24(9):1085–1087.

[423] Diaz Lanza AM, Abad Martinez MJA, Matellano L, Fernandez Carretero CR, Villaescusa Castillo L, Silvan Sen AM, et al. Lignan and phenylpropanoid glycosides from *Phillyrea latifolia* and their *in vitro* anti-inflammatory activity. Planta Med. 2001;67(3):219–223.

[424] Piccinelli AL, Arana S, Caceres A, d'Emmanuele di Villa Bianca R, Sorrentino R, Rastrelli L. New lignans from the roots of *Valeriana prionophylla* with antioxidative and vasorelaxant activities. J Nat Prod. 2004;67(7):1135–1140.

[425] Prasad AK, Kumar V, Arya P, Kumar S, Dabur R, Singh N, et al. Investigations toward new lead compounds from medicinally important plants. Pure Appl Chem. 2005;77(1):25–40.

[426] Kamikado T, Chang CF, Murakoshi S, Sakurai A, Tamura S. Isolation and structure elucidation of growth inhibitors on silkworm larvae from *Magnolia kobus* DC. Agric Biol Chem. 1975;39(4):833–836.

[427] Ghosal S, Banerjee S, Jaiswal DK. Chemical constituents of justica. Part 2. New furofurano lignans from *Justicia simplex*. Phytochemistry. 1980;19(2):332–334.

[428] Ghosal S, Banerjee S, Srivastava RS. Simplexolin, a new lignan from *Justicia simplex*. Phytochemistry. 1979;18(3):503–505.

[429] Hevey M, Donehower LA. Complementation of human immunodeficiency virus type 1 vif mutants in some CD4+ T-cell lines. Virus Res. 1994;33(3):269–280.

[430] Katsuzaki H, Kawakishi S, Osawa T. Structure of novel antioxidative lignan triglucoside isolated from sesame seed. Heterocycles. 1993;36(5):933–936.

[431] Sih CJ, Ravikumar PR, Huang FC, Buckner C, Whitlock H Jr. Isolation and synthesis of pinoresinol diglucoside, a major antihypertensive principle of tu-chung (*Eucommia ulmoides*, oliver). J Am Chem Soc. 1976;98(17):5412–5413.

[432] Kikuzaki H, Kayano S, Fukutsuka N, Aoki A, Kasamatsu K, Yamasaki Y, *et al.* Abscisic acid-related compounds and lignans in prunes (*Prunus domestica L.*) and their oxygen radical absorbance capacity (ORAC). J Agric Food Chem. 2004;52(2):344–349.

[433] Duan H, Takaishi Y, Momota H, Ohmoto Y, Taki T. Immunosuppressive constituents from *Saussurea medusa*. Phytochemistry. 2002;59(1):85–90.

[434] Kita M, Ohmoto Y, Hirai Y, Yamaguchi N, Imanishi J. Induction of cytokines in human peripheral blood mononuclear cells by mycoplasmas. Microbiol Immunol. 1992;36(5):507–516.

[435] Okuyama E, Suzumura K, Yamazaki M. Pharmacologically active components of todopon puok (*Fagraea racemosa*), a medicinal plant from Borneo. Chem Pharm Bull. 1995;43(12):2200–2204.

[436] Li YS, Wang ZT, Zhang M, Luo SD, Chen JJ. A new pinoresinol-type lignan from *Ligularia kanaitizensis*. Nat Prod Res. 2005;19(2):125–129.

[437] Gu JQ, Wang Y, Franzblau SG, Montenegro G, Yang D, Timmermann BN. Antitubercular constituents of valeriana laxiflora. Planta Med. 2004;70(6):509–514.

[438] Nikaido T, Ohmoto T, Kinoshita T, Sankawa U, Nishibe S, Hisada S. Inhibitors of cyclic AMP phosphodiesterase in medicinal plants. II. Inhibition of cyclic AMP phosphodiesterase by lignans. Chem Pharm Bull. 1981;29(12):3586–3592.

[439] Deyama T, Nishibe S, Kitagawa S, Ogihara Y, Takeda T, Omoto T, *et al.* Inhibitors of cyclic amp phosphodiesterase in medicinal plants. XIV. Inhibition of adenosine 3′,5′-cyclic monophosphate phosphodiesterase by lignan glucosides of *Eucommia bark*. Chem Pharm Bull. 1988;36(1):435–439.

[440] Oshima Y, Takata S, Hikino H, Deyama T, Kinoshita G. Anticomplementary activity of the constituents of *Eucommia ulmoides* bark. J Ethnopharmacol. 1988;23(2–3):159–164.

[441] Wu YC, Chang GY, Ko FN, Teng CM. Bioactive constituents from the stems of *Annona montana*. Planta Med. 1995;61(2):146–149.

[442] Kaneda N, Chai H, Pezzuto JM, Kinghorn AD, Farnsworth NR, Tuchinda P, *et al.* Cytotoxic activity of cardenolides from *Beaumontia brevituba* stems. Planta Med. 1992;58(5):429–431.

[443] Kardono LBS, Tsauri S, Padmawinata K, Pezzuto JM, Kinghorn AD. Studies on indonesian medicinal plants. Part II. Cytotoxic constituents of the bark of *Plumeria rubra* collected in Indonesia. J Nat Prod. 1990;53(6):1447–1455.

[444] Bowers WS. Juvenile hormone: activity of natural and synthetic synergists. Science. 1968;161(3844):895–897.

[445] Brekhman II, Dardymov IV. Pharmacological investigation of glycosides from ginseng and *Eleutherococcus*. Lloydia. 1969;32(1):46–51.

[446] Deyama T. The constituents of *Eucommia ulmoides* oliv. I. Isolation of (+)-medioresinol di-O-β-D-glucopyranoside. Chem Pharm Bull. 1983;31(9):2993–2997.

[447] Nishibe S, Kinoshita H, Takeda H, Okano G. Phenolic compounds from stem bark of *Acanthopanax senticosus* and their pharmacological effect in chronic swimming stressed rats. Chem Pharm Bull. 1990;38(6):1763–1765.

[448] Darias V, Bravo L, Rabanal R, Sanchez-Mateo CC, Martin-Herrera DA. Cytostatic and antibacterial activity of some compounds isolated from several lamiaceae species from the Canary Islands. Planta Med. 1990;56(1):70–72.

[449] Hirose N, Doi F, Ueki T, Akazawa K, Chijiiwa K, Sugano M, *et al.* Suppressive effect of sesamin against 7,12-dimethylbenz[a]anthracene induced rat mammary carcinogenesis. Anticancer Res. 1992;12(4):1259–1265.

[450] Wu JHY, Hodgson JM, Clarke MW, Indrawan AP, Barden AE, Puddey IB, et al. Inhibition of 20-hydroxyeicosatetraenoic acid synthesis using specific plant lignans: *in vitro* and human studies. Hypertension. 2009;54(5):1151–1158.

[451] Kiso Y. Antioxidative roles of sesamin, a functional lignan in sesame seed, and its effect on lipid- and alcohol-metabolism in the liver: a DNA microarray study. Biofactors. 2004;21(1-4):191–196.

[452] Fukuda Y, Nagata M, Osawa T, Namiki M. Contribution of lignan analogs to antioxidative activity of refined unroasted sesame seed oil. J Am Oil Chem Soc. 1986;63(8):1027–1031.

[453] Fukuda Y, Namiki M. Recent studies on sesame seed and oil. Nippon Shokuhin Kogyo Gakkaishi. 1988;35(8):552–562.

[454] Katsuzaki H, Kawakishi S, Osawa T. Sesaminol glucosides in sesame seeds. Phytochemistry. 1994;35(3):773–776.

[455] Miyake Y, Fukumoto S, Okada M, Sakaida K, Nakamura Y, Osawa T. Antioxidative catechol lignans converted from sesamin and sesaminol triglucoside by culturing with Aspergillus. J Agric Food Chem. 2005;53(1):22–27.

[456] Osawa T, Nagata M, Namiki M, Fukuda Y. Sesamolinol, a novel antioxidant isolated from sesame seeds. Agric Biol Chem. 1985;49(11):3351–3352.

[457] Cui G, Duan H, Ji L. Antioxidant furofuran lignans identified from chinese propolis. Shipin Kexue. 2002;23(12):117–120.

[458] San Feliciano A, Medarde M, Pelaez Lamamie de Clairac R, Lopez JL, Puebla P, Garcia Gravalos MD, et al. Synthesis and biological activity of bromolignans and cyclolignans. Arch Pharm. 1993;326(7):421–426.

[459] San Feliciano A, Gordaliza M, del Corral JMM, Castro MA, Garcia-Gravalos MD, Ruiz-Lazaro P. Antineoplastic and antiviral activities of some cyclolignans. Planta Med. 1993;59(3):246–249.

[460] Gordaliza M, Gastro MA, Garcia-Gravalos MD, Ruiz P, del Corral JMM, San Feliciano A. Antineoplastic and antiviral activities of podophyllotoxin related lignans. Arch Pharm. 1994;327(3):175–179.

[461] Schöttner M, Gansser D, Spiteller G. Lignans from the roots of *Urtica dioica* and their metabolites bind to human sex hormone binding globulin (SHBG). Planta Med. 1997;63(6):529–532.

[462] Kuepeli E, Erdemoglu N, Yesilada E, Sener B. Anti-inflammatory and antinociceptive activity of taxoids and lignans from the heartwood of *Taxus baccata* L. J Ethnopharmacol. 2003;89(2-3):265–270.

[463] Pullela SV, Takamatsu S, Khan SI, Khan IA. Isolation of lignans and biological activity studies of *Ephedra viridis*. Planta Med. 2005;71(8):789–791.

[464] Takamatsu S, Galal AM, Ross SA, Ferreira D, El-Sohly MA, Ibrahim ARS, et al. Antioxidant effect of flavonoids on DCF production in HL-60 cells. Phytother Res. 2003;17(8):963–966.

[465] Hasegawa M, Shirato T. Abnormal constituents of prunus wood. isoolivil from *P. jamasakura* wood. Nippon Rin Gakkaishi. 1959;41:1–3.

[466] Yang Kuo LM, Zhang LJ, Huang HT, Lin ZH, Liaw CC, Cheng HL, et al. Antioxidant lignans and chromone glycosides from *Eurya japonica*. J Nat Prod. 2013;76(4):580–587.

[467] Konuklugil B. Lignans with anticancer activity. Ankara Univ Eczacilik Fak Derg. 1994;23(1-2):64–75.

[468] ur Rehman A, Malik A, Riaz N, Ahmad H, Nawaz SA, Choudhary MI. Lipoxygenase inhibiting constituents from *Indigofera herantha*. Chem Pharm Bull. 2005;53(3):263–266.

[469] Li HR, Feng YL, Yang ZG, Wang J, Daikonya A, Kitanaka S, et al. New lignans from *Kadsura coccinea* and their nitric oxide inhibitory activities. Chem Pharm Bull. 2006;54(7):1022–1025.

[470] Li H, Wang L, Yang Z, Kitanaka S. Kadsuralignans H-K from *Kadsura coccinea* and their nitric oxide production inhibitory effects. J Nat Prod. 2007;70(12):1999–2002.

[471] Ohashi K, Winarno H, Mukai M, Inoue M, Prana MS, Simanjuntak P, *et al*. Indonesian medicinal plants. XXV. Cancer cell invasion inhibitory effects of chemical constituents in the parasitic plant *Scurrula atropurpurea* (loranthaceae). Chem Pharm Bull. 2003;51(3):343–345.

[472] Thongphasuk P, Suttisri R, Bavovada R, Verpoorte R. Antioxidant lignan glucosides from *Strychnos vanprukii*. Fitoterapia. 2004;75(7-8):623–628.

[473] Favela-Hernández JMJ, García A, Garza-González E, Rivas-Galindo VM, Camacho-Corona MR. Antibacterial and antimycobacterial lignans and flavonoids from *Larrea tridentata*. Phytother Res. 2012;26(12):1957–1960.

[474] Rajasekhar D, Subbaraju GV. Jusmicranthin, a new arylnaphthalide lignan from *Justicia neesii*. Fitoterapia. 2000;71(5):598–599.

[475] Fukamiya N, Lee KH. Antitumor agents. 81. Justicidin A and diphyllin, two cytotoxic principles from *Justicia procumbens*. J Nat Prod. 1986;49(2):348–350.

[476] Hui YH, Chang CJ, McLaughlin JL, Powell RG. Justicidin B, a bioactive trace lignan from the seeds of *Sesbania drummondii*. J Nat Prod. 1986;49(6):1175–1176.

[477] Vasilev NP, Ionkova I. Cytotoxic activity of extracts from Linum cell cultures. Fitoterapia. 2005;76(1):50–53.

[478] Asano J, Chiba K, Tada M, Yoshi T. Antiviral activity of lignans and their glycosides from *Justicia procumbens*. Phytochemistry. 1996;42(3):713–717.

[479] Bedoya LM, Alvarez A, Bermejo M, Gonzalez N, Beltran M, Sanchez-Palomino S, *et al*. Guatemalan plants extracts as virucides against HIV-1 infection. Phytomedicine. 2008;15(6-7):520–524.

[480] Ghosal S, Srivastava AK, Srivastava RS, Chattopadhyay S, Maitra M. Chemical constituents of Justicia. Part 4. Justicisaponin-I, a new triterpenoid saponin from *Justicia simplex*. Planta Med. 1981;42(3):279–283.

[481] Chen CC, Hsin WC, Ko FN, Huang YL, Ou JC, Teng CM. Antiplatelet arylnaphthalide lignans from *Justicia procumbens*. J Nat Prod. 1996;59(12):1149–1150.

[482] Navarro E, Alonso SJ, Trujillo J, Jorge E, Perez C. Central nervous activity of elenoside. Phytomedicine. 2004;11(6):498–503.

[483] Day SH, Chiu NY, Tsao LT, Wang JP, Lin CN. New lignan glycosides with potent antiinflammatory effect, isolated from *Justicia ciliata*. J Nat Prod. 2000;63(11):1560–1562.

[484] Sanmugapriya E, Shanmugasundaram P, Venkataraman S. Anti-inflammatory activity of *Justicia prostrata* gamble in acute and sub-acute models of inflammation. Inflammopharmacology. 2005;13(5-6):493–500.

[485] Correa GM, Alcantara AFdC. Chemical constituents and biological activities of species of Justicia - a review. Rev Bras Farmacogn. 2012;22(1):220–238.

[486] Chang ST, Wang DSY, Wu CL, Shiah SG, Kuo YH, Chang CJ. Cytotoxicity of extractives from *Taiwania cryptomerioides* heartwood. Phytochemistry. 2000;55(3):227–232.

[487] Tseng YP, Kuo YH, Hu CP, Jeng KS, Janmanchi D, Lin CH, *et al*. The role of helioxanthin in inhibiting human hepatitis B viral replication and gene expression by interfering with the host transcriptional machinery of viral promoters. Antiviral Res. 2008;77(3):206–214.

[488] Kavitha J, Gopalaiah K, Rajasekhar D, Subbaraju GV. Juspurpurin, an unusual secolignan glycoside from *Justicia purpurea*. J Nat Prod. 2003;66(8):1113–1115.

[489] Day SH, Chiu NY, Won SJ, Lin CN. Cytotoxic lignans of *Justicia ciliata*. J Nat Prod. 1999;62(7):1056–1058.

[490] Therien M, Fitzsimmons BJ, Scheigetz J, Macdonald D, Choo LY, Guay J, *et al*. Justicidin E: a new leukotriene biosynthesis inhibitor. Bioorg Med Chem Lett. 1993;3(10):2063–2066.

[491] Wu CM, Wu SC, Chung WJ, Lin HC, Chen KT, Chen YC, *et al*. Antiplatelet effect and selective binding to cyclooxygenase (COX) by molecular docking analysis of flavonoids and lignans. Int J Mol Sci. 2007;8(8):830–841.

[492] Baba A, Kawamura N, Makino H, Ohta Y, Taketomi S, Sohda T. Studies on disease-modifying antirheumatic drugs: synthesis of novel quinoline and quinazoline derivatives and their anti-inflammatory effect. J Med Chem. 1996;39(26):5176–5182.
[493] Gertsch J, Tobler RT, Brun R, Sticher O, Heilmann J. Antifungal, antiprotozoal, cytotoxic and piscicidal properties of justicidin b and a new arylnaphthalide lignan from *Phyllanthus piscatorum*. Planta Med. 2003;69(5):420–424.
[494] Rao YK, Fang SH, Tzeng YM. Anti-inflammatory activities of constituents isolated from *Phyllanthus polyphyllus*. J Ethnopharmacol. 2006;103(2):181–186.
[495] Kaur K, Jain M, Kaur T, Jain R. Antimalarials from nature. Bioorg Med Chem. 2009;17(9):3229–3256.
[496] Wang CLJ, Ripka WC. Total synthesis of (±)-justicidin P. A new lignan lactone from *Justicia extensa*. J Org Chem. 1983;48(15):2555–2557.
[497] Munakata K, Marumo S, Ota K, Chen YL. Justicidin A and B, fish-killing components of *Justicia hayatai*. Tetrahedron Lett. 1965;(47):4167–4170.
[498] Su CL, Huang LLH, Huang LM, Lee JC, Lin CN, Won SJ. Caspase-8 acts as a key upstream executor of mitochondria during justicidin A-induced apoptosis in human hepatoma cells. FEBS Lett. 2006;580(13):3185–3191.
[499] Williams RB, Hoch J, Glass TE, Evans R, Miller JS, Wisse JH, et al. A novel cytotoxic guttiferone analogue from *Garcinia macrophylla* from the *Surinam rainforest*. Planta Med. 2003;69(9):864–866.
[500] Pradheepkumar CP, Panneerselvam N, Shanmugam G. Cleistanthin A causes DNA strand breaks and induces apoptosis in cultured cells. Mutat Res, Fundam Mol Mech Mutagen. 2000;464(2):185–193.
[501] Susplugas S, Nguyen VH, Bignon J, Thoison O, Kruczynski A, Sevenet T, et al. Cytotoxic arylnaphthalene lignans from a Vietnamese *Acanthaceae*, *Justicia patentiflora*. J Nat Prod. 2005;68(5):734–738.
[502] Lu YH, Wei BL, Ko HH, Lin CN. DNA strand-scission by phloroglucinols and lignans from heartwood of *Garcinia subelliptica* merr. and Justicia plants. Phytochemistry. 2008;69(1):225–233.
[503] Navarro E, Alonso SJ, Navarro R, Trujillo J, Jorge E. Elenoside increases intestinal motility. World J Gastroenterol. 2006;12(44):7143–7148.
[504] Navarro E, Alonso SJ, Trujillo J, Jorge E, Perez C. General behavior, toxicity, and cytotoxic activity of elenoside, a lignan from *Justicia hyssopifolia*. J Nat Prod. 2001;64(1):134–135.
[505] Navarro E, Alonso SJ, Alonso PJ, Trujillo J, Jorge E, Perez C. Pharmacological effects of elenoside, an arylnaphthalene lignan. Biol Pharm Bull. 2001;24(3):254–258.
[506] Day SH, Lin YC, Tsai ML, Tsao LT, Ko HH, Chung MI, et al. Potent cytotoxic lignans from justicia procumbens and their effects on nitric oxide and tumor necrosis factor-α production in mouse macrophages. J Nat Prod. 2002;65(3):379–381.
[507] Weng JR, Ko HH, Yeh TL, Lin HC, Lin CN. Two new arylnaphthalide lignans and antiplatelet constituents from *Justicia procumbens*. Arch Pharm. 2004;337(4):207–212.
[508] Chung BH, Lee JJ, Kim JD, Jeoung D, Lee H, Choe J, et al. Angiogenic activity of sesamin through the activation of multiple signal pathways. Biochem Biophys Res Commun. 2010;391(1):254–260.
[509] Meckes M, David-Rivera AD, Nava-Aguilar V, Jimenez A. Activity of some Mexican medicinal plant extracts on carrageenan-induced rat paw edema. Phytomedicine. 2004;11(5):446–451.
[510] Cow C, Leung C, Charlton JL. Antiviral activity of arylnaphthalene and aryldihydronaphthalene lignans. Can J Chem. 2000;78(5):553–561.
[511] Pettit GR, Schaufelberger DE. Isolation and structure of the cytostatic lignan glycoside phyllanthostatin A. J Nat Prod. 1988;51(6):1104–1112.

[512] Iwasaki T, Kondo K, Nishitani T, Kuroda T, Hirakoso K, Ohtani A, et al. Arylnaphthalene lignans as novel series of hypolipidemic agents raising high-density lipoprotein level. Chem Pharm Bull. 1995;43(10):1701–1705.

[513] Kuroda T, Kondo K, Iwasaki T, Ohtani A, Takashima K. Synthesis and hypolipidemic activity of diesters of arylnaphthalene lignan and their heteroaromatic analogs. Chem Pharm Bull. 1997;45(4):678–684.

[514] Prieto JM, Recio MC, Giner RM, Manez S, Massmanian A, Waterman PG, et al. Topical anti-inflammatory lignans from *Haplophyllum hispanicum*. Z Naturforsch, C: J Biosci. 1996;51(9-10):618–622.

[515] Prieto JM, Giner RM, Recio MC, Schinella G, Manez S, Rios JL. Diphyllin acetylapioside, a 5-lipoxygenase inhibitor from *Haplophyllum hispanicum*. Planta Med. 2002;68(4):359–360.

[516] Gan LS, Yang SP, Fan CQ, Yue JM. Lignans and their degraded derivatives from *Sarcostemma acidum*. J Nat Prod. 2005;68(2):221–225.

[517] Wu SJ, Wu TS. Cytotoxic arylnaphthalene lignans from *Phyllanthus oligospermus*. Chem Pharm Bull. 2006;54(8):1223–1225.

[518] Tuchinda P, Kumkao A, Pohmakotr M, Sophasan S, Santisuk T, Reutrakul V. Cytotoxic aryl-naphthalide lignan glycosides from the aerial parts of *Phyllanthus taxodiifolius*. Planta Med. 2006;72(1):60–62.

[519] Yamasaki T, Kawabata T, Masuoka C, Kinjo J, Ikeda T, Nohara T, et al. Two new lignan glucosides from the fruit of *Vitex cannabifolia*. J Nat Med. 2008;62(1):47–51.

[520] Chang CW, Lin MT, Lee SS, Liu KCSC, Hsu FL, Lin JY. Differential inhibition of reverse transcriptase and cellular DNA polymerase-α activities by lignans isolated from Chinese herbs, *Phyllanthus myrtifolius* Moon, and tannins from *Lonicera japonica* Thunb and *Castanopsis hystrix*. Antiviral Res. 1995;27(4):367–374.

[521] Ghosal S, Banerjee S, Frahm AW. Chemical constituents of justicia. III. Prostalidins A, B, C and retrochinensin: a new antidepressant: 4-aryl-2,3-naphthalide lignans from *Justicia prostata*. Chem Ind. 1979;(23):854–855.

[522] Kimura M, Suzuki J, Yamada T, Yoshizaki M, Kikuchi T, Kadota S, et al. Anti-inflammatory effect of neolignans newly isolated from the crude drug 'shin-i' (*Flos magnoliae*). Planta Med. 1985;51(4):291–293.

[523] Kadota S, Tsubono K, Makino K, Takeshita M, Kikuchi T. Convenient synthesis of magnoshinin, an anti-inflammatory neolignan. Tetrahedron Lett. 1987;28(25):2857–2860.

[524] Chambers HF, Sachdeva M. Binding of β-lactam antibiotics to penicillin-binding proteins in methicillin-resistant *Staphylococcus aureus*. J Infect Dis. 1990;161(6):1170–1176.

[525] Ono M, Nishida Y, Masuoka C, Li JC, Okawa M, Ikeda T, et al. Lignan derivatives and a norditerpene from the seeds of *Vitex negundo*. J Nat Prod. 2004;67(12):2073–2075.

[526] Delorme D, Ducharme Y, Brideau C, Chan CC, Chauret N, Desmarais S, et al. Dioxabicyclooctanyl naphthalenenitriles as nonredox 5-lipoxygenase inhibitors: structure-activity relationship study directed toward the improvement of metabolic stability. J Med Chem. 1996;39(20):3951–3970.

[527] Iwasaki T, Kondo K, Kuroda T, Moritani Y, Yamagata S, Sugiura M, et al. Novel selective PDE IV inhibitors as antiasthmatic agents. Synthesis and biological activities of a series of 1-aryl-2,3-bis(hydroxymethyl)naphthalene lignans. J Med Chem. 1996;39(14):2696–2704.

[528] Gonzalez AG, Darias V, Alonso G. Cytostatic lignans isolated from *Haplophyllum hispanicum*. Planta Med. 1979;36(3):200–203.

[529] Liu KCSC, Lee SS, Lin MT, Chang CW, Liu CL, Lin JY, et al. Lignans and tannins as inhibitors of viral reverse transcriptase and human DNA polymerase-α: QSAR analysis and molecular modeling. Med Chem Res. 1997;7(3):168–179.

[530] Kashiwada Y, Nishizawa M, Yamagishi T, Tanaka T, Nonaka GI, Cosentino LM, et al. Anti-AIDS agents, 18. Sodium and potassium salts of caffeic acid tetramers from *Arnebia euchroma* as anti-HIV agents. J Nat Prod. 1995;58(3):392–400.

[531] Kashiwada Y, Bastow KF, Lee KH. Novel lignan derivatives as selective inhibitors of DNA topoisomerase II. Bioorg Med Chem Lett. 1995;5(8):905–908.

[532] Joseph H, Gleye J, Moulis C, Mensah LJ, Roussakis C, Gratas C. Justicidin B, a cytotoxic principle from *Justicia pectoralis*. J Nat Prod. 1988;51(3):599–600.

[533] Anjaneyulu ASR, Ramaiah PA, Row LR, Venkateswarlu R, Pelter A, Ward RS. New lignans from the heartwood of *Cleistanthus collinus*. Tetrahedron. 1981;37(21):3641–3652.

[534] Murakami T, Matsushima A. Studies on the constituents of Japanese *Podophylaceae* plants. I. On the constituent of the root of *Diphylleia grayi*. Yakugaku Zasshi. 1961;81:1596–1600.

[535] Ueng YF, Chen CC, Chen CF. Inhibition of benzo[a]pyrene hydroxylation by lignans isolated from *Justicia procumbens*. Yaowu Shipin Fenxi. 2000;8(4):309–314.

[536] Herz W, Gordon W, Falk H, Hunek S. Progress in the Chemistry of Organic Natural Products 81, Fortschritte der Chemie Organischer Naturstoffe / Progress in the Chemistry of Organic Natural Products Series. Springer-Verlag; 2001.

[537] Hwang DY. Therapeutic effects of lignans and blend isolated from schisandra chinesis on hepatic carcinoma. In: Kuang H, editor. Recent Advances in Theories and Practice of Chinese Medicine. InTech; 2012. p. 390–406.

[538] Richardson MD, Peterson JR, Clark AM. Bioactivity screenings of plants selected on the basis of folkloric use or presence of lignans in a family. Phytother Res. 1992;6(5):274–278.

[539] Chen YG, Wu ZC, Gui SH, Lv YP, Liao XR, Halaweish F. Lignans from *Schisandra henryi* with DNA cleaving activity and cytotoxic effect on leukemia and hela cells in vitro. Fitoterapia. 2005;76(3-4):370–373.

[540] Pao TT, Hsu KF, Liu KT, Chang LY, Chuang CH, Sung CY. Protective action of schizandrin B on hepatic injury in mice. Chin Med J. 1977;3(3):173–179.

[541] Fang SD, Xu RS, Gao YS. Some recent advances in the chemical studies of Chinese herbal medicine. Am J Bot. 1981;68(2):300–303.

[542] Wickramaratne DBM, Pengsuparp T, Mar W, Chai HB, Chagwedera TE, Beecher CWW, et al. Novel antimitotic dibenzocyclo-octadiene lignan constituents of the stem bark of *Steganotaenia araliacea*. J Nat Prod. 1993;56(12):2083–2090.

[543] Kupchan SM, Britton RW, Ziegler MF, Gilmore CJ, Restivo RJ, Bryan RF. Tumor inhibitors. LXXXX. Steganacin and steganangin, novel antileukemic lignan lactones from *Steganotaenia araliacea*. J Am Chem Soc. 1973;95(4):1335–1336.

[544] Hikino H, Kiso Y, Taguchi H, Ikeya Y. Validity of the oriental medicines. 60. Liver-protective drugs. 11. Antihepatotoxic actions of lignoids from schizandra chinensis fruits. Planta Med. 1984;50(3):213–218.

[545] Ikeya Y, Taguchi H, Mitsuhashi H, Sasaki H, Matsuzaki T, Aburada M, et al. Studies on the metabolism of gomisin a (TJN-101). I. Oxidative products of gomisin a formed by rat liver S9 mix. Chem Pharm Bull. 1988;36(6):2061–2069.

[546] Kiso Y, Tohkin M, Hikino H, Ikeya Y, Taguchi H. Mechanism of antihepatotoxic activity of wuweizisu C and gomisin A1. Planta Med. 1985;51(4):331–334.

[547] Sun QZ, Chen DF, Ding PL, Ma CM, Kakuda H, Nakamura N, et al. Three new lignans, longipedunins A–C, from *Kadsura longipedunculata* and their inhibitory activity against HIV-1 protease. Chem Pharm Bull. 2006;54(1):129–132.

[548] Xu LJ, Huang F, Chen SB, Zhang QX, Li LN, Chen SL, et al. New lignans and cytotoxic constituents from *Schisandra propinqua*. Planta Med. 2006;72(2):169–174.

[549] Xu L, Huang F, Chen S, Chen S, Xiao P. A new triterpene and dibenzocyclooctadiene lignans from *Schisandra propinqua* (WALL.) BAILL. Chem Pharm Bull. 2006;54(4):542–545.

[550] Shen YC, Liaw CC, Cheng YB, Ahmed AF, Lai MC, Liou SS, et al. C18 dibenzocyclooctadiene lignans from *Kadsura philippinensis*. J Nat Prod. 2006;69(6):963–966.

[551] Shen YC, Lin YC, Ahmed AF, Cheng YB, Liaw CC, Kuo YH. Four new nonaoxygenated C_{18} dibenzocylcooctadiene lignans from *Kadsura philippinensis*. Chem Pharm Bull. 2007;55(2):280–283.

[552] Chiu PY, Mak DHF, Poon MKT, Ko KM. In vivo antioxidant action of a lignan-enriched extract of *Schisandra fruit* and an anthraquinone-containing extract of *Polygonum* root in comparison with schisandrin b and emodin. Planta Med. 2002;68(11):951–956.

[553] Liu Z, Zhang B, Liu K, Ding Z, Hu X. Schisandrin B attenuates cancer invasion and metastasis *via* inhibiting epithelial-mesenchymal transition. PLoS One. 2012;7(7):e40480–.

[554] Liu CS, Fang SD, Huang MF, Kao YL, Hsu JS. Studies on the active principles of *Schisandra sphenanthera* Rehd. *et* Wils. The structures of schisantherin A, B, C, D, E, and related compounds. Sci Sin. 1978;21(4):483–502.

[555] Yang XW, Miyashiro H, Hattori M, Namba T, Tezuka Y, Kikuchi T, et al. Isolation of novel lignans, heteroclitins F and G, from the stems of *Kadsura heteroclita*, and anti-lipid peroxidative actions of heteroclitins A–G and related compounds in the *in vitro* rat liver homogenate system. Chem Pharm Bull. 1992;40(6):1510–1516.

[556] Yang X, Hattori M, Namba T, Chen D, Xu G. Antilipid peroxidative effect of an extract of the stems of *Kadsura heteroclita* and its major constituent, kadsurin, in mice. Chem Pharm Bull. 1992;40(2):406–409.

[557] Yang GY, Li YK, Wang RR, Li XN, Xiao WL, Yang LM, et al. Dibenzocyclooctadiene lignans from *Schisandra wilsoniana* and their anti-HIV-1 activities. J Nat Prod. 2010;73(5):915–919.

[558] Lei C, Huang SX, Chen JJ, Pu JX, Yang LB, Zhao Y, et al. Lignans from *Schisandra propinqua* var. propinqua. Chem Pharm Bull. 2007;55(8):1281–1283.

[559] Ikeya Y, Ookawa N, Taguchi H, Yosioka I. The constituents of *Schisandra chinensis* Baill. XI. The structures of three new lignans, angeloylgomisin O, and angeloyl- and benzoylisogomisin O. Chem Pharm Bull. 1982;30(9):3202–3206.

[560] Ikeya Y, Taguchi H, Yosioka I, Kobayashi H. The constituents of *Schizandra chinensis* Baill. V. The structures of four new lignans, gomisin N, gomisin O, epigomisin O and gomisin E, and transformation of gomisin N to deangeloylgomisin B. Chem Pharm Bull. 1979;27(11):2695–2709.

[561] Ikeya Y, Taguchi H, Yosioka I, Kobayashi H. The constituents of *Schizandra chinensis* Baill. IV. the structures of two new lignans, pre-gomisin and gomisin J. Chem Pharm Bull. 1979;27(7):1583–1588.

[562] Ikeya Y, Taguchi H, Yosioka I, Iitaka Y, Kobayashi H. The constituents of *Schizandra chinensis* Baill. II. The structure of a new lignan, gomisin D. Chem Pharm Bull. 1979;27(6):1395–1401.

[563] Ikeya Y, Taguchi H, Yosioka I, Kobayashi H. The constituents of *Schizandra chinensis* Baill. VIII. The structures of two new lignans, tigloylgomisin P and angeloylgomisin P. Chem Pharm Bull. 1980;28(11):3357–3361.

[564] Ikeya Y, Miki E, Okada M, Mitsuhashi H, Chai JG. Benzoylgomisin Q and benzoylgomisin P, two new lignans from *Schisandra sphenanthera* Rehd. *et* Wils. Chem Pharm Bull. 1990;38(5):1408–1411.

[565] Ikeya Y, Sugama K, Okada M, Mitsuhashi H. The constituents of schisandra species. Part 17. Two lignans from *Schisandra sphenanthera*. Phytochemistry. 1991;30(3):975–980.

[566] Ikeya Y, Taguchi H, Yosioka I, Kobayashi H. The constituents of *Schizandra chinensis* Baill. III. The structures of four new lignans, gomisin H and its derivatives, angeloyl-, tigloyl- and benzoyl-gomisin H. Chem Pharm Bull. 1979;27(7):1576–1582.

[567] Ikeya Y, Taguchi H, Yosioka I, Kobayashi H. The constituents of *Schisandra chinensis* Baill. I. Isolation and structure determination of five new lignans, gomisin A, B, C, F and G, and the absolute structure of schizandrin. Chem Pharm Bull. 1979;27(6):1383–1394.

[568] Ikeya Y, Taguchi H, Yosioka I. The constituents of *Schizandra chinensis* Baill. X. The structures of γ-schizandrin and four new lignans, (−)-gomisins L1 and L2 (±)-gomisin M1 and (+)-gomisin M2. Chem Pharm Bull. 1982;30(1):132–139.

[569] Kuo YH, Li SY, Huang RL, Wu MD, Huang HC, Lee KH. Schizanrin B, C, D, and E, four new lignans from *Kadsura matsudai* and their antihepatitis activities. J Nat Prod. 2001;64(4):487–490.

[570] Kuo YH, Huang HC, Kuo LMY, Chen CF. Novel C_{19} homolignans, taiwanschirin A, B, and cytotoxic taiwanschirin C, and a new C_{18} lignan, schizanrin A, from *Schizandra arisanensis*. J Org Chem. 1999;64(19):7023–7027.

[571] Peng HL, Chen DF, Lan HX, Zhang XM, Gu Z, Jiang MH. Anti-lipid peroxidation of gomisin J on liver mitochondria and cultured myocardial cells. Acta Pharmacol Sin. 1996;17(6):538–541.

[572] Zhang XM, Chen DF, He XJ, Yang S, Zheng P, Jiang MH. Blocking effects of heteroclitin D and gomisin J on L-type calcium channels in ventricular cells of guinea pig. Acta Pharmacol Sin. 2000;21(4):373–376.

[573] Chang J, Reiner J, Xie J. Progress on the chemistry of dibenzocyclooctadiene lignans. Chem Rev. 2005;105(12):4581–4609.

[574] Fujihashi T, Hara H, Sakata T, Mori K, Higuchi H, Tanaka A, *et al*. Anti-human immunodeficiency virus (HIV) activities of halogenated gomisin J derivatives, new nonnucleoside inhibitors of HIV type 1 reverse transcriptase. Antimicrob Agents Chemother. 1995;39(9):2000–2007.

[575] Ohtaki Y, Nomura M, Hida T, Miyamoto Ki, Kanitani M, Aizawa T, *et al*. Inhibition by gomisin A, a lignan compound, of hepatocarcinogenesis by 3′-methyl-4-dimethylaminoazobenzene in rats. Biol Pharm Bull. 1994;17(6):808–814.

[576] Nomura M, Nakachiyama M, Hida T, Ohtaki Y, Sudo K, Aizawa T, *et al*. Gomisin A, a lignan component of *Schizandra fruits*, inhibits development of preoplastic lesions in rat liver by 3′-methyl-4-dimethylamino-azobenzene. Cancer Lett. 1994;76(1):11–18.

[577] Miyamoto K, Wakusawa S, Nomura M, Sanae F, Sakai R, Sudo K, *et al*. Effects of gomisin A on hepatocarcinogenesis by 3′-methyl-4-dimethylaminoazobenzene in rats. Jpn J Pharmacol. 1991;57(1):71–77.

[578] Yasukawa K, Ikeya Y, Mitsuhashi H, Iwasaki M, Aburada M, Nakagawa S, *et al*. Gomisin A inhibits tumor promotion by 12-O-tetradecanoylphorbol-13-acetate in two-stage carcinogenesis in mouse skin. Oncology. 1992;49(1):68–71.

[579] Yamada S, Murawaki Y, Kawasaki H. Preventive effect of gomisin A, a lignan component of shizandra fruits, on acetaminophen-induced hepatotoxicity in rats. Biochem Pharmacol. 1993;46(6):1081–1085.

[580] Kubo S, Matsui-Yuasa I, Otani S, Morisawa S, Kinoshita H, Sakai K. Effect of splenectomy on liver regeneration and polyamine metabolism after partial hepatectomy. J Surg Res. 1986;41(4):401–409.

[581] Kubo S, Ohkura Y, Mizoguchi Y, Matsui-Yuasa I, Otani S, Morisawa S, *et al*. Effect of gomisin A (TJN-101) on liver regeneration. Planta Med. 1992;58(6):489–492.

[582] Wang JP, Raung SL, Hsu MF, Chen CC. Inhibition by gomisin C (a lignan from *Schizandra chinensis*) of the respiratory burst of rat neutrophils. Br J Pharmacol. 1994;113(3): 945–953.

[583] Lee KH, Xiao Z. Lignans in treatment of cancer and other diseases. Phytochem Rev. 2003;2(3):341–362.

[584] Chen DF, Zhang SX, Chen K, Zhou BN, Wang P, Cosentino LM, et al. Two new lignans, interiotherins A and B, as anti-HIV principles from *Kadsura interior*. J Nat Prod. 1996;59(11):1066–1068.

[585] Jin X, Gu Z, Hu T, Wang M, Chen D. Effects of lignan gomisin J from kadsura interior on liver mitochondria lipid peroxidation and on the superoxide anion radical. Zhongguo Yaolixue Tongbao. 2000;16(1):26–28.

[586] Toda S, Kimura M, Ohnishi M, Nakashima K, Ikeya Y, Taguchi H, et al. Natural antioxidants (IV). Antioxidative components isolated from *Schisandra fruit*. Shoyakugaku Zasshi. 1988;42(2):156–159.

[587] Tan R, Li L, Fang Q. Studies on the chemical constituents of *Kadsura longipedunculata*: isolation and structure elucidation of five new lignans. Planta Med. 1984;50(5):414–417.

[588] Ikeya Y, Kanatani H, Hakozaki M, Taguchi H, Mitsuhashi H. The constituents of *Schizandra chinensis* Baill. XV. Isolation and structure determination of two new lignans, gomisin S and gomisin T. Chem Pharm Bull. 1988;36(10):3974–3979.

[589] Ikeya Y, Taguchi H, Yosioka I. The constituents of *Schizandra chinensis* Baill. The cleavage of the methylenedioxy moiety with lead tetraacetate in benzene, and the structure of angeloylgomisin Q. Chem Pharm Bull. 1979;27(10):2536–2538.

[590] Ikeya Y, Taguchi H, Yosioka I. The constituents of *Schizandra chinensis* Baill. VII. The structures of three new lignans, (−)-gomisin K1 and (+)-gomisins K2 and K3. Chem Pharm Bull. 1980;28(8):2422–2427.

[591] Ikeya Y, Taguchi H, Yosioka I. The constituents of *Schizandra chinensis* Baill. The structures of three new lignans, angeloylgomisin H, tigloylgomisin H and benzoylgomisin H, and the absolute structure of schizandrin. Chem Pharm Bull. 1978;26(1):328–331.

[592] Su D, Tang W, Hu Y, Liu Y, Yu S, Ma S, et al. Lignan glycosides from *Neoalsomitra integrifoliola*. J Nat Prod. 2008;71(5):784–788.

[593] Chen JJ, Wang TY, Hwang TL. Neolignans, a coumarinolignan, lignan derivatives, and a chromene: anti-inflammatory constituents from *Zanthoxylum avicennae*. J Nat Prod. 2008;71(2):212–217.

[594] Sarker SD. Biological activity of magnolol: a review. Fitoterapia. 1997;68(1):3–8.

[595] Chiou T, Po A, Lin M, Romanuik T, Shum A. The antimicrobial effects of magnolol on the intracellular pH of *Bacillus subtilis* WB746 and *Escherichia coli* B23. J Exp Microbiol Immunol (JEMI). 2003;3:41–49.

[596] Schühly W, Khan SI, Fischer NH. Neolignans from North American *Magnolia* species with cyclooxygenase 2 inhibitory activity. Inflammopharmacology. 2009;17(2):106–110.

[597] Ai J, Wang X, Nielsen M. Honokiol and magnolol selectively interact with gabaa receptor subtypes in vitro. Pharmacology. 2001;63(1):34–41.

[598] Hasegawa S, Yonezawa T, Ahn JY, Cha BY, Teruya T, Takami M, et al. Honokiol inhibits osteoclast differentiation and function in vitro. Biol Pharm Bull. 2010;33(3):487–492.

[599] Begum S, Sahai M, Ray A. Non-conventional lignans: coumarinolignans, flavonolignans, and stilbenolignans. In: Kinghorn AD, Falk H, Kobayashi J, editors. Progress in the Chemistry of Organic Natural Products, Fortschritte der Chemie organischer Naturstoffe / Progress in the Chemistry of Organic Natural Products. vol. 93. Vienna: Springer-Verlag GmbH; 2010. p. 1–70.

Part VI
Outcome and Challenges

12

Summary, Conclusions, and Perspectives on Lignin Chemistry

12.1 Sources of Lignin

As it is explained throughout this book, lignin has been known since the 19th century and has been studied for decades by botanists, chemists, and engineers. It provides rigidity to plants, depending on its abundance, which varies widely according to the species, as is evident for herbaceous plants, softwood, or hardwood species.

As has also been mentioned, the term "lignin" applies to different types of biopolymers, better called lignins. Among them, the so-called "native lignin" directly renders the biomass. Nevertheless, even with regard to native lignin, a great diversity is found depending on the plant, and lignins from herbaceous plant species, hardwood trees, or softwood trees differ in composition and structure, with evident consequences for the rigidity and properties of plants.

Other issue to clarify regarding types of lignin is the method used for isolation (see Chapter 5). Most of the procedures used for isolation of lignin from biomass are quite aggressive, and therefore the resulting material is frequently modified with respect to the original polymer. Probably the least aggressive method is the one using organic solvents together with some chemicals, yielding the so-called organosolv lignin. In this particular case, the degree of polymer degradation is not very high, which makes organosolv lignin one of the most desirable types of lignin for further applications. The use of other new procedures such as ionic liquids is also desirable, but still very limited because of the high price of ionic liquids, which hampers the current extraction of lignin in bulk amounts.

In addition, another type of lignin can be derived as a by-product of the paper industry. The pulping process for the separation of cellulose from lignin when making paper is quite aggressive, and lignin remains for longer time at high temperatures and in the presence of reactive chemicals. Therefore, the isolation of this kind of lignin yields chemically modified polymers, whose modification, as with native lignin, depends on the procedure (Kraft, alkali). Despite such chemical modification of the lignin, this method can be scaled up to bulk production.

Lignin and Lignans as Renewable Raw Materials: Chemistry, Technology and Applications, First Edition.
Francisco G. Calvo-Flores, José A. Dobado, Joaquín Isac-García and Francisco J. Martín-Martínez.
© 2015 John Wiley & Sons, Ltd. Published 2015 by John Wiley & Sons, Ltd.

Finally, in recent years, a new source of lignin has arisen with the development of biorefineries. Biofuels development is a promising research field to produce fuels from renewable sources. In an initial stage, alcohols were made by fermenting starch-rich plants such as corn, but this approach gravely distorts food markets and feedstock. Accordingly, different efforts are being put forward for using alternative plants not related to food. As such, herbaceous or wood plants are being used as alternative sources for producing biofuels, and the processes employed produce large amounts of lignin as a side product, which represents an increasing source of lignin for diverse applications.

12.2 Structure of Lignin

The structure of lignin has been the object of research and controversy for decades, due mainly to the wide diversity of sources and isolation methods that make lignin structure highly variable. Thus, several models have been proposed to provide a reasonable description of lignin's main structural features. Hence, lignin models described by scientists over the years have offered a partial vision of the structure and have shown some remarkable aspects that have helped to explain lignin's behavior. A constant during this research process has been effort to formulate a reasonable model on lignin's structure, which, as we mentioned is a challenge in practice. Early work on this subject was based on the study of fragments produced by several hydrolysis methods. This provides valuable data, but addressed only a part of the model. Later, as spectroscopic methods were improved, a fuller idea of the real structure of lignin was provided. Over these years, it has been relatively easy to describe the main units that form lignin's polymer and also the way that these units are bonded. Thus, four main monomers have been identified (main monolignols), although other monomers slightly different from these have also been found. These monolignols and the latter characterize monomers from a list of lignin building blocks. The composition, percentage, and linking characteristics of monolignols vary, defining the nonuniform polymer with a great structural diversity known as lignin.

Several conclusions can be drawn from this. First of all, there is no real model of lignin that represents all different types of linkages, connectivity, frequency, degree of branching, and other characteristics of such a complex polymer. Depending on the origin of the plant species, and even the period of the year in which native lignin is collected, many differences can be found in the structure of lignin. Moreover, the method used for the isolation of lignin (see Chapter 5) also defines its composition. Most of the isolation procedures used for native lignin from biomass are aggressive, and therefore the resulting material is modified with respect to the original polymer. There are distinct models that represent different lignins from diverse sources through different isolation methods. It would thus be more correct to use the term "lignins" because of this great diversity in the origin and method of isolating lignin.

On the other hand, bulk amounts of lignin are currently available from the paper industry and from biorefineries, which are specifically designed for making biofuels from biomass. Lignin from the paper industry is an especially modified material compared to native lignin, where molecular mass and functional groups differ. The degradation of original lignin and the introduction of new functional groups, especially sulfur, are some of the chemical transformations that may occur. In the case of lignin from biorefineries, differences to native lignin also arise, depending mainly on the pretreatment procedure for separating lignin from cellulose or other plant components.

For all these reasons, we emphasize once more that there is no complete and single model of lignin that represents this polymer in only one way but rather different models that represent different lignins from different sources, produced through different isolation methods.

12.3 Biosynthesis and Biological Function

Another remarkable issue related to lignin that has been addressed for many years is the way that lignin is formed in plants. Lignin is produced by the oxidative coupling of the three basic H, G, and S monolignols (*p*-hydroxyphenyl, guaiacyl, and syringyl, respectively) [1] in the lignification process. In this chemical process, lignin is first synthesized and later deposited in the secondary cell wall of some specialized cells, such as xylem vessels, tracheids, and fibers. It is also deposited in minor amounts in the periderm [2].

From the extensive work already done in this regard, two major hypotheses have been formulated to explain the mechanism of lignin biosynthesis, both being frequently debated in the literature. The first one is the combinatorial random coupling [2, 3] between monolignol radicals according to an oxidative radical process. This hypothesis is the most widely accepted mechanism for the lignification process. The second hypothesis is based on a protein-directed synthesis that involves the so-called dirigent proteins (DIR) [4, 5].

The biological function of lignin can be summarized in the following considerations [6]: Lignin gives stiffness to the cell structure, working as a binder and fixating polymer in the cell walls of the woody plant. It interacts closely with polysaccharides, to make the fibers relatively stiff and rigid, and it offers mechanical support that builds up the stem and branches. In addition, lignin acts as a glue that keeps different cells together in woody tissues. In wood, the middle lamella (ML) consists mainly of lignin, which works as an efficient and resistant adhesive that again keeps the cells together. It inhibits cell walls from swelling in water, and therefore water leaks from a woody cell wall, making it waterproof. In fact, in nonwoody plants, this is the main function of lignin [7]. Finally, lignin efficiently protects against microbial degradation of wood.

12.4 Applications of Lignin

The vast diversity of lignin makes it difficult to produce a homogeneous material, but also provides the advantage of having lignins with diverse properties for very different applications, for example, water-soluble or insoluble polymers and very polar or with a moderate polarity.

As mentioned several times, this diversity depends on the lignin source, and therefore it also determines the application of lignin. Within this framework, lignins from paper industry are probably the best available for practical applications. The huge amounts of lignins that are produced in the cooking process of paper industry to extract cellulose have been considered for years to be only a by-product. Sometimes, this lignin has been used directly as fuel in the same factory, but in most of the cases it has been underutilized. In fact, this lignin is the cheapest and more easily available. Another source of lignins is that directly isolated from biomass, in which a key factor is to determine the origin and isolation method since these define the structural differences.

Deserving mention is the case of biorefineries, where even if the so-called carbohydrate platform has been extensively developed to produce not only alcohols, such as biofuel, but many others chemicals that are very close to compete with petrochemical industry, the lignin platform is still at a very early stage (Figure 12.1).

Building on these modern concepts, a promising general model for a lignocellulosic biorefinery is based on the sugar–lignin platform in which five-carbon (C_5) and six-carbon (C_6) sugars, issued from lignocellulosic matrix fractionation, are converted into fuels and building block chemicals by biotechnological and chemical pathways, respectively. Lignin output can be used as solid

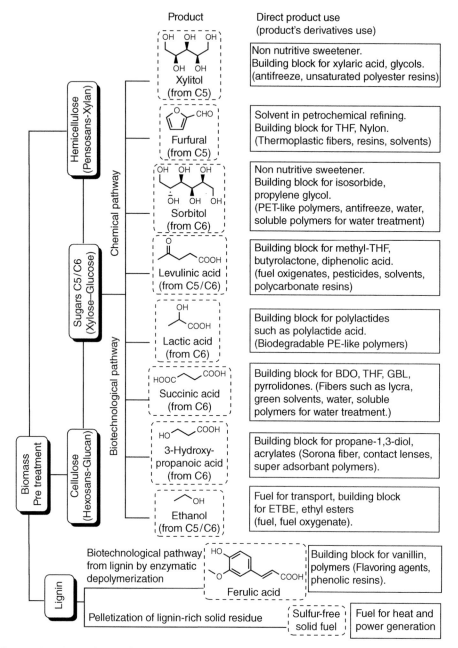

Figure 12.1 Sugar–lignin platform potential for value-added chemical production. Redrawn from http://bpe.epfl.ch (accessed December 10, 2014).

fuel for cogeneration, with or without previous pelletization, or it can be upgraded into value-added chemicals.

Concerning applications, an enormous amount of work has been carried out for many years, although there is still a relatively modest range of practical applications on the market, in contrast to all the academic research and patents being licensed. Nevertheless, there are several remarkable items related to applications and uses of lignins. Lignins can be used directly without transformation for the preparation of binders, polymers, different additives, and industrial goods, and it can be used as food for livestock. It should be pointed out that most of these applications are based on lignin from the paper industry.

Another option is to chemically transform lignin in order to improve its reactivity or to provide modifications that target new applications. The problem for this use of lignin, such as macromonomers in the polymer industry, is the relatively low reactivity due to its very cross-linked structure, which hides reactive functional groups within the polymer matrix. Demethylation and phenolation are processes that improve lignin reactivity due to the implementation of free hydroxyl groups of phenolic moieties. Also, inserting spacers with the appropriate functionalization is possible. These procedures prepare modified lignins with new functional groups not present in previous structures, while the spacers avoid the steric hindrance of reactive sites introduced in the 3D structure of the polymer. The development of these techniques in a very controlled manner is an endeavor for coming years of research on the applications of lignin.

12.5 Lignans

Lignans are naturally occurring molecules found in plants and animals, but especially widespread throughout the plant kingdom [8–10]. These are considered to be secondary metabolites, and are commonly included in the human diet. Their biological role in plants remains unclear, although it is thought to be related to defensive mechanisms against external agents [11, 12]. In fact, lignans and related compounds may play an autoprotective role for plants because they induce response to a fungal and bacterial attack.

Lignans can be found in more than 60 families of vascular plants [13] and have been isolated from different parts of the plant: roots and rhizomes, woody parts, stems, leaves, fruits, and seeds, as well as from exudates and resins [14–17]. They have also been detected in the urine of humans and other mammals, and are known to be produced by several *Streptomyces* species [18].

According to their structure, lignans can be classified into five families of compounds: proper lignans, neolignans, sesquilignans and dineolignans, norlignans, and hybrid lignans. Lignans and neolignans are natural products derived from the same phenylpropane (C_6-C_3) unit that makes up the lignin polymer. In total, lignans might account for up to about 900 compounds.

Lignan biosynthesis presents some common features with lignin biosynthesis. In fact, they are biosynthesized through the same pathway in the earlier steps, starting from hydroxycinnamyl alcohols.

Lignans and related compounds present biological and pharmacological properties. Many of them are components of orally administered herbal medicines and when metabolized they are often converted to pharmacologically active compounds by intestinal flora. Some lignans in plants are even considered to be vegetal hormones, and hence they are called phytoestrogens[1] because they are transformed by intestinal microflora into molecules such as enterodiol and enterolactone, which show estrogen-like biological activity [19–22].

---

[1] Compounds structurally related to mammalian estrogen 17β-estradiol (E2), originated in plants, which increase plant immunity to disease, and act as estrogen agonists or antagonists in animals.

Podophyllotoxin is probably the most widely known and studied lignan because of its proven antitumoral properties. This fact has promoted extensive studies on this compound that have given rise to a great number of publications. For the same reason, much work has been done on the chemical synthesis of derivatives and similar compounds that could result in new drugs against cancer. Because of the importance of this issue, concerted effort has been made in this book to summarize and synthesize different aspects of this subject.

12.6 Perspectives

It should be emphasized from the outset that this book provides the state of the art of science and technology on lignin. It does not, however, rule out perspectives on other tasks and tendencies to be confronted in the coming years. There are many unsolved problems and several fields that draw our attention for the middle and long term.

Lignin is being considered a valuable material and no longer simply a by-product for burning or a waste in paper industry and biorefineries. In particular, biorefineries have developed mainly the carbohydrate platform as a waste and have made considerable effort in terms of energy, chemicals, and time to separate lignin from other biomass components. In addition, it would be useful to develop genetically modified species that enable the separation of lignin under milder reaction conditions. Another point to consider in this area is the biotechnology needed to prepare new enzymes capable of separating lignin from biomass in a more specific and efficient way.

Furthermore, some applications described over the course of this book require the chemical functionalization of lignin in order to improve its reactivity. One example is the field of polymer synthesis. However, the current methods for introducing new reactive sites into the polymer matrix in a controlled way need to be improved. This field could open the window to the development of new applications related to the reactivity of the functional groups in lignin.

Some of the most promising fields are depolymerization processes. It has become clear that lignin is the major source of aromatic compounds in nature. It is even competitive with oil-related compounds and with an added value, because many of them show complex functionalizations difficult to introduce even with petrochemical technology. Nevertheless, the results available in the literature are still too modest to have an industrial application. Perhaps the problem could be addressed sequentially in two phases: first, breaking lignin into smaller pieces and second, producing simpler chemicals. The most challenging goal will be to improve selectivity, which implies the development of new catalysts.

Finally, substantial research needs to be devoted to finding new applications in emerging fields such as nanotechnology, new materials, organocatalysts, and biosciences, to mention just a few. Nature gives us a sophisticated and complex material. It would be a colossal oversight not to use this highly accessible renewable substance and an expensive one in these times as raw materials become scarce.

References

[1] Boerjan W, Ralph J, Baucher M. Lignin biosynthesis. Annu Rev Plant Biol. 2003;54:519–546.
[2] Sarkanen KV, Ludwig CH. Lignins: Occurrence, Formation, Structure and Reactions. Wiley-Interscience; 1971.
[3] Neish AC. Monomeric intermediates in the biosynthesis of lignin. In: Freudenberg K, Neish AC, editors. Constitution and Biosynthesis of Lignin, Molecular Biology, Biochemistry and Biophysics. vol. 2. Berlin: Springer-Verlag; 1968. p. 1–43.

[4] Davin LB, Wang HB, Crowell AL, Bedgar DL, Martin DM, Sarkanen S, et al. Stereoselective bimolecular phenoxy radical coupling by an auxiliary (dirigent) protein without an active center. Science. 1997;275(5298):362–367.
[5] Davin LB, Lewis NG. Lignin primary structures and dirigent sites. Curr Opin Biotechnol. 2005;16(4):407–415.
[6] Henriksson G. Lignin. In: Ek M, Gellerstedt G, Henriksson G, editors. Wood Chemistry and Biotechnology, Pulp and Paper Chemistry and Technology. vol. 1. Walter de Gruyer; 2009. p. 121–145.
[7] Falkehag SI. Lignin in materials. Appl Polym Symp. 1975;28(Proc. Cellul. Conf., 8th, 1975, Vol. 1):247–257.
[8] Umezawa T. Diversity in lignan biosynthesis. Phytochem Rev. 2003;2(3):371–390.
[9] Milder IEJ, Feskens EJM, Arts ICW, Bueno de Mesquita HB, Hollman PCH, Kromhout D. Intake of the plant lignans secoisolariciresinol, matairesinol, lariciresinol, and pinoresinol in Dutch men and women. J Nutr. 2005;135(5):1202–1207.
[10] Penalvo JL, Adlercreutz H, Uehara M, Ristimaki A, Watanabe S. Lignan content of selected foods from japan. J Agric Food Chem. 2008;56(2):401–409.
[11] Harmatha J, Dinan L. Biological activities of lignans and stilbenoids associated with plant-insect chemical interactions. Phytochem Rev. 2003;2(3):321–330.
[12] Schroeder FC, Del Campo ML, Grant JB, Weibel DB, Smedley SR, Bolton KL, et al. Pinoresinol: a lignol of plant origin serving for defense in a caterpillar. Proc Natl Acad Sci U S A. 2006;103(42):15497–15501.
[13] Gordaliza M, Garcia PA, del Corral JMM, Castro MA, Gomez-Zurita MA. Podophyllotoxin: distribution, sources, applications and new cytotoxic derivatives. Toxicon. 2004;44(4):441–459.
[14] Rao CBS. Chemistry of Lignans, Andhra University Series. Andhra University Press; 1978.
[15] Landete JM. Plant and mammalian lignans: a review of source, intake, metabolism, intestinal bacteria and health. Food Res Int. 2012;46(1):410–424.
[16] Castro MA, Gordaliza M, Del Corral JMM, Feliciano AS. The distribution of lignanoids in the order coniferae. Phytochemistry. 1996;41(4):995–1011.
[17] Cunha WR, Andrade e Silva ML, Sola Veneziani RC, Ambrósio SR, Kenupp Bastos J. Lignans: chemical and biological properties. In: Rao V, editor. Phytochemicals - A Global Perspective of their Role in Nutrition and Health. vol. 33. InTech; 2012. p. 213–234.
[18] Chiung YM, Hayashi H, Matsumoto H, Otani T, Yoshida K, Huang MY, et al. New metabolites, tetrahydrofuran lignans, produced by Streptomyces sp. IT-44. J Antibiot. 1994;47(4):487–491.
[19] Fletcher RJ. Food sources of phyto-oestrogens and their precursors in Europe. Br J Nutr. 2003;89(Suppl. 1):S39–S43.
[20] Liggins J, Grimwood R, Bingham SA. Extraction and quantification of lignan phytoestrogens in food and human samples. Anal Biochem. 2000;287(1):102–109.
[21] Nurmi T, Heinonen S, Mazur W, Deyama T, Nishibe S, Adlercreutz H. Lignans in selected wines. Food Chem. 2003;83(2):303–309.
[22] Heinonen S, Nurmi T, Liukkonen K, Poutanen K, Wähälä K, Deyama T, et al. In Vitro metabolism of plant lignans: new precursors of mammalian lignans enterolactone and enterodiol. J Agric Food Chem. 2001;49(7):3178–3186.

Glossary

Admixtures	A material other than water, aggregates, hydraulic cement, and fiber reinforcement, used as an ingredient of a cementitious mixture to modify its freshly mixed, setting, or hardening properties and that is added to the batch before or during its mixing.
Anthracite fines	Small pieces of material from an anthracite coal preparation plant, usually below 3 mm diameter.
Black liquor	Aqueous solution composed of lignin, hemicellulose, and inorganic chemicals that is used in the Kraft pulping process.
Bleaching	Removal of colored residual lignin from chemical pulp to increase its brightness, cleanliness, and other desirable properties, while preserving the strength (cellulose integrity) and carbohydrate yield (cellulose and hemicellulose) of the unbleached fiber, with due regard for potential effects on the environment.
Chelation	IUPAC defines chelation as "the formation or presence of bonds (or other attractive interactions) between two or more separate binding sites within the same ligand and a single central atom."
Crops	Refer to plants that are grown on a large scale for food, clothing, and other human uses. They are nonanimal species or varieties grown to be harvested as food, livestock fodder, fuel, or for any other economic purpose (e.g., for use as dyes, medicine, and cosmetics). Major crops include sugarcane, pumpkin, maize (corn), wheat, rice, cassava, soybeans, hay, potatoes, and cotton. While the term "crop" most commonly refers to plants, it can also include species from other biological kingdoms. For example, mushrooms like shiitake, which are in the Fungi kingdom, can be referred to as "crops." In addition, certain species of algae are also cultivated, although they are also harvested from the wild. By contrast, animal species that are raised by humans are called livestock, except those that are kept as pets. Microbial species, such as bacteria and viruses, are referred to as cultures. Microbes are not typically grown for food, but are rather used to alter food. For example, bacteria are used to ferment milk to produce yogurt.
Gymnosperms	A group of seed-producing plants that includes conifers, cycads, Ginkgo, and Gnetales.

Hardwood	Botanically hardwoods are angiosperms. Anatomically, hardwoods are porous; that is, they contain vessel elements. Typically, hardwoods are plants with broad leaves that fall in the autumn or winter.
Heartwood	The dead, inner wood, of a woody stem or branch that often comprises the majority of a stem's cross section.
Laccase	Copper-containing oxidase enzyme (EC 1.10.3.2) that is found in many plants, fungi, and microorganisms.
Lignans	Plant products of low molecular weight formed primarily from oxidative coupling of two p-propylphenol moieties at their β carbon atoms.
Molasse	A viscous by-product of the refining of sugarcane or sugar beets into sugar.
Monolignols	Phenolic alcohols acting as source materials in the biosynthesis of lignans and lignins.
Native lignin	A lignin isolated in such a way that the solvent does not react with the lignin or alter it in any way.
Neolignans	Products with two p-propylphenol moiety units coupled in other ways that lignans defined functionally are substances that promote estrogenic.
Phytoestrogen	Compounds structurally related to mammalian estrogen 17β-estradiol (E2), originated in plants, which increase plant immunity to disease and act as estrogen agonists or antagonists in animals.
Podophyllum peltatum	Podophyllum means foot-shaped leaf and peltatum means shield-like.
Protolignin	Immature form of lignin that can be extracted from the plant cell wall with EtOH or dioxane.
Pulp	Lignocellulosic fibrous material prepared by chemically or mechanically separating cellulose fibers from wood, fiber crops, or wastepaper.
Sapwood	The living, outermost portion of a woody stem or branch.
Softwood	Botanically softwoods are gymnosperms or conifers. Anatomically, softwoods are nonporous and do not contain vessel. Typically, softwoods are cone-bearing plants with needle- or scale-like evergreen leaves.
Supercritical fluid	Any substance at a temperature and pressure above its critical point, where distinct liquid and gas phases do not exist. It can effuse through solids like a gas, and dissolve materials like a liquid.

Index

3-Glycidyloxypropyltrimethoxysilane,
  3-(triethoxysilyl)propylisocyanate
  (IPTES) 265

A. gemmatalis 386
Abaca 28, 56, 58
Absorbent(s) 264
Acanthomax senticosus 411
Acanthopanax chiisanensis 401, 415
Acacia mollissima 56
Acetic anhydride 217
Acetosolv 134
Acetyl bromide 62, 66, 220
Acetyl bromide/acetic acid 60, 62
Acetylation 217
Acid detergent fiber (ADF) 59
Acid media depolymerization 296
Acidolysis 11, 116, 160, 190
Acridine 52, 53
Acriflavine 52, 53
Actaea racemosa 401
Actinidia deliciosa 329
Active carbon 306, 308
Acylation 216
Adhesive(s) 253, 260
Adipic acid 262
Adsorption 255
Adsorption kinetic(s) 255
Aedes aegypti 386
Aegdops ouata 386
Aglicone(s) 393
Aglycone(s) 84, 88, 390, 409
Agrochemical 258

Agropyron lignocellulose(s) 228
AIDS 256
Aiphanes aculeata 319
Aiphanol 319
Alcell© 134
Alcohol(s)
  amyl 125
  benzyl 96, 154, 162
  butyl 125
  cinnamyl 76, 86, 90, 316
  coniferyl 12, 18, 28, 49, 75, 81, 95,
    336, 371
  p-coumaryl 12, 18, 28, 49, 75, 79, 81
  dehydrodiconiferyl 90
  furfuryl 260
  hydroxycinnamyl 12, 81, 370, 461
  p-hydroxycinnamyl 12
  sinapyl 12, 18, 28, 31, 49, 75, 81
  tert-butyl 195
  tetrahydrofurfuryl 123
Alcoholysis 125
Aldehyde(s)
  coniferyl 51, 87
  hydroxycinnamyl 51
  phenolic 192
Alkoxysilane(s) 265
Alkylation 218, 251, 333, 343, 351
Allium sativum 329
Allium schoenoprasum 329
p-Aminodiphenylamine 52
Aminolysis, method of 153
Aminophenol 51
Ammonia fiber explosion (AFEX) 136

Lignin and Lignans as Renewable Raw Materials: Chemistry, Technology and Applications, First Edition.
Francisco G. Calvo-Flores, José A. Dobado, Joaquín Isac-García and Francisco J. Martín-Martínez.
© 2015 John Wiley & Sons, Ltd. Published 2015 by John Wiley & Sons, Ltd.

Ammonium nitrate 258
Anacardium occidentale 329
Ananas comosus 329
Angiosperm(s) 12, 16, 84, 101, 102, 118
Aniline 265
Anthracite 254
Anthriscus sylvestris 393
Antibacterial 257
Antibiotic 254, 257
Antibodie(s) 102
Anticarcinogenic 256
Antifeeding 383
Antifungal 383
Antimicrobial 256, 257, 383
Antioxidant 383
Antioxidant(s) 254, 256–258, 263, 400, 401, 409, 411–413
Antitumor 256
Antiviral 256, 383
AOAC International 49, 57, 66
Apoplast 85, 88
Aquasolv 127
Arabidopsis 88
Arabinose 38
Arachis hypogaea 329
Arboform 264
Arboform® 264
Arctigenin 402–404
Arctium lappa 402
Aromatic amine(s) 51
Aryldihydronaphthalenes 332
Asparagus officinalis 329
Aspergillus flavus 417
Aspergillus fumigates 417
Aspergillus usamii 411
Atropoisomery 324
Avena sativa 329
AVIDEL, method of 30

B hepatitis 256
Bacillus subtilis 384
Bacteria 76, 77
Base media depolymerization 296
Battelle-Geneva process 135
Battery(ies) 265
  lead-acid 267

Benzidine 52
Benzoyl chloride 218
Benzylium ion 220
Betula pendula 39
Betula verrucosa 56
BHT 199
Bile acid 257
Binder(s) 253, 258, 259, 461
Biodegradation 251, 254, 260
Bioethanol 115, 116
Biofluid(s) 331
Biofuel(s) 252, 458
Biomass 115, 116, 120, 123, 133, 136, 249–252, 457, 459, 462
Biomedical sciences 264
Bioplastic 263, 264
Biopolymer(s) 145, 260, 265
Biorefinery(ies) 30, 290, 458, 459, 462
Biphenyl 89
Bis(trimethoxysilyl)hexane 265
Bisphenol A 274, 275
Black liquor(s) 57, 128, 129, 132
  Kraft 253, 254
  Palm 257
Blastoschizomyces capitatus 417
Bleached chemical pulp 57
Bleaching 59
Bombyx mori 409
Bond(s)
  cross-linked 41
  glycosidic 97
  hydrogen 39, 40
Borregaard 305
  lignotech 129, 133
Brake friction material 260
Briquette(s) 252
  coal 253
Bryophyta 16, 326
BTX 290, 292
Bupleurum scorzonerifolium 401
Bursera fagaroides 393, 401
Bursera graveolens 393
Bursera permollis 393
Bursera schlechtendalii 401
Bursera simaruba 393

1-Butyl-3-methylimidazolium chloride
([BMIM][Cl]) 124

Cadensin G 319
Cadoxene (cadmium
    oxide/ethylenediamine) 60, 62
Caffeic acid 79
Caffeoyl-CoA 79
Calcium chloride 268
Calcium lignosulfonate 261
Candida albicans 256, 257, 417
Candida tropicalis 257
Carbamate 261
Carbazolo 52
Carbohydrate(s) 11, 12, 31, 38, 39, 55,
    57, 59, 64, 76, 97, 99, 118–121, 125,
    172, 256, 257
Carbon fiber(s) 308
Carcinoma verrucosus 391
Carum carvi 329
Caryolanemagnolol 319
Catechol(s) 51, 209, 227, 289, 296, 297,
    304, 404
Cathode 265
Cedrus deodara 404
Cellulose 13, 30, 38–40, 55, 59, 61, 75,
    97, 116, 118, 119, 124, 126, 128, 249,
    257, 263, 264, 457–459
Cement 269
Ceriporiopsis subvermispora 228
Cetyltrimethylammonium bromide 59
Chameacyparis obtusa 326
Chelate(s) 254, 255
Chelating effect 258
Chemical depolymerization 295
Chinese propolis 411
Chipboard 258
ChipSep 327
Chlamydomonas reinhardtii 267
Chloride salt(s) 268
Chlorine 64
Chlorine dioxide 299
Chlorine number 61, 66
2-Chloro-4,4,5,5-tetramethyl-1,3,
    2-dioxaphospholane (TDMP) 154
Chlorolignin(s) 219

p-Chloromercuribenzoate 87
Chloromethoxycatechol 54
Chloronium ion(s) 299
Chlorophyll 258
Cholesterol 257
Chorismate 77
Chrysomya megacephala 386
Cicer arietinum 329
Cinnamaldehyde 24, 51
Cinnamate-4-hydroxylase/5-hydroxylase
    (C4H::F5H) 31
Cinnamic acid(s) 78, 79, 316, 370
Circular dichroism (CD) 331
Citrus limon 329
Citrus sinensis 329
Cleome viscosa 319
Cleomiscosin D 319
click Chemistry 394
Clovanemagnolol 319
CO_2 explosion, method of 136
Coagulation 128
Coefficient(s)
    extinction 57
    sedimentation 42
Coffea arabica 329
Coke 254
Colchicine 392
Collagen 254
Colorimetric 63
Combustion heat 252
Commiphora erlangerian 396
Composite(s) 265
Concrete 269
Conducting polymer(s) 265
Conductivity 265
Condyloma acuminatum 388
Conifer(s) 100
Coniferous 18
Coniferyl acetate 83
Coniferyl alcohol oxidase(s) 85
Constant(s)
    diffusion 42
Copolymerization 261
Corchorus capsularis 56
Corylus avellana 329
Cosmetic 256

COSY 168
Cotton 56
p-Coumarate 28
p-Coumaric acid 79, 101
Coumarin(s) 80
Coumarino-lignan(s) 319
Coupling
  DCC 218
  oxidative 459
  radical 91
  random 459
β-4 Coupling 88
β-5 Coupling 88
CP/MAS 172
Cresol(s) 289
  p- 51, 198
  m- 51
  o- 51
Crop(s) 17, 254
Cross-linked 264, 265
Cross-linking 12, 260, 265, 461
Cryoscopy 150
Culex pipiens 386
cumene 290
Cuminum cymicum 329
Cuprammonium hydroxide 118
Curaua 28
Curcumin 304
Curing process 260
Cyclophosphamide 257
CYP81Q1 374
Cytochrome C 397
Cytochrome P450 375
Cytokinin (CK) 386
Cytoplasm 83
Cytosol 13, 84
Cytotoxic effects 256

Daphne mucronata 319
Daphne tangutica 370
Daphnecin 319
DDQ 156, 223, 225, 351
Deacetylation 153
Deamination 79
Decanedioyl dichloride 261
Decarboxylation 304

Deciduous 56, 256
Decoupling 168
Degradation method(s) 125, 145, 189, 457, 459
  acidolysis 198, 201
  DFRC 206, 207
  Hydrogenolysis 203
  hydrolysis with dioxane/water 198
  hydrolysis with thioglycolic acid 200
  hydrolysis with water 196
  mercaptolysis 200
  NE 209
  oxidation with cupric oxide 190, 193, 194
  oxidation with nitrobenzene 190, 192
  oxidation with permanganate 195, 196
  pyrolysis 215
  solvolysis 198
  thioacetolysis 200, 201
  thioacidolysis 201
Dehydrogenation 80
Dehydrogenation polymer (DHP) 97
3-Dehydroshikimate 77
Dehydroxylation 378
Delignification 118, 123, 125, 126, 128, 136, 154, 198
Demethylation 272, 378, 461
Dengue 386
Depolymerization 134, 290, 298, 462
DEPT 169
Determination method(s)
  acetyl bromide 193
DFRC 228
Diacid(s) 261
Dialdehyde(s) 119
Diarylpropane 89
Diarylpropanediol(s) 160
Diazoethane 218
Diazomethane 218
Dibenzodioxocin(s) 20, 22, 24, 25, 96, 164
Dicotyledonous 101
Dielectric 40
Dietary fiber 257
Digestion 76, 257
Dihydrodimerization 34

Dihydroxynaphthalene 51
Dilatometry 40
Dilignol(s) 88, 90, 154, 196
Dimerization 88
Dimethyl sulfide (DMS) 306
Dimethyl sulfoxide (DMSO) 306
Dimethyl-*p*-phenylenediamine 52
Dineolignan(s) 315, 316, 461
Dinitrofluorobenzene 153
Diol(s) 261
    diarylpropane 196
Diphenyl ether 89
1,1-Diphenyl-2-picrylhydrazyl (DPPH) 256
Diphenylamine 52
Diphenylpentane 316
Diphenylpicrylhydrazyl (DPPH) 405
Dirigent protein (DIR) 75, 370, 459
Dispersant 258, 269
Dispersity 129, 256 see Index
Distillation 123
DMF 153
Dopant(s) 265
Drosophila melanogaster 386
Dust control 269
Dye(s) 263

Ebers papyrus 398
Ebulliometry 150
Eco-friendly 261
EDXA 62
Elastomers 263
Electron microscopy 174
Electron spin resonance (ESR) 256
Elemental analysis 146
Emulsifier 269
Enantiomer(s) 316, 370
Encrusting
    material 3
Energy storage 265
Enterodiol 331, 461
Enterolactone 331, 461
Enterolactone (ENL) 402
Enterolignan(s) 331, 377
Enzymatic 116
Enzymatic hydrolysis 126

Epicatechin 256
Epichlorohydrin 226, 274, 275, 276
Epoxidation 226
Epoxy resins 274
Epstein–Barr virus (EBV) 401, 407
Erythrose-4-phosphate 77
ESR 174
Ethanolysis 199
Ethylation 154
Ethylene glycol 123, 125, 267
Eucalyptus globulus 56
Eucalyptus grandis 56
Eucalyptus hemiphloia 326
Eudeshonokiol A 319
Eudesmagnolol 319
Eudesobovatol A 319
Eudicotyledonous 18
Expander(s) 267
Exudate(s) 326

Factor
    contraction (g), 42
Fagopyrum esculentum 329
Feed(s) 254
Fermentation 115, 126
Fertilizer 257, 258
Ferulate ester(s) 80
Ferulic acid 79, 101
Feruloyl-CoA 79–81
Fiberboard(s) 259
Fire log(s) 253
Fitzroyia cuppresssoides 326
Flame 253
Flavonolignan(s) 319
Flax 56
Flocculant(s) 261, 279
Flovonoid(s) 80
Foam(s) 261, 263, 265, 267
Fomes annosus 383
Forage 76
Formacell 134, 135
Formaldehyde 251, 260
Forsythia intermedia 97, 370
Fragaria anamassa 329
Free-radical scavenger 256
Fremy's salt 54, 163, 223, 225

Freundlich adsorption isotherm 255
Friedel–Crafts acylation 251
FTIR 164, 165
Fuel(s) 249, 252, 459
Fuming HCl 118
Fungi 76, 115
  red-rot 229
  soft-rot 229
  white-rot 193, 228
Fungus 256
Furan(s) 52
Furfural 57, 134

γ-rays 174
Galactoglucomannan 97
Galactose 38
GC-MS 192, 199, 292, 295, 328
GC/MS 383
Genoprotective 257
Geocomposite(s) 265
Globodera pallida 386
Globodera rostochiensis 386
D-Glucopyranose 97
Glucopyranoside(s) 388
β-Glucosidases 88
Glucoside(s) 84
  4-O-β-D- 12
  coniferin alcohol 12
  p-glucocoumaryl 12
  monolignol(s) 13
  syringin alcohol 12
Glucosyl transferase(s) 88
Glucuronoxylan(s) 99
Glutaraldehyde 260
Glycerol 24, 154, 274, 275
Glycine max 329
Glycoconjugate(s) 84
Glycol(s) 123
Glycoprotein 87
Glycoside(s) 84, 330
Gmelina leichardtii 326
Gnetucleistol F 319
Gnetum cleistostachyum 319
Golgi apparatus 99
Gomisin(s) 421–422
GPC/DV 219

GPC/LALLS 219
Granulate 264
Grass 16
Gravimetry 55, 60, 66, 118
Green electricity 253
Greenhouse gas (GHG) 252
Grignard reagent(s) 335
Grinding 121
GS-MS 174
Guaiacol(s) 289, 297, 304
Guaiacun officinale 326
Guaiacyl (G) 16, 459
Gymnosperm(s) 12, 16, 21, 56, 84, 101–103, 118
Gymnospermophyta 326

Halogenation 219
Heavy metal(s) 254, 255
HeLa 397, 405, 407, 411, 418
Helianthus annuus 329
Hemicellulase(s) 38
Hemicellulose(s) 13, 38, 61, 97, 100, 122, 125, 127, 257
Hemipinic acid 163
Hemoprotein(s) 87
Hemp 56
Hepatoprotective effect 256
Hernandia nymphaeifolia 401
Hernandia ovigera 401, 407
Herpes simplex 256, 394
Heteropolymer 38
Hexamethylenetetramine (HMTA) 260
Hibiscus cannabinus 56, 83
High-boiling-point solvent(s) (HBS) 123
High-density polyethylene (HDPE) 260
Himantandra baccata 326
HIV 391, 392, 397, 398, 401, 404, 412
HMQC 169
Holocellulose 36
Holstein calve 257
Hordeum vulgare 329
Hot water process 136
HPLC 174, 192, 199, 328, 331
HPLC/MS 383
HSQC 172
Hydrodeoxygenation (HDO) 294

Hydrogel(s) 264
Hydrogen abstraction 90
Hydrogen fluoride 119
Hydrogen peroxide 87, 223, 300, 372
Hydrogenolysis 22, 190, 294
Hydrolysis 115, 116, 118, 119, 126
  alkaline 195
  mild 196, 198
Hydrolysis, method of 458
Hydrolytic enzymes 122
Hydroperoxide anion 223
Hydrophilicity 261
Hydroquinone 51
Hydrothermal hydrolysis 295
Hydrothermolysis 127
p-Hydroxybenzaldehyde(s) 11, 51, 191
p-Hydroxybenzoate(s) 28
p-hydroxybenzoic acid 153
Hydroxycinnamic acid(s) 82
p-Hydroxycinnamoyl-CoA 76
Hydroxylation 79
Hydroxymatairesinol (HMR) 327, 331
Hydroxymethylation 226, 273
Hydroxymethylfurfural 57
p-Hydroxyphenyl 16, 459
Hydroxyproline 99
Hymenolepis nana 256
Hypo number 64, 66
Hypochlorite
  calcium 64
  sodium 64

IBICOM biorefinery 253
Immunocytochemistry 102
INADEQUATE 169
Index
  dispersity (*Đ*) 40, 128
  dispersity (DJ) 256
  refractive 149
Indole 52
Indophenol(s) 54
Indulin AT (IAT) 253
Influenza virus 256
Insect(s) 49
Insecticidal 383

Insulation 265
Interference microscopy (IM) 174
Ionic liquid(s) 60, 115, 124, 125, 457
Ionic nutrient(s) 254
Iron-TAML 225
Isocyanate 251, 261, 262, 265
Isocyanate(s) 265
Isoeugenol 191, 304
Isolation method(s) 457
Isolation, method of 115
  fuming HCl 118
  Willstätter 118
IUPAC 319, 369

Juglans nigra 329
Juniperus phoenicea 393
Juniperus sabina 385
Justicia hayatai 417
Justicia pectoralis 413
Justicia procumbens 417
Justicia prostrata 413
Justicidin(s) 416–418
Jute 33, 56

Kadsura coccinea 419
Kadsura interior 422
Kadsura longipedunculata 419
Kadsura philippinensis 419
Kalundborg 253
Kappa number 64–66
  rapid 66
Karatex process 259
Kenaf 28, 33, 56, 58, 83
Kiesegel G 330
Klebsiella pneumoniae 256
Knots 327

Laccase enzyme 383
Laccase(s) 38, 85, 99, 229, 279
Lamella
  middle 76, 99, 100, 459
  middle (ML) 13
Langmuir adsorption isotherm 255
Lariciresinol 327
Larix decidua 326
Larrea tridentata 398, 400, 405

Layer(s)
  S1 13
  S2 13
  S3 13
Leguminosae 39
Leishmania donovani 398
Lentinus edodes mycelia (LEM) 256
Lepidoptera larvae 386
Lewis acid 201, 276
Libocedrus plumose 393
Lignan(s) 461
  arylnaphthalene 316
  aryltetralin 316
  biosynthesis 370, 461
  dibenzocyclooctadiene 316, 331, 418–421
  dibenzylbutane 316
  dibenzylbutyrolactol 316
  dibenzylbutyrolactone 316
  furanoid 316
  furofuranoid 316
  hybrid 315, 319, 461
  metabolism 377–385
  nomenclature 319
Lignan(s)Dibenzocyclooctadiene 377
Lignicolous fungi 126
Lignification 12, 34, 75, 79, 87, 89, 90, 93, 99, 100, 102, 174, 459
Lignin carbohydrate complex (LCC) 256
Lignin(s) 127
  Fredenhagen 119
  acetosol 259
  acetylated 28, 41
  acid soluble 57
  acid sulfite 118
  acrylamide 261
  Alcell 257, 263
  alkali 42, 125, 133, 457
  alkali sulfite (ASL) 219
  alkaline 265
  ammoxidized 281
  biochemical transformation(s) 227
  biodegradation 228
  biosynthesis 75, 76, 459, 461
  bisulfite 118
  Brauns 28, 117, 120, 126, 219
  brown-rot 126
  cellulolytic enzyme (CEL) 39, 117, 122
  commercial 127
  cuoxam 118
  cuproxam 118
  detection 50
  determination method(s)
    acetyl bromide 62, 68
    acid detergent 59, 68
    alkali 60
    Benedikt and Bamberger 61
    chemical 61
    chlorine 64
    Cross, Bevan, and Briggs 61
    Ellis 57
    fluorescence 60
    fuming HCl 59
    hypo number 64
    hypochlorite 64
    König and Rump 59
    Kappa number 64
    Klason 68
    Mehta 60, 61
    methanol number 64
    Morrison 63
    nitrosation 63
    oxidant consumption 64
    Pearl and Benson 63
    permanganate 68
    permanganate number 64
    precipitate formation 59
    Roe chlorine number 64
    Schulze 61
    Seidel 61
    solution 59
    spectrophotometric 59, 61
    thioglycolate 62
    total organic carbon (TOC) 66
    Uppsala 59
    Waentig and Gierisch 61
  Dioxane acidolysis 123
  Enzymatic mild acidolysis (EMAL) 123
  Fengel 118
  Freudenberg 118

glycol 221
Halse 118
hardwood 28, 30, 57, 79, 96, 159
hydrogenolysis 154
hydrolysis 154, 251
hydrolyzed 259
hydroxymetylated 273
hydroxypropyl (HPL) 219
hydroxypropylated 151
indirect method(s) 60
Indulin AT 263
isolation 115, 116
Klason 57, 64, 68, 118, 119, 199
Kraft 42, 57, 118, 127–129, 134, 151, 154, 218, 250, 251, 254, 255, 257–261, 265, 272, 279, 292, 294, 295, 306, 457
Kraft pine 265
lignosulfonate 250, 251, 259
liquefaction 219
low-sulfonated 267
milled wood (MWL) 117, 147, 121, 126, 153
modified 252
native 11, 21, 41, 51, 115, 116, 120, 122, 123, 145, 219, 249, 256, 257, 289, 457, 458
nonsulfur 127
organosolv 41, 123, 127, 219, 254–257, 259, 262, 295, 457
periodate 119, 120
phenol 125
pulping 40, 42, 249
purification 115
Purves 119, 120
pyrolysis 136
pyrolytic 274
Runkel 118
Schubert-Nord 126
soda 127, 133
softwood 20, 21, 28, 30, 57, 79, 96, 159
solubility 41, 154
spruce 87, 158
standard 60, 63
steam-exploded 125, 136
structure 145

sulfate *see* Lignin(s)
 Kraft 128
sulfite 257
sulfur 264
sulfur-containing 127
sulfur-free 257, 260
sulfuric acid 118
Swelled enzyme (CEL) 122
unmodified 252, 267
Urban 218
wheat straw 35
Willstatter 217, 219
Lignin(s) determination
 direct method(s) 55
 fuming HCl method(s) 55
 indirect method(s) 55, 60
 Klason method(s) 57
 Metha method(s) 55
 sulfuric acid method(s) 55, 57
Lignin(s) hardwood 17
Lignin-modified phenolic (LPF) resin 260
Lignin-carbohydrate complex (LCC) 34, 38, 40
Ligninsulfonate(s)
 calcium 218
Ligninsulfonic acid
 α- 222
 β- 222
Ligno-pani™ 265
LignoBond 253
LignoBoost 129
Lignocellulose 126, 268
Lignocellulosic(s) 55, 60, 62, 91, 116, 119, 459
Lignoglucoside 14
Lignols 99
Lignosulfonate 258, 259, 265
Lignosulfonate(s) 42, 60, 63, 127, 131, 132, 151, 193, 253, 254, 257, 267–269, 305
 calcium 253, 254, 269
 magnesium 270
 sodium 265, 269
Lignosulfonic acid(s) 218, 265
Linkage(s) 17
 radical coupling 24

Linum flavum 393
Linum nodiflorum 376
Linum usitatissimum 56, 329, 399
Lipid(s) 328
    metabolism 257
Lipoprotein 257
Liquefaction 295
Liquid hot water (LHW) 127
Liquid wood see wood(s) 263
Liquid(s)
    ionic 298
Liquor(s)
    black 252, 255
    pulping 249
Lizard's tail 316
Low-boiling-point solvent(s) (LBS) 123
Low-density polyethylene (LDPE) 260
Lumen 99

Maackia amurensis 319
Macrofibril(s) 13
Macromonomer 261, 263, 268
Macronutrient(s) 254
Magnolia biondii 407
Magnolia grandiflora 384
Magnoliales 316
Magnoliophyta 326
Manassantin 316
Mannan(s) 39, 100
Mannose 38
Matairesinol 327
Matairesinol (MAT) 402
MeadWestvaco 129
Medioresinol 328
Medium-density fiberboard (MDF) 258
Membrane
    cell 83
    plasma 84
Mesomeric form(s) 89
Mesophase pitch (MPP) 308
Metabolite(s)
    secondary 315, 318
Metahemipinic acid 163
Metal ion(s) 254
Metaperiodate (IO_4^-) 225
Methanol number 64

Methanolysis 160
Method(s)
    isolation 458
Methoxy-*o*-quinone 54
1-Methyl-2,5-diaminobenzene 52
1-Methyl-2-amino-6-nitrobenzene 52
Methylation 195
O-Methylation 89
Methylene diphenyl isocyanate (MDI) 262
Methylene quinonium 220
4-Methylquinoline 51
Methyolation 259
Metso Corporation 129
Meyloid leukemia (Ml) 402
Micellar electrokinetic chromatography (MEKC) 328
Michael addition 128, 338
Microalgae 267
Microfibril(s) 13, 16, 100
Micronutrient(s) 254, 258
Microorganism(s) 4, 76, 97
Microwave 292
Microwave-assisted extraction (MAE) 330
Milling 121
Milling ball 68, 115, 121, 261
Milling Wiley 68
Milox procedure 135
MOF 225, 302
Moisture 253
Molasses 265
Molecular weight (M_w) 13, 21, 39–42, 128, 132, 134–136, 291, 292, 294, 295
Molluscum contagiosum 390
Monochloroimide 156
Monocot(s) 82, 102, 103
Monocotyledon(s) 316
Monocotyledonous 18
Monoelectronic oxidation 87
Monoglucuronide 331
Monolignol(s) 11–13, 17, 18, 21, 24, 31, 49, 63, 75, 76, 154, 316, 458, 459
    biosynthesis 80
    dimerization 89
    transport 83

Monosaccharide(s) 36, 119
Morphology 265
Mosher's ester 331
Mucilage 257
Muconic acid 120
Multi-walled carbon nanotube (MWCNT) 268
Musca domestica 386
M_w 296
Mycobacterium avium 404
Mycobacterium smegmatic 384
Mycobacterium tuberculosis 404

N,*N*-Dimethylacetamide (DMAc) 261
Nanoparticle(s) (NPs) 267, 279, 281
Nanotechnology 462
β-Naphthol 51
β-Naphtylamine 52
Neoalsomitra integrifoliola 422
Neolignan(s) 88, 315, 316, 370, 377, 422–423, 461
Nitration 221
Nitric acid 60, 61
Nitrite
  sodium 63
Nitroaniline 52
Nitrobenzene 11
Nitrobenzene oxidation 190
Nitrolignin(s) 221
Nitrophenol 51
1-Nitroso-2-naphthol 153
Nitrosodisulfonate
  potassium 54
Nitrosophenol 63
NMR 127, 145, 156, 168, 172, 219, 331
  ^{27}Al 281
  ^{13}C 153, 169, 171, 172, 228, 273, 292
  ^{13}C 60
  1D 331
  2D 22, 100, 168, 172, 331
  ^{19}F 172
  ^{1}H 153, 159
  ^{1}H 163, 168, 172, 292
  ^{17}O 212
  ^{31}P 164, 172, 292
  ^{31}P 85

^{29}Si 172
NOESY 331
Noguchi–Crown–Zellerbach process 294
Nontransgenic 33
Nordihydroguaiaretic acid (NDGA) 386, 400–401
Norlignan(s) 315, 318, 370, 377, 461
Norway spruce 327
Novolac resin 259, 260
Nucleophilic attack 89, 370
Number-average molar mass (M_n) 39, 40, 128
Nutrient(s) 4, 255

Olea cunninghamii 326
Olea europa 326
Olea europaea 398
Oligolignol(s) 88, 100
Oligosaccharide(s) 257
Orcinol 51
Organocell process 135
Organosolv 115
Oriented strand board (OSB) 258
Oriza 56
OSB 259
Osmotic pressure 150
Oxalic acid 227
Oxidation 222
Oxidative coupling 75, 370
Oxidative polymerization 75
Ozone 300

Packing 254
Palliation 319
Panicum miliaceum 329
Parameter
  branching 42
Paraperiodate
  sodium 119
Pathogen(s) 49, 257
Pectin(s) 100, 257
Pellet(s) 252, 253, 461
Pellia epiphylla 319
Pentose phosphate pathway 77
Performic acid 135
Periderm 76

Periodic acid 120
Permanganate
  potassium 61, 64
Permanganate number 64, 66
Permanganate oxidation 59
Peroxidase(s) 38, 85, 99, 229
Peroxyacid 302
Persulfate (salt),$K_2S_2O_8$ 261
Pesticide(s) 258, 279
Petrochemical 264
Petrochemical industry 459
pH 126, 128, 129, 163, 195, 255, 267, 281, 295
Phenol 51
Phenol formaldehyde (PF) 258, 260
Phenol–formaldehyde resin 259
Phenolation 209, 251, 271, 461
Phenolic acid(s) 76
Phenolic resin 259
Phenolysis 271
4-O-Phenoxy radical(s) 77
L-Phenylalanine 76, 78, 81, 370
Phenylcoumaran(s) 20, 25, 34, 89, 90, 160
Phenylenediamine 52
Phenylpropane 40
Phenylpropanoid(s) 11, 12, 49, 76, 78, 89, 371
Phloroglucinol 51, 52, 61, 158
Phosphate 257
Phosphitylation 154
Phosphoenol pyruvate 77
phospholipase C γ1(PLCG1) 398
Phosphomolybdic acid 61
Phosphoric acid 60, 61, 115, 118
Phosphotungstic acid 61, 330
Photooxidative degradation 260
Phryma leptostachya 386
Phyllanthus 328
Phyllanthus oligospermus 413
Phyllantus myrtifolius 413
textslPhyrma leptostachya 386
Phytoestrogen(s) 377, 400, 402, 409–410, 461
Picea abies 56, 327, 383
Picea glauca 328

Picea sp. 326
Pieris rapae 385, 386
Piezoelectric 40
Pinoresinol 95, 163, 328
Pinoresinol (PIN) 409–410
Pinoresinol synthase 370
Pinus radiata 56
Pinus silvestris 56
Pinus taeda 39
P. mullesua 386
Piper solmsianum 386
Piperales 316
Piperitylmagnolol 319
Pisum sativum 329
Plant tissue(s) 50, 55
Plasticizer 262, 263
Plastid 77
platelet activating factor (PAF) 405
Plywood 258, 259
Poales 39
Podocarpus spicatus 326
Podophyllin 387, 388
Podophyllotoxin 374, 375, 384, 386
Podophyllotoxin (PPT) 386–397, 462
Podophyllum emodi 338, 388, 391
Podophyllum peltatum 387, 388, 393
Podophyllum rhizome 338, 388
Pollutant(s) 254, 255, 258
Poly (ε-caprolactone) 261
Poly (butylene succinate) 261
Poly (propylene glycol) 268
Polyacrylonitrile (PAN) 308
Polyalkene 258
Polyamide 264
Polydextran 148
Polydispersity 40, 42
Polyester(s) 258, 260, 261
Polyether polyol(s) 265
Polyethylene (PE) 260
Polyethylene glycol (PEG) 262
Polyethylene glycol diglycidyl ether 260
Polygalacturonase enzyme 383
Polymer industry 461
Polymerization 251, 265
  bulk 89
  cross-linked 13

random 90
  step-growth 268
Polyol 261, 262
Polyolefin 260
Polyoxometalate 279
Polyphenol oxidase(s) 85
Polyphenol(s) 75, 264
Polypropylene (PP) 260
Polypropylene glycol (PPG) 262
Polypyrrole 265
Polysaccharide(s) 42, 49, 60, 76, 93, 97, 99, 101, 118, 119, 122, 127, 256, 459
Polystyrene (PS) 42, 260
Polyurethane (PU) 261, 275
Populus tremula 56
Populus trichocarpa 79
Potassium chlorate 61
Prebiotic 257
Precipitation(s) 255
Propane-1,3-diol 253
Prostaglandin E2 401
Protocatechuic acid 227
Protolignin(s) 12, 25, 26, 42, 116
Protoplast 99
Pseudomonas aeruginosa 256, 400
pseudo second-order kinetic 255
Pseudotsuga menziesii 56
Pseudotsuganol 319
Psidium guajava 329
Psoriasis vulgaris 391
Pteridophyta 326
Pteridophyte(s) 39, 103
PTSA 219
Pulp(s)
  unbleached 64
Pulping 3, 127, 457
  organosolv 134
PVC 270
Py-GC/MS 175
Pyridine 217
Pyridinium chloride/[D_6]DMSO 68
Pyrogallol 51
Pyrolysis 66, 126, 175, 227, 290–292, 294
Pyrrole 52, 265
Pyrrolidine 153

Quinomethilide(s) 54
Quinone(s) 265, 289
  methide(s) 89, 90, 93, 96, 97, 99, 128, 134, 273
  mono chlorine 54
  o- 54, 120, 163, 219

Racemate(s) 316
Radical 85
Radical coupling 88
Radical scavenger 258
Rami 56
Raney nickel 201
Ratio(s)
  G/S 28, 42, 101
  H/G/S 154
  S/G 68, 97, 102, 172, 199, 225
Ray parenchyma 102
Reaction(s)
  C–C coupling 268
  Cannizzaro 273
  color 49–51, 53
    blue autofluorescence 50
    Mäule 50, 54
    Wiesner 50, 51, 158
    Wiesner test 51
    with toluidine blue 50
  condensation 198
  Diels–Alder 339, 340
  Heck 340
  Mannich 226, 279
  McMurry 346
  Simmons–Smith 348
Resin(s) 116, 326
  ion-exchange 255
Resinol(s) 25, 34, 89
Resorcinol(s) 51, 289
Reverse osmosis 255
Rheumatoid arthritis 391
Rhoda technique 305
Rhodnius prolixus 386
Roe chlorine number 64
Rosin 258
RP-HPLC 328
Rubbers 258
Ruminants 76

Saccharum 56
Saucerneol D 316
Saururus cernuus L. 316
Saururus chinensis 316, 401
Sawdust 40
Scattering
  light 149
Scavenger power 256
Schisandra chinensis 328, 398
Schisandra hernyi 418
Schisandra propinqua 419
Schisandra wilsoniana 420
Schizandraceae 343
Schweizer's reagent 118
Sebacoyl chloride 261
Secale cereale 329
Secoisolariciresinol 327
Secoisolariciresinol (SECO) 399–400
Secoisolariciresinol dehydrogenase
  (SIRD) 372
Secondary metabolite(s) 369, 461
SEM 62
Semiempirical C_9 formula 146
Sensoring 268
Sephadex 148
Sesame 328
Sesamin 328, 411
Sesamolin 328
Sesamum indicum 329, 411
Sesquilignan(s) 315, 316, 461
Shikimate pathway 76
Shikimic acid 76, 79
Silibinin 319
Sinapate ester(s) 80
Sinapyl acetate 83
Sinapyl aldehyde 51
Sisal 28, 56, 58
Skatole 52
Sodium chlorite 60, 62
Sodium lignosulfonate 268
Solanum melongena 329
Sorption 255
Spectroscopic method 458
Spectroscopy
  IR 331
  UV 57, 63, 330, 331

UV/vis 60, 229
X-ray 331
Spirodienone(s) 34
Sporobolomyces roseus 257
Stabilizer(s) 260, 267, 268
Staphylococcus aureus 384
Starch 265
Stauntonia chinensis 405, 412
Steam 252
Steam explosion, method of 261
Stereocenter(s) 324
Steric hindrance 260
Stilbene 157
Stilbenolignan(s) 319
Streptomyces 326
Streptomyces setonii 228
Streptomyces viridosporus 228
Sucrose 82
Sugarcane bagasse 122, 133
Sulfonation 222, 251
Sulfurous acid 222
Supercritical 136, 206, 227, 295,
  298, 328
Supercritical) 330
Superoxide dismutase (SOD) 405
Surface tension 269
Symplocos lucida 326
Syngas 290, 291
Syringaldehyde 11, 51, 191, 192, 193,
  194
Syringaresinol 163, 196, 328
Syringaresinol (SYR) 410–411
Syringyl (S) 16, 50, 459
Syringylpropane 193
Systematic name 369

Tannin(s) 63
TAPPI 64, 66
TAPPI, method of 57
Tarenna attenuata 422
Tarenna attenuate 422
Taxaceae 328
Taxus 328, 400
Taxus baccata 326
TEM 62
Temperature

glass transition (T_g) 40, 41, 276
Tensile strength 251
Terminalia bellerica 398
Tetraamine copper(II) 118
1,2,3,5-Tetrahydroxybenzene 51
Tetralignol(s) 154, 196
Tetralin 294
Thalline 52
Thermal
   analysis 40
   decomposition 40
   stability 251, 261, 265, 270, 279
Thermolysis 227, 291
Thermoplastic 261, 276, 278
Thin film 268
Thioacetic acid 200
Thioacetolysis 190
Thioacidolysis 154
Thioether(s) 62
Thioglycolate 63
Thioglycolic acid 60, 62, 68
Thiolignin see Lignin(s)
   Kraft 128
Thiosulfate 66
Thuja plicata 398, 412
Thymol 51
Tinction(s) 53
   acridine orange 53
   acriflavine 53
Titration, method of 152
TLC 330
TMDP 172
Toluene 121
toluene-2,4-diisocyanate 268
Toluidine 52
Topoisomerase 398, 405
Topoisomerase(s) 404
Torreya jackii 328
Tracheid(s) 76, 102, 459
Transesterified 79
Transgenic 30, 33, 37
Trichosporon cutaneum 257
Trifluoroacetic acid 118
Trigonotins A–C 332
Trilignol(s) 154, 196
Tripterygium wilfordii 405

Triticum aestivum 329
Trivial mame 369
Trypanosoma brucei 417
Trypanosoma cruzi 404, 417
Turbidimetry 60
Type
   -G 16
   -G-S 16
   -H-G 16
   -H-G-S 16, 17
L-Tyrosine 76, 81, 370

UDP-glycosyltransferase 84
Ultrafiltrated 268
Ultrafiltration (UF) 151
Ultrasonic 123
Ultra thin film 268
Unit(s)
   cinnamyl 316
   guaiacylpropane 28
   phenylpropane 315
   syringylpropane 28
Urea 257, 258
Urea formaldehyde 258
Urethane 261

Vaccine(s) 267
Vaccinium macrocarpon 329
Vacuole(s) 13, 84, 88
Vanilla planifolia 302
Vanillin 11, 190–193, 223, 289, 297, 302, 304
Vanillylmandelic acid 304
Vapor pressure 124
Vapor-pressure osmometry (VPO) 150
Vinblastine 392
Viscometry 149
Viscosity(ies) 42
Vitis vinifera 329
VTT 254

Wall(s)
   cell 12, 13, 34, 38, 49, 75, 76, 83
   primary (P) 13
   secondary (S) 13, 99
   secondary cell 459

Wax 264
Waxe(s) 253
Weight-average molar mass (M_w) 40
Weyerhaeuser NR Company 253
White liquor 128
Wilms tumor 391
Wood(s)
  early 118
  heart 118
  late 118
  liquefaction 219
  liquid 263
  milled 39, 123
  powder 40
  sap 118
  sweetgum 192

Wooten, method of 259
Workability 269

Xantholignan(s) 319
Xylan(s) 38, 62, 66, 99, 100
Xylem vessel(s) 76
Xylem(s) 81, 99, 102, 459
Xyloglucan(s) 38
Xylose 38

Yeast(s) 77, 267

Zanthoxylum avicennae 422
Zeolite 292
Zulauf, method of 97
Zutropf, method of 97

Printed and bound by CPI Group (UK) Ltd, Croydon, CR0 4YY
06/12/2023